Advance Praise for *Sustainable Transportation Systems Engineering*

"This book can be used for an entirely new approach to teaching transportation engineering, one where environmental considerations are no longer of secondary concern. It is comprehensive in its coverage of modes and of current issues, and thus also makes a great reference to practitioners who are trying to quickly expand their knowledge."

Eric Bruun
Visiting Professor, Aalto University, Espoo, Finland
Author, Better Public Transit Systems: Analyzing Investment and Performance

"This much-needed new book fills a large gap in existing transportation texts. While others deal tangentially with the concept of sustainability, this book takes a systems approach to cover all traditional transportation engineering concepts—from environmental impacts, to traffic engineering, to public and active transport modes—through the sustainability lens. A special strength of the book is the critical analyses of emerging technologies' impacts on sustainability in both passenger and freight transportation. The concepts in the book are reinforced with exhaustive international examples such that the evolution and current state of global transportation are fully explained. The strongest part of the text is the balance of data-driven, quantitative modeling approaches with an assessment of how these techniques influence policy. These presentations make this a perfect text for multiple engineering disciplines, but also for planners or for teaching in social sciences."

Jeff Casello
Associate Professor, School of Planning and Department of Civil and Environmental Engineering
University of Waterloo, Canada

"This extremely comprehensive book offers a *tour de force* of the major issues and dimensions of the relationship between transportation and environmental sustainability. Pulls together the various strands that make up transportation, from population to GDP to technology, and presents their interrelationships in a highly coherent and meaningful manner. The book covers a broad range of topics around which there is considerable discussion among experts as well as in the informed citizenry at large. Provides an excellent exposition of the many challenges facing countries who are not making sufficient efforts to reduce their transportation 'footprints' as well as the need for many of them, especially the United States, to increase their investment in environmentally and socially equitable better modes and the infrastructure, policies, and planning that is needed for these to succeed. The book is filled with much valuable up-to-date information and many well-designed highly informative graphics. Each chapter includes helpful exercises for students to undertake. This book should be welcomed by the transportation engineering profession, from study to teaching and practice, as it searches for ways to help transportation meet sustainability goals."

Preston Schiller
Adjunct Lecturer, Queens University, Canada
Author, An Introduction to Sustainable Transportation: Policy, Planning, and Implementation

"This text provides a rich tool kit for students of sustainable transportation, embracing a systems approach. The authors aptly blend engineering, economics, and environmental impact analysis approaches. Best practices and innovative technologies for passenger vehicles; public transportation; shared-use mobility services, such as carsharing and bikesharing; and freight transportation are examined. Transportation demand management and choice are featured early in the discussion. Technological strategies range from vehicles and fuels to infrastructure and information technology. Exercises at the end of each chapter offer valuable opportunities for further understanding."

SUSAN SHAHEEN
Professor, Department of Civil and Environmental Engineering, and
Co-Director, Transportation Sustainability Research Center
University of California, Berkeley

"This sophisticated book addresses one of the great challenges of the 21st century—how to transform our resource-intensive passenger and freight transportation system into a set of low-carbon, economically efficient, and socially equitable set of services. The authors provide a broad array of tools to analyze economics, vehicles, and infrastructure. The sweep of this book is awesome, covering virtually every mode and aspect of transportation across the world. This book is an excellent text for students and a great resource for anyone interested in learning the details of transportation."

DAN SPERLING
Professor and Director, Institute of Transportation Studies
University of California, Davis
Author, Two Billion Cars: Driving toward Sustainability

"This is a very welcome addition to the literature on transport. It adopts a systems approach and makes explicit links between engineering, ecological, environmental, and quality-of-life perspectives, and it is focused on solutions. The emphasis on solutions linked to improving sustainability is a significant contribution to the transport debate as is the rigorous approach to technology, energy security, and climate change. The emphasis on solutions, linkages, and systems fills a longstanding gap in transport science and does the job with the clarity and insight that will benefit all transport professionals and students and enrich our understanding of sustainability."

JOHN WHITELEGG
Editor-in-Chief, World Transport Policy & Practice
Author, Transport for a Sustainable Future: The Case for Europe

Sustainable Transportation Systems Engineering

Francis M. Vanek, Ph.D.
Largus T. Angenent, Ph.D.
James H. Banks, Ph.D.
Ricardo A. Daziano, Ph.D.
Mark A. Turnquist, Ph.D.

New York Chicago San Francisco
Athens London Madrid
Mexico City Milan New Delhi
Singapore Sydney Toronto

Sustainable Transportation Systems Engineering

1 2 3 4 5 6 7 8 9 0 DOC/DOC 1 2 0 9 8 7 6 5 4

ISBN 978-0-07-180012-9
MHID 0-07-180012-3

The pages within this book were printed on acid-free paper.

Sponsoring Editor
Robert Argentieri

Editorial Supervisor
Stephen M. Smith

Production Supervisor
Pamela A. Pelton

Acquisitions Coordinator
Amy Stonebraker

Project Manager
Yashmita Hota, Cenveo® Publisher Services

Copy Editor
Ragini Pandey

Proofreader
Amy Rodriguez

Indexer
Cenveo Publisher Services

Art Director, Cover
Jeff Weeks

Composition
Cenveo Publisher Services

To my wife Catherine, my children Mira and Ray,
and my parents Wilda and Jaroslav

—Francis Vanek

To my wife and son, Ruth Ley and Miles Ley Angenent

—Lars Angenent

To Denis, Marcela, Rolf, Gloria, and Jaime

—Ricardo Daziano

To my wife Lynn

—Mark Turnquist

About the Authors

Francis M. Vanek, Ph.D., is Senior Lecturer and Research Associate in the School of Civil and Environmental Engineering at Cornell University, where he specializes in energy efficiency, alternative energy, and energy for transportation. He is the lead author of *Energy Systems Engineering: Evaluation and Implementation*, Second Edition.

Largus T. Angenent, Ph.D., is Professor of Biological and Environmental Engineering at Cornell University, where he specializes in waste-to-energy conversion technologies. He is an editor of *Bioelectrochemical Systems: From Extracellular Electron Transfer to Biotechnological Application*.

James H. Banks, Ph.D., is Professor Emeritus of Civil, Construction and Environmental Engineering at San Diego State University. He is the author of *Introduction to Transportation Engineering*, Second Edition.

Ricardo A. Daziano, Ph.D., is the David Croll Fellow Assistant Professor in Civil and Environmental Engineering at Cornell University. His research focuses on engineering decision making, specifically on econometrics of consumer behavior and discrete choice models applied to technological innovation in transportation and energy.

Mark A. Turnquist, Ph.D., is Professor of Civil and Environmental Engineering at Cornell University. His research focuses on large-scale network optimization models for use in transportation, logistics, manufacturing systems, and critical infrastructure security.

Contents at a Glance

Contents

Unit 2 Tools and Techniques

Preface

The goal of this book is to introduce professionals with an interest in transportation as well as upper-class and first-year graduate students to contemporary transportation systems and options for improving their sustainability. Another goal, and one very much connected with the objective of achieving system-wide sustainability, is to provide context for the study of transportation, in terms of trends in population, economic activity, ecological impact, and other important factors.

Both the title and the focus of the book are a response to the many challenges we face with the current situation of our transportation system. One of these challenges is the need to create more *livable cities and communities.* At a time when growing wealth around the world is creating new demand for motorized mobility, we need to adapt communities of all population sizes so that they have less congestion and easier access to amenities for all ages and walks of life, including those for whom age, income, or disability make travel challenging. Another challenge is *energy security and climate change.* As one of the major consumers of world energy, the global transportation system must move toward energy sources that are available for the long term and that do not contribute to the net increase in atmospheric greenhouse gases. A third challenge is the creation and maintenance of *sound transportation infrastructure.* For industrializing countries, rapid expansion of transportation demand leaves governments struggling to develop sufficient infrastructure to keep pace. At the same time, industrialized countries are faced with the challenge of maintaining and renovating aging infrastructure systems at a time when governments are under financial stress. The response to challenges like these as offered in the book is to provide a wide array of solutions for sustainable transportation systems, covering passenger and freight transportation, as well as alternative energy sources that might eventually be made available on land, sea, and air.

Much of the material presented is taken from two engineering courses taught on transportation systems, whose subject matter has moved in recent years toward addressing concerns about sustainability. One course, *Introduction to Transportation Engineering*, introduces students to some of the building blocks, including the capacity and energy requirements of transportation systems, or the function of public transportation or freight transportation networks. The second course, *Future Transportation Technologies and Systems*, goes into greater depth on the emerging alternative solutions for contributing to livable cities or incorporating alternative energy sources. Additional material is taken from courses on discrete choice modeling or economics of environmental protection, which are taught at the upper-class and first-year graduate level.

Topics in the book move from providing a framework for discussing transportation systems in the early chapters to providing solutions in the later ones. The first two units consider motivations and tools. After that, passenger and freight transportation each has a separate unit, with an overview chapter for each followed by several chapters focused on details. The final unit covers system-wide considerations, largely focused on alternative energy sources and systems integration, as well as the concluding discussion in the final chapter.

The assumed background of the reader is that of an engineering student with two or three years of course work completed, and training in mathematics (including some basic calculus skills) as well as physics and chemistry. Readers with training in physical sciences (such as physics or chemistry majors) should find the material accessible as well. Some of the topics draw on a basic knowledge of probability and statistics, and readers who have taken a college-level course in this subject should find this material accessible. Technologies or operations are described from first principles and the book does not assume previous technical knowledge of devices such as motor vehicles powered by internal combustion engines, rail vehicles (in either an urban or intercity setting), aircraft, marine vessels, or other technologies.

In general, we have taken a systems approach to the structure of the book. This approach includes the following steps: (1) identifying goals under the general heading of "sustainable transportation" that should guide the development of the book, (2) consciously drawing both geographic and topical boundaries around the content of the book, (3) giving structure to the content judged to be within the scope of the book, and (4) considering both short- and long-term perspectives on solutions for sustainable transportation.

The goals of the systems approach are already given above in the form of the three focal points of sustainable transportation. In terms of geographic boundaries, we are largely focused on the United States and North America, but we bring in examples from a number of other countries as well. As authors based in the United States, North American examples of specific systems, urban regions, or national policy are the ones most familiar to us. However, we look to Europe, Asia, and other parts of the world for additional examples to make the book more useful for readers outside of North America, and also to provide an opportunity for readers from our region to learn about models from other parts of the world that may be reproducible here. In terms of topical boundaries, some of the topics that are given relatively shorter treatment in the book include aviation and marine transportation, the transportation and air-quality connection, and the social sustainability aspects of transportation systems. A number of topics that are outside of the scope are given shorter mention, however, to point the reader in the direction of further resources. A diagram of the book's content is given in Chap. 1 in Fig. 1-8.

Turning to the topics that are within the scope of the book, these topics are given structure at first by looking at them from several different perspectives; namely, technological, economic, and environmental. From a technological perspective, devices and systems should be explained from first principles, where possible using quantitative analysis to illustrate capacity and efficiency. From an economic perspective, the economic life of the device or system and its capital and operating costs are important. From an ecological perspective, the primary metrics of concern are energy consumption as proxy for the general category of resource consumption and CO_2 as a proxy for emissions.

Structure is also given, on both the passenger and freight sides of transportation, by identifying the dominant mode and considering how it can be adapted to advance the goals of sustainability. In the case of passenger transportation, this mode is the private automobile, which is the largest generator of overall passenger-miles and consumer of energy. For freight, the truck (also called *road freight* or *highway freight* mode) is the focal point in terms of impact on the transportation system, value as an economic activity, or consumer of freight energy. In each case, an appropriate response is to better understand the impact of these critical modes on congestion or infrastructure wear, and then to improve their prospects for smooth flow in the transportation system. Also, means of improving energy efficiency or introducing alternative energy sources should be developed for these two modes.

A third aspect of providing structure is to show how other modal options besides private cars and trucks should be developed to encourage a more diverse, multimodal solution. This is especially true because improving the flow or changing the energy source of private cars or freight trucks will not in a single move completely achieve a sustainable outcome. For passenger transportation, a wide range of alternative modes can relieve some of the pressure on the dominant mode, including public transportation, commercial intercity transportation (e.g., high-speed rail or more efficient aviation), nonmotorized modes, carsharing, and carpooling. For freight transportation, road and rail modes can be combined into intermodal truck-rail systems where shipments move smoothly between the road network for pickup and delivery, and the rail network for long-haul movements. Multimodalism of this type is a necessary part of the solution because even if congestion is alleviated and road modes are shifted to sustainable energy supplies, livable cities and communities require a mix of modes beyond motor vehicles.

In terms of temporal aspects of the content, the sustainable development principle of *intertemporality* dictates that both short- and long-term options should be considered. Existing vehicle technology can be used now to improve the design of internal combustion engine vehicles, and best practices allow us to use them in a more efficient way. Energy efficiency of delivering transportation services is an important theme throughout the book, and is therefore considered not only in the discussion of alternative energy sources for transportation in Unit 5 but also as part of the discussion of passenger and freight transportation in Units 3 and 4. Turning to longer-term issues, it is the goal of the book not only to introduce vehicles technologies such as electric- or plug-in hybrid-electric vehicles, but also to explore at a systems level where the primary energy for these vehicles will be sourced, what mixture of electricity, biofuels, or hydrogen might be used as energy carriers, and how the fuels will be supplied to the vehicles. Fossil fuels dominate the primary energy supply of the transportation system today, but in the long run it is not clear what mix of zero-carbon energy sources and carriers will emerge to power the vehicles of land, sea, and air. Therefore, electricity, biofuels, and hydrogen are all given treatment in Unit 5, rather than picking one source a priori.

We turn now to some practical considerations. Units of measure are generally U.S. customary units. For certain international examples, metric units are used instead when the transportation community has set the precedent. Also, in some cases in the United States as well as in other countries the convention has been established to use metric instead of U.S. units, such as the measure of hydrogen in kilograms rather than pounds when discussing its use as a transportation fuel. Where it is useful to do so, a selection of worked in-chapter examples are presented in both U.S. and metric units.

Images taken with permission from other sources are accompanied by their attribution. Where no attribution is given, the source is the authors.

Financial figures are generally given in U.S. dollars; the reader can of course adapt the examples by translating financial figures into their own local currency. Although an attempt is made to give financial values that are representative of actual values at present, cost figures contained in the book should not be used as a basis for decision-making about actual transportation systems. Similarly, specific products or systems are at times identified in the book, sometimes with approximate cost values, but this information is provided solely as a service to the reader and does not imply any endorsement of the product in question. The authors were not compensated in any way by the vendor of the product or system for providing this information in the book.

In conclusion, global transportation is currently at a crossroads. Those who have access to the modern transportation systems derive many benefits from it, in terms of quality of life or economic opportunities. The project of making the system truly sustainable for the long term is incomplete, however. A large part of the world population does not yet have the benefit of access to this system. Also, it is threatened by system-wide challenges, such as the lack of well-maintained infrastructure or the lack of consensus on a secure source of energy that does not contribute to climate change. The latter challenge is particularly complex, because greenhouse gas emissions from transportation contribute to climate change, while simultaneously climate change threatens effective transportation by jeopardizing infrastructure through extreme weather events and rising sea levels. Meeting all of these challenges is a stimulating opportunity for creative collaboration between many different types of professionals. We wish each of you much success in your part of this great endeavor.

Acknowledgments

From Francis Vanek: I would like to first thank my wife Catherine Johnson and my children Ray and Mira for their support for me during this project and patience while I spent so much time working on writing. I would also like to appreciate their interest in the topic of sustainable transportation, which included both discussions of our transportation experiences while on family trips and relevant content we found on the Internet having to do with everything from livable communities to energy and climate change. I would like to thank my parents Jaroslav and Wilda Vanek for their unfailing support and encouragement as well.

I would like to remember the late Edward Morlok who was professor of transportation systems at the University of Pennsylvania and passed away in 2009. Ed was my dissertation supervisor and also the author of the McGraw-Hill textbook *Introduction to Transportation Engineering and Planning* (1978), which provided an example of a book with broad content that we could follow. I would also like to thank Sam Landsberger of Cal State Los Angeles and formerly of Cornell University who provided some early guidance in the direction of sustainability while I was an undergraduate at Cornell.

I wish to dedicate a "thank you!" to the many students with whom I interacted in the years leading up to the publication of this book. A special thank you is due to Scott Cloutier, Mike Hyland, Bingyan Huang, and Hyun Park for their role as production assistants on the project. In addition, students who participated in individual and team research projects or the role of TA or grader for relevant classes that contributed to the content of the book include Hamzah Al-Jefri, Jeffrey Bernstein, Kim Campbell, Alexandra Wai-Sung Cheng, Tyler Coatney, Ryan Cummiskey, Sandeep George, Dan Grew, Nicole Gumbs, Christina Hoerig, Bingyan Huang, Mike Hyland, Ben Kemper, Bhavna Kolakaluri, Tim Komsa, Richard Larin, Sunjeet Matharu, Happiness Munedzimwe, Rachel Philipson, Jason Ryu, Julie Schwartz, Tao Shi, Karl Smolenski, Kunrawee Tangmitpracha, Selin Un, Jun Wan, and Ema Yamamoto. They have all earned my grateful thanks.

Several chapters in the book benefitted from outside reviews. I would like to thank Eric Bruun of Aalto University for reviews of Chaps. 8 and 12, Susan Shaheen of the University of California at Berkeley for review of Chap. 9, and John DeCicco of the University of Michigan for reviews of earlier versions of Chaps. 14 and 15 that appeared in the textbook *Energy Systems Engineering*. Thanks also to Judy Eda and my wife Catherine for editorial feedback that improved the overall clarity of the writing. I would further like to thank colleagues in the School of Civil and Environmental Engineering, the College of Engineering, and elsewhere at Cornell for their encouragement during this process, and in particular Linda Nozick and Oliver Gao for their input. While all of this input is much appreciated, responsibility for any and all errors rests with the authors.

Heartfelt thanks are due to the team at McGraw-Hill that helped make this book possible, including Robert Argentieri, Steve Chapman, Bettina Faltermeier, David Fogarty, Larry Hager, Mike Penn, Richard Ruzycka, Lauren Sapira, Amy Stonebraker, Pamela Pelton, Stephen Smith, and Bridget Thoreson. They have worked tirelessly to answer questions and support the development of the book. Thanks also go out to the production team at Cenveo Publisher Services for a job well done, including Yashmita Hota, Ragini Pandey, Amy Rodriguez, and Vastavikta Sharma.

I would like to thank Richard Bausch for his essay "Ten Commandments for writing" and in particular commandments #5 "be patient" and #6 "be willing"; although this piece was probably not written with an engineering textbook author in mind, the advice proved very valuable over the years of developing the content, as it hung on the wall next to my desk. Thanks to Ludwig van Beethoven for his *Piano Concerto in C Minor*, which energized my writing again and again. Thanks also go to my neighbors at the Ecovillage at Ithaca cohousing community where I have lived for the past 11 years, who have inspired me to keep working for the environment. An additional thank you goes to the yoga community at Cornell and in Ithaca, and to the teachers and students who together have created a space to restore mind and body, where I was able to "keep calm and carry on," and also keep from throwing my back out.

Lastly, to anyone else who should be thanked for their contribution or support but who I may have overlooked, please know that the oversight was unintentional and that you are thanked as well.

From Lars Angenent: I would have not been able to prepare the materials for this book without the courses that I have taught over the years to many groups of students. Thanks, undergraduate students, for helping me to develop as a teacher. I thank academic colleagues, who have introduced me to other energy systems through collaborations. I want to especially thank Prof. Adrianus van Haandel, who invited me to tour an ethanol plant in Brazil, and Johannes Lehmann. I am thankful to my graduate students Catherine Spirito, Joe Usack, Devin Doud, Dr. Michaela TerAvest, and Dr. Elliot Friedman for their assistance in developing problem sets, figures, and text.

Note to Instructors

For instructors who adopt this book for use as a course textbook, the authors provide a suite of instructor materials, including a sample course syllabus, solutions to end-of-chapter exercises, sample exam problems with solutions, and PowerPoint slides. These materials are available for download from the book's Web page at www.mhprofessional.com/STSE; please contact the publisher for the username and password.

Please note that the exercises appearing at the end of the chapters are only a subset of the complete suite of exercises and exam problems that are available for download. We warmly encourage instructors to access these materials and download the full range of materials. Instructors should feel free to copy the content and then modify and adapt the problems to suit their needs, including elaboration on the questions asked or updating of numerical figures to local and regional circumstances, or else to the most up-to-date values.

Sustainable Transportation Systems Engineering

UNIT 1

Motivations and Drivers

CHAPTER 1

Introduction

1-1 Overview

This chapter introduces the study of transportation systems by first looking from several different perspectives at the function and contribution of transportation in the economy and society. Next, we introduce the relationship between transportation and sustainability, and review the history of transportation systems from preindustrial times up to the present. We also discuss the ways in which transportation can contribute to the ecological, social, and economic goals of sustainable development, with a particular focus on the relationship between national wealth and access to transportation. After considering a "laundry list" of obstacles and challenges that stand in the way of fulfilling transportation's role in sustainability, we give an overview of the remaining topics in the book.

1-2 Introduction: Dimensions of Transportation

At the Massachusetts Division of Marine Fisheries in the United States, marine biologist Greg Skomal studies the Greenland shark, a strange, reclusive Arctic Ocean inhabitant that is adapted to the extremely cold water temperatures below the ice. Skomal and his support team create temporary encampments above the ice, then break through it to scuba dive in search of the shark. When they encounter a shark, they attach an acoustic telemetry device (also called a "pinger") to it so they can track its movements over time. Among other questions, Skomal and his colleagues are interested in the shark's feeding habits, since a changing Arctic environment may affect its ability to find food (Gellerman, 2012).[1]

The ability of this unusual shark to stay alive in such extreme conditions and still successfully hunt for prey is a fascinating story in itself. Equally fascinating from a transportation perspective is the way Skomal's research project uses the capabilities of the modern global transportation system. This system can carry people and supplies to and from the harshest environments on the planet: scorching deserts, icy mountains, or, in this case, from the East Coast of the United States to beneath the sea ice at the North Pole, several thousand miles away. Such journeys are by no means comfortable or free from the risk of injury or death, but thanks to specialized vehicles of land, sea, and air, and the systems that support them, travel to these distant and forbidding regions is no longer considered foolhardy.

[1]As reported to weekly environmental newsmagazine *Living on Earth* from Public Radio International, on a show titled "Greenland sharks: the apex predator," which aired August 10, 2012.

This vignette about a remote project points to the first of several *dimensions* of transportation, namely that it is a vast network stretching around the globe, reaching and connecting communities as diverse as the earth itself. Part of the influence of this transportation system stems from its geographic reach, which by the twenty-first century has become nearly universal. Almost all cities of any size, rich and poor alike, are served by mechanized vehicles using internal combustion engines: buses, taxis, private cars, and two-wheeled vehicles. Larger, wealthier cities also use electrified transportation, including streetcars, commuter rail and light rail systems, and subways, so as to reduce emissions and cut congestion. For intercity and international travel, a passenger can start at almost any airport with scheduled flights and reach any other airport in any other place in the world, no matter how remote, completing the journey within four or five "legs" or less (each leg being a single flight in an itinerary made up of several flights to intermediate airports). Rural inhabitants in the poorest countries may use only the simplest forms of transportation on a day-to-day basis (bicycles or their own two legs), but even they have occasional contact with mechanized transport in the form of four-wheel-drive vehicles, helicopters, or aircraft making drops (e.g., deliveries of food and relief supplies to a drought-stricken region). A very small fraction of indigenous tribes in remote rain forests may have no concept whatsoever of mechanized transportation, but researchers can reach even these regions, thanks to the modern transportation system.

A second dimension of transportation is its role as a major human endeavor, ranked with agriculture, manufacturing, and the various commercial activities that make up the service sector. When research organizations such as the Oak Ridge National Laboratories of the U.S. Department of Energy divide up the entire U.S. annual energy budget into major categories, transportation constitutes one of them, alongside industrial, commercial, and residential energy use. For a typical resident of an industrialized country, transportation is a major expense, comparable to housing, food, and health care. In many U.S. households, combined expenditures on owning one or more vehicles may constitute the second-largest expense after payments for housing. In some ways, the influence of transportation extends beyond a single sector to being a dominating factor in the modern way of life and human culture as a whole. When MIT researchers James Womack, Daniel Roos, and Daniel Jones published the findings of their acclaimed research on the automobile industry in a book for the popular press, they titled it *The Machine That Changed the World* (Womack et al., 1990), considering the transformative effect the automobile has had on the way the industrial, residential, and commercial sectors are organized. In short, transportation is not only a vital activity in and of itself but also has profoundly impacted all the other major areas of endeavor.

A third dimension of transportation is its capacity as an employer of workers from all backgrounds and walks of life. In the first instance, the transportation industry employs a vast cadre of service personnel to keep vehicles and systems operating, including drivers, conductors, railroad engineers, and network supervisors and administrators. Design, construction, and maintenance of both infrastructure and vehicles employ many more professionals. Another large group of engineers, economists, sociologists, and other professionals in related disciplines analyze and manage the transportation network to address problems and improve performance.

A fourth and final dimension of transportation is its role in consuming resources and generating ecological impact. In the United States alone, the need for transportation is responsible for the existence of tens of thousands of miles of interstate expressways and millions of additional miles of paved highways, rural feeder roads, and city streets and boulevards, along with more than 100,000 miles of railroads. Although turnover varies with the health of the economy, in a typical year some 10 million cars and light trucks (along with additional large road vehicles and vehicles for other modes) are replaced with newly manufactured ones as they reach the end of their life cycle. Over 200 billion gallons of gasoline, diesel, and aviation fuel are consumed each year to power the national fleet of road vehicles, trains, boats, and aircraft, and electrified transportation consumes an additional quantity of energy on the order of hundreds of trillions of British thermal units (Btu). Oak Ridge National Labs (Davis et al., 2012) reports a figure of 313 trillion Btu for 2009 consumed by a combination of pipelines, railroads, and road vehicles. (Note: See App. A for common conversions, such as between metric and U.S. standard units.) This level of energy consumption leads to the release of nearly 2 gigatonnes (Gt) of CO_2 into the atmosphere, out of approximately 5.5 Gt total from the United States and 32 Gt total from the world economy (2009 figures).[2] Transportation also contributes to the failure to meet federally mandated air quality standards in a large fraction of U.S. counties; it endangers wildlife; and it creates noise, vibration, and safety hazards for numerous communities.

1-3 Transportation and Sustainability: Historical and Contemporary Aspects

Almost since the dawn of the earliest forms of conveyance, the classical goals of transportation have been to find some balance between performance and affordability. The goal of performance includes several aspects:

- *Safety or security*: The chosen mode of transportation should be able to complete the journey and deliver passengers or goods safely to their destination.
- *Comfort or lack of damage*: In the case of passenger transportation, the traveler should not be subject to excessive nuisance or discomfort. For freight, the goods carried should not be physically harmed by the journey.
- *Speed and reliability:* The mode of transportation should be able to complete the journey within an acceptable time and arrive acceptably close to the predicted hour or date. Customers of the transportation system require both adequate speed and reliability: a transportation service may be rapid, but if its arrival time is unpredictable, it creates a nuisance for the customer. By the same token, a transportation service may be relatively slow in terms of miles per hour, but if its arrival time is predictable to the customer, it can be entirely acceptable.

Travelers and shippers of goods have throughout history endured a great deal of unpredictability on some of these points. On transatlantic crossings in the colonial era, for example, ships faced great risk of sinking. Passengers set out from one side of the ocean or the other simply hoping for the best, accepting that they would arrive whenever they arrived, if at all.

Against the desire for performance in its various dimensions, the human traveler (or shipper of freight) has always had to weigh the ability to pay. In ancient times,

---

[2]1 gigatonne = 1×10^9 tonnes. The spelling "tonne" denotes a metric ton, equivalent to 1,000 kg.

travelers might have desired the protection of an armed escort when traveling over remote pathways. If this was not affordable, then they would forgo the expense and instead risk being robbed. Today, a traveler might desire the extra space of a first-class car on a passenger train or first-class seat in an aircraft but forgo these amenities to hold down cost. Collective decision-making about transportation is similar: large cities in wealthy countries may be able to afford the investment needed for a high-speed urban rail system (e.g., subway, underground, or elevated "people movers"), whereas large cities in poor countries must instead rely on street transit, with its tendency toward overcrowding and congestion-related delays.

The advent of a new goal marks the transition from a focus on safety, speed, and reliability that persisted throughout history to the urgent concern of today: sustainability. As long as the number of people living on the planet was relatively small and the per-capita ecological impact was low, both providers and users of transportation services did not need to think about their impact on ecosystems. For most of human history, the world's population was less than 1 billion, or less than one-seventh the number today; this threshold was surpassed only in the year 1800. As population grew during the industrial revolution and beyond, modern transportation became much more widespread, and the effects of resource extraction and emissions into air, water, and soils became a more serious problem.

1-3-1 Definition and Interpretation of the Term "Sustainability"

When speaking of sustainability, we should consider its true meaning. The dictionary defines a process or activity as sustainable if it is "able to be maintained at a certain rate or level."[3] From an ecological perspective, the implication is that the process or activity can be maintained at a certain rate or level indefinitely without causing environmental harm or exhausting finite resources. In everyday life, however, many products, services, or practices are labeled "sustainable" even if they cannot truly be maintained indefinitely. This usage of the term has become acceptable in some cases because the practice in question has taken steps toward true sustainability and is seeking to make improvements in areas where there is a shortfall. Take *sustainable agriculture*, for example; in some locations, these practices are truly maintaining or even improving the health of the soil for the long term, thus fulfilling the definition of sustainability in one dimension. However, these operations rely on the fossil fuel industry, with its concomitant emissions of greenhouse gases (GHGs) to the atmosphere, for transportation, operating equipment, and/or plastic packaging, so on this point they fall short.

The use of the term *sustainability* in this way has at times led to controversy. Some people object to what is effectively a watering down of the term because the activities in question are not fully sustainable in all respects. For others, an activity that has taken some steps and adopted full sustainability as a goal does merit the use of the label. In any case, we must be aware of how the term is being used when first introduced to an activity, and study its performance in several dimensions, rather than categorically assuming that the activity in question has fully solved all sustainability-related problems pertaining to it.

---

[3]Oxford online dictionary, *www.oxforddictionaries.com*.

1-3-2 Historical Evolution of Transportation through Three Phases

The history of the development of transportation can be divided into three phases: preindustrial, industrial revolution, and contemporary. The preindustrial phase extends for thousands of years before the year 1800, nearly to the beginning of recorded history. In this earliest phase, one of the primary purposes of transportation was to advance human civilization from subsistence farming, relying on foods grown in the immediate surroundings of dwellings, to carrying grains and other goods over distances that made it possible for communities to survive at least in part on trade.

Trade during this time relied primarily on ships and boats of various types, since they were easier to construct in a size larger than any land vehicle available. Vessels could use wind, river currents, human power (oars), or animal power in the case of boats towed from land. To a lesser extent, animal-drawn carts were used to move people and goods over primitive roads. Towns and cities could then increase in population, since they might trade with several different regions for essential food. If one agricultural region did not fare well in a growing season, the town or city might trade with another. Even today, one of the characteristics of the most impoverished regions of the world is their inability to move foodstuffs in and out: if they have a shortfall in local production, they cannot easily bring in necessary foods using the usual channels of trade, nor can they raise themselves economically in times of excess harvest by selling the surplus to the outside, all for lack of adequate transportation.

Along with transportation of goods, another preindustrial application of transportation systems was the development of road networks for communications. The Roman Empire exemplified this: a network of roads paved with stones allowed messengers to move between different parts of the empire relatively efficiently, and companies of soldiers traveled between different regions to defend against hostile forces, especially along the perimeter of their empire. The expression "all roads lead to Rome" comes from this period. With similar intent, the Inca Empire of South America built a network of roads stretching from present-day Ecuador to Chile. Several empires in China developed advanced road networks; also, the Grand Canal, which connected Beijing in the north to Hangzhou in the south, was constructed in stages between approximately AD 600 and 800, and renovated in the 1400s.

Lastly, although preindustrial systems were modest in terms of their total throughput of passengers and goods, they were exemplary in one sense: they were powered entirely by renewable energy. Wind energy for sails, river flow brought about by the hydrological cycle and rainfall in high elevations, and biomass energy in the form of draft animals eating fodder are all examples of transportation energy sources that can be sustained indefinitely thanks to the ongoing availability of solar energy and natural ecosystems.

Transportation in the Age of Comparative Advantage and Industrial Revolution

The second phase of transportation history accompanied the industrial revolution and its many advances in technology, including those used in transportation systems. This was an age of scientific and political innovation. On the technological side, the steam engine developed by Newcomen and Watt in the eighteenth century set the stage for steam-powered propulsion. Early steam engines were fired either by wood or coal, but as time passed and its advantages became apparent, coal became the dominant energy source. On the political side, one of the most influential ideas that shaped the growth of transportation was the concept of *comparative advantage*, advanced by the Englishman

David Ricardo in his treatise *On the Principles of Political Economy and Taxation*, first published in 1817 (Ricardo, 1817). Ricardo proposed that each country or region was best served by focusing its efforts on products to which it was best suited, and trading surplus production with other regions to obtain the full range of goods needed to meet the needs of the population. The implementation of trade driven by comparative advantage assumed that the cost of moving goods was not so steep as to preclude the benefits of trading with other producers rather than seeking total regional self-sufficiency. At the same time, it provided a catalyst for the further modernization of transportation technologies and systems: the more advanced and efficient they became, the more barriers to trade would decline, and the more growing demand for trade would pay for additional investments in transportation.

Advances in transportation in Europe and North America began with the expansion of canal systems in the 1700s to facilitate inland movement of goods. Following the development of mobile steam engines that could be placed on wheels to pull railway trains, and steel rails of sufficient quality to support their weight, the age of canal building gave way to railway network construction. On the seas and oceans, steam propulsion-superseded sail power. At first, the fossil fuel of choice for propulsion in locomotives and ships was coal, since manufacturing techniques were not yet advanced enough to allow combustion inside the cylinders of engines. Around the turn of the twentieth century, this engineering challenge was overcome, and for reasons of greater efficiency, internal combustion engines on roads, rails, and waterways gradually replaced steam engines. Discoveries in surface transportation eventually led to the creation of propeller planes and jet aviation. Each new development answered both the call to overcome technical challenges and the promise of economic reward for new modes of transportation that could travel farther, faster, and safer.

Contemporary Transportation: Globalization in an Age of Resource and Ecological Constraints

The third phase of transportation development finds the industrial revolution continuing to spread to industrializing countries in Asia, Latin America, and Africa. Meanwhile, the transportation system taken as a global whole begins to confront limits to the planet's ability to provide sufficient resources and to absorb the ecological impact of this activity without suffering adverse consequences. In the industrialized countries of North America and Western Europe, as well as in Japan, between the 1960s and 1990s, networks of limited-access expressways, freeways, and toll roads are largely built out, with all major metropolitan areas either adjacent to or near nodes on these networks. To varying degrees, between the 1960s and the year 2010, many of these countries are expanding their high-speed rail (HSR) networks, to enable the rail mode to capture a larger share of the passenger transportation market. Higher prices for raw materials, more complex planning requirements, and increasingly scarce and costly land space combine to slow the growth of these countries' highway infrastructure, with current projects limited to completion of remaining key links in networks or strategic expansion of capacity (e.g., adding lanes to freeways) where it is merited. During this time, growth in population and demand for transportation outpace that of physical capacity of the infrastructure, so congestion grows.

Laying information technology networks over physical transportation infrastructure to increase efficiency and capacity has become a popular option, because IT is relatively inexpensive to deploy compared to the per-mile cost of building roads or rails. Aviation is in a similar phase, with major cities all operating domestic and international airports, and periodically renovating and expanding terminals and runways. Occasionally, all-new airports are built, such as Denver International Airport, which opened in 1995, replacing Stapleton Airport. Part of the slowdown in infrastructure expansion is due to the impact of stricter environmental regulations, and debate has emerged about whether this is appropriate. Some see this slowdown as desirable for ecological protection, since it means that fewer, more carefully planned projects will be built, whereas others see it as an unnecessary intrusion into the ability of business to solve transportation problems.

In the United States, for example, in recent years the slowdown in infrastructure expansion has begun to dampen the demand for growth in transportation. As shown in Fig. 1-1, from 1970 to 2005 demand for passenger-miles and ton-miles continued to grow faster than population but not as fast as total GDP. Also, between 2005 and 2008 transportation demand stopped growing, at least temporarily. The high pace of GDP growth compared to transportation demand suggests that the economy is looking to add value through activities that are not as transportation-intensive.

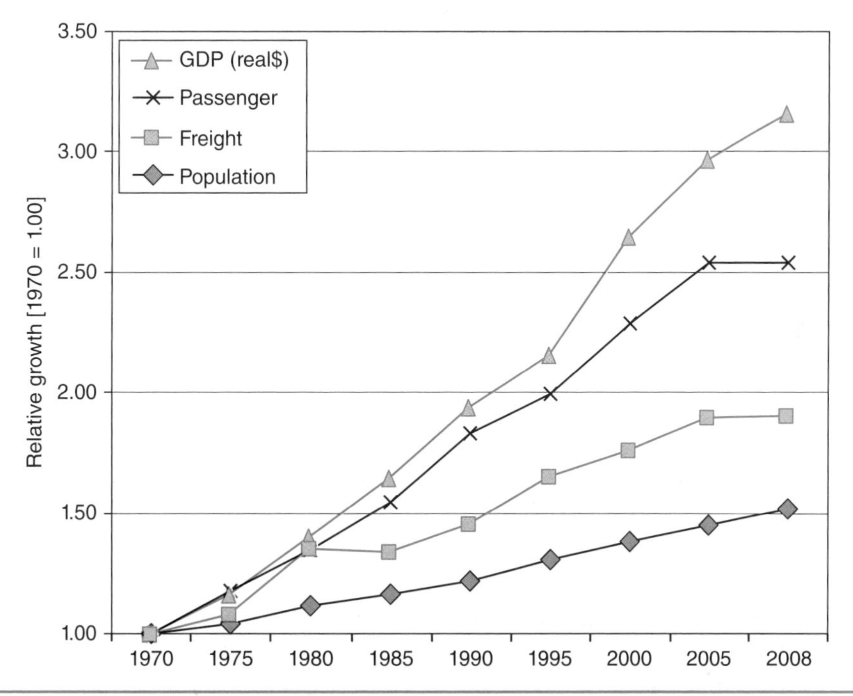

FIGURE 1-1 Relative growth from 1970 to 2008 of U.S. freight ton-miles, passenger-miles (p-mi), economic activity, and population, indexed to 1970 = 1.00.

Source: U.S. Bureau of the Census (2009), for population and GDP; own calculations based on Bureau of Transportation Statistics and Eno Foundation for freight and passenger transportation.

Notes: In 1970, population was 204 million, freight activity was 2,206 billion ton-miles, passenger activity was 2,196 billion passenger-miles, and GDP was $3.7 trillion (all GDP values in 2000 dollars).

The situation in the industrialized countries, represented here by the U.S. data in Fig. 1-1, provides a striking contrast to that of the industrializing countries, where transportation infrastructure and vehicles that accompany it are expanding at a robust rate. These countries, including high-growth economies such as China, India, and Brazil, seek the benefits of modern transportation systems for their citizens, including access to private automobiles, modern urban rapid transit systems, or the rapid transportation available from HSR. These countries also desire the economic opportunity that comes from transportation modernization: these changes create new markets for infrastructure development as well as vehicle sales for domestic companies. Major vendors of infrastructure and vehicles from the industrialized countries are eager to join in, ranging from the major car makers of North America, Europe, and Japan to aircraft makers such as Boeing and Airbus.

Several developments in the Chinese transportation market epitomize these changes. First was the rapid pace of modernization of rail technology: as recently as the early 1990s, steam locomotives were in ordinary commercial service on the Chinese railways. By 2004, these had all been replaced, and China leapfrogged past peer countries to become the first country in the world to put into service a magnetically levitated, or Maglev, train, between the Shanghai city center and airport (Fig. 1-2). Just 17 years separate Fig. 1-2(*a*) and Fig. 1-2(*b*). Although the adoption of modern technology

(*a*)

Figure 1-2 An example of rapid technological transformation in China: (*a*) Steam locomotive in regular service, Guangdong Province, southern China, 1989; (*b*) Maglev train in Shanghai, 2006. *Source:* Photograph (*b*): Jian Shuo Wang, *wangjianshuo.com*. Reprinted with permission.

(b)

FIGURE 1-2 (Continued)

has generally increased transportation intensity in China, it has also increased efficiency in some cases, as happened with the phasing out of inefficient steam locomotives.

Secondly, in 2009, China surpassed the United States to become the world's largest automobile market, with 13.6 million units sold in that year, compared to 10.2 million units in the United States. As an example of the interest of industrialized countries in these developing markets, in that year, the largest market in the world for the U.S. make Buick was not the United States but China.

As the industrializing countries have adopted modern transportation systems similar to those in the industrialized countries, they have not been able to avoid many of the same social and ecological impacts that already challenge the latter. As countries such as China and India have adopted the construction of limited access highways and large arterial streets around cities, along with increased sales of private automobiles, their cities have come to suffer similar levels of congestion with accompanying noise and air pollution. For those members of the middle- and upper classes that can afford to own these vehicles, time spent driving or number of miles traveled has grown substantially as well. Thus many of the solutions to transportation problems discussed in this book apply as well in the industrializing countries as they do in Europe, North America, or Japan.

1-4 Transportation in the Context of Sustainable Development

At present one of the best-known concepts that considers multiple goals for the various economic activities that underpin human society, among them the management of transportation systems, is *sustainable development*. The sustainable development movement is the result of efforts by the World Commission on Environment and Development (WCED) in the 1980s to protect the environment and at the same time eradicate poverty (World Commission on Environment and Development, 1987). The report that resulted from this work is often referred to as the Brundtland report, after its chair, Norwegian prime minister and diplomat Gro Harlem Brundtland. According to the report:

> Sustainable development is development that meets the needs of the present without compromising the ability of future generations to meet their own needs.
>
> (Page 43)

Seen by this standard, the transportation sector, along with the industrial, commercial, and residential sectors, runs the risk of degrading the ability of future generations to meet their needs. Therefore, development should proceed carefully. Both the extraction of resources and the disposal of infrastructure, such as transportation infrastructure along with the vehicles and products it carries, have the potential to harm the ecological health of the planet. The human population is included in this ecological health because human beings are ultimately biological organisms, subject to the same needs for safe air, clean water, and adequate nutrition. Furthermore, while some resources such as solar energy are in practical terms inexhaustible over the time frame that society must consider in its planning, other resources such as fossil fuels are nonrenewable and should therefore be treated with special care. The consideration of resource extraction, end-of-life disposal, and non-renewability of certain resources all have an impact on the development of transportation systems.

The Brundtland report has influenced the way the community of nations approaches environmental problems in the years since its publication. Many of the goals considered in its pages have parallels with the goals for sustainable transportation: satisfactory performance to meet human needs, economic viability, and environmental protection. Three of the report's goals stand out as particularly important:

1. *Protection of the environment:* The Brundtland report recognizes that the growing world population and increased resource and energy consumption was adversely affecting the natural environment in a number of ways that needed to be addressed by the community of nations.
2. *The right of poor countries to improve the well-being of their citizens:* This goal is implied by the words "to meet present needs" in the quote above. The report recognized that all humans have certain basic needs, and that it is the poor whose needs are most often not being met. The report then outlined a number of concrete steps for addressing poverty and the quality of life in poor countries.
3. *The rights of future generations:* According to the report, not only do all humans currently living have the right to a basic quality of life, the future generations of humanity also have the right to meet their basic needs. Therefore, a truly sustainable solution to the challenge of environmentally responsible economic development cannot come at the expense of our descendants.

These three components are sometimes given the names *environment, equity,* and *intertemporality.* The effect of the sustainable development movement was to raise a new level of awareness toward environmentalism. Advocates of environmental protection, especially in the wealthy countries, recognized that the environment could not be protected at the cost of human suffering in poor countries. These countries were guaranteed some level of access to resources and technology, including transportation technology, which could help them to improve quality of life and economic opportunity, even if this technology generated some undesirable side effects. It was agreed that the rich countries would not be allowed to "kick the ladder down behind themselves"; that is, having used polluting transportation technologies to develop their own quality of life, they could not deny other countries access to them. On the other hand, the Brundtland report sees both poor and rich countries as having a responsibility to protect the environment, so the poor countries do not have unfettered access to polluting technologies. The report does not resolve the question of what is the optimal balance between protecting the environment and lifting poor countries out of poverty, but it does assert that a balance must be struck between the two, and this assertion in itself was an advance.

The rise of the sustainable development movement has boosted efforts to protect and enhance the natural environment by invoking a specific objective, namely, that the environment should be maintained in a good enough condition to allow future generations to thrive on planet Earth. National governments were encouraged to continue to monitor present-time measures of environmental quality, such as the concentration of pollutants in air or water, but also to look beyond these efforts in order to consider long-term processes whose consequences might not be apparent within the time frame of a few months or years, but might lead to a drastic outcome over the long term. Indeed, many of the possible consequences of increased concentration of CO_2 in the atmosphere are precisely this type of long-term, profound effects. These efforts to limit long-term consequences affect the transportation sector as well as the others.

Since its inception in the 1980s, many people have come to divide the term sustainable development into three "dimensions" of sustainability:

1. *Environmental sustainability:* Managing the effects of human activities so that they do not permanently harm the natural environment.
2. *Economic sustainability:* Managing the financial transactions associated with human activities so that they can be sustained over the long term without incurring unacceptable human hardship.
3. *Social/cultural sustainability:* Allowing human activity to proceed in such a way that social relationships between people and the many different cultures around the world are not adversely affected or irreversibly degraded.

Some people use variations on this system of dimensions, for example, by modifying the names of the dimensions, or splitting social and cultural sustainability into two separate dimensions. In all cases, the intent of the multidimensional approach to sustainable development is that in order for a technology, business, community, or nations to be truly sustainable, they must succeed in all three dimensions. For example, applied to transportation, this requirement means that a transportation system that eliminates all types of environmental impacts but requires large subsidies from the government to maintain its cash flow is not sustainable, because there is no guarantee that the payments can be maintained indefinitely into the future. Similarly, a program that converts a country to completely pollution-free transportation by confiscating all private assets, imposing martial law, and banning all types of political expression might succeed regarding environmental sustainability, but clearly fails regarding social sustainability (!).

The business world has responded to the sustainable development movement by creating the *triple bottom line* for corporate strategy. In the traditional corporation, the single objective has been the financial *bottom line* of generating profits that can be returned to investors. The triple bottom line concept keeps this economic objective, and adds an environmental objective of contributing to ecological recovery and enhancement, as well as a social objective of good corporate citizenship with regard to employees and the community. Thus the triple bottom line of economics, environment, and society is analogous to the three dimensions of sustainable development. A growing number of businesses have adopted the triple bottom line as part of their core practice.

Transportation affects all three dimensions of sustainability, from both a government and enterprise perspective. Businesses are interested in keeping their transportation costs for both products and employees at a level that does not place an excessive financial burden and at the same governments wish to avoid both excessive cost for maintaining transportation infrastructure and excessive time or fuel loss due to congestion for the traveling public. Individual citizens push both government and businesses to reduce the environmental burden of transportation, for example on air quality or climate change. The negative effect of noise and other nuisances from transportation activities on ecological sustainability are of concern from a social sustainability perspective as well.

1-4-1 Relationship between Prosperity and Access to Modern Transportation

It is a premise of modern, technologically-based societies that the development of modern mechanized transportation both creates and reflects prosperity, assuming that protection of the environment and of social well-being is maintained. Modern transportation networks give businesses access to new markets for their products and services, and at the same time give individuals access to new employment opportunities and other amenities. As nations grow wealthy, they can afford to spend more on transportation, such as more advanced modes including high-speed mass transit, high-speed rail, or limited-access highways. These expenditures allow both individuals and private enterprise to increase their level of economic activity, which can support further expansion of the transportation network. This cause-and-effect relationship assumes that transportation development is conducted in a responsible way that does not eventually end up undercutting economic activity, for example, by creating social or ecological problems.

The relationship between transportation and economic output can be observed in national statistics for personal wealth and transportation use for different countries. To this end, the transport intensity and per-capita income of a group of countries, including Australia, Canada, China, Germany, India, Japan, and the United States, is compared in Fig. 1-3. Data on economic output and population are obtained from the World Trade Organization and the United Nations, respectively. Data on transportation volume are compiled by the departments of transportation of the various national governments, and in this case gathered by NationMaster (Nation Master, 2012), a data-tracking Web site. Because consistent time series data for transportation volume across all modes for these seven countries was not available, passenger transportation by the road mode was used as an approximation of total transportation volume. This substitution is reasonable,

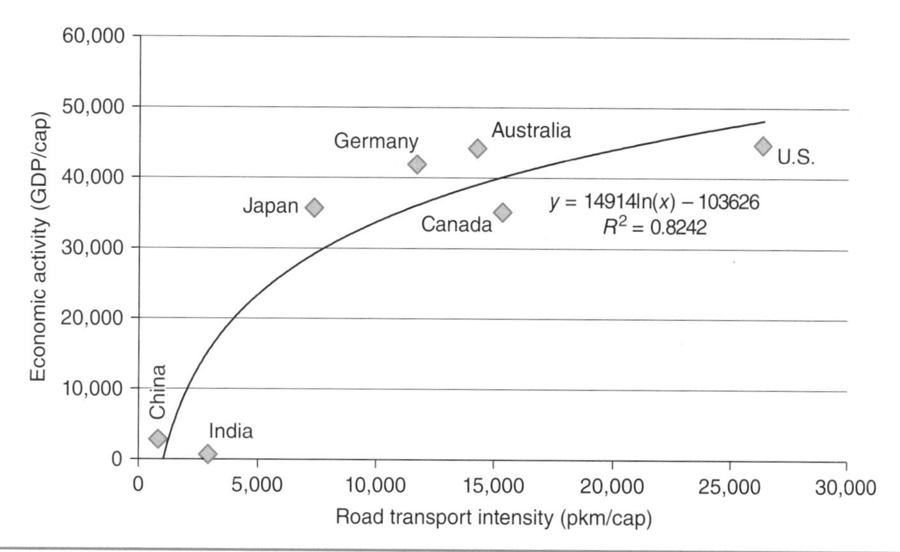

Figure 1-3 Scatter chart of passenger-kilometers per capita versus GDP per capita for seven representative countries, with best-fit logarithmic curve.

Sources: World Trade Organization (2012), for GDP data; United Nations (2012), for population data; NationMaster (2012), for transportation volume data.

since the road mode (including private cars, taxis, buses, and motorcycles) comprises the majority or most of the passenger transportation volume in middle-income or wealthy countries.[4] Data are obtained for countries for years from 2002 to 2007, depending on when they were made available. The countries shown represent approximately half of the world's population and two-thirds of world GDP.[5]

The scatter plot in Fig. 1-3 shows that the difference between middle-income countries like China or India and the wealthy countries is one of both lower transportation and economic activity per person. The wealthy countries vary considerably in their per-capita transportation volume, from Japan at 7,400 pkm/capita (passenger-kilometers per capita) to the United States at 26,400 pkm/person. In general, these countries can be divided into two groups, with higher-population density countries such as Japan and Germany having lower per-capita transportation volumes, while low-density countries such as the United States, Canada, and Australia have higher volumes. Their level of economic activity, however, lies in a fairly close range, between $35,200/person for Canada to $44,800/person for the United States. Thus the anticipated trajectory for China and India is that they will "climb up" the curve shown into the band of countries with mature economies, with economic output on the order of $40,000/person, and road transport capacity in a range from 7,500 to 25,000 km/person (with some additional capacity for rail and air passenger transport, although the scatter chart does not

[4]Among the seven countries, one exception is China, where road and rail passenger volumes measured in passenger-kilometers are of a similar size.

[5]GDP/capita as a measure of national development has been criticized for overemphasis on economic output at the expense of other potential measures such as life expectancy or educational levels. See Chap. 3 ("Systems tools") for a discussion of a competing measure called Human Development Index, or HDI, which incorporates life expectancy and education as well as per-capita wealth.

include these volumes). Given the relatively high population density of China and India, especially in the urban areas, it might also be expected that their eventual transportation intensity value would fall closer to that of Germany or Japan than of the other three countries. In any case, according to this vision of the future all seven countries would have in common characteristics of sufficiently high economic output and access to transportation capacity, with stable transportation energy supplies and concerns about environmental impact being met.

1-4-2 Discussion: Contrasting Mainstream and Deep Ecologic Perspectives

Underlying the previous discussion, economic wealth and access to transportation is a fundamental assumption that the model presented by the wealthy countries (the United States, Japan, etc.) is the right one, and that therefore they should maintain current levels, and other countries, notably China and India, should emulate that model. In other words, a fundamental premise for addressing world poverty is the development of modern transportation systems.

Figure 1-3 suggests that as poor countries increase their access to modern transportation their quality of life should improve as well. By this logic, the wealthy countries of the world should not reduce their per-capita transportation activity, since this would imply a loss of quality of life. Instead, these countries should neutralize the negative environmental effects from transportation systems, for example, impact from resource extraction to build vehicles, impact from CO_2 emissions on climate change, and so on. As long as this is done, the environment is protected at whatever level of per-capita or total transportation activity emerges, and the mix of activities in which their citizens engage (manufacturing, commerce, retailing, tourism, mass media entertainment, and the like) is no longer a concern. The technologies and systems thus developed can be made available to other countries so that they too can have greater access to transportation activity and achieve a high quality of life in a sustainable way.

This mainstream perspective has been widely adopted by economists, engineers, and political leaders, but it is not the only one. Another approach is to fundamentally rethink the activities in which humans engage, including mechanized transportation along with industrial mass production, the construction and inhabitation of modern residential and commercial structures, and the like. Instead of allowing the mix of activities to continue in its present form, activities and lifestyles are chosen to reduce energy and raw material requirements as well as waste outputs so that the effect on the environment is reduced to a sustainable level. This approach is one of the fundamental tools of *deep ecology*, an alternative philosophy for society's relationship with the natural world that has emerged in recent decades. By implication, it criticizes the mainstream approach for being one of shallow ecology, which only addresses the ecological crisis at a superficial level by making technical changes to resource extraction and by-product disposal, but does not address the core problem of excessive interference in the natural systems of the world.

For example, in the case of transportation systems, instead of generating per capita passenger-miles or ton-miles of transportation at the same rate but in a more environment-friendly way, the deep ecologist seeks to change society's choice of built environment, land use patterns, and trip destinations so that much less transportation is required. A primary target of this effort is *consumer culture*, that is, the purchase especially by middle- and upper-class citizens of resource-intensive goods and services that lead to large homes located far from city centers, trips to numerous retail outlets for various food products and durable goods, vacations in distant locations, and other choices of this nature. These activities are beyond what is necessary for basic existence, but are thought to improve quality of life. Deep ecologists seek to reorient purchases and consumption away from consumer culture toward focusing on *essential* goods and services. In the long term, the philosophy of deep ecology also advocates gradually reducing the total human population so as to reduce demand for resource-consuming activities, among them transportation.

As an example, consider the relative passenger transportation intensity of the United States, the United Kingdom, and India, as shown in Fig. 1-4. In this figure, intensity is calculated using passenger-kilometers across road, rail, and air modes, not just road modes, unlike in Fig. 1-3. The United States, as a relatively transportation-intensive country, has risen from 25,000 pkm/year in 1990 and reached a plateau in the range of 29,000 to 30,000pkm/year in the period 2004 to 2008. For the United Kingdom, being more densely populated and having more densely inhabited cities, a plateau has occurred in the same time period around 13,000 km/year. For India the situation is different: having started at 1,492 pkm/year and been at or below 2,500 pkm/year up until the year 2000, industrialization has brought opportunities for economic growth, and the figure has reached 5,980 pkm/year in 2008.

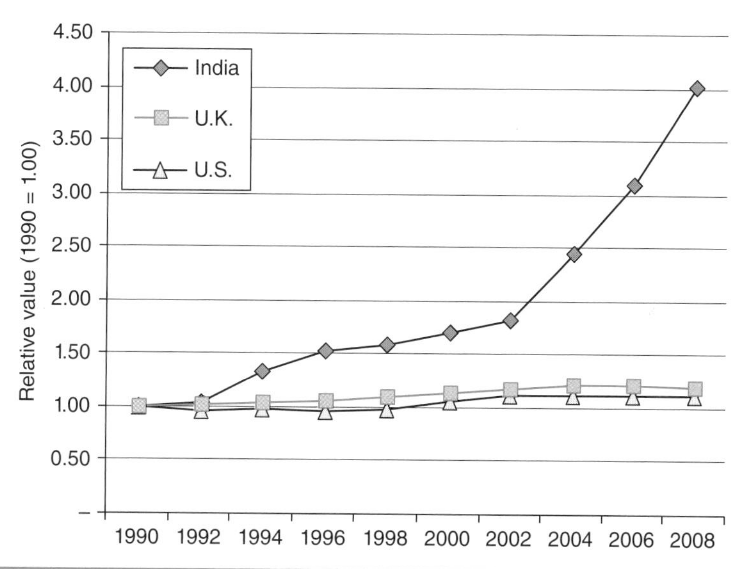

FIGURE 1-4 Relative value of passenger-kilometers per capita for India, the U.K., and the United States, 1990 to 2008, indexed to a value of 1990 = 1.00.

Notes: India, U.K., and the United States values for 1990 are 1,492, 12,063, and 24,739 pkm/capita, respectively. U.K. data point for 2008 is for 2007.

Sources: Ministry of Road Transport and Highways (2009), and Ministry of Railways (2009), for India; Transport Statistics Great Britain, for U.K. (Her Majesty's Stationery Office, 2009); U.S. Bureau of Transportation Statistics (2010), for the United States.[6]

[6]Acknowledgment: Transportation data in India were compiled by Bhavna Kolakaluri, student in the Master of Transportation Systems Engineering Program at Cornell University, 2009–2010. Her contribution is gratefully acknowledged.

The question for the debate between mainstream and deep ecologic perspectives in this case is the level at which any of these three countries should stabilize. From a mainstream perspective, the United States should be maintained at a level on the order of 30,000 pkm/year; this level might be viewed as too high for the United Kingdom or India, because of their high population density and older urban form, but a level of 13,000 pkm/year for the United Kingdom might be seen as appropriate, and India could also increase its transportation intensity to this level. From a deep ecologic perspective, however, India may already be at a desirable level at 6,000 pkm/year, and the United Kingdom should reduce transportation intensity so that it achieves this level as well. The United States might aspire to this level too, or to a level of 10,000 pkm/year due to the large distances involved in travel between major U.S. cities. Even this amount would represent a major reduction of around two-thirds of transportation intensity in the United States, while reduction to 6,000 pkm/year would require a cut of more than 50% for the United Kingdom.

The deep ecologic alternative incorporates some attractive advantages but also some significant challenges. The reduction of resource consumption obtained by following the deep ecological path can yield real ecological benefits. With less consumption of transportation services as well as other goods and services, the rate of resources being extracted, energy being consumed, CO_2 being emitted to the atmosphere, and so on, is reduced. In turn, reducing the throughput of resources, from extraction to the management of the resulting by-products, can simplify the challenge of reconciling human systems with the environment. This problem is made more difficult in the mainstream case, where the throughput rate is higher.

However, in the short to medium term, the deep ecologic approach would profoundly challenge society, because many of the activities in a modern service economy that appear not to be *essential* from a deep ecologic perspective are nevertheless vital to the economic well-being of individuals who earn their living by providing them. Transportation activity in a transportation-intensive country like the United States generates great numbers of jobs in areas such as vehicle repairs, sales of fuel and other supplies to travelers, commercial travel services such as airlines or train lines, activities related to the tourism industry, and numerous other pursuits. If demand for these services disappeared suddenly, it might reduce resource consumption and emissions but also create economic hardship for many people. Therefore, the adoption of a deep ecologic approach to solving the environmental problem would require a transformation of the way in which economies in wealthy countries provide work for their citizens, which could only be carried out over a period measured in years or decades.

The number of persons fully committed to a deep ecologic path is relatively small at the present time, especially compared to the numbers adhering to the more mainstream idea of continued high levels of economic activity with the intention of upholding adequate environmental safeguards. However, a related approach to personal behavior of giving environmental considerations priority in lifestyle and consumption decisions has started to influence the choices of some individual consumers. Understanding the environmental impact of choices concerning size of vehicle to purchase, choice of travel mode, destination for discretionary trips, and other transportation decisions, as well as other consumer decisions, these consumers choose to forego purchases that are economically available to them so as to reduce their personal environmental footprint. This shift has in turn led to some curbing of growth in transportation activity in the industrialized countries among certain segments of society, for example, those who identify themselves as *environmentalists*. It is likely that its influence will grow in the future, as will its impact on transportation demand.

1-5 An Overview of Challenges for Sustainable Transportation

While governments and businesses, and to some extent individuals, may have adopted the goals of sustainability, including sustainable transportation, the current status of the transportation system is in many instances far from having achieved these goals. There are in fact few aspects of daily life in both modern and developing countries that present so many multifaceted challenges: perhaps just education, health care, and access to sustainable energy. The following list of sustainable transportation systems goals includes economic, social, and ecological challenges, on both a local and global scale. Since the majority of the human population, and the great majority of economic activity, exists in and around urban areas, many of the problems in this list concern urban transportation problems. Other problems occur at an intercity or international level, for example, many of the problems dealing with freight transportation, since most freight activity occurs at an intercity level rather than within a single urban area.

1. *Sustainable maintenance of infrastructure:* For a transportation system to be economically sustainable, the cost of maintenance must be in line with the available funding that is derived from various revenue streams. This has not been the case in the United States for many years, and as a result, deficiencies in the overall quality of infrastructure have been growing. The American Society of Civil Engineers (ASCE), keeps a report card on the status of U.S. infrastructure, and in 2009 gave it an overall grade of "D" on a scale from A to F, with an estimated repair bill over 5 years of $2.2 trillion dollars to make essential repairs, upgrades, and improvements so that all infrastructure would be either in good or excellent condition (American Society of Civil Engineers, 2009).
2. *Providing access to amenities and overcoming congestion:* One important goal for transportation systems, especially in urban areas, is for users to be able reach the amenities they need: workplace, education, retail outlets, leisure, and so on. When the system does not function well and this access is not provided adequately, the quality of life in the region suffers. Closely related to providing access is the challenge of excessive congestion on the transportation network. When system users are slowed or halted in their travels due to congested conditions, both their productivity as contributing members of the economy and quality of life as citizens suffer.
3. *Equity and access to economic opportunity:* For economically deprived communities in both urban and rural settings, lack of mobility (point #2) is often a barrier to jobs and education that can lift these communities out of poverty. The particular need of this challenge goes beyond making the transportation network more efficient: often the barrier is access, from affordable motor vehicles to affordable public transportation. For other members of society such as the young, old, and those with disabilities, there may be the need for alternatives not requiring drivers' licenses or the physical ability to use the traditional public transportation network.
4. *Local quality of life impacts (noise, vibration, water pollution, habitat impact):* Transportation systems may have a negative impact on their immediate surroundings, affecting both human and nonhuman life. Noise and vibrations from vehicles of various modes affect quality of life for both human and nonhuman residents. Runoff from vehicles, such as lubricants and other chemical fluids,

find their way into local waterways, causing harm. Transportation infrastructure may also divide natural habitat into sections that are too small to safely or comfortably support animal and bird species.

5. *Air quality impact:* Tailpipe emissions from motorized transportation modes can accumulate in the atmosphere, making air harmful to breathe. Great strides have been made in the past several decades to reduce tailpipe emissions, especially from light-duty vehicles, and today the contribution from the transportation sector to air-quality problems is much less than it was. Still, many urban areas in the United States do not comply with national air quality standards. In 2009, 41 states in the United States had one or more counties that counted some degree of air quality nonattainment for one or more air pollutants, according to the U.S. Environmental Protection Agency (USEPA). Further improvements can be made, especially in regard to older vehicles that may lack emissions control equipment, or for which the equipment is no longer working properly.
6. *Security challenges for the transportation system:* Various challenges having to do with protecting the system, managing elevated-risk transportation activities, and continued operation of the system under duress fall under this heading. Examples abound. Arguably the highest profile topic in this area is the prevention of malicious attack exemplified by the terrorist attacks of September 11, 2001. Management of transportation systems during extreme weather events such as floods and hurricanes is another important challenge, involving activities ranging from prepositioning of supplies and emergency responders to evacuation of residents from affected areas. A third area is the transportation of hazardous or radioactive waste through the transportation network.
7. *Upward pressure on energy use and adequacy of energy supply:* Both the United States and the rest of the world use a large amount of energy on an annual basis, compared to levels in the early stages of the industrial revolution, or even those of 20 or 30 years ago. The transportation sector is one of the major energy users, and contributes to the growing demand for energy, which will pose challenges to be met in the future. Along with the general problem of needing to maintain access to large amounts of energy to support the transportation system is the particular problem of heavy reliance on nonrenewable petroleum. Not only is the resource finite but the peaking of world conventional crude oil may arrive quite soon [Fig. 1-5(*a*)]. The figure shows that even if the value of estimated ultimate recovery (EUR) is increased by 1 or 2 trillion barrels, the year of peak output might be postponed by only 15 or 20 years. Peaking of output will force a transition either to nontraditional petroleum sources (e.g., oil shale, tar sands) or to nonpetroleum sources (e.g., electricity from renewable, nuclear, or carbon-free coal or gas sources used to power electric vehicles or plug-in hybrid electric vehicles).
8. *Contribution of transportation to climate change:* Because transportation energy consumption is dominated by petroleum products (gasoline, diesel, and jet fuel) that release CO_2 to the atmosphere when combusted, it is a major contributor to climate change. Transportation along with other major fossil fuel consuming activities, chief among them electricity generation, are rapidly driving up world annual CO_2 emissions [Fig. 1-5(*b*)]. The challenge is either to

substitute low- or non-carbon fuels for traditional petroleum-based ones, or to improve energy efficiency per unit of transportation service delivered to reduce CO_2. It may also be possible to make offsetting carbon reductions in other sectors (industrial, residential, commercial) to compensate for emissions from transportation, but this option may encounter criticism since it fails to address the carbon-intensive nature of transportation at present.

Challenges and solutions discussed in the body of this book can be broadly divided into two groups. The first group covers Challenges 1-4 and contains elements more local or regional in nature. These solutions often involve changing the balance of travel modes or modifying transportation infrastructure, and are therefore a particular focus of urban passenger transportation. [More details about these challenges are detailed elsewhere in books dealing with sustainable transportation and urban planning, for example, Vuchic (1999), Schiller et al. (2010).] Challenges 7 and 8 are more

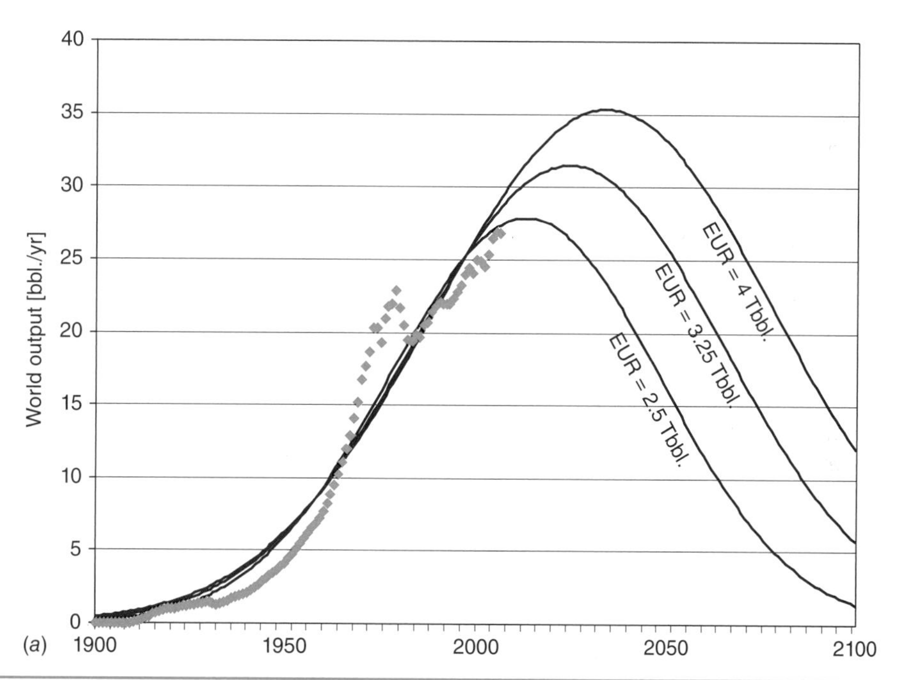

FIGURE 1-5 Twin global challenges from (*a*) peaking of petroleum production; and (*b*) rapid rise in annual CO_2 emissions.

Notes: Fig. 1-5(*a*) shows actual global oil production up to 2006, and possible pathways for peaking and subsequent decline of oil output depending on estimated ultimate recovery (EUR). Annual output is measured in billion barrels, and EUR is measured in trillion barrels (Tbbl). Representative values of 2.5 Tbbl, 3.25 Tbbl, or 4 Tbbl are shown. Fig. 1-5(*b*) shows the total emissions of carbon from combustion of fossil fuels for energy measured in million tonnes (i.e., metric tonnes) of carbon. The "Total" curve is the sum of oil, coal, and gas curves in the figure. Conversion: 1 tonne carbon = ~2.7 tonnes CO_2.

Sources: U.S. Energy Information Administration (2012), for oil production data; Carbon Dioxide Information and Analysis Center (2012) at Oak Ridge National Laboratories, for carbon emissions data.

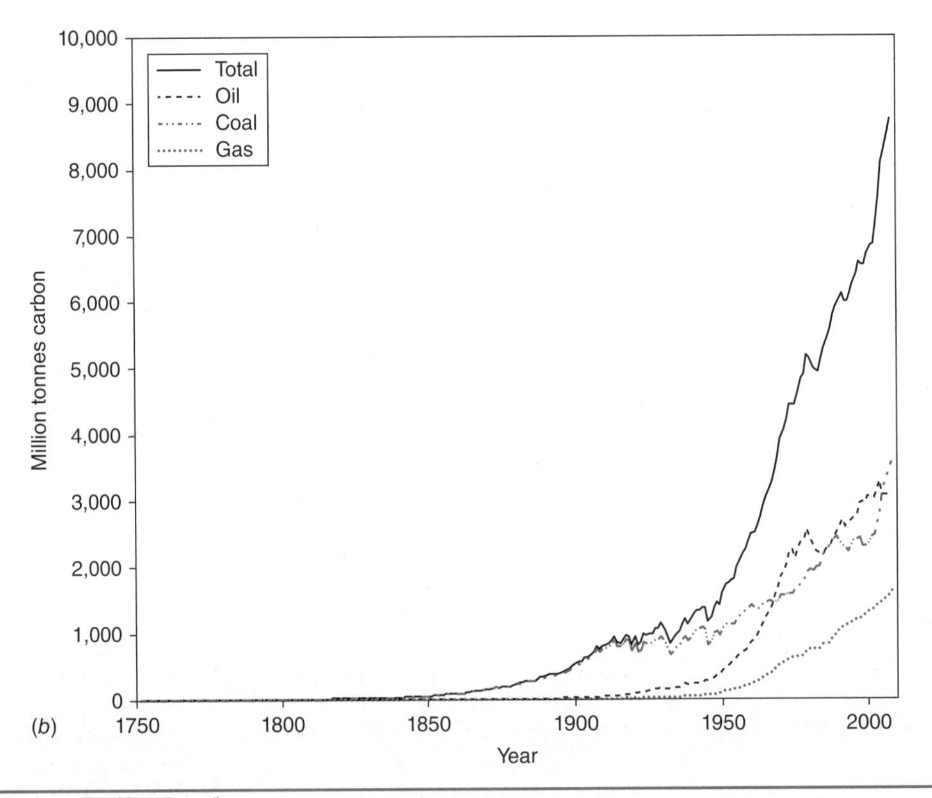

Figure 1-5 *(Continued)*

national and global in scope and focus on improvement in vehicle technology, modal share for eco-friendly modes, or development of cleaner energy sources to reduce pressures on the energy supply, improve air quality, or decrease GHG emissions. Passenger and freight activities of all types have role in meeting these challenges. Challenge 5 and 6 fall in the middle, since they have both global and regional components, and also depend on a mixture of technological and operational or behavioral remedies. Because Challenges 7 and 8 on energy security and climate change are particularly important in terms of their economic, political, and social implications at this time, they are given further treatment in Chap. 2.

1-5-1 A Scenario for Sustainable Transportation in the Twenty-First Century

The challenge of achieving sustainable transportation will require a commitment that is global in its scope because true sustainability can only be achieved when the benefit of having basic transportation needs met is extended to the entire global population. Pursuing this goal requires a commitment for the very long term as well. Therefore, as part of introducing the topic in this first chapter, it is useful to create a scenario of what this transition to sustainability might look like, that encompasses the entire globe and the duration of the twenty-first century.

We begin with several assumptions related to economic activity, population, and per-capita transportation demand:

1. Global population growth will slow and reach a plateau of around 9 billion people around the year 2060, and will remain approximately constant at that level from 2060 to 2100. This scenario is based on a United Nations midrange projection of population growth in the twenty-first century.
2. The global economy will function normally through the twenty-first century, encountering typical challenges that arise related to cycles of growth and recession, resource scarcity, frictions between nations, and effects of climate change, and navigate them with a reasonable degree of success. As a result, wealthy countries remain wealthy, and low- and middle-income nations grow in wealth.
3. Thanks to relative economic stability, wealthy countries maintain current rates of per-capita passenger and freight transportation intensity, and other countries have their levels of transportation intensity grow to resemble those of the wealthy countries.

The world's population is next divided into four groups as a basis for creating a whole-world scenario:

1. *High-income high-volume:* These wealthy countries are relatively transportation-intensive due to long distances of travel required and relatively low density of urban areas. They therefore have relatively high values of 25,000 pkm/person/year and 17,000 tkm/person/year for passenger and freight. Their total population is approximately 500 million, which remains constant for the duration of the twenty-first century.
2. *High-income low-volume:* This group of countries is also wealthy but relatively less transportation-intensive due to more compact country size and denser urban form. Per-capita transportation values are therefore lower, at 15,000 pkm/person/year and 8,000 tkm/person/year for passenger and freight. These countries have an estimated total population of 800 million, which remains constant for the duration of the twenty-first century.
3. *Middle-income:* This group includes the countries in the world where presently the most rapid economic growth is occurring, in BRIC has not been defined countries such as Brazil, India, and China. The population is 2.5 billion in 2000, and grows to 4.1 billion in 2050, where it remains until 2100. Transportation intensity is currently growing the most rapidly among these countries, so on the passenger side it increases from 5,000 pkm/person/year in 2000 to 15,000 in the year 2050, and remains at that level until 2100. On the freight side, it grows from 4,000 tkm/person/year in 2000 to 8,000 tkm/person/year in 2050 and again remains constant at that level after that until 2100.
4. *Low-income:* This group includes the remaining countries of the world that have the lowest per-capita income. The population is 2.2 billion in 2000 and reaches 3.6 billion in 2050, where it remains until 2100. Like the middle-income countries, transportation intensity is projected to grow over the twenty-first century, but more slowly and from a lower starting point. On the passenger side,

it increases from 2,000 pkm/person/year in 2000 to 15,000 in the year 2100, thus growing continually throughout the century before leveling out at the end. On the freight side, it grows from 2,000 tkm/person/year in 2000 to 8,000 tkm/person/year in 2100, again growing gradually and leveling out at the end of the century.

The results of the scenario development for passenger transportation are shown in Fig. 1-6. The dominant feature is the impact of the growth of both population and transportation intensity in the low- and middle-income countries. As the goals of sustainable development are carried out and these countries reach parity with the high-income, low-volume countries by the end of the century, total world demand increases from 40 to 140 trillion pkm/year, a factor of increase of approximately 3.5. By mid-century, population in these two groups of countries has reached 7.7 billion, where it stabilizes, compared to 1.3 billion in the high-income countries. This increase in transportation volume assumes that a sustainable energy supply can be developed for all of this new transportation volume, and that the ecological impacts of the additional volume are managed successfully.

The impact of the growth of freight transportation is similar, as shown in Fig. 1-7. By the end of the century, per-capita freight activity in low- and middle-income countries

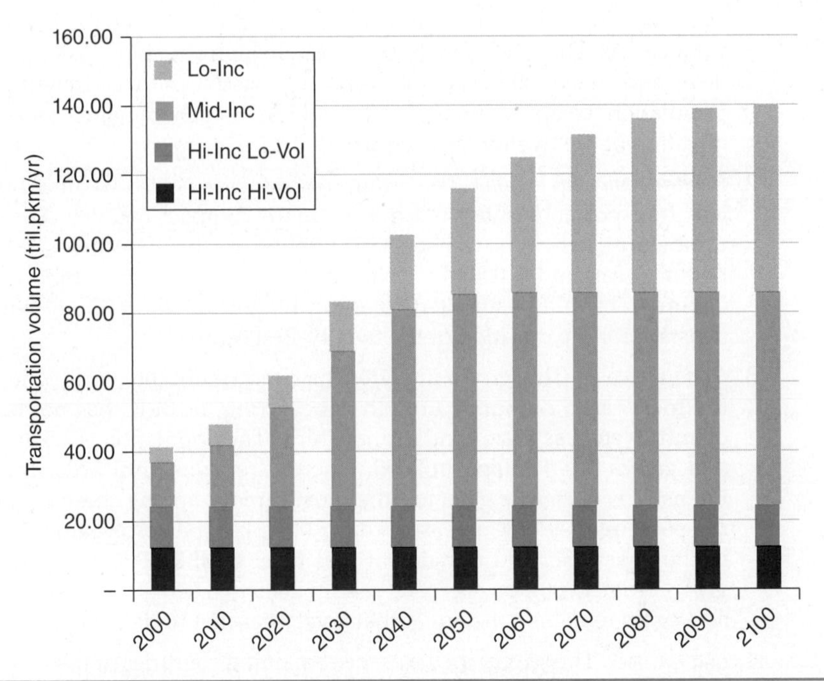

FIGURE 1-6 A possible scenario for passenger-km demand growth in the twenty-first century, based on population and per-capita demand growth projections.

Notes: "Lo-inc" = lowest income countries; "Mid-Inc" = industrializing countries with midrange incomes; "Hi-Inc Lo-Vol" = high-income countries with relatively low passenger-km per capita; "Hi-Inc Hi-Vol" = high-income countries with relatively high passenger-km per capita. See text.

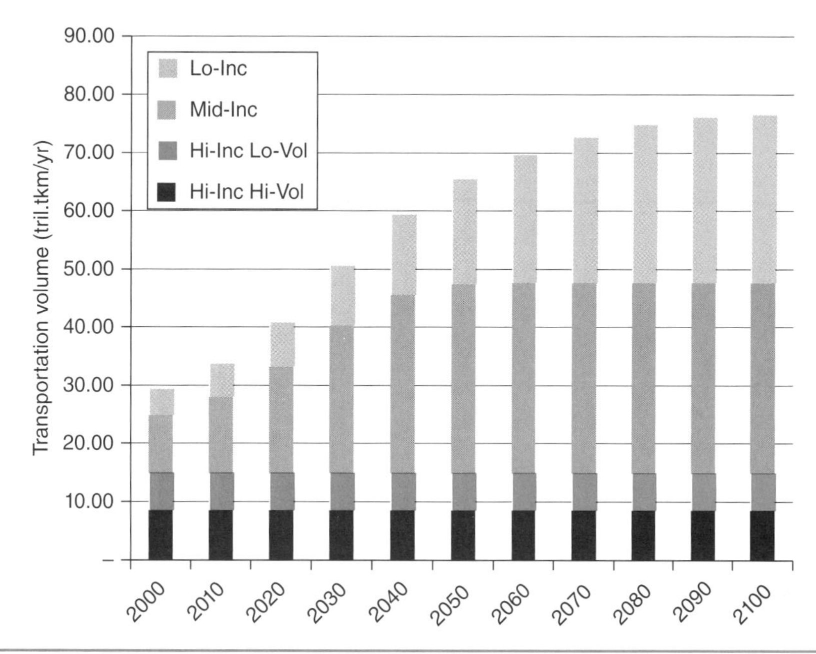

FIGURE 1-7 A possible scenario for freight transportation tonne-kilometers growth in the twenty-first century, based on population and per-capita demand growth projections.

Notes: "Lo-inc" = lowest income countries; "Mid-Inc" = industrializing countries with midrange incomes; "Hi-Inc Lo-Vol" = high-income countries with relatively low tonne-kilometers per capita; "Hi-Inc Hi-Vol" = high-income countries with relatively high tonne-km per capita. See text.

reaches levels equal to that of the high-income low-volume countries. Total freight demand worldwide increases from 29 to 76 trillion tonne-km,[7] a factor of increase of 2.6.

For brevity, only one of several possible scenarios for passenger and freight transportation has been presented here. Many variations could be developed. For instance, the assumption that the high-income, high-volume countries must maintain higher per-capita passenger and freight intensity levels could be relaxed; instead, these countries could use changes in land-use patterns and other adjustments to gradually reduce their levels to parity with the other three groups in the year 2100. This outcome would have the benefit of putting the entire world population on an equal footing in terms of transportation usage, reducing political tensions that sometimes arise at present between countries that use relatively different amounts of transportation per capita. This change would have only a small impact on the total volumes observed in 2100 in Figs. 1-6 and 1-7, however, because the population affected is small (500 million out of 9 billion). Total world passenger transportation would decline from 140 to 135 trillion pkm/year (–3.6%), and freight would decline from 76 to 72 trillion tkm/year (–5.9%).

Assumed Endpoints for Transportation in the Year 2100

The scenario in Figs. 1-6 and 1-7 suggests that the outcome for economies and transportation systems will fall along the lines of the mainstream vision described above.

[7]Conversion: 1 tonne = 1 metric ton = 1,000 kg = 1.1 standard tons. See App. A.

Notwithstanding the arguments made for a less transportation-intensive human society from the deep ecologic perspective, the main tenets of the vision are:

1. Stable global population of approximately 9 billion people in 2100.
2. Passenger transportation demand between 15,000 and 25,000 pkm/person/year in each region of the world, depending on regional factors.
3. Freight transportation demand between 8,000 and 17,000 tkm/person/year in each region of the world, depending on regional factors.
4. Stable supply of energy and other resources to allow the transportation system to function sustainably into the indefinite future, with problems of ecological impact solved.

The vision of sustainable transportation systems outlined by these points is used as a paradigm for many of the topics and examples in subsequent chapters.

1-6 Contents and Organization of This Book

The contents of this book can be divided into three parts: motivations and drivers (covered in part in Secs. 1-3 to 1-5), tools, and applications. The flowchart in Fig. 1-8 graphically represents the logical connection among these topics.

Four major challenges make up the motivations and drivers for implementing sustainable transportation systems, and are covered primarily in Unit 1 (Chaps. 1 and 2), with additional consideration in Chaps. 3 to 6. The first is *livable communities* (covered in this chapter and in Chap. 6, "Infrastructure Tools"), meaning continued development and maintenance of transportation systems in a way that does not adversely affect quality of life. The second is *sustainable system economics*, meaning successfully paying for adequate maintenance of both vehicles and infrastructure. This topic is covered in this chapter and Chap. 3 ("Economic Tools"). The third major challenge is that of environmental concerns, chiefly climate change (Chap. 2), which requires that society stabilize (and possibly someday reduce) the amount of CO_2 and other greenhouse gases (GHGs) in the atmosphere, through decisions about transportation systems, among other things. Last, as the world continues to consume nonrenewable fossil fuels, oil and gas production will peak sometime in the next several decades, requiring the development of alternative energy sources (Chap. 2).

Techniques and tools included in the book appear in Unit 2 (Chaps. 3 to 6). These include systems tools (Chap. 3) that enable the modeling of transportation activity as a system with interacting parts. Economic tools (Chap. 4) are also included, to make the connection between transportation systems as technologies and as financial investments that include both upfront capital costs and ongoing maintenance and operating costs that must be repaid over time through revenues (transportation system revenues, financial benefit of energy savings, and so on). Next, vehicle design tools (Chap. 5) consider means of evaluating a vehicle's performance as a function of physical and operating characteristics. Lastly, infrastructure design tools (Chap. 6) provide inform the creation of transportation infrastructure that supports the movement of vehicles. Note that "hard" engineering tools are the assumed background that the transportation student or practitioner brings to the study of transportation systems—algebra and calculus, Newtonian mechanics, electricity and magnetism, chemistry, and probability and statistics.

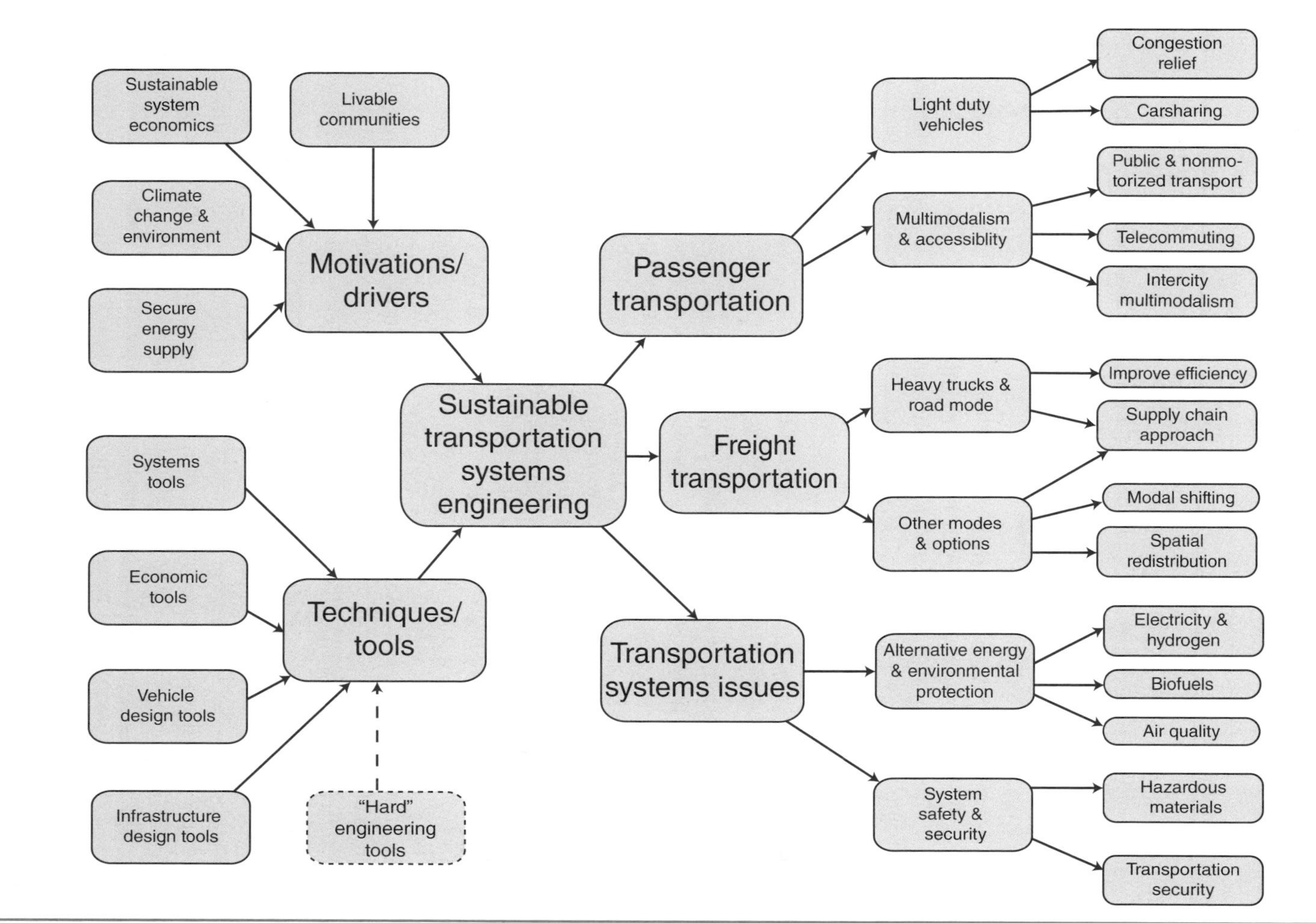

FIGURE 1-8 Flowchart of major topics, including motivations or drivers, tools or techniques, and applications.

The remaining elements on the right-hand side of Fig. 1-8 are dedicated to the range of transportation systems applications included in the book. These applications are divided into three general areas: passenger (Unit 3, Chaps. 7 to 10), freight (Unit 4, Chaps. 11 to 13), and overarching transportation issues (Unit 5, Chaps. 14 to 17). Passenger transportation topics include an overview (Chap. 7), urban public transportation and multimodal solutions (Chap. 8), personal mobility and accessibility (Chap. 9), and multimodal intercity transportation (Chap. 10). The coverage of freight transportation includes an overview (Chap. 11), efficient technology and alternative modes (Chap. 12), and supply chain and spatial analysis of freight transportation patterns (Chap. 13). The final chapters in the book consider a range of issues that affect transportation systems across the board, both passenger and freight, both urban and intercity, and so on. These include an overview of principles for incorporating alternative energy (Chap. 14); electricity and hydrogen options (Chap. 15); biofuel and bio-energy options (Chap. 16); transportation and air quality, hazardous materials transportation, and securing transportation systems against malicious attacks or extreme weather events (Chap. 17). Unit 5 concludes with suggestions for achieving a sustainable transportation system in the twenty-first century (also in Chap. 17). The appendices include units of measure used in transportation systems, lists conversions between metric and U.S. standard units, and appendices on infrastructure design and engineering economics.

1-7 Summary

Transportation systems and technologies play many roles in society: as a universal means of moving people and goods, as a major segment of world economic activity, as major employers, and as a major consumer of resources and origin of various ecological impacts. Both historically and in recent years, transportation providers have sought to provide systems that functioned reliably, safely, and affordably. Within the last several centuries, these efforts started with simple conveyances such as ships and horse-drawn carriages and extended to newer and faster modes, starting with canals, then railroads, then on to automobiles and other highway modes, and eventually to aviation. From the present time forward into the future, the overriding goal will be to integrate existing modes into society and into the natural environment that surrounds the human-built environment so that transportation becomes truly sustainable.

Although criticized in some quarters for generating excessive throughput of energy and resources, the mainstream transportation profession generally views access to modern transportation capacity, whether measured in passenger-kilometers per capita for passenger transportation or tonne-kilometers per capita for freight, as desirable for the development and maintenance of prosperous countries. From this perspective, negative consequences can be addressed by reducing the amount of impact per passenger-kilometer or tonne-kilometer. Regardless of the pathway that eventually emerges for the volumes of transportation eventually delivered by the system, there are several challenges that can be grouped into four basic categories: (1) making sure the economics for operating and maintaining transportation systems are sound; (2) making sure transportation systems function efficiently with minimal congestion, delays, accidents, and security risks; (3) addressing air quality concerns and other local and regional impacts; and (4) addressing global concerns such long-term transportation energy security and climate change. This book focuses on solutions to these problems in the twenty-first century.

References

American Society of Civil Engineers (2009). *2009 Report Card for America's Infrastructure.* ASCE, Reston, VA.

Carbon Dioxide Information Analysis Center (2012). CDIAC: Carbon Dioxide Information Analysis Center of the U.S. Department of Energy. Informational Web site. Available at: *cdiac.ornl.gov*. Accessed Sept. 30, 2012.

Davis, S., S. Diegel, and R. Boundy (2012). *Transportation Energy Data Book, 31 Ed.* Oak Ridge National Laboratories, Oak Ridge, TN.

Gellerman, B. (2012). "Greenland sharks: the apex predator." *Living on Earth* radio show, August 10, 2012. Public Radio International, Boston, MA.

Her Majesty's Stationery Office (2010). *Transport Statistics Great Britain.* HMSO, London.

Ministry of Railways. (2009) *Indian Railways.* Ministry of Railways, Government of India, New Delhi.

Ministry of Road Transport and Highways (2009). *Road Transport Year Book.* MORTH, Government of India, New Delhi.

NationMaster (2012). Passengers carried by roads in million passenger-km by country. Web resource, available at *www.nationmaster.com*. Accessed Sep. 29, 2012.

Ricardo, D. (1817). *On the Principles of Political Economy and Taxation.* John Murray, London.

Schiller, P., E. Bruun, and J. Kenworthy (2010). *An Introduction to Sustainable Transportation: Policy, Planning, and Implementation*. Earthscan, London.

United Nations (2012). *Population and Vital Statistics Report.* Electronic resource, available at *unstats.un.org.* Accessed November Nov. 9, 2012.

U.S. Bureau of the Census (2009). *Statistical Abstract of the United States.* USBOC, Washington, DC.

U.S. Bureau of Transportation Statistics (2010). *Transportation Facts.* U.S. Dept of Transportation, Washington, DC.

U.S. Energy Information Administration (2012). Main informational website. USEIA, US Department of Energy, Washington, DC. Electronic resource, available at *www.eia.doe.gov*.

Vuchic, V. (1999). *Transportation for Livable Cities*. Prentice-Hall, Englewood Cliffs, NJ.

Womack, J., D. Jones, and D. Roos (1990). *The Machine that Changed the World: The Story of Lean Production.* Harper-Perennial, New York, NY.

World Commission on Environment and Development (1987). *Our Common Future.* Oxford University Press, Oxford, UK.

World Trade Organization (2012). *Statistics Database.* Electronic resource, available at *www.wto.org*. Accessed Nov. 9, 2012.

Further Readings

Bullard, R., and G. Johnson (1997). *Just Transportation: Dismantling Race and Class Barriers to Mobility.* New Society Publishers, Gabriola Island, BC.

Devall, W., and G. Sessions (2001). *Deep Ecology: Living as if Nature Mattered.* Gibbs Smith, Layton, UT.

Merchant, C. (1992). *Radical Ecology: The Search for a Livable World.* Psychology Press, Abingdon, Oxfordshire, UK.

Owen, W. (1987). *Transportation and World Development: Mobility and the Global Economy.* Johns Hopkins Press, Baltimore, MD.

Sale, K. (1985). *Dwellers in the Land: The Bioregional Vision.* Sierra Club Books, San Francisco, CA.

Sperling, D., and D. Gordon (2009). *Two Billion Cars: Driving Toward Sustainability.* Oxford University Press, Oxford, UK.

Temple, R., and J. Needham (2007). *The Genius of China: 3,000 Years of Science, Discovery, and Invention.* Inner Traditions Press, Rochester, VT.

World Business Council on Sustainable Development (2010). *Mobility 2030: Meeting the Challenges to Sustainability.* WBCSD, Geneva.

Exercises

1-1. Write one paragraph describing a transportation technology or system with which you have personal experience in a specific metropolitan area or region, in either the United States or a foreign country. Was it successful? Why or why not? To answer this question, write a second paragraph evaluating its success (or lack thereof). Your answer may include qualitative or anecdotal information, quantitative data (including numbers such as cost, capacity, and so on), or a mixture of the two.

1-2. The underlying population trajectories for the four country groups ("low income," etc.) discussed in the future scenarios presented in this chapter are given below for the period 2000 to 2100, with population given in millions of people throughout the table. Create your own trajectories for world passenger and/or freight volumes by changing any or all of the following values: (*a*) assignment of population to each of the four country groups, (*b*) size of total world population, (*c*) per capita transportation volumes in different years. Explain why you made the changes you did, and discuss how total world transportation and its division between country groups would be different from the values presented in the body of the chapter.

	High income				
Year	**High volume**	**Low volume**	**Middle income**	**Low income**	**Total**
2000	500	800	2,500	2,200	6,000
2010	500	800	3,032	2,668	7,000
2020	500	800	3,457	3,043	7,800
2030	500	800	3,777	3,323	8,400
2040	500	800	3,989	3,511	8,800
2050	500	800	4,069	3,581	8,950
2060	500	800	4,096	3,604	9,000
2070	500	800	4,096	3,604	9,000
2080	500	800	4,096	3,604	9,000
2090	500	800	4,096	3,604	9,000
2100	500	800	4,096	3,604	9,000

CHAPTER 2

Background on Energy Security and Climate Change

2-1 Chapter Overview

This chapter continues the discussion of two problems highlighted in Chap. 1, namely, energy security and climate change. Regarding energy security, as world demand for transportation continues to grow, especially in the emerging economies of Asia, Latin America, and Africa, sufficient energy will be required on an ongoing basis to meet this need. Regarding climate change, growing levels of carbon dioxide (CO_2) and other greenhouse gases (GHGs) in the atmosphere are increasingly leading to climate change, with negative consequences for both the natural and human-built environments. Since transportation is a major emitter of CO_2, efforts are already underway to reduce emissions so as to slow down climatic changes. The goal of this chapter is to provide more background on these topics.

2-2 Introduction: The Role of Transportation in Energy Consumption

In both industrial and emerging countries across the globe today, the transportation sector is one of the major consumers of energy. Changes in levels of energy consumption, both for individual countries and for the world as a whole, are in a symbiotic relationship with levels of both population and wealth. That is, increasing access to energy makes it possible for human society to support larger populations and also increasing levels of wealth and, at the same time, a growing population and increasing wealth will spur the purchase of energy for all aspects of daily life, including transportation services. In the first part of this chapter we focus on energy consumption and energy security (i.e., having a sufficient supply of energy to avoid disruptions or damaging fluctuations in price). Transportation is also responsible for large quantities of GHG emissions, chief among them CO_2, and we discuss these emissions in a later part of the chapter.

The large global requirement for transportation energy is a function of both the large number of people who participate in the system and the high expectations that have been created for the rapid, reliable movement of passengers and freight.

During preindustrial times before the year 1800, in most countries, only a handful of elite, wealthy members of the population who could afford to travel any distance; most of the population remained close to the place of their birth for most of their lives. Today, by contrast, millions of citizens of industrial countries, including those in the poorest demographic strata, have access to mechanized transportation. Furthermore, a large number of middle- and upper-class travelers within this group expect to be able to travel quickly, by private car on an expressway, by high speed rail, or by air. Therefore, it is inevitable that this type of transportation will be energy-intensive on a per-person basis. Lastly, as consumers, these citizens require large volumes of freight all around the world as products are brought into being through modern supply chains and delivered to retail outlets or consumers' doorsteps.

World energy consumption in 2009 totaled 483 quadrillion Btu, or quads [510 EJ(EJ is exajoule)], according to the U.S. Energy Information Administration (USEIA). Energy data sources, such as those published by the USEIA, separate end-use energy consumption between transportation and nontransportation sectors for some countries but not others. Therefore, it is not possible to present a comprehensive comparison of transportation versus nontransportation energy use for all countries. Nonetheless, in the next sections on energy versus wealth and energy consumption growth, we look at energy consumption for different countries across all end uses, with the understanding that transportation energy consumption strongly influences total consumption. As an indication of transportation's role, Table 2-1 gives the percentage of total end-use energy consumption for the transportation sector for select countries. Values range from 11% for China and India to 40% for the United States. The eight countries in the table together

Country	Transportation	Total	Percent
Brazil	2.48	7.57	33%
China	6.38	56.87	11%
Germany	2.14	8.89	24%
India	2.04	17.83	11%
Japan	3.02	12.44	24%
Russia	3.56	16.78	21%
UK	1.66	5.24	32%
USA	22.93	58.04	40%
Sum of 8	44.21	183.66	24%

Note: Both transportation and total energy figures exclude energy losses in conversion and distribution. For example, for 2009, USEIA reports 94.5 quads for the United States in 2009, thus 36.5 quads or 39% of gross energy consumption are lost in conversion and distribution, and the remaining ~58 quads or 61% delivered for end use.

Source: International Energy Agency (2012)

TABLE 2-1 Percentage of 2009 End-Use Energy Consumed by Transportation Sector for Select Countries (quads)

consume 44 quads for transportation out of the total 184 quads, so that transportation consumes 2% of this total.

2-2-1 Relationship between Energy Consumption and Wealth

Access to energy brings advantages not only to the transportation sector, but to all sectors across the economy of any given country. Therefore, in this section we study the relationship between energy consumption in a country and level of wealth in that country, measured in terms of gross domestic product (GDP) per capita. Wealthy countries have GDP per-capita levels in the range of $35,000 to $50,000 per person per year, although some small (by population) and very wealthy countries such as Luxemburg, Norway, and Switzerland have levels that are even higher. Per-capita energy consumption varies more than per capita GDP between wealthy countries, and the reasons are discussed further in this section.

By definition, GDP is the sum of the monetary value of all goods and services produced in a country in a given year. For purposes of international comparisons, these values are usually converted to a common currency such as U.S. dollars, using an average exchange rate for the year. In some instances, the GDP value may be adjusted to reflect purchasing power parity (PPP), since even when one takes into account exchange rates, a dollar equivalent earned toward GDP in one country may not buy as much as in another country (e.g., in the decade between the years 2000 and 2010, dollar equivalents typically had more purchasing power in the United States than in Japan or Scandinavia).

Though other factors play a role, the wealth of a country measured in terms of GDP per capita is, to a fair extent, correlated with the energy use per capita of that country, notwithstanding variability in per-capita energy between countries of similar wealth. Table 2-2 presents a subset of the world's countries, selected to represent varying degrees of wealth as well as different continents, which is used to illustrate graphically the connection between wealth and energy. Using data in the table from the year 2008, both the GDP and per-capita energy consumption vary by around two orders of magnitude between the countries with the lowest and highest values. For GDP, the values range from $230 per person for Zimbabwe at the low end to $47,400 per person for the United States at the high end; for energy, they range from a low of 12.6 GJ (Gigajoules) (11.9 million Btu/person) for Zimbabwe to a high of 725 GJ/person (687 million Btu/person) for Bahrain.

Plotting these countries' GDP per capita as a function of energy use per capita (see Fig. 2-1) shows that GDP per capita rises with energy per capita up to a point, especially if one excludes countries such as Russia and Bahrain, which may fall outside the curve due to their status as major oil producers or due to extreme climates. Gabon represents a large group of emerging countries whose per-capita energy consumption (36 GJ/person) and GDP ($6,430/person) are not yet among the wealthy countries. Once the band including the most prosperous countries in the figure is reached, namely, those with a per-capita GDP at or above $35,000 (Germany, Australia, Japan, Canada, and the United States), there is a wide variation in energy use per capita, with Canadian and U.S. citizens using roughly twice as much energy as those of Japan or Germany. A detailed explanation of this difference is beyond the scope of this chapter, but factors such as average energy cost, extent of policy to increase energy efficiency, density of urban areas and intercity travel patterns, or harshness of climate (especially in the case of Canada) may all play a role.

	Population [millions]		Energy [EJ]		GDP [billion US$]	
	2004	2008	2004	2008	2004	2008
Australia	20.2	21.3	5.98	6.07	699.7	1013
Bahrain	0.7	0.8	0.46	0.58	11.7	21.2
Brazil	186.4	195.1	10.56	11.21	804.5	1573
Canada	32.3	33.3	15.05	14.79	1131.5	1499.5
China	1,315.80	1324.7	47.73	89.74	2240.9	4327.5
Gabon	1.4	1.4	0.05	0.05	9	14.5
Germany	82.7	82.2	17.26	15.15	2805	3673.1
India	1,103.40	1149.3	15.4	21.05	787.8	1206.9
Israel	6.7	7.5	0.95	0.91	122.8	202.1
Japan	128.1	127.7	26.38	23.07	4583.8	4910.7
Poland	38.5	38.1	4.26	4.08	303.4	527.9
Portugal	10.5	10.6	1.31	1.12	183	244.6
Russia	143.2	141.9	33.93	32.1	768.8	1676.6
Thailand	64.2	66.1	3.49	4.18	165.5	273.3
United States	298.2	304.5	116.76	106.11	12555.1	14441.4
Venezuela	26.7	27.9	3.54	3.37	134.4	319.4
Zimbabwe	13	13.5	0.28	0.17	3.9	3.1

Sources: UN Department of Economics and Social Affairs (2006) and Population Reference Bureau (2011), for population; U.S. Energy Information Administration (2011), for energy production; International Monetary Fund (2011), for economic data.

TABLE 2-2 Population, Energy Use, and GDP of Selected Countries, 2004 and 2008

2-2-2 Historic Growth in World Energy Consumption

Having observed the desirability of greater per-capita access to energy in the preceding section, it follows that, over time, as more countries succeed in raising their standard of living, world energy usage will grow. This trend is reflected in the estimates of world population and energy production from 1850 to 2010 shown in Fig. 2-2. The growth in world energy consumption per capita is also shown, measured in either GJ or million Btu per capita. The values shown are the total energy production figures divided by the population for each year. While population growth was unprecedented in human history over this period, growing nearly fivefold to approximately 7 billion in the year 2010, growth in energy consumption was much greater, growing more than twenty fold over the same period.

From observation of Fig. 2-2, the energy growth trend can be broken into five periods, each period reflecting events in worldwide technological evolution and social change. From 1850 to 1900, industrialization and the construction of railroad networks was

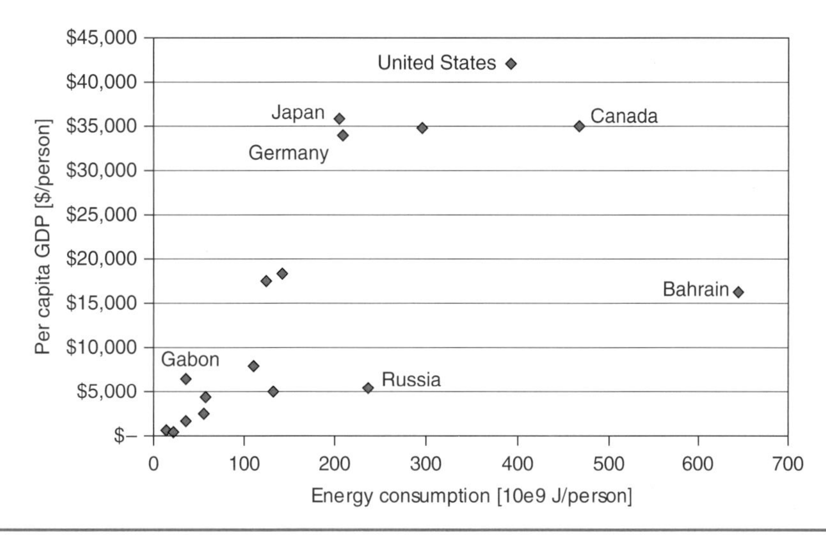

FIGURE 2-1 Per-capita GDP as a function of per-capita energy consumption in gigajoules (GJ) per person for selected countries, 2004.

Note: Conversion: 1 million Btu = 1.055 GJ.

Sources: UN Department of Economics and Social Affairs (2006), for population; U.S. Energy Information Administration (2006), for energy consumption; International Monetary Fund (2005), for GDP. Countries are taken from list in Table 2-2, with highlighted countries labeled to provide examples for comparison.

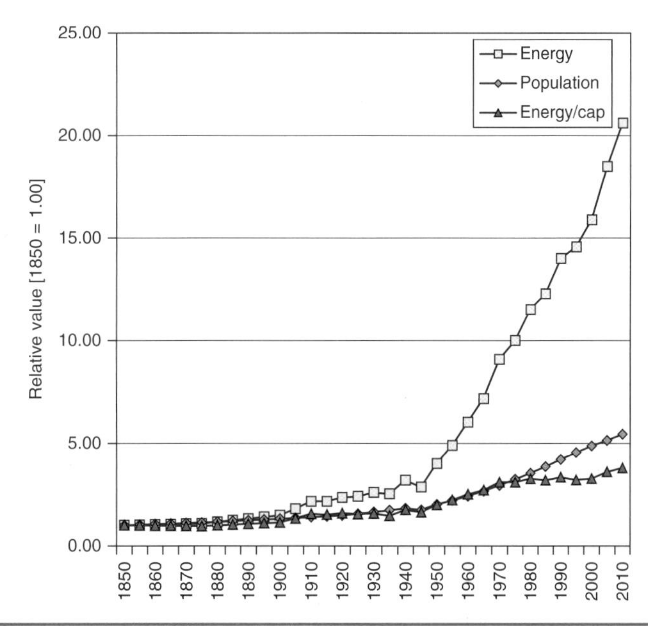

FIGURE 2-2 Relative growth in world population, world energy consumption, and average energy consumption per capita 1850 to 2010, indexed to 1850 = 1.00.

Notes: In 1850, population was 1.26 billion, energy was 25EJ or 23.7 quads, and per-capita energy use was 19.8GJ/person or 18.8 million Btu/person. For all data points, energy per person is the value of the energy curve divided by the value of the population curve; see text.

Source: Own calculations based on energy data from U.S. Energy Information Administration (2012) and population data from U.N. Population Reference Bureau (2011).

underway in several parts of the world, but much of the human population did not yet have the financial means to access manufactured goods or travel by rail, so the effect of industrialization on energy use per capita was modest, and energy consumption grew roughly in line with population. From 1900 to 1950, both the part of the population using modern energy supplies and the diversity of supplies (including oil and gas as well as coal) grew, so that energy consumption began to outpace population, and energy intensity per capita doubled by 1950 compared to 1850. From 1950 to 1975, energy consumption and energy intensity grew rapidly in the post-World War II period of economic expansion in Europe, North America, and Japan. From 1975 to 2000, both energy and population continued to grow, but limitations on output of some resources for energy, notably crude oil, as well as higher prices, encouraged more efficient use of energy around the world, so that energy consumption per capita remained roughly constant. Lastly, from 2000 to 2010, industrialization and economic growth in emerging countries led to a further increase in world average energy use per capita from 68 GJ or 65 million Btu to 75 GJ or 71 million Btu per person, and overall energy consumption from 417 EJ or 395 quads to 514 EJ or 488 quads.

It is this last development since 2000 that is currently putting significant upward pressure on world energy use, in which the rapid economic advance of the so-called "BRIC" countries (Brazil, Russia, India, and China) is leading the way. These four countries account for roughly 40% of the world's population, so movement toward U.S. levels of per-capita energy consumption would pose a tremendous challenge for energy security and climate change. Led by China's very rapid increase of overall gross energy consumption since the year 2000, the BRIC countries have surged past the United States and the European countries to become the single largest energy consumer in the world (Fig. 2-3). Also, despite being the largest overall consumer, the BRIC countries have relatively low per-capita energy consumption: approximately 60 million Btus per person per year in 2006, compared to 360 million Btus per person in the United States.

The energy growth path of China can be viewed from a different perspective by comparing it to that of the United States in terms of the annual growth since reaching a threshold of 40 quads of gross energy consumption, as shown in Fig. 2-4. The United States reached this threshold in 1955, and China in 2001. The figure shows the subsequent growth pathway, where the horizontal axis gives the number of years elapsed since 1955 or 2001, depending on the country, and the vertical axis gives the total energy consumption in that year. For the United States, growth from 40 to 90 quads required 40 years (1955–1995). In the case of China, this growth took place over a period of just 8 years (2001–2009). Also, the economic position of BRIC countries such as China leaves open the possibility for even more growth in the future. Whereas the industrial countries typically grow their GDP 2% to 3% per year, China's economy has been growing at 7% to 10% per year in recent years. The values in 2008 of $4.4 trillion GDP and $3,266 per capita for China, compared to $14.4 trillion and $47,400, respectively, for the United States, indicate that there is much room for GDP and GDP per capita in China to grow. This growth is likely to continue to put upward pressure on energy consumption in China. The other BRIC countries, and to a varying extent other emerging economies outside of BRIC, are in a similar situation.

2-2-3 Transportation Energy Consumption in the United States

We now turn to the specific case of the United States and its transportation energy consumption levels, with consideration of trends over time and comparisons to other sectors.

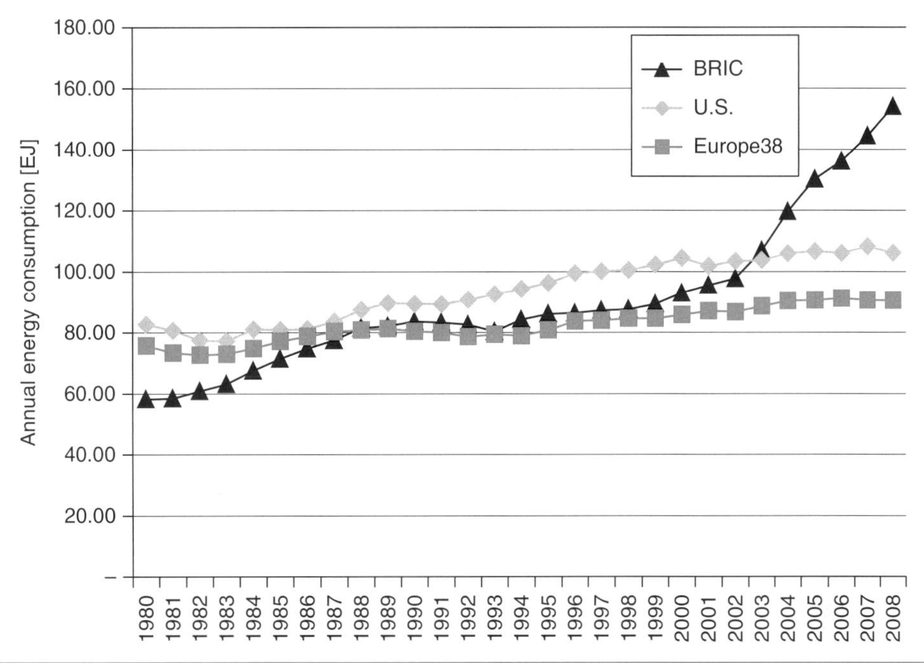

FIGURE 2-3 Energy consumption of U.S., European Union, and BRIC (Brazil/Russia/India/China) countries in EJ, 1980 to 2008.

Notes: The countries shown in the Europe group include the European Union plus various smaller countries that are part of the European land mass, such as Switzerland, Lichtenstein, etc. The energy consumption of other countries besides United States/Europe/BRIC is not included to avoid cluttering the figure; however, the countries shown represent approximately 75% of the world's energy consumption.

Source: U.S. Energy Information Administration.

This detailed case study of the United States is somewhat representative of the trend in other industrialized countries, although per-capita transportation energy use is higher in the United States than in peer countries in Europe or Asia, due to factors such as more energy-intensive private automobiles or higher average passenger-miles per capita per year. For example, using the International Energy Agency (IEA) figures in Table 2-1, U.S. per-capita transportation energy consumption stood at 75 million Btu per person in 2009, compared to 26 and 24 million Btu/person for Germany and Japan, respectively.

The transportation sector in the United States consumes a measurable share of not only the United States but also world energy (6.8% of the world total in 2003 and 5.9% in 2008), as shown in Fig. 2-5; the remaining sectors in the United States consume roughly 1%, with the remaining 8% of all energy split between the other industrialized and emerging countries. Although the U.S. transportation energy share of the world total is declining due to the increasing share of energy consumption occurring in emerging economies, its size remains significant at present.

Both passenger and freight transportation have contributed to energy consumption growth in the United States. On the passenger transportation side, the combination of more private vehicles on the road, larger average size of vehicles, and increase in distance driven

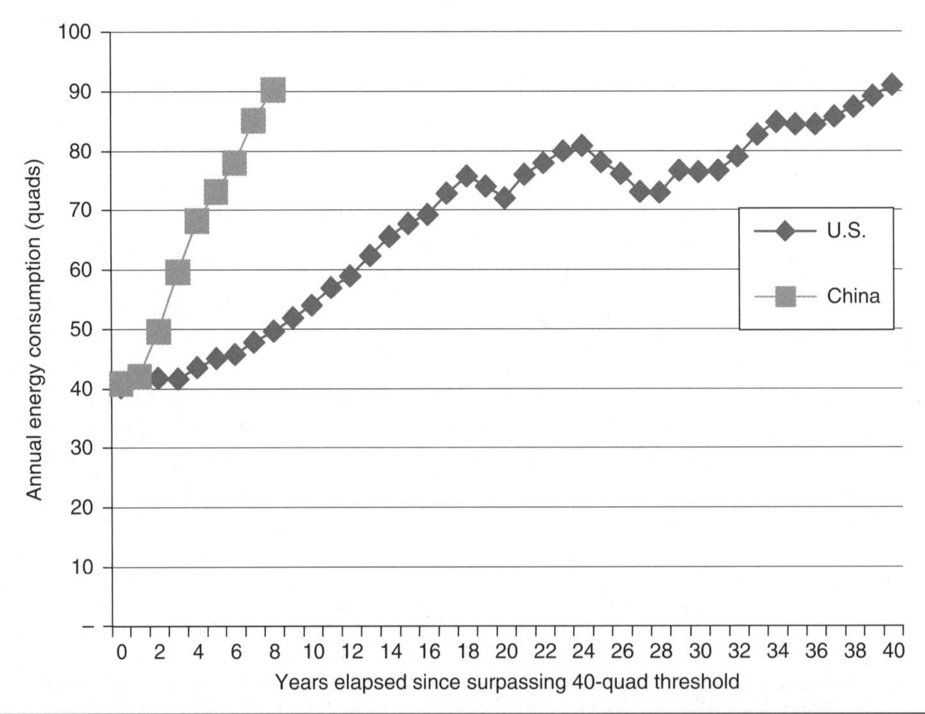

FIGURE 2-4 40 to 90 quads in 8 years? Figure of annual energy consumption for China and the United States for the number of years required to grow from 40 to 90 quads of total energy. Year 0 in the figure is equivalent to the years 2001 and 1955 for China and the United States, respectively.

Source: U.S. Energy Information Administration (2012).

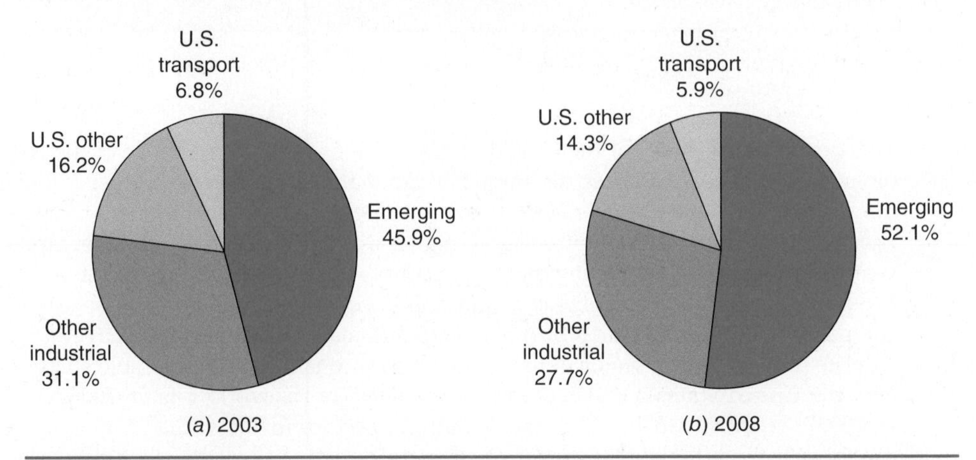

FIGURE 2-5 Comparison for 2003 and 2008 of U.S. transportation energy consumption to other U.S. energy consumption ("U.S. other"), total energy consumption of other industrial countries ("Other industrial"), and total energy consumption for emerging countries ("Emerging").

Note: Total value 520 EJ (493 quads) for 2008 and 426 EJ (403 quads) for 2003. Note that total energy consumption for all industrial countries is U.S. transport + U.S. other + other industrial, for example, 47.9% of the world total in 2008.

Source: U.S. Energy Information Administration.

per vehicle per year, has led to a 44% increase in energy consumption from 1970 to 2009. Freight transportation energy consumption is growing due both to the move toward a more global economy and changes to domestic distribution systems. In the United States, the freight sector energy consumption has been increasing by more than 2% each decade since 1970. Although the absolute value of 7.8 quads in 2009 is smaller than some other sectors, it is a large amount of energy in absolute terms, and the fact that it is almost entirely derived from petroleum and that its rate of consumption is on a rapidly increasing trend is cause for concern. During the same period, the passenger sector was subject to the noticeable impact of the Corporate Average Fuel Economy (CAFE) standards for automobiles, residential or commercial energy use was helped by improvements in lighting and insulation, and industrial energy use benefited from the transition away from smokestack industries. For lack of such mitigating circumstances, freight outpaced all other sectors in terms of growth in energy use over the period from 1970 to 2009, growing by 11%. The result, combining passenger and freight, is a total increase of 6% over 39 years from 1970 to 2009, from approximately 17 to 27 quads, as shown in Fig. 2-6, making the transportation sector

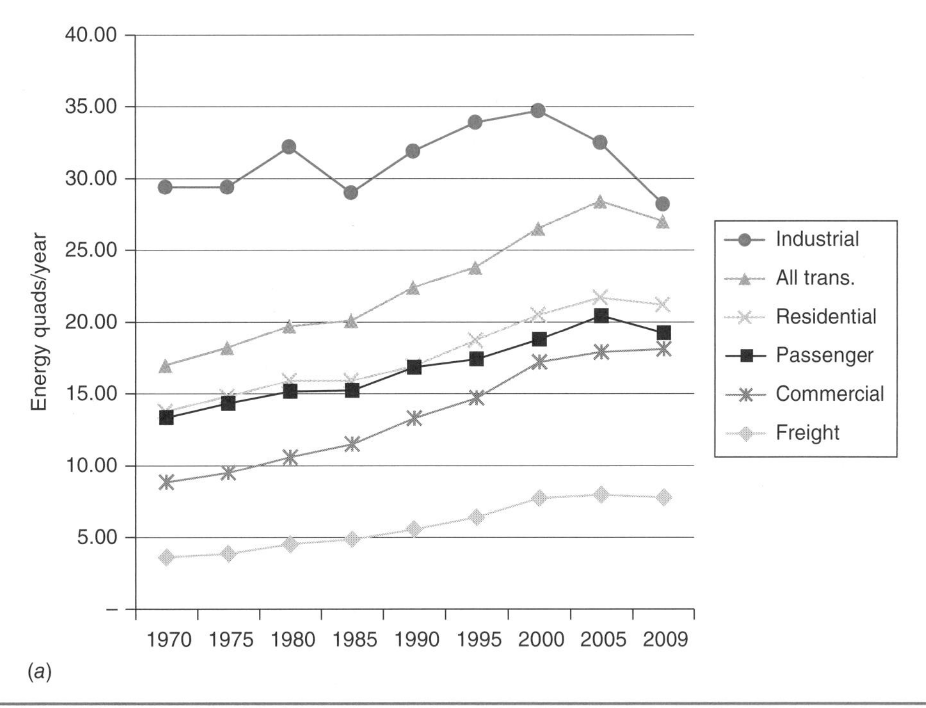

Figure 2-6 Relative growth of U.S. energy consumption: (*a*) absolute growth by sector; (*b*) indexed to 1970 = 1.00.

Source: Own calculations for freight and passenger breakdown, all other data from U.S. Energy Information Administration. 1970 values: freight transportation = 3.6 quads (3.8 EJ), passenger transportation = 13.4 quads (14.1 EJ), all transportation = 17.0 quad (17.9 EJ), commercial = 8.8 quad (9.3 EJ), residential = 13.8 quad (14.5 EJ), industrial = 29.4 quad (31.0 EJ), all energy = 69.0 quad (72.8 EJ).

Notes: (i) "Total Energy" trend is not included in Fig. 2-6(*a*) to make other trends more visible; (ii) 2005 and 2009 breakdown of transportation energy between passenger and freight is preliminary, due to limitations of source data.

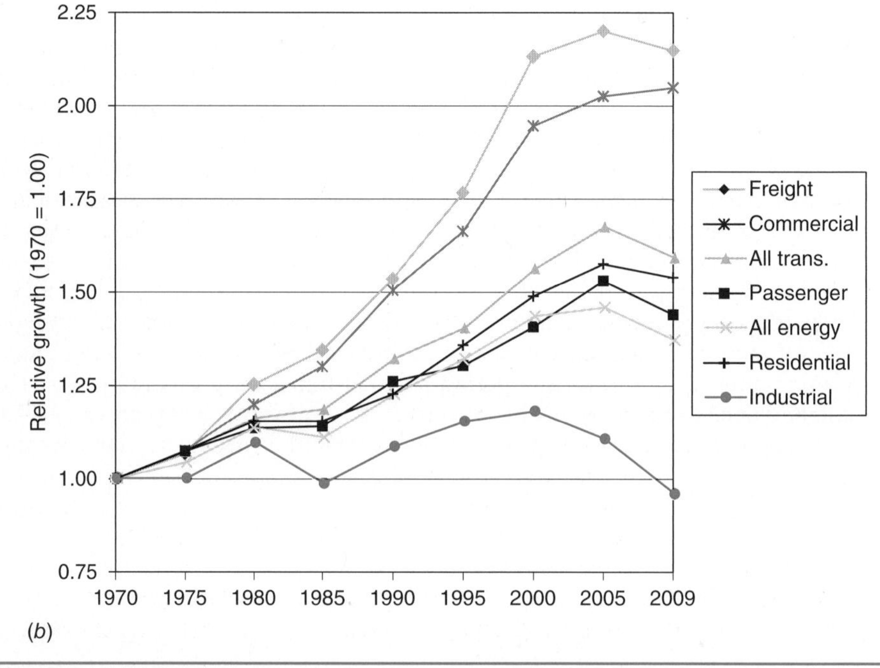

FIGURE 2-6 *(Continued)*

second only to the industrial sector at 28 quads. Note that the percent increase for transportation was second only to that of the commercial sector over this period.

Given the importance of transportation to energy use, emissions, and other issues, its growth in response to rising population and economic activity is of concern. As shown in Fig. 1-1 in Chap. 1, during the period from 1970 to 2008 the growth of both passenger and freight transportation as measured in passenger-km and tonne-km, respectively, outpaced the growth of the U.S. population by a substantial margin, increasing by 154% and 91%, whereas population increased by 52%. In addition, international freight, in the form of container shipments to and from the United States, also grew rapidly—time series data prior to 1995 were not available, but in the period 1995 to 2007, total United States 20-foot equivalent units, or TEUs, grew 101% from 22 to 45 million, before declining to 38 million in 2008 due to the recession. In an increasingly productive and affluent society the increased movement of passengers and goods does not come as any surprise; however, without an effective energy efficiency policy, increased population combined with increasing transportation activity per capita will drive up energy use in freight, all other things being equal.

2-3 Fuel Supplies for Meeting Transportation Energy Requirements

In this section, we turn from energy requirements measured in quads or exajoules to fuel supplies, which, in the case of the transportation sector, are typically derived from petroleum, also called crude oil. This oil is extracted from beneath the surface of the earth and is typically refined into three possible liquid fuels: (1) gasoline, for use in reciprocating (i.e., cylinder-and-piston) engines in light-duty vehicles; (2) diesel, for use in reciprocating engines in heavy-duty vehicles (heavy trucks, trains, ships) and some light-duty vehicles, and (3) jet fuel, for use in jets and turboprop-propelled aircrafts, which typically use rotating turbine rather than cylinder-and-piston technology. Additionally, a small fraction of the total volume of crude oil extracted is refined into lubricants used in transportation as well as other applications.

Lack of diversity of energy sources due to petroleum dominance poses a significant challenge for the transportation sector, as it becomes vulnerable to supply disruption or price volatility. Taking the case of the United States in the year 2010, 86% of the total 27.6 quads consumed in the transportation sector came in the form of either gasoline or diesel, with an additional 8% in the form of jet fuel. Thus, only 6% of the energy supply came from nonpetroleum sources (e.g., electricity generated from a mix of sources and delivered to electrified rail systems). By contrast, the U.S. electric grid in that same year was much more diverse: the largest source, coal, represented 45% of the market for primary energy sources, with natural gas at 24%, nuclear at 20%, and a small fraction of oil (<1%) and a remaining mixture of renewable sources (hydro, wind, solar, etc.) making up the remaining 11%. As an example of the variability in oil costs, Fig. 2-7 shows the price per barrel for oil from 1990 to 2011, including much greater volatility from 2005 onward and a peak value of $147 per barrel on July 9, 2008.

Along with problems raised by excessive dependence on a single resource, use of petroleum products as a transportation fuel has the further disadvantage that the underlying resource is finite and nonrenewable, at least on a nongeologic time scale.

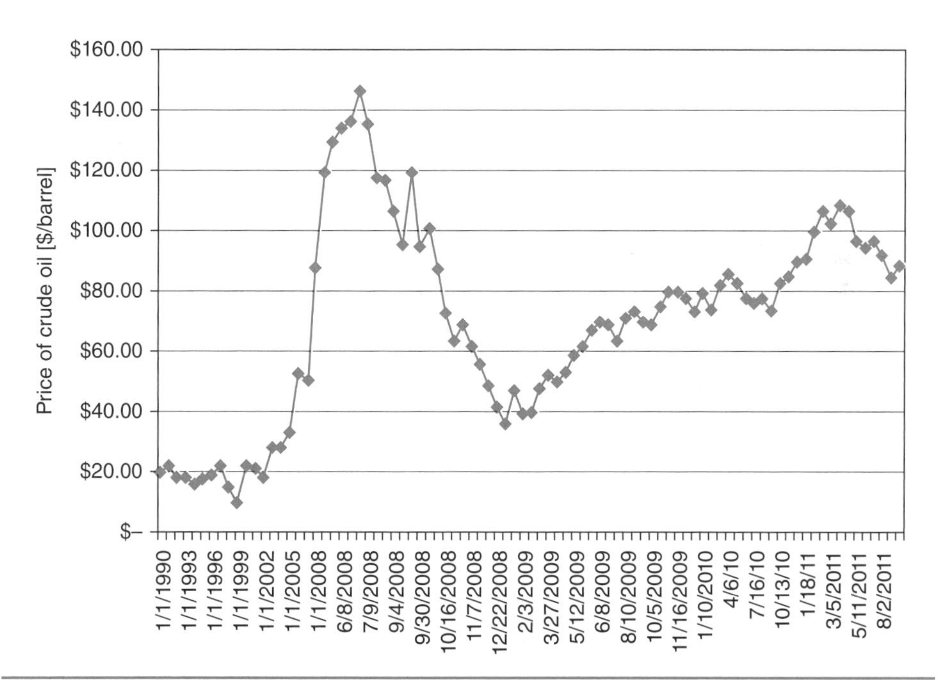

FIGURE 2-7 World price of crude oil, annual average from 1990 to 2006 and occasionally sampled value from January 2006 to September 2011.

Note: Change of time scale used from 2005 to 2011 to illustrate wide swings in price during this period.

Source: U.S. Energy Information Administration (1990–2006) and occasional personal sampling of publicized values (2006 to present).

Resources of this type exhibit a peak-and-decline production pattern, in which the resource reaches peak production at or near the point where 50% of the estimated ultimate recovery (EUR) has been extracted. After this point, maximum output of the resource becomes restricted, complicating efforts to maintain an adequate supply to the transportation fuels market. The rest of this section contains a quantitative treatment of resource-depletion pathways (i.e., alternative scenarios for the year-by-year quantities of consumption that lead to the eventual exhaustion of the resource), and a discussion of alternatives for conventional petroleum as supplies are exhausted.

2-3-1 Pathways for Annual Production of Nonrenewable Energy Sources

We now consider the pathway for nonrenewable resource production, in which annual output first increases as production is expanded and then declines as the remaining resource dwindles. This pathway for the life of the resource as a whole reflects that of individual oil fields, coal mines, and so on, which typically follow a life cycle in which productivity grows at first, eventually peaks, and then begins to decline as wells or mines gradually become less productive. Because total output is the sum of outputs from individual locations, near the end of the resource's worldwide lifetime, older wells or mines, where productivity is declining, outnumber those that are newly discovered, so that overall production declines. Hereafter, we focus on the particular case of conventional crude oil, but similar methods could be applied to nonconventional crude oil, natural gas, or coal.

Estimation of Annual Output Using Gaussian Curve

The use of a "bell-shaped" curve to model the lifetime of a nonrenewable resource is sometimes called a "Hubbert curve" after the geologist M. King Hubbert, who in the 1950s first fitted such a curve to the historical pattern of U.S. petroleum output in order to predict the future peaking of domestic U.S. conventional oil output. Here we use the Gaussian curve formula to fit an estimated production curve to observed annual production, but other mathematical curves with a similar bell shape could be used. Later we use a logistics curve to fit an estimated curve to the observed "S-shaped" cumulative production curve.

The Gaussian curve has the following functional form:

$$P = \frac{Q_{\text{inf}}}{S\sqrt{2\pi}} \exp\left[-(t_m - t)^2/(2S^2)\right] \tag{2-1}$$

Here P is output of oil (typically measured in barrels) in year t; Q_{inf} is the EUR, or the amount of oil that will ultimately be recovered; S is a width parameter for the Gaussian curve, measured in years; and t_m is the year in which the peak output of oil occurs. To fit the Gaussian curve to a time series of oil output data, one must first obtain a value for Q_{inf}. Thereafter, the values of t_m and S can be adjusted either iteratively by hand or using a software-based solver in order to find the values that minimize the deviation between observed values of production P_{actual} and values P_{est} predicted by Eq. (2-1). For this purpose, we can use the *root mean squared deviation* (RMSD), which has the following formula:

$$\text{RMSD} = \sqrt{\frac{1}{n}\sum_{i=1}^{n}(P_{\text{actual},\,i} - P_{\text{est},\,i})^2} \qquad (2\text{-}2)$$

Here n is the number of years for which output data are available, and i is the number of the year in the data set, from $i = 1$ to n. Example 2-1 uses a small number of data points from historical U.S. petroleum output figures to illustrate the use of the technique.

Example 2-1 The U.S. domestic output of petroleum for the years 1910, 1940, 1970, and 2000 is measured at 210, 1503, 3517, and 2131 million barrels, respectively, according to the U.S. Energy Information Administration. Suppose the predicted ultimate recovery of oil is estimated at 223 billion barrels: (a) If we arbitrarily guess values of $S = 30$ years and $t_m = 1970$, what is the value of RMSD? (b) By trial and error, choose an improved value of S and t_m and show that the value of RMSD is reduced. (c) Use an electronic solver to optimize the values of S and t_m to estimate the point in time at which oil output peaks, to the nearest year, based on the given four data points. Also, calculate the predicted output in that year.

Solution

(a) As a starting point, assuming $S = 30$ years and $t_m = 1970$, for each year t, the value of P_{est} can be calculated using Eq. (2-1). For example, for the year 1910:

$$P(1910) = \frac{2.23\times10^{11}}{30\sqrt{2\pi}}\exp[-(1970-1910)^2/(2(30)^2)] = 4.01\times10^{8}\,\text{bbl}$$

The appropriate values for calculating RMSD that result from repeating this process are given in the following table, leading to sum value of the $(P_{actual} - P_{est})^2$ terms in the right column of 5.38×10^{17}:

Year	P_{actual} (10^6 barrel)	P_{est} (10^6 barrel)	$(P_{actual} - P_{est})^2$
1910	210	401	3.665E+16
1940	1,503	1,799	8.772E+16
1970	3,517	2,966	3.032E+17
2000	2,131	1,799	1.101E+17
		Total	5.377E+17

The value of the RMSD is then calculated using Eq. (2-2):

$$\text{RMSD} = \sqrt{\frac{1}{4}(5.377\times10^{17})} = 3.67\times10^{8}$$

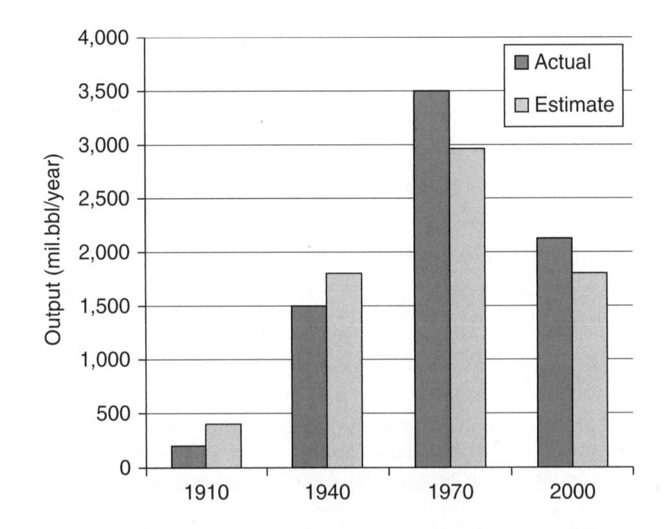

(b) To find an improved fit, the values of S and t_m are systematically changed by hand in order to reduce RMSD. The following figure shows the actual and estimated values from part (*a*):

From inspection of the figure, it appears that the estimated peak value is too low, and therefore the chosen value of S is too high. We can therefore keep t_m fixed and change only S to reduce the value of RMSD. Repeating the calculation with $t_m = 1970$ retained, S changed to $S = 25$, and creating a second table gives:

	P_{actual}	P_{est}	
Year	**10^6 barrel**	**10^6 barrel**	**$(P_{actual} - P_{est})^2$**
1910	200	210	1.050E+14
1940	1,732	1,503	5.250E+16
1970	3,559	3,517	1.728E+15
2000	1,732	2,131	1.591E+17
		Total:	2.134E+17

The value of the sum of error column of 2.134×10^{17} leads to a reduced value of RMSD = 2.31×10^8, due to making this change. Repeating the process can further reduce RMSD.

(c) To optimize the fit, we have used an electronic solver to find values $S = 25.3$ years and $t_m = 1974$ that minimize RMSD for the given data. Using the solver in Microsoft® Excel, the "target cell" is the "Total," that is, the bottom cell in the rightmost column in the table below, and the "changing cells" are the values of S and t_m (not shown), which the solver manipulates to minimize the value of the target cell. Results are:

T	P_{actual} (10^6 barrel)	P_{est} (10^6 barrel)	$(P_{actual} - P_{est})^2$
1910	210	146	4.028E+15
1940	1,503	1,442	3.671E+15
1970	3,517	3,479	1.456E+15
2000	2,131	2,056	5.659E+15
		Total	1.481E+16

The sum of values in the right column of 1.481×10^{16} leads to the RMSD being reduced to a minimum value of RMSD = 6.09×10^7.

To calculate output in the peak year, since the exponential term in Eq. (2-1) has value equal to one in the peak year, substituting Q_{inf} = 223 billion, S = 25.3 years, and t_m = 1974 into Eq. (2-1) gives

$$P = \frac{2.23 \times 10^{11}}{25.3\sqrt{2\pi}} \cdot 1 = \frac{2.23 \times 10^{11}}{25.3\sqrt{2\pi}} = 3.518 \times 10^9 \text{ bbl}$$

This value is quite close to the observed maximum for U.S. petroleum output of 3.520 billion barrels in 1972.

Application to Complete U.S. Annual Output Data Set 1900 to 2005

The Gaussian curve can be fitted to the complete set of annual output data from 1900 to 2005, as shown in Fig. 2-8. Assuming an EUR value of Q_{inf} = 225 billion, the optimum fit to the observed data is found with values S = 27.8 years and t_m = 1976. Therefore, the predicted peak output value from the curve for the United States occurs in 1976, 6 years after the actual peak, at 3.22 billion barrels/year. By 2005, actual and predicted curves have declined to 1.89 billion and 1.86 billion barrels/year, respectively. If the Gaussian curve prediction were to prove accurate, the output would decline to 93 million barrels/year in 2050, with 99.3% of the ultimately recovered resource consumed by that time. The Gaussian curve fits the data for the period from 1900 to 2005 well, with an average error between actual and estimated output for each year of 6%. Note that this figure does not include the impact of nonconventional oil resources, whose growing production has led to an increase to 2.06 billion barrels per year between 2005 and 2011. Because nonconventional oil is a new and separate resource that does not fall within the original EUR of 223 billion barrels, it is treated separately as shown in Fig. 2-8.

Application to Peaking of World Conventional Oil Production

The same techniques applied to U.S. oil production in the preceding section can be applied to worldwide production time series data for the years 1900 to 2005 to predict the peak year and output value. A range of possible projections for the future pathway of world annual oil output can be projected, depending on the total amount of oil that

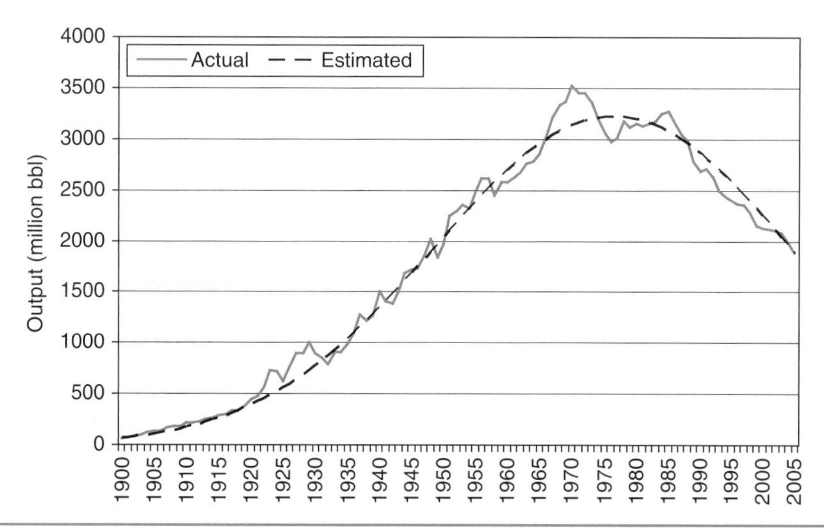

FIGURE 2-8 United States petroleum output including all states, from 1900 to 2005.
Source: U.S. Energy Information Administration.

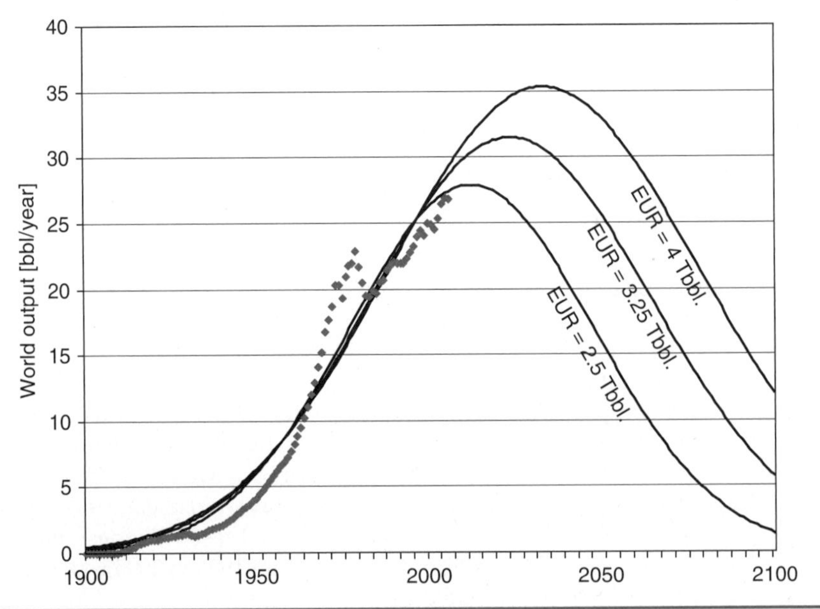

FIGURE 2-9 World oil production data from 1900 to 2006, with three best-fit projections as a function of EUR values of 2.5 trillion, 3.25 trillion, and 4 trillion barrels.

Source: Own calculation of production trend and projection using Eqs. (2-1) and (2-2), using historical production data from U.S. Energy Information Administration.

is eventually recovered, as shown in Fig. 2-9. Since the world total reserves and resources are not known with certainty at present, a wide range of EUR values can be assumed, such as values from 2.5 trillion to 4 trillion barrels, as shown. For each value of EUR, an electronic solver of the type explained in Example 2-1 has been used to find values of t_m and S that minimize RMSD for the years with known value of world oil output. The curve is then extrapolated forward to the year 2100 to project a possible pathway for year-on-year future oil production.

Predicting when the conventional oil peak will occur is of great interest from an energy policy perspective, because from the point of peaking onward, society will come under increased pressure to meet ever greater demand for the services that oil provides, with a fixed or declining available resource. This phenomenon has been given the name *peak oil.* Figure 2-9 suggests that the timing of the peak year for oil production is not very sensitive to the value of the EUR. Increasing the EUR from 2.5 trillion to 4 trillion barrels delays the peak by just 21 years, from 2013 to 2034. In the case of EUR = 3.25 trillion, output peaks in 2024. Also, once past the peak, the decline in output may be rapid. For the EUR = 4 trillion barrel case, after peaking at 35.3 billion barrels/year in 2034, output declines to 25.6 billion barrel/year in 2070, a reduction of 28% in 36 years.

Not all analysts agree that the future conventional oil output curve will have the pronounced peak followed by steady decline shown in the figure. Some believe that the curve will instead reach a plateau, where oil producers will not be able to increase annual output, but will be able to hold it steady for 2 or 3 decades. Most agree, however, that the total output will enter an absolute decline by the year 2050, if not before. Here the potential role of nonconventional fossil fuels in the future is important, as discussed in Sec. 2-3-3.

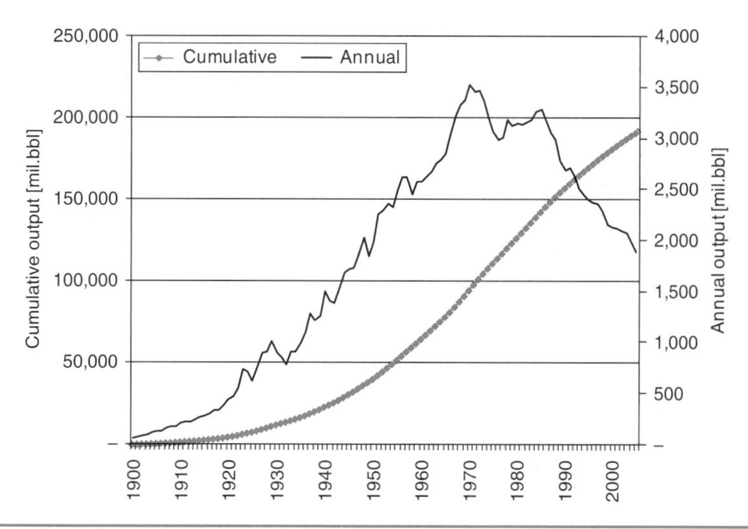

FIGURE 2-10 Annual and cumulative U.S. petroleum output including all states, 1900–2005. *Source:* U.S. Energy Information Administration

2-3-2 Modeling Conventional Oil Resource Pathway Using Cumulative Production Figures

Along with time series data for annual oil production analyzed above, data for cumulative oil production from the start year of production onward are of interest as well, where cumulative production for year t is simply the sum of annual output from the start year t_0 to t. Since the annual pattern is one of rise, peak, and decline, the cumulative value will have the "s-curve" shape characteristic of many penetration or depletion processes. In Fig. 2-10, the annual output from Fig. 2-8 above has been superimposed on the cumulative production curve for the period 1900 to 2005. Although the annual curve undulates substantially from year to year, the cumulative s-curve shows little impact, with a maximum rate of increase achieved sometime between 1970 and 1980, and the rate of growth declining thereafter. The large magnitude of total extraction in the first 70 to 80 years explains the relative insensitivity of the cumulative curve. From 1970 to 1980, output values varied between 3 and 3.5 billion barrels per year, but by that time, cumulative amounts were growing from approximately 100 billion barrels in 1970 to 130 billion barrels in 1980.

A logistics curve can be fit to the observed values for cumulative production from 1900 to 2005 in Fig. 2-10 to project forward the pathway of production and eventual depletion. First, we define t' as the number of years elapsed between t_0 and current year t, that is,

$$t' = t - t_0$$

Then, if we define F as the ultimate level of penetration and $f(t')$ as the level of penetration achieved after t' years have elapsed, the following curve can be fit to the observed data by adjusting parameters c_1 and c_2 to minimize RMSD, similar to the fitting of the Gaussian curve (Fig. 2-1):

$$f(t') = \frac{F \cdot e^{(c_1 + c_2 t')}}{1 + e^{(c_1 + c_2 t')}} \tag{2-3}$$

Values F and f can be given in terms of either percentage of ultimate penetration, for example, $F = 100\%$ if the penetration is to be complete or resource is to be completely used and f equal to some fraction of F. These values can also be absolute amounts, for example, $F = 2.23 \times 10^{11}$ barrels in the case of the U.S. ultimate amount of oil recovered. Example 2-2 illustrates fitting of a cumulative production curve to observed U.S. production for the period 1900 to 2005.

Example 2-2 For an estimate ultimate recovery value of $EUR = 2.23 \times 10^{11}$ bbl and the U.S. production values for the years 1900 to 2005, model cumulative oil production using the logistics function by finding parameters c_1 and c_2 that minimize RMSD.

Solution Set up a spreadsheet or computational software table in which the cells in the "modeled" column are written formulaically in terms of F, c_1, and c_2 to provide an estimated value for each year, for comparison with the "observed" values. A table with 106 rows, one for each year in the time series, results. An abbreviated version of the table with values for 1900, 1901, 1950, and 2005 is shown here, including calculations based on both the absolute numbers of production ("absolute approach") and percent of EUR consumed ("percent approach"). Note that absolute value of cumulative oil production is reported in millions of barrels.

Version of table in unformatted Excel format:

	Cumulative Barrels (millions)			Cumulative Barrels (millions)		
Year	**Observed**	**Modeled**	**(Obs – Mod)²**	**Observed**	**Modeled**	**(Obs – Mod)²**
1900	6.36E+01	2.26E+03	4.84E+06	0.03%	1.01%	9.73E-05
1901	1.33E+02	2.40E+03	5.16E+06	0.06%	1.08%	1.04E-04
etc.	etc.	etc.	etc.	etc.	etc.	etc.
1950	4.00E+05	3.93E+05	4.16E+05	17.92%	17.63%	8.38E-06
etc.	etc.	etc.	etc.	etc.	etc.	etc.
2005	1.91E+05	1.914E+05	2.91E+03	85.80%	85.82%	5.89E-08
	Sum of error terms:		2.73E+08	Sum of error terms:		5.50E-03

The solver minimizes the sum of RMSD by changing c_1 and c_2 in the model, resulting in values $c_1 = -4.58$ and $c_2 = 0.06077$ for either absolute or percent approach, giving the figures shown in the "modeled" columns in the table. Substituting the sum of error terms at the bottom of the table into Eq. (2-2), we get the following values for RMSD, respectively:

$$\text{RMSD\_Absolute} = \sqrt{(1/106)(2.73 \times 10^8)} = 1606$$

$$\text{RMSD\_Percent} = \sqrt{(1/106)(5.497 \times 10^{-3})} = 0.0072$$

Thus RMSD values are different, but both are minimal for their respective approaches.

Next, we can illustrate the calculation of the modeled value in a specific year from the table by taking the year 1950 as an example, as follows:

$$t' = 1950 - 1900 = 50$$

$$f(50) = \frac{\text{EUR} \cdot e^{(c_1 + c_2 t')}}{1 + e^{(c_1 + c_2 t')}} = \frac{2.23 \times 10^{11} \cdot e^{(-4.58 + 0.06077 \cdot 50)}}{1 + e^{(-4.58 + 0.06077 \cdot 50)}} = 3.931 \times 10^{10}$$

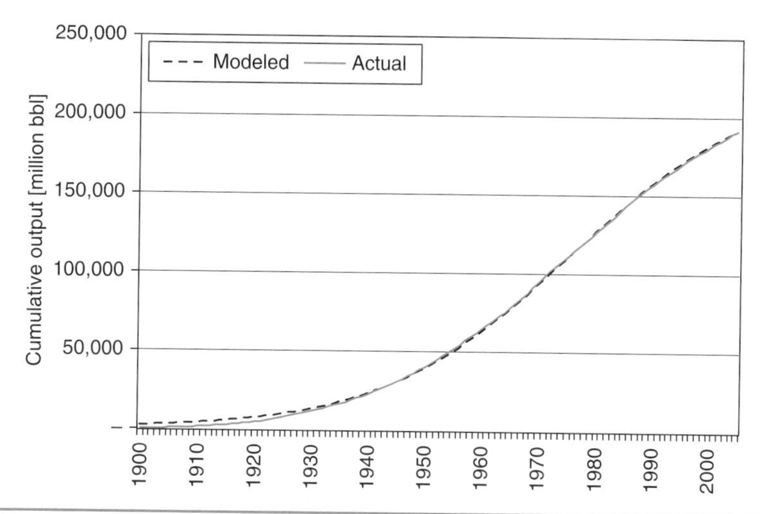

FIGURE 2-11 Comparison of observed and estimated cumulative consumption curves for U.S. conventional petroleum production, 1900 to 2005.

Source: U.S. Energy Information Administration, for observed cumulative consumption.

that is, estimated value of 39.31 billion barrels, which can be compared with the observed value of 39.96 billion barrels, a difference of 650 million barrels or 1.6% of the actual value. The overall agreement between the estimated and observed curves is quite good, as can be seen from the comparison of the two curves in Fig. 2-11.

One of the benefits of fitting the penetration curve to the observed cumulative consumption is that it provides another way to project future extent of depletion. For example, if one extrapolates the curve with $c_1 = -4.580$ and $c_2 = 0.06077$ out to the year 2025 (i.e., $t' = 125$), approximately 213 billion barrels, or 95%, of the original total amount available have been depleted, leaving 10 billion barrels to be consumed.

2-3-3 Nonconventional Oil and Other Nonconventional Fossil Resources

As mentioned in the preceding section, nonconventional fossil fuels that are derived from the same geologic process but that are not currently in widespread use provide a possible alternative to conventional petroleum. There has been interest in these nonconventional petroleum alternatives for several decades, since these resources have the potential to provide a lower-cost alternative at times when oil prices on the international market are high, especially for countries that import large amounts of fossil fuels and have nonconventional resources as a potential alternative, such as the United States. In other cases, nonconventional resources may make up for shortfalls of conventional resources if the latter dwindle in availability. In general, the conversion of nonconventional resources to a usable form is more complex and expensive than for conventional oil, which has hindered their development up until now. However, with the rapid rise in the price of oil in recent years, output from existing nonconventional sources is growing, and new locations are being explored.

There are three main options for nonconventional alternatives to crude oil that can provide a primary energy source for transportation, as follows:

1. *Oil shale:* Oil shale is composed of fossil organic matter mixed into sedimentary rock. When heated, oil shale releases a fossil liquid similar to petroleum.

Where the concentration of oil is high enough, the energy content of the oil released exceeds the input energy requirement, so that oil becomes available for refining into end-use products.

2. *Tar sands:* Tar sands are composed of sands mixed with a highly viscous hydrocarbon tar. As with oil shale, tar sands can be heated to release the tar, which can then be refined.
3. *Gaseous or liquid transportation fuel products from gas or coal:* Unlike the first two resources, the creation of this resource entails transforming a nonoil resource, either gas or coal, into an oil substitute. A variety of resources are possible. For example, techniques similar to those used to extract oil shale can be used to extract shale gas, which can then be liquefied to create liquid petroleum gas (LPG); vehicles can also be adapted to run on compressed natural gas (CNG). Coal can also be used as a primary energy source for transportation: Part of the energy content of the coal can be used in the transformation process, and the potential resulting products include a natural gas substitute or a synthetic liquid diesel fuel, either of which can be used as transportation fuel.

These nonconventional fossil resources are already being extracted, converted, and sold on the world market at the present time. For example, the oil sands of Alberta, Canada, produced approximately 1.3 million barrels of crude oil equivalent per day in 2008, up from 1 million barrels per day in 2005, according to the Province of Alberta's Energy Ministry. At a global level, economically recoverable tar sand deposits are concentrated in Canada and Venezuela, and are estimated at approximately 3.6 trillion barrels.

For the United States, one of the most significant developments for nonconventional fossil resources in recent years is the rise of shale oil extraction as a growing contributor to overall domestic crude oil production. Among the states, North Dakota is leading the growth of extraction of this resource: output in this state grew from 40 million barrels in 2006 to 153 million barrels in 2011, an almost fourfold increase in 5 years. The rise of shale oil accounts for the upturn in general in total U.S. oil production since 2005, which had been in decline, as shown Fig. 2-8. From a value of 1.89 billion barrels in 2005, output increased to 2.06 billion barrels in 2011, an increase of 9%. The total oil shale resource is estimated at 2.6 trillion barrels, with large resources in the United States and Brazil, and smaller deposits in several countries in Asia, Europe, and the Middle East. Taken together, world tar sands and oil shale resources are estimated to be of a similar order of magnitude to conventional oil resources, in terms of energy content.

Along with extraction and processing of nonconventional resources, conversion of coal to a synthetic gas or oil substitute is another means of displacing conventional gas and oil. Since 1984, the Great Plains Synfuels Plant in North Dakota, United States, has been converting coal to a synthetic natural gas equivalent that is suitable for transmission in the national gas pipeline grid, and therefore available to CNG-equipped vehicles. Processes are also available to convert coal into liquid fuel for transportation, although these have relatively high costs at present and are therefore not in commercial use.

Based on possible rate of expansion and the total availability of reserves, nonconventional resources show the potential to address the resource sufficiency side of the energy challenge, at least for an interim period of several decades. For example, one scenario from Cambridge Energy Associates suggests that, although conventional oil production would not keep pace with demand going forward and would peak in 2040, nonconventional oil would make up the difference and postpone any pronounced decline in total availability until 2060 or later (Fig. 2-12).

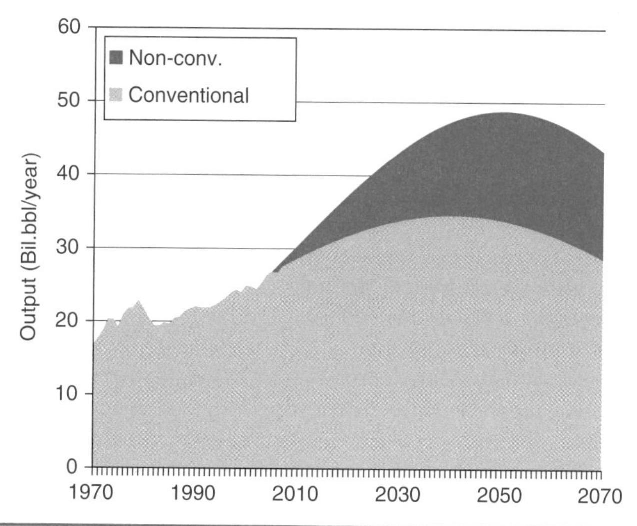

Figure 2-12 Projection of annual world oil output to 2070, showing impact of unconventional oil on total production over time period.

Source: Adapted from Cambridge Energy Research Associates (2006), as presented in Sperling & Gordon [2009].

Ecological Concerns with Nonconventional Oil Resource Development

Development of nonconventional oil resources or other nonconventionals raises two major ecological concerns. The first are the local and regional impacts specific to nonconventionals based on their extraction characteristics. Nonconventionals may require larger amounts of energy or water for extraction and processing, such as hydraulic fracturing that requires injection of water into the ground. They may also require the use of various industrial additives, for example in fracturing fluids used to extract shale oil or gas, whose use increases risk of groundwater contamination. Equipment requirements at extraction sites may lead to intensive traffic of industrial trucks and other vehicles needed to carry equipment in and out, which increase CO_2 emissions and contribute to compaction of soils and destruction of local habitats. Lastly, there may be tailings left behind by the extraction process that must be properly managed.

The second ecological concern is the continued emissions of CO_2 to the atmosphere that result when petroleum products are combusted by end users in vehicles large and small. This concern applies to conventional and nonconventional production alike, although nonconventionals typically release more CO_2 per unit of energy delivered to vehicles due to the extra energy involved in the extraction and refining process. Going forward, development of nonconventionals implies that the long-term trajectory of annual emissions levels would be extended well into the future, since nonconventionals might displace other sources such as renewables that might otherwise be required to meet transportation energy needs. Conversely, efforts to reduce CO_2 emissions might curtail development of world oil production, or negative consequences of climate change might slow world economic activity and hence oil consumption in the future. These considerations are covered in the Sec. 2-4.

2-4 Transportation Energy Demand, Greenhouse Gas Emissions, and Climate Change

Climate change is the process by which increased concentration of CO_2 and other GHGs in the atmosphere leads to higher average world temperatures and changes in climate patterns that have largely detrimental effects on systems in both the built and natural environment. Just as the peaking of world conventional petroleum production is the most important problem for energy security, responding to climate change is arguably the most pressing ecological challenge of the present time.

Conversion of fossil fuels (oil, gas, and coal) into various types of usable energy is the single most important contributor to climate change. While other human activities such as forestry and agriculture also play a role, and changes in these activities can reduce the threat of climate change, the single highest priority action society can take to combat climate change is to alter the way we generate and consume energy. Transportation is one of the main contributing sectors to the release of CO_2 from fossil fuel combustion, as it consumes approximately one-fourth of total world fossil fuel usage, so it is to be expected that the transportation sector must contribute to the reduction of CO_2 emissions to the atmosphere.

2-4-1 Growth in CO_2 Emissions and Impacts from Climate Change

As already illustrated in Fig. 1-5 in Chap. 1, both the total quantity of CO_2 present in the atmosphere and the annual rate at which CO_2 emissions occur are increasing over time. Values in Fig. 1-5 are given in carbon equivalents; on this basis, emissions increased from 2004 to 2008 by 9%, from 7.5 to 8.3 gigatonnes of carbon equivalent. Measured in mass of CO_2, this change is an increase from 27.4 to 30.3 gigatonnes per year. From the period of the 1990s onward, concern about climate change has grown steadily and, in response, some countries have stabilized or even decreased their rate of annual CO_2 emissions. The positive impact of these countries is more than offset, however, by countries in which year-on-year emissions are growing and, as a result (except for the year 2009 when the global economic downturn was in full effect), the emissions rate each year since 1998 has been higher than the year before.

The three groups of countries (BRIC, United States, and Europe) in Fig. 2-13 illustrate the underlying drivers of the global trend. In 2010, these countries combined represented approximately 70% of world CO_2 emissions. Emissions for the United States and Europe remained roughly constant over the period 1980 to 2010, but because BRIC emissions increased substantially after 2000, the sum of emissions from these three regions has increased overall since 1980.

Negative impacts from climate change come in various forms, as shown in Table 2-3. They range in scope from widespread decline in the amount of surface ice in polar regions to changes in the behavior of plants and animals as they respond to changing climatic conditions. Some changes are the result of generally increased average temperatures, such as extreme heat events in both tropical regions (e.g., Pakistan) and temperate ones (e.g., Russia). Also, some impacts are already being felt currently, such as the increased cost of damage from extreme weather events that are made stronger by greater availability of thermal energy from a warmer planet. Other impacts may only be felt in the future, even though trends are currently in motion that will result in those impacts. For instance, glaciers such as the one that supplies drinking water to the city of La Paz in Bolivia may be in decline, but the point at which their productivity begins to slow and supplies to the population become restricted may be some way off.

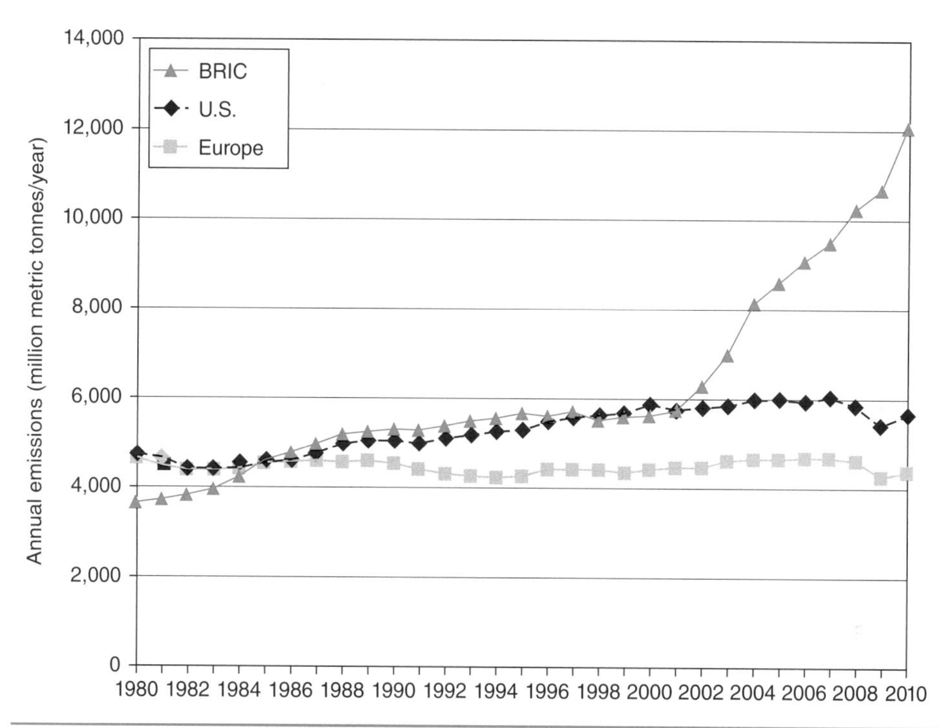

FIGURE 2-13 Annual CO_2 emissions from Europe, the United States, and the BRIC countries in million tonnes per year, 1980 to 2010.

Note: "Europe" group includes all western European countries, small nation states (e.g., Monaco, Lichtenstein, etc., and former East Bloc countries (Poland, Estonia, etc.), but not member states of the former Soviet Union (Ukraine, Belarus, etc.).

Source: U.S. Energy Information Administration.)

Contribution of the Transportation Sector to GHG Emissions and Climate Change

Transportation has been found to be one of the fastest-growing contributors across all sectors to GHGs in general and CO_2 emissions in particular since 1990, measured in percentage terms. As shown in Fig. 2-14, in the United States the emissions for heavy-duty highway vehicles (mostly heavy-duty trucks, with a small fraction of full-length buses included) rose 63% over the period from 1990 to 2005. Light trucks, which are primarily used for passenger transportation, increased 64% over the same period. Passenger transportation contributes to other types of emissions in the figure, including passenger cars (the single largest source in 2005), aviation, and rail transportation of passengers. Freight transportation contributes to emissions from aircraft (air freight in either dedicated freight aircraft or in the freight compartment of passenger aircraft) and other nonroad modes in the figure, which include rail freight, pipeline movements, and marine shipping. A rapid increase in transportation emissions, outpacing most other sectors, has been observed in other industrialized countries as well. In the United Kingdom, for example, the Department for Transport found that between 1990 and 2010 total GHG emissions from the transport sector increased by 11%.[1]

[1]U.K. Department for Transport (2012).

Effect	Description
Receding glaciers and loss of polar ice	Glaciers are receding in most mountain regions around the world. Notable locations with loss of glaciers and sheet ice include Mt. Kilimanjaro in Africa, Glacier National Park in North America, La Mer de Glace in the Alps in Europe, and the glaciers of the Himalayas. Also, the summer ice cover on the Arctic Ocean surrounding the North Pole is contracting each year. Loss of mountain glaciers may jeopardize water supply systems for major urban areas that depend on them (e.g., La Paz, Bolivia) and lack of summer ice will increase absorption of incoming solar radiation in the Arctic, further stimulating climate change. Glaciers also serve as tourist attractions in various parts of the world and their eventual disappearance would have a negative impact on the tourist industry.
Extreme heat events and drought events	In the 1990s and 2000s, extreme heat waves have affected regions such as Europe, North America, and South Asia. A heat wave in Moscow, Russia, drove temperatures above 100°F five times during the summer of 2010, where previously they had not exceeded this level in modern record keeping. On June 1, 2011, the highest temperature ever recorded in Asia of 53.5°C (128.3°F) was reached in Pakistan. Drought events are also more probable, such as the 2012 drought that affected much of the grain-growing region of the central United States, reducing agricultural productivity.
Change in animal and plant ranges, including invasives and pest species	Due to milder average temperatures in temperate regions, flora and fauna have pushed the limit of their range northward in the northern hemisphere and southward in the southern hemisphere, or to higher elevations in tropical regions. This includes pest species, such as insects that carry diseases, entering areas in which human and nonhuman populations previously were not affected by them. It also includes invasive plants that can crowd out native species or interfere with crops.
Interference with oceanic life	Climate change leads to both higher average ocean temperatures and acidification due to increased concentration of CO_2 in the water, both of which can negatively harm oceanic life. For instance, coral reef bleaching, in which corals turn white due to declining health and eventually die, is widespread in the world's seas and oceans. Along with other human pressures such as overfishing and chemical runoff, increased seawater temperature due to global warming is a leading cause of bleaching. Acidification also harms small and microscopic oceanic life, as is the case with certain phytoplankton that are negatively affected by higher levels of acidity. Since these small species are the foundation of oceanic food chains, this change can have repercussions for larger species such as fish or marine mammals.
Rising sea levels	Both the heating of the upper levels of water in the oceans, resulting in expansion of the affected water (warmer water is less dense than colder water), and runoff from melting glaciers in regions such as Antarctica or Greenland, raise ocean levels. Sea level rise makes regions more vulnerable to flooding during extreme weather events, such as occurred with Hurricane Sandy in late October and early November 2012, where an unusually high storm surge along with high winds destroyed areas of the coastal built environment in the states of New Jersey and New York in the United States, and flooded road and train tunnels around New York City.
Growing strength of extreme weather events	The presence of increased thermal energy in tropical and temperate oceans can contribute to stronger hurricanes, cyclones, and other extreme weather events. Hurricane Sandy mentioned above was judged to be unusually strong by historical standards for a hurricane making landfall in the late autumn along the upper reaches of the U.S. Atlantic coast (at around 40° north latitude).

Table 2-3 Current and Ongoing Negative Impacts from Climate Change

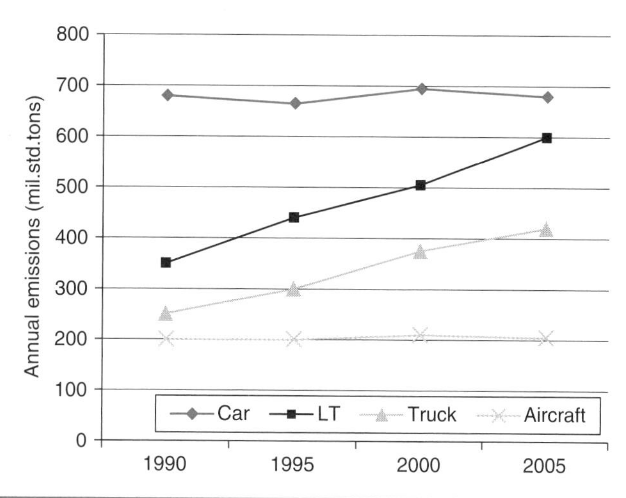

FIGURE 2-14 CO_2 emissions from U.S. transportation modes 1990-2005 in million tons of CO_2

Note: "LT" = light truck (i.e., pickup trucks, minivans, and sport utility vehicles); "Truck" = heavy duty truck, typically greater than 3.5 tons and used for hauling freight.

Source: U.S. Environmental Protection Agency (2007). Attribution: Thanks to J Bernstein, S Matharu, S Tao, R Larin, H Al-Jefri, and S Un for compiling this data as part of their Masters of Engineering project.

Challenges with Preventing CO_2 Emissions from the Transportation Sector

From an engineering perspective, climate change poses some particularly complex and difficult problems, due to the nature of the fossil fuel combustion and CO_2 emissions process. Unlike air pollutants such as nitrogen oxide (NO*x*), CO_2 cannot be converted into a more benign by-product using a catalytic converter in an exhaust stream. Also, due to the volume of CO_2 generated during the combustion of fossil fuels, it is not easy to capture and dispose of it in the way that particulate traps or baghouses capture particles in diesel and power plant exhausts (although such a system is being developed for CO_2 emissions from power plants, in the form of *carbon capture and sequestration* (CCS), which often involves long-term underground storage of CO_2). At the same time, CO_2 has an equal impact on climate change regardless of where it is emitted to the atmosphere, unlike emissions of air pollutants such as NO*x*, whose impact is location and time dependent. Therefore, a unit of CO_2 emission reduced anywhere helps to reduce climate change everywhere.

Threat Posed by Climate Change to Transportation Infrastructure and Systems

Just as the normal function of the transportation system at the present time can threaten the stability of the world's climate, climate change can in turn threaten the viability of the transportation infrastructure. Certain facets of climate change have already arrived and can be expected to have an impact for the foreseeable future, such as the increased ability of extreme weather events to affect low-lying areas. It is therefore essential that, in addition to investing in energy systems that address the root causes of climate change, society also increases investment in the protection of vulnerable transportation infrastructure from storms and flooding.

For example, as a result of Hurricane Sandy in October 2012, numerous tunnels and rail links in the greater New York City area and the state of New Jersey were damaged by wind and flooding. The resulting loss of use and repair cost both led to significant economic losses, estimated at $60 billion for the affected region as of December 2012. Effective investments in strengthening of this infrastructure so that it could survive a similar event in the future, with negligible damage and no loss of days of operation once the storm had passed, might yield a very positive return on investment. These investments might include changes to tunnel mouths and other orifices leading to below sea-level tunnels to make them less likely to flood.

2-4-2 Role of GHGs in Greenhouse Effect and Climate Change

The problem of climate change is directly related to the "greenhouse effect," by which the atmosphere admits visible light from the sun, but traps reradiated heat from the surface of the earth in much the same way that a glass greenhouse heats up by admitting light-waves but trapping infrared radiation. Gases that contribute to the greenhouse effect are called *greenhouse gases* (GHGs). Some level of GHGs are necessary for life on earth as we know it, since without their presence the earth's temperature would be on average below the freezing temperature of water. However, increasing concentration of GHGs in the atmosphere may lead to widespread negative consequences from climate change.[2]

Among the GHGs emitted to the atmosphere, some are *biogenic,* or resulting from interactions in the natural world, while others are *anthropogenic,* or resulting from human activity other than human respiration. The most important biogenic GHG is water vapor, which constitutes approximately 0.2% of the atmosphere by mass, and is in a state of quasi-equilibrium in which the exact amount of atmospheric water vapor varies up and down slightly around a long-term average, due to the hydrologic cycle (evaporation and precipitation). The most important anthropogenic GHG is CO_2, primarily from the combustion of fossil fuels but also from human activities that reduce forest cover and release carbon previously contained within trees and plants to the atmosphere. Other important GHGs include methane (CH_4), nitrous oxide (N_2O), and various human-manufactured chlorofluorocarbons (CFCs). While the contribution to the greenhouse effect of these other GHGs per molecule is greater than that of CO_2, their concentration is much lower, so overall they make a smaller contribution to the warming of the atmosphere. The contribution of water vapor to the greenhouse effect is also changing: as the average temperature of the planet warms, the amount of moisture held in the atmosphere increases, which acts to trap heat but also deflect some incoming solar radiation away from the planet, so that the net effect is difficult to assess with accuracy.

Both short-term and long-term cycles are at work bringing CO_2 into and out of the atmosphere. The collection of plants and animals living on the surface of the earth exchange CO_2 with the atmosphere through photosynthesis and respiration, while carbon retained more permanently in trees and forests eventually returns to the atmosphere through decomposition. In a healthy forest, this transfer of CO_2 is balanced by CO_2 removed by new trees that grow in place of the old. The oceans also absorb carbon from and emit carbon to the atmosphere, and some of this carbon ends up accumulated at the bottom of the ocean. Lastly, human activity contributes carbon to the atmosphere, primarily from the extraction of fossil fuels from the earth's crust, which then result in CO_2 emissions due to combustion.

---

[2]Both the calculation of total energy available from the sun and the difference between surface temperature of the earth with and without the greenhouse effect are left as end-of-chapter exercises.

The problem of increasing atmospheric levels of CO_2 stems from an imbalance in emissions to the atmosphere that is relatively small compared to the ongoing exchange between terrestrial and oceanic reservoirs on the one hand and atmospheric reservoirs on the other. In other words, exchange of CO_2 to and from the atmosphere is on the order of 600 Gt CO_2 in each direction each year, with an excess going to the atmosphere compared to coming from it on the order of just 4 Gt CO_2. Nevertheless, because this imbalance is sustained year on year and, in fact, is increasing slightly due to rising annual CO_2 emissions from the combustion of fossil fuels, the concentration of CO_2 in the atmosphere measured in parts per million volume (ppmv, or the number of CO_2 molecules per million molecules in the atmosphere of all types) is also rising. This gradual and slightly accelerating growth in CO_2 concentration is shown in Fig. 2-15, where the average ppmv value for every second year is graphed for the years 1958 to 2010. Also shown is the deviation between the global average temperature for every second year and the 30-year average for the period 1951 to 1980. As observed in the figure, from 1958 to 1982 the annual average temperature fluctuated around 0°C, but after 1982 it has climbed steadily into positive territory, surpassing + 0.6° in 2010.

2-4-3 Steps toward Climate Stabilization through CO_2 Reduction

The possible negative consequences from climate change both now and into the future provide strong motivation for the community of nations to *stabilize* atmospheric CO_2 concentration at a certain level beyond which concentration would rise no further. The Intergovernmental Panel on Climate Change (IPCC) is the main international body charged with creating and enforcing policies that address climate change. According to the IPCC, the increase over the period from 1900 to 2100 might range from 1.1 to 6.4°C. Values at the high end of this range would lead to very difficult changes that would be worthy of the effort required to avoid them. Because we cannot know with certainty

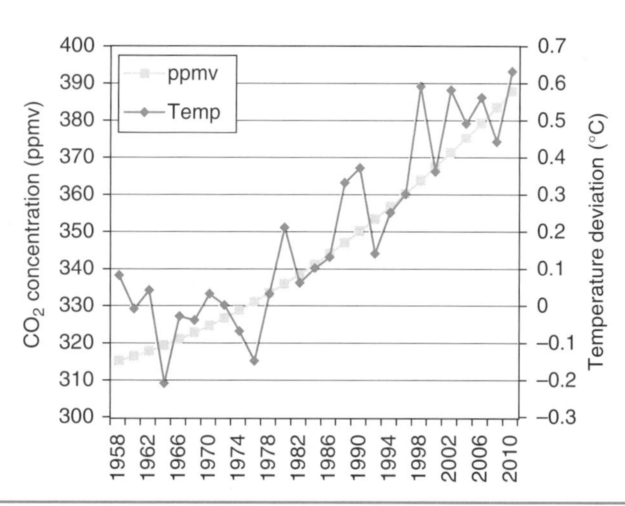

FIGURE 2-15 Comparison of concentration of CO_2 in the atmosphere in parts per million, 1958 to 2010, and temperature deviation in degrees Celsius compared to 1951 to 1980 average temperature.

Source: National Ocean & Atmosphere Administration (NOAA), for temperature; Carbon Dioxide Information & Analysis Center (CDIAC), for CO_2.

how average global temperature will respond to continued emissions of GHGs, and because the possibility of increases of as much as 5 or 6°C at the high end have a real chance of occurring if no action is taken, it is in our best interest to take sustained, committed steps toward transforming our energy systems so that they no longer contribute to this problem. Working through the IPCC, the community of nations has in fact agreed on a goal of holding global temperature increase to no more than 2°C, an outcome of the climate conference in Copenhagen in 2009.

To guide their work and also explain climate-change policy to constituent governments and individuals, the IPCC has created CO_2 stabilization scenarios, as shown in Fig. 2-16. Each scenario begins in Fig. 2-16(*a*) with a pathway for annual CO_2 emissions, at first increasing in the period out to at least the year 2025 (or beyond in many of the scenarios), but eventually peaking and declining so that by the year 2300 most scenarios converge to levels at or below 3 Gt carbon per year, so that concentration levels in Fig. 2-16(*b*) have stabilized at various possible levels of ppmv CO_2, as illustrated. Also, there is a band of shaded color around the emissions pathways, indicating that there is uncertainty about which exact pathway might lead to the CO_2 concentration level shown. Economic pathways and their consequences for CO_2 emissions and concentration are also included in the figures and given labels such as "A1," "A1B," "B1," and so on. Note that the rates of emissions shown in Fig. 2-16(*a*) are total flows from the surface to the atmosphere and not net emissions after taking account of amounts that might have been reabsorbed by plants or oceans, and are therefore larger than the figure of approximately 4 Gt CO_2 net outflow to the atmosphere quoted above.

The differences between annual emissions scenarios, and the resulting difference in stabilization levels of CO_2, are quite substantial at the extremes. For instance, the lowest scenario in Fig. 2-2-16(*b*), which stabilizes at 450 ppmv, shows annual emissions peaking already around the year 2015 at ~9.5 GtC/year before declining to around 2 GtC/year in the year 2100. At the other extreme, the highest scenario, which is shown with a solid line in Fig. 2-16 and stabilizes at around 1000 ppmv, increases to ~15 GtC/year in the year 2075, or nearly twice current levels, before decreasing to 4 GtC/year in the year 2300. This level of variation in the outcome of efforts to rein in carbon emissions could clearly have a profound impact on the average global temperature level that is eventually reached. To make the connection between pathways in the figure and underlying demographic and technological factors more tangible, Example 2-3 illustrates the estimation of levels in the near future based on current trends.

Example 2-3 Suppose an emissions pathway is created from a beginning position in the year 2000 of total emissions at 8 GtC/year, consistent with Fig. 2-15(*a*), and world population at the historically observed value of 6.08 billion. (The figure of 8 Gt of carbon per year is equivalent to 29.3 Gt of CO_2 per year on the basis of 12 weight units of carbon per 44 weight units of CO_2, but the remainder of this example will be calculated on the basis of GtC, to facilitate comparison with Fig. 2-15.) Thus the world average per capita emissions in 2000 are $(8 \times 10^9$ tonne C$)/(6.08 \times 10^9$ persons$) = 1.316$ tonne C/person/year. Also assume the population growth rate in the years 2000 and 2020 are the UN's observed and projected values of 76.5 and 68.4 million people per year by 2020, respectively. Assume that the number of people added each year declines linearly in the intervening years, for example, in 2010, the average of the 2000 and 2020 values or 72.45 million is added. Lastly, assume that high per-capita emissions countries become less carbon-intensive while industrializing countries become more so, so that the two factors balance each other and world per-capita emissions remain constant at 2000 levels. What is the resulting emissions rate in 2020, and how does it compare to the various curves in Fig. 2-15(*a*) over the 2000 to 2020 time period?

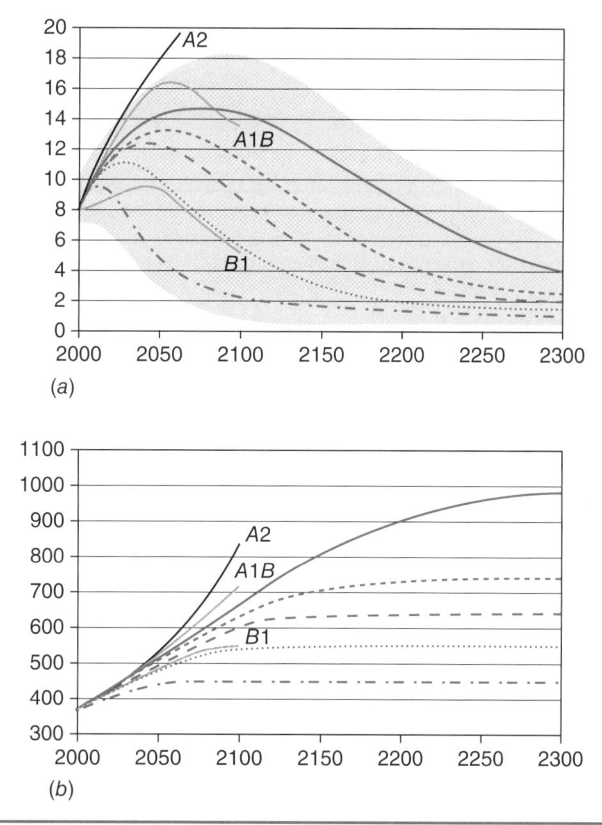

Figure 2-16 Possible pathways for CO_2 emissions (*a*) and atmospheric concentration (*b*) for a range of stabilization values from 450 to 1000 ppm.

Note: The numbered curves are IPCC economic growth scenarios, for example, A1B = rapid economic growth and peaking population with balance of energy sources, A2 = continued population growth with widely divergent economic growth between regions, B1 = similar to A1B but transition to less materially intensive economy, cleaner energy sources.

Source: Intergovernmental Panel on Climate Change (2001). Reprinted with permission.

Solution

Since the per-capita emissions rate is constant and the population grows nearly but not quite linearly as well (11% decrease in annual population added over 20 years), the emissions rate grows nearly linearly to 9.91 GtC/year. Emissions at 5-year intervals are shown in the following table. From visual inspection, the level of 9.91 GtC/year puts the pathway somewhere between the 450 ppmv and 550 ppmv pathways in Fig. 2-15.

Year	Population (million)	Emissions (GtC/year)
2000	6080	8.00
2005	6458	8.50
2010	6827	8.98
2015	7185	9.45
2020	7533	9.91

The rise to nearly 10 GtC/year represents an almost 25% increase in the annual emissions rate, which implies nonlinearly rising cumulative emissions of CO_2 to the atmosphere. In other words, if the emissions rate had remained flat at 8 GtC/year for 20 years, the total emissions would be 160 GtC. Thanks to increasing emissions rates, the actual emissions from the scenario are 188 GtC.

The effect of the proposed scenario on CO_2 concentration requires knowledge of the uptake of CO_2 by plants and oceans and is therefore beyond the scope of this example (a calculation involving both emissions and concentration has been left as an exercise at the end of the chapter). Notice, however, that the concentration value for the various scenarios in Fig. 2-15 only begin to level out once the emissions rates begin to drop below 2000 levels, for example, emissions levels in the 450 ppmv drop below 8 GtC/year in approximately 2030, and at this point the concentration begins to approach 450 ppmv asymptotically. Since the proposed scenario has annual emissions rates growing almost linearly, it is not yet at the point where emissions rates are falling and a final stabilized concentration level can be projected.

For comparison, world carbon emissions in the year 2000, according to the USEIA solely from use of fossil fuels for energy conversion were at 6.51 GtC/year, or 1.07 tonnes/person. By 2010, they had risen to 8.67 GtC/year, which is faster than the rate of population growth, so that the per capita emissions rate increased to 1.27 tonnes/year. Thus the assumption that the per-capita emissions rate could be held constant as presented in Example 2-3 may prove optimistic between now and the year 2020, since per-capita emissions may continue to increase beyond the 2010 rate and surpass the 1.32 tonnes/year used in the example.

In any case, the purpose of the various IPCC scenarios is to illustrate to world decision makers and citizens alike the tradeoffs between taking action on climate change either sooner or later. Both physical pathways, exemplified by the stabilization levels between 450 and 1000 ppmv, and political/economic pathways, in which the actions and performance of the world economy drives emissions and concentration levels, are presented. The advantage of peaking the growth of annual CO_2 emissions more rapidly and thereafter reducing them below 2 GtC/year earlier is that the resulting stabilization level is lower and arrives sooner; on the other hand, such an outcome requires rapid action in the near future to reduce the amount of carbon released per unit of energy consumed. In each scenario, once emissions reach a level of 2 GtC/year or less, emissions to the atmosphere from human activity are assumed to be in balance with absorption by oceans or land-based plant life, so that the concentration level remains constant.

Efforts to Actively Reduce CO_2 Concentration Levels toward Preindustrial Levels

The IPCC scenarios in the preceding section consider pathways by which atmospheric CO_2 might reach stabilization, but they do not consider what might happen after that point is reached. In terms of maintaining the healthy function of the global climate system and the natural environment in general, it is desirable not only to stabilize CO_2 but to reduce it back to current levels (e.g., 400 ppmv in 2013) or below, if possible. This outcome might arise from natural forces as CO_2 gradually settles out of the atmosphere, leading to lower concentration levels, but such a transfer of carbon back to the surface might also require centuries to run its course. CO_2 added to the atmosphere at present is thought to have an expected lifetime of approximately 400 years. Allowing CO_2 to remain in the atmosphere at high concentrations in the meantime, even after concentration levels stop increasing, may prove unacceptable due to heavy damage caused by climate change.

Motivated by a desire to reduce atmospheric CO_2 at a faster rate, a number of governments, academic researchers, and nongovernmental organizations are working to find solutions to actively remove CO_2 from the atmosphere. For example, the U.S.-based nongovernmental organization (NGO) 350.org campaigns for accelerated steps to reduce the rate of CO_2 emissions rise as a step toward actively reducing atmospheric concentration back to 350 ppmv or below, since there is a consensus among climate scientists that this is the maximum allowable safe level for the long term. As shown in Fig. 2-14, the atmosphere reached this level in 1990; in the subsequent 22 years up to 2012, it increased by an additional 42 ppmv. In a similar vein, in 2007, entrepreneur and philanthropist Richard Branson offered a $25 million prize for a device or system that could win approval from a panel of international judges for its ability to remove significant quantities of CO_2 from the atmosphere on an ongoing basis.

These campaigns notwithstanding, efforts to develop reliable and affordable technologies and systems for large-scale removal of CO_2 from the atmosphere for purposes of CCS stand at a preliminary level at this time. There are two general approaches:

- *Biological uptake:* Biological organisms including oceanic plant life and trees and plants on land might be enhanced or accelerated so that they absorb carbon at a substantial rate. Oceanic CO_2 removal leads to acidification and other negative consequences from higher CO_2 levels in sea water, so these efforts are currently focused on land-based sequestration. Carbon might be stored long term in forests that are allowed to grow back. It might also be stored in a product called *biochar*, which is produced from plant matter and creates an inert carbonaceous substance that can be plowed into the ground to sequester carbon and simultaneously improve agriculture yield. The largest problem with biological methods known at the present time is that their maximum global yield in any realistic configuration is nowhere near the rate of emissions on the order of ~9 GtC/year. There are academic discussions of biological systems that could sequester carbon much more rapidly, but these are in a conceptual stage.
- *Mechanical removal:* This approach entails the deployment of devices that circulate air through a chamber where CO_2 is selectively removed and then the remaining air passes out. The energy supply for such a device would need to be carbon-free, or at least release significantly less carbon to the atmosphere than the amount of carbon captured. Thus, the device might use renewable or nuclear energy, or else use fossil energy coupled directly to a CCS system, since the site would in any case require a means of removing CO_2 once it had been captured by the device and transporting it to some long-term CCS operation. CCS coupled to mechanical removal could happen in the form of *geological sequestration*, in which the CO_2 is pumped underground into saline aquifers, abandoned mines, or other locations from where it would be unable to escape. It could also occur in the form of *capture in carbonaceous materials*, in which the carbon is reacted with sources of, for example, calcium or magnesium to form calcium carbonate or magnesium carbonate. Of the stages in the mechanical removal system, the device for capturing CO_2 from the air is the least developed. Initial analyses suggest such a carbon-removal device is possible, but prototypes have not yet been built. More generally, both the great volume of CO_2 capture required to significantly alter atmospheric concentration levels and the requirement for cost-effectiveness of the process pose significant challenges to its development.

To summarize, both biological and mechanical options for carbon sequestration face challenges for scaling of output. Nevertheless, some pieces of potential systems are taking shape. As an indication of steps already taken to develop a comprehensive

Figure 2-17 Sleipner gas platform sequestration process.

Source: StatoilHydro. Reprinted with permission.

system, Fig. 2-17 shows the system already at work in the Sleipner gas field in the North Sea. The Norwegian company StatoilHydro separates about 1 million tonnes of CO_2 per year from gas being extracted from the field and reinjects it into a saline aquifer formation above the gas field but below the caprock, successfully retaining the CO_2 below the seabed without leakage to date.

2-5 Summary

Two of the most pressing issues facing the global community are the challenge of meeting rising demand for energy and that of counteracting global climate change caused by increasing levels of CO_2 and other GHGs in the atmosphere. Transportation is a major consumer of nonrenewable resources, and is in fact among the four major sectors of economic activity (industry, residential, and commercial, along with transportation) the sector that is the most dependent on petroleum for all but a small percentage of primary energy requirements. Transportation is also a major emitter of CO_2, as it is the second largest emitting economic activity after electricity generation. Thus any discussion of sustainable transportation must include these two topics.

The situation with transportation energy and CO_2 emissions poses a dilemma: access to transportation services (and the resulting energy use and emissions) is strongly correlated with a strong economy and high quality of life, and yet energy as it is currently extracted and used heavily burdens the natural environment. To solve this problem, the world must develop sources of energy for transportation that are plentiful and accessible to both developed and developing countries, but that do not harm the environment. Conventional petroleum supplies have already peaked in some countries (such as the United States) and are at or near their peak globally. The fitting of either Gaussian or logistics curves fit to historical production patterns and then extrapolation into the future can be used to predict possible future scenarios for conventional oil production. Nonconventional sources such as shale oil or tar sands can potentially extend the world transportation energy supplies for several decades, but do not address the concerns around GHG emissions in their current form. A complete solution to both the energy security and climate-change challenge might involve continued use of fossil fuels with some form of carbon capture and sequestration (CCS) alongside expanded use of renewables and continued use of nuclear energy. CCS could take place either during the conversion of fossil fuels to a carbon-free transportation fuel, or as part of a system for removing CO_2 already in the atmosphere.

References

Cambridge Energy Research Associates (2006). Press Release 60907-9. Cambridge Energy Research Associates, Cambridge, MA.

Hubbert, M. K. (1956). *Nuclear Energy and the Fossil Fuels.* American Petroleum Institute, Drilling and Production Practices: 7–25.

Intergovernmental Panel on Climate Change (IPCC) (2001). *Climate Change 2001: Synthesis Report.* IPCC, Geneva.

International Energy Agency (2012). *IEA Energy Statistics.* Electronic resource, available from *www.iea.org.* Accessed Nov.12, 2012.

Sperling, D. and D. Gordon (2009). *Two Billion Cars: Driving Toward Sustainability.* Oxford University Press, Oxford, UK.

U.K. Department for Transport (2012). *Transport Statistics Great Britain: 2012.* DfT, London.

U.S. Energy Information Administration (2012). *International CO_2 Emissions Trends.* Electronic resource, available at *www.eia.gov.* Accessed Nov. 25, 2012.

Further Readings

350.org. (2012) 350 Science. Web resource, available at *www.350.org.* Accessed Dec. 4, 2012.

Brohan, P., J.J. Kennedy, I. Harris, et al. (2006). "Uncertainty Estimates in Regional and Global Observed Temperature Changes: A New Dataset from 1850." *Journal of Geophysical Research* 111.

Global CCS Institute. (2012) "Sleipner CO_2 Injection." Web resource, available at *http://www.globalccsinstitute.com/projects/12401.* Accessed Dec. 5, 2012.

Marland, G., T. Boden, and R. Andres (2003). *Global, Regional, and National Fossil Fuel CO_2 Emissions.* Report, Carbon Dioxide Information Analysis Center, Oak Ridge National Laboratory, U.S. Department of Energy, available online at *cdiac.esd.ornl .gov.* Accessed Jan. 22, 2014

McKibben, B. (2012) "Global Warming's Terrifying New Math." *Rolling Stone,* Aug. 2, 2012.

Pickrell, J. (2004) "Oceans Found to Absorb Half of All Man-Made Carbon dioxide." *National Geographic News,* July 15, 2004. Electronic resources, available at *www.news.nationalgeographic.com.* Accessed Apr. 19, 2004.

Oreskes, N. (2004). "Beyond the Ivory Tower: The Scientific Consensus on Climate Change." *Science* 306(5702):1686.

Robinson, A., N. Robinson, and W. Soon (2007). "Environmental Effects of Increased Atmospheric Carbon Dioxide." *Journal of American Physicians and Surgeons* 12(3): 7990.

U.S. EPA (2007) *Trends in Greenhouse Gas Emissions.* USEPA, Washington, DC.

Exercises

2-1. Suppose the EUR for world conventional oil resources is 3.5 trillion barrels. Find on the internet or other source data on the historical growth in world oil production from 1900 to the present. Then use the Gaussian curve technique applied to annual production to predict: (a) the year in which the consumption peaks, (b) the world output in that year, and (c) the year following the peak in which the output has fallen by 90% compared to the peak.

2-2. Repeat Prob. 2-1 using cumulative production data and the logistics function, also known as the technological substitution function.

2-3. It is a well-known fact that because of the greenhouse effect caused by greenhouse gases (GHGs) such as CO_2, the surface temperature of the planet earth is noticeably higher than if there were no layer of GHGs in our atmosphere. The current average world temperature is approximately 288K or 15°C, before taking account of ongoing impact from climate change. In this problem, you will estimate the impact of GHGs by considering the case of our planet without them. Solar energy arrives from the sun at the average rate of 1372 W/m^2, of which 30% is reflected back into space and 70% penetrates the atmosphere and provides energy to the earth. The earth's radius is, on average, approximately 6,400 km. Use the following additional information: with no greenhouse effect, the surface temperature of the planet T is the value where the amount of energy coming from the sun is in equilibrium with the amount of energy leaving due to black body radiation. The flux of black body radiation $F_{BB} = \sigma T^4$, where $\sigma = 5.67 \times 10^{-8}$ W/m^2·°K^4 and T is measured in units of degrees Kelvin, or K.

Hint: In order to solve correctly, take into account the fact that the flux of energy from the sun is spread over the surface of the planet earth, which is spherical, while the radiation intercepted by the earth is in the shape of a disc.

a. What is the predicted value of T with no greenhouse effect in degrees K?

b. What is the value of T in degrees centigrade? What is one practical impact of such a temperature for life on the planet?

2-4. Adequacy of energy for the world: In 2008, world total energy consumption was estimated at 520 exajoules (EJ) 1 EJ = 10^{18} joule; the amount of energy is equivalent to 493 quadrillion Btu, or quads, although the rest of this problem is solved in metric units. Solar energy arrives from the sun at the average rate of 1372 W/m^2, of which 30% is reflected back into space and 70% penetrates the atmosphere and provides energy to the earth. The earth's radius is, on average, approximately 6,400 km. Recall that 1 W = 1 joule/second.

a. What is the ratio of the energy reaching the planet to the total world energy consumption?

b. Without revealing the exact answer, it is safe to say that the amount of energy reaching the earth from the sun is orders of magnitude larger than the amount used around the world. Since this is the case, why is only a small fraction of energy used by humans extracted directly from arriving sunlight? Give one possible reason, short answer form, in one or two sentences.

Hint: There are many possible correct reasons.

UNIT 2

Tools and Techniques

CHAPTER 3

Systems Tools

3-1 Overview

The goal of this chapter is to present and explore a number of systems tools that are useful for understanding and improving transportation systems. We first introduce the terminology for understanding systems and the systems approach to implementing a transportation technology or system. The remainder of the chapter reviews specific systems tools. Some are qualitative or "soft" tools, such as conceptual models for describing systems or the interactions in systems. Other tools are quantitative, including life-cycle analysis, multicriteria analysis, optimization, and Divisia analysis.

3-2 Introduction

Transportation systems are complex, often involving combinations of fixed infrastructure systems, mechanically powered vehicles, and electrical or electronic systems for energy supply and digital communications. These systems are designed to achieve a range of possible goals, such as the movement of passengers and/or freight; operation on land (including roads, railways, or pipelines), on water, or through the air; and the fulfillment of transportation requirements over short segments or long distances. Any transportation system must interact with input from human operators (both professionals that control them and passengers that use them) and with other systems that are connected to it (for example, telecommunications systems); these inputs may come from sources that are distributed over a wide geographic expanse. Transportation systems therefore lend themselves to the use of a systems approach to problem solving.

When transportation professionals make decisions about transportation systems, in many instances they must take into account a number of goals or criteria that are local, regional, or global in nature. These goals can be grouped into three categories:

1. *Physical goals:* Meeting physical requirements that make it possible for the system to operate. For example, many transportation systems require some combination of a network within which to operate; a type of vehicle, vessel, or craft; and a terminal facility (stop, station, or port) at which passengers and goods can enter or leave the system. They also require energy resources for power, which in most cases involves mechanical power derived from either electricity or a liquid fuel of some sort. Any type of transportation system must function efficiently, reliably, and safely.

2. *Financial goals:* Meeting monetary objectives related to the transportation system. For a privately held system (collection of delivery vehicles, fleet of shuttle buses belong to company, etc.), the goal may be to meet internal transportation requirements at minimum cost. For a commercial transportation system, the goal may be to earn a profit based on total revenues from the system exceeding total cost.
3. *Environmental goals:* Satisfying objectives related to the way the transportation system impacts the natural environment. Regional or global impacts include the emissions of greenhouse gases that contribute to climate change, air pollutants that degrade air quality, and physical effects from extracting resources used either for materials (e.g., metals used in fabrication of vehicles or vessels) or energy (e.g., crude oil, uranium). Goals may also address impacts in the physical vicinity of the transportation system, such as noise and vibrations, pollution of surrounding bodies of water, the land footprint of the system, or disruption of natural habitats.

Depending on the size of the transportation system, not all objectives will be considered in all cases. For example, a private individual or small private firm might only consider financial goals in detail when choosing a transportation system or adjusting their current choice of transportation system. In such a case, they might assume that physical goals are already met by the vendors of the various types of equipment involved, such as vehicles, and that environmental goals are outside the scope of their decision. On the other hand, a large-scale transportation system project may be required by law to consider a wide range of environmental impacts, and in addition be concerned with the choice of basic infrastructure and vehicle or vessel technologies, as well as the projections of costs and revenues over a project lifetime that lasts for decades.

When considering steps to meet environmental goals, the distinction between *eco-efficiency* and *eco-sufficiency* is insightful. Eco-efficiency is the pursuit of technologies or practices that deliver more of the service that is desired per unit of impact; for example, a vehicle that drives further per unit of energy consumed or kilogram of CO_2 emitted. Eco-sufficiency is the pursuit of the combination of technologies and their amount of use that achieves some overall target for maximum allowable environmental impact. For example, a region might determine a maximum sustainable amount of CO_2 emissions from driving, and then find the combination of more efficient vehicles and programs to reduce kilometers driven that achieves this overall goal. In general, eco-sufficient solutions are also eco-efficient, but eco-efficient solutions may or may not be eco-sufficient.

3-3 Fundamentals of the Systems Approach

In the preceding section, we discussed the existence of multiple goals for transportation systems. In Chaps. 1 and 2, we also identified the presence of different stakeholders that each wish to meet their transportation needs. These stakeholders may be separated by space or socioeconomic status, such as the rich countries and the poor countries, or by time, such as the current generation and future generations. The presence of both multiple goals and multiple stakeholders favors thinking about transportation from a systems perspective.

3-3-1 Initial Definitions

First, what is a *system*? A system is a group of interacting *components* that work together to achieve some common *purpose*. Each system has a *boundary*, either physical or conceptual, and entities or objects that are not part of system lie outside the boundary. The area outside the boundary is called the *environment*, and *inputs* and *outputs* flow across the boundary between the system and its environment. The inputs and outputs may be physical, as in raw materials and waste products, or they may be virtual, as in information or data. Also, each component within a system can (usually) itself be treated as a system (called a *subsystem*), unless it is a fundamental element that cannot be further divided into components. Furthermore, each system is (usually) a component in a larger system.

Figure 3-1 shows a conceptual example of the human-built infrastructure system relevant to sustainable transportation systems engineering, since it illustrates the importance of understanding the interaction between human-made and natural elements. The built infrastructure system consists of all physical infrastructure built and controlled by humans. It has two components, institutional infrastructure and physical infrastructure. The institutional infrastructure represents the functioning of human knowledge in a conceptual sense; it receives information flows from the environment and the physical infrastructure about the current condition of each, and then transmits information to the physical infrastructure, usually in the form of commands meant to control the function of the physical infrastructure. The boundary shown is conceptual in nature; in physical space, the human physical infrastructure and natural environment are interwoven with each other.

The highway network in a country, which we might rename the "highway system," represents a more concrete example of a transportation-related system. Each vehicle (car, truck, bicycle, etc.) is a subsystem in the grid system, and if we focus on motor vehicles specifically, each motor vehicle can be further subdivided into supporting subsystems, such as body, engine or motor, transmission, pollution control apparatus, etc. Going in the other direction, the broadest possible view is that the national highway system is a subsystem of the complete set of all national highway systems around the globe, which form the single *global highway system*. Even if they are not physically

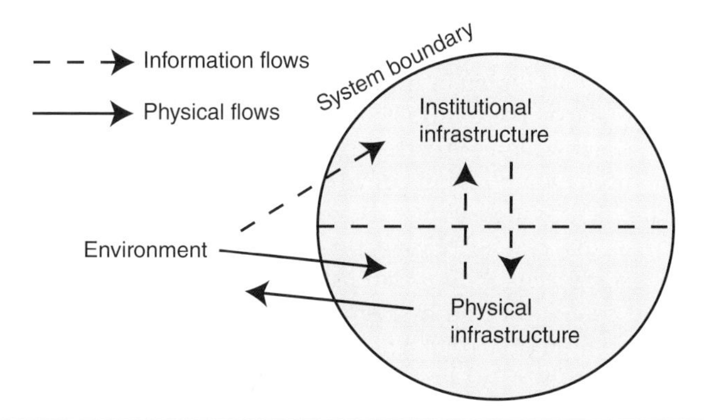

FIGURE 3-1 Conceptual model of built infrastructure system, surrounded by natural environment. *Source:* Vanek (2002).

connected, the various highway systems of the world are related to one another in the world because they draw from the same world fossil fuel resource for the majority of the energy required for propulsion. One could conceivably argue that the world highway system is a subsystem in the system of all highway systems in the universe, including those on other planets in other galaxies that support life forms that have also developed highway systems, but such a definition of system and subsystem would serve no practical purpose (!). Failing that, the world highway system is the broadest possible systems view.

In the field of systems engineering, a distinction is made between the term *system* as a concept and the *systems approach*, which is the process of conceptualizing, modeling, and analyzing objects from a systems perspective in order to achieve some desired outcome. When implementing the systems approach, the purpose, components, and boundary of a system are usually not preordained, but rather defined by the person undertaking the analysis. This definition process can be applied to physical systems, both in a single physical location (e.g., power plant) or distributed over a large distance (the electrical grid). Systems can also be purely conceptual, or mixtures of physical and conceptual; for example, the transportation services marketplace system with its physical assets, energy companies, government regulators, and consumers, where both physical transportation assets (vehicles, vessels) and logical commands interact with each other.

A system can be *abstracted* to whatever level is appropriate for analysis. For example, a regional transportation system may consist of road and rail infrastructure, various types of moving vehicles, energy-delivery systems, and dedicated monitoring and control telecommunications systems. If an engineer needs to solve a problem that concerns primarily just one of these subsystems, it may be a legitimate and useful simplification to create an abstract model of the transportation system as being a single system in isolation. In other words, for electrical and electronics monitoring and control systems, the transportation network could be treated as consisting only of an electrical network and computer network, where computers gather information about the condition of the system and then transmit commands or updates by either hard-wired or wireless networks. This abstracting of the system can be enacted provided that by doing so no critical details are lost that might distort the results of the analysis.

In the transportation network example, two different meanings for the term *subsystem* are possible. One is the definition just presented, that is, isolating the entire electrical subsystem in all geographic parts of the transportation system. On the other hand, each major geographic region of the system is a subsystem, each containing all the elements described above: fixed infrastructure, vehicles, energy supply, and controls. Neither definition of subsystems is correct or incorrect; in the right context, either definition can be used to understand and improve the function of the system as a whole.

In regard to identifying a boundary, components, and purpose of a system, there are no absolute rules that can be followed in order to implement the systems approach the right way. This approach requires the engineer to develop a sense of judgment based on experience and trial-and-error. In many situations, it is in fact beneficial to apply the systems model in more than one way (e.g., first setting a narrow boundary around the system, then setting a broad boundary) in order to generate insight and knowledge from a comparison of the different outcomes. It may also be useful to vary the number and layers of components recognized as being part of the system, experimenting with both simpler and more complex definitions of the system.

3-3-2 Steps in the Application of the Systems Approach

The engineer can apply the systems approach to a wide range of transportation topics, ranging from specific projects (e.g., a new public transportation line or local alternative fuel supply network) to broader ones (e.g., implementing a policy to improve the optimal mix of the various transportation modes by improving the interchange points between their respective networks, for example, where the highway and rail networks intersect). The following four steps are typical:

1. *Determine stakeholders:* Who are the individuals or groups of people who have a stake in how a project or program unfolds? Some possible answers are customers, employees, shareholders, community members, representatives of nongovernmental organizations (NGOs), or government officials. What does each of the stakeholders want? Possible answers include a clean environment, a population with adequate access to transportation services, a return on financial investment, and so on.
2. *Determine goals:* Based on the stakeholders and their objectives, determine a list of goals, such as expected performance, cost, or environmental protection. Note that not all of the objectives of all the stakeholders need to be addressed by the goals chosen. The goals may be synergistic with each other, as is often the case with technical performance and environmental protection—technologies that perform the best often use fuel the most efficiently and create the smallest waste stream. The goals may also be in conflict with each other, such as level of technical performance and cost, where a high-performing technology often costs more.
3. *Determine scope and boundaries:* In the planning stage, decide how deeply you will analyze the situation in planning a project or program. For example, for the economic benefits of a project, will you consider only direct savings from reduced fuel use, or will you also consider secondary economic benefits to the community of cleaner air stemming from your system upgrade? It may be difficult to decide the scope and boundaries with certainty at the beginning of the project, and the engineer should be prepared to make revisions during the course of the project, if needed (see Step 4).
4. *Iterate and revise plans as the project unfolds:* Any approach to project or program management will usually incorporate basic steps such as initial high-level planning, detail planning, construction or deployment, and launch. For a large fixed asset such as a transportation infrastructure project, "launch" implies full-scale operation of the asset over its revenue lifetime; for a discrete product such as a rail vehicle or hybrid car, "launch" implies mass production. Specific to the systems approach is the deliberate effort to regularly evaluate progress at each stage and, *if it is judged necessary,* return to previous stages in order to make corrections that will help the project at the end. For example, difficulties encountered in the detail planning stage may reflect incorrect assumptions at the initial planning stage, so according to the systems approach one should revisit the work of the initial stage and make corrections before continuing. This type of *feedback* and *iteration* is key in the systems approach: in many cases, the extra commitment of time and financial resources in the early stages of the project is justified by the increased success in the operational or production stages of the life cycle.

The exact details of these four steps are flexible, depending in part on the application in question and the previous experience of the practitioner responsible for implementing the systems approach. Other steps may be added as well.

The Systems Approach Contrasted with the Conventional Engineering Approach

The opposite of the systems approach is sometimes called the *unit approach*. Its central tendency is to identify one component and one criterion at the core of a project or product, generate a design solution in which the component satisfies the minimum requirement for the criterion, and then allow the component to dictate all other physical and economic characteristics of the project or product. For a vehicle, the engineer would choose the key component (e.g., the power train, composed of engine and transmission), design the power train so that it meets the criterion, which in this case is likely to be technical performance (maximum power output or mechanical efficiency), and then design all other components around the power train.

In practice, for larger projects it may be impossible to apply the unit approach in its pure form, because other criteria may interfere at the end of the design stage. Continuing the example of a transportation infrastructure system project, the full design may be presented to the customer only to be rejected on the grounds of high cost. At that point, the designer iterates by going back to the high-level or detail design stages and making revisions. From a systems perspective, this type of iteration is undesirable because the unproductive work embedded in the detail design from Round 1 could have been avoided with sufficient thinking about high-level design versus cost at an earlier stage. In practice, this type of pitfall is well-recognized by major engineering organizations, which are certain to carry out some type of systems approach at the early stages in order to avoid such a financial risk.

While the systems approach may have the advantages discussed, it is not appropriate in all situations involving transportation. The systems approach requires a substantial commitment of *administrative overhead* beyond the unit approach in the early stages in gathering additional information and considering tradeoffs. Especially in smaller projects, this commitment may not be merited. Take the case of a contractor who is asked to repair a small section of a minor road to address problems with pavement quality or other issues. The paving requirements may be specified by codes in many countries, so the contractor has limited latitude to optimize the overall design of the roadway as part of the repairs. She or he might reduce the amount of concrete or asphalt needed by using a systems approach and discussing the arrangement of grades, culverts, sidewalks, and intersections so as to reduce overall cost. However, the local government agency that bid on the project has by this time already decided that they would like to keep the geometric design of the roadway in its current form, and they are likely not interested in spending much extra time revisiting this plan to save a relatively small amount of money on paving material. In short, there may be little to learn from a systems approach to this problem. A *linear* process may be perfectly adequate, in which the contractor looks at the segment of road, acquires the necessary materials, carries out all necessary paving, and bills the customer at the end of the job, based either on a fixed contract price or the cost of time and materials.

3-3-3 Examples of the Systems Approach in Action

The following examples from past and current transportation problem solving reflect a systems approach. In some cases, the players involved may have explicitly applied systems engineering, while in others they may have used systems skills without even being aware of the concept (!).

Example 1: Transportation demand management (TDM)

According to the Victoria Transportation Policy Institute, transportation demand management (TDM) is "a general term for strategies that result in more efficient use of transportation resources."[1] As such, TDM is an approach to the management of transportation services in which transportation decision-makers actively manage the total demand for transportation measured in passenger-miles or trips rather than assuming that the demand is beyond their control and that the only policy tool at their disposal is to respond to increasing demand by increasing capacity (measured in highway lane-miles or size of public transportation system).

The distinction in transportation decision-making with and without TDM is illustrated in Fig. 3-2. The shortcoming for the transportation network in the original case is that the system boundary is drawn in such a way that customer demand for transportation is outside it. Not only must the infrastructure provider (often the government)

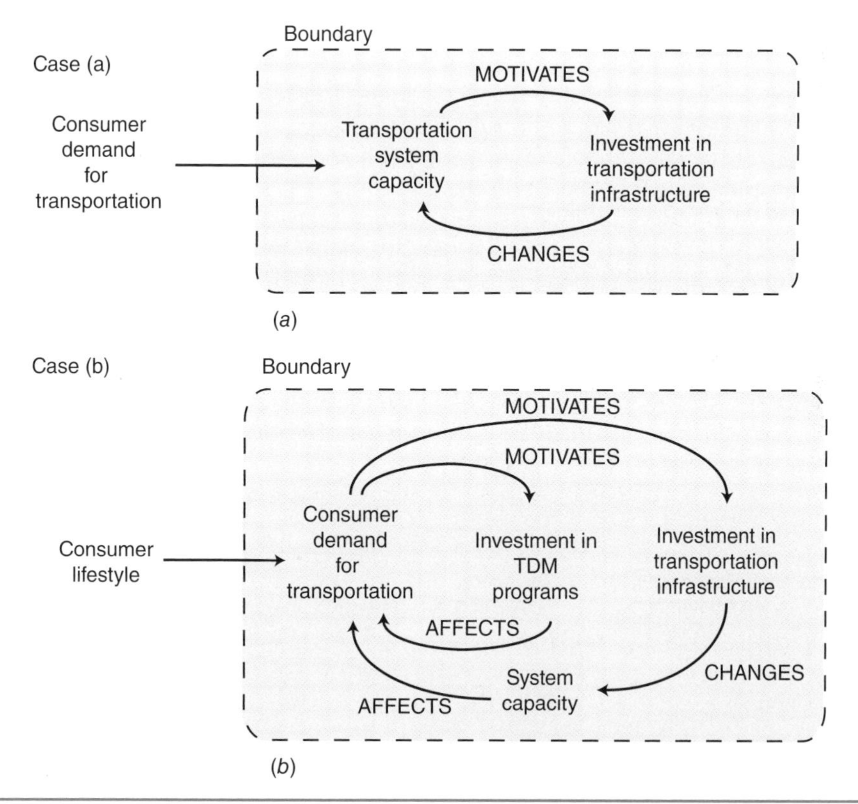

Figure 3-2 System boundaries and function: (*a*) conventional case, (*b*) case with TDM.

[1]See Victoria Transportation Policy Institute (2012). VTPI documents examples of TDM applications in locations such as Vancouver, Canada; Bellingham, WA the United States; and Darlington, Peterborough, and Worcester, England.

respond to the customer by building expensive new infrastructure when consumer demand increases, but consumer demand may be highly peaked, so that the network's capital-intensive capacity is only well utilized at peak demand times and is underutilized at others, and therefore is not very cost-effective.

In the system that uses TDM, the boundary has been redrawn to incorporate consumer demand for transportation as an endogenous part of an interacting system. The new input acting from outside the boundary is labeled *consumer lifestyle*, which can represent the goal of traveling to the workplace, to shopping, to education, or to other amenities. In the previous system this input did not appear, since the only aspect of the consumer seen by the system was her/his demand for mobility. The effect of consumer goals on the system is important, because it means that the consumer has the right to expect a certain level of access to amenities that the transportation system can provide, and that the system cannot meet its needs by arbitrarily cutting the consumer off from access. Within the system, however, a new path has been added between demand and TDM, representing the decision-maker's ability to influence demand through various TDM measures, for example, varying prices charged with time of day, encouraging more efficient use of public transportation or bicycling, and so on. The option of *investment* continues to exist alongside TDM, but now the decision-maker can determine the optimal combination of expanding supply and managing demand.

Example 2: Combined energy supply for transportation and electric grid

This example involves a comparison of the refining of liquid fuels (gasoline, diesel, and jet fuel) for transportation and the generation of electricity for distribution through the electric grid. Currently, these two energy supply systems are mostly separate, although the electric grid does provide a small fraction of the energy that the transportation sector consumes (for electric subways, intercity trains, and so on).

With the current demand for new sources of both electricity and energy for transportation, engineers are taking an interest in the possibilities of developing systems that are capable of supporting both systems. This transition is represented in Fig. 3-3. The *original* boundaries represent the two systems in isolation from each other. When a new boundary is drawn that encompasses both systems, many new solutions become available that were not previously possible when the two systems were treated in isolation.

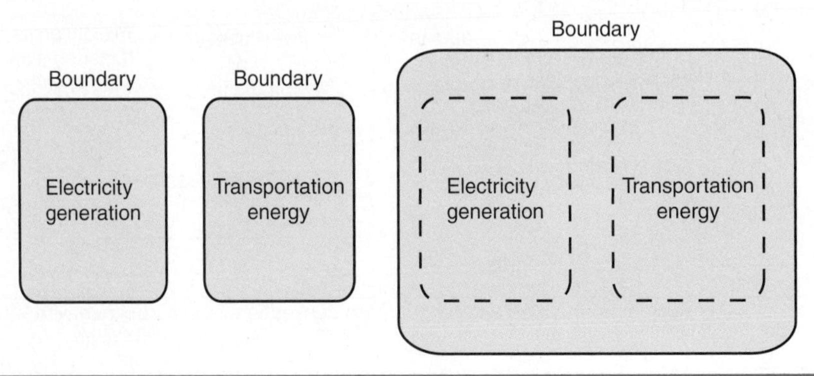

Figure 3-3 Electricity and transportation energy as two systems in isolation and combined into a single large system.

For example, an intermittent energy source such as a wind farm might sell electricity to the grid whenever possible, and at times that there is excess production, store energy for use in the transportation energy system that can be dispensed to vehicles later. Plug-in hybrid vehicles might take electricity from the grid at night during off-peak generating times, thereby increasing utilization of generating assets and at the same time reducing total demand for petroleum-based liquid fuels. Also, new vehicle-based power sources such as fuel cells might increase their utilization if they could plug in to the grid and produce electricity during the day when they are stationary and not in use.

Example 3: Modal shifting among freight modes

In the freight arena, movers of goods can reduce energy consumption by shifting them to more energy-efficient modes of transportation. (A similar approach applies to passenger transportation.) This policy goal is known as *modal shifting*. In the case of land-based freight transportation, rail systems have a number of technical advantages that allow them to move a given mass of goods over a given distance using less energy than movement by large trucks.

How much energy might be saved by shifting a given volume of freight from truck to rail? We can make a simple projection by calculating the average energy consumption per unit of freight for a given country, or the total amount of freight moved divided by the total amount of energy consumed, based on government statistics. Subtracting energy per unit freight for rail from that of truck gives the savings per unit of modal shifting achieved. Figure 3-4(*a*) represents this view of modal shifting.

This projection, however, overstates the value of modal shifting because it does not take into account differences between the movement of high-value, finished goods and

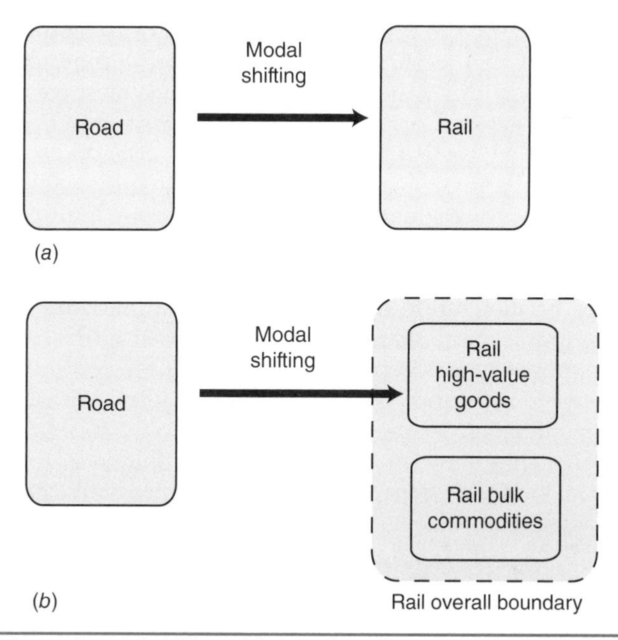

FIGURE 3-4 Two views of modal shifting of freight: (*a*) rail treated as uniform system, (*b*) rail divided into two subsystems.

bulk commodities in the rail system. The latter can be packed more densely and moved more slowly than high-value products, which customers typically expect delivered rapidly and in good condition. The movement of high-value goods also often requires the use of trucking to make connections to and from the rail network, further increasing energy use. In Fig. 3-4(*b*), the rail system is divided into two subsystems, one for moving high-value goods and one for moving bulk commodities. Connecting the modal shifting arrow directly to the high-value goods subsystem then makes the estimation of energy savings from modal shifting more accurate. A new value for average energy consumption of high-value rail can be calculated based on total movements and energy consumption of this subsystem by itself, and the difference between truck and rail recalculated. In most cases, there will still be energy savings to be had from modal shifting of freight, but the savings are not as great as in the initial calculation—and also not as overoptimistic.

3-4 Systems Tools Focused on Interactions between System Elements

In this section, we turn from the systems approach in general to specific systems engineering or systems analysis tools that can be applied when taking a systems approach to solving a problem. One commonality between the tools in this section is that they focus on the interaction between elements in the system, in either a qualitative or quantitative way.

3-4-1 Stories, Scenarios, and Models

The terms stories, scenarios, and models describe perspectives on problem solving that entail varying degrees of data requirements and predictive ability. Although these terms are widely used in many different research contexts as well as in everyday conversation, their definitions in a systems context are as follows. A *story* is a descriptive device used to qualitatively capture a cause-and-effect relationship that is observed in the real world in certain situations, without attempting to establish quantitative connections. A story may also convey a message regarding the solution of transportation systems problems outside of cause-and-effect relationships. A *scenario* is a projection about the relationship between inputs and outcomes, which incorporates quantitative data, without proving that the values used are the best fit for the current or future situation. It is common to use multiple scenarios to bracket a range of possible outcomes in cases where the most likely pathway to the future is too difficult to accurately predict. Lastly, a *model* is a quantitative device that can be calibrated to predict future outcomes with some measure of accuracy, based on quantitative inputs. Note that the term model used in this context is a *quantitative* model, as opposed to a *conceptual* model such as the conceptual model of the relationship between institutional infrastructure, physical infrastructure, and the natural environment in Fig. 3-1.

The relationship between the three terms is shown in Fig. 3-5. As the analysis proceeds from stories to scenarios to models, the predictive power increases, but the underlying data requirements also increase. In some situations, it may not be possible to provide more than a story due to data limitations. In others, the full range of options may be available, but it may nevertheless be valuable to convey a concept in a story to provide an image for an audience that can then help them to understand a scenario or model. In fact, in some situations, a modeling exercise fails because the underlying story was not clear or was never considered in the first place.

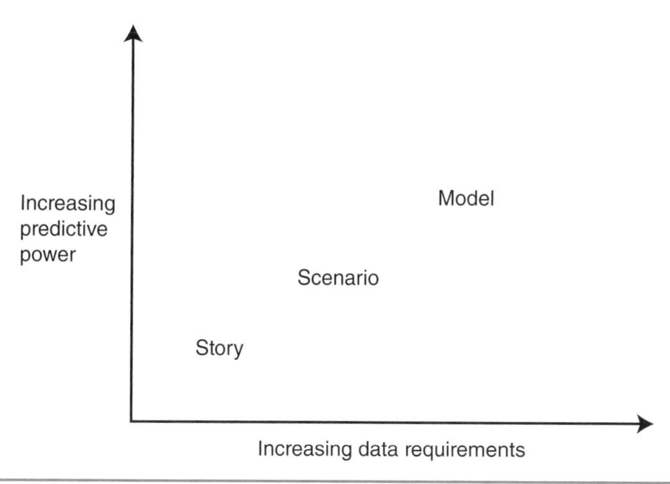

FIGURE 3-5 Predictive ability versus data requirements for stories, scenarios, and models.

Relationship between Stories, Scenarios, and Models: Example of "Rising Tide"

The following example illustrates how a common concept can be characterized using either a story, a scenario, or a model. In this case, the stories, scenarios, or models help us to visualize situations and problems, but do not necessarily generate the specific solutions to the problems themselves.

The story in this instance is the simple adage from economics that "a rising tide lifts all boats." The image used in this case is that of a harbor with boats moored; as the high tide moves in and the water level rises, all the boats in the harbor inevitably rise with it. The story suggests that, in the same way, improvements in the national or global economy will improve the situation for all people, and also make possible investments in more advanced transportation systems that enhance environmental protection or improve urban livability (by reducing congestion, improving average travel speed, reducing noise and other nuisances, etc.). The story by itself does not prove that the improvements in well-being will be spread sufficiently, that all boats will truly be "lifted" rather than some boats being left behind entirely. However, it describes with some accuracy the way in which economic resources of the wealthy countries (North America, Europe, and Japan) created markets for imports that lifted newly industrialized countries such as Taiwan or Korea to the point where they have recently been spending more on upgrading their national transportation systems.

Naturally, it is desirable to move beyond qualitative descriptions of the behavior suggested by the story to some level of quantitative analysis. At the same time, in some situations the connections between interactions between elements in a system (the different countries, demand for goods, financial flows, and policies in this case) may be too complex to capture in the form of a definitive mathematical function that relates one to the other. As a fallback, scenarios can be used to create a range of plausible outcomes that might transpire. Suppose a modeler wishes to explore the future impact of rapid economic expansion in certain major world economies on world economic growth in general and the resulting investment in transportation infrastructure. Each scenario starts from the present time and then projects forward the variables of interest, such as total economic growth, demand for transportation services, and type of transportation technology used. A mathematical relationship may still exist between variables, but the modeler is free to vary the relationship from scenario to scenario, using her or his subjective judgment, for example, to adjust the rate at which transportation demand grows in response to economic growth, across a likely range of values. Some scenarios will be more optimistic or pessimistic in their outcome, but each should be plausible, so that, when the analysis is complete, the full range of scenarios brackets the complete range of likely outcomes for the variables of interest. In the rising tide example, if one of the outcomes of interest were to estimate future transportation greenhouse gas (GHG) emissions, a range of scenarios might be developed that incorporate varying rates of economic growth among the emerging economies, the economic partners that import goods from these economies, and rates of improvement in clean transportation technology, in order to predict in each scenario a possible outcome for total emissions in some future year. A distinction might also be made between scenarios with and without the *policy intervention* to encourage greener transportation. Chapter 1 presented a possible scenario for pathways for global passenger and freight demand in the twenty-first century based on economic activity and population growth, in such a way that the same underlying data inputs (e.g., projected population levels by decade from 2000 to 2100) could be manipulated so as to construct alternative scenarios with different quantitative outcomes.

Going a step beyond scenario building in terms of levels of gathering and using data, the modeler might build an economic model of the relationship between the elements in the story. In this model, as wealth grows in the rich countries, imports also grow, leading to increased wealth in the industrializing countries and the upgrading of transportation systems. The model makes connections using mathematical functions, and these are calibrated by adjusting numerical parameters in the model so that they correctly predict performance from the past up to the present. It is then possible to extrapolate forward by continuing to run the model into the future. If the rate of adoption of clean transportation technology falls short of some desired goal, the modeler may introduce the effect of *policy variables* that use taxation or economic incentives to move both the rich and industrializing countries more quickly in that direction.

3-4-2 Systems Dynamics Models: Exponential Growth, Saturation, and Causal Loops

The concepts of stories, scenarios, and models illustrate the difference that increasing availability of data can make in allowing more quantitative representation of a system, but they do not explicitly incorporate any temporal dimension by which the changing state of the system could be measured over time, nor do they explicitly state what method should be used to incorporate feedback between outputs from one element in a system and inputs to another. This section presents time-series and causal loop models to serve these purposes.

Exponential Growth

In the case of transportation systems, exponential growth occurs during the early stages of periods of influx of a new technology, or due to a human population that is growing at an accelerating rate. Exponential growth is well known from compound interest on investments or bank accounts, in which an increasing amount of principal accelerates the growth from one period to the next. Exponential growth over time in demand for some activity, such as transportation demand, can be expressed as follows:

$$E(t) = a \cdot \exp(bt) \tag{3-1}$$

where t is the time in years

$E(t)$ is the annual energy consumption in year t

a and b are parameters that fit the growth to a historic data trend.

Note that a may be constrained to be the value of energy consumption in year 0, or it may be allowed other values to achieve the best possible fit to the observed data. Example 3-1 illustrates the application of the exponential curve to historical data, where a and b are allowed to take best-fit values.

Example 3-1 Car sharing, or shared use of cars among members in a membership organization designed to increase vehicle utilization rates and reduce driving cost (see Chap. 9), is an emerging transportation option in the United States and other countries. The following table provides annual membership levels for the United States from 2000 to 2008. Use the exponential function to fit a minimum-squares error curve to the data, using data points from the period 2000 to 2007 only. Then use the curve to calculate a projected value for the year 2008, and calculate the difference in energy between the projected and actual value.

Year	Members	Year	Members
2000	422	2005	76,420
2001	5,377	2006	102,993
2002	12,098	2007	184,292
2003	25,640	2008	279,174
2004	52,347		

Solution For each year, the estimated value of the number of car-sharing members $N_{est}(t)$ is calculated based on the exponential growth function and the parameters a and b. The values of a and b that best fit the modeled exponential curve to the observed data are found using a numerical solver. Let y be the year of a given data point. Then $N_{est}(y)$ is calculated as

$$N_{est}(y) = a \cdot \exp[b(y - 2000)]$$

The value of the squared error term $(ERR_y)^2$ in each year based on observed value $N_{obs,y}$ and modeled value $N_{est,y}$ is

$$(ERR_y)^2 = (N_{obs,y} - N_{est,y})^2$$

Finding the values of a and b that minimize the sum of the errors for the years 2000 through 2007 gives $a = 6329$ and $b = 0.4799$. The following table shows the observed and estimated values along with the error squared term. For example, for the year 2001, we have $N_{obs,2001} = 5{,}377$ and

$$N_{est}(2001) = 6329 \cdot \exp[0.4799(2001 - 2000)] = 10{,}227$$
$$(ERR)^2 = (10{,}227 - 5{,}377) = 2.35 \times 10^7$$

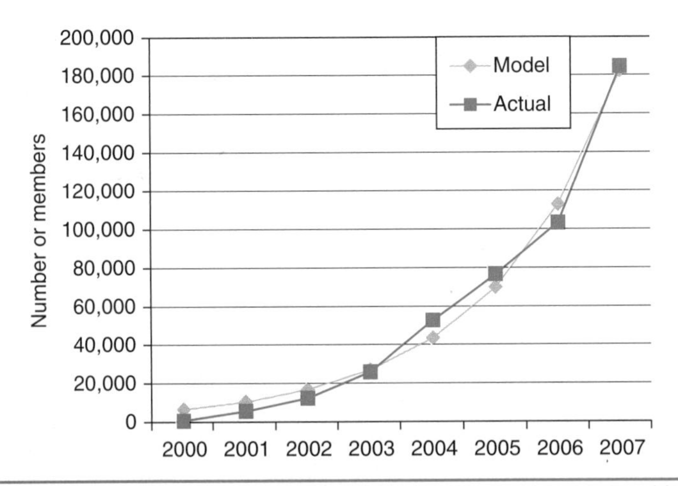

FIGURE 3-6 Observed versus modeled exponential growth of car-share membership, 2000 to 2008.

The results in tabular form are as follows:

Year	Actual	Year t	Model	Error	ERR^2
2000	422	0	6,329	5907	3.49E+07
2001	5,377	1	10,227	4850	2.35E+07
2002	12,098	2	16,525	4427	1.96E+07
2003	25,640	3	26,702	1062	1.13E+06
2004	52,347	4	43,146	-9201	8.47E+07
2005	76,420	5	69,717	-6703	4.49E+07
2006	102,993	6	112,652	9659	9.33E+07
2007	184,292	7	182,027	-2265	5.13E+06

Finally, solving for year = 2008 using Eq. (3-1) and parameter values $a = 6329$ and $b = 0.4799$ gives the following:

$$N_{est}(2008) = 6329 \cdot \exp[0.4799(2008 - 2000)] = 294{,}126$$

Thus the modeled value is quite close to the observed value of 279,174. The actual and estimated curves can be plotted as shown in Fig. 3-6.

Asymptotic Growth, Logistics Function, and Triangle Function

The growth of car-sharing membership discussed in Example 2-1 is typical of a successful new product or service that is taking root in a marketplace and experiencing exponential growth. Such growth cannot continue indefinitely. In many applications, ranging from growth of micro-organisms in a laboratory experiment, to spread of an infectious disease, to penetration of a new product or service into a marketplace, exponential growth gives way to a period of constant growth followed by tapering off of growth. Since in the case of a new service like car sharing there is a finite pool of persons who can join the market, and since there is a finite capacity for each consumer to use car-sharing services, it makes sense that there is also a limit on total market penetration that is possible. This trajectory is called *asymptotic* growth, in which the growth rate slows as it approaches some absolute ceiling in terms of magnitude.

Just as the exponential function models exponential growth, two other mathematical functions, the *logistics function* (also known as the *technological substitution function*) and the *triangle function*, model asymptotic growth. Taking the first of these, the logistics function is an empirically derived formula that has been found to fit many observed real-world phenomena ranging from the penetration of a disease into a population to the conversion of a market over to a new technology (e.g., the penetration of personal computers into homes in the industrialized countries). Although it can be written in different ways, one possible form is the following:

$$f(t) = \frac{F \cdot e^{(c_1 + c_2 t)}}{1 + e^{(c_1 + c_2 t)}} \tag{3-2}$$

where t is the value of time in appropriate units, years/months

$f(t)$ is either the population, in absolute units or the percentage (out of the entire population) in time t

c_1 and c_2 are parameters used to shape the function

F = maximum population or penetration

The values of c_1 and c_2 shape the rate at which $f(t)$ grows along the s-curve toward saturation at the final level of penetration F. As the value of t grows large, the impact of the value of the constant in the denominator becomes negligible and the ratio of numerator and denominator approaches unity times the value of F.

Unlike the logistics function, where the value of $f(t)$ approaches but never actually achieves the limiting value F, the triangle function depends on the user to set in advance the time period at which 100% penetration or the cessation of all further growth will occur. In the triangle function, the value of $p(t)$ is the percentage relative to the completion of growth, which can be multiplied by an absolute quantity (e.g., maximum number of vehicle owners or passengers using a given transportation technology or service) to compute the absolute value of penetration. Let a and b be the start and end times of the transition to be modeled. The triangle function then has the following form:

$$p(t) = 2\left(\frac{t-a}{b-a}\right)^2 \quad \text{if } a \le t \le \frac{a+b}{2}$$

$$p(t) = 1 - 2\left(\frac{b-t}{b-a}\right)^2 \quad \text{if } \frac{a+b}{2} \le t \le b \tag{3-3}$$

For times less than a, $p(t) = 0$, and for times greater than b, $p(t) = 1$.

In terms of the difference between the two functions, the logistics curve can be fit to the initial values to project when the population will get within some increment of full saturation, in forward-looking cases where the penetration is still relatively low (e.g., less than 10 or 15%) and the goal is to predict penetration in the future. It is also possible to use the logistics curve retrospectively to fit a curve to a technology penetration by adjusting the parameters c_1 and c_2 to minimize error. With the triangle function, the values of a and b are set at the endpoints (for instance, 30 years apart) to estimate the likely population at intermediate stages. Example 3-2 illustrates the application of both the logistics and triangle functions.

Example 3-2 Automobile production in a fictitious country amounts to 1 million vehicles per year, which does not change over time. Starting in the year 1980, automobiles begin to incorporate fuel injection instead of carburetors in order to reduce energy consumption. Compute future penetration of fuel injection as follows:

1. *Using the logistics function:* Suppose we treat 1980 as year 0, and that 5,000 vehicles are produced with fuel injection in this year. In 1981, the number rises to 20,000 vehicles. The expected outcome is 100% saturation of the production of vehicles with fuel injection. How many vehicles are produced with fuel injection in 1983?
2. *Using the triangle function:* Suppose now that only the total time required for a full transition to fuel injection is known, and that this transition will be fully achieved in the year 1990. How many vehicles are produced with fuel injection in 1983, according to the triangle model?

Solution

(a) *Logistics function*: Since the new technology is expected to fully penetrate the production of vehicles, we can set $F = 1 = 100\%$. In general, with several years in a historical curve we would use a numerical solver to find values of c_1 and c_2 that minimize error, but in this illustrative example there are two equations (for years 1980 and 1981) along with two unknown values of c_1 and c_2, so we can solve exactly. We first use an equation for 1980 to solve for c_1, and then solve for c_2 using an equation for 1981. In 1980, or year $t = 0$, since the penetration is 5,000 out of 1 million, we have $f(0) = 0.5\%$. We can therefore solve the following equation for c_1. (Note that since $t = 0$, c_2 cancels in the equation.):

$$f(0) = \frac{1 \cdot \exp(c_1 + c_2 \cdot 0)}{1 + \exp(c_1 + c_2 \cdot 0)} = \frac{1 \cdot \exp(c_1)}{1 + \exp(c_1)} = 0.5\%$$

Solving gives $c_1 = -5.29$. In 1981, at time $t = 1$, we have $f(1) = 2 \times 104/1 \times 106 = 2.0\%$, so we can now solve for c_2 in a similar manner:

$$f(1) = \frac{1 \cdot \exp(c_1 + c_2 \cdot t)}{1 + \exp(c_1 + c_2 \cdot t)} = \frac{1 \cdot \exp(-5.29 + c_2 \cdot 1)}{1 + \exp(-5.29 + c_2 \cdot 1)} = 2.0\%$$

Solving gives $c_2 = 1.40$. Lastly, we solve for the case of 1983, or $t = 3$:

$$f(3) = \frac{1 \cdot \exp(c_1 + c_2 \cdot t)}{1 + \exp(c_1 + c_2 \cdot t)} = \frac{1 \cdot \exp(-5.29 + 1.4 \cdot 3)}{1 + \exp(-5.29 + 1.4 \cdot 3)} = 25.2\%$$

Since $f(3) = 0.252$, the prediction is that $0.252(1 \times 106) = 252{,}000$ cars will be made with fuel injection in 1983.

(b) *Triangle function:* We adopt the same convention as in part (a), namely, $t = 0$ in 1980, $t = 1$ in 1981, and so on. Therefore $a = 0$, $b = 10$, and $(a + b)/2 = 5$. Since $t = 3 < (a + b)/2$, we use $p(t) = [(t - a)/(b - a)]^2$, which gives the following:

$$p(3) = \left(\frac{3 - 0}{10 - 3}\right)^2 = 0.18$$

$$(0.18)(1 \times 10^6 \text{ veh}) = 180{,}000 \text{ veh}$$

Therefore, the triangle function predicts that 180,000 cars are built with fuel injection in 1983.

The two curves from Example 3-2 are plotted in Fig. 3-7. Notice that the logistics function in this case has reached 85% penetration in the year 1985, when the triangle function is at 50%, despite having very similar curves for 1980 to 1982. A triangle curve starting in 1980 and ending in 1987 ($a = 0$, $b = 7$) would have a shape more similar to the logistics function.

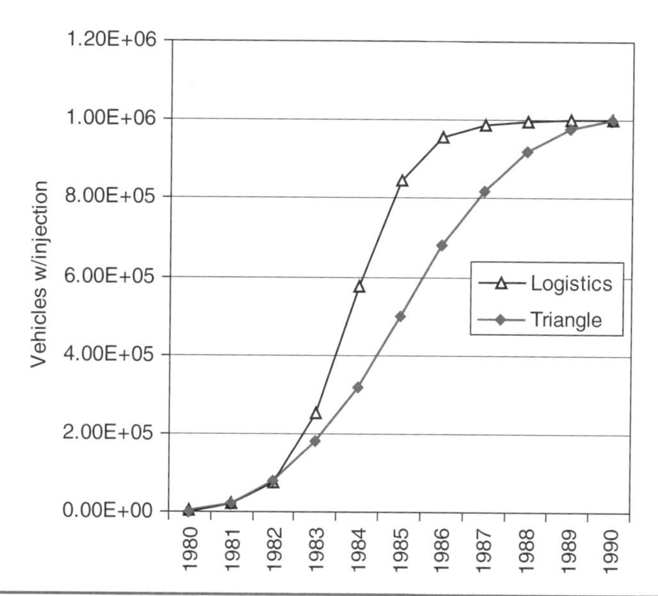

Figure 3-7 Projection of influx of fuel injection technology using logistics and triangle functions.

Causal Loop Diagrams

The cause-and-effect relationships underlying the trend in Example 3-2 can be described as follows. After the initial startup period (years 1980 and 1981), the industry gains experience and confidence with the technology, and resistance to its adoption lessens. Therefore, in the middle part of either of the curves used, the rate of adoption accelerates (1982 to 1985). As the end of the influx approaches, however, there may be certain models that are produced in limited numbers or that have other complications, such that they take slightly longer to complete the transition. Thus the rate of change slows down toward the end. Note that the transition from carburetors to fuel injection is not required to take the transition path shown; it is also possible that that transition could end with a complete transition of all the remaining vehicles in the last year; for example, in the year 1986 or 1987 in this example. However, experience shows that the "S-shaped" curve with tapering off at the end is a very plausible alternative.

These types of relationships between and catalysts for and constraints on growth can be captured graphically in *causal loop diagrams*, which map the connections between components that interact in a system and where each component is characterized by some quantifiable value (e.g., temperature of the component, amount of money possessed, and the like). As illustrated in Fig. 3-8, linkages are shown with arrows leading from sending to receiving components and, in most cases, either a positive or negative sign. When an increase in the value of the sending component leads to an increase in the value of the receiving component, the sign is positive, and vice versa. Often the sending component receives either positive or negative feedback from a loop leading back to it from the receiving component, so that the systems interactions can be modeled. A component can influence, and be influenced by, more than one other component in the model.

Figure 3-8 shows the general relationship between atmospheric CO_2 concentration, temperature, demand for heating and cooling, and CO_2 emissions rate, to which much

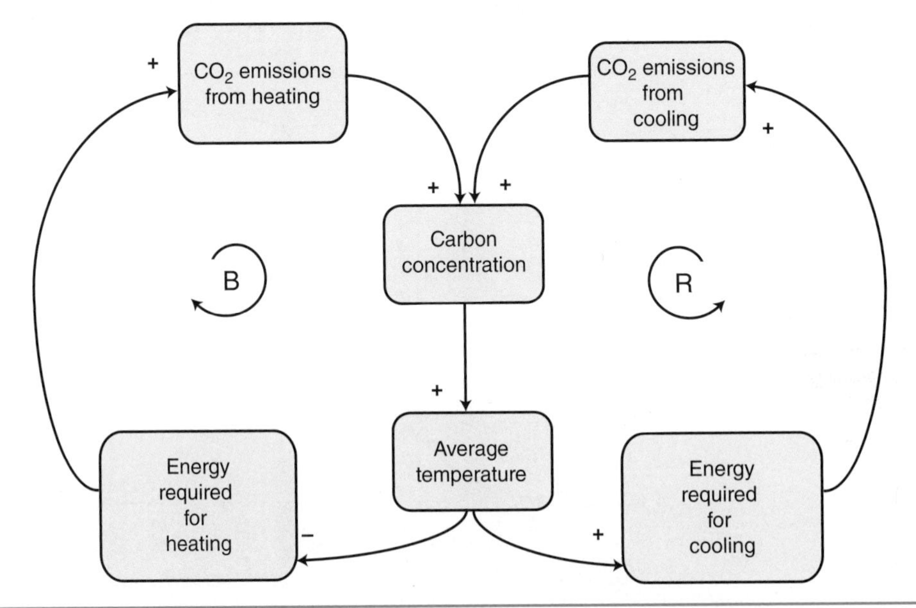

FIGURE 3-8 Causal loop diagram relating outdoor climate and indoor use of climate control devices.

of the built environment including transportation infrastructure is subject. In the figure, the symbol in the middle of the loop ("B" for balancing or "R" for reinforcing) indicates the nature of the feedback relationship around the loop. On the left side, increasing average temperature leads to reduced emissions of CO_2 from the combustion of fossil fuels for heating space, so the loop is balancing. On the right, increased CO_2 increases the need for the use of air conditioning, which further increases CO_2 concentration. The diagram does not tell us which loop is the more dominant, and hence whether the net carbon concentration will go up or down. Also, it only represents two of the many factors affecting average CO_2 concentration in the atmosphere; more factors would be required to give a complete assessment of the expected direction of CO_2 concentration. Thus it is not possible to make direct numerical calculations from a causal loop diagram. Instead, its value lies in identifying all relevant connections to make sure none are overlooked, and also in developing a qualitative understanding of the connections as a first step toward quantitative modeling.

The causal loop diagram in Fig. 3-9 incorporates two new elements, namely, a discrepancy relative to a goal and a delay function. In this case, a goal for the target annual CO_2 emissions is set outside the interactions of the model; there are no components in the model that influence the value of this goal, and its value does not change. The *discrepancy* component is then the difference between the goal and the actual CO_2 emissions from all sorts of energy production, including conversion of fossil fuels for transportation energy as well as other forms; for example, stationary power plants that produce electricity. Here we are assuming that these emissions initially exceed the goal, so as they increase, the discrepancy also increases. Since the target does not change, it does not have any ability to positively or negatively influence the discrepancy, so there is no sign attached to the arrow linking the two. The discrepancy then leads to the introduction of policies that reduce CO_2 emissions toward the goal. As the policies take hold,

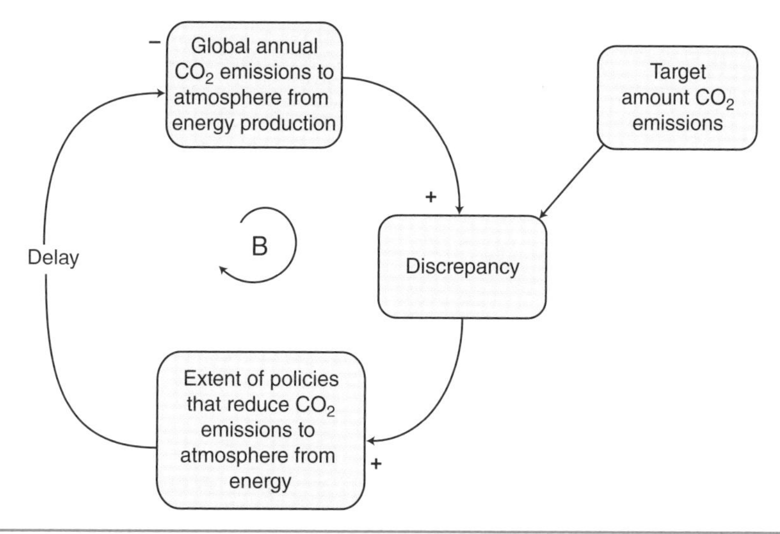

FIGURE 3-9 Causal loop diagram incorporating a goal, a discrepancy, and a delay: the relationship between CO_2 emissions and energy policy.

emissions decline in the direction of the target and the discrepancy decreases, so that eventually the extent of emissions reduction policies can be reduced as well. The link between policy and actual emissions is indicated with a *delay*, in that reducing emissions often takes a long lead time in order to implement new policies, develop new technologies, or build new infrastructures.

It is useful to explicitly include delays in the causal loop diagram since they can often lead to situations of overshoot and oscillation. Since the result of the policy may only be seen after some months or years, the adjustment to the policy may only occur after the condition of the system has overshot the desired target. Reversing the policy may again lead to a result that is observed only after the system has overshot the target in the other direction; continued overshooting in this way results in oscillation around the target. This behavior is analogous to a person adjusting the temperature of a shower or bath, where the delay between changing the rate of flow of hot and cold water at the faucet and the change in the water temperature coming out of the spout leads to oscillation between overshooting and undershooting the optimal temperature.

The simple examples of causal loop diagrams presented here can be incorporated into larger, more complete diagrams that are useful for understanding interactions among many components. The individual components and causal links themselves are not complex, but the structures they make possible allow the modeler to understand interactions that are quite complex.

3-5 Other Systems Tools for Transportation Systems

The remainder of this chapter discusses a range of tools used in analyzing transportation systems and making decisions that best meet the goal of developing transportation systems that are economical and sustainable. Some of the tools, such as life-cycle analysis (LCA), multicriteria analysis, optimization, Divisia analysis, or Monte Carlo simulation are more general in nature and can be applied to answering questions about ecological impact, system capacity, financial decision-making, and the like. Other tools, such as the Kaya equation and energy return on investment, deal specifically with energy and CO_2 emissions because these are such an important concern in achieving sustainable transportation.

3-5-1 Life-Cycle Analysis

When analyzing the performance of a transportation system in terms of its capacity, cost-effectiveness, ecological impact, or other metric, it is often a necessary simplification to focus on a single stage of its life cycle to make the performance analysis tractable. For instance, we might focus only on the *operation* or *end use* phase of the life of a transportation system or vehicle, and consider measures such as daily or yearly net revenue, fuel economy, and the like. In reality, the life of the system or vehicle has several stages, such as the construction or manufacturing stage that occurs *upstream* from end use. *Life-cycle analysis* is a technique that attempts to take a broader view of the impact of a system by including all parts of the life cycle. In this section, we focus on LCA applied to a single criterion, namely energy consumption. In Sec. 3-5-2, we incorporate multiple criteria so as to compare the effects on energy with effects on other objectives.

LCA can help the practitioner overcome at least two important problems in measuring performance and comparing alternatives, in this case in regard to energy consumption:

- *Incomplete information about the true impact of a technology on energy efficiency*: If only one stage of the life cycle, or some subset of the total number of stages, is used to make claims about energy savings, the analysis may overlook losses or quantities of energy consumption at other stages that wash out projected savings. For example, in the case of comparing vehicle-propulsion systems, one propulsion system may appear very efficient in terms of converting stored energy on board the vehicle into mechanical propulsion. However, if the process that delivers the energy to the vehicle is very energy intensive, the overall energy efficiency of the entire process from the original source may be poor.
- *Savings made at one stage may be offset by losses at another stage*: This effect can occur in the case of a manufactured product that has one or more layers of both manufacturing and shipping before it is delivered to a consumer as a final product. It is often the case that concentrating manufacturing operations in a single large facility may improve manufacturing efficiency, since large versions of the machinery may operate more efficiently, or the facility may require less heating and cooling per square foot of floor space. However, bringing raw materials to and finished goods from the facility will, on average, incur a larger amount of transportation than if the materials were converted to products, distributed, and sold in several smaller regional systems. In this case, gains in the manufacturing stage may be offset, at least in part, by losses in the transportation stage.

Table 3-1 shows eight life-cycle steps that can be applied to transportation systems in various ways, including both physical systems or products, such as fixed

Life Cycle Stage	Typical Activities
Resource extraction	Machinery used for mining, quarrying, etc., of coal, metallic ores, stone resources Energy consumed in pumping crude oil and gas Energy consumption in forestry and agriculture to maintain and extract crops for bio-energy resources, including machinery, chemical additives, etc.
Transportation of raw materials	Movement of bulk resources (coal, grain, bulk wood products) by rail, barge, truck Movement of oil and gas by pipeline Intermediate storage of raw materials
Conversion of raw materials into semi-finished materials and energy products	Refining of crude oil into gasoline, diesel, jet fuel, other petroleum products Creation of chemical products from oil, gas, and other feedstocks Extraction of vegetable oils from grain Production of bulk metals (steel, aluminum, etc.) from ores Creation of bulk materials (concrete, glass, fabric, paper, etc.) from various feedstocks
Manufacturing and final production	Mass production of energy-consumption-related devices (steam and gas turbines, wind turbines, motorized vehicles of all types, etc.) Mass production of consumer goods Mass production of information technology used to control energy systems
Transportation of finished products	Transportation by truck, aircraft, rail, or ship Intermediate storage and handling of finished products in warehouses, regional distribution centers, air- and marine ports, etc.
Commercial "overhead" for system management	Energy consumed in management of production/distribution system (office space) Energy consumed in retail space for retail sale of individual products (e.g., vehicle dealerships)
Infrastructure construction and maintenance	Construction of key components in transportation system (intercity road or rail corridor, station or terminal, port facility) Construction of resource extraction, conversion, and manufacturing plants Construction of commercial infrastructure (office space, shopping malls, car dealerships, etc.) Energy consumed in maintaining infrastructure
Demolition/disposal/recycling	Disposal of consumer products and vehicles at end of lifetime Dismantling and renovation of infrastructure Recycling of raw materials from disposal/demolition back into inputs to resource-extraction stage

Note: In approximately the order in which they occur in the life cycle of a transportation system, mass-produced transportation system component (e.g., vehicle), or transportation energy product.

Table 3-1 Examples of Categories in Life-Cycle Analysis

infrastructure systems (bridges, rail lines, and so on) or vehicles, and transportation energy products, such as electricity or motor fuels. The list of examples of each stage shown in the right column is not exhaustive, but does provide a representative list of the type of activities that consume energy in each stage. For transportation energy products, resources such as coal may be converted to electricity after the raw material stage, while other resources such as crude oil remain in a liquid form as they are refined, distributed, sold, and consumed in vehicles. Consumer products generally depend on several *chains* of conversion of raw materials to components before final assembly, such as a motor vehicle, which incorporates metal resources and wires, plastics, fabrics, and other inputs, all arising from distinct forms of resource extraction. Energy is consumed to greater or lesser degrees in all of these stages, as well as in the building of fixed infrastructure and in commercial enterprises needed to control the system.

In theory, one could estimate energy-consumption values for each stage of the life cycle, allocate these quantities among different products where multiple products rely on a single activity (e.g., energy consumed in office used to administer product), and divide by the number or volume of products produced to estimate an energy consumption value per unit. In practice, it may prove extremely laborious to accurately estimate the energy consumed in each activity of each stage. Therefore, as a practical simplification, the engineer can inspect stages for which energy consumption values are not easily obtained. If they can make a defensible argument that these stages cannot, for any range of plausible values, significantly affect the life-cycle total energy consumption, then they can either include a representative fixed value resulting from an educated guess of the actual value or explicitly exclude the component from the life-cycle analysis.

3-5-2 Multicriteria Analysis of Energy Systems Decisions

The preceding section presented the use of LCA using a single measure, namely that of energy consumption at the various stages of the life cycle. The focus on energy consumption as a sole determinant of the life-cycle value of a transportation system or component may be necessary to complete an LCA with available resources, but care must be taken that other factors are not overlooked. For example, a project may appear attractive on an energy LCA basis, but if some other aspect that causes extensive environmental damage is overlooked, for example, the destruction of vital habitat, the natural environment may end up worse off.

Multicriteria analysis of decisions provides a more complete way to incorporate competing objectives and evaluate them in a single framework. In this approach, the decision maker identifies the most important criteria for choosing the best alternative among competing projects or products. These criteria may include air pollution, water pollution, greenhouse gas emissions, or release of toxic materials, as well as direct economic value for money. The criteria are then evaluated for each alternative on some common basis, such as the dollar value of economic benefit or economic cost of different types of environmental damage, or a qualitative score based on the value for a given alternative relative to the best or worst competitor among the complete set of alternatives. An overall multicriteria *score* can then be calculated for alternative i as follows:

$$\text{Score}_i = \sum_c w_c x_{ic} \ , \forall i \tag{3-4}$$

where Score_i is the total score for the alternative in dimensionless units

w_c is the weight of criterion c relative to other criteria

x_{ic} is the score for alternative i in terms of criterion c.

Category	EPA Weight	Harvard Weight
Global warming	0.27	0.28
Acidification	0.13	0.17
Nutrification	0.13	0.18
Natural-resource depletion	0.13	0.15
Indoor air quality	0.27	0.12
Solid waste	0.07	0.10

Source: USEPA (1990) and Norberg-Bohm (1992), as quoted in Lippiatt (1999).

TABLE 3-2 Comparison of USEPA and Harvard Weights for Important Environmental Criteria

At the end of the multicriteria analysis, the alternative with the highest or lowest score is selected depending on whether a high or low score is desirable.

Naturally, how one weights the various criteria will have an important impact on the outcome of the analysis, and it is not acceptable to arbitrarily or simplistically assign weights, since the outcome of the decision will then appear arbitrary, or worse, designed to reinforce a pre-selected outcome. One approach is therefore to use as weights sets of values that have been developed using previous research, and which those in the field already widely accept. Table 3-2 gives relative weights from two widely used studies from the U.S. Environmental Protection Agency (EPA) and from Harvard University. In the tables, the category most closely related to energy consumption is global warming, since energy consumption is closely correlated with CO_2 emissions. This category has the highest weight in both studies (tied with indoor air quality in the case of EPA), but other types of impacts figure prominently as well. In both studies, acidification (the buildup of acid in bodies of water), nutrification (the buildup of agricultural nutrients in bodies of water), natural resource depletion, and indoor air quality all have weights of at least 0.10. Only solid waste appears to be a less pressing concern, according to the values from the two studies. In order to use either set of weights from the table correctly, the scoring system for the alternatives should assign a higher score to an inferior alternative, so that the value of Score_i will correctly select the alternative with the lowest overall score as the most attractive.

In one possible approach, the decision-maker can adopt either EPA or Harvard values as a starting point for weights and then modify values using some internal process that can justify quantitative changes to weights. In Table 3-2, the weights shown add up to 1.00 in both cases. As long as the value of weights relative to one another correctly reflects the relative importance of different criteria, it is not a requirement for multi criteria analysis that they add up to 1. However, it may be simplest when adjusting weights to add to some while taking away from others, so that the sum of all the weights remains 1.

3-5-3 Choosing among Alternative Solutions Using Optimization

In many calculations related to transportation systems, we evaluate a single measure of performance, either by solving a single equation or by solving a system of equations such that the number of equations equals the number of unknown variables, and the value of each unknown is therefore uniquely identified. In some situations, there may be a *solution space* in which an infinite number of possible combinations of *decision variable* values, or *solutions*, can satisfy the system requirements, but some solutions are better than others in terms of achieving the system objective, such as maximizing financial earnings or minimizing pollution. *Optimization* is a technique that can be used to solve such problems.[2]

We first look at optimization in an abstract form before considering examples that incorporate quantitative values for components of transportation systems. In the most basic optimization problem relevant to transportation systems, the decision variables regarding service provided by facilities can be written as a vector of values $x_1, x_2, \ldots x_n$ from 1 to n. These facilities are constrained to meet demand for transportation service, that is, the sum of the output from the facilities must be greater than or equal to some demand value a. There are then constraints on the values of x_i, for example, if the ith facility cannot provide more than a certain amount of service b_i in a given time period, this might be written $x_i \leq b_i$. Then the entire optimization problem can be written as

$$
\begin{aligned}
&\textit{Minimize } Z = \sum_{i=1}^{n} c_i x_i \\
&\text{subject to} \\
&\sum_i x_i \geq a \\
&x_i \leq b_i, \forall i \\
&x_i \geq 0, \forall i
\end{aligned}
\tag{3-5}
$$

In Eq. (3-5), the top line is called the *objective function*, and the term "minimize" indicates that the goal is to find the value of Z that is as low as possible. The vector of parameters c_i are the cost coefficients associated with each decision variable x_i. For example, if the objective is to minimize expenditures, then cost parameters with units of "cost per unit of x_i" are necessary so that Z will be evaluated in terms of monetary cost and not physical units. The equations below the line "subject to" are the constraints for this optimization problem. Note that in the objective function, we could also stipulate that the value of Z be maximized, again subject to the equations in the constraint space. For example, the goal might be to maximize profits, and instead of costs per unit of production, the values c_i might represent the profits per unit of production from each plant. Also, the *nonnegativity constraint* $x_i > 0$ does not appear in all optimization problems; however, in many problems, negative values of decision variables do not have any useful meaning (for example, negative values of output from a transportation facility) so this constraint is required in order to avoid spurious solutions.

Example 3-3 illustrates the application of optimization to transportation problem that is simple enough to be solved by inspection and also to be represented in two-dimensional space, so that the function of the optimization problem can be illustrated graphically.

---

[2]Various textbooks on operations research or related topics can be referenced for more background on optimization techniques; for example, Hillier and Lieberman (2002).

Example 3-3 A company that produces and ships products has two product sources (e.g., factories or processing centers) available to meet demand for product. Source 1 has a capacity of six units of product, and Source 2 a capacity of four units of product. The cost of transportation is \$10 per mile, and Source 1 and Source 2 are located 1,500 miles and 3,000 miles, respectively, from the market for the product, where demand is 8 units. The cost of producing the product (as opposed to shipping it) is the same at both sources, and is therefore left out of this calculation. What is the minimum cost at which the company can meet a demand for eight units of product?

Solution Given the above data, the cost per unit to ship from Source 1 and Source 2 is \$15,000 and \$30,000 per unit, respectively. It is convenient to write costs in the form c_i, that is, c_1 = \$15,000/unit and c_2 = \$30,000/unit.

The solution to this simple problem can be found by inspection: the company will source 6 units from Source 1 and the remaining 2 units with Source 2. The minimum total cost is then \$150,000. To arrive at the same solution using a formal optimization approach, Eq. (3-5) can now be rewritten for the specific parameters of the problem as follows:

$$\begin{aligned}&\textit{Minimize } Z = 15000x_1 + 30000x_2\\&\text{subject to}\\&x_1 \leq 6\\&x_2 \leq 4\\&x_1 + x_2 \geq 8\\&x_1, x_2 \geq 0\end{aligned}$$

Note that the third constraint equation says that the output from the two sources combined must be at least eight units.

Next, it is useful to present the problem graphically by plotting the feasible region for the solution space on a two-dimensional graph with x_1 and x_2 on the respective axes (Fig. 3-10). The constraint $x_1 + x_2 > 8$ is superimposed on the graph as well, and only combinations of x_1 and x_2 above and to the right of this line are feasible, that is, able to meet the product requirements. Two *isocost lines* are visible in the figure as well, and combination of x_1 and x_2 on each line will have the same total cost; for example, \$150,000 or \$180,000 in this case. The intersection of the constraint line with the isocost line $c_1x_1 + c_2x_2$ = \$150,000 gives the solution to the problem, namely

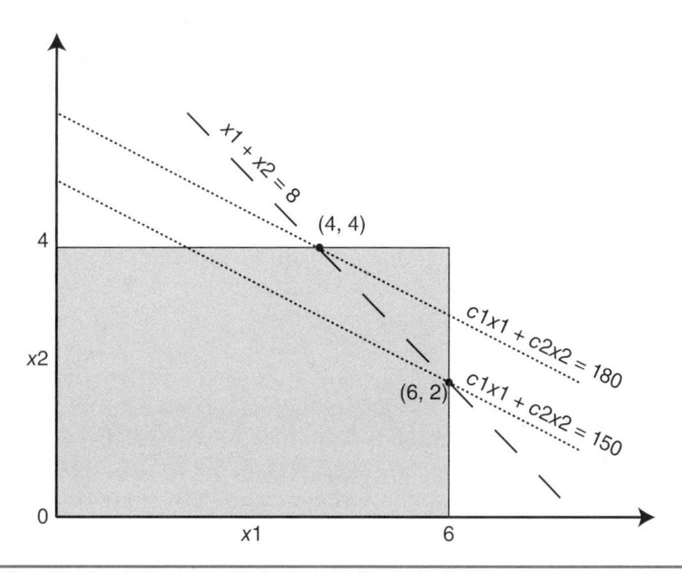

FIGURE 3-10 Graphical solution to optimization in Example 3-3.

that since one cannot go to a lower isocost line without leaving the feasible space, the solution is at the point ($x_1 = 6$, $x_2 = 2$).

Although it is possible to solve for a larger optimization problem by hand, it is usually more convenient to use a computer software algorithm to find the solution. The general goal of any such algorithm is to first find a feasible solution (i.e., a combination of decision variables that is in the decision space) and then move efficiently to consecutively improved solutions until reaching an optimum where the value of Z cannot be improved. At each feasible solution, a *search direction* is chosen such that movement in that direction is likely to yield the largest improvement in Z compared to any other direction. The transportation systems analyst can accelerate the solution process by providing an initial feasible solution. For instance, suppose the initial feasible solution given was ($x_1 = 4$, $x_2 = 4$) in this example. The algorithm might detect that there were two search directions from this point, either along the curve $x_1 + x_2 = 8$ or the curve $x_2 = 4$. Of the two directions, the former decreases the value of Z while the latter increases it, so the algorithm chooses the former direction and moves along it until reaching the constraint $x_1 = 6$, which is the optimum point. If no initial solution is provided, the algorithm may start at the origin (i.e., all decision variables equal to zero) and first test for feasibility of this solution. If it is feasible, the algorithm moves toward consecutively improved solutions until finding the optimal one. If not, it moves in consecutive search directions until arriving at the boundary of the feasible space, and then continuing the optimization.

The problem presented in Example 3-3 is small enough that it can be presented in a visual form. Typical optimization problems are usually much larger than can be presented in two-dimensional or three-dimensional space. For example, suppose the problem consisted of *i* product sources, and output from these sources needed to be assigned to *j* demand destinations in an optimal way. (By convention, the subscript *i* is usually used for origins in transportation problems, and the subscript *j* for destinations.) The total number of dimensions in this problem is $i \times j$, which is a much larger number than three, the largest number that can be represented graphically.

Optimization can be applied to a wide array of transportation systems problems. As presented in Example 3-3, one possibility is the optimal allocation of production from different plants, where the cost per unit of transportation may be different between plants. Another is the allocation of capital resources to different transportation infrastructure projects so as to maximize the expected profit from operation of the projects once completed. Optimization may also be applied to the efficient use of energy in networks, such as the optimal routing of a fleet of vehicles so as to minimize the projected amount of fuel that they consume.

Lastly, one of the biggest challenges with successful use of optimization is the need to create an objective function and set of constraints that reproduce the real-world system accurately enough for the recommendations output from the problem to be meaningful. An optimization problem formulation that is overly simplistic may easily converge to a set of decision variables that constitute a recommendation, but the analyst can see in a single glance that the result cannot in practice be applied to the real-world system. In some cases, adding constraints may solve this problem. In other cases, it may not be possible to capture in the system of objective function and mathematical constraints all of the real-world factors that influence the problem in a way that the problem is both solvable and leads to a realistic, useful optimal solution. If so, the analyst may use the optimization tool for *decision support*, that is, generating several *promising* solutions to the problem, and then using one or more of them as a basis for creating by hand a *best* solution that takes into account subtle factors that are missing from the mathematical optimization.

3-5-4 Understanding Contributing Factors to Time-Series Trends Using Divisia Analysis

A common problem in the analysis of the performance of transportation systems is the need to understand trends over time up to the most recent year available to evaluate the effectiveness of previous policies or discern what is likely to happen in the future. Specifically, because the performance of alternative technologies that compete in a market (e.g., road, rail, air, and marine transportation) and the share of each technology in the overall market are bound to change over time, the translation of change in demand into change in other metrics (total cost, energy consumption) is not a simple linear conversion. (In other words, an $X\%$ increase in demand for transportation services does not automatically result in an $X\%$ increase in total expenditure or energy consumption.) In such situations, the transportation systems analyst would like to know the contribution of the factors to the quantitative difference between the actual performance value and the *trended* value, which is the performance that would have been observed if factors had remained constant. *Divisia analysis*, also known as *Divisia decomposition*, is a tool that serves this purpose. *LaSpeyres decomposition* is a related tool; for reasons of brevity, it is not presented here, but the reader may be interested in learning about it from other sources (for example, Schipper et al., 1997).

A typical application of Divisia analysis is to energy consumption of a transportation system. In such an application, there might be multiple modes that contribute to overall energy consumption, each with a quantitative value of conversion efficiency or some other measure of performance that changes with time. Hereafter consumption is referred to as *activity*, denoted A, and the various contributors are given the subscript i. The share of activity allocated to the contributors also changes with time. Thus a contributor may be desirable from a sustainability point of view because of its high energy efficiency, but if its share of A declines over time, its contribution to improving overall system performance may be reduced.

The contribution of both efficiency and share can be mathematically *decomposed* in the following way. Let E_t be the total energy consumption in a given time period t. The relationship of E_t to A_t is then the following:

$$E_t = \left(\frac{E}{A}\right)_t \cdot A_t = e_t \cdot A_t \tag{3-6}$$

where e_t is the intensity, measured in units of energy consumption per unit of activity.

We can now define the share s for a given mode to be the fraction or percentage of total activity attributed to that mode, and divide the aggregate energy intensity and energy use into the different contributors i from 1 to n:

$$e_t = \left(\frac{E}{A}\right)_t = \sum_{i=1}^{n}\left(\frac{E_{it}}{A_{it}}\right)\left(\frac{A_{it}}{A_t}\right) = \sum_{i=1}^{n} e_{it}s_{it} \tag{3-7}$$

$$E_t = \left(\frac{E}{A}\right)_t = \sum_{i=1}^{n}\left(\frac{E_{it}}{A_{it}}\right)\left(\frac{A_{it}}{A_t}\right)A_t = \sum_{i=1}^{n} e_{it}s_{it}A_t \tag{3-8}$$

We have now divided overall intensity into the intensity of contributors e_{it}, and also defined the share of contributor i in period t as s_{it}, which is calculated as follows:

$$s_{it} = \frac{A_{it}}{A_t} \tag{3-9}$$

Next, we consider the effect of changes in intensity and share between periods $t - 1$ and t. By the product rule of differential calculus, we can differentiate the right-hand side of Eq. (3-7):

$$\frac{de_t}{dt} = \sum_{i=1}^{n}\left[\left(\frac{de_{it}}{dt}\right)s_{it} + \left(\frac{ds_{it}}{dt}\right)e_{it}\right] \tag{3-10}$$

The differential in Eq. (3-10) can be replaced with the approximation of the change occurring at time t, written $\Delta e_t/\Delta t$, which gives the following:

$$\begin{aligned}\Delta e_t &= \Delta t\frac{de_t}{dt}\\ &= \Delta t\sum_{i=1}^{n}\left[\left(\frac{\Delta e_{it}}{\Delta t}\right)s_{it} + \left(\frac{\Delta s_{it}}{\Delta t}\right)e_{it}\right]\\ &= \sum_{i=1}^{n}[(\Delta e_{it})s_{it} + (\Delta s_{it})e_{it}]\end{aligned} \tag{3-11}$$

The change in contributors' intensity and share is now rewritten as the difference in value between periods $t-1$ and t, and the current intensity and share values are rewritten as the average between the two periods, giving

$$\begin{aligned}\Delta e_t &= \sum_{i=1}^{n}\left[(e_t - e_{t-1})_i\frac{(s_t + s_{t-1})_i}{2} + (s_t - s_{t-1})_i\frac{(e_t + e_{t-1})_i}{2}\right]\\ &= \sum_{i=1}^{n}\left[(e_t - e_{t-1})_i\frac{(s_t + s_{t-1})_i}{2}\right] + \sum_{i=1}^{n}\left[(s_t - s_{t-1})_i\frac{(e_t + e_{t-1})_i}{2}\right]\end{aligned} \tag{3-12}$$

It is the value of the two separate summations in Eq. (3-12) that is of particular interest. The first summation is called the *intensity term,* since it considers the change in intensity for each i between $t - 1$ and t, and the second term is called the *structure term,* since it has to do with the relative share of the various contributors.

The steps in the Divisia analysis process are (1) for each time period t (except the first), calculate the values of the summations for the intensity and structure terms, and (2) for each time period (except the first and second), to calculate the cumulative value of the structure term by adding together values of all previous intensity and structure terms. The cumulative values in each time period are then used to evaluate the contribution of each term to the difference between the actual and trended energy consumption value in that period, by multiplying the cumulative value by the total activity observed. Example 3-4 demonstrates the application of

Divisia analysis to a representative problem with intensity and structure data for three time periods.

Example 3-4 The activity levels and energy consumption values in the years 1995, 2000, and 2005 for two alternative modes for a hypothetical activity are given in the following table. The amount of the activity is given in arbitrary units; actual activities that could fit in this example include passenger-kilometers or passenger-miles for different passenger modes, tonne-kilometers or ton-miles for different freight modes, distance driven by two different models of vehicles, and the like. Energy consumption is given in giga watt-hours (GWh) of electricity (i.e., million kWh). Construct a graph of the actual and trended energy consumption, showing the contribution of changes in structure and intensity to the difference between the two.

	Year	Activity (million units)	Energy (GWh)
1	1995	15	300
1	2000	14.5	305
1	2005	14.3	298
2	1995	22	350
2	2000	25.5	320
2	2005	26.5	310

Solution The first step in the analysis process is to calculate total energy consumption and activity, since these values form a basis for later calculations. The following table results:

Year	Activity (million units)	Energy (GWh)
1995	37	650
2000	40	625
2005	40.8	608

From this table, the average energy intensity in 1995 is $(650 \text{ GWh})/(3.7 \times 10^7) = 17.6$ GWh per million units = 17.6 kWh/unit, which is used to calculate the trended energy consumption values of 703 GWh and 717 GWh for the years 2000 and 2005, respectively; that is:

$$(4 \times 10^7 \text{ unit})(17.6 \text{ kWh/unit}) = 703 \text{ GWh}$$
$$(4.08 \times 10^7 \text{ unit})(17.6 \text{ kWh/unit}) = 717 \text{ GWh}$$

The energy intensity e_{it} is calculated next using the energy consumption and activity data in the first table, and the share s_{it} is calculated from the first and second table by comparing activity for *i* to overall activity. Both factors are presented for the two sources as follows:

Year	Source 1 Intensity	Source 1 Share	Source 2 Intensity	Source 2 Share
1995	20.00	40.54%	15.91	59.46%
2000	21.03	36.25%	12.55	63.75%
2005	20.84	35.05%	11.70	64.95%

The intensity and structure terms for the years 2000 and 2005 are now calculated for each source using Eq. (3-12). It is convenient to create a table for each source, as follows:

For Source 1:

	e_{it}	s_{it}	$(e_t - e_{t-1})_i$	$(s_t - s_{t-1})_i$	$(e_t + e_{t-1})_i/2$	$(s_t + s_{t-1})_i/2$	Intensity	Structure
1995	20.00	40.54%	—	—	—	—	—	—
2000	21.03	36.25%	1.03	–4.29%	20.52	38.40%	0.397	(0.880)
2005	20.84	35.05%	(0.20)	–1.20%	20.94	35.65%	(0.070)	(0.251)

For Source 2:

	e_{it}	s_{it}	$(e_t - e_{t-1})_i$	$(s_t - s_{t-1})_i$	$(e_t + e_{t-1})_i/2$	$(s_t + s_{t-1})_i/2$	Intensity	Structure
1995	15.91	59.46%	—	—	—	—	—	—
2000	12.55	63.75%	(3.36)	4.29%	14.23	61.60%	(2.070)	0.611
2005	11.70	64.95%	(0.85)	1.20%	12.12	64.35%	(0.548)	0.146

Next, the overall intensity and structure term for each year is calculated by adding the values for sources 1 and 2. For example, for 2000, the intensity term is 0.397 – 2.070 = –1.673. The cumulative value for 2005 is the sum of the values for 2000 and 2005 for each term. A table of the incremental and cumulative values gives the following:

	Incremental Values		Cumulative Values	
Year	**Intensity**	**Structure**	**Intensity**	**Structure**
2000	(1.673)	(0.270)	(1.673)	(0.270)
2005	(0.617)	(0.106)	(2.290)	(0.376)

The effect on energy consumption in the intensity and structure of the sources can now be measured by multiplying activity in 2000 or 2005 by the value of the respective terms. Units for both intensity and structure terms are in kWh/unit. Therefore, the effect of intensity in 2000 and 2005 is calculated as

$$\text{Year 2000: } (4 \times 10^7 \text{ unit})(-1.673 \text{ kWh/unit})(10^{-6} \text{ GWh/KWh}) = -66.9 \text{ GWh}$$

$$\text{Year 2005: } (4.08 \times 10^7 \text{ unit})(-2.29 \text{ kWh/unit})(10^{-6} \text{ GWh/KWh}) = -93.4 \text{ GWh}$$

Using a similar calculation for the effect of structural changes, we can now show that the cumulative effect of both changes accounts for the difference in actual and trended energy consumption, as shown in the following table:

Factor	2000	2005
E(trended), GWh	703	717
Intensity, GWh	(67)	(93)
Structure, GWh	(11)	(15)
E(actual), GWh	625	608

The accuracy of the Divisia analysis calculation can be verified by confirming that the trended, intensity, and structure terms add up to the actual value observed:

$$E_{\text{Trended}} + E_{\text{Intensity}} + E_{\text{Structure}} = E_{\text{Actual}}$$

$$703 - 67 - 11 = 625$$

$$717 - 93 - 15 = 608$$

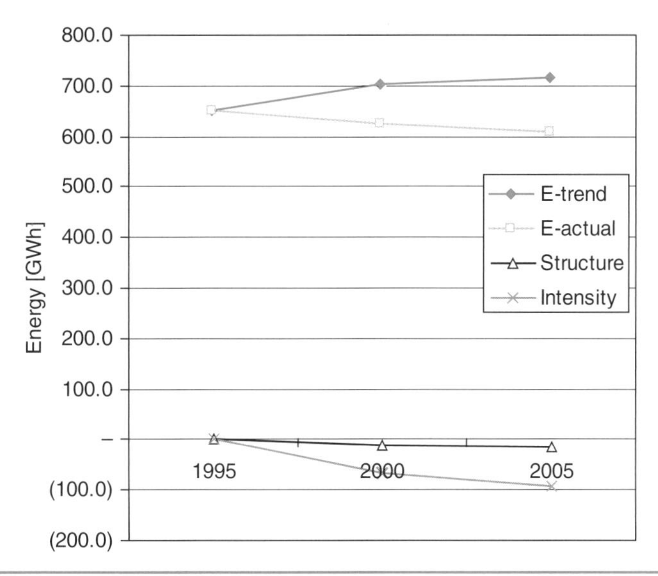

FIGURE 3-11 Graphical representation of Divisia analysis.

This information can also be presented in graphical form, as in Fig. 3-11. In conclusion, from the preceding table and figure, both intensity and structure factors contribute to the reduction in energy consumption relative to the trended value, with the intensity factor accounting for most of the change, and the structure factor accounting for a smaller amount.

One practical example of the use of Divisia analysis is the decomposition of the contribution of different modes of transportation to total CO_2 emissions, with activity measured in kWh generated, and CO_2 emissions by source used in place of energy consumption in Example 3-4. Another example is the contribution of different types of vehicles to the total value of the transportation market in a country. A wide variety of other applications are also possible.

3-5-5 Incorporating Uncertainty into Analysis Using Probabilistic Approaches and Monte Carlo Simulation

When we solve forward-looking problems such as that of Example 3-3, where the goal is to meet a certain level of energy demand for the lowest possible cost, it is often assumed for simplicity that the exogenous values (such as the cost per kWh of production) are known with certainty, and the problem is solved based on knowing the values completely accurately, leading to a deterministic solution. In reality, there may be uncertainty surrounding these values such that if values in the future prove to be different from expectation, the actual cost will be different from what was predicted. Furthermore, in the case of optimization, in some situations the optimal answer might even change if the actual exogenous values change sufficiently.

One way to compensate for this potential weakness in the deterministic approach is to simply describe qualitatively (e.g., in the form of a list of caveats) the level of uncertainty surrounding input values. Another is to create multiple scenarios, as described in Sec. 3-4-1, to calculate different possible outcomes based on a range of possible input values. Both of these approaches have drawbacks. Listing caveats may describe qualitatively the possible changes to the outcome, but does not quantify alternative outcomes. Use of scenarios is limited by the total number of scenarios that can be created and the ability to judge what values are appropriate in place of the expected value.

Where data about the statistical distribution of specific inputs are available (e.g., the expected value and standard deviation of the cost of some required resource), we can treat uncertainty in a more sophisticated way by incorporating the distribution data directly into the solution to the problem (see for example Devore, 2011, for a review of statistical distributions). Instead of solving the problem once as in the deterministic case, we repeat the calculation many times, each time generating new random values for independent input values (such as the cost per unit of energy), and in the end studying the distribution of answers returned. Each answer returned is called an iterate. This technique is called Monte Carlo simulation or alternatively static simulation. The latter name distinguishes it from, for example, discrete-event simulation, since the former does not capture the passage of time within the calculation of each iterate, whereas the latter focuses on how a system changes over time in response to random inputs. This section covers Monte Carlo (static) simulation only.

The steps in Monte Carlo simulation are the following:

1. *Preprocess:* Calculate the deterministic solution to the stated problem using expected values for each input. For instance, if a value is normally distributed, the expected value is the mean of the distribution, while if it is uniformly distributed, the expected value is the average of the two endpoint values. This first step helps to organize the calculations that occur in the simulation and also provides an expected value around which the iterate values will fall. Probabilistic values in Step 2 falling in a range significantly away from the deterministic solution usually means an error in the calculation, so this step has troubleshooting value as well.
2. *Simulation process:* Calculate the desired number of probabilistic iterates. Typically 1,000 or more iterates are calculated. Also, for each iterate, random numbers should be generated for each input value. For example, if a simulation involves fuel economy and the price of gasoline to calculate expected fuel expenses, the value for fuel economy should be generated separately from that of gasoline.
3. *Postprocess evaluation:* Tabulate the results of the simulation in a table or histogram and analyze the results. Output values can be put into *bins* of a given range and then the number of iterate responses in each bin can be tabulated in either a table or graphically in a histogram. The analyst can then evaluate the amount of variability in the histogram around the expected value from Step 1, and also assess whether the likelihood of a value falling above a critical threshold is acceptable (e.g., expenditure exceeding the amount in the budget plus contingency, which may have serious repercussions on the function of a transportation business or government transportation agency).

Monte Carlo simulation at its core depends heavily on the ability to generate random variates from any statistical distributions that might be necessary to model uncertainty in an input value, including not only the uniform and normal distributions mentioned above but also the many others available from the field of probability and statistics. As an example, the steps for generating from a uniform distribution a normally distributed value with known mean μ and standard deviation σ are the following:

1. Generate a random variate u from uniformly distributed $U \sim (0,1)$ (i.e., between 0 and 1). These functions are commonly built into a spreadsheet or engineering calculation software packages.
2. Find a value z such that for the standard normally distributed random variable Z, $P(Z \leq z) = u$. Again, z can be evaluated using built-in functions, or the standard normal table that appears in any statistics textbook.
3. The random variate x that is used in the simulation is then evaluated:

$$x = \mu + z \cdot \sigma \tag{3-13}$$

Alternatively, for a uniformly distributed x between a and b, the calculation given u from $U \sim (0,1)$ is simpler:

$$x = a + u \cdot (b - a) \tag{3-14}$$

The entire simulation can be built in a spreadsheet using built-in functions, or the process can be streamlined using specialist Monte Carlo simulation packages such as @Risk® or Crystal Ball®. Example 3-5 illustrates the deterministic solution preprocess, the evaluation of random variates, and the presentation of results from a model Monte Carlo simulation.

Example 3-5 A common problem in transportation systems is the effect of budgetary uncertainty on decision-making for the future. Consider a municipality that must prepare a budget for fuel expenditures for the operation of snow-removal equipment (SRE) on streets and highways in the upcoming winter season. The simulation will consider three normally distributed random inputs: severity of winter storms (as represented by the total number of miles driven by the SRE fleet), fuel economy delivered by the SREs in terms of average miles per gallon, and the cost of diesel fuel per gallon. Mean and standard deviation values are given in the following table, as is the coefficient of variation (CV), that is, ratio of SD to mean, which indicates the amount of variability in the input value. For this simulation, (a) calculate the deterministic solution, (b) calculate the value of a single iterate if three samples of u are generated from the $U(0,1)$ distribution with values 0.823, 0.204, and 0.020 for the three required inputs, and (c) produce a histogram from a simulation of 100 iterates and discuss its implications if the municipality has a hard budget constraint of $140,000 for the season for SRE fuel expenditures. Note that the relatively small simulation size of 100 iterates is chosen here to make the example more transparent, although in general at least 1,000 iterates are desirable to make the results more robust.

Variables	Units	Mean	SD	CV
Demand	Miles driven	80,000	24,000	30%
Fuel economy	MPG	3.5	0.525	15%
Fuel cost	$/gallon	$3.10	$0.775	25%

Solution

(a) The deterministic value is calculated from the mean values of demand, fuel economy, and fuel cost:

$$\text{Expense} = \frac{\text{demand}}{\text{fuel economy}}(\text{fuel cost}) = \frac{80,000}{3.5}(\$3.10) = \$70,857$$

(b) For this particular iterate, we will calculate the value for demand in detail, with the assumption that the steps can be repeated for fuel economy and fuel cost. Given $u = 0.823$, first solve for z using the standard normal probability table or a built-in spreadsheet or math software function. From the standard normal table, $P(Z \leq 0.93) = 0.8238$, so for the purposes of the calculation, it is sufficiently accurate to set $z = 0.93$. Using Eq. (3-13) then gives

$$x = \mu + z \cdot \sigma = 80,000 + 0.93 \cdot 24,000 = 102,320$$

The reader can verify that starting with random values 0.204 and 0.020 instead of 0.823 and repeating these steps for fuel economy and cost gives, to the nearest hundredth, 3.07 MPG and \$1.51/gal, respectively. Thus the value returned for this iterate is

$$\text{Expense} = \frac{\text{demand}}{\text{fuel economy}}(\text{fuel cost}) = \frac{102,320}{3.07}(\$1.51) = \$50,327$$

Note that the municipality has "gotten lucky" in this iterate because they have experienced an unusually low fuel cost of \$1.51/gal, which is more than two standard deviations below the mean cost of \$3.10/gal. The result is an annual cost of approximately \$50 K, well below the expected value of approximately \$71 K from part (a). Returning such a low-valued variate in a Monte Carlo simulation is possible, but unusual.

(c) In principal, the simulation is carried out by repeating the steps in part (b) in an automated way. No two runs of the Monte Carlo simulation will be identical, but the results for a particular run of 100 iterates are tabulated in Table 3-3 and Fig. 3-12. No iterate returned a value lower than \$20,000, and none higher than \$200,000; however, within this range there was wide

Bin Values (\$1000s)		
Minimum	**Maximum**	**Number**
<\$40	\$40	2
\$40	\$60	12
\$60	\$80	24
\$80	\$100	29
\$100	\$120	12
\$120	\$140	12
\$140	\$160	6
\$160	\$180	1
\$180	>\$180	2
	Total	100

Table 3-3 Distribution of 100 Iterates for Representative Monte Carlo Simulation Run

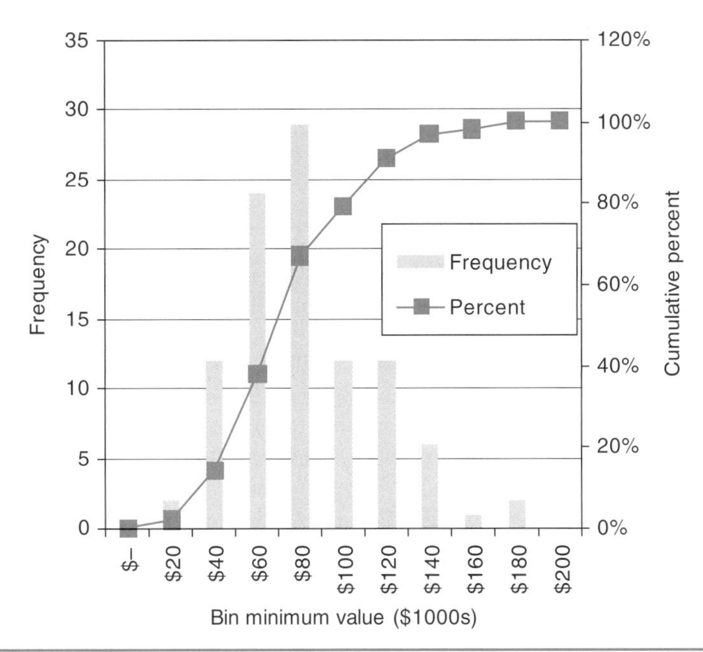

FIGURE 3-12 Graphical representation of distribution of 100 iterates into bins (left *y*-axis) and cumulative percentage of iterate values below bin upper value (right *y*-axis).

Note: Bin values shown are the minimum for a given bin; the maximum for that bin is the next value to the right (e.g., the bin marked "$20" is actually the $20 K to $40 K bin).

variability in returned values, reflecting relatively high CVs both for miles driven and the cost of fuel. This high level of variability may pose a concern for the municipality. In particular, 91% of the values fall in the bins $120 to $140 K or below, suggesting that there is a 9% probability that the budget constraint of $140 K will be exceeded. This result may motivate the municipality to consider various possible responses, such as investments in improving SRE fuel economy, policy changes that might reduce the level of commitment to snow removal (to lower the expected value of demand), or increasing the size of the SRE budget by reducing commitments elsewhere. In any case, the example shows the power of the Monte Carlo simulation: if we only calculate the deterministic solution, then we know the expected value of total expenses for the winter but we do not obtain any quantitative information about the risk that the upper bound will be exceeded.

Note: The values are distributed with mean = $72,687 and CV = 45%. Interpretation of the table: For example, 2 out of 100 iterates returned total expenditure values less than $40,000, 12 returned values between $40,000 and $60,000, and so on.

3-5-6 Kaya Equation: Economic Activity, Energy Consumption, and CO_2 Emissions

With the introduction of the Kaya equation we turn to systems tools that are specific to energy and climate change, as motivated by the discussion in Chap. 2. The Kaya equation was popularized by the economist Yoichi Kaya as a way of disaggregating the various factors that contribute to the overall emissions of CO_2 of an individual country or of the whole world. Since transportation is such a significant user of energy and generator of CO_2 and other GHG emissions, it is desirable to analyze the effect of transportation alongside other sectors of the economy in shaping emissions levels.

The Kaya equation incorporates population, level of economic activity, level of energy consumption, and carbon intensity, as follows:

$$CO_{2\text{Emit}} = (P) \times \left(\frac{\text{GDP}}{P}\right) \times \left(\frac{E}{\text{GDP}}\right) \times \left(\frac{CO_2}{E}\right) \tag{3-15}$$

If this equation is applied to an individual country, then P is the population of the country, GDP/P is the population per capita, E/GDP is the energy intensity per unit of GDP, and CO_2/E is the carbon emissions per unit of energy consumed. Thus the latter three terms in the right-hand side of the equation are performance measures, and in particular the last two terms are metrics that should be lowered in order to benefit the environment. The measure of CO_2 in the equation can be interpreted as CO_2 emitted to the atmosphere, so as to exclude CO_2 released by energy consumption but prevented from entering the atmosphere, a process known as *sequestration*. Because the terms in the right-hand side of the equation multiplied together equal CO_2 emitted, it is sometimes also called the *Kaya identity*.

The key point of the Kaya equation is that a country must carefully control all the measures in the equation in order to reduce total CO_2 emissions or keep them at current levels. As shown in Table 3-4, values for five of the major players in determining global CO_2 emissions, namely, China, Germany, India, Japan, and the United States, vary widely between countries. Although it would be unrealistic to expect countries with the largest populations currently, namely China and India, to reduce their populations to be in line with the United States or Japan, all countries can contribute to reducing CO_2 emissions by slowing population growth or stabilizing population. China and India have low GDP/capita values compared to the other three countries in the table, so if their GDP/capita values are to grow without greatly increasing overall CO_2 emissions, changes must be made in either E/GDP, or CO_2/E, or both. As for Germany, Japan, and the United States, their values for E/GDP and CO_2/E are low relative to China and India, but because their GDP/P values are high, they have high overall CO_2 emissions relative to what is targeted for sustainability. This is true in particular of the United States, which has the highest GDP/P in the table and also an E/GDP value that is somewhat higher than that of Japan or Germany. Using these comparisons as an example, individual countries can benchmark themselves against other countries or also track

	P (million)	**GDP/P ($1000/person)**	**E/GDP (GJ/$1000 GDP]**	**CO_2/E [tonnes/GJ]**	**CO_2 [million tonne]**
China	1298.8	$1,725	28.05	0.075	4,707
Germany	82.4	$34,031	5.53	0.056	862
India	1065.1	$740	20.65	0.068	1,113
Japan	127.3	$35,998	5.21	0.053	1,262
USA	293.0	$42,846	8.44	0.056	5,912

TABLE 3-4 Kaya Equation Values for Selection of Countries, 2004

changes in each measure over time in order to make progress toward sustainable levels of CO_2 emissions.

From a systems perspective, one issue with the use of these measures is that they may mask a historic transfer of energy consumption and CO_2 emissions from rich to industrializing countries. To the extent that some of the products previously produced in the United States or Europe but now produced in China and India are energy-intensive, their transfer may have helped to make the industrial countries "look good" and the industrializing countries "look bad" in terms of E/GDP. Arguably, however, the rich countries bear some responsibility for the energy consumption and CO_2 emissions from these products, since it is ultimately their citizens who consume them.

Human Development Index (HDI): An Alternative to Gross Domestic Product (GDP)

In the Kaya identity as in many other measures of economic performance, either GDP or GDP per capita is used as a measure of the overall developmental success of a country. Gross domestic product as a measure of development has been criticized in some quarters for overlooking other facets of the quality of life of a nation, such as health, education, or overall citizen satisfaction with quality of life based on survey research. To create a measure of prosperity that better reflects broad national goals beyond the performance of the economy, the United Nations has since the early 1990s tracked the value of the human development index (HDI). The HDI is measured on a scale from 0 (worst) to 1 (best), and is an average of the following three general indices for life expectancy, education, and GDP per capita:

$$
\begin{aligned}
\text{Life expectancy index} &= (\text{LE} - 25)/(85 - 25) \\
\text{Education index} &= 2/3(\text{ALI}) + 1/3(\text{CGER}) \\
\text{CGER} &= 1/3(\text{GER}_{\text{Prim}} + \text{GER}_{\text{Second}} + \text{GER}_{\text{Terti}}) \\
\text{GDP index} &= \frac{[\log(\text{GDPpc}) - \log(100)]}{[\log(40000) - \log(100)]}
\end{aligned}
\tag{3-16}
$$

where LE = the average life expectancy, years
ALI = the adult literacy index, or the percentage of adults that are literate
CGER = the combined gross enrollment rate, or average of the primary, secondary, and tertiary gross enrollment rates (i.e., ratio of actual enrollment at each of three educational levels compared to expected enrollment for that level based on population in the relevant age group)
GER = the gross enrollment rate;
GDPpc = the GDP per capita.

As was the case with GDP per capita, countries with high HDI values have higher values of energy use per capita than those with a low value (Fig. 3-13). For example, Zimbabwe, with a life expectancy of 38 years and an HDI value of 0.491, has an energy per capita value of 22 GJ/capita or 21.0 million Btu/capita, whereas for Canada, the corresponding values are 0.950, 470, and 442, respectively. Also, among countries with a high value of HDI (> 0.85), there is a wide range of energy intensity values, with Bahrain consuming 6.5 × 1011 J/capita (612 million Btu/capita) but having an HDI value of 0.859, which is somewhat lower than that of Canada.

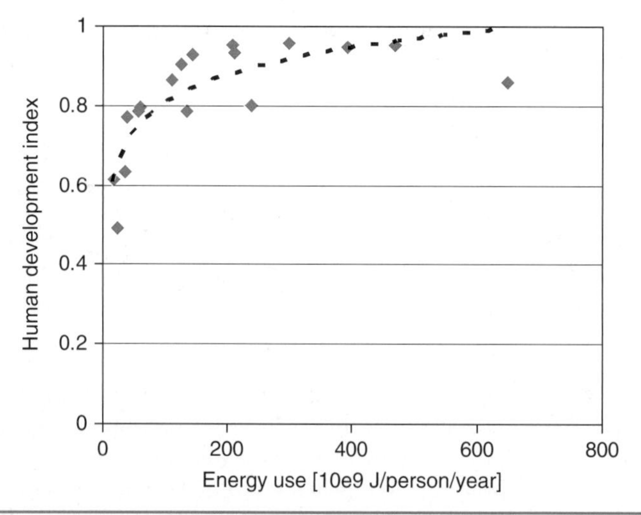

FIGURE 3-13 Human development index as a function of per-capita energy consumption in gigajoule per person for selected countries, 2004.

3-5-7 Energy Return on Investment

When investing in changes to transportation systems that are intended to reduce overall energy consumption, it is important not only to understand the life-cycle energy consumption, but also to verify that reductions in energy consumption at the end use or operational stage justify energy expended upstream in its extraction, conversion, and delivery. On this basis, we can calculate the *energy return on investment* (EROI), which is the energy delivered at the end of the life cycle divided by the energy used in earlier stages. For example, it would not make sense to expend more energy on the manufacture and deployment of some energy-efficient transportation system that requires more energy in resource extraction, manufacturing, and installation than it can save in its usable lifetime. Practitioners may also use the term *energy payback*, meaning the amount of net energy savings returned from the use of a transportation system after expending energy on manufacture, installation, and so on.

3-6 Summary

During the course of the twenty-first century, transportation systems will be required to meet several important goals, including conformance with the environmental, economic, and social goals of sustainable development. The existence of multiple goals, multiple stakeholders, and numerous available technologies lends itself to the use of a "systems approach" to solving transportation problems, which emphasizes additional effort in the early conceptual and design stages in order to achieve greater success during the operational lifetime of a product or project. Qualitative systems tools available to the decision-maker include the use of *stories* or causal loop diagrams to describe systems interactions. Quantitative tools include exponential, logistics, and triangle functions to project rates of growth or market penetration of transportation technologies. They also include scenario analysis and quantitative modeling to estimate the future impact of changes to transportation systems on consumption or pollution, and systems tools such as life-cycle analysis, multicriteria analysis, optimization, Divisia analysis, and Monte Carlo simulation to determine whether a particular technology is acceptable or to choose among several competing technologies or solutions. Lastly, energy- and climate-specific tools such as the Kaya equation and EROI provide a means of understanding the contribution of underlying factors to overall energy consumption or CO_2 emissions trends.

References

Devore, J. (2011). *Probability and Statistics for Engineering and the Sciences: 8th ed.* Duxbury Press, London.

Hillier, F. and G. Lieberman (2002). *Introduction to Operations Research: 7th ed.* McGraw-Hill, New York.

Norberg-Bohm, V. (1992). *International Comparisons of Environmental Hazards: Development and Evaluation of a Method for Linking Environmental Data with the Strategic Debate Management Priorities for Risk Management.* Report from Center for Scientific and International Affairs, John F. Kennedy School of Government, Harvard University, Cambridge, MA.

Schipper, L., L. Scholl, and L. Price (1997). *Energy Use and Carbon Emissions from Freight in 10 Industrialized Countries: An Analysis of Trends from 1973 to 1992.* Transportation Research Part D 2(1):57–75.

U.S. Environmental Protection Agency (1990). *Reducing Risk: Setting Priorities and Strategies for Environmental Protection.* Report SAB-EC-90-021. Scientific Advisory Board, Washington, DC.

Vanek, F. (2002). "The Sector-Stream Matrix: Introducing a New Framework for the Analysis of Environmental Performance." *Sustainable Development* 10(1):12–24.

Victoria Transportation Policy Institute (2012). *Online TDM Encyclopedia.* Electronic Resource, available at *www.vtpi.org/tdm/*. Accessed Dec. 12, 2012.

Further Readings

Ackoff, R. (1978). *The Art of Problem Solving.* Wiley & Sons, New York.

H.H. Goode and R.E. Machol (1957). *Systems Engineering: An Introduction to the Design of Large-Scale Systems.* McGraw-Hill, New York.

Greene, D. and Y. Fan (1995). "Transportation Energy Intensity Trends: 1972–1992." *Transportation Research Record* (1475):10–19.

Hawken, P., A. Lovins, and L. Lovins (1999). *Natural Capitalism: Creating the Next Industrial Revolution.* Little, Brown, Boston, MA.

Jackson, P. (2010). *Getting Design Right: A Systems Approach.* CRC Press, Boca Raton, FL.

Jenkins, G. (1969). "The Systems Approach." *Journal of Systems Engineering,* 1:3–49.

Kossiakoff, A. (2003). *Systems Engineering: Principles and Practice.* Wiley & Sons, New York.

Lippiatt, B. (1999). "Selecting Cost-Effective Green Building Products: BEES Approach." *Journal of Construction Engineering and Management* (Nov-Dec):448–455.

Sterman, J. (2002). "System Dynamics Modeling: Tools for Learning in a Complex World." *IEEE Engineering Management Review* 30(1):42–52.

Vanek, F., P. Jackson, and R. Grzybowski (2008). "Systems Engineering Metrics and Applications in Product Development: A Critical Literature Review and Agenda for Further Research."*Systems Engineering*. 11(2): 107–124

World Commission on Environment and Development (1987). *Our Common Future*. Oxford University Press, Oxford, UK.

Exercises

3-1. Consider a technical project with which you have been involved, such as an engineering design project, a research project, or a term project for an engineering course. The project can be related to transportation systems or to some other field. Consider whether or not the project used the systems approach. If it did use the approach, describe the ways in which it was applied, and state whether the use of the approach was justified or not. If it did not use the approach, do you think it would have made a difference if it had been used? Explain.

3-2. Causal loop diagrams: In recent years, one phenomenon in the private automobile market of many countries, especially the wealthier ones, has been the growth in the number of very large sport utility vehicles and light trucks. Create two causal loop diagrams, one *reinforcing* and one *balancing*, starting with *number of large private vehicles* as the initial component.

3-3. The values of world energy consumption in 25-year increments for the period 1850 to 1975 are given as below. Fit a curve to the data points using the exponential growth function, and then use this curve to predict the value of energy consumption in 2000. If the actual value in 2000 is 419 EJ, by how many EJ does the projected answer differ from the actual value in units of EJ?

1850	25 EJ
1875	27 EJ
1900	37 EJ
1925	60 EJ
1950	100 EJ
1975	295 EJ

3-4. Current global energy consumption trends combined with per-capita energy consumption of some of the most affluent countries, such as the United States, can be used to project the future course of global energy demand in the long term.

a. Using energy consumption and population data from Chap. 2, calculate per-capita energy consumption in the Unites States in 2008, in units of GJ/person.

b. If the world population ultimately stabilizes at 10 billion people, each with the energy intensity of the average U.S. citizen in 2008, at what level would the ultimate total annual energy consumption stabilize, in EJ/year?

c. From data introduced earlier, it can be seen that the world reached the 37 EJ/year mark in 1900, and the 100 EJ/year mark in 1950. Using these data points and the logistics formula, calculate the number of years from 2008 until the year in which the world will reach 97% of the value calculated in part (b).

d. Using the triangle formula, pick appropriate values of the parameters a and b such that the triangle formula has approximately the same shape as the logistics formula.

e. Plot the logistics and triangle functions from parts (c) and (d) on the same axes, with year on the x-axis and total energy consumption on the y-axis.

f. Discussion: The use of the United States as a benchmark for the target for world energy consumption has been chosen arbitrarily for this problem. If you were to choose a different country to project future energy consumption, would you choose one with a higher or lower per-capita energy intensity than the United States? What are the effects of choosing a different country?

3-5. Multicriteria decision making: A construction firm that is building a passenger transportation terminal facility is evaluating two materials to choose the one that has the least effect on the environment. The alternatives are a traditional *natural* material and a more recently developed *synthetic* alternative. Energy consumption in manufacturing is used as one criterion, as are the effect of the material on natural resource depletion and on indoor air quality, in the form of off-gassing of the material into the indoor space. Cost per unit is included in the analysis as well. The scoring for each criterion is calculated by assigning the inferior material a score of 100, and then assigning the superior material a score that is the ratio of the superior to inferior raw score multiplied by 100. For example, if material A emits half as much of a pollutant as material B, then material B scores 100 and material A scores 50. Using the following data, choose the material that has the overall lower weighted score. (Note that numbers are provided only to illustrate the technique and do not indicate the true relative worth of synthetic or natural materials.)

Data for Exercise 3-5:

	Units	Natural	Synthetic	Weight
Energy	MJ/unit	0.22	0.19	0.24
Natural resource	kg/unit	1,000	890	0.15
Indoor AQ	g/unit	0.1	0.9	0.21
Cost	$/unit	$1.40	$1.10	0.40

CHAPTER 4

Individual Choices and Transportation Demand

4-1 Overview

Understanding choice behavior of individuals is fundamental for forecasting travel demand. This chapter provides an overview of modeling decision making of transportation systems users through discrete choice models. Discrete choice models represent the cognitive process of economic decisions based on a probabilistic representation of neoclassical consumer theory. Discrete choice analysis is a common tool in transportation engineering, applied economics, marketing, and urban planning. This chapter will assume that the reader has a working knowledge of standard regression analysis.

4-2 Introduction: Why Are We Interested in Understanding Behavior?

A successful engineering solution must not only be technically adequate but also be one that society is ready to adopt. Suppose that a research group develops a novel propulsion technology for cars with important energy efficiency gains. Profitability of the new product will depend not only on production costs and the use of resources but also on sales. The impact of the new technology on achieving sustainability goals also depends on the number of consumers that adopt the new product. Engineering analysis for policy-making and firm decisions must then address consumer *preferences* and *demand.* For example, to analyze and design any transportation project, one needs to understand current demand levels (vehicle flow) as well as predict demand in the new scenario set by the project. If the project is a toll bridge, transportation engineers need to predict the amount of traffic passing over, which will be determined by the number of users willing to pay the toll.

Among the different fields in civil and environmental engineering, transportation system analysis stands out with a long tradition of integrating supply and demand. In fact, the tools used by a transportation engineer have a strong behavioral component. Both travel demand and supply reflect a decision-making process by different agents, including users of the transportation system, planners, and firms.

To forecast demand we need a quantitative understanding of the tradeoffs among product or service attributes, such as the toll in the bridge example above. Demand estimation also serves to assess monetary valuations that are needed for cost-benefit analysis of infrastructure projects and policies. In sustainable engineering, the determination of demand needs special tools. For example, environmental costs and benefits are not transmitted through prices. We need then an indirect mechanism to understand consumer response to energy efficiency and to determine the monetary valuation of environmental costs and benefits. In the case of transportation, drivers do not pay for the emissions they produce, which not only affect other users of the transportation system but also affect the whole community. Drivers do not pay either for the additional delays they cause other users of the system. These two—pollution and congestion—are externalities that distort costs (drivers perceive a lower cost of using their vehicles), leading to indiscriminate use of cars.

Describing and predicting the behavior of agents is extremely challenging. Sophisticated mathematical models are required to better represent individuals' decisions among mutually exclusive alternatives.

4-3 Travel Behavior: Demand for Trips and Transportation Choices

Demand represents the amount of a good or service desired by consumers. Demand is motivated by needs, determined by the *willingness to pay* a price for consumption, and constrained by available income. The standard *inverse demand curve* in economics is given by the willingness to pay for a specific good or service as a function of quantity. Consumers extract benefits from consumption due to satisfaction of needs, but in a market economy consumers are paying a price for consumption. Those individuals who are willing to pay more than the actual price of the good or service extract a benefit called *consumer surplus*.

A typical representation of quantity in aggregate travel demand models is given by a measure of number of trips (counts or flows). Other standard aggregate measures of travel demand are:

1. *VMT (vehicle miles traveled):* Number of miles that residential vehicles are driven (See Fig. 4-1, where the inverse demand curve is exemplified using VMT.)
2. *Ton-mile:* A ton of freight moved over 1 mile
3. *Passenger-mile:* A passenger moved over 1 mile

However, in the case of travel demand, there are several additional considerations that need to be accounted for. First, transportation is a *service*, meaning that trips are not

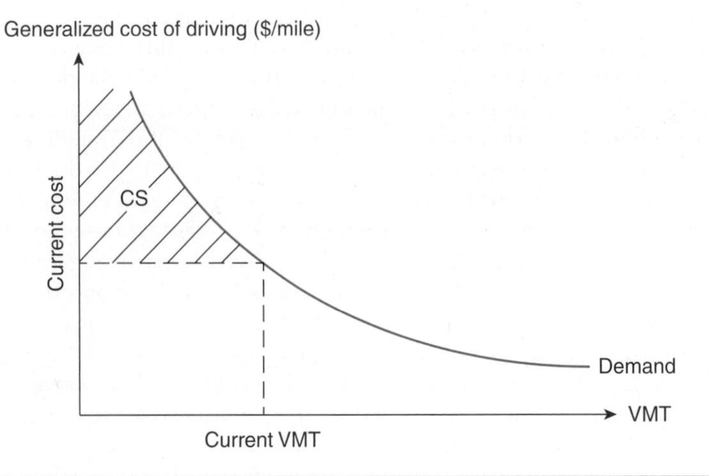

FIGURE 4-1 Inverse demand curve and consumer surplus.

commodities that can be exchanged or stored. In addition, travel demand is dynamic and requires a physical network, meaning that trips have both temporal and spatial dimensions. The aggregate measures presented above lose relevant information about trips. For example, 1,000 ton-miles could refer to 1,000 tons moved 1 mile or 1 ton moved 1,000 miles. In addition, we don't know when movement took place or over which route. Transportation has *temporal* and *spatial* dimensions that are translated in transportation demand being dynamic and having interactions with the built environment. A user of the transportation system shares the time and space dimensions with other users of the system, creating complex *network effects.* For instance, our transportation decisions affect other users. The clearest example is *congestion*, where the decision of others of driving their cars creates delays for our own travel time, and our decision of driving creates delays for others. Furthermore, users not only interact among themselves but transportation also interacts with other *systems.* For example, characteristics of the *built environment* clearly affect mobility (*New Yorkers ride the subway, whereas people in LA drive everywhere*). These interactions create complex *equilibrium effects* as well as energy and environmental impacts.

Travel is not an end in itself: users of transportation systems *consume* trips, but do not extract a direct benefit from transportation services. You may be satisfied with the decision of taking a taxi because you arrived on time at your destination, but this satisfaction is associated with the *activity* that created the need for the trip. This is why demand in transportation is a *derived demand.*

4-3-1 Energy Requirements for Transportation and Environmental Sustainability

Current mobility patterns in the United States are characterized by automobile dependence, creating several societal, economic, geographic, energy, and environmental problems. The negative externalities of automobile-dependent societies range from congestion and high levels of pollution to health issues due to lack of physical activity. Transportation consumes 28% of the total energy demanded in the United States. Road transportation explains about 80% of the energy (mainly fossil fuel) demanded by the transportation sector, with passenger transportation consuming about 70%.

Automobile dependence is a serious problem in the United States. According to the 2009 National Household Travel Survey, 72% of all trips of 3 miles or less is actually taken by car. A sustainable future requires that society at large adopt low-carbon consumption behavior as a result of a consumer shift to energy-efficient technologies. In the following sections, we will discuss consumer demand models for representing the behavioral response to sustainable energy and for evaluating the changes that are needed for promoting decarbonization of transportation.

4-4 Discrete Choice Models

Discrete choice models are by far the most widely used behavioral models in transportation analysis (McFadden, 2001). In fact, developments in discrete choice models of transportation rank among the most relevant recent advances in econometrics. Transportation researchers, firms, and policy-makers use discrete choice to predict demand for new alternatives and infrastructure (e.g., a new light rail, a new highway, alternative fuels), to analyze the market impact of certain firm decisions (e.g., merger of two airline companies), to set pricing strategies (e.g., congestion pricing, toll roads, emission pricing, revenue management), to prioritize research and development decisions (e.g., automotive industry looking at the introduction of new technologies), and to perform engineering economy analyses of projects (e.g., construction of a tunnel). Discrete choice theory is also a relevant input for traffic assignment models (route choice problems).

Beyond transportation, discrete choice theory is widely used in applied economics (health and labor economics, as well as environmental economic studies of energy, water, and conservation), marketing, political science (voting preferences), and urban planning. There is a growing literature on the use of discrete choice for all sorts of environmental problems, especially on the valuation of environmental externalities. This literature looks at the use of discrete choice theory to analyze consumers' willingness to pay for renewable energy, energy-efficiency gains, and emission reductions. In addition, there is interest in using discrete choice to better regulate energy providers (pricing, terms of contracts, energy audits, mix of renewable power), resource abuse (emission taxes, energy-efficiency mandates, subsidies for carbon dioxide abatement), and new infrastructure developments (building-emission standards).

Discrete choice models are a probabilistic representation of choice among a finite (discrete) group of differentiated products. Differentiated products are heterogeneous goods that fulfill the same need. For example, different travel modes (car, bus, subway) cover the same commuting needs. Even more, if a household decides to buy a car, the possible options of covering the need of traveling by car are copious. Just in model-year 2008, the 2009 National Household Travel Survey (NHTS) contains 242 makes and models. If trim levels are considered (options for specific makes and models summarized in names such as LX, LS, and EX), then there are more than 1,000 options for a new car. And the household may also consider buying a used vehicle!

4-4-1 Differentiated Products as Mutually Exclusive, Discrete Alternatives

Neoclassical consumption theory considers the problem of continuous demand functions. A standard demand curve for a continuous, homogenous good represents the amount consumers are willing to pay for each unit of consumption. However, there are several examples of individual choices that have a discrete nature, because of either discreteness of the quantities demanded, discreteness of the feasible alternatives (choice set), or both. Choice of a specific brand of a product at a supermarket, personal choice of entering the labor market or not, a household's heating fuel choice, and a household's decision of which vehicle to buy, are all examples of discrete choices.

Whereas continuous demand assumes homogeneity of goods, discrete demand appears for differentiated goods that have different qualitative characteristics (Fig. 4-2). This is a relevant fact, as demand for differentiated products will not depend only on price (which is the case of a standard demand function). Discrete choice models follow a hedonic approach, where consumer preferences are derived from the characteristics or *attributes* of a good. These attributes characterize quality of the discrete goods. Think of the value of a property depending on attributes such as square footage, number of bedrooms, number of bathrooms, location, and so on. Note that discrete alternatives are mutually exclusive.

4-4-2 Some Basic Definitions

1. ***Decision maker,*** *i:* Individual (individual consumer, household, firm, agency) who faces a decision-making problem characterized by a discrete choice. Characteristics of the decision maker are summarized in the vector $\mathbf{w}_i$, which may contain variables such as age, gender, education, and the like.

FIGURE 4-2 Different modes can satisfy the same transportation needs.

2. Choice set $C_i = \{1, \ldots, j, \ldots, J\}$: Exhaustive polytomous set of feasible, mutually exclusive *alternatives* that are available in the market. For example, for an individual the choice of alternative transportation modes for commuting can be given by {car, bus, subway}. Combinations of modes can be included as a separate alternative (e.g., {car, bus, subway, park&ride}, where park&ride represents a trip with a first portion using car and a second portion using subway). Note that the choice set may be specific to each individual. For example, if an individual does not have access to subway where she lives, then subway does not belong to her choice set.
3. *Alternative-specific attributes* q_{ij} *of quality:* Characteristics of the discrete alternatives. For example, in a travel-mode choice context, a set of relevant attributes is given by in-vehicle travel time, access time, and waiting time. In general, we will assume that $\mathbf{q}_{ij}$ contains K attributes.
4. *Price* p_{ij}: Cost of the discrete good (as experienced by individual i). (Sometimes the alternative attribute vector is extended to include price.)

4-4-3 Preferences and Individual Choice Behavior

A theory of choice requires identification of the decision maker, the alternatives, the alternative attributes, price of the alternatives, and decision rules. Neoclassical economic behavior is represented as a decision-making problem where individuals make choices by *maximizing the satisfaction* (payoffs) they get from consumption (behavioral action), subject to a budget constraint. Microeconomic theory of consumer choice is operationalized using consumer *preferences*. Preferences are characterized axiomatically and the preference relation can be summarized using the concept of *utility function*. Utility is a continuous real-valued function that assigns higher numbers to preferred actions. Utility works as an index of attractiveness of actions that can be used to rank different combinations of consumption. Note that utility is never unique, as there exist infinite ways of deriving the same ranking.

Whereas for continuous goods satisfaction or utility extracted from consumption depends on quantity, for discrete goods utility depends on quality as measured by the attributes that characterize the discrete alternative. For example, consider two consumers looking at the possibility of buying an electric car. Suppose that the two consumers have different preferences for driving range (maximum distance allowed by a single charge of the electric battery) and power, assuming the other characteristics of the cars they are evaluating are equal (purchase price, efficiency, space, etc.). The first potential buyer is really worried about the possibility of running out of power while driving and does not care much about power. In contrast, the second potential buyer has a short commute (range is less of an issue for this consumer) and desires a car with good power. When different consumers have different preferences for the same concepts, then we can introduce the concept of *consumer heterogeneity*.

Figure 4-3 shows heterogeneous preferences for performance and driving range. Each *indifference curve* in the graph represents the combinations of performance and driving range that provide the exact same of satisfaction (utility) to the consumer. Consumers on

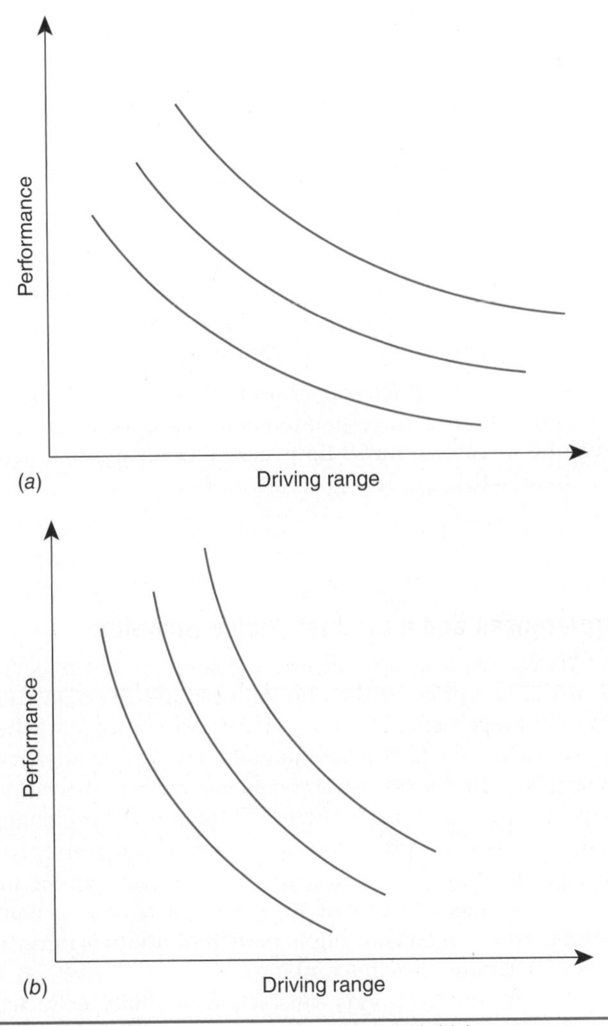

FIGURE 4-3 Heterogeneous preferences for performance and driving range.

Fig. 4-3*a* prefer cars with better performance and are willing to give up little performance in exchange for improvements in driving range. Consumers on Fig. 4-3*b* are willing to exchange a large amount of performance for even a small gain in driving range.

Discrete choice analysis is based on a reformulation of the standard consumer problem in microeconomics. Demand for a discrete good can be represented as a decision process that involves consumption of a set of standard homogenous goods (or a composite good that summarizes all other goods) in addition to the problem of which discrete alternative to select.

Consider individual i who, in a given purchase situation, is simultaneously choosing between alternatives in C_i as well as how much to consume of a composite good $c \in \mathbb{R}_+$. Preferences are indexed by the utility function $u(\mathbf{q}_{ij}, c)$. If we assume that price of the composite good has been normalized to one, the consumer problem is given by

$$\max_{j \in C_i,\ c \in \mathbb{R}_+} u(\mathbf{q}_{ij}, c)$$
$$\text{s. t. } p_{ij} + c \leq I_i$$

where I_i is the income of the individual and $p_{ij} + c \leq I_i$ is the budget constraint.

This problem can be solved in two stages. One problem is associated with determining the optimal consumption of $c \in \mathbb{R}_+$ conditional on the discrete good. The solution to this first problem is c^*. The second problem is characterized by the mutually exclusive decision that involves discrete maximization of the *conditional indirect utility* $v_j(\mathbf{q}_{ij}, I_i - p_{ij}) \equiv u(\mathbf{q}_{ij}, c^*(I_i - p_{ij}, \mathbf{q}_{ij}))$. Assuming that alternative j is chosen, this indirect utility represents the maximal utility that the consumer can achieve given the available income $I_i - p_{ij}$ and the attributes $\mathbf{q}_{ij}$. Note that $v_j(\mathbf{q}_{ij}, I_i - p_{ij}) > v_l(\mathbf{q}_{il}, I_i - p_{il}) \Leftrightarrow j \succ l$, meaning that alternative j is preferred to ($\succ$ is the symbol used in economics to denote a preference relation) alternative l if and only if the indirect utility v_j is greater than v_l.

Example 4-1 Consider a commuter deciding which mode to use to get to work. After evaluating trip cost, travel time, waiting time, and walking time of the different alternatives for a given day, the commuter managed to construct a conditional indirect utility for each mode. Given the utility values in the following table, which mode does the commuter choose? What happens if we add 10 to all values of the indirect utility?

Alternative	Conditional Indirect Utility	Conditional Indirect Utility (+10)
Car	$V_{car} = -4$	$V_{car} = 6$
Bus	$V_{bus} = -6$	$V_{bus} = 4$
Subway	$V_{subway} = -3.5$	$V_{subway} = 6.5$
Park&ride	$V_{park\&ride} = -4.7$	$V_{park\&ride} = 5.3$

Solution The alternative that maximizes the conditional indirect utility is subway. Hence, the commuter chooses subway as her commuting mode. Given the information, we can actually conclude that subway $\succ$ car $\succ$ park&ride $\succ$ bus. Note that choice is determined by a maximization process that is discrete, and that results in a ranking of the alternatives. To determine the preferred alternative, we can simply compare which one offers a higher value of utility relative to the others (i.e., the alternative that is ranked first in terms of *incremental utility*).

Note that if we add the same value (either positive or negative) to the conditional indirect utility of all alternatives, the preference relation is not changed. What would happen if we multiply the conditional indirect utility of all alternatives by the same positive constant?

A Revealed Preference Mechanism

Suppose that we observe that individual i chooses alternative $j_i \in C_i$. Then we know that $j_i = \arg\max_j v_j(\mathbf{q}_{ij}, I_i - p_{ij})$. In layman's terms, the chosen alternative is revealed to be preferred to the other feasible alternatives in the choice set. Note that j_i is such that $v_{j_i}(\mathbf{q}_{ij_i}, I_i - p_{ij_i}) - v_j(\mathbf{q}_{ij}, I_i - p_{ij}) \geq 0,\ \forall j \neq j_i$. This inequality establishes a constraint to the possible values of the indirect utility function. The revealed preference mechanism can be exploited to infer which alternative will be chosen after a change in the values of the attributes or available income.

In Fig. 4-4 a vehicle choice situation is represented. In Fig. 4-4*a*, both performance and driving range are attributes that are desired by the consumer. The straight lines

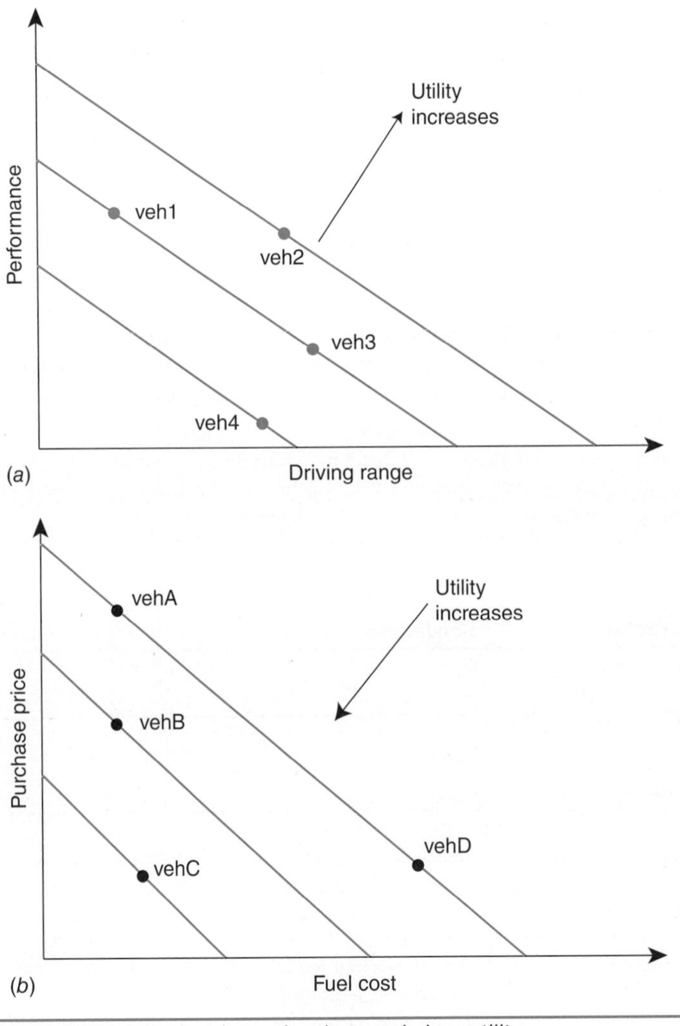

FIGURE 4-4 Individuals choose the alternative that maximizes utility.

represent combinations of the attributes that make the individual indifferent (*indifference curves* that provide the same utility level). Vehicles 1 and 3 are then perceived as being equally attractive in terms of performance and driving range. Vehicle 2 is the one that maximizes utility. In Fig. 4-4*b*, purchase price and fuel cost are attributes that are not desired. In this case, vehicle C is the one that maximized utility. Finally, the slope of the indifference curves is related to the concepts of marginal rate of substitution and willingness to pay, which will be discussed later.

Choice Microdata

Suppose that we observe choice behavior in a given market at a given purchase situation. This can be, for example, travel-mode choice at the morning peak period of a given workday at a specific city. What we observe is a sample of size N of individuals ($i \in \{1, \ldots, N\}$). For each individual we collect the following information:

1. Individual choice via a suitable *choice indicator*. If individual i chose alternative j_i, then we can define a choice indicator $y_i = j_i$. Possible values for y_i are all feasible alternatives in the choice set. Another possibility is a series of binary choice indicators $y_{ij} = \mathbb{I}[j = \arg\max_j v_j(\mathbf{q}_{ij}, I_i - p_{ij})]$, that is, $y_{ij} = 1$ if j was chosen and $y_{ij} = 0$ otherwise.
2. Attributes $\mathbf{q}_{ij}$ and price p_{ij} not only for the chosen alternative but for the whole set of available alternatives in the choice set of each individual.

Additional information that may be part of a choice data set is:

- Indicators of availability
- Identifier of the individual (ID), identifier of the choice situation (we could have *cross-sectional data*, where each individual is observed once, or *panel data*, where each individual is observed at different purchase situations)
- Characteristics of the individual $\mathbf{w}_i$

There are two standard formats for choice microdata. In the *wide* shape, information for each individual appears as a single row. For the wide shape, the choice indicator y_i is standard. In the *long* shape, for each individual there are as many rows as available alternatives in the choice set. For the long shape, the binary choice indicator y_{ij} is standard.

Example 4-2 Greene (2012, sec. 18. 2. 9) analyzes travel mode choice between Sydney and Melbourne. The sample contains observations for 210 individuals. All individuals have the same choice set composed of air, train, bus, and car. Tables 4-1 and 4-2 present information for three individuals (ID, mode, choice, travel time, trip cost, household income) in the sample in both the wide and long shapes.

Note that values of the utility function are not part of a choice dataset. From the point of view of a researcher trying to explain the observed choice behavior, the conditional indirect utility of each alternative is not known. To make inference on choices, a statistical model is needed. In Sec. 4-5, basic concepts of econometric modeling are summarized. These concepts are needed to build the microeconometric model of choice that we will use.

Individual	Mode	Choice	Travel Time	Trip Cost	Income
1	Air	0	100	59	35
1	Train	0	372	31	35
1	Bus	0	417	25	35
1	Car	1	180	10	35
2	Air	0	68	58	30
2	Train	0	354	31	30
2	Bus	0	399	25	30
2	Car	1	255	11	30
3	Air	0	125	115	40
3	Train	0	892	98	40
3	Bus	0	882	53	40
3	Car	1	720	23	40

Source: http://pages.stern.nyu.edu/%7Ewgreene/Text/tables/tablelist5.htm. Also available as the TravelMode data frame of the Applied Econometrics with R package (AER).

TABLE 4-1 Travel Mode Choice in Australia—Long Shape

Individual	Mode	Choice	Time Air	Cost Air	Time Train	Cost Train	Time Bus	Cost Bus	Time Car	Cost Car	Income
1	car	4	100	59	372	31	417	25	180	10	35
1	car	4	68	58	354	31	399	25	255	11	30
1	car	4	125	115	892	98	882	53	720	23	40

http://pages.stern.nyu.edu/%7Ewgreene/Text/tables/tablelist5.htm. Also available as the TravelMode data frame of the Applied Econometrics with R package (AER).

TABLE 4-2 Travel Mode Choice in Australia—Wide Shape

4-5 Overview of Econometric Modeling

Discrete choice theory is a special case of microeconometric modeling. Econometrics is the field of statistics that uses empirical data to analyze, elucidate, and test economic principles—including models of economic decision making that lead to consumer demand. Microeconometrics is concerned with causal inference on economic behavior of microagents (individuals or firms) using microdata (agent-based data). Choice datasets are microdata because information is provided at the individual level. The structural approach of microeconometric models of consumer behavior aims at estimating fundamental parameters that characterize individual preferences. Microeconometrics was recognized as an official subfield of econometrics with the 2000 Nobel Prize in Economics to James Heckman (for his work in censored and truncated variable models) and Daniel McFadden (for his work in discrete choice analysis).

In econometrics we work with a statistical model that establishes a causal relationship between the vector $\mathbf{y}$ of possible outcomes and both the matrix of exogenous variables $\mathbf{X}$ and the vector of unknown parameters $\boldsymbol{\beta}$. The standard model in econometrics is the regression $y_i = \mathbf{x}_i'\boldsymbol{\beta} + \varepsilon_i$, where i denotes an individual observation, $\mathbf{x}_i'$ is row i of matrix $\mathbf{X}$, and ε_i is an error term. If we assume that the error term is normally distributed $\varepsilon_i \sim \mathcal{N}(0, \sigma^2)$, then we can define the vector of unknown parameters of the model $\boldsymbol{\theta} = (\boldsymbol{\beta}', \sigma^2)'$. The *likelihood function* is the mapping that takes the parameters of a model and evaluates the probability of observing the data, that is, $\ell(\boldsymbol{\theta};\mathbf{y}) : \boldsymbol{\theta} \mapsto f(\mathbf{y}|\,\boldsymbol{\theta})$, where $f(\mathbf{y}|\,\boldsymbol{\theta})$ is the density of the sample $\mathbf{y}$ given the parameters $\boldsymbol{\theta}$ (cf. the probability density function, which is the mapping $\mathbf{y} \mapsto f(\mathbf{y}|\,\boldsymbol{\theta})$). A *conditional likelihood function* is defined as $\ell(\boldsymbol{\theta};\mathbf{y}\,|\,\mathbf{X})$.

Consider that the data we observe was generated by the parameter $\boldsymbol{\theta}_0$. The statistical estimation problem is to propose a value to the true but unknown parameter $\boldsymbol{\theta}_0$. There are different rules to decide which guess to propose as a solution to the estimation problem. A particular decision rule is called the *estimator*. When the estimator is evaluated at a particular observed sample, we obtain the *estimate.*

A common decision rule for proposing a good guess about the true but unknown parameter is the *maximum likelihood estimator* (MLE). Once the sample $\mathbf{y}$ is observed for a conditional model, the MLE keeps the value that maximizes the conditional likelihood function, i.e. $\hat{\boldsymbol{\theta}}(\mathbf{y}|\,\mathbf{X}) = \arg\max_{\boldsymbol{\theta}} \ell(\boldsymbol{\theta};\mathbf{y}\,|\,\mathbf{X})$. It can be shown that the true parameter $\boldsymbol{\theta}_0$ maximizes the expectation of the likelihood function. Thus, MLE searches for the value that maximizes the empirical counterpart of the expectation, which is the sample average likelihood function in a sampling model.

There are five steps that give a general structure for econometric modeling:

1. *Specification:* A structural model is built following a specific theory and available data. When some variables are relevant in theory but are missing in practice, *latent variables* that account for the unobserved information can be specified. Latent variables require an indirect way of measurement, usually provided by means of a measurement equation.

2. *Identification:* Before estimating the model one needs to check whether the degrees of freedom in the model allow the researcher to *identify* a unique set of values for the estimates. Formally, the parameter $\boldsymbol{\theta}$ is identified if $\ell(\boldsymbol{\theta}^1;\mathbf{y}\,|\,\mathbf{X}) = \ell(\boldsymbol{\theta}^2;\mathbf{y}\,|\,\mathbf{X}) \Rightarrow \boldsymbol{\theta}^1 = \boldsymbol{\theta}^2,\ \forall \boldsymbol{\theta}^1, \boldsymbol{\theta}^2$. Weak identification occurs when there are relatively flat areas in the likelihood function.

3. *Estimation:* The statistical estimation problem is to propose a value for the true but unknown parameter $\boldsymbol{\theta}$. If a single value is proposed, then we are solving the *point estimation* problem. To account for uncertainty in the determination of the parameters, it is possible to solve the *interval estimation* problem that finds a set of possible values for $\boldsymbol{\theta}$.

4. *Testing:* The first step in statistical inference is to draw valid conclusions about the model. Probably the most standard test regards statistical significance of individual parameter estimates. Goodness of fit and statistics for comparing competing models are additional examples of statistical testing. After testing the model, specification may be revisited.

5. *Forecasting:* Because econometric models are constructed following specific economic theories, the relationships that are modeled represent causal effects. Changes in the explanatory variables can be used to infer changes in the dependent variable. Forecasting also includes *welfare analysis* of *counterfactual*

scenarios, that is, the analysis of changes in economic surplus (satisfaction) coming from controlled changes in the explanatory variables that represent a situation that does not correspond to the conditions used for estimation.

4-6 Additive Random Utility Maximization

Consider the problem of a researcher trying to explain choice behavior. The goal is to propose a structural model of utility-maximizing behavior. We will assume that the conditional indirect utility is not fully known to the researcher. First, the researcher does not have knowledge about how the individual weighs each attribute to calculate the utility index. Second, the researcher may have incomplete information about all the different factors that influence consumers' decisions. In general, the first problem can be addressed by assuming a vector of unknown parameters ($\boldsymbol{\theta}$). The second problem can be handled by introducing an error term (ε_{ij}). The error term ε_{ij} is sometimes called a *taste shock* and represents both unobserved alternative attributes and consumer characteristics. Thus, the modeler postulates a model where choice is determined by maximization of the *random utility* $U_{ij} = v_j(\mathbf{q}_{ij}, I_i - p_{ij}, \varepsilon_{ij} \mid \boldsymbol{\theta}), \forall j$.

In additive random utility maximization (ARUM) models, the random utility U_{ij} is assumed to be separable into a deterministic component V_{ij} and the random term ε_{ij}, such that $U_{ij} = V_{ij} + \varepsilon_{ij}, \forall j$. The deterministic component of utility depends on factors that the modeler can observe and measure, as well as on a subset of the unknown parameters $\boldsymbol{\theta}$. The standard assumption is to consider a linear specification of the type:

$$U_{ij} = \mathbf{q}'_{ij}\boldsymbol{\beta}_{\mathbf{q}} - \alpha p_{ij} + \mathbf{w}'_i \gamma_j + \varepsilon_{ij} \tag{4-1}$$

where $\boldsymbol{\beta}_{\mathrm{q}}$ is a vector of marginal utilities or taste parameters (the weights assigned to each attribute to calculate the utility index of each alternative)

$-\alpha$ represents the marginal (dis)utility of cost of the discrete alternative

γ_j is a vector of parameters that account for observed consumer heterogeneity.

Example 4-3 Specify an ARUM model for the travel-mode choice between Sydney and Melbourne shown in Example 4-2.

Solution There are four alternatives. This means that for each individual we will need to specify four equations (one utility for each alternative). We will consider the variables shown in the example (the actual dataset contains more attributes). Travel time and cost are *alternative-specific* and *individual-specific* variables. Income is an individual-specific variable. If we assume that there are mean unobserved effects, we can introduce alternative-specific constants to our model. So, for individual *i* we have

$$U_{i\text{car}} = \text{ASC}_{\text{car}} + \beta_{\text{time}}\,\text{time}_{\text{car}} - \alpha\,\text{cost}_{\text{car}} + \gamma_{\text{male, car}}\,\text{male}_i + \varepsilon_{i\text{car}}$$

Due to convenience of notation, the ARUM model is usually rewritten as

$$U_{ij} = \mathbf{x}'_{ij}\boldsymbol{\beta} + \mathbf{w}'_i \gamma_j + \varepsilon_{ij}$$

where $\mathbf{x}_{ij}$ represents a vector of extended attributes that includes price (i.e., $\mathbf{x}_{ij} = (\mathbf{q}'_{ij}, p_{ij})'$).

4-6-1 Willingness to Pay

A key concept in consumer behavior is the *marginal utility of income* (MUI). Choices are constrained by the availability of resources. If a consumer receives an additional dollar, she will have more options for consumption. Assuming that "more is always better," that additional dollar will increase her utility (MUI > 0). The MUI measures precisely the marginal increase in utility after a marginal increase in income. Note that

$$\text{MUI} = \frac{\partial v_j(\mathbf{q}_{ij}, I_i - p_{ij}, \varepsilon_{ij} \mid \boldsymbol{\theta})}{\partial I_i} = -\frac{\partial v_j(\mathbf{q}_{ij}, I_i - p_{ij}, \varepsilon_{ij} \mid \boldsymbol{\theta})}{\partial p_{ij}} = \alpha$$

where $-\alpha$ not only represents the marginal (dis)utility of cost of the discrete alternative, but α is also the Lagrange multiplier of the budget constraint $p_{ij} + c \leq I_i$. The equality holds because a dollar saved on the cost of the discrete good—after a drop in its price—equals an additional dollar to spend on something else (the marginal increase of income).

Following the same concept, it is also possible to define the marginal utility of any attribute. Consider attribute k of alternative j (i.e., q_{kij}). The marginal utility of that particular attribute is $\partial v_j(\mathbf{q}_{ij}, I_i - p_{ij}, \varepsilon_{ij} \mid \boldsymbol{\theta}) / \partial q_{kij}$. This derivative measures the marginal change in utility after a marginal change in attribute q_{kij}. In a linear utility model $U_{ij} = \mathbf{x}'_{ij}\boldsymbol{\beta} + \mathbf{w}'_i \boldsymbol{\gamma}_j + \varepsilon_{ij}$, the marginal utility of attribute k equals the parameter β_k. Thus, $\boldsymbol{\beta}$ is a vector of marginal utilities. If $\beta_k > 0$, then utility increases after an improvement in x_{kij} (meaning that x_{kij} is desired by the consumer). If $\beta_k < 0$, then utility decreases after an increase in x_{kij}.

The marginal rate of substitution between attribute k and cost of the discrete good (at constant utility) is the *willingness to pay* for a marginal improvement of an attribute that provides utility:

$$\text{WTP}_{\Delta q_{kij}} = \frac{\partial v_j(\mathbf{q}_{ij}, I_i - p_{ij}, \varepsilon_{ij} \mid \boldsymbol{\theta}) / \partial q_{kij}}{\partial v_j(\mathbf{q}_{ij}, I_i - p_{ij}, \varepsilon_{ij} \mid \boldsymbol{\theta}) / \partial I_i} = -\frac{\partial v_j(\mathbf{q}_{ij}, I_i - p_{ij}, \varepsilon_{ij} \mid \boldsymbol{\theta}) / \partial q_{kij}}{\partial v_j(\mathbf{q}_{ij}, I_i - p_{ij}, \varepsilon_{ij} \mid \boldsymbol{\theta}) / \partial p_{ij}} \tag{4-2}$$

Unlike marginal utilities, which lack a measurement unit, $\text{WTP}_{\Delta q_{kij}}$ is a monetary valuation that can be easily interpreted.

4-6-2 The Value of Travel Time Savings

In transportation analysis, we are especially interested in the willingness to pay for *reducing* travel time $\text{WTP}_{\Delta \text{time}}$. This measure is sometimes called the *subjective value of travel time savings* or, more simply, the *value of time* (VOT) (Jara-Díaz, 2010). Because travel is a derived demand, users of the transportation system extract benefits from reductions in time. For example, a commuter would appreciate her morning commute being reduced by 5 minutes because she could use that time to sleep 5 minutes more. Another commuter may appreciate those extra 5 minutes to do additional work, and earn more income. From this example it is clear that the value of time measures the *opportunity cost* of travel time savings.

For a commuting mode choice model $U_{ij} = \beta_c \text{cost}_{ij} + \beta_t \text{time}_{ij} + \varepsilon_{ij}$, the value of time equals the ratio of the parameters of time and cost $\text{VOT} = \beta_t / \beta_c$. If travel time is measured in minutes, and travel cost in dollars per trip, then the unit of the value of time is dollar per minute ($/min).

4-7 Vehicle Purchase Choices

The automotive industry is a perfect example of firms producing highly differentiated products that consumers choose from. A single automaker offers multiple classes of vehicles, with several models and trims within each class. In addition, incumbent

FIGURE 4-5 Cars are highly differentiated goods that fulfill the mobility needs of heterogeneous consumers.

automakers are numerous and the market for used vehicles is big. Product differentiation in the automotive market responds to consumers' preference heterogeneity and differing budget constraints. Each maker-year-model-trim features a large set of attributes, including price, fuel economy, technical characteristics, interior and exterior features, and even color (see Fig. 4-5).

The analysis of vehicle purchase decisions is one of the most established subfields in discrete choice modeling. Each consumer is modeled as weighing the attributes of the feasible alternatives and choosing the vehicle that has the largest conditional indirect utility. Understanding the process of vehicle purchase decisions is key for firms and policymakers. Consumer response to vehicle features not only informs engineering design of the cars, but also marketing decisions (showcasing the most desired attributes according to consumer preferences). In addition, models of consumer demand can be used to determine reasons for the market success or failure of specific models or brands. For example, using the results of a discrete choice model, Train and Winston (2007) analyzed the reasons for the decline in the market share of U.S. automakers with respect to that of foreign manufacturers. The authors conclude that to regain their position in the market domestic automakers need to make up for the gap in basic attributes such as price, size, power, operating cost, transmission type, and reliability. In these basic attributes, Japanese and European automakers offer more attractive features.

The automotive market is highly competitive. As we mentioned above, each firm produces several products to respond to the differing tastes of heterogeneous consumers. Different degrees of competition are expected among makers and models. Additionally, demand for vehicles is motivated not only by mobility needs. Cars are associated with freedom, are viewed as an extension of the personality of the owner, and are perceived as symbols of lifestyle and status. We will discuss the fact that new, energy-efficient vehicle technologies are associated with symbols of environmentalism. All these special characteristics make vehicle purchase decisions a difficult process to model.

4-7-1 Extended Example: Binary Model of Vehicle Choice

Suppose that we collected data among consumers choosing between an internal combustion vehicle (ICV) and a battery electric vehicle (BEV). The microdata contains the following information for 1,000 consumers:

1. Choice: The vehicle actually purchased by each consumer
2. ICV and BEV purchase price($)
3. ICV and BEV fuel cost($/100 miles)

We will now follow the five steps of econometric modeling to specify, identify, estimate, test, and forecast with the model.

Specification of the Model

- Consumer $i \in \{1, \ldots, N\}$ chooses among the two alternatives, ICV and BEV
- ICV considered by i is characterized by $\text{price}_{\text{ICV},i}$ and $\text{fuel cost}_{\text{ICV},i}$
- BEV considered by i is characterized by the attributes $\text{price}_{\text{BEV},i}$ and $\text{fuel cost}_{\text{BEV},i}$
- There may be other factors that the buyer considered when deciding which vehicle to purchase that are not part of the dataset. These factors are represented by error terms $\varepsilon_{\text{ICV},i}$ and $\varepsilon_{\text{BEV},i}$
- Satisfaction (utility) when buying ICV:

$$U_{\text{ICV},i} = \beta_{\text{price}}\,\text{price}_{\text{ICV},i} + \beta_{\text{fuelcost}}\,\text{fuelcost}_{\text{ICV},i} + \varepsilon_{\text{ICV},i}$$

- Satisfaction (utility) when buying BEV:

$$U_{\text{BEV},i} = \beta_{\text{price}}\,\text{price}_{\text{BEV},i} + \beta_{\text{fuelcost}}\,\text{fuelcost}_{\text{BEV},i} + \varepsilon_{\text{BEV},i}$$

- Buyer i is a utility maximizer and exhibits *compensatory behavior.* That individuals are compensatory utility maximizers is a *behavioral assumption.* Other decision rules could be hypothesized. (For instance, the researcher may assume *bounded rationality.*)
- Formulation of the decision rule: chosen alternative (indicated by y_i) maximizes utility.

  If $y_i = \text{ICV}$, then $U_{\text{ICV},i} > U_{\text{BEV},i}$, and if $y_i = \text{BEV}$, then $U_{\text{BEV},i} > U_{\text{ICV},i}$
- Because utility is random, the model is completed with *statistical assumptions* over the error term. Suppose that $\varepsilon_{\text{BEV},i}$ and $\varepsilon_{\text{ICV},i}$ are distributed in such a way that $\varepsilon_{\text{BEV},i} - \varepsilon_{\text{ICV},i}$ has a probability density function f and a cumulative distribution function F.

Some notes about specification of the model:

1. The dataset is a *sample* from a population of interest that contains both observed choices and a set of characteristics of the alternatives and the decision maker.

2. The vector $\boldsymbol{\beta} = (\beta_{\text{price}}, \beta_{\text{fuelcost}})'$ contains *marginal utilities* that measure the change in satisfaction with a given alternative after a change of one unit in a specific attribute. These marginal utilities are unknown to the researcher and are the *parameters* of interest of the econometric model. Note that the hypothesized specification assumes parameters that are *generic*, representing preferences that are *homogeneous* among consumers (all individuals in the sample are assumed to have the same preferences) and among alternatives (preferences do not vary for different alternatives). The homogeneity assumption can be contested. For example, one could specify a model in which preferences for fuel cost are different for ICV and BEV. *Alternative-specific* marginal utilities for fuel cost $\beta_{\text{fuelcost, ICV}}$ and $\beta_{\text{fuelcost, BEV}}$ can be introduced to measure the sensitivity of the buyers to the energy source being used to refuel the vehicles. In addition, if one thinks that different individuals have different sensitivities for price, then it is possible to introduce individual-specific parameters $\beta_{\text{price}, i}$.

Econometric Derivation of the Model

Suppose that consumer i chose to buy an ICV. We then know that $U_{\text{ICV}, i} > U_{\text{BEV}, i}$, which is equivalent to $U_{\text{ICV}, i} - U_{\text{BEV}, i} > 0$. Arriving to this last inequality is trivial, but it contains a very important reinterpretation of the decision rule. *Only differences in utility are relevant* for determining choice. ICV is chosen because attributes of ICV compared to those of BEV (combined and weighted by preferences) are more attractive to the buyer. Based on the model in difference it is possible to rewrite utility as one structural equation:

$$
\begin{aligned}
y_i^* &= U_{\text{ICV}, i} - U_{\text{BEV}, i} \\
&= \beta_{\text{price}}(\text{price}_{\text{ICV}, i} - \text{price}_{\text{BEV}, i}) + \beta_{\text{fuel cost}}(\text{fuelcost}_{\text{ICV}, i} - \text{fuelcost}_{\text{BEV}, i}) \\
&\quad + (\varepsilon_{\text{ICV}, i} - \varepsilon_{\text{BEV}, i}) = \mathbf{x}'_{\Delta i}\boldsymbol{\beta} + \epsilon_i
\end{aligned}
$$

where $\mathbf{x}'_{\Delta i} = (\text{price}_{\text{ICV}, i} - \text{price}_{\text{BEV}, i}, \text{fuelcost}_{\text{ICV}, i} - \text{fuelcost}_{\text{BEV}, i})$

$$\boldsymbol{\beta} = (\beta_{\text{price}}, \beta_{\text{fuelcost}})'$$

$$\epsilon_i = \varepsilon_{\text{ICV}, i} - \varepsilon_{\text{BEV}, i}.$$

For the whole sample we can write the model as

$$
\underbrace{\begin{bmatrix} y_1^* \\ \vdots \\ y_i^* \\ \vdots \\ y_N^* \end{bmatrix}}_{\mathbf{y}^*} = \underbrace{\begin{bmatrix} \text{price}_{\text{ICV}, 1} - \text{price}_{\text{BEV}, 1} & \text{fuelcost}_{\text{ICV}, 1} - \text{fuelcost}_{\text{BEV}, 1} \\ \vdots & \vdots \\ \text{price}_{\text{ICV}, i} - \text{price}_{\text{BEV}, i} & \text{fuelcost}_{\text{ICV}, i} - \text{fuelcost}_{\text{BEV}, i} \\ \vdots & \vdots \\ \text{price}_{\text{ICV}, N} - \text{price}_{\text{BEV}, N} & \text{fuelcost}_{\text{ICV}, N} - \text{fuelcost}_{\text{BEV}, N} \end{bmatrix}}_{\mathbf{X}_\Delta} \underbrace{\begin{bmatrix} \beta_{\text{price}} \\ \beta_{\text{fuelcost}} \end{bmatrix}}_{\boldsymbol{\beta}} + \underbrace{\begin{bmatrix} \epsilon_1 \\ \vdots \\ \epsilon_i \\ \vdots \\ \epsilon_N \end{bmatrix}}_{\epsilon}.
$$

Note that this model, now rewritten as $\mathbf{y}^* = \mathbf{X}_\Delta\boldsymbol{\beta} + \epsilon$, seems to be a standard regression problem. So can we apply the normal equations of a standard regression to propose a value for β (see Prob. 2 in the exercise section)? The answer is no, because we don't observe the dependent variable $\mathbf{y}^*$. (Utility is not part of the data. What we do observe is the choice indicator $y_i = \mathbb{I}(U_{\text{ICV},i} - U_{\text{BEV},i} > 0)$, that is, $y_i = 1$ when $U_{\text{ICV},i} - U_{\text{BEV},i} > 0$. Can we use the vector $\mathbf{y} = (y_1, \ldots, y_N)'$ as dependent variable for a regression? We could, but we would encounter several problems in doing so. The choice indicator is not continuous. In fact, choice in this case is a dichotomous variable (see Fig. 4-6). However, a standard linear regression would treat $\mathbf{y}$ as a vector of continuous outcomes. Predictions of such a model would produce unbounded values for $\mathbf{y}$. Even more, a standard regression will try to fit a linear relationship between choice and the explanatory variables (such as the price difference).

Looking at Fig. 4-6 it is clear that a *nonlinear* relationship is necessary for explaining the impact of the attributes on choice; that is, we need a *link* to connect $\mathbf{y}$ and $\mathbf{y}^* = \mathbf{X}_\Delta \boldsymbol{\beta} + \epsilon$. Note that elements in $\mathbf{y}$ are random variables that have a Bernoulli distribution. The Bernoulli is a parametric distribution that is completely determined by the probability of producing an outcome equal to 1. So, from a statistical point of view, the vector $\mathbf{y}$ is a sample of Bernoulli random variables, where each element y_i has an unknown probability $P_{\text{ICV},i} = \Pr(y_i = 1 \mid \mathbf{x}'_{\Delta i}\boldsymbol{\beta})$. The probability that consumer i will choose ICV depends on the attribute levels that they experience and their preferences. Note that this conditional *choice probability* is individual-specific. It is the choice probability that connects both $\mathbf{y}$ and $\mathbf{y}^*$. Effectively, the probability of choosing ICV is equivalent to the probability of the utility of ICV being larger than that of BEV:

$$P_{\text{ICV},i} = \Pr(y_i = 1 \mid \mathbf{x}'_{\Delta i}\boldsymbol{\beta}) = \Pr(y_i^* > 0) = \Pr(\mathbf{x}'_{\Delta i}\boldsymbol{\beta} + \epsilon_i > 0) = \Pr(-\epsilon_i < \mathbf{x}'_{\Delta i}\boldsymbol{\beta})$$

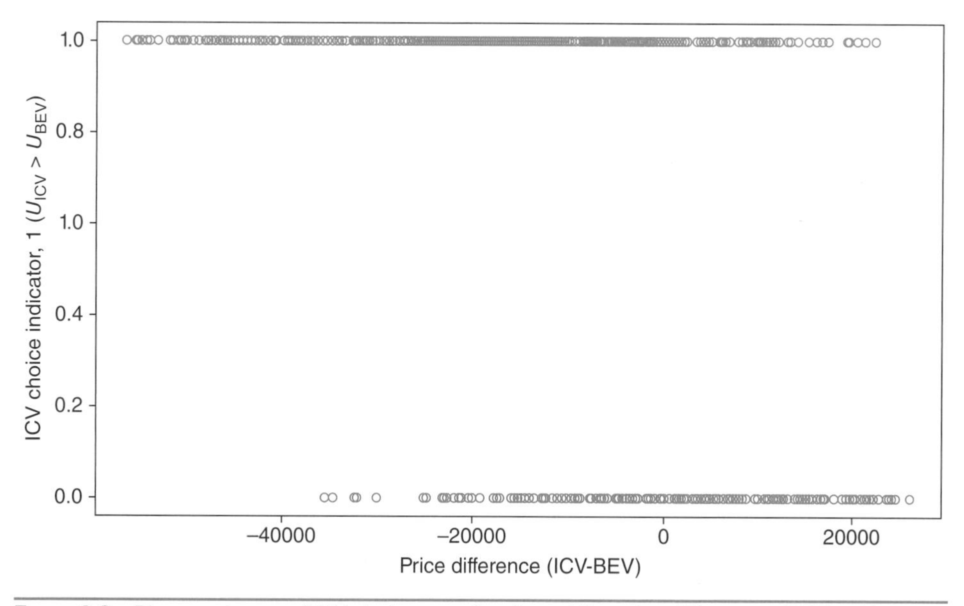

FIGURE 4-6 Binary outcome of ICV choice as a function of the price difference of ICVs with respect to BEVs. When the price difference is positive it means that ICVs are more expensive than BEVs.

which in this specific case is equal to

$$P_{\text{ICV},i} = \Pr[\varepsilon_{\text{BEV},i} - \varepsilon_{\text{ICV},i} < \beta_{\text{price}}(\text{price}_{\text{ICV},i} - \text{price}_{\text{BEV},i}) + \beta_{\text{fuelcost}}(\text{fuelcost}_{\text{ICV},i} - \text{fuelcost}_{\text{BEV},i})].$$

From the equation above, the individual choice probability is the probability of a random variable $(\varepsilon_{\text{BEV},i} - \varepsilon_{\text{ICV},i})$ being less than a deterministic value $[\beta_{\text{price}}(\text{price}_{\text{ICV},i} - \text{price}_{\text{BEV},i}) + \beta_{\text{fuelcost}}(\text{fuelcost}_{\text{ICV},i} - \text{fuelcost}_{\text{BEV},i})]$. This probability is, by definition, a cumulative distribution function. According to our assumptions, $P_{\text{ICV},i} = F(\mathbf{x}'_{\Delta i}\boldsymbol{\beta})$.[1]

The derivation above may seem rather technical, but it contains a relevant practical implication. Due to uncertainty in individual choices, random utility maximization leads to a *probabilistic model of choice*. In other words, a researcher is able to make an inference on the probability of a consumer performing a certain action, but would not be able to determine with certainty the actual choice.

Identification

When we introduced the notion of a utility function, we said that it was an index summarizing preferences and that it was never unique. The same choice probability $P_{\text{ICV},i}$ can be derived from an infinite number of utility functions that preserve the same preference order. In particular, suppose that the distribution $F(\mathbf{x}'_{\Delta i}\boldsymbol{\beta})$ has location (mean) μ and scale (standard deviation) s. Consider that the shape of $F(\mathbf{x}'_{\Delta i}\boldsymbol{\beta})$ is fully determined by the location and scale parameters, which are unknown.[2] The whole set of unknown parameters of the model is now $(\boldsymbol{\beta}', \mu, s)$. The problem is that the conditional choice probability remains the same, independent of the values of the nuisance parameters μ and s:

$$P_{\text{ICV},i} = \Pr\left(\frac{-\epsilon_i - \mu}{s} < \frac{\mathbf{x}'_{\Delta i}\boldsymbol{\beta} - \mu}{s}\right)$$

As a result (μ, s) cannot be identified, and identification of $\boldsymbol{\beta}$ requires assuming specific values of the nuisance parameters. The most common identification restriction is to assume that we are working with a standardized model $(\mu = 0,\ s = 1)$, ensuring identification of $\boldsymbol{\beta}$.

Estimation: Deriving the Estimator

The estimation problem is to estimate the identified parameters of the model. Because the parameters are unknown, we would like to propose a value $\hat{\boldsymbol{\beta}}$ to the true parameters. As explained when first introducing econometric models in this chapter, maximum likelihood estimates are obtained by finding the parameters that maximize the likelihood of having sampled the particular set of observations $\mathbf{y}$ in the dataset. The conditional likelihood is the joint density of observing the sample as a function of the parameters of the model, conditional on the alternative attributes and individual characteristics. Since each choice indicator is Bernoulli distributed, the probability density function of outcome y_i is $f_y(y_i | x'_{\Delta i}, \boldsymbol{\beta}) = (P_{\text{ICV},i})^{y_i}(1 - P_{\text{ICV},i})^{1-y_i}$. (Why?)

---

[1]What we do is to parameterize the probability of the Bernoulli distribution to depend on the differences of the deterministic component of utility.

[2]In econometrics, $\boldsymbol{\beta}$ is considered the parameter of interest, and (μ, s) are considered nuisance parameters.

Assuming that observations in the sample are independent, the likelihood function of the sample is

$$\ell(\boldsymbol{\beta};\mathbf{y}|\mathbf{X}_\Delta)=\prod_{i=1}^{N}[F(\mathbf{x}'_{\Delta i}\boldsymbol{\beta})]^{y_i}[1-F(\mathbf{x}'_{\Delta i}\boldsymbol{\beta})]^{1-y_i}$$

Taking logs, we derive the log-likelihood function:

$$\mathcal{L}(\boldsymbol{\beta};\mathbf{y}|\mathbf{X}_\Delta)=\sum_{i=1}^{N}y_i\ln[F(\mathbf{x}'_{\Delta i}\boldsymbol{\beta})]+(1-y_i)\ln[1-F(\mathbf{x}'_{\Delta i}\boldsymbol{\beta})]$$

The first-order condition for maximizing the log-likelihood function—known as the *likelihood equation*—is

$$\frac{\partial\mathcal{L}(\hat{\boldsymbol{\beta}};\mathbf{y}|\mathbf{X}_\Delta)}{\partial\boldsymbol{\beta}}=\sum_{i=1}^{N}\frac{y_i-F(\mathbf{x}'_{\Delta i}\hat{\boldsymbol{\beta}})}{F(\mathbf{x}'_{\Delta i}\hat{\boldsymbol{\beta}})[1-F(\mathbf{x}'_{\Delta i}\hat{\boldsymbol{\beta}})]}f(\mathbf{x}'_{\Delta i}\hat{\boldsymbol{\beta}})=0$$

which is an implicit equation. [Finding $\hat{\boldsymbol{\beta}}_{\text{MLE}}(\mathbf{y}|\mathbf{X}_\Delta)$ requires numerical optimization methods.] Note that even with a linear specification of utility, what enters the likelihood function are the choice probabilities that are nonlinear. Thus, discrete choice theory is an example of nonlinear models.

MLE Properties

MLE has several large-sample properties:

1. $\hat{\boldsymbol{\beta}}_{\text{MLE}}(\mathbf{y}|\mathbf{X}_\Delta)$ is *consistent* if the choice probabilities are correctly specified. Consistency is desired, because it means that the estimator converges in probability to the true parameter.
2. $\hat{\boldsymbol{\beta}}_{\text{MLE}}(\mathbf{y}|\mathbf{X}_\Delta)$ is asymptotically *efficient*, meaning that no other consistent estimator has lower asymptotic mean squared error. MLE is efficient because its asymptotic variance is given by the inverse of the information matrix:

$$\mathcal{J}^{-1}(\boldsymbol{\beta})=\left[-\mathbb{E}_{\beta}\left(\frac{\partial^2\mathcal{L}(\boldsymbol{\beta};y|x_\Delta)}{\partial\boldsymbol{\beta}\partial\boldsymbol{\beta}'}\right)\right]^{-1}$$

3. Under regularity, the MLE is asymptotically normal, that is, when the sample is large enough $[\hat{\boldsymbol{\beta}}_{\text{MLE}}(\mathbf{y}|\mathbf{X}_\Delta)\xrightarrow{a}\mathcal{N}(\boldsymbol{\beta},\mathcal{J}^{-1}(\boldsymbol{\beta})]$.

The Binary Logit Model

To make the model operational, we need an additional assumption to complete the specification of the model. The general likelihood expression depends on the general link $P_{\text{ICV},i}=F(\mathbf{x}'_{\Delta i}\boldsymbol{\beta})$, but a specific shape is needed for the cumulative distribution. There are several possibilities. The most common in applied work is the *logit* link. The binary *logit model* specifies

$$P_{\text{ICV},i}=\Lambda(\mathbf{x}'_{\Delta i}\boldsymbol{\beta})=\frac{\exp(\mathbf{x}'_{\Delta i}\boldsymbol{\beta})}{1+\exp(\mathbf{x}'_{\Delta i}\boldsymbol{\beta})} \tag{4-3}$$

where $\Lambda(\cdot)$ is actually the cumulative distribution function of the *logistic distribution*. This means that we can derive the logit model by assuming that the differences $\varepsilon_{\text{BEV},i} - \varepsilon_{\text{ICV},i}$ are independent and identically distributed standard logistic. Note that

$$P_{\text{ICV},i} = \frac{\exp[\beta_{\text{price}} (\text{price}_{\text{ICV},i} - \text{price}_{\text{BEV},i}) + \beta_{\text{fuelcost}} (\text{fuelcost}_{\text{ICV},i} - \text{fuelcost}_{\text{BEV},i})]}{1 + \exp[\beta_{\text{price}} (\text{price}_{\text{ICV},i} - \text{price}_{\text{BEV},i}) + \beta_{\text{fuelcost}} (\text{fuelcost}_{\text{ICV},i} - \text{fuelcost}_{\text{BEV},i})]}$$

$$= \frac{\exp(\beta_{\text{price}}\,\text{price}_{\text{ICV},i} + \beta_{\text{fuelcost}}\,\text{fuelcost}_{\text{ICV},i})}{\exp(\beta_{\text{price}}\,\text{price}_{\text{ICV},i} + \beta_{\text{fuelcost}}\,\text{fuelcost}_{\text{ICV},i}) + \exp(\beta_{\text{price}}\,\text{price}_{\text{BEV},i} + \beta_{\text{fuelcost}}\,\text{fuelcost}_{\text{BEV},i})}.$$

For the binary logit model, the likelihood equation becomes $\Sigma_{i=1}^{N}[y_i - \Lambda(\mathbf{x}'_{\Delta i}\hat{\boldsymbol{\beta}})]\mathbf{x}_{\Delta i} = 0$. Furthermore, the Hessian is negative definite, meaning that the log-likelihood function is globally concave. From the point of view of optimization, finding the maximum likelihood estimates is easy. When the number of attributes is low, even grid search methods can easily be implemented. However, for more sophisticated models, iterative methods such as Newton-Raphson work best.

Estimation: Obtaining the Estimates

If specification leads to a standard model, such as the binary probit or logit, then the researcher can use a variety of commercial and freeware statistical software with ready-to-use procedures for limited-dependent variables. If the model is nonstandard, then derivation of the estimator is necessary and the researcher will need to use an optimization package for maximizing the likelihood function.

In order to give a better idea about MLE implementation, we present below a code in *R* that maximizes the likelihood function of the binary logit using Newton-Raphson. Estimation results are shown in Table 4-3.

```
logl<- function(beta, y, Xdiff)
{
 # Loglikelihood of the binary logit (in differences)
return(sum(y*log(plogis(Xdiff%*%beta)) + (1-y)*log(1-plogis(Xdiff%*%beta))))
}

# Maximizing the loglikelihood using Newton Raphson
MLE.nr <- maxNR(logl, y=choice, Xdiff=cbind(price, fuelcost), start=c(0,0))

# Computing standard errors from the Hessian
MLE.se <-sqrt(diag(solve(-MLE. nr$hessian)))

MLE.nr$estimate
MLE.se
```

Attribute	Point Estimate	Standard Error (s.e.)	z-Value	p-Value
Price (β_{price})	−0.000105	0.000007	−14.494	0.000
Fuel Cost (β_{fuelcost})	0.002717	0.005080	0.535	0.593

TABLE 4-3 Binary Vehicle Choice—Model 1

Testing

Before any statistical testing, you should remember that the parameters of discrete choice models are structural; that is, the values can be interpreted as marginal utilities that are connected with preferences. How would you feel about an increase in price in the car that you were considering buying? Increases in price should have a negative impact on the consumer's satisfaction. The estimate of the price parameter being negative ($\hat{\beta}_{\text{price}} = -0.000105$) is then compatible with a rational consumer. The estimate of the parameter of fuel cost is positive ($\hat{\beta}_{\text{fuelcost}} = 0.002717$). Fuel cost is another component of price, so following the same rationale we should expect a negative parameter. So, should we conclude that consumers are irrational? Before doing so, we should take a look at the standard error. With respect to the point estimate, the standard error of price is small, but the standard error of fuel cost is large. This observation leads us to the implementation of a *statistical test of significance.*

When testing significance of a single parameter, the null hypothesis is that the parameter has no effect on the dependent variable $H_0 : \beta_k = 0$ (vs. $H_a : \beta_k \neq 0$). For performing the test, we can calculate the ratio $z = \hat{\beta}_k / s.e.(\hat{\beta}_k)$. Under the null hypothesis, $z \sim \mathcal{N}(0, 1)$. In our example, we can reject $\beta_{\text{price}} = 0$, but we cannot reject $\beta_{\text{fuelcost}} = 0$.

Consumers in the sample may be insensitive to fuel costs, but we should check if there is a problem with the model. When analyzing identification, we assumed that both location and scale of the difference of the error term were normalized to the values of a standard distribution. Let us see what happens if we add a constant to the model. Consider now a model with *alternative specific constants* in order to capture the average effect of the unobserved attributes:

$$U_{\text{ICV},\,i} = \beta_{\text{ICV}} + \beta_{\text{price}}\,\text{price}_{\text{ICV},\,i} + \beta_{\text{fuel cost}}\,\text{fuel cost}_{\text{ICV},\,i} + \varepsilon_{\text{ICV},\,i}$$

$$U_{\text{BEV},\,i} = \beta_{\text{BEV}} + \beta_{\text{price}}\,\text{price}_{\text{BEV},\,i} + \beta_{\text{fuel cost}}\,\text{fuel cost}_{\text{BEV},\,i} + \varepsilon_{\text{BEV},\,i}$$

We now know that what enter the likelihood function are choice probabilities, which depend on utility differences. So, for the model with constants

$$\begin{aligned} y_i^* &= U_{\text{ICV},\,i} - U_{\text{BEV},\,i} \\ &= (\beta_{\text{ICV}} - \beta_{\text{BEV}}) + \beta_{\text{price}}(\text{price}_{\text{ICV},\,i} - \text{price}_{\text{BEV},\,i}) + \beta_{\text{fuelcost}}(\text{fuelcost}_{\text{ICV},\,i} \\ &\quad - \text{fuelcost}_{\text{BEV},\,i}) + (\varepsilon_{\text{ICV},\,i} - \varepsilon_{\text{BEV},\,i}) \end{aligned}$$

Note that in this model the difference ($\beta_{\text{ICV}} - \beta_{\text{BEV}}$) can be identified as a single constant, but β_{ICV} and β_{BEV} cannot be separately identified. A usual normalization restriction is to normalize any of the two to zero. What the remaining constant will estimate is the average effect of the unobserved attributes with respect to the alternative. If we set $\beta_{\text{BEV}} = 0$ as identification restriction, the model becomes

$$\begin{aligned} y_i^* &= \beta_{\text{ICV}} + \beta_{\text{price}}(\text{price}_{\text{ICV},\,i} - \text{price}_{\text{BEV},\,i}) + \beta_{\text{fuelcost}}(\text{fuelcost}_{\text{ICV},\,i} - \text{fuelcost}_{\text{BEV},\,i}) + (\varepsilon_{\text{ICV},\,i} \\ &\quad - \varepsilon_{\text{BEV},\,i}) = \mathbf{x}'_{\Delta i}\boldsymbol{\beta} + \epsilon_i, \end{aligned}$$

where $\mathbf{x}'_{\Delta i} = (1, \text{price}_{\text{ICV},\,i} - \text{price}_{\text{BEV},\,i}, \text{fuelcost}_{\text{ICV},\,i} - \text{fuelcost}_{\text{BEV},\,i})$ and $\boldsymbol{\beta} = (\beta_{\text{ICV}}, \beta_{\text{price}}, \beta_{\text{fuelcost}})'$.

Attribute	Point Estimate	Standard Error	*z*Value	*p*Value
Constant (β_{ICV})	2.711000	0.290400	9.335	0.000
Price (β_{price})	−0.000108	0.000008	−13.362	0.000
Fuel cost ($\beta_{fuelcost}$)	−0.124600	0.014560	−8.557	0.000

Table 4-4 Binary Vehicle Choice—Model 2

Table 4-4 shows the estimation results of the model with the alternative-specific constant (see also Fig. 4-7). Note that all parameters are significantly different from 0. When including the constant, both the marginal utilities of price and fuel cost have the expected negative sign.

The positive constant indicates that ICVs are preferred to BEVs. Even with identical price and fuel costs the consumer will be more likely to choose the ICV because of the constant. This constant actually measures the average effect of the attributes that the consumer evaluates but we do not observe as modelers. For instance, the consumer

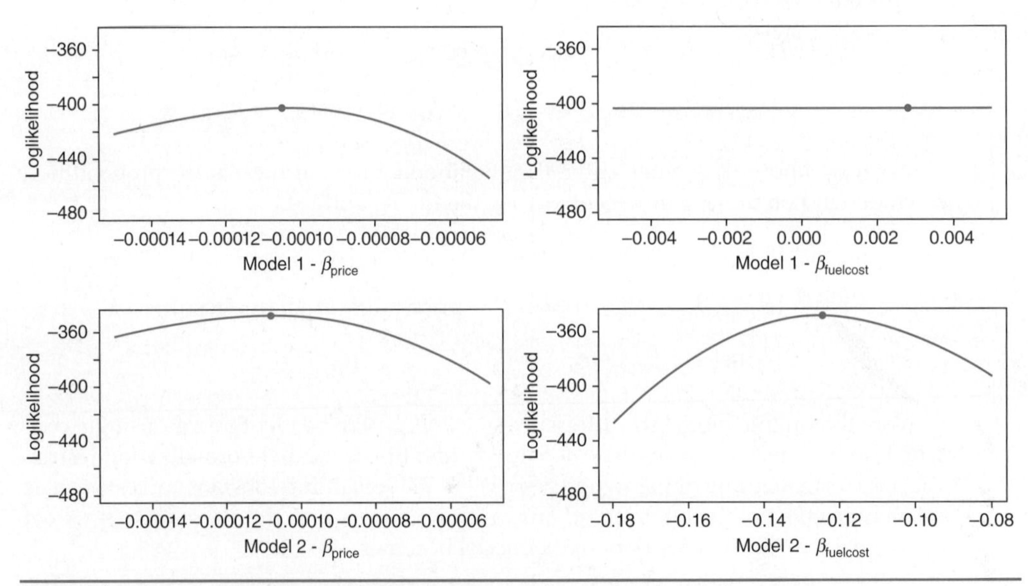

Figure 4-7 Finding the maximum likelihood estimates for models 1 and 2.

Note how flat the loglikelihood function is for $\beta_{fuelcost}$ in the case of model 1. In that model, very little information about the true parameter yielded a large standard error, and one could not reject $\beta_{fuelcost} = 0$. Note that a better specification, as in model 2, allows the researcher to produce a more efficient estimate.

may prefer ICVs because these do not exhibit problems with limited driving range and lengthy recharging.

Forecasting

Once the parameters have been estimated, we can produce estimates of the choice probabilities using the choice probability equation.

$$\hat{P}_{\text{ICV},\,i} = \Lambda(\mathbf{x}'_{\Delta i}\hat{\boldsymbol{\beta}}) = \frac{\exp(\mathbf{x}'_{\Delta i}\hat{\boldsymbol{\beta}})}{1+\exp(\mathbf{x}'_{\Delta i}\hat{\boldsymbol{\beta}})} \tag{4-4}$$

As we explained earlier, although the data—the choice indicators—are discrete, the researcher's best solution to explain the data is choice probabilities. Figure 4-8 shows the choice probability estimates for choosing ICVs as a function of the price difference between ICVs and BEVs. The blue dots represent the choice probabilities of each observation in the sample. The black line represents the choice probability for a representative individual, assuming there were no differences in fuel costs.

The curvature of the choice probability as a function of relative attributes is determined by how random or deterministic the choice process is. For example, for a segment of consumers who exhibit high sensitivity to price, the choice probability would look almost like a step function and would be immediately equal to 1 when ICVs are cheaper than BEVs even by 1 cent. For a segment of consumers who are inattentive to price, choice with respect to price differences would be completely random (a flat function with $P_{\text{ICV}} = 0.\,5$ for any price combination).

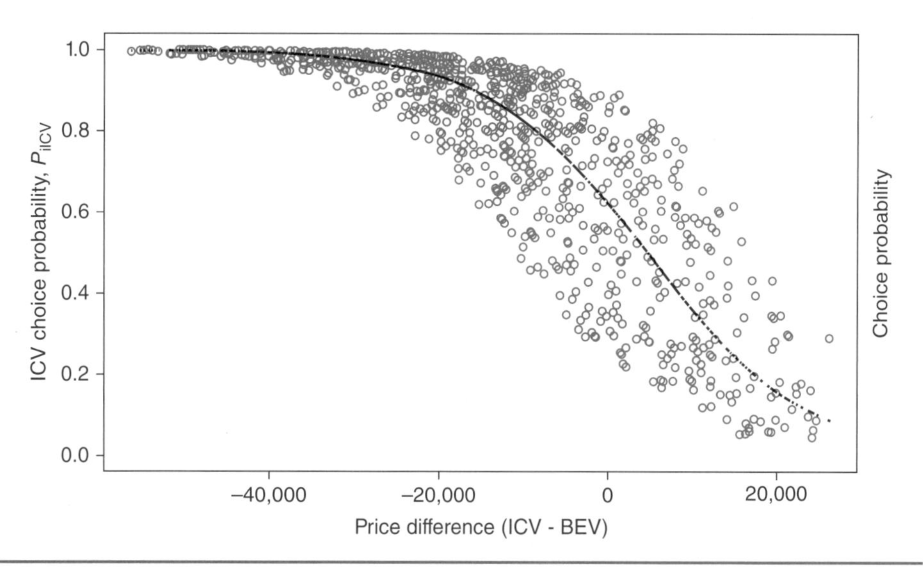

Figure 4-8 ICV choice probability estimates. As ICVs become more attractive (cheaper) relative to BEVs, the probability of choosing an ICV approaches 1.

4-8 Multinomial Discrete Choice[3]

Choice among differentiated products often requires a comparison of conditional utilities of several alternatives. Suppose that consumer i, who is looking to purchase a new car, is considering a compact sedan (1), a small SUV (2), a compact hybrid (3), and a hybrid SUV (4). When we first introduced the choice indicator, we said that there are two standard ways of manifesting the chosen alternative. We can use the polytomous choice indicator $y_i = j_i$. For example, if the consumer bought the compact hybrid, then $y_i = \text{compact hybrid}$, or $y_i = 3$. The second possibility is a series of dichotomous choice indicators $y_{ij} = \mathbb{I}[j = \arg\max_j v_j(\mathbf{q}_{ij}, I - p_{ij})]$. For example, $y_{i1} = 0, y_{i,2} = 0, y_{i,3} = 1, y_{i,4} = 0$. Note that when using the y_{ij} notation, only J_{i-1} choice indicators are needed (for example, $y_{i,1} = 0, y_{i,2} = 0, y_{i,3} = 1$). (Can you explain why?)

Multinomial response models are needed for choice situations with more than two alternatives. For multinomial response sampling models, the likelihood of observing the sample is

$$\ell(\boldsymbol{\beta}; \mathbf{y} | \mathbf{X}_\Delta) = \prod_{i=1}^{N} P_{i1}^{y_{i1}} \cdots P_{ij}^{y_{ij}} = \prod_{i=1}^{N} P_{ij_i} \tag{4-5}$$

where P_{ij} is the choice probability of individual i choosing alternative j (P_{ij_i} is the probability of the consumer choosing the actually chosen alternative).

Assuming a linear specification of the conditional utilities $U_{ij} = \mathbf{x}'_{ij}\boldsymbol{\beta} + \varepsilon_{ij}$, the choice probability can be written as

$$\begin{aligned} P_{ij} &= \Pr(U_{ij} > U_{il}, \forall\, l \in C_i \setminus \{j\}) \\ &= \Pr(\mathbf{x}'_{ij}\boldsymbol{\beta} + \varepsilon_{ij} > \mathbf{x}'_{il}\boldsymbol{\beta} + \varepsilon_{il}, \forall\, l \in C_i \setminus \{j\}) \\ &= \Pr(\varepsilon_{il} - \varepsilon_{ij} < (\mathbf{x}_{ij} - \mathbf{x}_{il})'\boldsymbol{\beta}, \forall\, l \in C_i \setminus \{j\}) \end{aligned}$$

The expression above simply means that the probability of choosing alternative j is defined as the probability of the utility of that alternative maximizing the satisfaction of the consumer.

4-8-1 Conditional Logit Model

For the conditional logit model, also known as *multinomial logit* (MNL) in transportation analysis, the choice probability P_{ij} has the following closed form:

$$P_{ij} = \frac{\exp(\mathbf{x}'_{ij}\boldsymbol{\beta})}{\sum_{l=1}^{J} \exp(\mathbf{x}'_{il}\boldsymbol{\beta})} \tag{4-6}$$

[3]This section summarizes the main elements of multinomial discrete choice models. Details, including seminal references and contributions, can be found in the textbooks that appear in the reference section at the end of this chapter.

When we derived the choice probability of the binary logit model we used the distribution of the error terms. It can be shown that the conditional logit choice probabilities are type 1 extreme value.

If individual-specific characteristics are added to the specification of utility ($U_{ij} = \mathbf{x}'_{ij}\boldsymbol{\beta} + \mathbf{w}'_i\gamma_j + \varepsilon_{ij}$), the choice probability becomes:

$$P_{ij} = \frac{\exp(\mathbf{x}'_{ij}\boldsymbol{\beta} + \mathbf{w}'_i\gamma_j)}{\sum_{l=1}^{J}\exp(\mathbf{x}'_{il}\boldsymbol{\beta} + \mathbf{w}'_i\gamma_l)}, \tag{4-7}$$

which can be rewritten as:

$$P_{ij} = \frac{\exp[(\mathbf{x}_{ij} - \mathbf{x}_{ib})'\boldsymbol{\beta} + \mathbf{w}'_i(\gamma_j - \gamma_b)]}{1 + \sum_{l \neq b}\exp[(\mathbf{x}_{il} - \mathbf{x}_{ib})'\boldsymbol{\beta} + \mathbf{w}'_i(\gamma_l - \gamma_b)]}$$

The expression above makes explicit the dependence of the choice probabilities on the differences of utility with respect to a base alternative. It is clear that one of the γ_j needs to be normalized to 0 for identification. In particular, only $J-1$ alternative specific constants can be estimated.

An interesting property of the conditional logit model is that the expected maximum utility (EMU) is equal to

$$EMU = \ln \sum_j \exp(\mathbf{x}'_{ij}\boldsymbol{\beta} + \mathbf{w}'_i\gamma_j)$$

which is known as *logsum* (or inclusive value). This property is a result of the distribution of the maximum of type 1 extreme value distributed variables. The logsum can be used to determine a measure of the *consumer surplus*, using the marginal utility of income in

$$\mathbb{E}(CS) = \frac{1}{\alpha}\ln \sum_j \exp(\mathbf{x}'_{ij}\boldsymbol{\beta} + \mathbf{w}'_i\gamma_j)$$

In addition, note that

$$\frac{P_{ij}}{P_{il}} = \frac{\exp(\mathbf{x}'_{ij}\boldsymbol{\beta} + \mathbf{w}'_i\gamma_j)}{\exp(\mathbf{x}'_{il}\boldsymbol{\beta} + \mathbf{w}'_i\gamma_l)} = \exp[(\mathbf{x}_{ij} - \mathbf{x}_{il})'\boldsymbol{\beta} + \mathbf{w}'_i(\gamma_j - \gamma_l)]$$

meaning that the relative odds between two alternatives do not depend on the presence or absence of, or changes in any other alternative. *Substitution patterns* are thus constant for logit models. This property is known as *independence of irrelevant alternatives* (IIA). If IIA holds, then it is possible to obtain consistent parameters using subsamples of alternatives.

MLE can be used for estimating the parameters of the model. The log likelihood function of the conditional logit model is globally concave.

Example 4-4 Bolduc et al. (2008) estimated a conditional logit model using data on 866 stated-preference interviews collected in 2008 among potential car buyers in Canada. Four hypothetical vehicles were presented to the respondents, based on fuel type: gasoline, compressed natural gas (CNG), hybrid (HEV), and hydrogen (HFC). The experimental attributes were the following: purchase price, fuel cost, fuel availability, express lane access, and power.

Attribute	Point Estimate	Standard Error
CNG Constant (β_{CNG})	−4.500	0.661
HEV Constant (β_{HEV})	−1.380	0.633
HFC Constant (β_{HFC})	−2.100	0.644
Price (β_{price})(10,000 CAD$)	−0.856	0.210
Fuel Cost ($\beta_{fuelcost}$)(100 CAD$/month)	−0.826	0.198
Fuel availability (β_{avail})(%)	1.360	0.186
Express lane access ($\beta_{express}$)	0.156	0.068
Power (β_{pow})(%)	2.700	0.655
Adjusted ρ^2	0.234	

4-8-2 Generalized Extreme Value Models

Although IIA is an attractive property, in practice it is likely that alternatives will exhibit different degrees of competition. For example, think of the introduction of the Tesla model S into the market. The conditional logit model will consider that all alternatives are equal substitutes to the new vehicle. However, one should expect higher substitution between the Tesla Model S and a luxury, gasoline-fueled sports car, and much lesser competition between the Model S and an economy compact car. In lay terms, a consumer considering buying an Audi A7 should be more likely to choose the Tesla Model S than a consumer considering buying a Ford Focus. The actual degree of competition among alternatives should be revealed by the data and not imposed by the model.

Generalized extreme value (GEV) models expand the conditional logit model by allowing for different degrees of competition among the alternatives. Subsets of alternatives may share unobserved attributes that introduce correlations in the taste shocks. Failing to account for this correlation will create a problem of misspecification, leading to inconsistent estimates. The classical example of GEV models is the *nested logit model*, which considers a partition of the alternatives, with *nests* defined according to the unobserved similarity of the alternatives. In a nested logit model, IIA holds within nests but does not hold between nests.

For a two-level nesting structure, the choice probability of a nested logit model is

$$P_{ij} = \frac{\exp\left(\frac{\mathbf{x}'_{ij}\boldsymbol{\beta}}{\lambda_m}\right)\left(\sum_{l\in \mathfrak{n}_m}\exp\left(\frac{\mathbf{x}'_{im}\boldsymbol{\beta}}{\lambda}\right)\right)^{\lambda_m-1}}{\sum_{m'=1}^{M}\left(\sum_{l\in \mathfrak{n}_{m'}}\exp\left(\frac{\mathbf{x}'_{il}\boldsymbol{\beta}}{\lambda_{m'}}\right)\right)^{\lambda_{m'}}} \tag{4-8}$$

where the J alternatives are partitioned into M nests, alternative j belongs to nest $\mathfrak{n}_{m'}$, and λ_m is an identified scale parameter of nest $\mathfrak{n}_m$. Note that when $\lambda_m = 1, \forall m$ (i.e., representing independence among all alternatives) the choice probability of the nested logit is the same as that of the conditional logit. The nested-logit parameters are $\boldsymbol{\theta} = (\boldsymbol{\beta}', \lambda_1, \ldots, \lambda_M)'$. The additional M parameters measure the degree of correlation of the alternatives belonging to the same nest. For instance, if $j, l \in \mathfrak{n}_m$ then $\mathrm{corr}(U_{ij}, U_{il}) = \sqrt{1-\lambda_m^2}$.

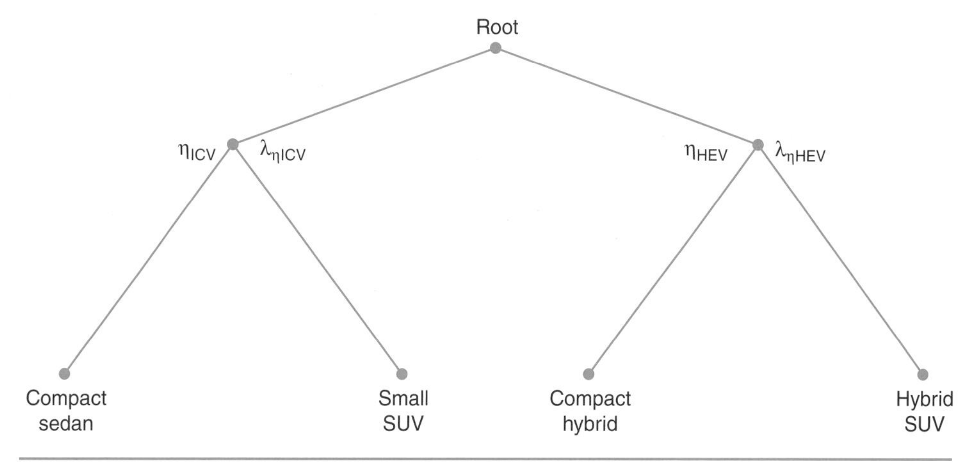

FIGURE 4-9 Example of (disjoint) nested structure.

Estimation of $\boldsymbol{\theta} = (\boldsymbol{\beta}, \lambda_1, \ldots, \lambda_M)'$ can be done using the MLE just as in the case of the conditional logit. However, the nested-logit likelihood function is not globally concave.

Example 4-5 *Nested logit:* A nested logit model for choice among a compact sedan (1), a small SUV (2), a compact hybrid (3), and a hybrid SUV (4) requires hypothesizing first a particular nesting structure (partition of the alternatives). For example, we can create the hypothesis that competition is stronger for vehicles with the same type of engine. In this case, we can have a first nest for vehicles with internal combustion engines ($\mathfrak{n}_{\text{ICV}}$ = {compact sedan, small SUV}), and a nest for the hybrid technology ($\mathfrak{n}_{\text{HEV}}$ = {compact hybrid, hybrid SUV}), as shown in Fig. 4-9. Using conditional probabilities, the nested-logit choice probability can be decomposed into the probability of choosing the nest and then choosing the alternative given that the specific nest is chosen. For example,

$$P_{i,\,\text{compact hybrid}} = P_{i,\,\mathfrak{n}_{\text{HEV}}} P_{i,\,\text{compact hybrid}|\mathfrak{n}_{\text{HEV}}}$$

Given that the consumer is choosing within the nest of hybrid technology, choice is only between the compact hybrid and the hybrid SUV. Because within a nest IIA holds, the conditional choice probability $P_{i,\,\text{compact hybrid}|\mathfrak{n}_{\text{HEV}}}$ then takes the MNL form:

$$P_{i,\,\text{compact hybrid}|\mathfrak{n}_{\text{HEV}}} = \frac{\exp(\mathbf{x}'_{i,\,\text{compact hybrid}}\boldsymbol{\beta}/\lambda_{\mathfrak{n}_{\text{HEV}}})}{\exp(\mathbf{x}'_{i,\,\text{compact hybrid}}\boldsymbol{\beta}/\lambda_{\mathfrak{n}_{\text{HEV}}}) + \exp(\mathbf{x}'_{i,\,\text{hybrid SUV}}\boldsymbol{\beta}/\lambda_{\mathfrak{n}_{\text{HEV}}})}.$$

If each nest is represented by the expected maximum utility that the consumer can extract in each case (for instance, $\text{EMU}_{\mathfrak{n}_{\text{HEV}}} = \lambda_{\mathfrak{n}_{\text{HEV}}} \ln[\exp(\mathbf{x}'_{i,\,\text{compact hybrid}}\boldsymbol{\beta}/\lambda_{\mathfrak{n}_{\text{HEV}}}) + \exp(\mathbf{x}'_{i,\,\text{hybrid SUV}}\boldsymbol{\beta}/\lambda_{\mathfrak{n}_{\text{HEV}}})]$), choice between the two nests can be represented by MNL probabilities of the form:

$$P_{i,\,\mathfrak{n}_{\text{HEV}}} = \frac{\exp(\text{EMU}_{\mathfrak{n}_{\text{HEV}}})}{\exp\left(\text{EMU}_{\mathfrak{n}_{\text{HEV}}}\right) + \exp(\text{EMU}_{\mathfrak{n}_{\text{ICV}}})}$$

$$= \frac{\left(e^{\mathbf{x}'_{i,\,\text{compact hybrid}}\boldsymbol{\beta}/\lambda_{\mathfrak{n}_{\text{HEV}}}} + e^{\mathbf{x}'_{i,\,\text{hybrid SUV}}\boldsymbol{\beta}/\lambda_{\mathfrak{n}_{\text{HEV}}}}\right)^{\lambda_{\mathfrak{n}_{\text{HEV}}}}}{\left(e^{\mathbf{x}'_{i,\,\text{compact hybrid}}\boldsymbol{\beta}/\lambda_{\mathfrak{n}_{\text{HEV}}}} + e^{\mathbf{x}'_{i,\,\text{hybrid SUV}}\boldsymbol{\beta}/\lambda_{\mathfrak{n}_{\text{HEV}}}}\right)^{\lambda_{\mathfrak{n}_{\text{HEV}}}} + \left(e^{\mathbf{x}'_{i,\,\text{compact sedan}}\boldsymbol{\beta}/\lambda_{\mathfrak{n}_{\text{ICV}}}} + e^{\mathbf{x}'_{i,\,\text{small SUV}}\boldsymbol{\beta}/\lambda_{\mathfrak{n}_{\text{ICV}}}}\right)^{\lambda_{\mathfrak{n}_{\text{ICV}}}}}$$

Then the joint probability is

$$P_{i,\text{ compact hybrid}}$$

$$= \frac{e^{\mathbf{x}'_{i,\text{ compact hybrid}}\beta/\lambda_{\mathrm{n_{HEV}}}} \left(e^{\mathbf{x}'_{i,\text{ compact hybrid}}\beta/\lambda_{\mathrm{n_{HEV}}}} + e^{\mathbf{x}'_{i,\text{ hybrid SUV}}\beta/\lambda_{\mathrm{n_{HEV}}}}\right)^{\lambda_{\mathrm{n_{HEV}}}-1}}{\left(e^{\mathbf{x}'_{i,\text{ compact hybrid}}\beta/\lambda_{\mathrm{n_{HEV}}}} + e^{\mathbf{x}'_{i,\text{ hybrid SUV}}\beta/\lambda_{\mathrm{n_{HEV}}}}\right)^{\lambda_{\mathrm{n_{HEV}}}} + \left(e^{\mathbf{x}'_{i,\text{ compact sedan}}\beta/\lambda_{\mathrm{n_{HEV}}}} + e^{\mathbf{x}'_{i,\text{ small SUV}}\beta/\lambda_{\mathrm{n_{ICV}}}}\right)^{\lambda_{\mathrm{n_{ICV}}}}}$$

which corresponds exactly to the choice probability of the nested logit as given in Eq. (4-8).

Because the scale parameters of each nest can be estimated, we can use $\hat{\lambda}_{\mathrm{n_{HEV}}}$ and $\hat{\lambda}_{\mathrm{n_{HEV}}}$ to test the hypothesis of the underlying structure being a conditional logit $(H_0 : \lambda_{\mathrm{n_{HEV}}} = 1, \lambda_{\mathrm{n_{ICV}}} = 1)$.

Other GEV models, such as the *cross-nested logit,* consider nests that are subsets of alternatives, not necessarily a partition. These models introduce the idea of allocation of an alternative to different nests.

Example 4-6 A cross-nested structure could consider nests of the form $\mathrm{n_{ICV}}$ = {compact sedan, small SUV}, $\mathrm{n_{HEV}}$ = {compact hybrid, hybrid SUV}, and $\mathrm{n_{SUV}}$ = {small SUV, hybrid SUV}. These nests are not disjoint and hence cannot be represented by a nested logit model.

4-8-3 Random Consumer Heterogeneity: Continuous Mixture Models

The marginal utilities β represent consumers' preferences (or tastes) for the different attributes. The parameter vector β may be the same for homogenous consumers, but heterogeneous consumers should have different preferences. Within the discrete choice modeling framework, several strategies for market segmentation can be applied. In particular,

1. Separate discrete choice models are specified for market segments that are constructed before estimation of the parameters. For example, three conditional logit models for three income classes are considered to account for income effects on choice. Estimates can then be used to test validity of the segmentation.
2. In a single model, interactions among socio-demographic variables and attributes account for *deterministic consumer heterogeneity* or observable taste variations. For example, suppose that we have reasons to believe that an individual with an annual income greater than \$100,000 exhibits a different sensitivity to purchase price of a product. Consider the indicator $I(\mathrm{Inc}_i > \$100\mathrm{K})$, which equals 1 if the income of individual i is greater than \$100,000 and 0 otherwise. Then in the expression $[\beta_{\text{price}} + \beta_{\text{price–high inc}} I(\mathrm{Inc}_i > \$100\mathrm{K})]\,\mathrm{price}_{ij}$, β_{price} represents the base marginal utility of price (for individuals with income lower than or equal to \$100,000), and $\beta_{\text{price–high inc}}$ measures the *taste variation* (with respect to β_{price}) for individuals with income greater than \$100,000. Note that for individuals with high income, the total marginal utility of price is $(\beta_{\text{price}} + \beta_{\text{price–high inc}})$.
3. In a single model, statistical distributions are used to account for *random consumer heterogeneity* or unobservable taste variations. In this strategy, preference differences cannot be explained in a fully systematic manner.

The most popular discrete choice model in current applied work, the random parameter logit or *mixed logit* model, considers heterogeneity distributions that are *continuous.* In the random parameter logit model,

$$U_{ij} = \mathbf{x}'_{ij}\boldsymbol{\beta}_i + \varepsilon_{ij}$$

where ε_{ij} are iid type 1 extreme value, and where the marginal utilities $\boldsymbol{\beta}_i$ are not only individual-specific but also random with heterogeneity distribution $\boldsymbol{\beta}_i \sim F(\boldsymbol{\beta}, \Sigma_\beta)$. Note that given $\boldsymbol{\beta}_i$, the conditional choice probability has the conditional logit form:

$$P_{ij|\boldsymbol{\beta}_i} = \frac{\exp(\mathbf{x}'_{ij}\boldsymbol{\beta}_i)}{\sum_{l=1}^{J}\exp(\mathbf{x}'_{il}\boldsymbol{\beta}_i)} \tag{4-9}$$

If $f(\boldsymbol{\beta}_i)$ is the probability density function of the random parameters, then the choice probability of a random parameter logit equals the unconditional probability:

$$P_{ij} = \int \frac{\exp(\mathbf{x}'_{ij}\boldsymbol{\beta}_i)}{\sum_{l=1}^{J}\exp(\mathbf{x}'_{ij}\boldsymbol{\beta}_i)} f(\boldsymbol{\beta}_i)d\boldsymbol{\beta}_i \tag{4-10}$$

which is a K-dimensional integral if all attributes are assumed to have random parameters. Under an appropriate specification, McFadden and Train (2000) show that the choice probabilities of a random parameter logit can approximate those of any RUM model to any degree of accuracy.

In applied work it is common to consider $\boldsymbol{\beta}_i \sim \mathcal{N}(\boldsymbol{\beta}, \Sigma_\beta)$. In fact, a diagonal covariance matrix is assumed. Under the normality assumption, $U_{ij} = \mathbf{x}'_{ij}\boldsymbol{\beta}_i + \varepsilon_{ij} = \mathbf{x}'_{ij}\boldsymbol{\beta} + \mathbf{x}'_{ij}\boldsymbol{\mu}_i + \varepsilon_{ij}$, where $\boldsymbol{\mu}_i \sim \mathcal{N}(0, \Sigma_\beta)$ represents a vector of *random taste variations* with respect to the population mean $\boldsymbol{\beta}$. Furthermore, if the taste variations are independent, then

$$U_{ij} = \mathbf{x}'_{ij}\boldsymbol{\beta} + \sum_{k=1}^{K}\sigma_k x_{ijk}\xi_{i,k} + \varepsilon_{ij}$$

where $\xi_{i,k}$ are iid standard normally distributed terms. In this latter model, the parameters to estimate are $\boldsymbol{\theta} = (\boldsymbol{\beta}', \boldsymbol{\sigma}')'$, that is, the mean and standard deviation of the random marginal utilities.

Although the choice probability in a random parameter logit does not have a closed form expression, it does take the form of an expectation: $P_{ij} = \mathbb{E}(P_{ij|\boldsymbol{\beta}_i})$. Thus, instead of working with the K-dimensional integral we can use its empirical counterpart:

$$\tilde{P}_{ij} = \frac{1}{S}\sum_{s=1}^{S}\frac{\exp\left(\mathbf{x}'_{ij}\boldsymbol{\beta}_i^{(s)}\right)}{\sum_{l=1}^{J}\exp\left(\mathbf{x}'_{il}\boldsymbol{\beta}_i^{(s)}\right)}$$

where $\boldsymbol{\beta}_i^{(s)}$ ($s \in \{1, \ldots, S\}$) are random draws from the density $f(\boldsymbol{\beta}_i; \boldsymbol{\beta}, \Sigma_\beta)$. $\tilde{P}_{ij}$ is a Monte Carlo simulator that is a valid estimator of the true choice probabilities and that is known to be unbiased, consistent (as $S \to \infty$), and smooth. Using the Monte Carlo simulator, we can derive the simulated log likelihood:

$$\tilde{\mathcal{L}}(\boldsymbol{\beta}, \Sigma_\beta; \mathbf{y}|\mathbf{X}) = \sum_{i=1}^{N}\sum_{j=1}^{J} y_{ij} \ln\left[\frac{1}{S}\sum_{s=1}^{S}\frac{\exp(\mathbf{x}'_{ij}\boldsymbol{\beta}_i^{(s)})}{\sum_{l=1}^{J}\exp(\mathbf{x}'_{il}\boldsymbol{\beta}_i^{(s)})}\right]$$

The *maximum simulated likelihood estimator* (MSLE) of the parameters of the model can be derived as $(\hat{\beta}, \hat{\Sigma})_{\mathrm{MSLE}} = \arg\max \sum_i \ln \tilde{P}_{ij_i}$, where j_i is the alternative actually chosen by individual i. Consistency of the estimator requires not only a large sample, but also a large number of replications. In fact, even though the Monte Carlo simulator is unbiased for the choice probabilities, for a finite number of repetitions the MSLE is biased.

Example 4-7 The work of Brownstone and Train (1999) is one of the earliest applications of a random parameter logit model estimated with microdata. 4,654 valid responses in an unlabeled discrete choice experiment were collected to analyze consumer response to alternative-fuel vehicles in California. The experimental attributes were: fuel type (gasoline, CNG, flexi-fuel methanol, and BEV), price (divided by the logarithm of income for estimation), driving range, acceleration, top speed, tailpipe emissions, vehicle size, body type, luggage space, operating cost (home recharging cost for BEV), refueling time, and fuel station availability. The authors specified an error component model of the type $U_{ij} = \mathbf{x}'_{ij}\boldsymbol{\beta} + \mathbf{z}'_{ij}\boldsymbol{\mu}_i + \varepsilon_{ij}$, where $\mathbf{z}_{ij}$ is not necessarily equal to the attribute vector. (If $\mathbf{z}_{ij} = \mathbf{x}_{ij}$, then as we discussed above, $\boldsymbol{\mu}_i$ represents random taste variations; if $\mathbf{z}_{ij}$ contains a subset of attributes, then $\boldsymbol{\mu}_i$ is the taste variation of those attributes that exhibit random heterogeneity; and if $\mathbf{z}_{ij}$ does not contain attributes, then $\mathbf{z}'_{ij}\boldsymbol{\mu}_i$ is a parametric representation of additive error terms that may introduce correlation or heteroscedasticity.) Selected estimates are shown in Table 4-5.

In the model presented in Table 4-5, both price/ln(income) and operating cost have fixed parameters. Size and luggage space were assumed to have normally distributed heterogeneity. So for these two attributes, the mean and standard deviation of the heterogeneity distribution are estimated. In addition, an additional error component nests the alternatives that are not a BEV, and a second additional error component nests the alternatives that are not a CNG-fueled car. The idea is that the term $\mu_{\text{non-BEV}}$ creates correlation (i. e., greater substitution) among gasoline, CNG, and methanol. The authors noted a problem with normally distributed parameters. The normal distribution is unbounded, meaning that a proportion of the population will exhibit a change of sign in the normally distributed marginal utility. For attributes such as vehicle size and luggage space this result can be justified: some consumers may prefer smaller cars. However, certain attributes need to have the same sign for the whole population. For example, β_{OC} needs to be negative since a positive marginal utility would indicate that the consumer would be willing to pay more for a less efficient car (everything else held constant). A solution is to use a one-side bounded distribution, such as the *lognormal*, to represent consumer heterogeneity for parameters that need to have a specific sign.

Attribute	Point Estimate	Standard Error
Price/ln(income) (β_{price})	−0.264	0.043
Operating cost (β_{OC})	−1.224	0.159
Vehicle size – mean (β_{size})	1.435	0.508
Vehicle size – std. dev ($\sigma_{\beta_{\text{size}}}$)	7.455	1.819
Luggage space – mean ($\beta_{\text{lug.space}}$)	1.702	0.482
Luggage space – std. dev ($\sigma_{\beta_{\text{lug.space}}}$)	5.994	1.248
Non-BEV – error component ($\sigma_{\mu_{\text{non-BEV}}}$)	2.464	0.541
Non-CNG – error component ($\sigma_{\mu_{\text{non-CNG}}}$)	1.072	0.377
Loglikelihood	−7375.34	

TABLE 4-5 Random Parameter Logit Example (Brownstone and Train, 1999)

Brownstone and Train (1999) also estimated a model with lognormally distributed parameters.

4-8-4 Random Consumer Heterogeneity: Discrete Mixture Models

Discrete, latent market segments can be identified using finite mixture models. The idea is the same as in the (continuous) random parameter logit or mixed logit model, but the random parameters $\boldsymbol{\beta}_i$ in $U_{ij} = \mathbf{x}'_{ij}\boldsymbol{\beta}_i + \varepsilon_{ij}$ are assumed to have a heterogeneity distribution that is *discrete*. The discrete possible values of the unobservable taste variations define *classes*. Consider a discrete latent partition of the consumers into Q classes. Given a class q, the conditional choice probability $P_{ij|q}$ has a known form (for instance, we can assume that $P_{ij|q}$ is given by a conditional logit model). Assignment to classes follows a random process. Let π_{iq} be the allocation probability of consumer i belonging to class q. Then, the unconditional choice probability is $P_{ij} = \sum_{q=1}^{Q} \pi_{iq} P_{ij|q}$. The loglikelihood of this latent class model is then

$$\mathcal{L}(\boldsymbol{\theta};\mathbf{y}|\mathbf{X}) = \sum_{i=1}^{N}\sum_{j=1}^{J} y_{ij} \ln \sum_{q=1}^{Q} \pi_{iq} P_{ij|q} \tag{4-11}$$

A common assumption in the latent class model is to assume a logit link for the allocation probabilities of the form:

$$\pi_{iq} = \frac{\exp(\mathbf{w}'_i\boldsymbol{\theta}_q)}{\sum_{r=1}^{Q} \exp(\mathbf{w}'_i\boldsymbol{\theta}_r)} \tag{4-12}$$

If the kernel $P_{ij|q}$ is a conditional logit, then

$$\mathcal{L}(\boldsymbol{\beta}, \boldsymbol{\theta};\mathbf{y}|\mathbf{X}) = \sum_{i=1}^{N}\sum_{j=1}^{J} y_{ij} \ln\left[\sum_{q=1}^{Q} \frac{\exp(\mathbf{w}'_i\boldsymbol{\theta}_q)}{\sum_{r=1}^{Q}\exp(\mathbf{w}'_i\boldsymbol{\theta}_r)} \frac{\exp(\mathbf{x}'_{ij}\boldsymbol{\beta}^{(q)})}{\sum_{l=1}^{J}\exp(\mathbf{x}'_{il}\boldsymbol{\beta}^{(q)})}\right]$$

where $\boldsymbol{\beta}^{(q)}$ is a vector of marginal utilities that are specific to class q.

Example 4-8 *Location decisions:* Modeling household location decisions are relevant for city planning. In particular, higher residential density is associated with lower car demand (Cervero and Kockelman, 1997; Brownstone and Golob, 2009). Walker and Li (2006) estimated a latent-class error-component logit model of residential choices. The data consisted of 611 individuals who answered to a stated-preference survey in Portland, Oregon. Five alternatives were presented in each of eight choice situations: buy a single-family home, buy multifamily, rent single-family, rent multifamily, and move out of the metro area. The experimental attributes were the following: price, residence and lot size, parking availability, school quality, safety, a set of community amenities (such as parks and bicycle paths), and measures of accessibility (such as travel time to work by car and walking time to shops). Based on different specifications, the authors found three latent classes of lifestyle: suburban households that are car-oriented, transit-oriented households, and car-oriented urban households. For the class allocation probabilities, socio-demographics were used. For example, affluent households are more likely to belong to the car-oriented, suburban class; younger families are more likely to be transit-riders; and non-family, professional households tend to have an urban, car-oriented lifestyle.

4-8-5 Multinomial Probit Model

The models that we have discussed thus far hypothesize specific substitution patterns through a particular covariance structure. However, exact competition among alternatives in a fully flexible model should be revealed from the data. The multinomial probit model assumes that the utility vector $\mathbf{U}_i = (U_{i1}, \ldots, U_{iJ})'$ is multivariate normally distributed with a full covariance matrix: $\mathbf{U}_i = \mathbf{X}_i\boldsymbol{\beta} + \varepsilon_i,\ \varepsilon_i \sim \mathcal{N}(\mathbf{0}, \Sigma)$, where $\mathbf{X}_i$ is a matrix of exogenous attributes with row j equal to $\mathbf{x}'_{ij}$ (such that $U_{ij} = \mathbf{x}'_{ij}\beta + \varepsilon_{ij}$), and where $\boldsymbol{\varepsilon}_i = (\varepsilon_{i1}, \ldots, \varepsilon_{iJ})'$. Because only utility differences can be identified, we derive the relative utility with respect to alternative j:

$$\Delta_j\mathbf{U}_i = \Delta_j\mathbf{X}_i\boldsymbol{\beta} + \Delta_j\varepsilon_i,\ \Delta_j\varepsilon_i \sim \mathcal{N}(\mathbf{0}_{(J-1)}, \Delta_j\Sigma\Delta'_j) \tag{4-13}$$

where Δ_j is a $(J-1\times J)$ matrix difference operator that takes the difference of each element of a vector or matrix with respect to the corresponding component of alternative j. For instance, $\Delta_j\mathbf{U}_i = (U_{i1} - U_{ij}, \ldots, U_{ij-1} - U_{ij}, U_{ij+1} - U_{ij}, \ldots, U_{iJ} - U_{ij})'$. Elements in Δ_j are defined as

$$[\Delta_j]_{lm} = \begin{cases} -1 & \text{if } m = j \\ 1 & \text{if } l = m \\ 0 & \text{otherwise} \end{cases}$$

The probit choice probabilities take the form

$$\begin{aligned} P_{ij} &= \Pr(U_{ij} \geq U_{ij'},\ \forall j' \neq j) \\ &= \Pr(\varepsilon_{ij'} - \varepsilon_{ij} \leq (\mathbf{x}_{ij'} - \mathbf{x}_{ij})'\beta,\ \forall j' \neq j) \\ &= \int_{-\infty}^{(\mathbf{x}_{iJ}-\mathbf{x}_{ij})'\beta} \cdots \int_{-\infty}^{(\mathbf{x}_{i1}-\mathbf{x}_{ij})'\beta} f(\Delta_j\varepsilon_i)\, d\Delta_j\varepsilon_i \end{aligned}$$

where $f(\Delta_j\varepsilon_i)$ is the probability density function of a multivariate normal:

$$f(\Delta_j\varepsilon_i) = \frac{1}{(2\pi)^{\frac{J-1}{2}} |\Delta_j\Sigma\Delta'_j|^{\frac{1}{2}}} \exp\left\{-\frac{1}{2}\varepsilon'_j\Delta'_j\Delta_j\Sigma\Delta'_j\Delta_j\varepsilon_j\right\}$$

The probit choice probability P_{ij} is an integral of dimension $J-1$ with an openform. Numerical integration is feasible for up to 3 dimensions (i.e., 4 alternatives). For larger-scale applications, parameters of probit models—marginal utilities $\boldsymbol{\beta}$ and elements of the covariance matrix $\Delta_j\Sigma\Delta'_j$—can be estimated using simulation-based inference, namely, the method of simulated moments, the method of simulated scores, or maximum simulated likelihood. For finding $(\hat{\beta}, \hat{\Sigma})_{\text{MSLE}} = \arg\max \sum_i \ln \tilde{P}_{ij_i}$ any choice probability simulator can be used for $\tilde{P}_{ij_i}$, but the smooth-recursive-conditioning GHK simulator is the most popular choice in empirical work. The GHK simulator exploits recursive truncation of a normal distribution and is continuous and differentiable, which is an advantage for finding the optimum. However, the probit log likelihood is not globally concave.

Example 4-9 Daziano and Achtnicht (2013) estimated a multinomial probit model using data on 600 stated-preference interviews collected in 2008 among potential car buyers in Germany. Seven hypothetical vehicles were presented to the respondents, based on fuel type: gasoline, diesel, hybrid, CNG, biofuel, hydrogen, and electric. The experimental attributes were the following: purchase price, fuel cost, fuel availability, engine power, and emissions. Using probit estimates the authors analyzed the impact on the market shares of increasing the network of service infrastructure for recharging electric vehicles as well as for refueling hydrogen. Forecasts indicate that if density of recharging becomes fully competitive, BEVs would experience a greater than threefold increase in market penetration.

Example 4-10 *Binary probit model of vehicle choice:* Specify and estimate a binary probit model for the binary vehicle choice data.

Solution *Specification:* For a binary probit model, consider

$$U_{ICV,i} = \beta_{ICV} + \beta_{price}\,\text{price}_{ICV,i} + \beta_{fuelcost}\,\text{fuelcost}_{ICV,i} + \varepsilon_{ICV,i}$$
$$U_{BEV,i} = \beta_{price}\,\text{price}_{BEV,i} + \beta_{fuelcost}\,\text{fuelcost}_{BEV,i} + \varepsilon_{BEV,i}$$

or

$$\mathbf{U}_i = \begin{bmatrix} U_{ICV,i} \\ U_{BEV,i} \end{bmatrix} = \begin{bmatrix} 1 & \text{price}_{ICV,i} & \text{fuelcost}_{ICV,i} \\ 0 & \text{price}_{BEV,i} & \text{fuelcost}_{BEV,i} \end{bmatrix} \begin{bmatrix} \beta_{ICV} \\ \beta_{price} \\ \beta_{fuelcost} \end{bmatrix} + \begin{bmatrix} \varepsilon_{ICV,i} \\ \varepsilon_{BEV,i} \end{bmatrix}$$

where

$$\varepsilon_i = \begin{bmatrix} \varepsilon_{ICV,i} \\ \varepsilon_{BEV,i} \end{bmatrix} \sim \mathcal{N}\left(\begin{bmatrix} 0 \\ 0 \end{bmatrix}, \begin{bmatrix} \sigma^2_{ICV} & \sigma_{ICV,BEV} \\ \sigma_{ICV,BEV} & \sigma^2_{BEV} \end{bmatrix} \right)$$

The ICV choice probability is

$$\begin{aligned} P_{ICV,i} &= \Pr(U_{ICV,i} > U_{BEV,i}) \\ &= \Pr[\varepsilon_{BEV,i} - \varepsilon_{ICV,i} < \beta_{ICV} + \beta_{price}(\text{price}_{ICV,i} - \text{price}_{BEV,i}) \\ &\quad + \beta_{fuel\ cost}(\text{fuel cost}_{ICV,i} - \text{fuel cost}_{BEV,i})] \end{aligned}$$

Using the properties of the normal distribution

$$\varepsilon_{BEV,i} - \varepsilon_{ICV,i} \sim \mathcal{N}(0, \sigma^2_{ICV} + \sigma^2_{BEV} - 2\sigma_{ICV,BEV})$$

or

$$\frac{\varepsilon_{BEV,i} - \varepsilon_{ICV,i}}{\sqrt{\sigma^2_{ICV} + \sigma^2_{BEV} - 2\sigma_{ICV,BEV}}} \sim \mathcal{N}(0, 1)$$

we can write $P_{ICV,i}$ as a recognizable cumulative distribution:

$$\begin{aligned} P_{ICV,i} &= \Pr\left(\frac{\varepsilon_{BEV,i} - \varepsilon_{ICV,i}}{\sqrt{\sigma^2_{ICV} + \sigma^2_{BEV} - 2\sigma_{ICV,BEV}}} < \frac{\mathbf{x}'_{\Delta i}\boldsymbol{\beta}}{\sqrt{\sigma^2_{ICV} + \sigma^2_{BEV} - 2\sigma_{ICV,BEV}}} \right) \\ &= \Phi\left(\frac{\mathbf{x}'_{\Delta i}\boldsymbol{\beta}}{\sqrt{\sigma^2_{ICV} + \sigma^2_{BEV} - 2\sigma_{ICV,BEV}}} \right) \end{aligned}$$

where $\mathbf{x}'_{\Delta i}$ is $(1, \text{price}_{ICV,i} - \text{price}_{BEV,i}, \text{fuelcost}_{ICV,i} - \text{fuelcost}_{BEV,i})$
$\boldsymbol{\beta}$ is $(\beta_{ICV}, \beta_{price}, \beta_{fuelcost})'$,
Φ is the cumulative distribution of a standard normal

The scale of the utility difference needs to be normalized for identification of $\boldsymbol{\beta}$. (*Note on identification:* in the choice probability, the marginal utilities $\boldsymbol{\beta}$ and the standard deviation of the error difference $\sqrt{\sigma^2_{ICV} + \sigma^2_{BEV} - 2\sigma_{ICV,BEV}}$ appear always together, meaning that it would not be possible to estimate both separately.) So, we set $\sigma^2_{ICV} + \sigma^2_{BEV} - 2\sigma_{ICV,BEV} = 1$.

Note that the same model can be derived from working directly with the model in differences:

$$\begin{aligned} y^*_i = \Delta_{BEV}\mathbf{U}_i &= \beta_{ICV} + \beta_{price}(\text{price}_{ICV,i} - \text{price}_{BEV,i}) + \beta_{fuel\ cost}(\text{fuelcost}_{ICV,i} - \text{fuelcost}_{BEV,i}) \\ &\quad + (\varepsilon_{ICV,i} - \varepsilon_{BEV,i}) = \mathbf{x}'_{\Delta i}\boldsymbol{\beta} + \epsilon_i,\ \epsilon_i \sim \mathcal{N}(0, 1) \end{aligned}$$

an expression that leads to a binary regression with a *probit link* $P_{\mathrm{ICV},i} = \Phi(\mathbf{x}'_{\Delta i}\boldsymbol{\beta})$ and loglikelihood:

$$\mathcal{L}(\boldsymbol{\beta};\mathbf{y}|\mathbf{X}_{\Delta}) = \sum_{i=1}^{N} y_i \ln\left[\Phi(\mathbf{x}'_{\Delta i}\boldsymbol{\beta})\right] + (1-y_i)\ln\left[1-\Phi(\mathbf{x}'_{\Delta i}\boldsymbol{\beta})\right]$$

The first order condition for maximizing the binary probit loglikelihood function is

$$\frac{\partial \mathcal{L}(\hat{\boldsymbol{\beta}};\mathbf{y}|\mathbf{X}_{\Delta})}{\partial \boldsymbol{\beta}} = \sum_{i=1}^{N} \frac{y_i - \Phi(\mathbf{x}'_{\Delta i}\hat{\boldsymbol{\beta}})}{\Phi(\mathbf{x}'_{\Delta i}\hat{\boldsymbol{\beta}})[1-\Phi(\mathbf{x}'_{\Delta i}\hat{\boldsymbol{\beta}})]}\phi(\mathbf{x}'_{\Delta i}\hat{\boldsymbol{\beta}}) = 0$$

where ϕ is the probability density function of a standard normal. Unlike the logit link, the probit choice probability does not have a closed-form expression.

Estimation: For a binary probit model, we do not need to use simulation. In fact, all statistical packages have integrated packages that evaluate both the pdf and CDF of a standard normal distribution. Furthermore, the loglikelihood of the binary probit is globally concave.

4-8-6 Addressing Endogeneity: The BLP Model

Endogeneity occurs when observable attributes are correlated with the unobserved factors absorbed by the error term. When this correlation happens, the estimates are no longer consistent. In discrete choice models, endogeneity problems are likely to occur. Attributes aim at representing quality of the discrete good. Because quality is hard to measure, and because there is an expected relationship between price and quality, price may easily be correlated with unobserved components in the error term.

Several methods for addressing endogeneity in discrete choice models have been proposed. Berry et al. (1995) introduced a structural model for market-level demand shocks that considers endogeneity in prices. The BLP model was originally applied to the automotive market and has become standard in the analysis of vehicle purchases with revealed-preference data. Consider that consumer behavior is observed in $t \in \{1, \ldots, T\}$ markets. The utility function is

$$U_{ijt} = \alpha_i \ln(I_i - p_{jt}) + \mathbf{x}'_{jt}\boldsymbol{\beta}_i + \xi_{jt} + \varepsilon_{ijt}$$

where income effects are made possible by considering $\ln(I_i - p_{jt})$, and where ξ_{jt} is an unobserved product attribute (demand shock) that is correlated with price p_{jt}. The possibility of no purchase is introduced through an outside good with utility $U_{i0t} = \varepsilon_{i0t}$. If ε_{ijt} is iid type 1 extreme value and $\boldsymbol{\beta}_i$ has a heterogeneity distribution $\mathrm{F}(\alpha, \boldsymbol{\beta})$, then the market share of the alternative j is

$$s_{jt} = \int_{\alpha,\beta} \frac{\exp(\alpha_i \ln(I_i - p_{jt}) + \mathbf{x}'_{jt}\boldsymbol{\beta}_i + \xi_{jt})}{1 + \sum_{l=1}^{J} \exp(\alpha_i \ln(I_i - p_{lt}) + \mathbf{x}'_{lt}\boldsymbol{\beta}_i + \xi_{lt})} d\mathrm{F}(\alpha, \boldsymbol{\beta})$$

To address the problem of endogeneity, the BLP model considers a vector of instrumental variables $\mathbf{z}_{jt}$. Instruments are such that they are not correlated with the demand shock $\mathbb{E}(\xi_{jt} | \mathbf{z}_{jt}) = 0$, but $\mathrm{cov}(\mathbf{z}_{jt}, [\mathbf{x}_{jt}, p_{jt}]) \neq 0$.

Suppose that the researcher observes market shares S_{jt} and market-level prices p_{jt}, as well as attributes $\mathbf{x}_{jt}$ and $\mathbf{z}_{jt}$. Estimation of the BLP model uses the moment equations $\mathbb{E}(S_{jt} - s_{jt}(\mathbf{p}_t, \mathbf{X}_t | \alpha, \boldsymbol{\beta}, \xi) | \mathbf{z}_{jt}) = 0$ in a GMM estimator with a contraction mapping that matches the observed and predicted shares.

4-8-7 Subjective Probabilities and Bayes Estimators

Finding the maximum likelihood estimator is not an easy task for flexible models that introduce unobserved heterogeneity. Global convergence problems can be experienced in the nested logit, the random parameter logit with independent log normally distributed parameters or with multivariate heterogeneity distributions, and the multinomial probit. MLE is not the only method for estimation of the parameters of a discrete choice model. In fact, parameters can be found using an approach that is not based on the classical concept of probability as a frequency. In Bayesian statistics, the true parameters are assumed random. Statistical inference is based on describing the subjective probabilities that measure the researcher's beliefs about the occurrence of an event. Before observing data, the researcher has prior beliefs $p(\boldsymbol{\theta})$ about the possible values of the parameters. After accessing the data, a *posterior distribution* $p(\boldsymbol{\theta} \mid \mathbf{y})$ can be derived. Beliefs are updated using Bayes' theorem: $p(\boldsymbol{\theta} \mid \mathbf{y}) = \ell(\boldsymbol{\theta};\mathbf{y})p(\boldsymbol{\theta})/p(\mathbf{y})$, where $\ell(\boldsymbol{\theta};\mathbf{y})$ is the likelihood function.

The goal of the Bayes estimation problem is to summarize the posterior distribution of interest. For some classes of models, the posterior distribution has a known parametric expression. Although discrete choice models do not have a closed-form posterior, Markov chain Monte Carlo (MCMC) methods can be used to approximate the desired distribution. Effectively, MCMC simulates $p(\boldsymbol{\theta} \mid \mathbf{y})$ using repeated sampling. There are several practical advantages to the Bayesian approach, namely, the derivation of estimators with exact properties (valid for small samples) that are both gradient and Hessian free. Bayes estimators are also useful for non-convex likelihood functions and weakly identified models, have the same asymptotic properties as those of MLE, and work particularly well for models that include latent variables.

Example 4-11 *Bayesian binary probit:* Consider the binary probit model for vehicle choice that was discussed in Section 4-7-1. We saw that we could summarize the model using the latent variable y_i^* of utility differences

$$y_i^* = \Delta_{\text{BEV}}\mathbf{U}_i = \beta_{\text{ICV}} + \beta_{\text{price}}(\text{price}_{\text{ICV},\,i} - \text{price}_{\text{BEV},\,i}) + \beta_{\text{fuelcost}}(\text{fuelcost}_{\text{ICV},\,i} - \text{fuelcost}_{\text{BEV},\,i})$$
$$+(\varepsilon_{\text{ICV},\,i} - \varepsilon_{\text{BEV},\,i}) = \mathbf{x}'_{\Delta i}\boldsymbol{\beta} + \epsilon_i,\ \epsilon_i \sim \mathcal{N}(0,\,1)$$

Note that $(y_i^* \mid \boldsymbol{\beta},\, \mathbf{x}'_{\Delta i}) \sim \mathcal{N}(\mathbf{x}'_{\Delta i}\boldsymbol{\beta},\, 1)$. However, we need to take into consideration that we observe the choice indicator $y_i = \mathbb{I}(y_i^* > 0) = \mathbb{I}(\Delta_{\text{BEV}}\mathbf{U}_i > 0)$. The choice indicator actually constrains the feasible space of the utility function such that

$$(y_i^* \mid y_i = 1,\ \boldsymbol{\beta},\ \mathbf{x}'_{\Delta i}) \sim \mathcal{N}(\mathbf{x}'_{\Delta i}\boldsymbol{\beta},\, 1)\mathbb{I}(y_i^* > 0) \text{ and } (y_i^* \mid y_i = 0,\ \boldsymbol{\beta},\ \mathbf{x}'_{\Delta i}) \sim \mathcal{N}(\mathbf{x}'_{\Delta i}\boldsymbol{\beta},\, 1)\mathbb{I}(y_i^* \le 0)$$

which means that $(y_i^* \mid y_i,\ \boldsymbol{\beta},\ \mathbf{x}'_{\Delta i})$ has a *truncated normal* distribution.

We know that if somehow we would observe each y_i^* in $\mathbf{y}^* = \mathbf{X}_\Delta\boldsymbol{\beta} + \epsilon$, then $\boldsymbol{\beta}$ could be estimated using standard regression techniques. We will use this idea, and the conditional distribution $\mathbf{y}^* \mid \mathbf{y},\ \boldsymbol{\beta}$ to derive a *Gibbs sampler* for the binary probit. Gibbs sampling is a numerical algorithm for creating an MCMC approximation of a multivariate distribution of interest. A Gibbs sampler is based on creating a Markov chain of draws of the conditional distributions of a partition of the parameter space. Before implementing the Gibbs sampler, we will introduce another concept that is useful for Bayesian inference on latent variables. Latent variables can be treated as additional parameters of the model in a process known as *data augmentation.* Thus, when augmenting the data, the joint posterior of interest is $(\mathbf{y}^*,\ \boldsymbol{\beta} \mid \mathbf{y},\ \mathbf{X}_\Delta)$. Instead of creating direct samples of this joint posterior, a

Gibbs sampler will generate draws from $\mathbf{y}^* \mid \boldsymbol{\beta}, \mathbf{y}, \mathbf{X}_\Delta)$ and $(\boldsymbol{\beta} \mid \mathbf{y}^*, \mathbf{y}, \mathbf{X}_\Delta)$. The steps for the binary probit Gibbs sampler are

- Start at any given point $\boldsymbol{\beta}^{(0)}$ and $\mathbf{y}^{*(0)}$ in the parameter space
- For iteration $g \in \{1, \ldots, G\}$
  1. If $y_i = 1$, draw a new $\mathbf{y}^{*(g)}$ from $(y_i^* \mid y_i = 1, \boldsymbol{\beta}^{(g-1)}, \mathbf{x}'_{\Delta i}) \sim \mathcal{N}(\mathbf{x}'_{\Delta i}\boldsymbol{\beta}^{(g-1)}, 1)\mathbb{I}(y_i^* > 0)$. If $y_i = 0$, update $\mathbf{y}^{*(g)}$ from $(y_i^* \mid y_i = 0, \boldsymbol{\beta}^{(g-1)}, \mathbf{x}'_{\Delta i}) \sim \mathcal{N}(\mathbf{x}'_{\Delta i}\boldsymbol{\beta}^{(g-1)}, 1)\mathbb{I}(y_i^* \le 0)$. This is the *data augmentation* step.
  2. Knowing $\mathbf{y}^{*(g)}$, $\mathbf{y}^{*(g)} = \mathbf{X}_\Delta \boldsymbol{\beta} + \epsilon$ becomes a regression problem (the data augmentation step creates observations for the latent variable). $\boldsymbol{\beta}^{(g)}$ is then updated using the Bayes estimator of a standard regression. With uninformative priors, $\boldsymbol{\beta}^{(g)}$ is drawn from $(\boldsymbol{\beta} \mid \mathbf{y}^{*(g)}, y, \mathbf{X}_\Delta) \sim \mathcal{N}((\mathbf{X}'_\Delta \mathbf{X}_\Delta)^{-1}\mathbf{X}'_\Delta \mathbf{y}^{*(g)}, (\mathbf{X}'_\Delta \mathbf{X}_\Delta)^{-1})$. (Do you recognize the normal equations?)
  3. Make $g = g + 1$ and go back to the data augmentation step until the desired number of repetitions G has been reached.
- The Bayes point estimate of $\boldsymbol{\beta}$ is the mean of the posterior samples $\hat{\boldsymbol{\beta}}_{\text{Bayes}} = \sum_{g=1}^{G} \boldsymbol{\beta}^{(g)}/G$. The standard deviation of the posterior samples can be used as a measure of uncertainty in the determination of the parameter.

4-9 Statistical Testing

For testing statistical significance of a single parameter we can use the *univariate Wald test* (also known as *asymptotic* t*-test or pseudo* t*-test*). In general, if variable $H_0 : \beta_k = \beta_0$ (vs. $H_a : \beta_k \ne \beta_0$), then

$$\frac{\hat{\beta}_k - \beta_0}{s.\,e.(\hat{\beta}_k)} \sim \mathcal{N}(0, 1) \tag{4-14}$$

Statistical packages by default calculate the z-value or (asymptotic) t-statistic against zero. The critical value of the test at the 5% significance level is $z_{\text{crit, 5\%}} = 1.95996$. When $\left|z = \hat{\beta}_k / s.\,e.(\hat{\beta}_k)\right| < z_{\text{crit, 5\%}}$ then we cannot reject $H_0 : \beta_k = 0$ at the 5% significance level.

For testing a hypothesis of the type $H_0 : g(\boldsymbol{\beta}) = 0$ (vs. $H_a : g(\boldsymbol{\beta}) \ne 0$), which includes comparing nested models, we can use any of three asymptotically equivalent tests, namely the Wald test, the likelihood ratio test, and the Lagrange multiplier test. For instance, the *likelihood ratio test* is defined as

$$\text{LR} = -2[\mathcal{L}_R(\hat{\boldsymbol{\beta}}; \mathbf{y} \mid \mathbf{X}_\Delta) - \mathcal{L}_U(\hat{\boldsymbol{\beta}}; \mathbf{y} \mid \mathbf{X}_\Delta)] \sim \chi^2_{d.f.} \tag{4-15}$$

where $\mathcal{L}_R$ is the loglikelihood of the restricted model (under the null), $\mathcal{L}_U$ is the loglikelihood of the unrestricted model, and the degrees of freedom $d.f.$ are equal to the number of restriction imposed by the null hypothesis. When $\text{LR} > \chi^2_{d.f., \text{crit, 5\%}}$ then the restricted model can be rejected.

Goodness of fit can be assessed using the likelihood ratio index

$$\rho^2 = 1 - \frac{\mathcal{L}(\hat{\boldsymbol{\beta}}; \mathbf{y} \mid \mathbf{X}_\Delta)}{\mathcal{L}(\mathbf{0}; \mathbf{y} \mid \mathbf{X}_\Delta)} \tag{4-16}$$

where $\mathcal{L}(\mathbf{0}; \mathbf{y} \mid \mathbf{X}_\Delta)$ is the loglikelihood of the equally likely model (completely random choice).

$$\mathcal{L}(\mathbf{0}; \mathbf{y} \mid \mathbf{X}_\Delta) = \sum_{i=1}^{N} \ln\left(\frac{1}{J}\right)$$

The index ρ^2 (McFadden's rho square) measures the percentage increase in the loglikelihood compared to the equally likely model. When $\rho^2 = 1$, then every choice is predicted perfectly. Unlike the index of goodness of fit R^2 for ordinary regressions that measures the percentage of explained variance, there is no clear interpretation for $0 < \rho^2 < 1$.

Another index of goodness of fit (sometimes called adjusted-ρ^2) is

$$\rho^2_{\text{ASC}} = 1 - \frac{\mathcal{L}(\hat{\boldsymbol{\beta}};\mathbf{y}|\,\mathbf{X}_\Delta)}{\mathcal{L}(\mathbf{ASC};\mathbf{y}|\,\mathbf{X}_\Delta)} \tag{4-17}$$

where $\mathcal{L}(\mathbf{ASC};\mathbf{y}|\,\mathbf{X}_\Delta)$ is the loglikelihood of a model considering only alternative-specific constants.

The Bayesian information criterion (BIC) is another tool to compare competing models. BIC is defined as

$$\text{BIC} = -2\mathcal{L}(\hat{\boldsymbol{\beta}};\mathbf{y}|\,\mathbf{X}_\Delta) + K\ln(N) \tag{4-18}$$

where K is the number of free parameters to be estimated and N is the sample size. The decision rule is to select the model with the lowest BIC.

4-10 Modeling Sustainable Choices

4-10-1 Adoption of Energy Efficiency in Transportation: Consumer Response to Ultra-Low-Emission Vehicles

The automotive industry is experiencing the emergence of new, energy efficient propulsion technologies that are competing with internal combustion engines. These new technologies include plug-in electric vehicles (PEVs), either pure battery electric (BEVs) or plug-in hybrids (PHEVs), and fuel cells that are fueled by hydrogen (HFCVs). Automakers also need to comply with energy-efficiency mandates set by the government (to attain environmental sustainability goals), and even the incumbent technologies are now exhibiting important energy efficiency gains.

Buying a car is an example of a durable good that is usable for many years. In the case of investing in energy efficient vehicles, consumers *pay more up front* for buying the technology but they experience *savings in the future*. Economies in operating costs related to fuel efficiency gains are among the most relevant benefits of adopting ultra-low and zero-emission vehicles. In addition, energy efficiency gains have other benefits such as reducing emissions. *Environmental preferences* thus become relevant in modeling the socio-technical transition to low carbon transportation.

Despite the benefits of alternative fuels in general, there are several challenges in making new technologies competitive. Higher price tags are obviously a barrier for adoption. Other barriers include limited recharging availability due to a lack of infrastructure (BEVs, HFCVs), lengthy recharging (BEVs), and limited driving range (BEVs). Figure 4-10 shows an example of a charging system.

Understanding the adoption of energy efficiency by consumers is critical for promoting sustainability.

FIGURE 4-10 Charging an electric car.

Example 4-12 *Driving range anxiety:* Limitations in electric driving range—the maximum distance that a BEV can go on a single charge—is one of the drawbacks preventing broad consumer acceptance of electric vehicles. In fact, the term *driving range anxiety* has been used to represent the consumer's fear that the electric battery will be depleted before arriving at the desired destination. Although automakers highlight that current driving range meets most of the average consumer's driving needs, electric vehicles will not represent true competition to ICVs unless driving range is increased and charging times reduced. The Tesla Model S (see Fig. 4-11) is an electric vehicle that offers a 265-mile range with the 85 kWh battery, but at a hefty cost ($72,400).

Sales of the Tesla show that consumers are willing to pay for improvements in driving range. In fact, the willingness to pay for marginal range improvements is expected to be decreasing in range and has been estimated at 83 to 134 $/mile for a vehicle offering 100 miles (Daziano, 2013).

We will now discuss several particularities that should be taken into account when modeling consumer response to energy efficiency and low carbon technologies.

The Energy Paradox or Energy-Efficiency Gap

Electric vehicles are more expensive to buy but they offer important savings in operating and maintenance costs. Investments in energy efficiency thus represent a tradeoff between a higher upfront price and *future energy savings.* A rational consumer should exhibit the same valuation for purchase price and for the *present value* of the life-cycle fuel costs. Therefore, an eventual difference between the marginal utility of purchase

Figure 4-11 The Tesla Model S, an electric vehicle with competitive driving range.

price and that of the *discounted* fuel cost will measure incorrect accounting of the benefits of energy efficiency. In fact, there is practical evidence suggesting that consumers underestimate future savings. At the limit, consumers may be inattentive to fuel cost. A lower elasticity to future energy costs than to out-of-pocket expenses at the time of purchase has been named *energy paradox* or *energy efficiency gap* (Jaffe and Stavins, 1994). Estimates of the energy paradox are critical for the evaluation of tighter *efficiency standards* versus other policies such as *emission pricing* or changes in *gasoline taxes.*

Suppose that we estimate a model where fuel costs enter as a monthly expense in a linear utility of the form: $U_{ij} = \beta_{\text{price}} \text{price}_{ij} + \beta_{\text{fuelcost}} \text{fuelcost}_{ij} + \cdots + \varepsilon_{ij}$. To make the comparison internally consistent, the parameter β_{fuelcost} adjusts the temporal dimension of future costs by implicitly accounting for the present value of fuel costs (PVFC):

$$\text{PVFC}_{ij} = \sum_{t=1}^{L_{ij}} \frac{\text{fuelcost}_{ij}}{(1+r)^t}$$

where r is an implicit discount rate that summarizes time preferences, and L is the lifespan of the vehicle. If L is large enough, $\text{PV}_{ij} \approx \text{fuelcost}_{ij}/r$. Thus, the ratio

$$r = \frac{\beta_{\text{price}}}{\beta_{\text{fuelcost}}} = \frac{1}{\text{WTP}_{\Delta\text{fuelcost}}}$$

can be used to infer the implicit discount rate (Hausman, 1979; Train, 1985). Annual implicit discount rates of 15% to 35% are very common in empirical work. Market rates for the automotive market are in the order of 6% to 7%. High discount rates make the present value of the future savings smaller. As a result, consumers with high discount rates will choose the cheapest, least efficient vehicles in their preferred class.

Because the ratio of the purchase price and operating costs may mask market failures other than underestimation of savings, specific instruments for elicitation of implicit discount rates can be used. If we determine r exogenously to the estimation of the discrete choice parameters, then it is possible to consider a utility function of the form:

$$U_{ij} = \beta_{\text{price}}\left[\text{price}_{ij} + \gamma_{\text{PVFC}}\text{PVFC}_{ij} + \mathbf{x}'_{ij}\boldsymbol{\omega}\right] + \varepsilon_{ij}$$

where γ_{PVFC} is the willingness to pay for a \$1 saving in discounted fuel costs. Note that in this specification of utility, the parameters are recast such that γ_{PVFC} and $\boldsymbol{\omega}$ represent willingness to pay. [This reparameterization, which works for any ARUM model, was proposed by Train and Weeks (2005) and is known as utility in *willingness to pay space.*] If the hypothesis $\gamma_{\text{PVFC}} < 1$ cannot be rejected then there is evidence in favor of the presence of an energy paradox.

Example 4-13 *Energy paradox estimate:* Using 86 million monthly vehicle transactions from 1999 to 2008, Allcott and Wozny (2012) estimated a discrete choice model of the type $U_{ij} = \beta_{\text{price}}[\text{price}_{ij} + \gamma_{\text{PVFC}}\text{PVFC}_{ij} + \mathbf{x}'_{ij}\boldsymbol{\omega}] + \varepsilon_{ij}$. The authors obtained $\hat{\gamma}_{\text{PVFC}} = 0.76$ and concluded that the energy paradox occurs.

The Energy Rebound Effect

Although the adoption of energy efficiency is desired, the provision of energy-efficient technologies may have unintended effects. Because driving is cheaper in a vehicle that offers important fuel savings, the owner may end up driving more. The energy rebound effect in transportation can be defined as the increase in VMT due to efficiency gains. A measure of the rebound effect is the additive inverse of the cost elasticity of VMT, that is, the percentage change in the distance driven divided by the percentage change in the cost of driving (p_{VMT}):

$$E_{\text{VMT},\, p_{\text{VMT}}} = \frac{\partial \text{VMT}}{\partial p_{\text{VMT}}} \frac{p_{\text{VMT}}}{\text{VMT}}$$

The interpretation of the rebound effect is as follows. If, for example, $E_{\text{VMT},\, p_{\text{VMT}}} = -0.1$, then the resulting rebound effect of 10% measures the *effectiveness loss of efficiency standards* (due to increased driving).

Example 4-14 *The declining rebound effect in the United States:* Using annual U. S. data by state from 1966 to 2004, Small and Dender (2007) found that the rebound effect is declining. For the full period 1966 to 2004, the short-run rebound effect was estimated at 4.1% and the long-run rebound effect at 21.0%. For the period 2000 to 2004, the short-run rebound effect is 1.1% and the long-run rebound effect is 5.7%. One of the reasons for the decline of the rebound effect is the increase in income. The main conclusion is that efficiency standards do translate into important fuel savings.

4-10-2 Vehicle Use: Discrete-Continuous Models

Sustainable driving not only depends on the vehicle that we buy, but even more on our use of that vehicle. (A person can have a gas-guzzler that she uses very infrequently, and most of her trips are done in a bike. This behavior is more sustainable than that of a person who drives her hybrid everywhere.) Vehicle use models try to explain how much a person will drive. VMT models can be specified jointly with vehicle ownership models. In this case, the discrete choice is the vehicle that the person will buy, and then there is a continuous demand function where the decision is mileage. Train (1986) and Bento et al. (2009) specified a discrete-continuous demand model, where conditional VMT is derived from the conditional utility of vehicle purchases using *Roy's identity*. Roy's identity is a microeconomic method for deriving a demand function from the indirect utility. In the case of vehicle choice, the indirect utility can be specified to depend explicitly on the per-mile cost of driving p_{ij}^{M} such that $U_{ij} = v_j(\mathbf{q}_{ij}, I_i - p_{ij}, p_{ij}^{M}, \varepsilon_{ij} \mid \boldsymbol{\theta})$. According to Roy's identity

$$\mathrm{VMT}_{ij}(\mathbf{q}_{ij}, I_i - p_{ij}, p_{ij}^{M}, \varepsilon_{ij} \mid \boldsymbol{\theta}) = -\frac{\partial v_j(\mathbf{q}_{ij}, I - p_{ij}, p_{ij}^{M}, \varepsilon_{ij} \mid \boldsymbol{\theta}) / \partial p_{ij}^{M}}{\partial v_j(\mathbf{q}_{ij}, I - p_{ij}, p_{ij}^{M}, \varepsilon_{ij} \mid \boldsymbol{\theta}) / \partial I_i}$$

To account for measurement error, the model is completed with $\widetilde{\mathrm{VMT}}_{ij} = \mathrm{VMT}_{ij}(\mathbf{q}_{ij}, I_i - p_{ij}, p_{ij}^{M}, \varepsilon_{ij} \mid \boldsymbol{\theta}) + \epsilon_{ij}$, where $\widetilde{\mathrm{VMT}}_{ij}$ is the observed (self-reported) VMT. The joint model of vehicle choice and use can be estimated using MLE. For the case of multiple discrete-continuous choices, Bhat (2008) proposed the multiple discrete-continuous extreme value (MDCEV) model.

The MPG Illusion

In the United States, fuel economy in miles per gallon (mpg) is the standard measure for communicating transportation energy efficiency. The problem with this fuel economy measure is that mpg improvements exhibit *diminishing returns in fuel cost*, but consumers assume *linear energy savings*. When shown mpg information, consumers underestimate improvements in vehicles with low mpg and overestimate improvements in vehicles with high mpg.

Consider the following utility function:

$$\mathrm{U}_{ij} = \beta_{\mathrm{price}}\left[price_{ij} + \gamma_{\mathrm{PVFC}}\sum_{t=1}^{L_{ij}}\frac{1}{(1+r)^t}\frac{\mathrm{VMT}_{ijt}\mathbb{E}(fp_{it})}{\mathrm{mpg}_j} + \mathbf{x}'_{ij}\boldsymbol{\omega}\right] + \varepsilon_{ij}$$

where $\mathbb{E}(fp_{it})$ is the expected fuel price of consumer i at period t. From this utility specification it is clear that the cost of driving is nonlinear in mpg, but it is linear in gallons per mile (GPM). (In the metric system, fuel economy is measured in liters per kilometer, thus avoiding the nonlinearity issue.)

Measurement of energy efficiency for BEVs adds to the fuel efficiency puzzle. For alternative fuels in general, the U. S. Environmental Protection Agency (EPA) has proposed the use of miles per gallon gasoline equivalent (mpge). However, it would be more effective to communicate fuel efficiency in terms of GPMe. In addition, efficiency of PHEVs depends on whether the vehicle is running on the gasoline or the electricity mode.

Willingness to Pay for Environmental Benefits

Environmental costs or benefits of transportation are not transmitted through prices. However, discrete choice models can be used to determine how individuals value environmental benefits of energy-efficient technologies such as reduced emission levels. The derivation of willingness to pay for renewable energy or for energy-efficiency gains is an indirect mechanism to determine the valuation of environmental impacts. Several authors have included carbon dioxide emissions as a vehicle attribute in modeling the decision of buying AFVs, and estimate a significant marginal utility that is negative. However, one of the problems is that emission levels are an engineering attribute usually measured in terms of mass of CO_2-equivalent (CO_2e) per unit of distance (*emission mileage*, such as pounds per mile) or per unit of time (such as pounds per year). It is thus not completely clear how consumers process this kind of measure.

Example 4-15 *Willingness to pay for CO_2 abatement:* Suppose that we collected data among consumers choosing between an internal combustion vehicle (ICV) and a battery electric vehicle (BEV). The microdata contains the following information for 1,000 consumers:

- Choice: the vehicle actually purchased by each consumer
- ICV and BEV purchase price ($)
- ICV and BEV fuel cost ($/100 miles)
- BEV recharge infrastructure as a percentage of the network of gas stations
- Power (HP)
- Carbon dioxide emissions (g/mile)
- Driving range (miles)
- Gender of the consumer

The following model with deterministic taste variations is estimated:

Attribute	Point Estimate	Standard Error	z-Value	p-Value
Constant (β_{ICV})	−0.729180	2.858300	−0.255	0.799
Price (β_{price})	−0.000202	0.000017	−12.023	0.000
Fuel Cost ($\beta_{fuelcost}$)	−0.224900	0.023805	−9.448	0.000
Recharge Infrastructure ($\beta_{Infr.}$)	0.079501	0.029274	2.716	0.007
Power (β_{POW})	0.028195	0.005902	4.777	0.000
CO_2 emissions (β_{CO2})	−0.026193	0.002366	−11.073	0.000
ln(range) ($\beta_{ln(range)}$)	1.687700	0.782800	2.156	0.031
Male × CO_2 emissions ($\beta_{male \times CO2}$)	0.003799	0.000971	3.913	0.000
Male × Power ($\beta_{male \times POW}$)	0.006182	0.006731	0.918	0.358
Loglikelihood		−184.49		

In this model, the ratio $\beta_{CO2}/\beta_{price} = 129.37$ $/g-mile represents the willingness to pay for reducing 1 g-mile of CO_2 emissions for the base segment, which in this case is the group of women. The willingness to pay for men is calculated as $(\beta_{CO2} + \beta_{male \times CO2})/\beta_{price} = 110.60$ $/g-mile. The fact that women have higher environmental preferences has been encountered in several studies.

4-10-3 Sustainable Mobility Choices

Beyond the introduction of energy-efficient cars, achieving a sustainable transportation system requires a set of socio-technical actions. Discrete choice models can be used to understand how to promote a shift toward sustainable transportation behavior. For example, there is the question of how to *make transit more attractive* to people living in a city. With adequate loading factors, public transportation systems can achieve interesting levels of energy efficiency. Additionally, transit makes a more efficient use of space. However, it is hard to convince drivers to leave their car at home and use public transportation instead. Understanding the tradeoffs for *mode choice,* it is possible to target the improvements in the sustainable mobility alternatives that are necessary to produce a behavioral change. For example, increasing the *level of service* (attributes) of public transit via dedicated lanes or imposing restrictions to the use of car (such as higher parking costs) can improve transit ridership. However, one lesson from the properties of discrete choice models is that big changes are needed to increase the choice probability of sustainable mobility options in cities where solo driving dominates the market.

Example 4-16 *Congestion pricing:* Drivers create negative externalities to users (delays and pollution) and non-users (pollution) of the system. The yearly cost of congestion has been estimated at $750 per commuter. An indiscriminate use of cars occurs because drivers do not perceive the full cost of driving. From the point of view of economic equilibrium, a solution to the problem of congestion is to charge drivers for the extra delays that they cause to the system. In fact, drivers perceive the average cost of driving, but not the marginal cost. Thus, the congestion charge should be equal to the difference between the marginal cost (extra delay) and the average cost of driving. If the cost of a mode (such as solo drivers) increases—due to higher fuel taxes, congestion charges, or higher parking prices—then the choice probability and market share of solo driving should decrease. Bhat and Castelar (2002) analyzed multiple congestion pricing scheme simulations in travel-mode choice models. The amount of the impact will depend on the resulting elasticity of demand of the mode to discourage. Congestion pricing is a demand-side management tool for any public good that experiences congestion (electricity transmission, airports, subway).

Nonmotorized Transportation

One of the solutions to the livability degradation caused by automobile dependency is the adoption of non-motorized alternatives (see Fig. 4-12). There are successful examples of

(*a*)

FIGURE 4-12 Shared bicycle systems are expanding in several cities around the world. (*a*) The NYC shared bicycle system. (*b*) Solar-powered bike-sharing stations in Santiago, Chile.

(b)

FIGURE 4-12 (Continued)

cities for which cycling is playing a major role in their paths toward sustainability. For example, 5.8% of commuters in Portland cycle to work. The percentage in New York City is only 0. 6%, despite 345 miles of bicycle routes being added in the last decade. To encourage the use of non-motorized alternatives we need to better understand the motives underlying demand. There are several challenges in applying choice modeling to nonmotorized options. Users of the transportation system may be motivated to cycle or walk not because of the tradeoff between cost and time, but because of *health and environmental benefits* of these alternatives. At the same time, there are several factors that may discourage the use of nonmotorized transportation, such as poor accessibility, safety concerns, and unfavorable route and weather conditions. For instance, it is often argued that the Northeast of the United States has a climate that discourages the use of biking. Accounting for factors beyond traditional compensatory attributes is not straightforward and requires a deep understanding of user behavior.

4-11 Forecasting and Welfare Analysis

Forecasting allows the researcher to perform what-if analyses of the resulting benefits of private or public investments to consumers. Effectively, estimates of discrete choice model can be used to determine choice probabilities, elasticities, market shares, and consumer surplus under *counterfactual scenarios* of qualitative improvement. Counterfactual scenarios compare the *welfare effects*—changes in consumer surplus—of different actions or policies to assess effectiveness of differing programs. The goal is to identify the most efficient welfare-improving actions. For example, Bento et al. (2013) analyze the policy of allowing solo-drivers of hybrids on HOV-lanes in California. The authors conclude that this policy, although popular, is much less efficient than a direct subsidy for purchase.

An alternative can become more attractive because of private investments (such as improvements in driving range) or public policies (such as tax credits for the purchase of ultra-low-emission vehicles). In a discrete choice model, if an alternative becomes more attractive its choice probability should increase (the consumer is more likely to choose the improved alternative). Because the choice probabilities for a consumer need to sum one, the choice probabilities of the competing alternative will decrease. These changes reflect substitution among the alternatives.

Marginal effects measure how much more attractive an alternative becomes if it is improved ($\partial P_{ij} / \partial x_{kij}$), or how unattractive the competition becomes after the improvement ($\partial P_{il} / \partial x_{kij}$). Choice *elasticities* measure the relative change in the choice probabilities after a change in a specific attribute of a given alternative. We can define two types of elasticities:

1. Own choice elasticity: the effect of the change is calculated on the choice probability of the alternative that is improved (or deteriorated)

$$E_{P_{ij}, x_{kij}} = \frac{\partial P_{ij}}{\partial x_{kij}} \frac{x_{kij}}{P_{ij}}$$

2. Cross-choice elasticity: the effect is calculated on a competing alternative

$$E_{P_{il}, x_{kij}} = \frac{\partial P_{il}}{\partial x_{kij}} \frac{x_{kij}}{P_{il}}$$

Because discrete choice models are nonlinear, the marginal effects and elasticities are not constant. In fact, the highest effect of improvements is found in situations where choice is more uncertain (meaning that the consumers seem to be more or less indifferent among the alternatives). However, when an alternative dominates the others, it is very hard to make the competition more attractive. This fact is relevant for the creation of policies that promote adoption of emerging technologies.

The steps for evaluating counterfactual scenarios are the following:

1. The base scenario is the one used for estimation: $(\mathbf{y}_i \mid \mathbf{X}_i^{(0)})$.
2. Model estimates $\hat{\boldsymbol{\theta}}(\mathbf{y}_i \mid \mathbf{X}_i^{(0)})$ are found, and preferences are assumed to be stable.
3. Choice probability estimates are calculated: $\hat{\mathbf{P}}_i^{(0)} = \mathbf{P}_i(\mathbf{X}_i^{(0)}, \hat{\boldsymbol{\theta}})$.
4. A new scenario is constructed for forecasting: $\mathbf{X}_i^{(1)}$. (Note that we do not observe new choices, and we do not re-estimate the model.)
5. Forecasted choice probabilities are: $\hat{\mathbf{P}}_i^{(1)} = \mathbf{P}_i(\mathbf{X}_i^{(1)}, \hat{\boldsymbol{\theta}})$.

Forecasting market demand is essential to plan the future of transportation systems. Changes in discrete choice models are evaluated first at the individual level. Sometimes a representative individual is chosen for performing a market analysis. However, note that the use of average attributes induces bias because the choice probabilities are nonlinear. Sample enumeration is an alternative for aggregation that considers the sample average of the individual choice probabilities. For example, the market share of alternative j (the proportion of consumers choosing that alternative) is calculated as

$$\hat{S}_j = \frac{1}{N}\sum_{i=1}^{N}\hat{P}_{ij}$$

Changes in the choice probabilities (and elasticities and market shares) are due to changes in utility. An improved alternative when consumed will provide a higher level of satisfaction and the consumer will be "happier" when choosing it. Improvements in utility provide higher benefits to consumption. However, improving an alternative requires resources. To evaluate the improvement we need a method for expressing changes in well-being in monetary terms, so that the monetized benefits can be contrasted with the cost of the improvement. The consumer surplus evaluated at the chosen alternative is given by the monetized utility:

$$\text{CS}_i = \frac{1}{\text{MUI}_i}\max_j U_{ij}$$

In the case of a conditional logit model, the expected consumer surplus is given by the logsum divided by the marginal utility of income

$$\mathbb{E}(\text{CS}_i) = \frac{1}{\text{MUI}_i}\ln\sum_j \exp(\mathbf{x}'_{ij}\boldsymbol{\beta})$$

In the case of welfare analysis of a counterfactual scenario, the change in expected consumer surplus is given by

$$\Delta\mathbb{E}(\text{CS}_i) = \frac{1}{\text{MUI}_i}\left[\ln\sum_j \exp(\mathbf{x}_{ij}^{(1)\prime}\boldsymbol{\beta}) - \ln\sum_j \exp(\mathbf{x}_{ij}^{(0)\prime}\boldsymbol{\beta})\right]$$

When the marginal utility of income is not a function of income, Small and Rosen (1981) proved that the expression above measures the *compensating variation* (CV), which is a formal benefit measure in economics defined as the solution to

$$\max_j v_j(\mathbf{q}_{ij}^{(0)}, I_i - p_{ij}^{(0)}, \varepsilon_{ij} \mid \boldsymbol{\theta}) = \max_j v_j(\mathbf{q}_{ij}^{(1)}, I_i - \text{CV} - p_{ij}^{(1)}, \varepsilon_{ij} \mid \boldsymbol{\theta}).$$

The CV is then the change in income that offsets the change in utility.

4-11-1 Energy and Environmental Policy

There are several policies that regulators can use as tools for reducing the impact of transportation demand and improve consumer welfare. These policies seek to create incentives for the adoption of technologies and behaviors that will promote sustainability. Forecasts of consumer response using discrete choice estimates can inform policymakers to design efficient incentives for deployment of energy-efficient technologies. Some of the policies that can be contrasted in terms of their impact on consumer welfare and sustainability goals include the following:

- *Price instruments* such as subsidies for purchase of new technology.
- *Tax instruments* such as gasoline taxes, VMT charges, emission pricing, and congestion pricing.

- *Energy efficiency standards* such as corporate average fuel economy (CAFE) standards mandate a minimum average fuel economy for the fleet produced by each car manufacturer. CAFE standards not only promote efficiency gains coming from alternative fuels (ethanol, electricity, or hydrogen), but create incentives for improving internal combustion vehicles. Some models are offering fuel economy levels that are very close to those of hybrids.
- *Renewable fuel standards,* which require a minimum volume of renewable fuel to be blended in transportation fuels. Note that flexible-fuel vehicles now can burn any proportion of gasoline and alternative fuel.
- *Feebates* establish a threshold of fuel efficiency. Vehicles above the efficiency threshold receive a rebate [Ultra Low Emission Vehicle (ULEV) subsidies], whereas vehicles below the threshold should pay a fee (gas guzzler tax). The amount of the rebates and fees can be a function of fuel efficiency ratings.
- *Labels* such as Monroney stickers that inform the consumer about the energy efficiency and environmental benefits of the vehicles offered in the market. The use of efficiency labels may help reduce the energy paradox.
- *Improve the level of service* of sustainable alternatives. For example, ultra-low emission vehicles may be granted access to HOV lanes ("fast lanes").

4-12 Integration of Consumer Demand into Engineering Design

4-12-1 Agent-Based Modeling

Agent-based modeling (ABM)—an approach frequently used in computational economics, social behavior, and epidemiology—has recently emerged as a computational tool for analyzing the market dynamics of the automotive industry. Agent-based models simulate the micro-leveled actions and interactions of several heterogeneous decision makers, including not only consumers, but also producers (car manufacturers, energy companies), distributors and dealers, and the government. The timing of the simulated decisions in ABM is discrete. At each period, each agent faces a stochastic decision-making problem subject to the current state of the simulated environment. For example, producers decide to manufacture a vehicle fleet with a certain mix of technical features, while they seek to maximize profit. Consumers seek to maximize their welfare and decide whether to keep their current vehicle (if they own one), to renew the existing vehicle, or to buy an additional car. The government acts as a regulator that imposes institutional constraints, such as fuel efficiency mandates, that can become more or less strict.

In terms of computational programming, the decision-making process is coded as modular, rule-based algorithms that exploit Monte Carlo methods for introducing uncertainty. ABM is especially good at simulating decisions that are shaped by different market forces, including both spatial and social network effects.

After identification and definition of the agents, links, parameters, data, and algorithms to be used, the usual steps in an ABM include:

1. Design of the conceptual model (simulation flowchart)
2. Coding
3. Initialization

4. Verification
5. Validation

ABMs of ultra-low-emission vehicles typically produce market penetration curves of the competing drivetrain technologies in a horizon of 15 to 30 years, under different scenarios that introduce elements such as differing incentives for BEV purchase. Because Monte Carlo methods are used, for each scenario multiple penetration curves are generated.

ABMs may use heuristics for determining agents' decisions. However, a discrete choice model can be used within an ABM. For example, an ABM can consider survey-based data to obtain estimates of the preference parameters that are used in initializing the autonomous agents.

Example 4-17 *ABM of electric vehicle deployment:* Eppstein et al. (2011) designed an ABM where at each period agents stochastically decide whether to buy a new vehicle or not. Then, agents who decided to renew their car choose which vehicle to buy (ICV, HEV, or BEV) based on maximization of a relative utility function that the authors label as "desirability":

$$D_{jl} = w_G \mathrm{RB}_{jl} + (1 - w_G)\mathrm{RC}_{jl}$$

D_{jl} is the desirability index of alternative j with respect to alternative l, and is a weighted average of the relative benefits RB_{jl} and relative costs RC_{jl}. Based on social and media influences, the weight w_G is updated for each agent following a heuristic. In addition, desirability of BEVs is adjusted according to an agent-specific threshold for the proportion of BEVs owned in union of the agent's social and spatial networks. The vehicle chosen is the one that maximizes the desirability index.

4-12-2 Decision-Based Engineering Design

When discussing the agents that can enter an ABM, we mentioned firms and their decisions that determine supply. Firms that produce engineered goods or services have an engineering team that creates new designs through research and development tasks. Engineering design is actually a decision-making process that can be modeled using the same concepts that we have discussed for the demand side.

The idea is that engineering solutions need to be designed with the goal of maximizing benefits. In the case of consumers, utility is maximized to find the best affordable alternative. In the case of firms, which can be viewed as design organizations, a clear example of an objective function is profit. Because benefits to the producers depend on demand, finding the optimal design of a marketable good or service requires a demand model. As you now know, discrete choice analysis is a powerful structural approach to the purpose of modeling consumer behavior. Thus, engineers can perform a market research analysis using discrete choice techniques to assess consumer preferences and forecast demand under competing designs. Note, however, that the way attributes are perceived by the consumers is not necessarily how attributes appear as decision variables from the point of view of the producer or designer. In fact, engineering attributes face technological constraints that may not match consumers' desires or needs.

The best engineering design will optimize an *expected utility function* (profit, life-cycle cost, system effectiveness) that depends on demand and is subject to performance and feasibility constraints.

4-13 Summary

Engineers have been working on developing clean technologies aiming at transforming the nation's energy supply by promoting the use of renewables and reducing our dependence on fossil fuels. On the policy side, federal and state governments have encouraged green-tech industry through special subsidies while at the same time imposing stricter energy-efficiency standards on manufacturers. However, since the energy industry is a consumer-driven business, an energy revolution will not happen if the energy demand does not change. It is thus up to final consumers to decide whether they will buy ultra-low-emission vehicles. In this chapter we introduced the concept of *utility maximizing behavior* as a tool to model the decision-making process that leads to transportation demand.

References

Allcott, H. and N. Wozny (2012). *Gasoline Prices, Fuel Economy, and the Energy Paradox.* Working Paper 18583, NBER, 2012.

Bento, A., L. Goulder, M. Jacobsen, and R. von Haefen (2009). "Efficiency and Distributional Impacts of Increased U. S. Gasoline Taxes." *American Economic Review* 99(3).

Bento, A., D. Kaffine, K. Roth, and M. Zaragoza (2013). *The Unintended Consequences of Regulation in the Presence of Competing Externalities: Evidence Form the Transportation Sector.* Working Paper, Charles H. Dyson School of Applied Economics and Management, Cornell University, Ithaca, N.Y.

Berry, S., J. Levinsohn, and A. Pakes (1995). "Automobile Prices in Market Equilibrium." *Econometrica* 63(4): 841–890.

Bhat, C. R. (2008). "The Multiple Discrete-Continuous Extreme Value (MDCEV) Model: Role of Utility Function Parameters, Identification Considerations, and Model Extensions. " *Transportation Research* B42(3): 274–303.

Bhat, C. R. and Castelar, S. (2002). "A Unified Mixed Logit Framework for Modeling Revealed and Stated Preferences: Formulation and Application to Congestion Pricing Analysis in the San Francisco Bay Area." *Transportation Research* B36(7): 593–616.

Bolduc, D., N. Boucher, and R. Alvarez Daziano (2008). "Hybrid Choice Modeling of New Technologies for Car Choice in Canada." *Transportation Research Record* 2082: 63–71.

Brownstone, D. and T. F. Golob (2009). "The Impact of Residential Density on Vehicle Usage and Energy Consumption." *Journal of Urban Economics* 65: 91–98.

Brownstone, D. and K. Train (1999). "Forecasting New Product Penetration with Flexible Substitution Patterns." *Journal of Econometrics* 89: 09–129.

Brownstone, D., D. S. Bunch, and K. Train (2000). "Joint Mixed Logit Models of Stated and Revealed Preferences for Alternative-Fuel Vehicles." *Transportation Research B* 34(5): 315–338.

Cervero, R. and K. Kockelman (1997). "Travel Demand and the 3Ds: Density, Diversity, and Design." *Transportation Research D* 2 (3): 199–219.

Daziano, R. A. (2013). "Conditional-Logit Bayes Estimators for Consumer Valuation of Electric Vehicle Driving Range." *Resource and Energy Economics* 35 (3): 429–450.

Daziano, R. A. and M. Achtnicht (2013). "Forecasting Adoption of Ultralow-Emission Vehicles Using Bayes Estimates of a Multinomial Probit Model and the GHK Simulator." *Transportation Science*, DOI 10. 1287/trsc. 2013. 0464.

Eppstein, M. J., D. K. Grover, J. S. Marshall, and D. M. Rizzo (2011). "An Agent-based Model to Study Market Penetration of Plug-in Hybrid Electric Vehicles." *Energy Policy* 39(6): 3789–3802.

Hausman, J. A. (1979). "Individual Discount Rates and the Purchase and Utilization of Energy-Using Durables. " *Bell Journal of Economics*, 10(1): 33–54.

Jaffe, A. and R. Stavins (1994). "The Energy Paradox and the Diffusion of Conservation Technology. " *Resource and Energy Economics* 16: 91–122.

McFadden, D. and K. Train (2000). "Mixed MNL Models for Discrete Response." *Journal of Applied Econometrics* 15(5): 447–470.

McFadden, D. (2001). "Economic Choices." *American Economic Review*, 91(3): 351-378.

Small, K. A. and K. vanDender (2007). "Fuel Efficiency and Motor Vehicle Travel: The Declining Rebound Effect." *Energy Journal* 28(1): 22–51.

Train, K. (1985). "Discount Rates in Consumers' Energy-related Decisions: A Review of the Literature." *Energy* 10: 1243–1253.

Train, K. (1986). *Qualitative Choice Analysis: Theory, Econometrics, and Application to Automobile Demand*. MIT Press, Cambridge, MA.

Train, K. and M. Weeks (2005). Discrete choice models in preference space and willingness to pay space, in R. Scarpa and A. Alberini, eds., *Applications of Simulation Methods in Environmental and Resource Economics*, Springer, Dordrecht, Chapter 1, 1-16.

Train, K. E. and C. Winston (2007). "Vehicle Choice Behavior and the Declining Market Share of U. S. Automakers." *International Economic Review* 48(4): 1469–1496.

Walker, J. L. and J. Li (2007). "Latent Lifestyle Preferences and Household Location Decisions." *Journal of Geographical Systems* 9(1): 77–101.

Further Reading

Anderson, S. P., A. de Palma, J. F. Thisse (1992). *Discrete Choice Theory of Product Differentiation*. The MIT Press, Cambridge, MA.

Ben-Akiva, M. and S. R. Lerman (1985). *Discrete Choice Analysis: Theory and Application to Travel Demand*. The MIT Press, Cambridge, MA.

Cameron, A. C. and P. K. Trivedi (2005). *Microeconometrics: Methods and Applications*. Cambridge University Press, New York, NY.

Domencich, T. and D. McFadden (1975). *Urban Travel Demand: A Behavioral Analysis*. North-Holland, Amsterdam.

Jara-Díaz, S. R. (2007). *Transport Economic Theory*. Elsevier Science, Amsterdam.

Greene, W. H. (2011). *Econometric Analysis*. Prentice Hall, Upper Saddle River, N. J.

Louviere, J. J., D. A. Hensher, and J. D. Swait (2000). *Stated Choice Methods: Analysis and Application*. Cambridge University Press, Cambridge, UK.

McCarthy, P. (2001). *Transportation Economics*. Blackwell, Malden, MA.

Small, K. and E. Verhoef (2007). *The Economics of Urban Transportation*. Routledge, New York, NY.

Ortúzar, J. de D. and L. G. Willumsen (2011). *Modelling Transport*. John Wiley and Sons, Chichester, UK.

Train, K. E. (2009). *Discrete Choice Methods with Simulation*. Cambridge University Press, Cambridge, UK.

Exercises

4-1. Preference essentials.

a. The conditional indirect utility function of a nonchosen alternative cannot be higher than the conditional indirect utility function of the actually chosen alternative. Explain.

b. Alternatives of discrete choice models need to be mutually exclusive. Does this mean that behavior such as choosing to use a combination of modes (park and ride) cannot be modeled using discrete choice?

c. In a binary choice context, if a consumer is indifferent between the two alternatives, what would the choice probabilities look like?

4-2. Show that the maximum likelihood estimator of β in the standard regression problem $y_i = \mathbf{x}_i'\beta + \varepsilon_i$, $\varepsilon_i \sim \mathcal{N}(0, \sigma^2)$, or $\mathbf{y} = \mathbf{X}\beta + \varepsilon$, $\varepsilon \sim \mathcal{N}(0, \sigma^2\mathbf{I})$ in matrix form, is $\hat{\beta}_{\text{MLE}} = (\mathbf{X}'\mathbf{X})^{-1}\mathbf{X}'\mathbf{y}$.

4-3. Consider $V_{in} = \beta_p p_{in} + \beta_q q_{in}$, where p_{in} is price of alternative i as experienced by individual n, and q_{in} is a qualitative attribute.

a. Suppose that we manage to find the prices as a function of q_{in}—$p_{in} = p(q_{in})$—that make the utility level constant, that is, $\beta_p p(q_{in}) + \beta_q q_{in} = V^0$, where V^0 is constant. What do you think is the economic interpretation of the pair $(p(q_{in}), q_{in})$?

b. Differentiate the expression $\beta_p p(q_{in}) + \beta_q q_{in} = V^0$ with respect to q_{in} and derive an expression for $p'(q_{in})$. What is the economic interpretation of your result?

c. Repeat (b) but consider now a nonlinear function for the utility $V_{in} = V[p(q_{in}), q_{in}]$.

4-4. Look at the point estimates shown in Example 4-4.

a. Comment on the sign and statistical significance of the parameters.

b. Write the general expression for the utility function of each of the four alternatives of the model: SGV, AFV, HEV, HFC.

c. Write the general expression for the choice probability of each of the four alternatives of the model: SGV, AFV, HEV, HFC.

d. Suppose that an individual faces a choice situation where the attributes take the following values:

Attribute	SGV	AFV	HEV	HFC
Purchase price (10,000$)	2.1	2.2	3.2	4.2
Fuel cost (100$/month)	1.3	1.5	0.7	0.9
Fuel network (density ratio, from 0: no stations to 1: current network)	1.0	0.6	1.0	0.2
HOV access (0: No,1:Yes)	0.0	0.0	0.0	1.0
Power (ratio compared to current car)	1.0	0.9	1.0	0.8

Calculate the choice probabilities for this individual.

e. Recalculate the choice probabilities of part (d) if hydrogen fuel cars receive a subsidy of $2,500.

4-5. Consider a conditional logit model with a linear specification of utility.

a. Show that the likelihood equation takes the form:

$$\frac{\partial \mathcal{L}(\hat{\boldsymbol{\beta}};\mathbf{y}|\mathbf{X})}{\partial \boldsymbol{\beta}} = \sum_{i=1}^{N}\sum_{j}(y_{ij} - P_{ij})\mathbf{x}_{in} = 0$$

b. Show that the choice elasticities are

$$E_{P_{ij},\, x_{kij}} = \beta_k x_{kij}(1 - P_{ij})$$

$$E_{P_{il},\, x_{kij}} = -\beta_k x_{kij} P_{ij}$$

c. Use the cross choice elasticity $E_{P_{il},\, x_{kij}} = -\beta_k x_{kij} P_{ij}$ to explain that the conditional logit imposes proportional substitution.

4-6. Consider individuals choosing among a compact sedan (1), a small SUV (2), a compact hybrid (3), and a hybrid SUV (4).

a. Derive the choice probability of a nested logit model with a hypothesized nesting structure with two sets based on vehicle class. The first nest contains compact vehicles (n_{Compact} = {compact sedan, compact hybrid}), while the second nest contains SUVs (n_{SUV} = {small SUV, hybrid SUV})

b. Propose a statistical test to validate the proposed nested structure.

c. Derive an expression of the consumer surplus.

d. How would you determine which model is best, the one with the nesting structure based on vehicle class of part (a), or the one with the nesting structure based on technology of Example 4-5?

4-7. Consider the estimates of Example 4-15.

a. Interpret the signs of the marginal utilities. Are they reasonable?

b. "The likelihood of consumers buying low-emission vehicles increases when the appropriate refueling/recharging network increases." Comment on this assertion.

c. Calculate the willingness to pay for a car that produces 1 gram/mile less of carbon dioxide.

d. A Volkswagen Jetta produces 193 grams/km of CO_2. The Prius produces 113 grams/km. Based on the results of the model, how much more could the Prius cost just based on its environmental benefits?

e. Find online the average price of a new Jetta versus the average price of a new Prius. Do you think that the result you found in (d) explains this actual price difference?

CHAPTER 5

Background on Transportation Systems and Vehicle Design

5-1 Overview

This chapter comprises the first of two background chapters on transportation technologies and systems. It provides background on transportation systems metrics along with aspects of vehicle design, and the next chapter (Chap. 6) considers transportation infrastructure design. In the first part of the chapter, system-wide measures of transportation activity and efficiency are discussed, including measures of transportation system capacity, productivity, and energy efficiency. Thereafter, a number of design factors are introduced and analyzed as a foundation for discussion of vehicle technology alternatives, with a primary focus on road vehicles but with applications to other vehicles (rail, aircraft) as well.

5-2 Introduction

The focus of this chapter is on both transportation systems and on vehicles that function within that system. Although the physical manifestation of a transportation system varies from one to the next, the components of those systems can generally be chosen from a menu of components that are common to all. Once assembled, systems can be evaluated in terms of their function through various measures of capacity, productivity, and efficiency. Similarly, the vehicles used in a transportation system can have their performance measured in terms of various metrics.

5-2-1 Components of a Transportation System

In Chap. 3, a system was defined as a group of components acting together to achieve some common purpose. In the case of transportation systems, the common purpose is to deliver transportation service (the movement of persons or goods), and the components generally draw from the following list:

- *Vehicles:* This category consists of surface transportation vehicles (e.g. automobiles, urban railcars, etc.), marine vessels, and aircraft that function within a transportation system. In common parlance the term *vehicle* is usually applied specifically to vehicles on roads or railways, and not to ships or airplanes, but in this context the term could be applied to any unit of conveyance large or small.

- *Infrastructure* (discussed in Chap. 6): The fixed component of a transportation system that provides physical support for and guidance to vehicles. Transportation infrastructure includes a network of *guideways* that connect locations in network, such as streets in a metropolitan area that are designed to physically support the weight of various road vehicles while at the same time providing them clear boundaries within which they should navigate. Another example is raillines in a rail network, or the network of navigable rivers and lakes plus artificially created canals that make up the inland waterway network. *Stations* are found at various points in a passenger transportation network, where individuals can enter or exit the network, along with *stops* that typically are smaller than stations. The corresponding component in the freight infrastructure network is the *terminal*, such as a port for transferring freight to and from the marine network.
- *Equipment:* Found in terminals or stations, this category consists of machine elements that serve a similar role to vehicles but over a shorter distance, usually for transferring freight or luggage between different types of vehicles. Cranes in container ports, forklifts in distribution centers (also known as warehouses), and baggage-handling systems in airports are all examples of equipment.
- *Power systems* (discussed in Unit 5 at the end of the book): Electrical systems that provide electrical power to certain types of vehicles that require electricity, or else to equipment and other electrical loads in terminals, stations, and stops, to allow them to function. Power systems function at the nexus between transportation infrastructure and the electrical grid infrastructure, which includes electricity-generating stations, high-voltage transmission lines, and lower-voltage distribution systems.
- *Fuel supply systems* (discussed in Unit 5 at the end of the book): Systems that provide liquid or gaseous fuels, usually in the form of some type of hydrocarbon, which serve as an energy supply for vehicles that do not use electricity as a primary source of energy. Similar to electric-power systems, fuel-supply systems exist at the intersection between transportation infrastructure and the fuel supply infrastructure, which includes oil and gas wells, pipelines, refineries, and retail outlets for motor fuels.
- *Control systems* (discussed in Chap. 7): Lastly, this "balance of system" component of transportation systems includes functions such as monitoring system operations to evaluate the function of the system in real time, controlling network operations by making changes in their operation through hardwired and wireless networks, informing transportation system users about operating conditions, and operation of financial networks that support monetary exchanges to keep the transportation system operating.

5-2-2 Ways of Categorizing Transportation Systems

Transportation systems can be categorized in a number of ways, and the category to which each system belongs influences how it functions. Using the following typology of four transportation categories, a classification on one level can be made independent of the other three.

- *Function: passenger or freight:* One general way to classify transportation is between *passenger* and *freight* transportation. Passenger transportation constitutes any movement of people (e.g., for work, errands, tourism, etc.), including all luggage or personal effects pertaining to their travel, and freight constitutes the unaccompanied movement of goods other than luggage (e.g., bulk commodities, finished products, livestock, mail and parcels, etc.). In general, vehicles are dedicated to one form or the other, although there are exceptions; for instance, in the case of aviation, large commercial airliners frequently carry passengers on the main deck and airfreight on the lower deck (also known as the *lower lobe*) of the aircraft.
- *Modes: road, rail, water, air,* and *pipeline:* Passenger and freight transportation can be further divided into one of these five major modes. For example, the road mode consists of cars, buses, and motorcycles on the passenger side, and vans and trucks on the freight side, which is sometimes called the *truck* mode. Broadly defined, the road mode can also include nonmotorized modes such as bicyclists and pedestrians. The water mode (also called the *marine* mode) includes movements of boats and ships on both inland bodies of water such as rivers, lakes, and canals, and on the open seas. All five modes exist in both a passenger and freight form, except for pipeline, which is used exclusively for freight, and primarily for the transport of energy products such as oil and natural gas. In addition to these major modes, a number of niche modes exist as well, such as an aerial tramway in a mountainous region that carries passengers and/or goods from the outside world to a remote community.[1] The five major modes are, however, the only ones that consume a significant fraction of the world's transportation energy budget.
- *Geographic scope: urban or intercity:* Both passenger and freight transportation can be divided between urban and intercity transportation movements. On the passenger side, the word *urban* is used as an umbrella term that covers movements in both large urban areas and small communities, but that in each case are characterized by movements for work, commerce, or recreation over a short distance and carried out on a daily or regular basis. Intercity passenger trips are typically of a distance of 50 to 100 km (approximately 30 to 60 miles) or more between two distinct towns or cities. On the freight side, intercity movements entail the long-distance *trunk* movement of goods between population centers, where the goods are often combined into a larger unit such as a tractor-trailer or a freight train with multiple cars. Urban movements are those at the endpoints of a trunk movement that are used to gather shipments together for long-distance movement, or distribute them once they have arrived at their destination. In some cases, smaller vehicles (such as delivery vans or light trucks)[2] are used to carry out urban distribution movements,

---

[1]Means of conveyance, such as aerial tramways, cruise ships, etc., in which the purpose is sightseeing or tourism, and the passenger travels in a circuit without the intention of reaching the endpoint of the circuit as a destination, are generally not counted in transportation statistics.

[2]In the remainder of this chapter, we use the term *light truck* to refer to pickup trucks, vans, minivans, and sport-utility vehicles (SUVs).

while in other cases, the same truck may be used for both intercity movement and urban collection and delivery activities.

- *Ownership: private or commercial:* Transportation activity can be divided between *private* transportation movements, which consist of any activity in which the driver or operator owns the vehicle, and *commercial* movements, which entails the selling of the transportation service to passengers or shippers of freight by professional transportation providers (taxi companies, airlines, for-hire trucking companies, etc.). In the case of freight transportation by road, a fleet of trucks that is owned by a company for movement of products it makes or sells itself is considered to be private transportation, even though the driver of the truck does not personally own the vehicle. For example, some large food retailers that operate chains of supermarkets may have their own private fleet, while others contract with for-hire firms to have this service carried out.

As examples of application of these categories, a driver in her/his own personal automobile to work represents passenger transportation using the road mode for urban transportation in a privately owned vehicle. A railroad moving a shipping container from a port to an inland city represents freight transportation using the rail mode for intercity transportation in a for-hire vehicle.

5-3 Units of Measure in Transportation

Metrics for evaluating transportation systems cover all major modes of transportation, for both passenger and freight applications. These metrics capture characteristics such as available capacity in the system; amount of throughput actually delivered in a given amount of time; productivity of the system (a comparison of service delivered to available capacity); and efficiency (amount of service delivered per unit of resource consumed, such as energy), or its inverse, intensity (resource consumed per unit of service delivered).

5-3-1 Measures of Capacity and Productivity

Capacity can first be measured in terms of overall vehicle-miles traveled (abbreviated "VMT" by convention) for road vehicles, aircraft-miles for aircraft, and train-miles for railroads.[3] Since railroad trains vary in length, an alternative measure for railroads called the car-mile exists, which signifies the movement of one car in a train a distance of 1 mile. Thus a longer train with more cars will have a higher capacity value in terms of car-miles than a shorter train traveling the same distance. For passenger transportation, the seat-mile comprises the movement of one seat within a bus, train, or aircraft a distance of 1 mile, and is used to reflect the different capacities provided by different sizes of vehicles. For instance, the movement of two traincars of different sizes a distance of 1 mile will both deliver 1 car-mile of capacity, but different numbers of seat-miles.

Actual service delivered by transportation systems is measured in a similar way. For passenger and freight transportation, the movement of a passenger or a ton

[3]Metric units can be used for any of the units in this section, for example, vehicle-kilometers (by convention abbreviated "VKT") in place of vehicle-miles, etc.

(or metric tonne[4]) a distance of 1 mile (or kilometer) is called a passenger-mile or ton-mile (tonne-kilometer), respectively. Comparison of service delivered to capacity provided serves as a measure of utilization, for example, the ratio of passenger-miles to seat-miles is equivalent to the percent of capacity used, with a higher value representing a more effective use of capacity. Alternatively, the volume of passengers or freight on board without reference to distance provides a different figure for delivered service, for example, passengers carried for passenger transportation, tons loaded (sometimes called "tons lifted") for freight transportation. In the case of container freight shipping, another measure is the *twenty-foot equivalent unit* (TEU), which represents a shipping container 20 ft in length. Thus a 40-ft container is equivalent to 2 TEUs, and so on. Where weight or volume of freight is not known, TEUs can provide a convenient measure of freight volume or productivity; for example, the TEU capacity of a container ship.

Transportation volume or productivity can be measured per unit of time as well, such as per hour or per year. For road transportation, highway volumes can be measured either in the number of cars per hour in one direction, or the number of cars per lane per hour, where the capacity per lane per hour multiplied by the number of lanes gives the capacity per hour. In the case of urban rail transportation, capacity per hour can be measured as well, both in terms of a theoretical maximum and a maximum observed flow. For intercity passenger and freight transportation the capacity per year is often of greater interest. Example measures include the number of passengers served by an airport per year (typically both arriving and departing passengers are counted, although this may not always be true), number of trips served by a passenger ferry boat or truck freight ferry boat per year, number of truckload equivalents served by a major rail or highway corridor per year, or the number of TEUs handled by a container port in a year. Hourly capacity of transfer facilities are sometimes published, such as a rail terminal that handles transfers of freight to and from trucks (e.g., number of truckloads per hour capacity) or number of cars and/or trucks handled per hour at an international border crossing. Example 5-1 illustrates the calculation of some representative transportation system metrics.

Example 5-1S A freight rail corridor is 1,000 miles in length and has a terminal at either end. Each day a total of 40 trains operate on the corridor, 20 in each direction, and this schedule is maintained 365 days per year. The average train is 80 cars in length, and the average loading is 44 tons per car. (a) Calculate the number of train-miles, car-miles, ton-miles, and tons loaded per year. (b) If the freight is divided evenly between the two terminals and each operates 24 hours per day, what is the average throughput of each terminal in tons loaded per hour?

Solution

(a) Multiplying 40 trains per day by 365 days per year gives 14,600 trains originated per year. Each train travels 1,000 miles, so the total is 14.6 million train-miles per year. Since each train pulls an average of 80 cars, the total car-miles is 1.168 billion car-miles per year. Furthermore, since each car-mile delivers 44 ton-miles, the total ton-miles are 51.39 billion ton-miles per year. Also, 14,600 trains per year with 80 cars per train and 44 tons per car results in 51.39 million tons loaded per year.

(b) Since there are 20 trains per day each with 80 cars per train and 44 tons per car, the total throughput is 70,400 tons per day, or 2,933 tons per hour.

---

[4]The spelling "tonne" is used to distinguish a metric tonne (1,000 kg) from a standard ton (2,000 lb).

Example 5-1M[5] A freight rail corridor is 1,500 km in length and has a terminal at either end. Each day a total of 40 trains operate on the corridor, 20 in each direction, and this schedule is maintained 365 days per year. The average train is 80 cars in length, and the average loading is 40 tonnes per car. (a) Calculate the number of train-miles, car-miles, tonne-miles, and tonnes loaded per year. (b) If the freight is divided evenly between the two terminals and each operates 24 hours per day, what is the average throughput of each terminal in tons loaded per hour?

Solution

(a) Multiplying 40 trains per day by 365 days per year gives 14,600 trains originated per year. Each train travels 1,500 km, so the total is 21.9 million train-km per year. Since each train pulls an average of 80 cars, the total car-km is 1.752 billion car-km per year. Furthermore, since each car-mile delivers 40 tonne-km, the total tonne-km is 70.08 billion tonne-km per year. Also, 14,600 trains per year with 80 cars per train and 40 tonnes per car results in 56.72 million tonnes loaded per year.

(b) Since there are 20 trains per day, each with 80 cars per train and 40 tonnes per car, the total throughput is 64,000 tonnes per day, or 2,667 tonnes per hour.

5-3-2 Units for Measuring Transportation Energy Efficiency

Transportation energy efficiency can be measured at various stages, from the testing of the equipment in the laboratory to the delivery of transportation services in the real world. Each successive stage introduces the possibility for ever greater numbers of intervening factors that can disrupt the smooth operation of a component, vehicle, or entire system, so that the potential for losses is increased, as shown in the following four stages:

1. *Technical efficiency of components:* A component of the drivetrain can be tested for its ability to transmit input to output energy. For example, the engine might be tested under steady-state, optimal conditions to determine what percent of the energy present in the fuel combusted is transferred to the rotation of the driveshaft. Engineers might evaluate losses in other drivetrain components in a similar way, that is, calculating efficiency on the basis of power out divided by power in.
2. *Laboratory vehicle fuel economy:* At this level of measurement of energy efficiency, vehicles are tested on a dynamometer to estimate fuel economy, where the dynamometer drive cycle is used to represent driving conditions in the real world. The results are given in units of city or highway miles per gallon or kilometers per liter. For metric measurement of fuel consumption, the measure of "liters per 100 kilometers" is commonly used. Laboratory testing recognizes that measuring technical efficiency of the drivetrain (Approach #1) does not capture the use of the component in the vehicle, or the effect of parts of the vehicle that do not directly consume energy (e.g., the vehicle body). Estimation of fuel economy also results in a measure that incorporates times when the vehicle is not operating at optimal energy efficiency, for example, stop-and-go driving conditions. Lastly, prospective buyers of the vehicle want to know the effect of fuel economy on operating cost of the vehicle. While a measure of technical efficiency of the drivetrain gives the buyer little information on this point, a measure of fuel economy can readily be translated into an

[5] For select problems throughout the book, metric and standard versions of examples are offered.

estimated cost per year for fuel, if the buyer knows how far she/he typically drives in a year.

3. *Real-world vehicle fuel efficiency or intensity:* Government agencies typically report overall fuel efficiency (also referred to as *energy intensity*) for different vehicle classes (passenger cars, light trucks, heavy duty vehicles, etc.), in terms of energy consumption per unit of distance traveled. Thus fuel efficiency or intensity is the inverse of fuel economy, which is distance per energy. (Liters per 100 km, introduced under Approach #2, is a measure of fuel intensity.) These agencies use vehicle counts on selected roadways and modeling techniques to estimate total vehicle-kilometers of travel, and the allocation of total annual transportation fuel sales to different transportation applications, as a basis for estimating actual kilojoule/vehicle-kilometers (kJ/vehicle-km, Btu/vehicle-mile in standard units).[6]
4. *Real-world transportation service efficiency or intensity:* Use of energy per vehicle-kilometers as a measure of technological progress is imperfect because the purpose of the transportation system is not the movement of vehicles but the delivery of transportation service, that is, the movement of passengers or freight. Movement of vehicles is a means to this end, but it is not the end itself. The quantity of transportation service is measured in units of passenger-kilometers and tonne-kilometers (e.g., the movement of one passenger or 1 tonne of freight for a distance of 1 kilometer, respectively) in order to incorporate the effect of distance on transportation intensity. In other words, the movement of 100 passengers for 100 kilometers will require more energy, incurs more wear and tear on transportation infrastructure, and so on, than the movement of 100 passengers for 1 kilometer. As with vehicle-km statistics, total passenger-km and tonne-km are published by governments through a mixture of sampling and modeling. Thus it is possible to publish measures of transportation energy intensity in terms of kJ/passenger-km and kJ/tonne-km, respectively. These measures capture the effect of changing transportation practices on energy consumption: for instance, if vehicles are not loaded fully and all other factors remain the same, energy intensity measured in energy per passenger-km or tonne-km will increase.

The roles of different measures in assessing the energy efficiency of the transportation system is summarized in Table 5-1. Comparison of laboratory vehicle fuel economy values (Approach #2) for an entire fleet of cars in a country and real-world fuel efficiency (Approach #3) typically reveals a fuel efficiency shortfall or gap between the predicted and actual fuel consumption. The laboratory fuel economy values can be used to predict the annual fuel consumption of each vehicle in the fleet, based on its rated fuel economy and estimated distance driven. The sum of all the vehicles in the national fleet gives one possible value for the amount of fuel consumed. When the real-world fuel efficiency is converted to units of fuel economy, it is not as high a value as the estimate based on laboratory fuel economy figures. The shortfall occurs because laboratory tests typically do not fully capture the negative effect of high-speed driving,

[6]For brevity, metric units are used in the remainder of this chapter. Conversion between standard and metric units is as follows: 1 passenger-mile = 1.6 passenger-km; 1 ton-mile = 1.45 tonne-km; 1 vehicle-mile = 1.6 vehicle-km.

Name	Units		Description
	Metric	Standard	
Technical efficiency of components	Percent efficiency; kJ out per kJ in	Percent efficiency; Btu out per Btu in	Laboratory testing of drivetrain components (engine, transmission, tires, etc.)
Laboratory vehicle fuel economy	Liters per 100 km, km per liter	Miles per gallon	Estimate of real-world energy consumption performance based on laboratory drive-cycle test
Real-world vehicle fuel efficiency or intensity	kJ/vehicle-km	Btu/vehicle-mile	Real-world energy efficiency based on estimates of actual energy consumption and vehicle distance traveled
Real-world transportation service efficiency or intensity	kJ/passenger-km, for passenger; kJ/tonne-km, for freight	Btu/passenger-mile, for passenger; Btu/ton-mile, for freight	Real-world energy efficiency based on actual energy consumption and passenger or freight distance traveled

TABLE 5-1 Levels of Measuring Transportation Energy Efficiency

traffic congestion, decline of fuel economy due to aging vehicles, inadequate owner maintenance practices (e.g., failure to keep tires sufficiently inflated), and other factors.

Influence of Transportation Type on Energy Requirements

The distinction between passenger and freight transportation is important because the movement of freight is strictly commercial in nature, while the movement of passengers entails more of a balance between competing factors including not only cost but also, in many cases, the comfort of the passenger, pleasure derived from the route of travel, or the image or status conveyed by the vehicle. When manufacturers, retailers, or other parties responsible for the overall cost of a product make decisions about freight, they determine the amount of protection needed for the product (e.g., protective packaging, refrigeration, etc.) as well as the desired speed and reliability of delivery, and then seek out the least-cost solution that meets these requirements. There is therefore no incentive to spend extra money on more expensive solutions, since the difference in cost will come directly out of the profitability of the product. Freight shippers may, on occasion, spend extra to use the vehicle (typically a truck or shipping container) as a means of promoting the product by decorating the exterior with information about the product or company, as a form of advertising similar to a stationary billboard, but this decision is a calculated investment that expects a return on the expenditure.

When passengers make transportation choices, on the other hand, they may be more likely to spend extra on a larger, more comfortable vehicle, or more distant vacation destinations, so long as they have the economic means to do so. Especially in the case of middle- and upper-class populations in the industrial countries, some of whom have in recent decades greatly increased purchasing power, the attraction of energy-intensive transportation choices has made curbing the growth in overall transportation energy use more challenging.

The effect of increasing wealth and more demand for transportation service spills over into the choice of mode as well. For example, in the case of freight, modes such as water and rail are on average more energy efficient because they allow the vehicle operator to consolidate more goods on a "vehicle," they move with less stopping and starting, and also water and rails create less rolling resistance than, e.g., rubber tires on roads. However, these modes also require greater coordination at the terminal points to consolidate and break apart shipments. From a transportation management point of view, shipment by road and air is a more "agile" option because the shipment reaches the destination more quickly and reliably, although the energy requirement is on average greater. An analogous argument can be made in the case of passenger transportation. Overall, the marketplace for both passenger and freight transportation has in recent years favored the higher service of road and air modes over the energy efficiency of rail and water, increasing total energy consumption.

Lastly, one of the main effects of the geographic scope of transportation is to limit energy source options. In the case of passenger transportation, the majority of transportation activity is generated in urban movements. For many of these trips, it would be possible for travelers to use alternative energy options such as electric vehicles with batteries or alternative liquid fuels that do not have as high of an energy density, because the distances between opportunities to recharge/refuel are not too great. It is also easier to connect vehicles such as buses or urban rail vehicles to a catenary grid (e.g., overhead wires), because the density of passenger demand is high enough to justify the cost. By contrast, the majority of freight transportation activity happens over long distances between cities, where the current expectation is that the vehicle or aircraft can travel for long periods between refueling stops. For the most densely traveled rail routes, electric catenary may be justified, but for other routes, rail locomotives must rely on liquid fuels stored onboard the vehicle between refueling stops in the same way that trucks, aircraft, or ships do. Also, the combination of long distances between refueling and large power requirements limits the practicality of battery-electric systems as an option for long distance air or surface freight.

5-4 Background on Vehicle Design

The first two measures in Table 5-1, technical efficiency of components and of vehicles, suggest that an important goal of any vehicle design is to create a vehicle that is as technically efficient as possible, all other factors being equal. Vehicle engineers in practice use a scientific approach and apply knowledge of physics, thermodynamics, fluid mechanics, and heat transfer to create a laboratory ideal of a vehicle against which actual prototypes can be empirically tested.

Regardless of the energy source or the propulsion technology used, design of any vehicle is based on meeting certain performance requirements, of which saving energy (and by extension reducing emissions such as CO_2) is just one. Based on the power, weight, and aerodynamic characteristics of a vehicle, it is possible to predict a number of its performance measures. These performance measures provide an indication of the success of the engineer in meeting the customer's desire for a vehicle that performs well, e.g. stopping, accelerating, climbing hills, etc. Vehicle owners, whether private individuals owning passenger cars or professional operators of heavy duty vehicles, are aware that there is usually a tradeoff between performance on the one hand and fuel economy/reduced cost of fuel on the other, so that vehicle performance may be only one of several

Vehicle	Drag Coefficient (C_D)
1970s or 1980s standard passenger cars	0.5–0.6
1970s or 1980s sports cars	0.4–0.5
Post-2000 high fuel economy passenger cars	0.25–0.3
Highly aerodynamic concept cars	0.15–0.18
Theoretical minimum drag (teardrop shape)	0.03

TABLE 5-2 Evolution of Drag Coefficient Values, 1970s to present

factors in choosing a vehicle for purchase. Also, aspects of driving style, including typical cruising speed or rates of braking and acceleration, affect delivered fuel economy, so even if an owner anticipates a certain level of fuel economy from a given vehicle, the way they drive the vehicle will affect the actual fuel consumption. Lastly, owner preferences change in response to outside information, so it is to be expected that as they learn more about the role of vehicle choices in affecting climate change and other environmental issues, at least some owners will give environmental concerns more weight in their decision-making.

For their part, vehicle manufacturers have responded to government requirements and consumers' desire to reduce fuel expenses by using technological innovations to improve energy efficiency of vehicles. For example, Table 5-2 shows how drag coefficients have improved over generations of vehicle design. The drag coefficient relates the velocity of the car to the amount of effort required to overcome aerodynamic drag, so as this coefficient is reduced, cars become more efficient, other things equal.

5-4-1 Criteria for Measuring Vehicle Performance

Typical performance measures used in vehicle design include the power requirement at cruising speed, maximum speed, maximum gradability, and maximum acceleration. Each is explained in turn below. The measures of performance can be applied to internal combustion engines (ICEs), electric motors, or vehicles which combine both (i.e., hybrids).

Power Requirement at Cruising Speed

The power requirement for a vehicle to maintain cruising speed on a level road is the power provided from the transmission that just equals the rolling and aerodynamic resistance of the vehicle, so that it neither accelerates nor decelerates. Let P_{TR} be the required tractive power, ρ be the density of air, A_F the frontal cross-sectional area of a vehicle, C_D the aerodynamic drag coefficient (as represented in Table 5-2), V the vehicle speed, and C_o the coefficient of rolling resistance. The relationship between P_{TR} and V is then

$$P_{TR} = 0.5\rho A_F C_D V^3 + mgV C_o \tag{5-1}$$

where m is the mass of the vehicle and g is the gravitational constant.

For a vehicle climbing a constant grade, Eq. (5-1) can be modified to include a term that incorporates the work done to move the mass of the vehicle against gravity (Albertus et al., 2008).[7]

[7] Alternatively, Albertus et al. (2008) use a variant of this model that includes the effect of acceleration on tractive power requirement.

Maximum Speed

By extension from Eq. (5-1), the maximum speed for a vehicle is the speed V when the transmission in an ICE vehicle is in highest gear, or motor output in an electric vehicle (EV) is at its maximum value, P_{TR} is in equilibrium with aerodynamic and rolling resistance, and an increase in engine or motor rotational speed, in revolutions per minute (RPM), would lead to a drop in P_{TR}, so that the vehicle cannot accelerate to a higher speed. For a given desired maximum speed V_{max}, Eq. (5-1) tells us the required tractive power that must be provided by the drivetrain. Alternatively, for a given amount of available tractive power, we can predict V_{max} for the vehicle.

Maximum Gradability

The maximum gradability is the grade of slope at which the gravitational force acting downward on the vehicle is just balanced by the maximum tractive force F_{TR} of the engine or motor acting upward, so that upward motion at an infinitesimal rate is just possible (Fig. 5-1). Here slope is the ratio of distance of rise to distance of run, measured in percent, that is, 10 m vertical rise over 100 m horizontal is a 10% slope. Maximum gradability GR_{max} is a function of F_{TR} and m, as follows:

$$GR_{max} = 100 \times \left[\frac{F_{TR}}{\sqrt{(mg)^2 - F_{TR}^2}} \right] \tag{5-2}$$

Maximum acceleration: The maximum acceleration A_{max} achievable on a level surface by the vehicle is based on its maximum available tractive force F_{TR} and its total mass. In a simple form, ignoring the effect of drag and assuming constant force across

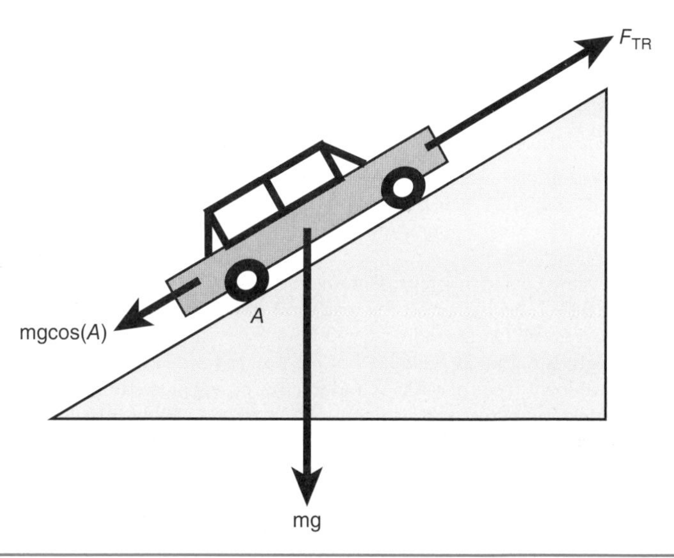

FIGURE 5-1 Force balance of gravitational and tractive forces acting on a vehicle on a grade, showing gravitational force acting vertically, angle A between the vertical and the slope of the grade, and vector component of force mgcos(A) acting opposite the tractive force F_{TR}.

the range of vehicle speeds, this relationship can be derived from Newtonian mechanics, as follows:

$$A_{max} = \frac{F_{TR}}{m} \tag{5-3}$$

Thus, in order to maximize acceleration, the engineer seeks to maximize force while at the same time reducing mass. In practice, increasing force tends to increase engine size, thereby increasing mass and lowering A_{max}, so a balance must be struck between the two factors. Examples 5-2 and 5-3 illustrate the use of these performance measures for evaluating vehicle designs at a basic level.

Example 5-2 A representative passenger car that is designed for fuel economy has a frontal area of 2.6 m^2, drag coefficient of 0.3, rolling resistance coefficient of 0.01, curb weight of 1,200 kg, maximum tractive force at low speeds of 3,000 N, and tractive power at maximum speed of 30 kW. The vehicle has a five-passenger capacity, plus rear end space behind the second seat for storage. Calculate the cruising power requirement at 96 km/h, the maximum speed, and the maximum gradability. (In standard units, the vehicle has a cross-sectional area of 28.1 ft^2, weighs 2,640 lb, and the desired cruising speed is 60 mph.)

Solution Assume air density of 1 kg/m^3. The cruising speed is equivalent to 26.7 m/s. Therefore, using Eq. (5-1), the power requirement is

$$P_{TR} = 0.5(1)(2.6)(0.3)(26.7)^3 + (1200)(9.8)(26.7)(0.01) = 10.5 \text{ kW}$$

Next, the tractive power at maximum speed of 30 kW will determine the maximum speed V_{max}. Plugging in known values gives

$$30 \text{ kW} = 0.5(1)(2.6)(0.3)V_{max}^3 + (1200)(9.8)V_{max}(0.01)$$

Solving using a numerical solver gives V_{max} = 40.2 m/s, or 145 km/h (90.4 mph).

Maximum gradability is determined using the maximum tractive force at low speed and the mass of the vehicle:

$$GR_{max} = 100 \times \left[\frac{3000}{\sqrt{(1200 \cdot 9.8)^2 - (3000)^2}}\right] = 26.4 = 26.4\%$$

Example 5-3 Now consider a representative sport-utility vehicle (SUV) also with five-passenger capacity. The SUV rides higher off the road, and so has a larger frontal area of 3.1 m^2 and a higher drag coefficient of 0.4. Due to more rugged tires, the rolling resistance coefficient increases to 0.015. The curb weight is higher at 1,500 kg, but the power train is also stronger, delivering a maximum tractive force at low speeds of 4,500 N, and tractive power at maximum speed of 50 kW. (In standard units, the vehicle has a cross-sectional area of 33.4 ft^2 and weighs 3,300 lb.) (a) Calculate the cruising power requirement at 96 km/h; the maximum speed; and the maximum gradability. (b) Suppose both the SUV and the economy car in Example 5-2 accelerate from a standstill at full force in a frictionless vacuum, and that they can maintain the maximum low-speed tractive force over the entire range of speeds. How fast will each of them reach the cruising speed of 96 km/h or 60 mph?

Solution

(a) repeating the calculations from Example 5-2 but using new parameters for the SUV, the answers are P_{TR} = 17.6 kW, V_{max} = 146 km/h (91.0 mph), and GR_{max} = 32.2%.

	Units	Compact	SUV	Change
Cruise	kW	10.5	17.6	67%
Maximum speed	Km/h	144.6	145.7	1%
	mph	90.4	91.0	1%
Gradability	percent	26.4%	32.2%	22%
Acceleration	seconds	10.7	8.9	–17%

TABLE 5-3 Comparison of Energy Requirement and Performance Measures for Representative Passenger Car and SUV

(b) Since there is no resistance and tractive force is constant, we can apply Eq. 5-3 to the case of both vehicles. For the car:

$$A_{max} = \frac{F_{TR}}{m} = \frac{3,000}{1,200} = 2.5\ m/s^2$$

In order to reach cruising speed of 26.7 m/s, the car must accelerate for $t = (26.7)/(2.5) = 10.7$ s. By similar calculation, for the SUV, $A_{max} = 3\ m/s^2$, $t = (26.7)/(3) = 8.9$ s. Note that because the assumptions about vehicle specifications are simplistic, the comparison is not transferable to actual vehicles having approximately the same dimensions.

A comparison of the results for the two vehicles from Examples 5-2 and 5-3 is given in Table 5-3. The percent change column gives the percent change up or down for the value for the SUV, relative to that of the car.

From the table, it is clear that the design features of the SUV give it the superior performance in the categories that consumers seek in such a vehicle, but also worsen its fuel economy. In practical terms, the maximum number of passengers is the same, but the SUV has presumably more cargo space, rides higher off the ground, and is heavier, giving it the impression of being a structurally stronger and hence safer car. However, cruising fuel consumption is 67% more; although it is not shown, the heavier mass will also lead to greater fuel consumption when accelerating. Increased drag and rolling resistance coefficients as well as mass lead to fuel consumption at constant speed being greater. On a per-unit of mass basis, the SUV can provide more tractive force when accelerating and more tractive power at maximum speed, so acceleration, maximum gradability, and maximum speed are superior, which are desirable features for this type of vehicle.

The results for the two generic vehicles presented in the above table can be compared to real-world vehicles to examine the validity of using engineering formuli to predict performance. For instance, the 2005 Toyota Corolla and Toyota RAV-4 small SUV are comparable to the example vehicles in terms of curb weight and drag coefficient (1150 kg/0.3 and 1448 kg/0.4, respectively). Using current estimates of highway fuel economy from the U.S. Environmental Protection Agency (USEPA), the increase in predicted fuel consumption for driving a RAV-4 in place of a Corolla is 37%. Though not as large as the predicted value in the table of 67%, the difference is nevertheless significant. Also, if we were to assume for the example SUV that the tires had the same rolling resistance as for the passenger car ($C_o = 0.1$), the fuel increase percentage in the table would be 49%, which is closer to that of real-world vehicles. Given that Eq. (5-1) is calculating power requirement at one speed and does not take into account many variables that affect overall fuel economy (highway driving cycle, power train efficiency in converting gasoline into power, effects of accelerating and braking, etc.), we can see that the use of Eqs. (5-1) and (5-2) as shown in this example is a reasonable way to make first-order predictions about differences in performance and energy efficiency.

Complicating Factors in Vehicle Design

It is possible to use Eqs. (5-1) to (5-3) to make the calculations necessary at a basic level to make broad comparisons between major groups of light-duty vehicles, such as compact cars versus SUVs and minivans, or hybrids versus ICE vehicles. In practice, complicating factors come into play, which make accurate comparisons between competing vehicle alternatives, or interpretation of experimental results from a vehicle test track, a much more complex enterprise.

As an example, let us focus on the question of acceleration. First, in real-world driving, the force acting to accelerate the vehicle is the net difference between tractive force from the power train and resistance acting on the vehicle. Since the resistance is a function of V, this component of the net force will change as the vehicle accelerates. Furthermore, F_{TR} changes with changing speed. Taking the example of the ICE, F_{TR} is related to the flywheel torque T_{FW}, measured in Newton-meters (N-m), of the engine, which is itself changing as a function of changing RPM, as follows:

$$T_{AX} = T_{FW} \times G_{TR} \times G_{FD} - L_{DR} \tag{5-4}$$

$$F_{TR} = T_{AX} \times r_{tire} \tag{5-5}$$

Where T_{AX} is the axle torque, N-m
G_{TR} is the transmission gear ratio (which varies depending on which gear the transmission is in)
G_{FD} is the final drive gear ratio (i.e., between the drive shaft and the transaxle)
L_{DR} are the various drivetrain losses
r_{tire} is the radius of the tire

This calculation assumes a manual transmission; additional losses are incurred due to slippage in an automatic transmission.

The practical result of Eqs. (5-4) and (5-5) is that since the engine has an RPM range where T_{FW} reaches a peak value, and then above or below that range T_{FW} falls off, the ability to contribute to F_{TR} will diminish once the RPM is above the range. Presence of a multispeed gearbox allows the driver to compensate by shifting into a higher gear where once again the engine will operate in ideal RPM range. However, with each higher gear, G_{TR} is decreased, so that F_{TR} and hence maximum acceleration decreases. The effect of changing F_{TR} with changing speed is shown in Fig. 5-2. The dashed line shows F_{TR} for the engine's highest gear, and the solid line shows the sum of aerodynamic and road resistance forces. At V_{max}, F_{TR} and resistance forces are in balance, so as V approaches V_{max}, resistance is increasing with the third power of V while F_{TR} is decreasing with increasing RPM, so that the vehicle will approach V_{max} asymptotically. This behavior is observed in vehicles traveling at very high speeds. A test driver can typically accelerate from a standstill to expressway speeds of 100 to 140 km/h relatively rapidly, depending on the vehicle in question, but thereafter finds that the increase of speed to the rated maximum speed of the vehicle on level ground happens much more slowly.

One practical outcome of the complex nature of relationship between design parameters and delivered performance is that manufacturers carry out performance testing

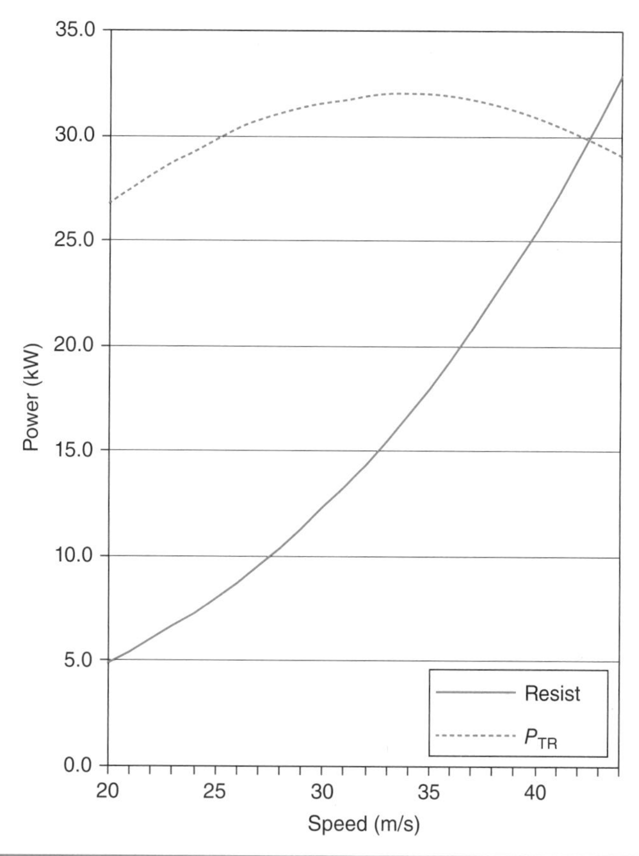

FIGURE 5-2 Tractive power P_{TR} versus resistance force as a function of velocity V for a hypothetical vehicle. Conversion: 10 m/s = 22.5 mph = 36 km/h.

using a mixture of theoretical modeling and empirical testing. As illustrated in Examples 5-2 and 5-3, theoretical evaluation can be used to make general predictions about the performance of classes of vehicles in terms of cruising fuel consumption, maximum gradability, and so on. On the other hand, where the goal is to create a transparent benchmark by which a discerning consumer will make choices between specific makes and models within a vehicle class, it is too complicated to create a defensible theoretical model that makes meaningful comparisons between vehicles, so the makers use empirical testing instead. For example, a vehicle's ability to accelerate is evaluated and published for promotional purposes in terms of "time from 0 to 30 MPH" (0 to 48 km/h) or "time from 0 to 60 MPH" (0 to 96 km/h), and the like, using professional drivers on a test course rather than through theoretical analysis.

5-4-2 Options for Improving Conventional Vehicle Efficiency

From the discussion in the preceding section, overall vehicle weight, maximum power, aerodynamic drag, and other parameters directly influence the energy consumption of gasoline and diesel internal combustion engine vehicles (ICEVs). It follows that making changes to the parameters, such as curb weight, aerodynamic drag, or rolling resistance, through improvements in vehicle design, provides a means to improve fuel economy. Some of the parameters are already evolving in this direction. As shown in Table 5-2, typical drag coefficient values have been decreasing steadily, and a further decrease to a value of $C_D = 0.25$ for many passenger cars is possible in the next 5 to 10 years. Makers have also reduced curb weight per unit of passenger compartment volume through more efficient use of space, advances in materials, and the abolition of the underbody chassis that was common to vehicles in the 1960s and before. On the other hand, in markets such as that of the United States, larger vehicles such as SUVs have become popular, putting upward pressure on the average curb weight of the light-duty vehicles in the fleet. Nevertheless, further improvements in ICEV weight should be possible, especially if higher fuel prices discourage buyers from purchasing the largest vehicles. Through these changes, some improvement in fuel economy is attainable without sacrificing performance or vehicle comfort. These incremental changes have a lower upfront cost to the maker than full-scale changes to alternative platforms, so they pose less of a financial risk.

Incremental improvements of this type have a limit, however. First, as time passes, it becomes harder to wring additional savings out of an ICEV platform that has already been substantially improved. Secondly, as long as the fuel for these vehicles remains gasoline or diesel derived from fossil fuel resources, the makers and car buyers cannot fully achieve the resource and climate goals discussed in Chap. 2 if only this option is pursued.

5-5 Adaptation of Vehicle Design Equations to Other Modes

Similar principles can be applied to the estimation of power requirements at cruising speed and maximum speed for other modes such as rail, marine, and air. For rail, the power requirement for a given train is a function of rolling and aerodynamic resistance, with rolling resistance dominant at low speeds and aerodynamic resistance dominant at higher ones. Parameter values for rail-rolling resistance reflect the advantage of the steel rail wheel on a steel rail compared to rubber tires on an asphalt road, and result in lower rolling resistance for a given speed, all other factors equal. On the other hand, rail vehicles on steel rails suffer from significantly lower maximum gradability than road vehicles; for this reason, some urban underground mass transit systems have adopted rubber-tired vehicles on concrete tracks. Rail vehicles also require a wider turning radius compared to road vehicles. Lastly, a fully detailed equation governing rail-power requirement considers not only cross-sectional area, speed, and coefficients, but also the length of the train, since as trains grow in length both aerodynamic and rolling resistance increase.

For marine vessels, the drag created by the friction between the hull of the vessel and the surrounding water dominate the generation of physical drag. Speed is measured in "knots," with 1 knot = 1.15 mph = 1.86 km/hour. A given hull design will have a specific "hull speed" to which the length of the hull contributes, with a longer hull being proportional to a higher hull speed. A significant discontinuity in the equation of drag as a function of speed occurs at the hull speed, with drag increasing greatly above this threshold.

Adaptation of design equations to aircraft introduces two complications. First, once the aircraft is airborne there is no rolling resistance, so the frictional drag comes in the form of aerodynamic drag as the surface of the aircraft passes through the air. A new component of power requirement is introduced; namely, the force required to generate the aerodynamic lift that keeps the craft aloft. Since drag increases with velocity and the power requirement for lift decreases, there is an optimal cruising speed at which power requirement is optimized (MacKay, 2009).[8] Secondly, in Eq. (5-1), the power requirement to accelerate any vehicle from standstill to cruising speed has generally been ignored, but in the case of aircraft the power requirement and hence energy consumption to accelerate the craft from zero speed on the runway to being airborne and eventually climbing to cruising altitude is a substantial component of the energy requirement for the entire flight. Therefore, a linear model of energy consumed as a function of distance traveled does not apply well to aviation. Instead, estimation of energy consumption for flying must take into account the relative proportion of the flight distance spent climbing to cruising altitude versus cruising.

5-6 Summary

In this chapter, we first reviewed the dimensions of a system for categorizing transportation, including function (passenger or freight), mode (road, rail, etc.), geographic scope (urban or intercity), and ownership (private or commercial). Next, we looked at units of measure for energy use in transportation, and also used a formulaic treatment of power requirement, acceleration, and gradability to characterize the basic design of vehicles, with a focus on highway vehicles. The content of this chapter is therefore applicable to the remaining parts of this book: (1) passenger transportation, (2) freight transportation, and (3) transportation energy.

References

Albertus, P, J. Coutsa, and V. Srinivasan (2008). "A Combined Model for Determining Capacity Usage and Battery Size for Hybrid and Plug-in Hybrid Electric Vehicles." *Journal of Power Sources* 183:771–782.

MacKay, D. (2009). *Sustainable Energy: Without the Hot Air.* UIT Press, Cambridge, UK.

Further Reading

Banks, J. (2001). *Introduction to Transportation Engineering, 2d ed.* McGraw-Hill, New York, NY.

Braess, H., and U. Seiffert, eds. (2004). *Handbook of Automotive Engineering.* Society of Automotive Engineers, Warrendale, PA.

Gillespie, T. (1999). *Fundamentals of Vehicle Dynamics.* Society of Automotive Engineers International, Warrendale, PA.

Khisty, C., and B. Lall (2002). *Transportation Engineering: an Introduction, 3d ed.* Prentice-Hall, New York.

Papacostas, C., and P. Prevedouros (2001). *Transportation Engineering & Planning.* Prentice-Hall, Upper Saddle River, NJ.

Stone, R., and J. Ball (2004). *Automotive Engineering Fundamentals.* Society of Automotive Engineers International, Warrendale, PA.

---

[8] For example, MacKay (2009, p. 273) estimates the optimal speed that minimizes total power requirement for a fully-loaded Boeing 747 at cruising altitude at 540 mph (220 m/s, or 874 km/h).

Exercises

5-1. Based on available data on train and commodity movements, the total tons and ton-miles of freight moved by U.S. railroads in 2011 for different commodity groups are estimated in the table below. During that year, it is estimated that the average railcar carried 62.9 tons of freight, and the average train had 39.1 cars besides the locomotive(s). Calculate total tons moved, total ton-miles moved, ton-miles per ton, total car-miles traveled, and total train-miles traveled. Note that the "other types" group captures all other commodities, including mixed freight, not captured in the top five most common commodities by tonnage.

Commodity	Million	Billion
Coal	816	748
Chemicals	194	178
Grain	157	144
Food products	107	98
Nonmetallic minerals	128	117
Other types	484	444

5-2. Use the balance of forces acting on a vehicle on a slope as shown in Fig. 5-1 and the definition of percent grade (i.e., ratio of vertical rise to horizontal run) to derive Eq. (5-2).

5-3. A high-efficiency passenger car has a frontal area of 2.4 m^2, a drag coefficient of 0.25, a rolling resistance value of 0.01, maximum tractive force at low speed of 2,800 N, maximum power at high speed of 32 kW, and a curb weight of 1,200 kg. Assume air density of 1.1 kg/m^3. Calculate: (a) power requirement at a cruising speed of 100 km/h, (b) maximum speed, and (c) maximum gradability.

5-4. A large sport-utility vehicle (SUV) has a frontal area of 3.8 m^2, a drag coefficient of 0.42, a rolling resistance value of 0.011, maximum tractive force at low speeds of 6,500 N, maximum power of 115 kW at high speed and in highest gear, and a curb weight of 2500 kg. Assume air density of 1.1 kg/m^3. Calculate (a) power requirement at a cruising speed of 100 km/h, (b) maximum speed, and (c) maximum gradability.

CHAPTER 6

Physical Design of Transportation Facilities

6-1 The Design Process

There are many ways to describe the design process for transportation facilities or transportation systems. The overall process of developing a transportation project is a mixture of technical, legal, and political elements. When it is carried out by a public agency, it must balance the interests of users, residents in the immediate vicinity of the project, and the general public. Many of its features will be spelled out by public laws and regulations or by agency policy. These laws, regulations, and policies are intended to ensure that the resulting facility is safe and economical, that its environmental impacts are reasonable, and that the interests of different political constituencies are adequately represented. Among the laws and regulations that govern the project development process are enabling legislation for transportation funding, which often establishes minimum design standards and requires compliance with other laws and regulations; federal and state planning regulations; environmental legislation [the National Environmental Policy Act (NEPA), the Clean Air Act, the Clean Water Act, etc.]; and the Americans with Disabilities Act.

In this process, there is no clear distinction between what is usually referred to as planning and the process known as design. *Planning* refers to the more general and abstract parts of the process and *design* to the more detailed and concrete, but both involve use of rational processes to decide how to use available resources to achieve goals. The overall design process is a coordinated process of information gathering, analysis, and decision making. In almost all cases, it is open-ended (that is, there is no one right answer, although some answers may be better than others in terms of particular goals) and iterative, so that various alternatives are proposed and evaluated before the final decision is made.

Figure 6-1 is one way of representing the overall transportation facility design process. In this representation, the overall process is divided into planning, traffic design, location, and physical design phases, and ultimately results in construction of the facility. These phases overlap to some extent, however, and some of them may be repeated several times. Specific steps (represented by the boxes in the diagram) include:

1. *Deciding generally what sort of system or facility is needed:* A highway, a mass transit route (or station), an airport, even a whole system. This step is normally considered to be part of the planning process, and is the responsibility

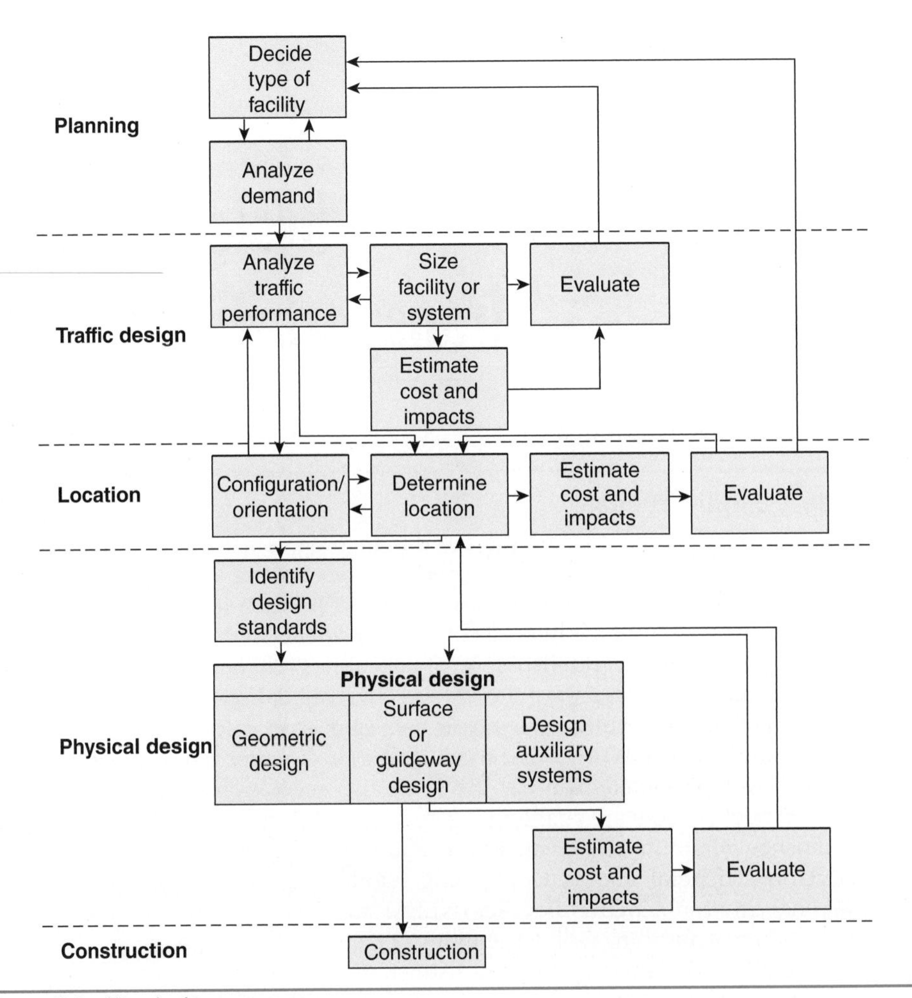

FIGURE 6-1 The design process.

of transportation planning officials and the political system as a whole; nevertheless, design engineers are key participants. Intelligent decisions depend on the ability to predict costs and impacts; these predictions, in turn, often depend on preliminary designs. In addition to cost and impact information, decisions at this point in the process often depend on transportation demand analysis for alternative facilities or systems.

2. *Demand analysis for the system or facility to be designed:* In this context, transportation demand analysis is an attempt to predict, as accurately as possible, the number and type of trips which will take place on a particular facility.
3. *Traffic performance analysis:* In this step, the designer establishes the relationship between anticipated demand and the design features of the facility or system. This step is often referred to as capacity analysis, although it usually involves

analysis of more than just capacity. Also, it is necessary to take into account some of the physical characteristics of the system in order to perform this step; consequently, it is usually necessary to reconsider preliminary performance analyses after the facility or system is located and certain features of the geometric design are decided.

4. *Size the facility or system, based on performance standards and the traffic analysis:* For a highway, for instance, this consists of deciding the number of lanes to be provided at various locations. For an airport, it involves determining whether several parallel runways will be required and, if so, how many. For a railroad, it involves decisions about whether to provide single or double track.
5. *Determine the location of the facility or system:* This step ordinarily requires consideration of several alternative locations. Deciding between them may further require preliminary designs, cost estimates, and environmental impact analyses, and will usually involve public hearings and other public decision processes. In order to carry out these analyses, detailed physical mapping, based on aerial photography and field surveys, may be required.
6. *Determine the configuration and/or orientation of the facility or system:* Orientation refers to such matters as the direction of an airport runway; configuration refers to things like transit system route structures or selection of highway interchange types.
7. *Identify physical design standards:* These are often a matter of policy within a given design organization, but the individual designer must judge the applicability of given design standards to particular situations.
8. *Geometric design:* Geometric design refers to establishment of horizontal and vertical alignments and cross sections, based on considerations such as operating characteristics of vehicles, design standards, and drainage.
9. *Design auxiliary systems:* These include drainage, lighting, traffic control, and power supply (for electrified rail lines).
10. *Design surface or guideway:* This refers to the design of pavement or track for land transportation facilities.
11. *Estimate construction costs and project impacts:* Major cost items in the design of a transportation facility include land (right-of-way), earthwork, structures, and control devices. Final cost estimates are necessary before jobs can go out to bid; it is good practice, however, for the designer to make rough cost estimates throughout the design process and to base design decisions on them. It is also necessary to identify environmental impacts and the cost of environmental mitigation.
12. *Evaluate design:* Designs should be evaluated continually throughout the design process. Evaluations are based on criteria such as physical feasibility; economy; and social, economic, and environmental impacts.

Feedback arrows in the diagram represent the process of redesign. This is the process of mutual adjustment of the various elements of the design. Like cost estimating and design evaluation, it goes on continuously throughout the design process.

6-2 Design Standards

Responsibility for the establishment of design standards varies, depending on the type of facility. Design standards for state highways are established by the state departments of transportation and are documented by design manuals such as the *Highway Design Manual* (Caltrans, 1995). These standards are usually based on the recommended standards of the American Association of State Highway and Transportation Officials (AASHTO). The principal source of recommended standards for the geometric design of highways is the AASHTO *A Policy on Geometric Design of Highways and Streets* (AASHTO, 1994), often referred to as the "Green Book." In addition, the Federal Highway Administration (FHWA) has established minimum design standards for the Interstate System. Design standards for local streets and roads are the responsibility of the local jurisdiction. These are usually based on AASHTO recommendations and state standards.

Establishment of design standards for rail facilities is the responsibility of each individual railroad company or transit authority. Recommended standards are published by the American Railway Engineering Association (AREA, 1996).

The Federal Aviation Administration (FAA) has established design standards for airport landing areas (runways, taxiways, etc.) constructed in the United States (FAA, 1989). These are mandatory for airports receiving federal funding for physical improvements. At the international level, recommended standards are published by the International Civil Aviation Organization (ICAO).

The physical performance of a transportation facility, including its comfort and safety, is a result of the interaction of vehicular characteristics, human characteristics, and the characteristics of the transportation facility. Physical design standards link physical performance to design elements such as horizontal alignment, vertical alignment, cross-section, and various design details.

Vehicular characteristics include physical dimensions such as length, width, height, and wheelbase; weight, including gross weight and wheel loads for various axle configurations; acceleration and deceleration characteristics; maximum speed; and (for aircraft only) lift. Table 6-1 summarizes transportation facility characteristics whose design standards are influenced by these vehicular characteristics. In some cases, the relationship between the vehicular characteristic and the design standard is straightforward, as in vehicle height and vertical clearance. In other cases, relationships between vehicular characteristics and facility design features are complicated. For instance, the relationship between vehicle height and minimum length of vertical curve for highways also depends on acceleration/deceleration characteristics, design speed, and human characteristics (see Sec. 6-3).

Human capabilities and characteristics important in setting design standards include visual ability, ability to hear, reaction times, gap-acceptance behavior, steering behavior, and comfort standards. In many cases, actual design standards are based on comfort. For instance, limitations on radial acceleration on horizontal curves for highways are normally based not on the coefficient of friction between the tires and the roadway but rather on the movement of the passenger's body about the seat. Similarly, limits on vertical acceleration in vertical curves are normally based not on the necessity of maintaining contact between the tires and the pavement, but rather on the feeling in the pit of the passenger's stomach. Other important interactions between human characteristics and design standards have to do with reaction times,

Vehicular Characteristic	Related Facility Characteristic
Length	Parking stall length Transit station platform length
Width	Lane width Parking stall width Lateral clearance
Height	Vertical clearance Minimum vertical curve length
Wheelbase (turning radius)	Lateral clearance on curves Intersection edge radii
Weight	Structural design of surface Structural design of guideway Structural design of bridges
Acceleration/deceleration	Maximum grade Minimum vertical curve length Horizontal curve radius
Speed	Horizontal curve radius Minimum vertical curve length Maximum superelevation
Lift	Runway length

TABLE 6-1 Relationships between Vehicular and Facility Characteristics

which are of major importance in determining stopping distances and hence sight distance requirements.

Transportation system characteristics (or design elements) to which design standards apply include the following:

- *Minimum radius of horizontal curve:* This standard applies to highways and railways. For a given design speed, minimum curve radius is limited by maximum allowable side friction, which is usually based on a comfort standard; maximum superelevation rate (or banking) for the curve, and the necessity to maintain stopping sight distance.
- *Maximum rate of superelevation:* This standard applies to highways and railways. For highways, maximum superelevation rate is limited by side friction and by presence of roadside features such as driveways. The major concern here is to prevent slow-moving vehicles from sliding to the inside of the curve under slippery conditions. For railways, it is limited by the need to limit imbalances in the loads on the rails.
- *Maximum grade:* This standard applies to highways, railways, and airport runways. Maximum upgrades are limited by vehicle power-weight ratios and vehicle traction. Maximum downgrades are also limited by stopping distances and sight distances. Maximum grade standards for certain classes of roadway or railway are also influenced by traffic levels and the need to maintain reasonable speeds on upgrades.
- *Minimum grades:* For some types of highway, these are limited by the need to provide drainage.

- *Minimum cross-slopes:* For highways, runways, and taxiways, these are also limited by the need to provide drainage.
- *Minimum length of vertical curve:* This standard applies to highways, railways, and airport runways and taxiways. For highways, minimum length of vertical curve is limited by stopping or passing sight-distance requirements, vertical acceleration, and appearance standards. For railways, minimum length of vertical curve is also limited by the need to prevent jerk on couplings in sag vertical curves. For runways and taxiways, minimum length of vertical curve is limited by sight distance requirements.
- *Edge radii in roadway and taxiway intersections:* These are limited by vehicle turning radii. These in turn are related to vehicle wheelbase dimensions.
- *Minimum intersection setbacks* (minimum distances to obstructions to vision): These are limited by stopping sight distance and driver gap-acceptance behavior.
- *Freeway ramp junction details:* These are limited by gap-acceptance behavior, steering behavior in entering or exiting lanes, and vehicle acceleration and deceleration capabilities.
- *Horizontal and vertical clearances:* These apply to all modes of transportation. They are limited by vehicle dimensions and, in the case of horizontal clearances for highways, by the need to provide clear recovery zones for vehicles that run off the road.

6-3 Design Speed

As an example of how design standards are developed, consider sight distances for highways. There are two types of sight distance. *Stopping sight distance* is the distance required to see an object 0.15 m high on the roadway. It is intended to allow drivers to stop safely after sighting an object on the roadway large enough to cause damage to the vehicle or loss of control. *Passing sight distance* is the distance required to see an oncoming vehicle of a certain minimum size. It is intended to ensure that a passing maneuver can be completed safely under certain assumptions as to vehicle speeds and acceleration capabilities. Passing sight distances are normally of concern only on two-lane roadways and need not be maintained everywhere on them; the usual consideration is that passing sight distance exists for a sufficient fraction of the highway's length to prevent driver impatience. Stopping sight distance, on the other hand, should be maintained at all points on the roadway.

Sight distance, like many other transportation facility design features, is related to the design speed of the facility. *Design speed* (for highways) is defined as *the maximum safe speed that can be maintained over a specified section of highway when conditions are so favorable that the design features of the highway govern.* Put another way, it is the maximum safe speed when weather conditions are favorable and traffic volumes are low enough that there is no significant interaction between vehicles. As a general rule, the stated design speed for a highway section establishes the minimum standard for design features related to it (such as horizontal curve radius and vertical curve length), and the same design speed will be used for each of these, so as to provide a "balanced" design. Design speeds vary depending on terrain and the anticipated level and character of use of the facility. For highways, AASHTO recommends the speeds given in Table 6-2.

Conditions	Design Speed, km/h
Limited Access Types	
Rural freeways in mountainous terrain	80–100
Freeways in urban areas	100–110
Rural freeways, level terrain	110
Unlimited Access Types	
Rural arterials	
Flat terrain	100–110
Rolling terrain	80–100
Mountainous terrain	60–80
Urban	
Arterial streets	60–100
Arterial streets, central business districts	50–60

Source: Based on *A Policy on Geometric Design of Highways and Streets*. Copyright 1994 by the American Association of State Highway and Transportation Officials, Washington, DC.

TABLE 6-2 Recommended Design Speeds

The stopping sight distance s depends on the reaction time of the driver (including both perception time and the time required to react physically) and the braking distance of the vehicle. That is,

$$s = d_r + d_b \tag{6-1}$$

where d_r is the distance traveled during the driver's reaction time and d_b is the braking distance. The distance traveled during the reaction time of the driver is just the speed of the vehicle times the reaction time, or

$$d_r = vt_r \tag{6-2}$$

where v is the design speed and t_r is the driver's reaction time (including perception time). The difficulty in evaluating this term is determining what the driver's reaction time will be. Reaction times vary widely within the driving population, depending on circumstances such as age, possibility of driver impairment (alcohol, drugs, etc.), and the extent to which the driver was anticipating the need to stop. As a result, the usual practice is to use a single, rather conservative value. AASHTO suggests a value of 2.5 s in its *Policy on Geometric Design of Highways and Streets.*

Braking distance also varies a great deal from vehicle to vehicle, and in theory is expressed by a rather complicated formula which involves braking force, vehicle mass, and vehicle speed. Since there is great variation in these features from vehicle to vehicle, braking distances for determining practical stopping distances are based on the simplified formula

$$d_b = \frac{v^2}{2g(f \pm G)} \tag{6-3}$$

Design Speed, km/h	Coefficient of Friction f	Stopping Sight Distance, m
30	0.40	29.6–29.6
40	0.38	44.4–44.4
50	0.35	57.4–62.8
60	0.33	74.3–84.6
70	0.31	94.1–110.8
80	0.30	112.8–139.4
90	0.30	131.2–168.7
100	0.29	157.0–205.0
110	0.28	179.5–246.4
120	0.28	202.9–285.6

Source: A Policy on Geometric Design of Highways and Streets. Copyright 1994 by the American Association of State Highway and Transportation Officials, Washington, DC. Used by permission.

Table 6-3 Coefficients of Friction and Stopping Sight Distances

where d_b is braking distance
g is acceleration of gravity
f is coefficient of friction between tires and pavement
G is average grade, dimensionless ratio (m/m).

For cases in which G varies (for instance, in a vertical curve), an average value for the entire brake reaction distance is used. AASHTO also gives the mixed unit formula

$$d_b = \frac{V^2}{254f} \tag{6-3a}$$

where V is in kilometers per hour, d_b is in meters, and the effect of grade is ignored. Values of f, like assumed reaction times, are chosen to be conservative and vary with design speed. Table 6-3 gives values of f recommended by AASHTO and minimum stopping sight distances recommended for design purposes by AASHTO. The lower bound of the range of stopping sight distances is calculated on the basis of actual average speeds, which tend to be less than design speeds; the upper bound is calculated by using the design speed. Also, the sight distances are calculated by assuming that the effect of grade is negligible. Where this is not the case, AASHTO suggests corrections which vary with the design speed and the length of the grade.

Example 6-1 Determine minimum stopping sight distance on a –3.5% grade for a design speed of 110 km/h.
Total required stopping sight distance:

$$s = d_r + d_b$$

Reaction distance:

$$d_r = vt_r = (110\text{ km/h})\left(\frac{1{,}000\text{ m/km}}{3{,}600\text{ s/h}}\right)(2.5\text{ s}) = 76.4\text{ m}$$

Braking distance:

$$f = 0.28 \text{ (Table 6-3)}$$
$$G = 0.035 \text{ (given)}$$

$$d_b = \frac{v^2}{2g(f \pm G)} = \frac{\left[(110 \text{ km/h})\left(\frac{1{,}000 \text{ m/km}}{3{,}600 \text{ s/h}}\right)\right]^2}{2(9.8 \text{ m/s}^2)(0.28 - 0.035)} = 194.4 \text{ m}$$

Total sight distance:

$$s = d_r + d_b = 76.4 + 194.4 = 270.8 \text{ m}$$

Calculation of passing sight distance is somewhat more complicated in that it depends on the relative speeds of leading, overtaking, and oncoming vehicles, and on the minimum gap between the oncoming vehicle and the vehicle being passed that the driver of the passing vehicle will accept. For purposes of analysis, AASHTO defines four distances:

d_1 = distance traversed during perception and reaction time and during the initial acceleration to the point of encroachment on the left lane
d_2 = distance traveled while the passing vehicle occupies left lane
d_3 = distance between the passing vehicle at the end of its maneuver and the opposing vehicle
d_4 = distance traversed by opposing vehicle for two-thirds of the time the passing vehicle occupies the left lane, or $\frac{2}{3}d_1$.

These are as shown in Fig. 6-2. Total passing sight distance is given by

$$s = d_1 + d_2 + d_3 + d_4 \tag{6-4}$$

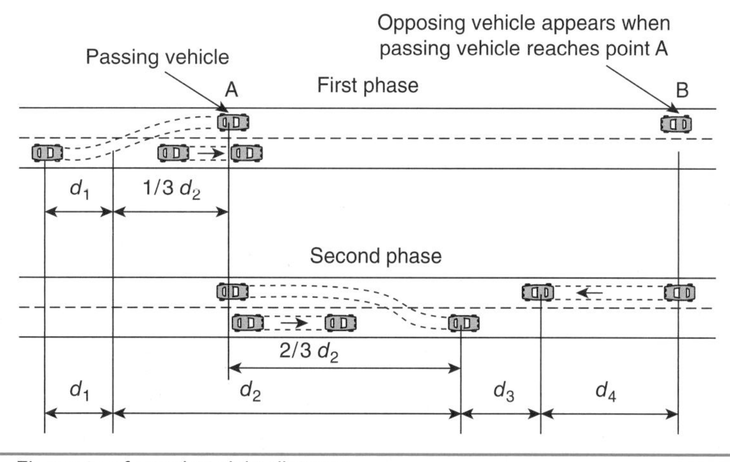

Figure 6-2 Elements of passing sight distance.

Source: A Policy on Geometric Design of Highways and Streets. Copyright 1994 by the American Association of State Highway and Transportation Officials, Washington, DC. Used by permission.

Design Speed, km/h	Passing Sight Distance, m
30	217
40	285
50	345
60	407
70	482
80	541
90	605
100	670
110	728
120	792

Source: A Policy on Geometric Design of Highways and Streets. Copyright 1994 by the American Association of State Highway and Transportation Officials, Washington, DC. Used by permission.

TABLE 6-4 Passing Sight Distances

Values d_1 through d_4 determined through studies of actual passing behavior form the basis for the AASHTO recommendations for passing sight distances on two-lane highways. These are given in Table 6-4.

6-4 Design Documents

Required design documents for transportation projects will vary somewhat, depending on the type of facility. In most cases, the agency owning the facility will contract out its construction. In order to solicit bids from potential contractors or to enter into a contract, it is necessary to document the design, allowable materials, and required construction techniques in detail. Bid documents will usually include *plans, specifications,* and *estimates. Plans* refers to drawings, usually accompanied by notes, of various aspects or components of the design. In the case of transportation projects, plans will document the basic geometric features of the facility as well as many details. *Specifications* are written instructions detailing how the facility is to be constructed. They include such things as allowable materials, allowable construction techniques, and performance standards for various components of the project. *Estimates* include cost estimates for various parts of the project and are used to evaluate the acceptability of bids and the financial feasibility of the project.

Construction plans for linear transportation projects (highways, railways, runways, etc.) consist of four basic elements, which together document the geometry of the facility. In addition, there will usually be plan sheets documenting various details. These detail sheets will often be reproduced from sets of *standard plans,* which are maintained by most design agencies. The four basic elements are:

1. The *plan view* (or simply "plan"): This is a drawing of the facility as it would look to an observer directly above it.

2. The *profile*: This drawing has elevation as its vertical axis, and horizontal distance, as measured along the centerline of the facility (or other recognized reference line), as its horizontal axis.
3. The *geometric cross section*: This view has elevation as its vertical axis and horizontal distance, measured perpendicular to the centerline, as its horizontal axis.
4. The *superelevation diagram:* This applies to curved facilities, such as highways or railways, only. It consists of a graph with roadway or railway cross-slope (vertical axis) versus horizontal distance (horizontal axis). The cross-slope is measured relative to the centerline or some other axis of rotation for the facility. An alternative version of the superelevation diagram plots the difference in elevation between the reference line and the edges versus horizontal distance.

Figures 6-3 and 6-4 are examples of a plan view and a profile, respectively. Figure 6-4 also includes a superelevation diagram. Geometric cross-sections and superelevation diagrams are covered in detail in the next chapter. These examples are of highway plans; however, similar construction plans are used for all linear transportation facilities. Note that *the plan view and the profile are not orthogonal views*—horizontal distance in the profile is measured along the centerline, as if the facility were "stretched out"—not along an arbitrary x axis.

The line representing the facility on the profile is called the *profile grade*. For most highway applications, and for runways, the profile grade represents the elevation of the pavement at the centerline of the facility. For multilane divided highways, it may represent the elevation of the inside edge of pavement. For railroads, it generally represents the top of the rail. In other words, unless otherwise specified, for a straight section, profile grade represents the highest point on the surface of the facility. It normally will not represent the highest point on the surface of a curved section.

Locations along the centerline are identified by *stations*. Depending on the design organization, stations may be either 1,000 m or 100 m apart. Throughout this book, 100-m stations will be used. Distances along the centerline may be measured either in meters or in stations. A distance of 1,024.5 m, for example, would be expressed as 10 + 24.5 stations (or 1 + 024.5, if 1000 m stations are used). Note that stations, like the horizontal axis of the profile, are measured along the centerline, not along an arbitrary axis. Generally, the horizontal alignment of a facility will first be established as a series of straight lines (or *tangents*). The initial stationing will be in terms of these tangents, but when curves are determined, the project will be restationed to reflect distances along the actual centerline. All distances, however, are measured as true horizontal distances; they do not represent distances along the surface of the facility where it is not level.

Elevations are in meters above some datum; usually this datum is mean sea level, but it may be any arbitrary scale. For instance, where projects are located below sea level, it is common to rescale elevations so that they will all be positive. Normally, profiles are drawn with an exaggerated vertical scale so that changes in elevation will be more obvious.

Grades (longitudinal slopes) are expressed as decimal fractions (meter/meter) or as percentages (meter/station).

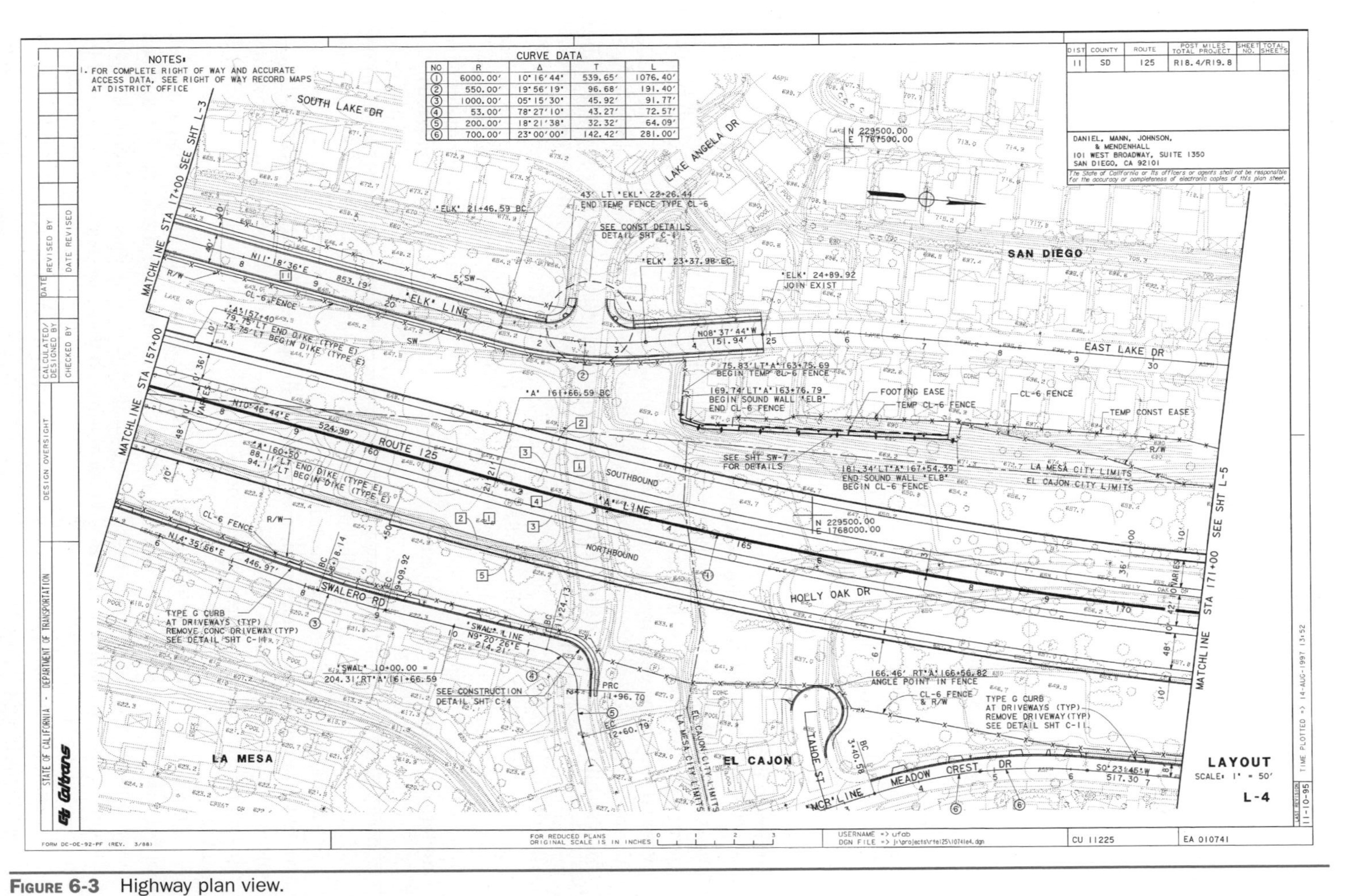

FIGURE 6-3 Highway plan view.

Source: California Department of Transportation and Daniel, Mann, Johnson, and Mendenhall, San Diego.

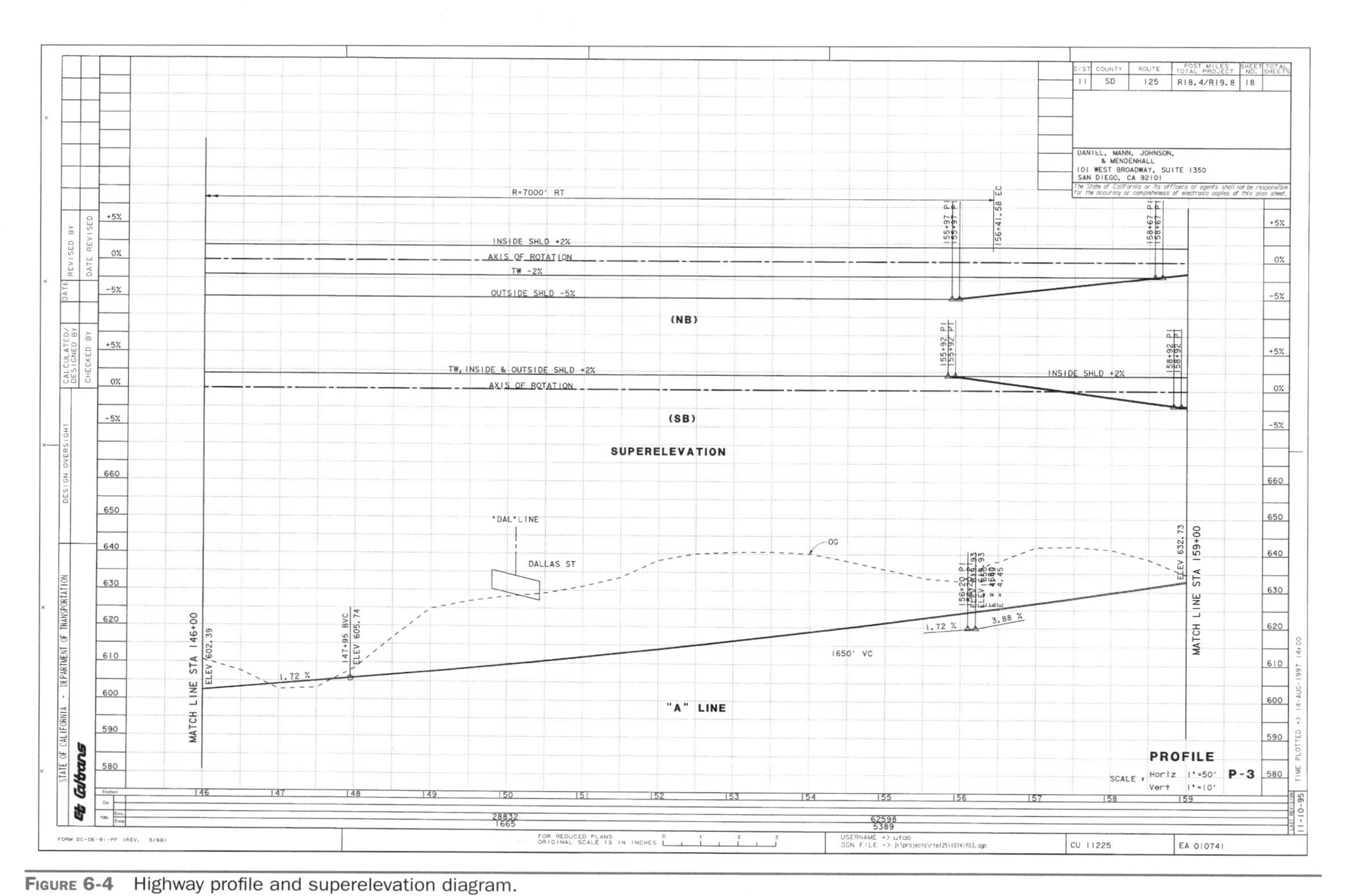

Figure 6-4 Highway profile and superelevation diagram.

Source: California Department of Transportation and Daniel, Mann, Johnson, and Mendenhall, San Diego.

Cross-slopes are normally specified in meter/meter or in percentages. In railway practice, cross-slopes may be expressed in millimeters (or inches) that the high rail is above the low rail, since there is a standard horizontal distance maintained between rails.

Specifications consist of *standard specifications,* which apply generally to all construction projects undertaken by a particular agency and *special provisions,* which apply to individual projects only. Most specifications applying to a particular project will be standard specifications, and will be incorporated in the contract documents by reference. In addition, specifications may be divided into *general clauses,* which deal with bidding procedures, award of the contract, execution of the work, scope of the project, control of work, and the like, and *other clauses* which deal with things like materials and the manner of executing the work, which may be more properly thought of as being part of the design of the project.

6-5 Summary

The design process for transportation facilities may be described in several ways. It involves a coordinated process of information gathering, analysis, and decision making, and is usually open-ended and iterative. Details of designs are based on design standards, which link the physical performance of the facility to various design elements. As an example, design standards for sight distances are derived from considerations of drivers' reaction times and brake reaction times. Major geometric design documents for linear transportation systems such as highways and railways include plan views, profiles, geometric cross sections, and superelevation diagrams.

References

AASHTO (1994). *A Policy on Geometric Design of Highways and Streets.* American Association of State Highway and Transportation Officials. Washington, DC.

AREA (1996). *Manual for Railway Engineering.* American Railway Engineering Association. Washington, DC.

Caltrans (1995). *Highway Design Manual,* 5th ed. California Department of Transportation. Sacramento, CA.

FAA (1989). *Airport Design: Advisory Circular,* 150:5300-5313. Federal Aviation Administration. Washington, DC.

Exercises

6-1. Determine the minimum stopping sight distance on a –2.5% grade at a design speed of 90 km/h.

6-2. Determine the minimum stopping sight distance on a +1.5% grade at a design speed of 100 km/h.

6-3. Determine the minimum stopping sight distance on a –4.0% grade at a design speed of 70 km/h.

6-4. Design standards link vehicle characteristics, human characteristics, and the characteristics of the transportation facility. What features of human and vehicle characteristics are important in the derivation of design standards?

6-5. List and briefly describe at least five transportation facility characteristics typically specified by design standards.

6-6. Four basic elements of facility plans document the geometry of linear transportation facilities such as highways and railways. List and briefly describe these four elements.

6-7. Use a spread sheet to construct a table of stopping sight distances for design speeds ranging from 30 to 120 km/h in increments of 10 km/h and grades ranging from 26% to + 6% in 2% increments.

UNIT 3

Passenger Transportation

CHAPTER 7

Overview of Passenger Transportation

7-1 Overview

This chapter introduces the unit of the book specifically focused on passenger transportation (as opposed to transportation in general), and covers overarching topics relevant to all types of passenger transportation, urban or intercity, public or private. The chapter first provides context for discussion of contemporary solutions by summarizing recent trends in passenger transportation activity, including concerns related to total transportation volume, congestion, and energy consumption. Next, the issue of congestion and measurement of roadway capacity as a function of vehicle speed and density is discussed, including the quantification of congestion impact of different types of disruptions. The chapter concludes with an overview of intelligent transportation systems (ITS), which have a pervasive role in improving all types of passenger transportation, including personal vehicles, public transportation, and intercity modes.

7-2 Introduction

In this unit, we focus on passenger transportation, which, in comparison to freight transportation, contributes the lion's share of vehicles to our road, highway, and airport infrastructure, and consumes the majority of transportation energy used. Urban and intercity passenger transportation are vital for a prosperous society, yet solutions to meet travelers' needs must be implemented in a way that also addresses ecological and social problems including pollution, climate change, congestion, and local nuisances.

Several important directions are highlighted for developing the passenger transportation systems of the future in this and subsequent chapters:

- *Passenger transportation and information technology:* (Chap. 7) We lay a foundation for understanding the modern relationship between transportation infrastructure and information technology infrastructure in ITS, including systems that monitor transportation system movement, relay information to first responders in emergency situations, and control system-wide operation. Much of the impetus for developing ITS is to improve transportation efficiency and reduce congestion, so part of this chapter is dedicated to explaining why congestion happens and measuring its impact.

- *Urban public transportation and multimodal solutions:* (Chap. 8) We cover the operation of public transportation by explaining planning and operating activities at the core of the work that transit agencies do, namely to determine what routes to serve, with what vehicles, and how frequently. We also discuss techniques and practices useful for improving system performance.
- *Personal mobility and accessibility options:* (Chap. 9) We cover options for improving "personal mobility," which is an umbrella term used to describe any type of transportation where the passenger travels independent of a public transportation operator, and includes walking, bicycling, driving in a private car, or driving in a shared car ("carsharing"). Topics covered include efforts to enhance nonmotorized modes, notably bicycling and walking; integrating access to carsharing as part of a strategy for personal mobility that encompasses nonmotorized travel, public transport, and carsharing travel; and substituting "accessibility" for mobility by encouraging telecommuting, where the need for amenities is met without requiring physical travel of the passenger.
- *Sustainable intercity transportation:* (Chap. 10) We focus on commercial intercity options, especially high-speed rail and aviation, to look at options for reducing ecological impact and potentially providing a less environmentally harmful alternative to passenger travel between cities in private cars. Conventional rail and bus modes are also considered to a lesser extent.

7-3 Recent Developments in Passenger Transportation

The development of modern passenger transportation begins with a classical, technology-based vision that established itself after World War II, first in North America and gradually in other countries. According to this vision, the transportation system would be built around private automobiles, which would be given wide access to workplaces, schools, shopping, and other amenities through the construction of a modern urban roadway infrastructure: expressways, arterials, and feeder streets. By providing enough lane-miles of roads and modern control systems (limited-access interchanges, coordinated traffic signals, etc.), a large fraction of passenger transportation needs in cities could be met with sufficient vehicles and roads.

Not all transportation needs could be met this way, so the road-and-car system would be complemented with a range of mechanized public transportation systems. Some travelers such as children and youth or the elderly, could not operate vehicles, so the road networks could be used by public buses as well as the yellow school buses that became ubiquitous in the latter half of the twentieth century in the United States. Also, for very concentrated travel routes, such as travel in and out of major cities like New York or Chicago, it would not be practical to replace highly space-efficient *heavy-rail systems* (an umbrella term for subways, elevated trains, commuter trains, and the like) with space-consuming road networks, so these systems would still play a role.

7-3-1 Limitations of the Automobile-Focused Urban Passenger Transportation System

Thus for a few decades transportation planning revolved around basically three types of conveyances: cars, buses, and heavy rail systems. In time, however, it became clear that the automobile-focused approach was not able to meet the demand for transportation with acceptable levels of congestion and local air quality impact for two reasons. First, in many cities private automobiles came to so dominate the share of passenger trips, and public transportation (including both buses and heavy rail) became so marginalized, that the latter could no longer contribute to the success of urban transportation in a meaningful way. Not only was service inadequate in terms of frequency and area coverage so that it became unappealing compared to driving a car, but the lack of ridership then meant that transit operators in many cities could not generate the revenue from passenger fares to win back the lost ridership.

The second reason was the inability of the road infrastructure to keep pace with burgeoning automobile usage. The flexibility of the automobile proved extremely popular, to allow middle-class individuals and families to live further out from the city center yet still have access to jobs, education, shopping, and other amenities. Despite ambitious road-building programs, road networks eventually became saturated with traffic and as time went by it became more and more difficult to find space to build more roads.

Figure 7-1 uses a causal loop diagram (CLD) to illustrate the impact that a growing urban highway network can have on public transit. The growing network initially creates two problems. First, it makes driving in a car relatively attractive. Second, it makes longer trips in particular more attractive, trips that in fact might not have been practical before new roads were opened, and these types of trips are especially not well served by transit. This phenomenon of unanticipated trips is known as *latent demand*, which

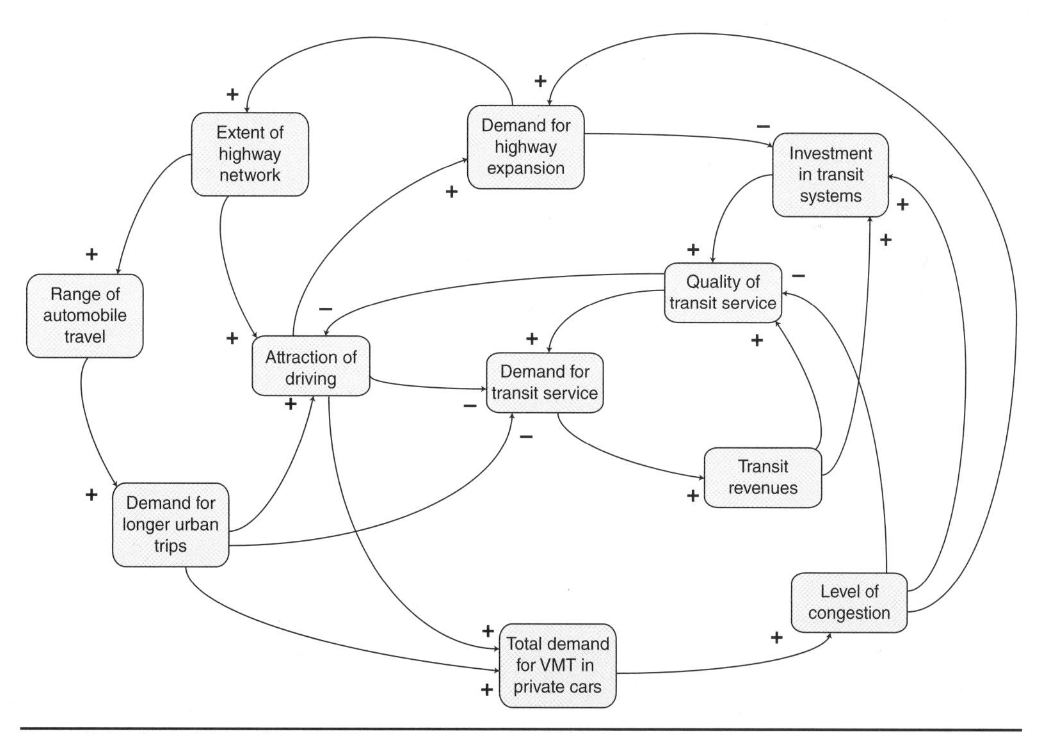

FIGURE 7-1 Causal loop diagram showing the relationship between urban highway expansion and public transit investment and performance.

Source: Adapted from Vuchic (1999, p. 8).

entails driving patterns that are not anticipated prior to completing a road project but that come into being when drivers realize that new destinations are now available to them. In any case, the net effect of the growing popularity of driving in general and long urban trips in particular is a vicious circle of underinvestment, declining service quality, and declining demand for transit.

Not all cause-and-effect relationships in the CLD are bad for transit. Starting with the "Level of Congestion" bubble, congestion can undercut transit service quality, but can also spur investment in new transit systems, which can in turn improve transit quality and reduce the relative attraction of driving in a private car. The result is a balancing loop within the CLD. Generally, though, the outcome for urban passenger transportation in the latter half of the twentieth century in the United States, as well as in a number of other industrialized countries, has been modal share concentration in private cars and decline for transit. It is only recently in some cities that, as available space for new roads has been exhausted and congestion on major roadways has remained endemic, these loops in the CLD have been saturated, leaving investment in transit as a more viable policy response to continued high congestion.

7-3-2 Emergence of an Expanded Approach to Urban Passenger Transportation

In response to the above limitations, an expanded approach to urban passenger transportation problems that seeks to reduce congestion and improve livability came into being. The new approach comprises a broad array of options, some building on the use of private automobiles and public transportation, and some consisting of entirely new directions:

- *Infrastructure—use information technology in the form of intelligent transportation systems (ITS) to manage highway, rail, and other components more effectively:* Rather than building ever more transportation infrastructure, ITS allows transportation system decision-makers and stakeholders to utilize existing systems more efficiently, at a fraction of the cost. From their beginnings as changeable roadside information boards in the 1960s, these systems have evolved into comprehensive IT networks that monitor transportation performance, take action to address problems, and inform the traveling public.
- *Public transit—diversify the offering of modes to be competitive with the private car in more markets:* The light-rail transit (LRT) system, with not as high a level of service as heavy rail, but at lower cost, and later the bus rapid transit (BRT) system came into being. Metropolitan areas faced with ongoing chronic congestion and dwindling space for new freeways have increasingly, in recent years, turned to the expanded menu of rapid transit options (LRT and BRT as well as heavy rail) as a solution. Examples in the United States include Houston and Denver, cities known for auto-dependency and extensive freeway networks that have recently stated their intention to expand LRT systems into comprehensive networks because continued reliance on private cars and freeways was becoming untenable. In addition, transit operators introduced new services such as circulator buses, minibuses for lightly traveled routes, and bus routes with route diversion that allow slight deviations from main routes to pick up and drop off passengers while holding to a fixed schedule, such as that of the Potomac and Rappahannock Agency in the suburbs of Washington, D.C.

- *Nonmotorized transportation (NMT)—rediscovering an overlooked mode:* For a time, if bicycling or walking were considered in urban planning at all, it was solely for recreational purposes, and not as a point-to-point means of conveyance. Therefore these modes were largely ignored in urban transportation planning, leading to roads built without sidewalks and intersections without safe crosswalks. Eventually, NMT reemerged as a viable alternative to other options, with many advantages: it is relatively inexpensive to build the infrastructure, it contributes to users' need for physical exercise, and it provides access to other modes such as transit.
- *Carsharing—adapting the car rental business to serve local needs:* Carsharing can be considered an adaptation of car rental systems with several changes, including: (1) vehicles are distributed around the community rather than at a central business, so that they are close to members' homes, (2) the sign-out procedure is highly streamlined to make it as easy as possible to use and return a vehicle, and (3) vehicles are rented by short time increments, so that several people might use a vehicle in a single day. As members adapt to using carsharing, they may find that they can discontinue owning a personal vehicle. Also, differences in the cost structure for the carshare participant compared to owning a private vehicle encourage modal choices other than driving.
- *Telecommuting—sometimes the best transportation option is no transportation:* Unlike other approaches which optimize the transportation network or offer alternative modes, telecommuting eliminates passenger travel entirely by allowing would-be travelers to meet their needs from home or another location near home. Local governments can work with employers to provide financial incentives or know-how for creating a home work space that provides a work experience equivalent to traveling to the office. Home shopping and distance learning play a complementary role for eliminating the need to travel to retail outlets and brick-and-mortar schools, respectively.
- *Land use planning—creating an urban form that encourages sustainable transportation habits:* Land use planning, the activity carried out by urban planners and other public servants to guide the development and maintenance of the urban built environment, can be directed toward patterns that encourage efficient transportation and discourage wasteful habits. For example, *transit-oriented development* (TOD) builds public transportation routes so as to encourage the growth of employment, retailing, and residences around transit stops, such as stations on a subway or light rail system.
- *Internalizing the full cost of transportation through techniques such as congestion pricing:* One of the major flaws of the transportation system as it currently functions is that many types of passenger services, such as driving in a private car on urban freeways or arterial streets, are underpriced, and travelers often do not see the full cost of their choices when they take a trip. Internalizing the full cost of trips gives the traveler realistic information about the true impact of their trip about problems such as congestion or pollution, and helps the system to function more efficiently. For example, congestion pricing, in use in cities such as Singapore and London, encourages alternative modal choices or time of day choices by charging a fee for driving a private car into the most congested parts of a city during business hours.

In the midst of all of these new options, the conventional options of private cars, heavy rail, and buses continue to evolve. Vehicle technology is continuously being updated to make it more city-friendly (such as the advent of micro-cars that take up less parking space), and in some cities new roadways can be added, or existing roadways widened, in a judicious way so as to smooth the flow of vehicles without leading to increased congestion. Bus and heavy rail systems can be added, where they are the right technology to serve a given demand. For example, in China, the cities of Beijing and Shanghai have rapidly added underground heavy rail systems along corridors with large numbers of riders that could not be served effectively by any other mode.

Even with all the tools, old and new, at our disposal, the problems of urban passenger transportation continue to be daunting. Congestion, petroleum dependence, and greenhouse gas (GHG) emissions remain perennial problems. But just as the problems are multifaceted, there is plenty of room at the table for the whole range of alternatives mentioned above, as well as new options that are only now emerging.

7-3-3 Contemporary Challenges from Travel Intensity and Energy Consumption

One of the most significant contemporary challenges for passenger transportation is the increasingly transportation-intensive nature of our society. Figure 7-2 compares light-duty vehicle miles traveled (VMT) and registrations (i.e., number of vehicles actively on the road) to growth in population. By the peak year in Fig. 7-2 (2006), population had risen to 299 million, an increase of 47%. VMT and registrations, by contrast, had risen much faster: VMT by 127% to 2.8 trillion and registrations by 150% to 234 million. Thus both the number of vehicles on the road and the number of vehicle-miles driven per year generated by each member of the population increased sharply. Between 2006 and 2010, both VMT and registrations declined, likely due to a combination of the effects of the global economic slowdown and perhaps as a correction since VMT and registrations may have excessively expanded up to 2006. Nevertheless the total growth between 1970 and 2010 remains very large compared to population growth.

Taking a snapshot of the modal distribution of passenger miles in a single recent year further shows how the light-duty vehicle mode along with air travel dominates the passenger transportation market (Fig. 7-3). The combination in 2010 of passenger cars and light trucks represents approximately 80% of the total market; if air travel is included, the total increases to 92% of the total market. Some modes have very small shares: all types of rail, including urban transit and intercity rail, constitute less than 1% in both 2007 and 2010. Rail's share of total passenger-miles over those three years actually increased even as total passenger-miles across all modes declined, so the change represents a 2.2% increase in the absolute number of passenger-miles; nevertheless, this growth would need to be sustained over many years for rail to recapture a significant modal share. Note also that for light-duty vehicles, the overall figures for passenger-miles are larger than those for vehicle-miles, because a significant fraction of vehicle-miles are driven with more than one passenger in the car. For example, for the year 2010, 3.6 trillion passenger-miles were observed for 2.6 trillion VMT, for a ratio of 1.38 passenger-miles per VMT. This ratio indicates that the most common condition for vehicles was that they were single occupant vehicles (SOVs), but that a measurable fraction of VMT were traveled by vehicles that were not.

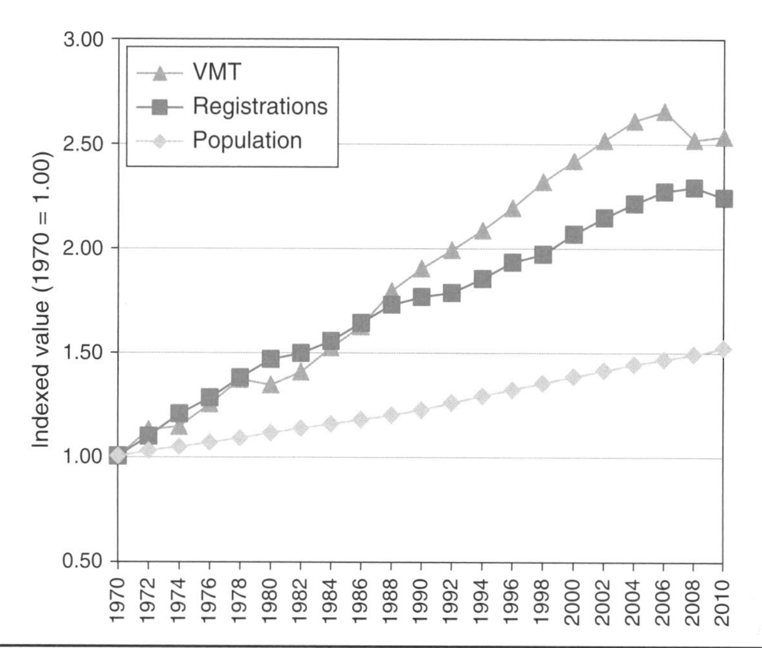

FIGURE 7-2 Indexed growth of U.S. light-duty VMT, vehicle registrations, and population 1970 to 2010, indexed to 1970 = 1.00.

Note: In 1970, VMT was 1,043 billion, registrations were 103 million, and population was 204 million. Light-duty vehicles in the figure include all passenger cars and light trucks.

Source: Oak Ridge National Laboratories (2011), for transportation data; U.S. Bureau of the Census, for population data.

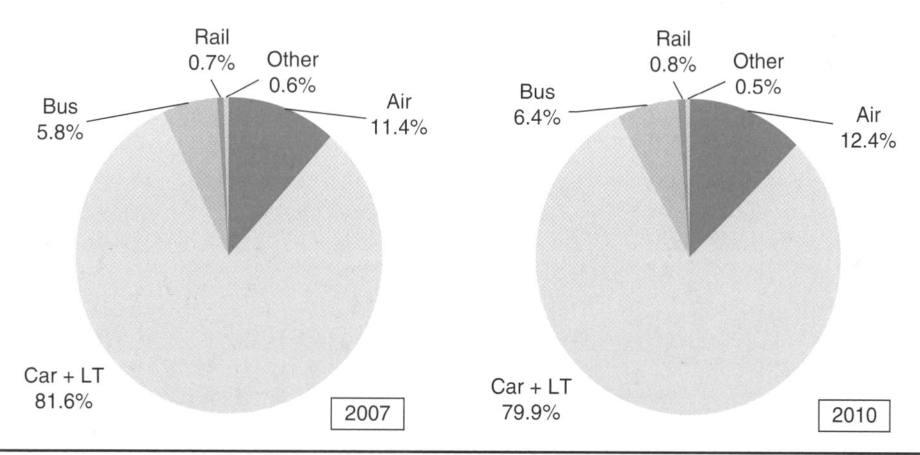

FIGURE 7-3 2007 and 2010 U.S. modal share of combined urban and intercity passenger miles.

Note: Total value across all modes = 5,322 billion passenger-miles in 2007 and 4,561 billion passenger-miles in 2010. "LT" = light truck. "Other" includes ferry, taxi, motorcycle, and other niche modes.

Source: U.S. Dept. of Transportation.

Comparison of U.S. Situation and Peer Countries

An examination of peer-industrialized countries shows that, like the United States, passenger cars in other countries have the largest modal share of passenger-miles, while at the same time the relative proportion of modal share for other modes can be quite different. Using the United Kingdom, Germany, and Japan as examples, the light-duty vehicle share of passenger-miles in 2010 was 85%, 84%, and 58%, respectively. Rail modal shares were much higher than those of the United States; however, the figures including intercity rail and all types of urban rail systems were 8.5%, 9.4%, and 30%, respectively. Note also that the total number of passenger-miles per capita is much lower in the selected countries: 7,756, 7,954, and 6,370 passenger-miles, respectively. This difference is partly due to geography, as the United Kingdom, Germany, and Japan are more compact and more densely populated countries. However, it is also partly due to transportation policy, as these countries have limited the growth of transportation demand by creating an environment where transportation costs more and infrastructure capacity is not as abundant.

Modal Contribution to Passenger Transportation Energy Consumption

Returning to the example of the United States, rapid growth in demand for passenger-miles and the dominance of an energy-consumptive mode such as the use of light-duty vehicles becomes apparent when looking at resulting trends such as annual energy consumption by mode. Figure 7-4 gives the growth in energy consumption from 1970 to 2008,

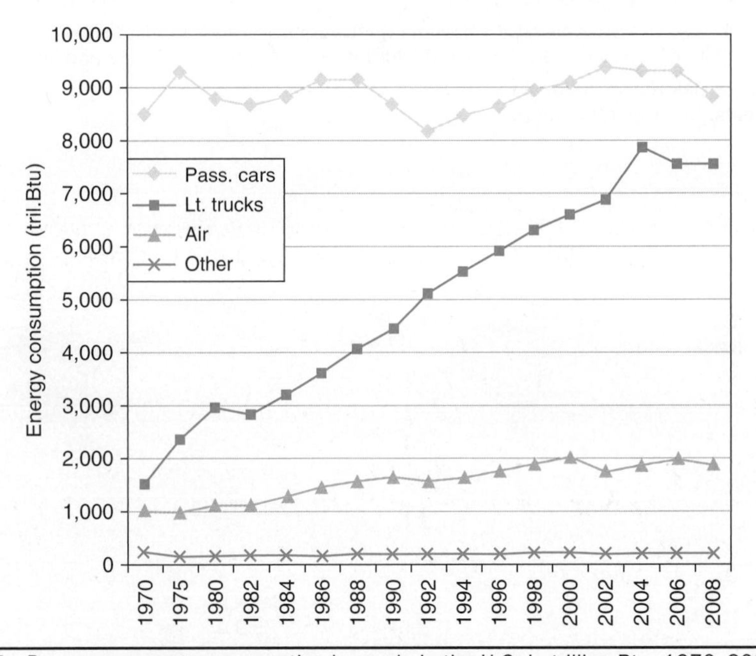

FIGURE 7-4 Passenger energy consumption by mode in the U.S. in trillion Btu, 1970–2008.

Note: "Other" includes intercity bus and rail, urban transit, ferry boats, motorcycles, and any other motorized mode not included. "Light truck" category includes all light truck movements, including those that were actually used to carry freight; no attempt has been made to isolate passenger-related light truck activity from freight-related. Note change of scale from 1970–80 to 1980–2008 to make recent trends more visible.

Source: Oak Ridge National Laboratories.

during which time the main driver of energy growth was light-duty trucks as opposed to passenger cars. First, as shown in Fig. 7-1, the annual VMT per capita of population increased substantially. For passenger-cars, improvements in fuel economy and hence reduction in energy consumed per mile offset VMT growth, so that total energy consumed by this mode hovered around 9 quads (i.e., 9,000 trillion Btu, as shown in Fig. 7-4) for the duration of this time period. On the other hand, light trucks [including sport utility vehicles (SUVs), minivans, and pickups] gained a much larger share of the overall light-duty vehicle market and experienced only a modest improvement in fuel economy, so that energy use by this mode increased approximately fivefold from 1,500 to 7,500 trillion Btu.

The remaining modes, including air and all other, constitute only about 10% of all energy consumed in 2008. Air passenger-miles in the United States grew more than any other in the time period shown in the figure, but total energy consumption did not grow as rapidly as, for example, that of light trucks, because the aviation industry took aggressive steps to improve engines and aircraft. The remaining modes (such as urban bus and rail) had relatively high percentage growth in demand, but because they remain a small fragment of total passenger-miles, their total energy consumption of approximately 250 billion Btu remains small in the context of the figure.

The large volume of U.S. passenger transportation demand partially explains the large total energy requirement represented in Fig. 7-4 of approximately 20 quads, or 65 million Btu per person. Another part of energy consumption is explained by the impact of traffic congestion on vehicles and aircraft in terms of requiring more fuel consumed to go a certain distance. Because of its impact on energy use and also on the quality of urban life, congestion is the topic of the following section.

7-4 Road Capacity and Roadway Congestion

The economic, social, and ecological consequences of roadway congestion on urban expressways and arterial streets pose a powerful motivation for improving the function of the transportation system. In 2011, according to the Texas Transportation Institute (TTI, 2012), the average U.S. urban automobile commuter living in a large metropolis lost 52 hours per year due to congestion, with the highest losses occurring in Washington, D.C. (67 hours/year), followed by Los Angeles and San Francisco (both at 61 hours/year). The total cost of congestion was estimated at $121 billion, between 5.5 billion hours wasted and 2.9 billion gallons of extra fuel consumed as vehicles sat in congested conditions. In addition, approximately 112 million tons of extra CO_2 was emitted as a result of congestion. Thus there is strong motivation to understand how congestion occurs and quantify its impact based on traffic conditions in the moment. Note that although congestion is introduced here in the context of passenger transportation, freight transportation (the subject of Chaps. 11–13) both contributes to and is affected by congestion, so many of the concepts discussed here apply to freight as well.

7-4-1 Relationship between Speed, Density, and Flow

To understand how congestion occurs, one must first study the relationship between the *velocity* at which vehicles travel, the *density* of vehicles per mile of street or highway lane, and the *flow* of vehicles per hour per lane. We initially assume steady-state conditions (continuous flow with no transition from off-peak to peak periods, interruptions by traffic signals or stop-and-go conditions, etc.) although a more complete and realistic analysis must consider these factors as well.

The density, k, of traffic flow is determined by the length of each individual vehicle plus its required *shadow*, or the space between the vehicle and the vehicle traveling in front of it. In congested conditions, drivers tend to decrease the size of the shadow by traveling closer to the vehicle in front. We can further call L_{veh} the length of the vehicle proper, L_{shadow} the length of the shadow, and denote L as the total length occupied by the vehicle. Since $L = L_{veh} + L_{shadow}$, using standard units and the conversion of 5,280 ft/mile, we can write the equation for $k$in units of vehicles per mile as

$$k = \frac{5280 \text{ ft/mile}}{L} = \frac{5280 \text{ ft/mile}}{L_{veh} + L_{shadow}} \tag{7-1}$$

The reader can verify that, as an example, vehicles with an average length of 16 ft and shadow length of 20 ft occupy a space of L = 32 ft and therefore a density of ~147 vehicles/mile.

Eq. (7-1) can be rewritten in metric units using the relationship of 1,000 m per kilometer:

$$k = \frac{1000 \text{ m/km}}{L} = \frac{1000 \text{ m/km}}{L_{veh} + L_{shadow}} \tag{7-1a}$$

To give an example in metric units, suppose we are given average values of $L_{veh} = 4.8$ m and $L_{shadow} = 6.1$ m; then applying Eq. (7-1a) gives $L = 10.9$ m and $k = 91.7$ vehicles/km.

We now assign variable u to speed and introduce variable q for flow, measured in vehicles per unit of time. Since velocity is measured in units of distance per unit of time, and density in vehicles per unit of distance, it holds that flow is the product of velocity and density:

$$q = uk \tag{7-2}$$

Because of its importance in understanding flow and congestion, this relationship is called the *fundamental equation of traffic flow.* As a further exploration of how u and k affect q, consider three possible conditions for flow on a roadway:

- *Free flow conditions:* Vehicles travel at the posted speed limit (and given typical traveler behavior, often at speeds above the posted limit) and at very low densities (very large shadow values relative to the average length of vehicles). Low density tends to outweigh high speed, so the value of q is relatively low.
- *Maximum flow conditions:* As the roadway becomes more congested, drivers respond by slowing down but also reducing shadow distance, which has the net effect of increasing q up to some maximum value q_{max}.
- *Severely congested conditions:* As the level of congestion continues to deteriorate past the maximum flow conditions, there are diminishing returns to reducing speed and shortening shadow distance, such that with increasing k the value of q declines from q_{max} to some lower value.

Example 7-1 illustrates some possible combinations of speed, shadow space, density, and flow.

Example 7-1 Suppose we choose three representative values of u, namely 65 mph (free flow), 35 mph (maximum flow), and 10 mph (severe congestion). Suppose furthermore that the average vehicle length is 16 ft, and that the density values for the three conditions are $k = 10$, $k = 65$, and $k = 112$ vehicles/mile, respectively. Create a table of k, L, L_{shadow}, u, and q values for each of the three conditions.

Solution As an illustration, we first calculate the missing values in the free-flow case. Since we are given $k = 10$ vehicles/mile and $v = 65$ mph, the remaining values L and q are:

$$L = \frac{5280 \text{ ft/mile}}{10 \text{ vehicles/mile}} = 528 \text{ ft}$$

$$L_{\text{shadow}} = 528 - 16 = 512 \text{ ft}$$

$$q = uk = (65)(10) = 650 \text{ vehicles/hour}$$

Repeating for the other two conditions results in the following table:

Condition	u (mph)	L (ft)	L_{shadow} (ft)	k (vehicles/mile)	Q (vehicles/hour)
Free flow	65	528	512	10	650
Maximum flow	35	81.2	65.2	65	2,275
Severe congestion	10	47.1	31.1	112	1,120

The critical insight from Example 7-1 is that as the roadway situation deteriorates from maximum flow to severe congestion conditions, drivers are able to travel closer together to increase k, but this does not make up for the reduction in u, so that q declines substantially. Hence the problem with congestion at peak periods (and increasingly in off-peak periods in cities such as Los Angeles, San Francisco, or Washington, D.C.): as roadways accumulate vehicles, and without adequate public transit, telecommuting, or nonmotorized options for many travelers, there is no option except to suffer through the extra time required to travel between points in the network, and the losses in productive time and fuel consumption cited above result.

Cities in other parts of the world suffer from similar levels of congestion loss. Some of the worst losses are in the major cities of industrializing countries such as China or India, where "motorization," or the increase in the number of light-duty vehicles per capita, is occurring rapidly. The need to provide information about delays to the traveling public, to suggest alternate routes if it is productive to do so, and to respond quickly and effectively to accidents and other incidents, all point to the use of ITS as a tool to combat congestion, as discussed later in Sec. 7-5.

7-4-2 Greenshields Model of Speed-Density Relationship

The causes and conditions that create either free flow or congestion can be further understood by exploring the mathematical relationship between u, k, and q. There are several possible functional forms for this relationship, either linear or nonlinear; one common linear model that is useful for a basic understanding is the Greenshields model, as shown in Fig. 7-5.

The starting point for the Greenshields model is the curve for speed as a function of density, also known as a "speed-density curve" (the lower right curve in the figure). At negligible values of k, traffic travels at the *free flow speed* u_f. Speed then declines linearly with increasing k from a point ($k = 0$, $u = u_f$) to the maximum possible value of k where $u = 0$. The value of density at this point is called the *jam density* k_j. In reality, both conditions $k = 0$ and $u = 0$ are lower bounds that cannot be attained in the real world: for the value of u to be meaningful, k must be nonzero, and even in the most congested conditions, $u = 0$ cannot be maintained indefinitely. Thus both endpoints of the Greenshields speed-density curve are bounding conditions.

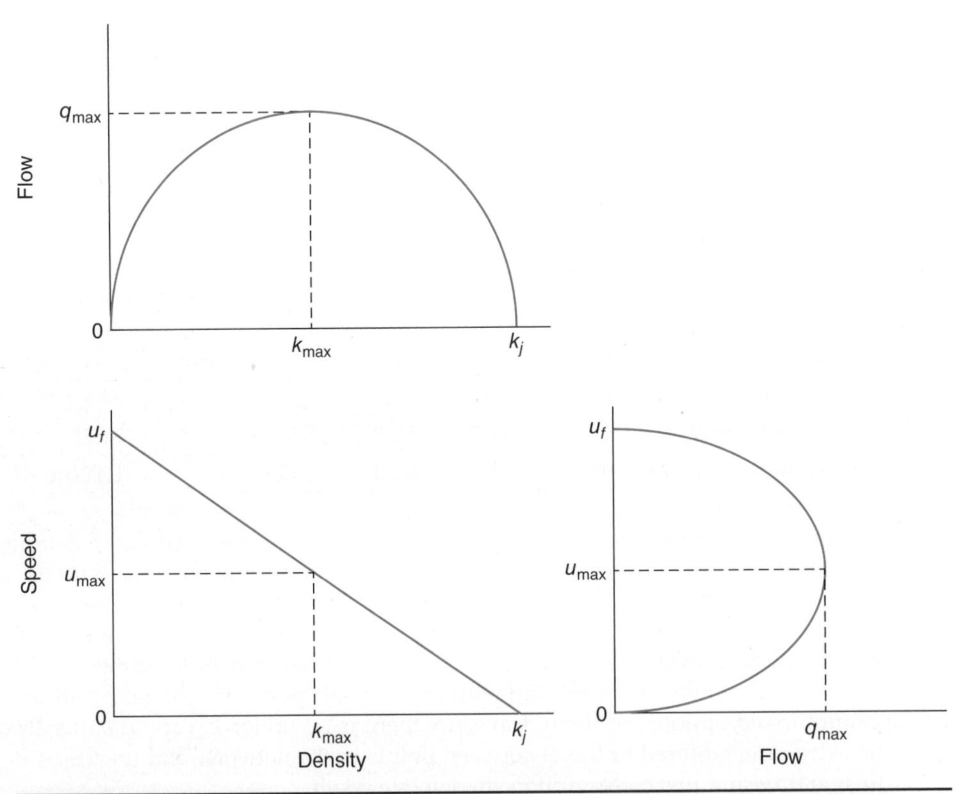

Figure 7-5 Greenshields model of speed-density, flow-density, and speed-flow relationships.

Values of k and u at specific points on the speed-density curve for known k_j (jam density) and u_f (free flow speed) can be calculated as follows:

$$u = u_f - \left(\frac{u_f}{k_j}\right)k \tag{7-3}$$

In some instances, it is useful to rearrange Eq. (7-3) to give k as a function of u, that is:

$$k = k_j - \left(\frac{k_j}{u_f}\right)u \tag{7-4}$$

Next, we consider flow as a function of density and speed using the Greenshields model. Equations. (7-3) and (7-4) can be substituted into Eq. (7-2) to derive q in terms of either k or u, respectively:

$$q = uk = u_f k - \left(\frac{u_f}{k_j}\right)k^2 \tag{7-5}$$

$$q = uk = uk_j - \left(\frac{k_j}{u_f}\right)u^2 \tag{7-6}$$

Eq. (7-5) is plotted in the upper set of axes in Fig. 7-5 (q as a function of k), while Eq. (7-6) is plotted in the set of axes on the right (q as a function of u).

As a function of either u or k, the curve for q has a parabolic shape, given the quadratic nature of Eqs. (7-5) and (7-6). In general, the value of k or u that maximizes q, or achieves $q = q_{max}$, can be found by differentiating and setting equal to zero, for any linear or nonlinear function that calculates q_{in} terms of u or k. In the particular case of the Greenshields model, the maximum value is achieved at the point where u or k is exactly half of the free-flow speed or jam density, respectively. Thus adopting the symbol $k = k_{max}$ for the value of k and $u = u_{max}$ for the value of u at q_{max}, we have

$$k_{max} = \frac{k_j}{2} \tag{7-7}$$

$$u_{max} = \frac{u_f}{2} \tag{7-8}$$

Note that the terms k_{max} and u_{max} should not be misinterpreted as the maximum possible values of k and u possible with the Greenshields model, which are in fact k_j and u_f, respectively.

Lastly, it follows from Eqs. (7-7) and (7-8) that the value of q_{max} can be calculated from Eq. (7-2):

$$q_{max} = u_{max}k_{max} = \left(\frac{u_f}{2}\right)\left(\frac{k_j}{2}\right) = \frac{u_f k_j}{4} \tag{7-9}$$

In Example 7-2, the value of q_{max} is calculated both using calculus and using Eq. (7-9) to show that they arrive at the same answer.

Example 7-2 Suppose for a particular highway, analysts have observed traffic conditions extensively at different times of day and found that k is related to u according to $u(k) = 70 - 0.538k$, where k is given in vehicles/mile/lane. Furthermore, the value of time for the particular region in question is \$12/hour. (a) What is the value of u_{max}, k_{max}, and q_{max}? (b) If a freeway with three lanes in the inbound direction has vehicles traveling under these flow-maximizing conditions, how many vehicles can travel over it in 1 hour? (c) If velocity is reduced to $u = 10$ mph, what is the new value of k and q? (d) Suppose the corridor over which the vehicles travel is the exact length such that it takes 1 hour to travel from end to end at u_{max}. What is the total value of lost time for the number of vehicles that previously traveled the corridor in 1 hour in part (b)?

Solution

(a) Calculus approach. Using the relationship $q = uk$, we can substitute for u, giving the equation:

$$q = (70 - 0.538k)k = 70k - 0.538k^2$$

To find q_{max}, we differentiate and set equal to zero to find the value that maximizes q:

$$\frac{d}{dk}q(k) = 70 - 1.076k$$

$$70 - 1.076k = 0$$

$$k_{max} = \frac{70}{1.076} = 65 \text{ vehicles/mile}$$

Substituting gives

$$u_{max} = 70 - 0.538k_{max} = 70 - 0.538(65) = 35 \text{ mph}$$

$$q_{max} = u_{max}k_{max} = (35)(65) = 2{,}275 \text{ vehicles/lane/hour}$$

Using the formuli in Eqs. (7-7) to (7-9) related to the Greenshields model gives the same outcome. From the relationship $u(k) = 70 - 0.538k$, it can be inferred that $u_f = 70$ mph and $k_j = 130$ veh/mi. This can be confirmed as follows:

$$\frac{u_f}{k_j} = \frac{70}{130} = 0.538$$

With known u_f and k_j, q_{max} is calculated as

$$q_{max} = \frac{u_f k_j}{4} = \frac{(70)\,(130)}{4} = \frac{9100}{4} = 2275 \text{ vehicles/hour}$$

(b) In 1 hour, with three lanes, the number that can travel is 2,275(3) = 6,825 vehicles/hour.

(c) First, the Greenshields model relationship should be rearranged to find k as a function of u:

$$u = 70 - 0.538k$$

$$k = k_j - \left(\frac{k_j}{u_f}\right)u = 130 - \left(\frac{130}{70}\right)u = 130 - 1.857u$$

Given $u = 10$ mph, k and q can now be calculated:

$$k = 130 - 1.857(10) = 111 \text{ vehicles/mile}$$

$$q = uk = (10)\,(111) = 1{,}110 \text{ vehicles/lane/hour}$$

(d) Since the vehicles in part (a) are traveling at 35 mph and they take 1 hour to traverse the corridor, the corridor length must be exactly 35 miles. The number of vehicles per hour was previously calculated as 6,825. The time required at $v = 10$ mph is then

$$35 \text{ miles} \left(\frac{1}{10 \text{ mph}}\right) = 3.5 \text{ hours} = 210 \text{ minutes}$$

Since the extra time required is 210 – 60 = 150 minutes = 2.5 hours, the total value of time lost is

$$(6{,}825)(2.5 \text{ hours}) \left(\frac{\$12}{\text{hour}}\right) = \$204{,}750$$

The situation in Example 7-2 represents an extreme situation in which a large number of vehicles are reduced to moving very slowly (10 mph) over a long distance, perhaps due to a severe incident, such as a major freeway accident that closes several lanes to traffic. Everyday *volume delays* (i.e., those delays routinely caused by too many vehicles trying to pass through a roadway with inadequate capacity) cause fewer lost minutes per vehicle per event, but because they are endemic, annual totals add up to millions of lost hours for major metropolitan areas in the United States.

Comparison of Greenshields Model to Observed Traffic Flow Values

Figure 7-6 shows modeled and empirical values for an actual roadway under both dry and rainy conditions, in metric rather than standard units. Although the values do not

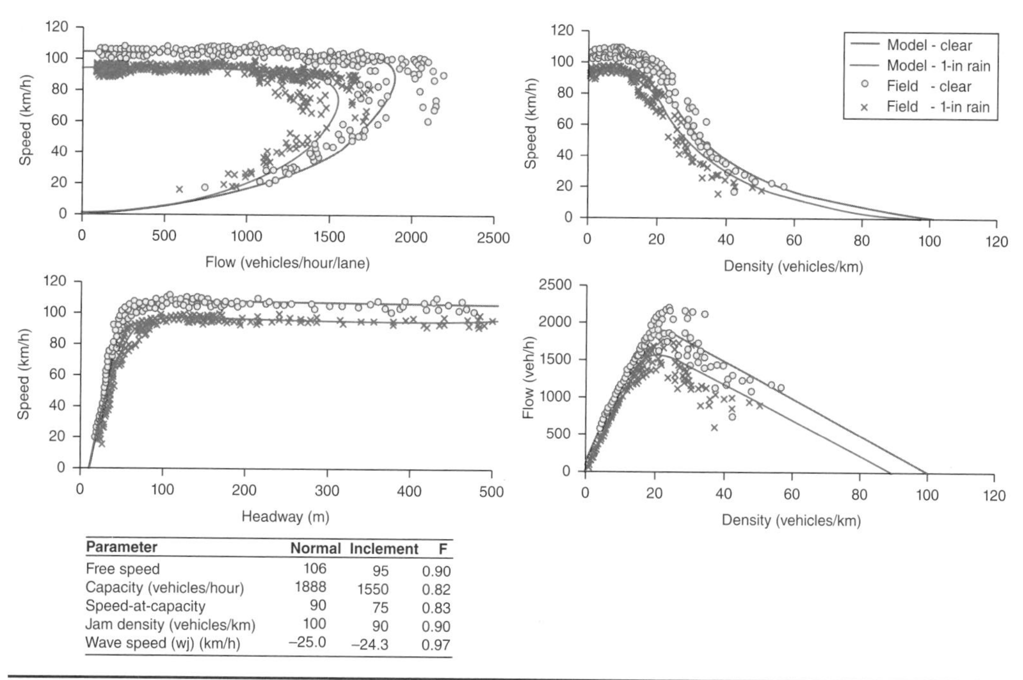

Parameter	Normal	Inclement	F
Free speed	106	95	0.90
Capacity (vehicles/hour)	1888	1550	0.82
Speed-at-capacity	90	75	0.83
Jam density (vehicles/km)	100	90	0.90
Wave speed (wj) (km/h)	−25.0	−24.3	0.97

Figure 7-6 Comparison of observed and modeled velocity, density, and flow values for dry and rainy conditions on an actual roadway.

Source: Bureau of Transportation Statistics (2011), public domain image.

have exactly the linear or semi-circular shapes observed in Fig. 7-5, they do show that the conceptual model is a reasonable first-order approximation of real-world conditions. For example, the figure on the upper right showing speed as a function of density has values on the order of 100 km/hour (= 62 mph) for density of 10 vehicles/km (= 16 vehicles/mile), and 20 km/hour for density values on the order of 60 vehicles/km. These values are similar to the results found in Example 7-2. The figure representing speed as a function of flow on the upper left is similar to the parabolic shape in the graph on the lower right in Fig. 7-5, and the figure representing flow as a function of density is similar to the upper graph in Fig. 7-5, although Fig. 7-6 appears to be more piecewise linear. Only the graph on the lower left (speed as a function of headway) has no equivalent in Fig. 7-5.

At the same time, Fig. 7-6 also suggests how in a more complex functional modeling of the speed-density-flow relationship, higher-order functions might be used to better capture the effect of one variable on another. For example, one limitation of the Greenshields model is that it assumes that as density begins to increase from $k = 0$, speed decreases, but in practice, the empirical speed-density curve in Fig. 7-6 shows that for the range $0 < k < 20$ vehicles/km, there is no appreciable decrease in observed speed. Therefore, a more plausible alternative is a piecewise linear function with constant u as a function of k over this range ($0 < k < 20$), and then linearly decreasing u for higher values of k up to k_j, where speed reaches zero. The relationship between flow and density or speed could also be refined with a higher-order function or piecewise function.

7-4-3 Effect of Slow-Moving Vehicles on Traffic Flow

The fundamental diagrams for basic traffic flow in the preceding section make a series of simplifying assumptions—vehicles have similar characteristics, drivers have similar desired speeds and car-following behavior, the physical characteristics of the roadway are uniform along the section under analysis, and so on. While these assumptions are not all strictly true in most cases, they are useful in developing practical models of traffic flow on the macro-scale.

However, drivers all experience conditions under which traffic is not flowing smoothly at a constant average speed. "Stop-and-go" traffic in the vicinity of major merge points along freeways, "back-ups" upstream from construction zones, and the "tail" following a slow-moving vehicle on a two-lane roadway with limited passing areas (such as many rural highways in the U.S.) are all illustrations of disturbances in the flow of traffic. When these disturbances occur, there are regions of rapid change in speed, density, and flow volume along the roadway. The locations of these rapid changes move over time, and this movement represents a "shock wave" through the traffic stream, much like waves moving through fluids.

Understanding how shock waves are generated, the speed with which they move, and other characteristics, is important to understanding the effects of disruptions and the delays that are caused. This understanding also creates the basis for assessing ways to minimize the occurrence and consequences of disruptions. We will not study disruptions, waves, and queues extensively, but we will look at two basic types of disruptions—a slow-moving vehicle on a two-lane road in this section, and a bottleneck (such as a lane-reduction for a construction zone) in Sec. 7-4-4.

Time-Space Diagram as a Tool for Visualizing Congestion

A visual aid called a *time-space diagram* is used to visualize how vehicles travel over time and how they respond to changes in travel conditions, such as a slow-moving vehicle in the travel lane in front of them, as shown in Fig. 7-7. In the figure, the distance moving to the right along the x-axis represents the passage of time, and the distance along the y-axis represents movement in a forward direction along the road. Each line in the figure represents the movement of an individual vehicle. Thus a vehicle that is at rest will be represented by a horizontal line, and the greater the slope of the line, the faster the vehicle is traveling. In the figure, a slow-moving vehicle (in this case a truck) enters the roadway at position $d = d_o$ and $t = t_o$. Because the density is different between the free-flowing vehicles that are not yet stuck at the end of the queue behind the slow vehicle and those that are, the sudden change in density is called a *shock wave*. The speed of the shock wave is denoted u_w and is the quotient of change in flow divided by change in density:

$$u_w = \frac{\Delta q}{\Delta k} = \frac{q_2 - q_1}{k_2 - k_1} \tag{7-10}$$

Equation (7-10) is derived later in this section. Example 7-3 shows the effect of a slow vehicle on traffic flow as well as the calculation of u_w.

Example 7-3 Suppose that a traffic stream is moving along a roadway (similar to the one shown in Fig. 7-7) with initial flow conditions $u = 40$ mph, $k = 25$ vehicles/mile, and $q = 1000$ vehicles/hour. At time t_0, a slow-moving truck (or piece of farm equipment, etc.) enters the roadway, moving at $u = 10$ mph. Because of the terrain, passing is not permitted and a platoon of vehicles begins to

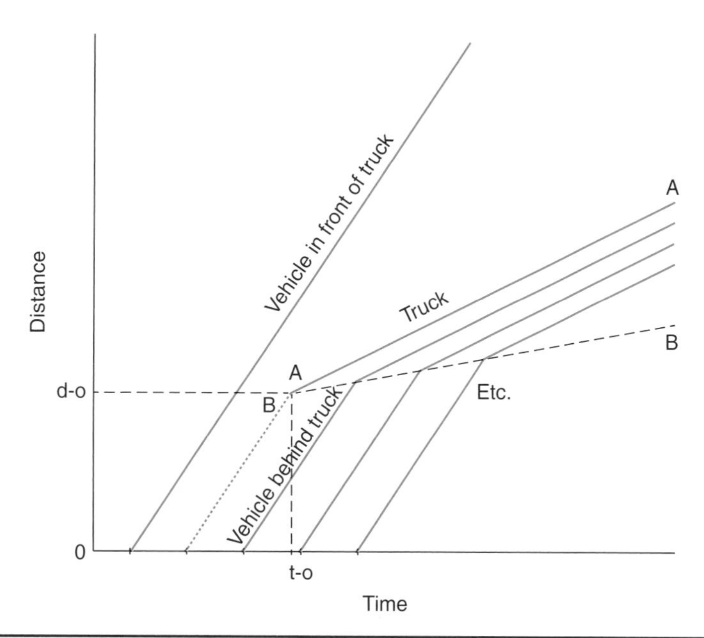

FIGURE 7-7 Time-space diagram of platoon formation behind a slow-moving vehicle.

form behind the truck, as illustrated by the series of lines to the right of time t_0. The dashed line prior to t_0 and d_0 represents the gap in the traffic moving at 40 mph into which the truck entered, under the simplifying assumption that it appeared instantaneously and ignoring time required to accelerate. The truck and the vehicles immediately behind it are moving at 10 mph, and at much higher density of k = 120 vehicles/mile than the other flow on the roadway. Unaffected vehicles continue to move at the free-flow speed represented by the more steeply sloped line of the vehicle ahead of the slow truck. In fact, ahead of the truck there is soon no traffic at all (density = 0 and speed—if there were any vehicles in this area—is at free-flow conditions), represented by the open space to the right of time t_0 between the free-flowing vehicle and the truck at the head of the queue. As time goes on, the platoon of vehicles following the truck gets larger as more vehicles catch up with the slow-moving group and join the end of the queue. At what speed is the shock wave at the end of the platoon moving?

Solution In Fig. 7-7 we use the letter A to note the position of the truck (the front of the platoon) and the letter B to note the position of the last vehicle in the platoon. The trajectory of the truck is indicated by the line A-A and the position of the rear of the platoon is shown by the line B-B. Both of these lines mark where sudden changes in density occur, and the location of these changes is moving over time.

The change in conditions can also be represented using a flow-density diagram as shown in Fig. 7-8. The initial conditions on the roadway correspond to point 1 on the diagram (k = 25 vehicles/mile, q = 1,000 vehicles/hour, and u = 40 mph). When the truck enters the roadway, the flow behind it is at a much higher density and lower speed (point 2 on the diagram). In this case, since u = 10 mph and k = 120 vehicles/mile, the flow q is

$$q = uk = (10)\ (120) = 1{,}200 \text{ vehicles/hour}$$

To a stationary observer, the flow rate behind the truck (i.e., the rate at which vehicles are passing a fixed point) is actually higher (q = 1200 vehicles/hour), but a driver stuck in the slow-moving platoon does not perceive that. Both flow rates are below the capacity of the roadway (q = 1,400 vehicles/hour at k = 70 vehicles/mile and u = 20 mph), indicated by point 3.

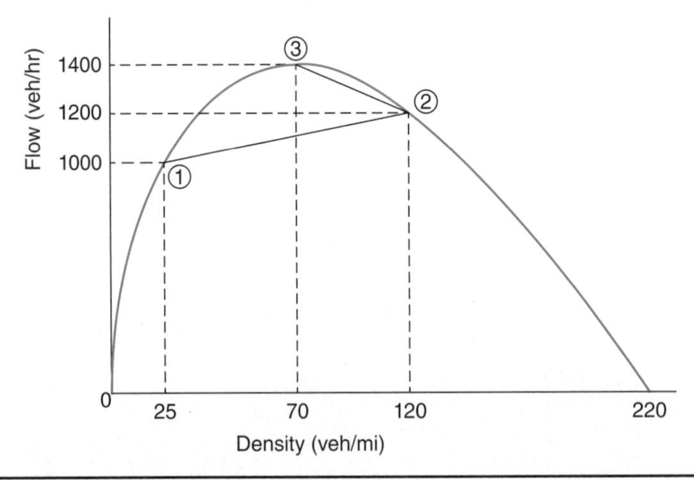

FIGURE 7-8 Flow-density diagram for Example 7-3.

The chord connecting points 1 and 2 describes the transition between the operating conditions that occurs at location B (the end of the platoon) as vehicles slow down and move at higher density. The slope of that chord is the speed at which the shock wave is moving, that is, u_w:

$$u_w = \frac{\Delta q}{\Delta k} = \frac{1{,}200 \text{ vehicles/hour} - 1{,}000 \text{ vehicles/hour}}{120 \text{ vehicles/mile} - 25 \text{ vehicles/mile}} = 2.1 \text{ mph}$$

Thus the shock wave is moving slowly in a forward direction along the road, at a speed of 2.1 mph.

The result from Example 7-3 is consistent with the picture in Fig. 7-7, where the back of the platoon (point B) is moving in a forward direction on the y axis as time progresses from left to right. Another indication is the small positive slope of the line B-B.

Characterization of Speed and Distance of Travel of Shock Wave

To get a better sense of why the chord on the flow-density diagram describes the movement of the shock wave, consider the diagram shown in Fig. 7-9, representing two regimes of flow (1 and 2) at different densities and speeds. Traffic is flowing from left to right, moving from a regime of lower density (k_1) into a regime of higher density (k_2). In the lower-density area, traffic is moving at speed u_1, and it slows to speed u_2 (i.e., $u_2 < u_1$) in the higher-density area. The location of the transition is denoted w, and this location is moving at speed u_w (which may be either positive, indicating w is moving left to right; or negative, indicating w is moving right to left).

Within area 1, the speed of vehicles relative to the line w is

$$u_1^r = u_1 - u_w$$

During a time period t, the number of vehicles crossing line w from area 1 is then

$$N_1 = u_1^r k_1 t$$

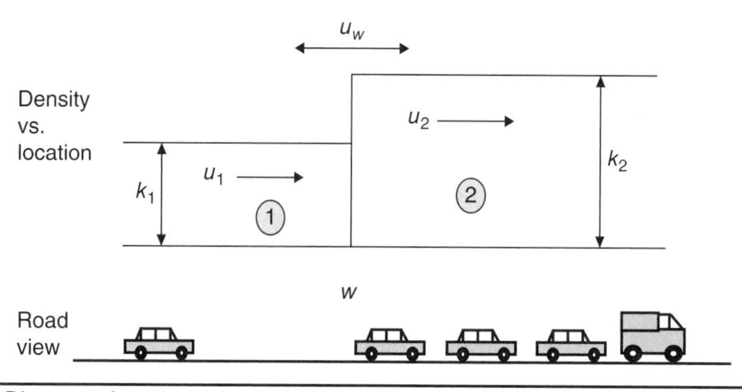

FIGURE 7-9 Diagram of roadway showing change in speed and density on either side of location *w* of a shock wave.

From the perspective of area 2, the speed within the area, relative to the line *w*, is

$$u_2^r = u_2 - u_w$$

The number of vehicles crossing line *w* during a time period *t* is

$$N_2 = u_2^r k_2 t$$

Vehicles must be conserved across the line *w*, so $N_1 = N_2$. This implies that

$$u_1^r k_1 t = u_2^r k_2 t$$

$$(u_1 - u_w)k_1 = (u_2 - u_w)k_2$$

$$u_w(k_2 - k_1) = u_2 k_2 - u_1 k_1$$

We know that the flow rates in sections 1 and 2 are $q_1 = u_1 k_1$ and $q_2 = u_2 k_2$. Thus we arrive at the value of u_w from Eq. (7-10):

$$u_w(k_2 - k_1) = q_2 - q_1$$

$$u_w = \frac{q_2 - q_1}{k_2 - k_1}$$

It is of interest to know how the value of u_w compares to the value of *u*. If we let the changes in *q* and *k* become small, we can convert the finite difference to a differential:

$$u_w = \frac{dq}{dk}$$

Since $q = uk$, this can be written as

$$u_w = \frac{d(uk)}{dk}$$

Speed is a function of density, so

$$u_w = u + k\frac{du}{dk}$$

Because speed decreases with increasing density (i.e., $\frac{du}{dk} < 0$), the speed of the wave is always less than the speed of the traffic itself.

Understanding the movement of the shock wave also gives us the ability to evaluate how the platoon behind the slow-moving vehicle grows over time, as well as the delays created by this disruption, as shown in Example 7-4.

Example 7-4 To continue with the example started above, suppose the truck travels a distance of 1.67 miles along the road and then turns off. How many vehicles accumulate behind the truck?

Solution Since the truck is traveling 10 mph and goes 1.67 miles it is on the road for 10 minutes (or 1/6 hour = 0.167 hour).During that time, the wave (B-B in Fig. 7-7 marking the back of the vehicle platoon is moving at 2.1 mph, so the location of the back of the platoon after 1/6 hour is (2.1)(0.167) = 0.35 miles from where the truck entered the roadway. The front of the platoon is clearly where the truck just turned off, so the platoon is 1.67 – 0.35 = 1.32-miles long. Within that platoon of vehicles, the density is 120 vehicles/mile, so there are 120 (1.32) = 158 vehicles that have accumulated, waiting to get by the truck.

Dissipation of Queue after Slow Vehicle Leaves Roadway

The queue or platoon of vehicles dissipates after the departure of the slow vehicle in the following way. With the slow vehicle gone, the vehicles at the front of the platoon are free to accelerate, and the speed-density-flow conditions begin to change rapidly. The first few vehicles are likely to reach free-flow speed quickly, but the whole platoon of vehicles does not suddenly accelerate as a unit to that speed. A reasonable assumption is that the *release conditions* at the front of the platoon are approximately the maximum flow point ($k = 70$, $q = 1{,}400$) on the flow-density curve.

The behavior of the queue both before and after the departure of the slow vehicle is shown in Fig. 7-10. After the truck exits, the queue continues to have vehicles added at the rear, but it is dispersing from the front. It will disappear when the two waves (going in opposite directions) meet. Thus the dispersal at the front of the queue creates a second *front* shock wave in addition to the original *back* shock wave that formed when the slow vehicle entered the roadway, as shown. Note that the front shock wave (representing dispersal) is moving backwards (i.e., right to left) in the traffic stream.

As shown in Fig. 7-10, the slow vehicle has reached position d_1 when it exits at time t_1. At that time, the end of the platoon has reached position d_1', so the length of the queue is represented by the distance between d_1 and d_1'. Later at time t_2, the two shock waves meet at position d_2, representing the complete dissipation of the platoon. Example 7-5 uses actual figures from the preceding examples to further illustrate the calculations.

Example 7-5 Once the truck in Example 7-4 has left the roadway, what is the speed of the front shock wave? How many minutes from the departure of the truck, and from its initial arrival in the road, does it take for the queue to completely disperse? At what distance from the initial arrival of the truck does this dispersal take place?

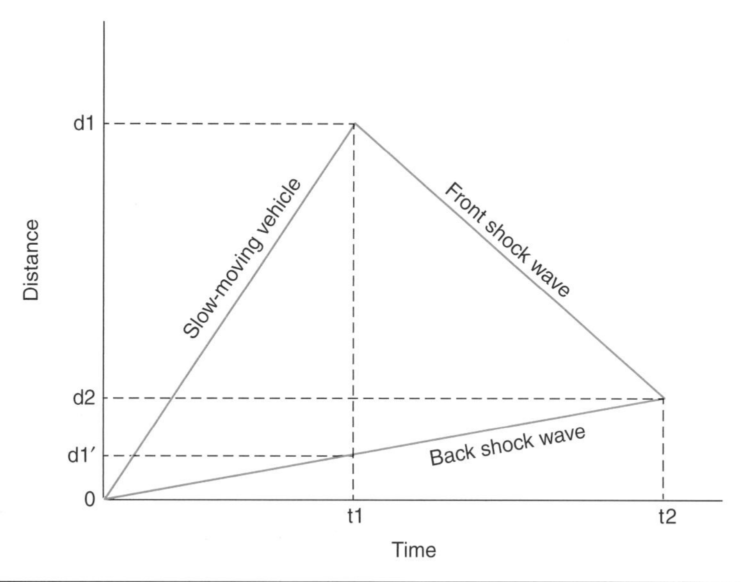

FIGURE 7-10 Time-distance diagram of platoon location, physical length, and duration.

Solution From point 3 on the diagram in Fig. 7-8 the flow of 1,400 vehicles/hour at a density of 70 vehicles/mile occurs at the front of the platoon, under the assumption that vehicles will travel at the speed that maximizes flow. The speed of this wave is

$$\frac{\Delta q}{\Delta k} = \frac{1,400-1,200}{70-120} = -4.0 \text{ mph}$$

When the truck exits, the platoon is 1.32 miles long, and the waves are approaching each other at 2.1 – (–4.0) = 6.1 mph, so the meeting will require the following amount of time:

$$\frac{1.32}{6.1}(60) = 13 \text{ minutes}$$

This is equivalent to 23 minutes after the truck entered the roadway. The distance from the truck's entry is then

$$(2.1)(23)\left(\frac{1 \text{ hour}}{60 \text{ minutes}}\right) = 0.805 \text{ mile}$$

Note in Example 7-5 that the point at which the platoon disappears (0.8 miles along the roadway from where the truck entered) is neither where the truck entered nor where it exits. Most drivers who enter the platoon may never see the truck that caused it—they simply experience an area of slow-moving traffic that has no apparent cause. Note also that the time for the platoon to dissipate is longer than it took for it to accumulate. This is also quite common—a disruption causes delays long after it has ended.

7-4-4 Effect of Roadway Bottlenecks on Traffic Flow

Another type of disruption to traffic flow occurs when a roadway transitions to a section of lower capacity, like a construction zone, the scene of a vehicle breakdown, or other obstacle. Often, this results in temporary loss of a lane and traffic has to squeeze into a portion of the roadway to pass the bottleneck, as shown in Fig. 7-11 for the case

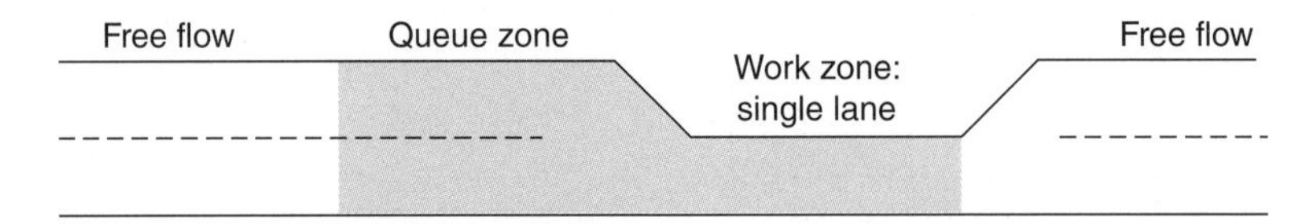

FIGURE 7-11 A bottleneck section due to construction. Shading indicates congested area.

of two lanes being reduced to one. In this case, there are two different flow-density curves that are relevant—one for the roadway before and after the bottleneck and the other (with lower capacity) for the bottleneck section itself, as shown in Fig. 7-12. The example of two lanes reduced to one could be generalized to other combinations (e.g., three lanes to two, etc.).

The roadway has a normal capacity of C_a (the maximum value of the upper curve in Fig. 7-12) at some density k_a (not shown). Note that in the figure, density is given in vehicles/lane/mile, but flow is given in total vehicle flow per hour. In the bottleneck, the capacity C_b is lower because there is only one lane available, hence the different maximum value of the two curves. In Fig. 7-12, the flow-density diagram for the bottleneck has been drawn so that the density in the bottleneck at capacity k_b (labeled k_3 in the figure) is lower than k_a, but this is not necessarily the case. The flow-density curve for the bottleneck could be drawn differently. What is important is that $C_b < C_a$, since if the flow q approaching the bottleneck is in the range $C_b < q < C_a$, a queue will form at the bottleneck.

One challenge with calculations underlying the bottleneck flow-density curve is that for situations where flow is given and density must be calculated, there are two solutions for k for flow values other than at the maximum flow. For instance, in Fig. 7-12,

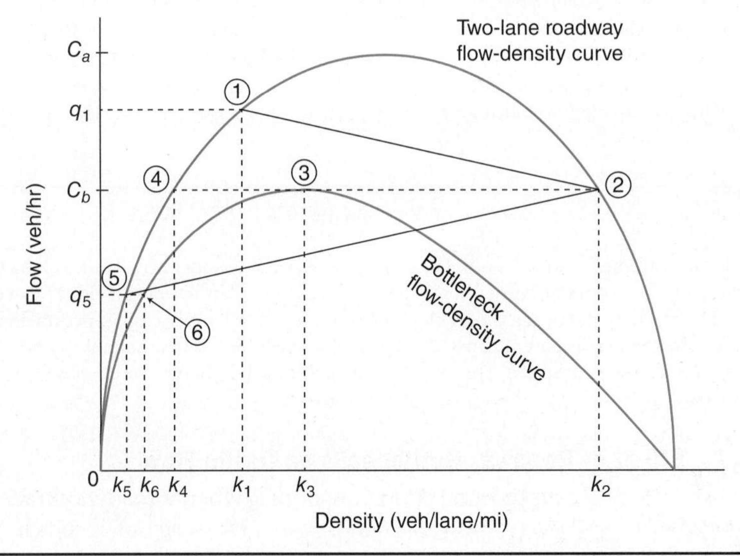

FIGURE 7-12 Flow-density diagram for a bottleneck, showing initial operating point at times when flow exceeds bottleneck capacity (Point 1), and when flow is less than bottleneck capacity (Point 5).

points 2 and 4 both satisfy the flow value $q = C_b$, but have different values of k. Therefore, a qualitative understanding of behavior around bottlenecks must be used to choose the correct value of k. In general, if traffic arrives under free-flow conditions at the end of a queue waiting to enter a bottleneck, its speed must decrease and density must increase. In the figure, however, $k_4 < k_1$, so in the case of calculating k and u, the only feasible choice is point 2, with $k = k_2$. The opposite logic applies at the end of a bottleneck: once vehicles have passed the obstruction, density must be low and speed must be high.

Suppose that the flow approaching the bottleneck is $q_1 > C_b$, as shown in Fig. 7-12. This flow is at density k_1 and speed $u_1 = q_1/k_1$. The flow through the bottleneck is limited to its capacity, C_b, so upstream of the bottleneck there is a transition from point 1 to point 2, reducing the flow rate to C_b at a much higher density k_2 and a much lower speed $u_2 = C_b/k_2$. This is the transition from unshaded to shaded in Fig. 7-11, indicating increasing density (the end of the queue forming to go through the bottleneck). The change in density creates a shock wave whose velocity is given by the slope of the line from point 1 to point 2 in Fig. 7-12. This slope is negative, indicating that the wave is moving right to left in Fig. 7-11—back up the roadway as the queue grows.

Within the bottleneck itself, flow must be on the inner flow-density curve, so at the entrance to the bottleneck there is a transition from point 2 to point 3. This is another change in density, so there is another shock wave here, but since the change in flow is zero, the wave velocity is zero [from Eq. (7-10), $u_w = 0$ if $\Delta q = 0$]. That is, it is a *standing wave* whose location stays at the entrance to the bottleneck section. As a driver, you experience this transition from density k_2 just before the bottleneck section to a lower density k_b inside the bottleneck as an increase in speed (from $u_2 = C_b/k_2$ to $u_3 = C_b/k_3$) as you leave the queue and enter the bottleneck itself.

At the end of the bottleneck section, there is a third transition (from point 3 to point 4 in Fig. 7-12) as vehicles re-enter the main roadway and return to the *outer* flow-density curve. The flow stays the same (C_b), but there is another reduction in density (from k_3 to k_4) and an associated increase in speed, to $u_4 = C_b/k_4$. This transition sets up another standing wave (velocity = 0 because there is no change in flow rate) at the exit from the bottleneck. As a driver, you experience this increase in speed and decrease in density, and often note that $u_4 > u_1$ (i.e., traffic is moving faster than it was before you came to the bottleneck). This is because the flow is being limited by the bottleneck capacity.

Recovery from Congested Conditions and Return to Uncongested Flow

Because the input flow rate q_1 is greater than the capacity of the bottleneck, the conditions illustrated in Figs. 7-11 and 7-12 cannot last forever. The queue waiting to get through the bottleneck section would just get longer and longer. Eventually, we must get to a situation where the incoming flow (shown as q_5 in Fig. 7-12) is less than C_b.

If the reduction of input flow rate occurs while there is a queue built up in front of the bottleneck, initially the situation is as depicted in Fig. 7-12 between points 5, 2, 3, and 4. The queue is still operating at point 2 (very high density and low speed) on the roadway upstream of the bottleneck, and the flow through the bottleneck section is at capacity (point 3). The transition at the exit from the bottleneck (point 3 to point 4) is the same as before. However, the input flow (point 5) is now at a flow rate that is less than the flow rate through the bottleneck, so the shock wave representing the transition from point 5 to point 2 has a positive slope. That is, the velocity of the wave is positive (left to right in Fig. 7-11), indicating that the back of the queue is moving toward the entrance to the bottleneck (the queue is dissipating).

After the queue dissipates, the situation shifts to a simpler one, shown by points 5 and 6 in the figure. Because the input flow q_5 is less than the capacity of the bottleneck, a transition occurs at the entrance to the bottleneck, where density increases from k_5 to k_6 (and speed decreases from $u_5 = q_5/k_5$ to $u_6 = q_5/k_6$) as conditions change from the *outer* flow-density curve to the *inner* one that is relevant for the bottleneck. There is a standing wave at the bottleneck entrance, representing the line between points 5 and 6. At the exit from the bottleneck, conditions transition back to point 5 on the main roadway. A driver notices this as a reduction in density and an increase back to the speed that existed before the bottleneck. Example 7-6 illustrates the calculation of flow, speed, and density values for conditions where the queue is growing, where it is dissipating, and where there is no queue.

Example 7-6 A work zone has reduced two lanes of a limited-access expressway to one lane in one direction. Flow in each lane behaves according to a Greenshields model with free-flow speed of 68 mph and jam density of 100 vehicles/lane/mile. Initially, vehicles approach the work zone with a total flow of 2,500 vehicles/hour across two lanes. Later the flow drops to 1,200 vehicles/hour across two lanes. (a) Under the initial conditions, what is the speed of the shock wave that propagates as vehicles queue to pass through the bottleneck? (b) What is the speed of the traffic upstream from the queue, waiting in the queue to enter the bottleneck, in the bottleneck, and downstream from the bottleneck? (c) If the flow conditions last for 20 minutes, how long is the queue, and how many vehicles does it contain? (d) Once the flow drops to the lower level, what is the new speed of the shock wave, and how long does it take for the queue to disperse? (e) Once the queue has dispersed and traffic is flowing through the bottleneck under uncongested conditions, what is the speed approaching the bottleneck and the speed in the bottleneck?

Solution

(a) It is useful to refer to Fig. 7-12 throughout this problem. The given values of $u_f = 68$ and $k_j = 100$ are first used to create equations for flow as a function of density, based on Eq. (7-3):

$$u = u_f - \left(\frac{u_f}{k_j}\right)k = 68 - \left(\frac{68}{100}\right)k = 68 - 0.68\,k$$

$$q = uk = k(68 - 0.68k) = 68k - 0.68k^2$$

For the flow in the roadway with two lanes, the total flow is doubled, thus

$$q = 2uk = 2(68k - 0.68k^2) = 136k - 1.36k^2$$

Next, C_b is calculated using Eq. (7-9):

$$q_{\max} = \frac{u_f k_j}{4} = \frac{68(100)}{4} = 1{,}700$$

Therefore since the incoming flow of $q_1 = 2{,}500$ exceeds $q_{\max}$ in a single lane, a queue will build backward. It is known that at point 2 the total flow across two lanes is $C_b = 1{,}700$, so we can use this information to solve for k_2:

$$q_2 = -1.36k^2 + 136k = 1{,}700$$

$$-1.36k^2 + 136k - 1{,}700 = 0$$

We now use the quadratic formula to solve for the values of k that satisfy the preceding equation:

$$k = \frac{-b \pm \sqrt{b^2 - 4ac}}{2a}$$

$$k = \frac{-136 \pm \sqrt{(136)^2 - 4(-1.36)(-1700)}}{2(-1.36)}$$

$$k = 14.6 \qquad k = 85.4$$

Because the vehicles are queuing to wait for their turn to enter the one-lane bottleneck, they are moving with high density and slow speed, so the correct choice is $k = 85.4$ vehicles/lane/mile. Density for the upstream uncongested traffic must be calculated as well by a similar method:

$$-1.36k^2 + 136k - 2500 = 0$$

$$k = \frac{-b \pm \sqrt{b^2 - 4ac}}{2a}$$

$$k = \frac{-136 \pm \sqrt{(136)^2 - 4(-1.36)(-2500)}}{2(-1.36)}$$

$$k = 24.3 \qquad k = 75.7$$

In this case, because traffic is arriving in uncongested conditions and a high rate of speed, the correct choice is $k = 24.3$ vehicles/lane/mile.

With known values for q and k, it is now possible to calculate the speed of the shock wave. Care should be taken to use the flow value on a *per-lane* basis rather than for two lanes, to be consistent with the value of density given on a per-lane basis. Thus the difference in flow is divided by 2 in the calculation:

$$\frac{\Delta q/2}{\Delta k} = \frac{(1700 - 2500)/2}{(85.4 - 24.3)} = -6.6 \text{ mph}$$

(b) The values of speed desired correspond with speeds at points 1 to 4, that is, u_1, u_2, u_3, and u_4. These values are calculated from the known values of q and k on a per-lane basis. Values at points 3 and 4 are $k_3 = k_{max} = k_j / 2 = 50$ and $k_4 = 14.6$ since this was the other root of the quadratic equation calculated above:

$$u_1 = \frac{q_1}{k_1} = \frac{1,250}{24.3} = 51.4 \text{ mph}$$

$$u_2 = \frac{q_2}{k_2} = \frac{850}{85.4} = 10.0 \text{ mph}$$

$$u_3 = \frac{q_3}{k_3} = \frac{850}{50} = 34.0 \text{ mph}$$

$$u_4 = \frac{q_4}{k_4} = \frac{850}{14.6} = 58.2 \text{ mph}$$

(c) The speed of the shock wave and the length of time determine the total length of the queue. Note that shock wave speed is presented as a positive number since the queue grows in length by propagating backward:

$$(6.6 \text{ mph})(20 \text{ minutes})\left(\frac{1 \text{ hour}}{60 \text{ minutes}}\right) = 2.2 \text{ miles}$$

Since the density is 85.4 vehicles/lane/mile and there are two lanes, the total number of vehicles is

$$(2.2 \text{ miles})\left(85.4\frac{\text{vehicles}}{\text{lane}-\text{mile}}\right)(2 \text{ lanes}) = 376 \text{ vehicles}$$

(d) Since the new flow value as the queue is dissipating is 1,200 vehicles/hour across two lanes, the density value is again solved using the quadratic formula:

$$-1.36k^2 + 136k - 1,200 = 0$$

$$k = \frac{-b \pm \sqrt{b^2 - 4ac}}{2a}$$

$$k = \frac{-136 \pm \sqrt{(136)^2 - 4(-1.36)(-1200)}}{2(-1.36)}$$

$$k = 9.8 \qquad k = 90.2$$

Again in this case because the flow is upstream from the congested portion, density is low, so the correct choice is $k = 9.8$ vehicles/lane/mile. Repeating the calculation of shock-wave speed with new values and then using the speed to calculate dissipation time gives

$$\frac{\Delta q/2}{\Delta k} = \frac{(1,700 - 1,200)/2}{(85.4 - 9.8)} = 3.3 \text{ mph}$$

$$(2.2 \text{ miles})\left(\frac{1}{3.3 \text{ mph}}\right)\left(\frac{60 \text{ min}}{1 \text{ hour}}\right) = 40 \text{min}$$

(e) From part (d), it is already known that $q_5 = 600$ vehicles/lane/hour and $k_5 = 9.8$ vehicles/lane/mile. The density value k_6 in the bottleneck is calculated using the quadratic formula on the one-lane flow-density equation:

$$-0.68k^2 + 68k - 1,200 = 0$$

$$k = \frac{-b \pm \sqrt{b^2 - 4ac}}{2a}$$

$$k = \frac{-68 \pm \sqrt{(68)^2 - 4(-0.68)(-1200)}}{2(-0.68)}$$

$$k = 22.9 \qquad k = 77.1$$

From Fig. 7-12, the density value inside the bottleneck must be less than $k_{max} = 50$, so the correct answer is $k = 22.9$. Using known values of q and k gives speeds u_5 and u_6:

$$u_5 = \frac{q_5}{k_5} = \frac{600}{9.8} = 61.2 \text{ mph}$$

$$u_6 = \frac{q_6}{k_6} = \frac{1200}{22.9} = 52.4 \text{ mph}$$

Several observations emerge from the results of Example 7-6. As was the case with the slow vehicle in the roadway previously, the congestion caused by the bottleneck took longer to dissipate once the cause was removed (i.e., flow levels about the capacity of the bottleneck in this case) than it did to form in the first place. Also, the driver navigating the bottleneck experiences the slowest driving speed while waiting to merge into the bottleneck (point 2), then faster speed in the bottleneck itself (point 3), then even faster speed after passing the bottleneck. In fact, a driver who at first loses time waiting in the queue to enter the bottleneck can reclaim some of the lost time because of the relatively high speeds that are possible downstream from the bottleneck. In the example, the overall flow value across two lanes downstream is q = 1,700 vehicles/hour, compared to q = 2,500 vehicles/hour upstream. On balance, however, once the driver has waited in the queue for any measurable length of time, their overall journey time will increase because of interference from the bottleneck.

7-5 Passenger Transportation and Intelligent Transportation Systems

We now turn from analyzing the causes of congestion and behavior of vehicles in congested conditions to one of the key tools for fighting congestion, namely ITS. As introduced earlier in the chapter, ITS is the umbrella term for the use of information technology, including sensing, communications, and analysis, to improve transportation system performance. As was the case with the discussion of congestion, the application in this section is primarily on the passenger side, but freight transportation can also benefit from ITS. (See Chowdhury and Sadek, 2003, for a full-length work on ITS.)

The motivation for ITS is both one of opportunities created by advances in computational power and wireless technology, and the constraints, both physical and monetary, on the expansion of physical transportation infrastructure. In many cases, the technology needed to wire a length of transportation infrastructure, whether road, rail, or waterway, and including sensors, transmission networks, and data centers, is much less expensive than growing the size of the infrastructure, in the form of adding highway lanes, rail tracks, bridges, and so on. It therefore makes sense to increase maximum throughput by using existing infrastructure more efficiently with ITS rather than increasing maximum throughput by expanding its physical size.

ITS began not as an overarching plan but rather as a collection of fragmented, piecemeal components. As early as the 1960s, transportation departments began using vehicle detectors to sense the presence of vehicles and electronic signage to provide information that was changeable depending on circumstances. This experience led eventually to the vision of a comprehensive system that could

- *Monitor* transportation system performance,
- *Intervene* in the control of the system to make it perform better, and
- *Inform* both passenger and freight system users from the general public of conditions in real time, and in particular warn them of disruptions.

In the late 1980s and early 1990s many national governments, including that of the United States, began to combine components that had already been developed into a comprehensive support system covering not only highway movements but also freight, public transportation, construction traffic, and emergency services. In 1993, the U.S. Department of Transportation began to invest heavily in national ITS architecture.

Today ITS is largely built out in the industrialized countries of the world, including the United States. It is deployed in all major U.S. cities, as well as in rural and statewide systems. Nevertheless, the traveling public is largely unaware of the ITS program, and is instead only familiar with some of its more visible components, such as roadside information signage, traffic reports, and public transportation interfaces.

7-5-1 ITS Architecture and ITS Design

To understand ITS, it is first necessary to understand the difference between *ITS architecture* and *ITS design*. ITS architecture is the framework for the development of ITS that the various stakeholders agree to uphold in the deployment of ITS. For example, the architecture dictates how different systems within ITS will interface with each other, such as the interface between systems that monitor street and highway conditions and those that support emergency response services. ITS design is then the development of specific systems that meet particular ITS needs within the framework provided by the architecture. Architecture is also important for smooth operation between regions within the overall national ITS network. As vehicles pass between regions, it is natural that ITS deployments should be able to pass information back and forth, so having a common architecture facilitates this function.

Given the complexity of the overall national transportation system and its many competing tasks (passenger versus freight, multiple modes, many tasks such as safety, road conditions, maintenance, or congestion) it is natural that the overall ITS architecture would have embedded within it many different specific functions, each interconnected with other functions and components of the system. Figure 7-13 therefore provides a highest-level perspective on the most basic management goals of ITS. Put in the simplest possible terms, the goal of ITS is to *manage transportation*, so this function appears at the center of the diagram. This goal is too general to be useful, however, so both the words *manage* and *transportation* can be made more specific, resulting in the two main bubbles on either side of the center of the diagram and the outer eight bubbles around the perimeter. The "general functions" category on the left is divided into subcategories relevant to all types of transportation affected by ITS: system monitoring services, payment services, traveler services (i.e., providing real-time information regarding travel conditions), and data gathering and analysis services. On the right are the "application-specific functions": management of maintenance/construction,

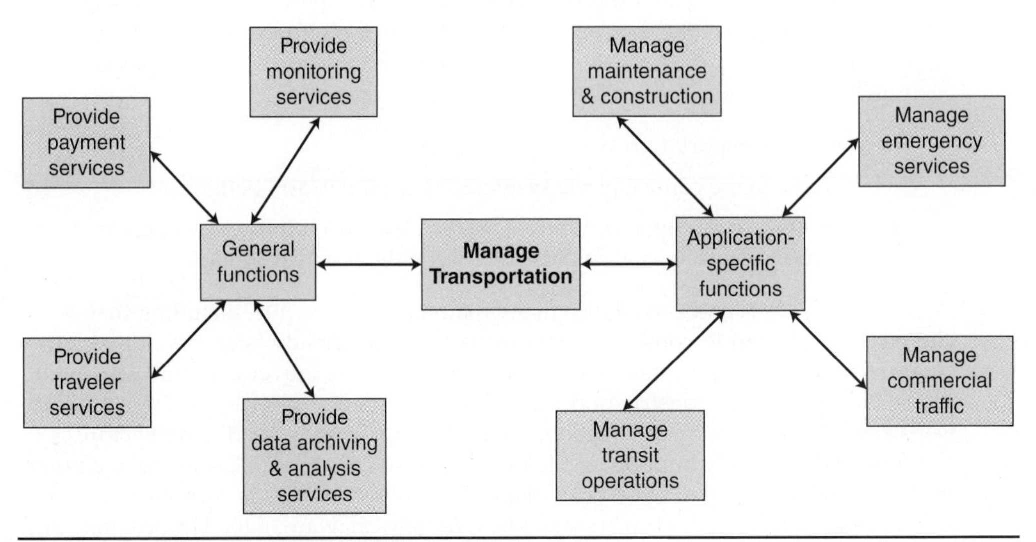

FIGURE 7-13 Highest-level diagram of ITS: functions view.
Source: Adapted from Chowdhury and Sadek (2003).

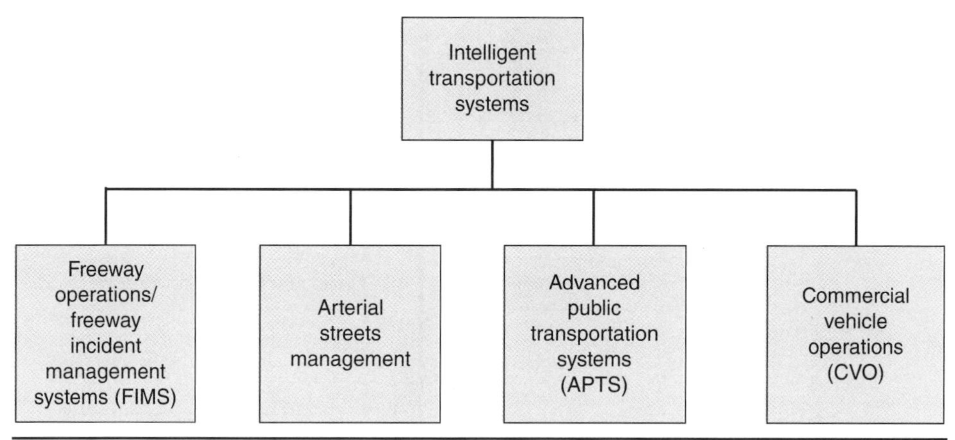

FIGURE 7-14 Highest-level diagram of ITS: applications view.

emergency vehicles, public transit, and freight (also referred to as "commercial") vehicles.[1]

Separate from the functions of ITS are the main applications shown in Fig. 7-14: Freeways, arterial streets, advanced public transportation systems, and commercial vehicle operations. Each application calls on multiple functions from Fig. 7-13. For example, freeway management requires monitoring, payment services, data archiving and analysis, traveler information dissemination, construction and maintenance, and emergency vehicle operations. Public transit and commercial vehicles have their own bubbles in Fig. 7-13, but they too call on many of the same functions as freeways or arterial streets, including monitoring, payment, data, and information services.

National and State Collaboration on ITS Architecture

Development of ITS in the United States has been a partnership between federal, state, and metropolitan area government agencies. At the outset, the federal government created a federal ITS architecture and supported the states in developing state-level architectures by providing resources such as ITS development software. This federal support helps to ensure that each state will not be *reinventing the wheel* as they develop and maintain ITS. Thereafter, states can add elements to their state-level architecture as their individual circumstances dictate. For example, Fig. 7-13 does not make any direct mention of bicycle and pedestrian safety, so a state may choose to develop these aspects beyond what is presented by the federal government. International collaboration on ITS is another possibility for countries such as the United States and Canada that share a land border.

7-5-2 ITS Components and Equipment

Figures 7-13 and 7-14 show the relationship between high-level objectives of ITS, but they do not introduce the specific physical components whose interaction makes up the

[1]In ITS parlance the term "commercial" or "commercial vehicle" refers specifically to freight as opposed to passenger transportation.

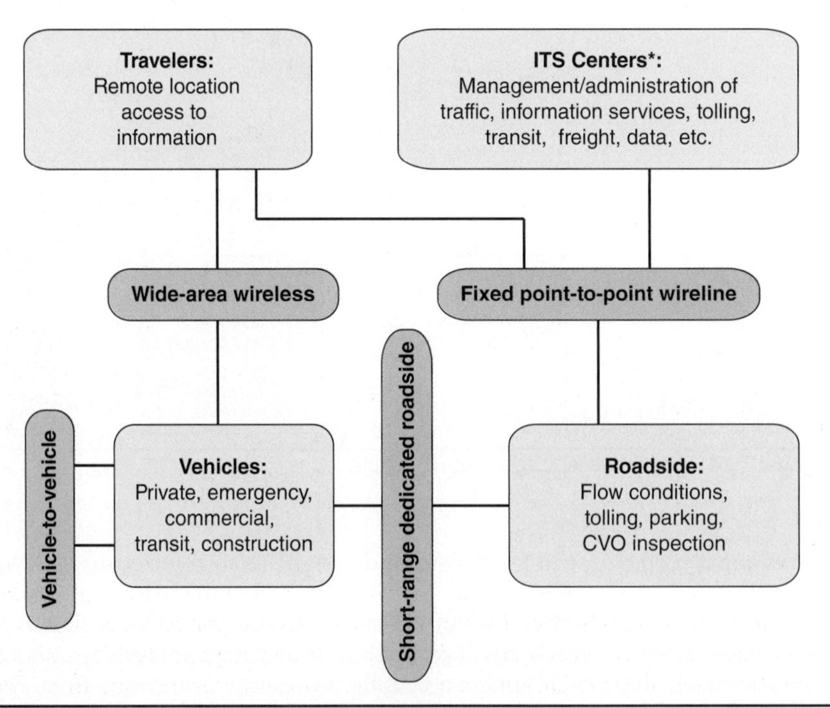

FIGURE 7-15 Connections between major interacting components of ITS: centers, travelers, vehicles, and roadside equipment.

Note: Shaded bubbles represent communications channels, including communication technology (wireless, wireline, etc.).

*Types of ITS centers roughly correspond with ITS functions given in Fig. 7-13.

Source: Adapted from Iteris (2013).

function of ITS. These are introduced in Fig. 7-15, where components are grouped into four primary categories:

1. *Centers:* These are focal points for administering some function within ITS, largely along the lines of purposes presented in Fig. 7-13. They may be housed in a single physical building or facility, or their function may be distributed among multiple locations, but either way the centers are the points to which other ITS components report in regard to the particular function in question. Thus information pertaining to traffic management originating from other components of the ITS system is reported to the "traffic management center," transit information to the "transit management center," and so on.
2. *Travelers:* These constitute either travelers moving through the passenger transportation networks or drivers moving freight through the freight network. For those in the passenger network, travelers in both private cars and public transportation modes are considered potential beneficiaries of ITS information. In the freight network, drivers of transportation vehicles, or freight professionals working on behalf of routing and planning for those drivers, are considered the customers of ITS.

3. *Vehicles:* The ITS structure divides vehicles into categories according to purpose, including private vehicles, commercial vehicles (another term for freight vehicles), emergency vehicles, transit vehicles, and construction or maintenance vehicles. General traffic vehicles include all cars and light trucks used for passenger transportation.
4. *Roadside equipment:* Roadside equipment comprises the collection of devices used in ITS to connect vehicles to centers. This category is divided up according to function. Many devices function to monitor the flow of vehicles. Other equipment handles the payment of tolls or the monitoring of parking activities. Still other equipment termed *commercial vehicles* is specific to the movement of freight vehicles.

Communications linkages between major ITS components require different types of technologies, and these are depicted in Fig. 7-15 as well. For instance, communication between roadside components and passing vehicles can be carried out either in the form of changeable signboards or short-range wireless communications. Some communication to and from vehicles may occur over longer distances, therefore requiring long-distance wireless communications through the cellular network. Since roadside components are generally in permanent positions and require more or less continuous communication with ITS centers, the logical communications solution is a wireline communication network. Travelers not already in the network (e.g., located in their homes or workplaces) may require remote access to real-time travel information, and this can be provided either through wireline or wireless communication.

Some of the equipment used in ITS is common to any modern system that relies on information technology and electronic communication. Included on this list are computer networks, data storage systems, and hard-wired and wireless communication networks. Some of these assets also exist in the form of public-private partnerships; for example, private cellular phone network providers may collaborate with government agencies to support ITS services on adjacent highways.

Some equipment is specific to the function of ITS. One of the most common and fundamental pieces of equipment is the *induction loop*, which is built into the roadway and uses the magnetic field induced by the vehicle passing over it as a way of recording and counting the passage of vehicles. Radio frequency identification (RFID) and closed-circuit TV can also be used, and these systems allow for the unique identification of specific vehicles, for example, for charging tolls in the case of toll highways. Induction loops, by contrast, can only detect the presence of a vehicle but not discern that vehicle's identity. Electronic signage in ITS provides information to vehicles and other travelers, in either changeable text or changeable graphic formats. Signal systems at intersections can also be linked to ITS centers so that they can remotely control timing and coordination, not only at intersections but also at freeway on-ramps.

Illustration of Interaction between ITS Components

A description of the interaction between components in a fictitious scenario can illustrate the function of ITS: (1) Travelers obtain real-time information about the condition of the transportation system that has been provided by the various ITS centers. (2) They then make a decision to take a trip in the system. Suppose the trip is taken in a private car. (3) Their car movements will be noted by roadside detection equipment as they pass through the network, and possibly by toll-collection or parking equipment if these services are required during the trip. In these instances, (4) information is flowing from the vehicle via the roadside detection equipment to the ITS centers. The individual identity of the vehicle is only taken into consideration when needed, for privacy reasons. In other words, roadside-detection equipment dedicated to monitoring overall levels of traffic flow needs only to know that a vehicle has passed a point at a given instant in time, and not the identity of the vehicle. Toll- or parking-collection equipment, on the other hand, must identify the individual vehicle in order to charge fees to the account of the vehicle owner. (5) If an incident arises that affects the route that the vehicle is taking, the ITS centers may broadcast information about possible delays or alternate routes to the vehicle via roadside devices, on-board navigation systems, or handheld wireless devices.

7-5-3 Applications of ITS: Examples from Passenger Transportation Systems

This section develops passenger transportation applications from Fig. 7-14: (1) freeway management, (2) arterial street management, and (3) transit management. Although it is not possible to present the complete range of possible specific ITS applications or technologies, the three examples developed in this section introduce the reader to the possible breadth of opportunities for improving transportation system performance through ITS.

Freeway Operations and Freeway Incident Management Systems

The most basic goal for management of freeways (also known as expressways or limited access highways, or turnpikes in cases where tolls are charged) is that all vehicles, whether general traffic, commercial vehicles, or transit, are able to travel safely and with minimal delays due to congestion. To accomplish this goal, roadside devices in ITS systems detect vehicle speed and highway temperature and moisture levels. From vehicle-detection devices, it is possible to use the rate at which vehicles pass a detection system to calculate both speed and density of traffic. In addition, sensors built into the road can detect temperature, and sensors next to the road can detect air temperature, precipitation, and airborne moisture levels.

Information can also flow outward from ITS centers to the traveling public. If data analysis reveals that road conditions are deteriorating due to weather, or traffic is slowing due to congestion, the ITS system may post warnings on roadside signage alerting drivers to either condition. In the case of congestion, ITS may inform drivers of the extent of congestion or suggest alternate routes that may be less congested. Using decision-support systems found at ITS centers, the system may also reduce demand on the system so that it is more in line with capacity using *ramp control*, or signals at freeway on-ramps that control the rate at which vehicles enter the freeway.

ITS developers have established *freeway incident management* and the freeway incident management system (FIMS) for the particular freeway conditions that result from traffic accidents and other one-off incidents that are distinct from ordinary volume congestion when demand exceeds capacity. Figure 7-16 shows how the freeway operations control center can involve various ITS elements to respond to an incident and keep overall delays and disruption to a minimum. First, the surveillance system can remotely detect the incident at the top of the figure by monitoring both vehicle flow and closed-circuit televisions. The freeway operations center at the center of the figure can then take action to aid an affected driver, divert traffic around an incident,

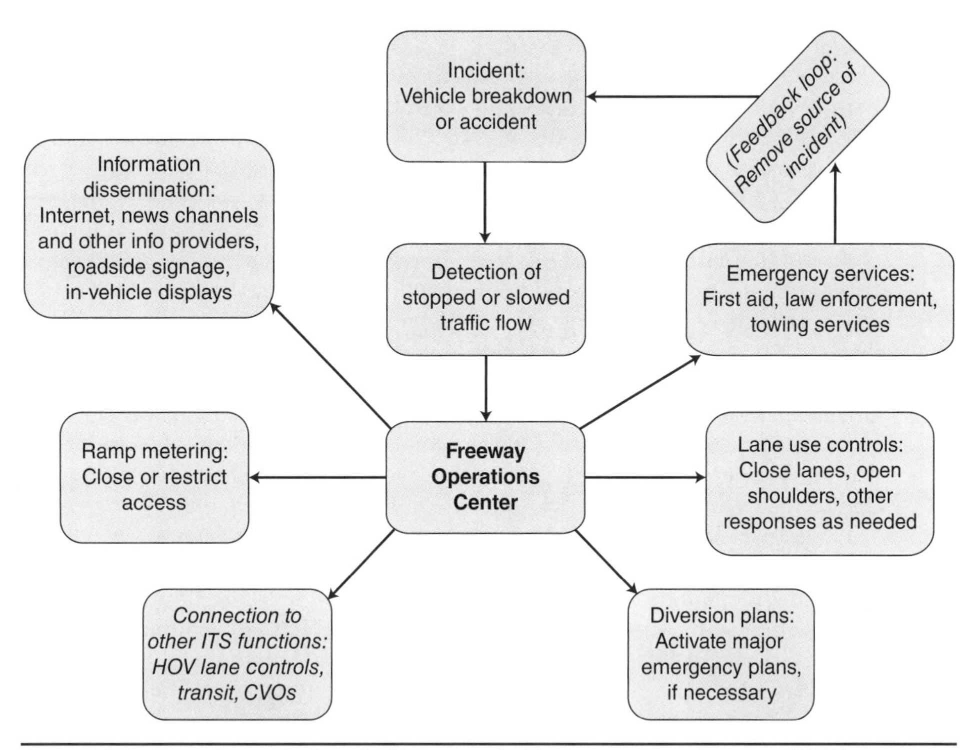

Figure 7-16 Example application of ITS: freeway incident management system (FIMS) response to disabled vehicle or vehicular accident.

Source: Adapted from Chowdhury and Sadek (2003).

restrict flow, and in general inform the public. Throughout this process, information dissemination (bottom of the figure) is key to informing both motorists at the scene (through lane closure signs and dynamic, or changeable, roadside signs) and elsewhere in the network (through internet communications, radio and television traffic reports, communication to onboard navigation systems, and the like). Lanes can be closed or opened remotely, or freeway ramp-metering systems adjusted. Other ITS applications not directly under the control of freeway operations but nevertheless affected by the incident, such as public transit and commercial vehicle operations, can be contacted as well.

Throughout response to an incident, the objective of FIMS is to create economic value by responding quickly and efficiently to disruptions in ordinary traffic flow. One could compare this response to a situation without any sort of detection and response system: Department of Transportation (DOT) officials and emergency responders might sense that something was amiss in the network, but they would need to deduce what had happened and where, and then attempt to reach the scene through an already congested system. ITS equipment helps these stakeholders quickly identify the problem and then respond in a coordinated, prepared way. With the potential for lost time due to traffic disruptions and large numbers of motorists affected, the investment in ITS is eminently sensible.

Arterial Management Systems

Arterial streets, avenues, and boulevards form a second network of main thoroughfares in an urban area that complements the role of the urban freeway network, where one exists. The primary difference from an ITS perspective between freeways and arterials is the presence of signals at intersections. ITS strategy for arterials therefore entails monitoring flow conditions and if needed making adjustments to signal timing to improve flow or respond to incidents.

At the simplest level, an ITS system for arterial streets can offer centralized adjustment of timing at intersections without an ability to detect and respond to events in real time. Even in this simplified role, it is useful for the ITS control center to provide *open-loop* adjustment of signal timing at signalized intersections in the network. In such a system, each signal is connected remotely to the traffic control center for the region. These centers can then make routine changes by time of day, for example, changing all signals to flashing red or flashing yellow during the early hours of the morning when traffic levels are very low. They may also make one-time changes to the system to update it to a new signal regimen that is thought to be an improvement over the existing one.

Open-loop, centrally controlled systems have an advantage over a more basic system where each signal is isolated from the network, since changes can be made centrally and with a single command. With an investment in traffic-flow detection systems and reporting to the operations center, however, the ITS application can move to the more advanced *closed-loop* system, where decisions to change signal timing (and possibly also lane and road closures) are taken *in response to actual, real-time conditions* in the network. Detection systems may consist of flow detectors at intersections or in the middle of blocks between intersections. Operations centers can continuously monitor flow conditions, and if some threshold for congestion is reached, put into place either (a) a predesigned plan from a library of such plans, or (b) a custom-developed plan generated in the moment based on observed conditions. In general, response (b) is more effective since each plan is optimized for current conditions, but it also costs more since it entails an investment in a decision-support system to quickly create a plan using some sort of computational algorithm.

Advanced Public Transportation Systems

Advanced public transportation systems (APTS) comprise a family of ITS applications that can be used for a range of public transportation applications. APTS uses some of the same equipment already seen in freeway and arterial management systems, but it also calls for some additional hardware and practices unique to public transportation. Four systems within APTS are of particular interest:

- *Automatic vehicle location (AVL):* AVL systems allow the public transportation agency to know where each vehicle is automatically in real time. In one possible application, global positioning system (GPS) devices on board the vehicle communicate wirelessly with the operations center and update the position of the vehicle. Note that, unlike roadside vehicle-detection systems in Fig. 7-15, which detect only that a vehicle is present but not the identity of the vehicle, AVL determines the unique identity of the individual vehicle. Agencies use AVL to determine whether a vehicle is delayed or has encountered an incident.
- *Transit operations software systems*: These software systems allow the agency to supervise the movements of both vehicles and drivers or other staff, and to

acquire and archive operational data. The support from ITS for software both helps the individual agency to perform its tasks better and helps state and federal government better understand the agencies they support, since almost all agencies receive both capital and operating support from government.

- *Transit information systems (TIS):* These systems play a complementary role to AVL. Where AVL channels information from the vehicles in to the agency so that it can monitor vehicle function in real time, TIS provides outgoing information from the agency to public transportation users in the form of pre-trip, wayside (i.e., at stops or stations), and in-vehicle information.
- *Fare payment systems:* Fare payment systems as part of ITS allow the agency to verify eligibility of the passenger at point of entry to the system and then, in systems where fare is based on time or distance, monitor the location and elapsed time of the passenger. In other systems that charge a flat fare, the system accumulates data on where and when users purchased passage in the system.

Relative to the high cost of vehicles, labor, fuel, and maintenance, and especially infrastructure costs for public transportation, cost of equipment to monitor, analyze, and announce public transportation system operations can be modest. Thus APTS represents a particularly appealing application. Example 7-7 illustrates the possible cost and benefit of implementing ITS in the form of APTS (Bruun, 2007):[2]

Example 7-7 Portsmouth, England, is a city with approximately 200,000 people located about 60 miles southwest of London. Current public transit usage is on the order of 12.4 million boardings per year, or 62 boardings per capita per year.

The transit agency is considering investment in a real-time passenger information system (RTPI) to be installed at all major stops in the network. Net cost to the agency after taking into account national government grants is 2 million British pounds (GBP). The investment lifetime of the system is 14 years, and the minimum attractive rate of return is 15%. (Note that this rate is higher than that of typical public works projects; the agency is applying a higher rate since the investment must compete with other money-saving investments that it could take internally.)

The fare paid by each rider in the system is 1.20 GBP. By what absolute and percentage amount would ridership need to increase for the investment in RTPI to break even?

Solution The approach to solving this problem is to calculate the number of additional boardings needed to pay for the additional cost of the system and then make some judgment on this basis. Using the notation (A/P, i%, N) to denote the annual capital cost factor and the given system cost and investment lifetime, the annual cost is

$$(A/P,\ 15\%,\ 14) = \frac{(0.15)\ (1+0.15)^{14}}{(1+0.15)^{14}-1} = 0.1747$$

$$A = (A/P,\ 15\%,\ 14)P = (0.1747)\ (2M) = 349.4\ K\ \text{GBP}$$

The number of additional boardings based on 1.20 GBP per boarding is then:

$$\frac{349.4K}{1.2} = 291.2\ K$$

---

[2]Example based on a longer case study of Portsmouth that appears in Bruun (2007). See the appendix for further background on the calculation of discounting factors using engineering economic techniques.

or approximately 291,200 new boardings. This represents a 2.3% increase in ridership, or in other words an increase from 62 to 63.5 boardings per capita of population per year. These boardings might come from existing patrons of the system riding the bus more frequently because its operation is more transparent, or from new patrons who begin to ride the bus for the first time because they have higher confidence in knowing the location of buses. In either case, the transit agency in conjunction with the national government authorities offering grants would decide whether or not the investment is merited based on whether the minimum increase in ridership is likely to be achieved, assuming the capital is available for the project.

Although it cannot be proven with given information that the investment would pay for itself, Example 7-7 illustrates the potential financial attraction of ITS. A system like that of Portsmouth need not own any of its own road or rail infrastructure, since the buses likely all operate on the street. However, just the fleet needed to carry more than 12 million boardings per year might be on the order of 200 vehicles. At the equivalent of 200,000 GBP per vehicle, the cost would be 40 million GBP, and this neither includes ongoing maintenance and labor costs, nor the cost of administrative and maintenance facilities that the agency requires. The RTPI is available at less than the cost of 10 buses.

Summary

For the latter half of the twentieth century and into the twenty-first century, passenger transportation systems have been in transition from being oriented toward the automobile with a complementary but limited role for public transportation, to a multifaceted and diverse mix of solutions: redesigned auto usage in the form of carsharing, diversified offerings in public transportation, nonmotorized modes, and changes to the relationship between transportation and land-use planning. Over this time period, the problem has grown more urgent as transportation activity measured in passenger-miles has grown much more quickly than available capacity, and passenger transportation energy use has grown to highly intensive levels on both an absolute and per-capita basis.

A pervasive problem facing all forms of passenger transportation, but especially on urban roadways, is congestion, which can be better understood through the relationship between average travel speed, density of vehicles per lane-mile or lane-kilometer, and throughput in vehicles per hour. Specific causes of congestion can be identified, such as slow-moving vehicles on single-lane highways or bottlenecks on multilane ones; flow conditions during congestion events dictate the quantitative extent of queues in terms of physical length, number of vehicles affected, and extra travel time spent. For both slow-moving vehicles and bottlenecks, we observed the pernicious effect of congestion that even once the factor that instigated the congestion in the first place has cleared, much additional time is required for the system to return to normal flow conditions.

Intelligent Transportation Systems constitute a tool that can assist many different types of transportation, including those found in an urban setting: freeways, arterial streets, and public transit. ITS functions include monitoring system performance and detecting incidents or other deterioration in flow; responding to incidents/deterioration by changing flow conditions, dispatching law enforcement or emergency services as needed, or other steps; and informing the traveling public with real-time information about travel conditions on the urban network.

References

Bruun, E. (2007). *Better Public Transit Systems: Analyzing Investments and Performance.* American Planning Association Press, Chicago.

Bureau of Transportation Statistics (2011). *Flow, Headway, and Density. Bureau of Transportation Statistics,* Electronic Resource. Available at: *http://ntl.bts.gov/lib/31000/31400/31419/14497_files/images/fig2_3.jpg.* Accessed Sep. 15, 2011.

Chowdhury, M and A. Sadek (2003). *Intelligent Transportation Systems: Planning Requirements for ITS.* Artech House, Norwood, MA.

Iteris (2013). *Overview of ITS Architecture.* Electronic resource, available at *http://www.iteris.com/itsarch/index.htm.* Accessed Jun. 7, 2013.

Oak Ridge National Laboratories (2011). *Transportation Energy Data Book, 30th ed.* ORNL, Oak Ridge, TN.

Texas Transportation Institute (2012). *2011 Annual Mobility Report.* TTI, Texas A&M University, College Station, TX.

U.S. Department of Transportation (2009). *ITS Strategic Research Plan 2010–2014.* Research and Innovative Technology Administration, Intelligent Transportation Systems Joint Program Office, Washington, DC.

Further Readings

Banks, J. (2002). *Introduction to Transportation Engineering. 2d ed.* McGraw-Hill, New York, NY.

Greenshields, B. (1935). "A Study of Traffic Capacity." *Highway Research Board Proceedings* **14**:448–477.

Institute of Transportation Engineers (2009). *Traffic Engineering Handbook, 6th ed.* ITE, Washington, DC.

Papacostas, C. and P. Prevedouros (2001). *Transportation Engineering & Planning.* Prentice-Hall, Upper Saddle River, NJ.

Sperling, D. and D. Gordon (2008). *Two Billion Cars: Driving Toward Sustainability.* Oxford: Oxford University Press.

Wardrop, J. (1952) "Some Theoretical Aspects of Road Traffic Research." *Proceedings of the Institute of Civil Engineering* **1**:325–362.

Wardrop, J. and G. Charlesworth (1954). "A Method of Estimating Speed and Flow of Traffic from a Moving Vehicle." *Proceedings of the Institute of Civil Engineering* **3**:158–171.

Exercises

7-1. Suppose that the relationship between speed and density for a particular highway facility is: $u = 6\sqrt{120 - k}$, where the speed, u, is in mph and the density, k, is in vehicles/mile. (a) What is the free-flow speed for this facility? (b) What is the jam density? (c) What is the capacity (q_{max}) for the facility? (d) At what speed is the maximum flow achieved?

7-2. Along a two-lane rural highway where there is no opportunity to pass because of curves and short sight distances, the speed-density relationship can be represented by a Greenshields model with $u_f = 62$ mph and $k_j = 100$ vehicles/mile. Traffic is moving at an average speed of 53 mph when a piece of farm equipment pulls onto the road and travels at 20 mph. (a) What is the speed of the shock wave at the end of the platoon of vehicles that builds up behind the farm implement as it travels? (b) After 5 minutes on the road, the farm implement turns off and leaves the road. At that time, how long (in miles) is the platoon that has built up behind it on the road? (c) At the implied density within the platoon, approximately how many vehicles are backed up in the platoon? (d) From the time the farm implement turns off, how many minutes does it take for the platoon to disperse? (e) How far has the queue traveled from the point where the implement entered the road when it disperses completely?

7-3. A construction zone on a freeway creates a bottleneck section where only one lane out of two is open. You may assume that traffic operates according to a Greenshields speed-density relationship, with $u_f = 120$ km/hour and $k_j = 70$ vehicles/lane/km. Approaching the construction zone, traffic flow is 2,700 vehicles/hour (or 1,350 vehicles/lane/hour). Inside the bottleneck zone, assume that the same speed-density relationship exists, but there is only one lane available. (a) What are the speed and density inside the queue that is building up in front of the construction zone? (b) What is the speed of the wave representing the back of the queue? (c) What is the speed and density of the traffic as it travels over the one-lane segment formed by the construction zone? (d) What is the speed of the traffic exiting the construction zone and resuming travel on the two-lane facility?

7-4. Below are data in standard units for U.S. cars and light trucks for the period 1970 to 2000. Use Divisia analysis to create a table and a graph for the period 1970 to 2005, showing four curves: (1) actual energy consumption, (2) trended energy consumption, and the contribution of (3) energy intensity, and (4) structural changes to the difference between actual and trended energy consumption.

	Car		Light Truck	
Year	Bil. vmt	Tril. Btu	Bil. vmt	Tril. Btu
1970	917	7,836	123	1,433
1975	1,034	8,537	201	2,287
1980	1,112	8,080	291	2,744
1985	1,247	8,262	391	3,171
1990	1,408	8,019	575	4,116
1995	1,438	7,866	790	5,275
2000	1,600	8,445	923	6,098

CHAPTER 8

Public Transportation and Multimodal Solutions

8-1 Overview

Urban public transportation systems, whether bus, rail, or ferry, provide a transportation solution that can, if used effectively, reduce congestion, improve ecological sustainability, and provide accessibility to urban travelers (intercity travel by public modes such as bus or rail is considered separately in Chap. 10). This solution assists travelers both with and without access to a private vehicle, since the former may benefit from having more convenient options for some trips, and the latter depend on local agencies to meet their mobility needs. In this chapter, we discuss the current status of public transportation and review fundamental requirements of public transportation system design. We also explain the transportation efficiency advantage of a robust public transportation system and illustrate several recent innovations that can improve public transportation performance. The primary focus of this chapter is on larger public transport vehicles that dominate markets in wealthier countries, from full-length buses (40 ft or 12 m) to the largest rail vehicles. Some concepts are, however, applicable to small- to medium-sized buses that are in widespread use in developing countries.[1]

8-2 Introduction

Public transportation constitutes any passenger transportation system that is open to the public and either operated or regulated by a public agency. Generally, public transportation involves the use of vehicles (buses, railcars of some sort, etc.) where a driver other than the passenger controls the movement of the vehicle, although it is possible to have public transportation vehicles that are controlled automatically with remote oversight. Also, public transportation systems usually operate on the principle of offsetting the total cost of the operation of the system (including staff and facilities) by consolidating

[1]For full-length works on urban public transportation that treat the subject in greater depth than can be covered in a single chapter, the reader is referred to Bruun (2014), Vuchic (2005), or Vuchic (2007). Note also that a thorough treatment of provision of semi-regulated markets found in developing countries, and the different vehicle types and operating practices that they engender, is beyond the scope of this chapter. For a more careful treatment of this topic, the reader is referred to, among other sources, the discussion of world "exemplars" in Schiller et al. (2009) or the writings of the Institute for Development and Development Policy (ITDP) in their periodical *Sustainable Transport.*

into the vehicle a larger number of passengers than would be found in a private vehicle (such as a passenger car, minivan, or sport utility vehicle [SUV]). Again, exceptions occur, such as taxi services or the concept of personal rapid transit. Because public transportation systems require consolidation of passengers, they are not as flexible as private vehicles in terms of the routes they must take or their ability to visit every street in an urban grid network. However, if public transportation systems are well-used with a sufficiently high average number of passengers on board each vehicle, they can reduce space consumption, energy use, air pollution, and greenhouse gas emissions as compared to private vehicles, all of which are important goals for urban areas.

Public transportation comes in several forms, each with specific characteristics in terms of cost and maximum capacity. Four major forms, widely deployed, are as follows:

- *Rail rapid transit:* These systems include subway (also called metros) and commuter trains that have few or no at-grade crossings with streets, and run underground or on elevated passageways. Subway systems in New York, London, Tokyo, etc. are examples.
- *Light-rail transit (LRT):* These systems comprise vehicles that run on rails but are typically smaller than those used in heavy-rail transit. They also use a mixture of guideways separated from street traffic and mixed in traffic, as well as occasional use of tunnels or overpasses to improve flow at key points. They are less expensive to build per kilometer of track than heavy rail. There are numerous examples around the world, including Manila, the Philippines; Manchester, England; Portland, Oregon, United States; or the example shown in Fig. 8-1*a* in Geneva, Switzerland. In recent years, the LRT railcar industry has pushed the maximum length of LRT vehicles while still allowing them to navigate street traffic where necessary. This development increases both maximum passenger capacity during peak periods and financial productivity of the asset, since the driver's wages are distributed among a larger passenger base, all other things equal.
- *Bus rapid transit (BRT):* These systems resemble light rail in their use of bus-only roadways separated from street traffic, but the use of buses instead of rail vehicles typically reduces total system cost. In some applications, BRT vehicles operate entirely on dedicated BRT rights-of-way and depend on off-board fare collection (similar to heavy-rail transit) while in others, BRT operates using conventional buses and routes extend from dedicated rights-of-way to and from local streets at the outer endpoints. Examples include Curitiba, Brazil; Bogota, Colombia; Cleveland, Ohio, United States (Fig. 8-1*c*); or Miami, Florida, United States (Fig. 8-1*d*). The BRT system in Fig. 8-1*c* features a dedicated right of way throughout its route, distinctive stations and platforms with off-board fare collection, signal priority at intersections, and articulated buses capable of boarding on either side. The busway for the system in Fig. 8-1*d* is exclusively for buses and street traffic crosses the route of the busway on an overpass, rather than at grade level. Bus service uses conventional buses with onboard fare collection that operate on city streets at the perimeter of the urban area and then use the busway to travel to the Central Business District (CBD).
- *Street transit:* These systems include motor buses (i.e., powered by internal combustion engines; hereafter simply referred to as *buses*), tramways and trolleys (i.e., electrically powered rail vehicles that operate using overhead catenary), and trolleybuses (also known as *trackless trolleys*), that operate

(a)

(b)

Figure 8-1 (a) Bombardier "Flexity" light rail transit (LRT) vehicle at a stop in Geneva, Switzerland; (b) conventional electric streetcar with high floor in mixed traffic, Toronto, Canada; (c) bus rapid transit, Cleveland, Ohio; (d) bus rapid transit, Miami, Florida. Typical public buses: (e) from an original equipment manufacturer (OEM) in the United States and (f) based on a U.S. school bus in Nicaragua.

Source for part (d): Photo: Jon Bell. Reprinted with permission.

(c)

(d)

Figure 8-1 *(Continued)*

(*e*)

(*f*)

Figure 8-1 (*Continued*)

entirely in the presence of street traffic. These systems are the least expensive to build and maintain, but they also provide the slowest service and are the most susceptible to congestion. Examples are shown in Fig. 8-1*e* and 8-1*f*. While operators in rich countries compete for ridership with other modes by purchasing modern vehicles equipped with low-floor access, and reduce headways to limit the maximum number of riders per vehicle (Fig. 8-1*e*), operators in poor countries hold costs down by using low-cost equipment (in this case a second-hand school bus from the United States (Fig. 8-1*f*) and by maintaining relatively long headways and maximizing occupancy, as many as 80 to 100 riders in a bus with 48 seats.

8-2-1 Brief History of Public Transportation

The earliest form of urban public transportation was the horse-drawn *omnibus* carriage of the nineteenth century, from which the name *bus* was eventually derived. In the latter half of the century omnibus operators added steel rails for the carriages to roll upon so that the ride would be smoother and encounter much lower rolling resistance, resulting in the first horse-drawn trams. The steam-powered railways of the 1820s and 1830s were adapted into the earliest urban rapid transportation systems in the form of the Metropolitan underground line, which opened in London in 1863 and connected the city center with Paddington Station in the West End. The availability of practical electric generation and transmission in the latter half of the nineteenth century made possible the use of electricity in place of horse power as a propulsion source, and in 1890 the first practical spring-loaded trolley pole with overhead electric wire was introduced by inventor Frank Sprague in Richmond, Virginia. Subsequently, cities around the world developed streetcar lines, and even new housing tracts called *streetcar suburbs* were built around these lines. With Sprague's further invention of the third rail, electrification was brought to underground subway systems in London as well as systems that first opened in Boston, Budapest, New York, Paris, and Istanbul by 1905.

In the early twentieth century, the internal combustion engine-powered bus emerged as an alternative to electric streetcars. By mid-century buses were in widespread use throughout the industrialized world, and proved to be more flexible in mixed traffic in terms of their ability to navigate obstacles or congested streets than the rail-bound streetcar. The paths of the various countries diverged after approximately 1950, with countries such as the United States, the United Kingdom, France, and Spain replacing most or all of their streetcars with buses, while other European countries such as Germany, the Netherlands, and Switzerland, and the countries of the Eastern Bloc continuing to operate large networks.

The reason for the emergence of LRT and BRT in recent years can be illustrated using a *cost versus level-of-service* (LOS) diagram, as shown in Fig. 8-2. Around 1950, public transportation primarily offered only heavy-rail and street transit systems (bus and/or streetcar, depending on the country). Between these two lay a gap in terms of service offering where for many cities, heavy-rail systems were too expensive and street transit was too slow to compete with the private automobile for passengers. LRT and BRT reduce the total system investment cost compared to rail rapid transit, but they also provide a measure of accelerated service, in which passengers travel faster than the stop-and-go speeds of street traffic. In this way, these modes are able to compete in market niches where previously only the car was present.

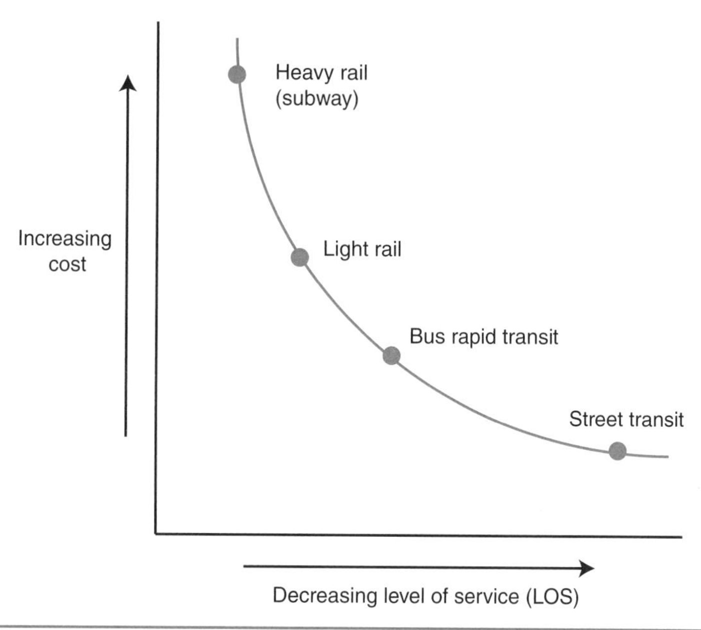

FIGURE 8-2 Cost versus level-of-service tradeoff for public transportation systems.

Since the 1960s, many cities of 1 million or less population that could not afford heavy-rail systems built LRT or BRT instead. Two early examples in the United States were the San Diego Trolley and the Blue Line in Los Angeles and Long Beach, California, which opened in 1981 and 1990, respectively. Today, the number of cities in the world that operate LRT and/or BRT has surpassed the number with heavy-rail systems. In the very largest cities as well, LRT and BRT can serve certain niches, for example, on the periphery of a city where demand is not high enough to justify a heavy-rail system. For instance, Mexico City uses primarily underground heavy rail but also uses some BRT and LRT operations. As an example of BRT used in the same role as a metro, the city of Bogota, Colombia, opened its Transmilenio system in the year 2000 and has since grown the network of BRT lines to 87 km as of 2012.

8-2-2 Challenges for Contemporary Public Transportation Systems

The primary challenge for public transportation systems, particularly in the United States but also to some degree in the other industrialized countries, is to increase their modal share. When implemented successfully, public transportation can improve the quality of life for its riders by providing a less stressful and more predictable means of conveyance in congested cities. For other riders, however, public transportation may be slow (especially taking into account full door-to-door time) or uncomfortable (due to crowding, poor-quality vehicles, and the like), and when incomes increase or road infrastructure improves, these riders may modal shift from public transportation to private car.

For many urban regions, if public transportation is to significantly improve overall urban travel efficiency, many more travelers must make use of it. Figure 8-3*a* gives non-intercity passenger miles for the United States in 2008, and clearly travel by light-duty vehicles dwarfs the other alternatives. The figure of 3.51 trillion passenger-miles constitutes more than 98% of the total. A small fraction of these movements are in taxis or carpools, but most are personal vehicles, often with a single occupant. The "Rail transit" figure includes urban movements of subways, commuter rail, LRT, and streetcar, while the "Bus transit" figure includes urban and commuter buses. The "Other" figure includes walking, bicycling, motorcycles, urban ferryboats, and other modes. Note that the "Car + LT" figure is approximated from the overall urban and intercity

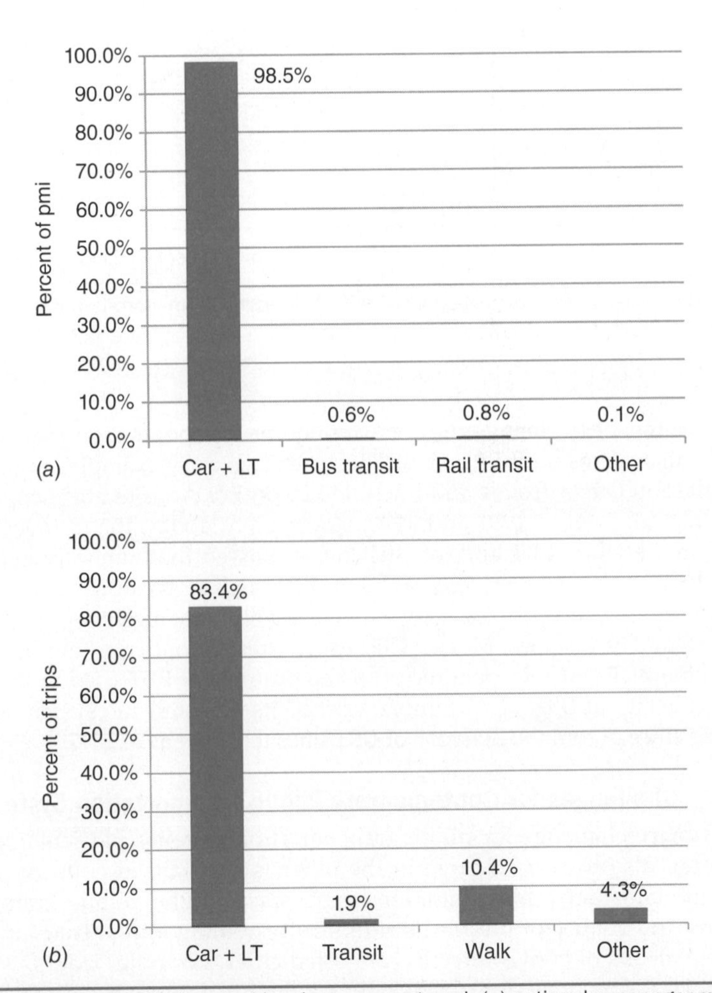

FIGURE 8-3 Modal share of urban and local passenger travel: (*a*) national passenger-miles (pmi) share—total: 3,565 billion pmi; (*b*) national passenger trips—total: 392 billion trips; (*c*) Manhattan intra-borough work trips; and (*d*) Cornell University graduate student commute to campus.

Note: "Car + LT" = car and light truck, that is, private light-duty vehicles.

Sources: U.S. Department of Transportation (2012); National Household Travel Survey (2011); New York City Government (2008); and Cornell University (2008), respectively.

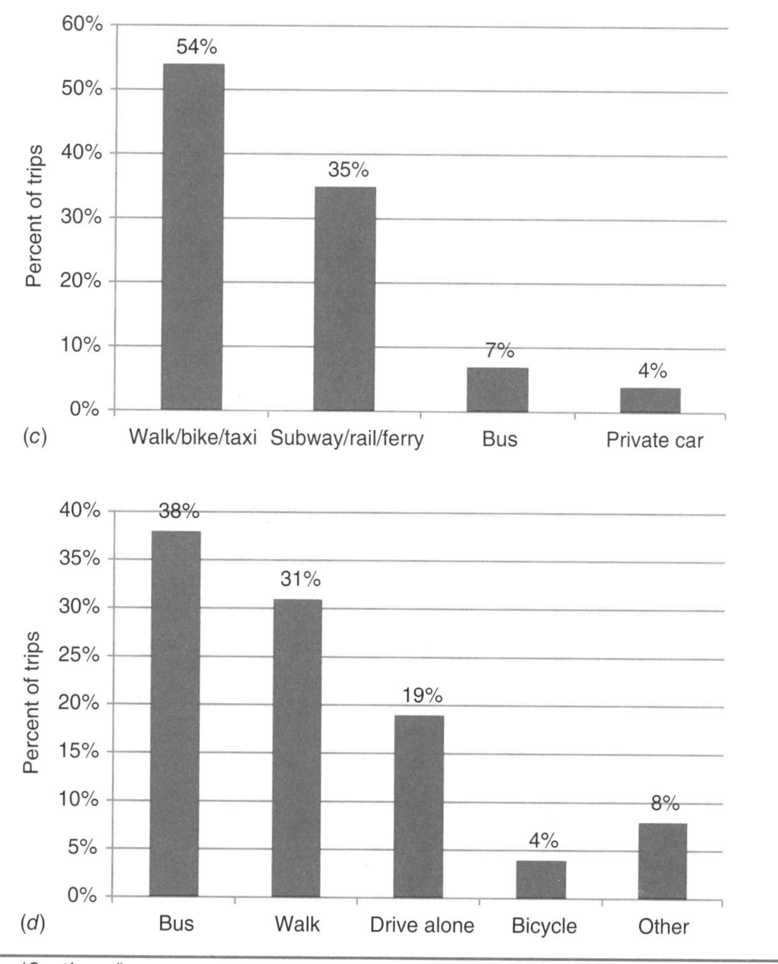

FIGURE 8-3 (*Continued*)

figure of 4.23 trillion passenger-miles published by the U.S. Department of Transportation multiplied by an estimated fraction of 83% of overall passenger-miles attributed to urban and local movements.

In addition use of passenger-miles as a measure of modal share somewhat overstates the importance of light-duty vehicles. Because of the convenience of this mode for traveling large distances (high speed, absence of wait time or transfers) it is relatively easy for light-duty vehicles to generate large numbers of passenger-miles compared to urban public transportation, urban nonmotorized modes, or other modes. Therefore, Fig. 8-3*b* gives modal share of individual passenger trips, where each trip counts equally regardless of length. The combined car and light truck mode remains dominant, though the percentage of the total is not as high.

Looking at more narrowly focused passenger transportation markets can reveal a different story. Figure 8-3*c* shows trip modal share for work trips within the Borough of Manhattan in New York City and Fig. 8-3*d* for graduate students traveling to campus at

Cornell University in and around Ithaca, N.Y. In each case, the private car is not the leading mode. In Manhattan, the private modes of walk, bike, or taxi lead (54% of trips), and in combination with the public modes subway, rail, or ferry comprise some 89% of work trips (Fig. 8-3*c*). Note that although taxi service relies on light-duty vehicles, these vehicles are not privately owned and avoid some of the burdens of private vehicles, such as space for parking while the occupant is at work. For graduate students at Cornell, thanks to extensive bus service and a highly walkable campus as well as the high cost of owning a car on a graduate student budget, the bus mode has the largest share of commute trips (38%) and bus and walk modes together have 69% of the overall share (Fig. 8-3*d*). Bicycling (4%) is also relatively high compared to the national average, although modest in absolute terms. As a counter-example, modal split can be very different the moment one looks at a slightly different demographic. Although not shown in Fig. 8-3*d*, the underlying data source reveals that for Cornell faculty and staff, the split is 55% driving alone (i.e., excluding carpools) versus 12% riding the bus, presumably due to longer average commute distance, more complex daily transportation needs, higher overall income, and other factors.

The Manhattan and Ithaca examples are offered as counter-evidence to the premise that urban passenger trips are uniformly dominated by light-duty vehicles. Permanent geographic factors determine some part of the modal share pattern; for example, in Manhattan high population density and relative lack of parking favor public transportation, taxis, bicycling, and walking. Modal-share patterns can also be affected by active policies to encourage a multimodal solution to transportation needs, such as efforts by Cornell University and the surrounding communities to expand bus service over the course of several decades, which led to the growth in graduate student use of buses. To the extent that other communities study and adopt these policies, modal share for public transportation (whether measured in passenger-miles or trips) can grow, leading to a larger impact on solving sustainable transportation problems.

8-3 Fundamental Calculations for Public Transportation Systems

A public transportation system can be understood in terms of three fundamental components. The fundamental building block is the *route*, which comprises the path between endpoints of a specific road, rail, or water vehicle along a particular path and with a designated number or name (e.g., "Route 23," "Blue Line," "Staten Island Ferry"). Routes may use different grades of rights of way, depending on the level of service that is required. The sum total of routes for a single mode or multiple modes constitute a *network,* which provides a certain breadth and intensity of service to an urban region taken as a whole. Finally, *network connectivity* takes into account the manner and degree to which intersections between routes in the network allow transfers between vehicles and facilitate origin-to-destination trips that cannot be completed using only a single route. This section covers these three components.

8-3-1 Grades of Right of Way and Types of Guideways for Routes

Public transportation routes operate over three possible grades of right of way (ROW), depending on the desired level of service, as follows:

- *ROW A:* Full separation of right of way from street traffic. Vehicles travel on a rail or road right of way that is laterally separated from ordinary street traffic.

In addition, where other traffic crosses, the right of way is separated vertically from traffic by an underpass, overpass, or tunnel. Examples include urban subway systems or BRT on elevated busways (or at-grade busways with traffic overpasses, as in Fig. 8-1*c*). One exception is suburban commuter rail, where at-grade crossings of the rail ROW and street traffic are possible. These are signalized in such a way, however, that rail vehicles passing through the crossing always have priority, so that the effect is the same as if they were vertically separate.

- *ROW B:* Lateral separation of right of way from street traffic. As in ROW A, vehicles travel laterally separated from ordinary street traffic but encounter cross-traffic at intersections where all vehicles may share priority, or transit may receive preferential treatment. In either case, ROW B has lower average speed than ROW A, but also lower construction cost.
- *ROW C:* No separation from street traffic, for example, vehicles such as buses or streetcars travel with traffic. Lowest cost and lowest average speed.

Rights of way can either have a fixed guideway, as in a track or monorail, or they can be nonfixed, as in a rubber-tired bus driving on an asphalt or concrete roadway. A tradeoff between fixed and nonfixed guideways exists in terms of the complexity of the combined vehicle-ROW system: a guideway places more complexity in the infrastructure design, but less complexity in the steering system, since the conductor or driver needs only to control the speed of the vehicle but not its lateral position in the guideway. Guideways also reduce the lateral space required for a given width of vehicle, thanks to this increased lateral control, and can save energy in the case of steel wheels on steel rails compared to rubber on pavement, as there is far less rolling resistance.

Example of BRT Right of Way: Lateral Separation without Guideway

As an example of design elements combined into a public transportation right of way, consider an urban BRT system with lateral separation such as the Euclid Avenue "Health Line" in Cleveland, Ohio. Since the system uses buses rather than railcars, the guideway is of course nonfixed. However, because it operates on separate ROW B infrastructure between termini in the Cleveland CBD and the eastern suburbs, the route never operates in mixed-city traffic, which facilitates the use of articulated vehicles for maximum capacity. The infrastructure is created by taking lanes away from the surrounding street, so that no additional lanes are built, nor is full separation (bridges, tunnels, etc.) provided at any point on the route.

The route installation incorporates several BRT features to improve performance and attract passengers, some of which are visible in Fig. 8-4. Off-board fare collection is offered at each station on the route, so that passengers can board and alight at any door on the bus and the driver need not manage fare collection while waiting at each stop. The stops incorporate uniformly designed platforms and shelters to *brand* the system and improve its image, while at the same time improving familiarity for passengers to make system access easier. Wheel-chair-accessible raised platforms accelerate the boarding and alighting process since passengers do not need to step up or down to enter the bus. Buses are equipped with doors on both sides of the vehicle so that they can approach a platform on either the right or left side. Lastly, a separate signal system at intersections reduces wait time for BRT vehicles by synchronizing green signals with bus arrivals.

(*a*)

(*b*)

Figure 8-4 BRT route and station design, Euclid Ave., Cleveland, Ohio, showing (*a*) separation of right of way, (*b*) standardized stop platform and shelter, and (*c*) separate intersection signal.

(c)

FIGURE 8-4 (Continued)

8-3-2 Design of Public Transportation Routes

A fundamental consideration of the design of an individual transportation route, independent of its specific geographic location, is in regard to the schedule to which it adheres. Let h be the *headway* or time elapsed between consecutive arrivals of vehicles at a stop. Let f be the frequency, or number of arrivals per hour; if h is given in minutes, which it typically is, then h and f are related by the equation

$$f = 60/h \tag{8-1}$$

For route design, a tradeoff exists between shorter and longer headways. Short headways will provide a more convenient service for the riding public but will also incur higher costs both for vehicle ownership and operating costs (wages, fuel, maintenance, etc.), so the headway must be matched to demand. To some extent, demand will respond to frequency of service, since higher frequency and shorter headways will encourage some travelers to switch to public transportation from other modes. In any case, for vehicles with a known passenger capacity C_v, the *capacity per hour* on a route can be measured in terms of maximum passengers per hour. For some types of public transportation, especially rail systems, the total capacity is a function of the capacity of individual cars multiplied by the number of cars in the train, which is given the symbol m, signifying multiples of the individual car capacity. Then total available line capacity Q measured in seats per hour is calculated as

$$Q = mC_v f \tag{8-2}$$

Example 8-1 illustrates the calculation of throughput for a representative route.

Example 8-1 A bus route uses buses with seating capacity for 38 and standing room for 28. The route headway is 15 minutes. What is the line capacity?

Solution Total capacity on each bus is 66 passengers. The headway of 15 minutes implies a frequency of 4 buses per hour. Since the service is a bus service, there is only one vehicle per arrival, that is, $m = 1$. The line capacity is therefore

$$Q = mC_v f = (1)(66)(4) = 264 \text{ passengers/hour}$$

Peaking of Demand and Measurement of Capacity Utilization

Ideally, public transportation routes would operate with nearly uniform occupancy of passengers on each vehicle from one end of a route to another. That is, passenger trips would be evenly distributed along the length of the geographic regions served by the route, so that most seated or standing places on the vehicle are occupied along the route, maximizing the possibility for earning revenue from the riding public.

In reality, typical urban form tends to undercut rather than support the goal of balancing occupancy on a transportation route. Many routes connect a suburban or rural *hinterland* with the CBD or other transportation hub, in which case vehicles tend to start a route in the hinterland with relatively low occupancy and gradually increase in occupancy until the destination is reached. In the opposite direction, the pattern is reversed: the vehicle starts out full and gradually empties. Routes may also start in the hinterland on one side of the urban area, pass through the CBD or hub, and continue to another hinterland in some other direction from the CBD or hub. In such a case, the peak demand will occur in the middle of the run, rather than at the end, because the largest number of passengers boarding the service on the run have trips that either originate or terminate in the CBD.

A second factor that leads to uneven occupancy is the difference in demand by time of day. The most common pattern is the morning and afternoon peaking of demand on weekdays due to work trips. Suppose we take the endpoint in the CBD or transportation hub as a specific location at which to measure peaking of demand over the course of a representative day. Not only does that location represent the point in space at which demand peaks along the length of the route, but if the location is observed for consecutive runs of the route arriving at that point, peaking will be observed during the morning peak period (sometimes called a rush hour), fall off in the middle of the day, and then peak again in the afternoon.

Therefore peaking occurs both in space along a route and over time for a fixed point on the route. It is desirable that the value of the maximum demand during the course of the day—that is, the point with maximum occupancy of the vehicle on the busiest or most congested run of the day—not be greatly in excess of the average occupancy observed on the route for all runs and in all locations along the route. The basic physics of passenger capacity implies that a high ratio of peak-to-average ridership will force the operator to provide a large capacity in the form of vehicle space, drivers, stop or station capacity, and potentially right of way, which will be underutilized much of the time. Some amount of imbalance is inevitable, but steps can be taken reduce it: a route may connect two or three hubs rather than passing through or ending in just one, or headways may be shortened during the peak and lengthened in the off-peak. These options are discussed further below.

Degree of peaking can be measured more carefully by considering the *Maximum Load Section* (MLS), which is the section of the route that on average has the highest number of passengers on board. Let P_{max} be the highest daily demand observed on the MLS, and P_{total} be the total demand observed past the MLS for all runs from the entire day. The peak-hour factor α_{peak} is the ratio of the highest demand to the total demand, that is,

$$\alpha_{peak} = \frac{P_{max}}{P_{total}} \tag{8-3}$$

Passenger counts at a given point on a run can be calculated by considering all boardings and alightings up to that point. Suppose there is a route with n sections for which boarding values b_j and alighting values a_j are known for each stop on the route. If we take P_j to be the accumulated number of passengers on the vehicle after stop j, then

$$P_j = B_j - A_j = \sum_{i=1}^{j} b_i - \sum_{i=1}^{j} a_i \tag{8-4}$$

where B_j is the cumulative boarding and A_j is the cumulative alighting up to stop j.

The value of P_j can in turn be used to evaluate utilization of capacity at a specific point on a route in the form of the *point load factor* α_j, where j is subscript of the number of the segment in the route:

$$\alpha_j = \frac{P_j}{mC_v} \tag{8-5}$$

It is natural to extend the calculation of utilization from a single point to the entire route. The variability in length of segments that make up a route complicates this calculation. To take this variability into account, the *space-averaged load factor*, LF_{avge}, is evaluated using occupancy P_j and segment length l_j.

$$LF_{avge} = \frac{\sum_j P_j l_j}{mC_v L} \tag{8-6}$$

Here L is the overall length of the route. Example 8-2 illustrates the calculation of MLS, point load factor, space-average load factor, and peak-hour factor.

Example 8-2 A representative bus route in Ithaca, N.Y., runs on the following itinerary between major stops, such that the route between consecutive major steps constitutes a single route segment on the route: Downtown, Collegetown, Central Campus, North Campus, Cayuga Heights, Community Corners, Kendal Place, and Ithaca Mall. Segment lengths and an origin-destination matrix of passengers boarding at each stop and their destinations are given below. Bus capacity is the same as in Example 8-1. Calculate (a) the MLS for this run, (b) the point load factor α_j on the MLS, and (c) the space-average load factor LF_{avge} for the run. Also, if the total demand for this route across all runs is 800 riders, calculate (d) the peak-hour factor α_{peak}.

Underlying data
Table of names of stops and distances for segments:

Stop No.	Stop Name	Next Stop	Distance to Next Stop (miles)
1	Downtown	Collegetown	1
2	Collegetown	Central Campus	0.6
3	Central Campus	North Campus	0.6
4	North Campus	Cayuga Heights	1
5	Cayuga Heights	Community Corners	0.7
6	Community Corners	Kendal Place	0.5
7	Kendal Place	Ithaca Mall	0.8

Origin-destination ridership pattern (stop number used to indicate start and end points):

To	1	2	3	4	5	6	7	8	Total
From									
1		10	20	6	4	3	1	4	48
2			8	12	0	1	3	2	26
3				1	1	0	2	1	5
4					3	2	0	0	5
5						1	3	1	5
6							2	2	4
7								6	6
8									
Total		10	28	19	8	7	11	16	

Solution The MLS is found by calculating how many passengers get on and off at each stop. For example, at the Downtown stop, the total number of boardings equals the sum of the boardings for each individual stop, or 48 riders, as shown in the table. Similarly, for the College Town stop, a total of 26 riders board. At the College Town stop, the only alightings possible are from the Downtown stop, of which there are 10 riders. Obviously, no passengers can alight at Downtown since the route originates there. Using the notation of segment 2 ($j = 2$) being Collegetown to Central Campus:

$$P_2 = B_2 - A_2 = \sum_{i=1}^{2} b_i - \sum_{i=1}^{2} a_i$$
$$= 48 + 26 - 0 - 10 = 64$$

By inspection, the number on board in Segment 1 is 48. Repeating the calculation above for Segments 3 to 7 gives 41, 27, 24, 21, and 16 riders, respectively. Thus the MLS is Segment 2, with a total number on board of $P_2 = 64$ passengers.

(b) Since the maximum capacity of the bus is 66 passengers, the point load factor is

$$\alpha_2 = \frac{P_2}{mC_v} = \frac{64}{(1)(66)} = 0.97 = 97\%$$

(c) To calculate the space-averaged load factor, we must first calculate the numerator of Eq. (8-6), taking into account the occupancy on each segment and the length of each segment, that is,

$$\sum_j P_j l_j = (48)(1) + (0.6)(64) + (0.6)(41) + (1)(27) + (0.7)(24) + (0.5)(21) + (0.8)(16)$$
$$= 178.1$$

The length of the route is in turn the sum of all the lengths of segments, or

$$L = \sum_j l_j = 1 + 0.6 + 0.6 + 1 + 0.7 + 0.5 + 0.8 = 5.2 \text{ miles}$$

$$\text{LF}_{\text{avge}} = \frac{\sum_j P_j l_j}{mC_v L} = \frac{178.1}{(1)(66)(5.2)} = \frac{178.1}{343.2} = 0.519 = 51.9\%$$

(d) There are 64 passengers on board in the peak segment, and 800 total boardings.

$$\alpha_{\text{peak}} = \frac{P_{\text{max}}}{P_{\text{total}}} = \frac{64}{800} = 0.08$$

As shown in Example 8-2, bus service is designed to meet peak demand on the MLS, and other segments inevitably have lower demand. Note that although the peak segment has 64 out of 66 nominal places on the bus occupied, by the end of the route the occupancy has dropped to 16 riders. This behavior tends to hold down average occupancy, even on a relatively well-occupied run such as this one. The lower value of LF_{avge} is preferred from a service point of view to either (1) exceeding the maximum capacity of the bus, and thus degrading comfort of individual passengers, or (2) having the driver judge the situation to be a "full bus," in which case passengers at individual stops will be bypassed, leading to customer annoyance and dissatisfaction. Thus, public transportation managers will tend to incur the extra cost of reducing headway rather than degrading service quality. Managers in low-income countries with limited resources may decide instead to keep headways longer and maximize occupancy on the bus (often well exceeding nominal capacity), thereby degrading customer service but maximizing fare revenue for a given incursion of operating cost. In any case, the implementation of a maximum capacity value is not precise: drivers almost never have access to technology that can track total occupancy on a bus in real time, and if the passengers are poorly distributed through the bus, the driver may judge it to be full before it reaches the nominal capacity (e.g., 66 passengers in this case).

A second point of discussion is the value of $\alpha_{\text{peak}} = 0.08$, which by itself does not convey the extent to which the ridership pattern on the route is "peaked" or not. In this example, the route is heavily peaked, with 64 passengers on board on the MLS compared to a total of 800 passengers per day on the route. Furthermore, 99 out of 800 boardings occur on this particular run. Suppose that through improvements in quality of service, advertising to the community, and the like, the daily boardings were increased to 1,600 passengers with no change in P_{max}. As a result, the value would change to $\alpha_{\text{peak}} = 0.04$, implying more even utilization of the bus capacity thanks to larger average numbers of passengers on board at other times of day.

Lastly, the evaluation of vehicle capacity is itself imprecise, because while the number of seats dictates a precise number of seated passengers, the standing component of the total capacity requires a judgment about how many passengers can stand per unit of floor area in the vehicle. This limitation applies to rail vehicles (rapid rail transit, LRT) as well as buses. The maximum capacity per area reflects regional or national expectations: in Europe and North America, a maximum of 4 passengers/m^2 is the assumed maximum, but in Asia and South America the figure is as high as 6 persons/m^2 in some locations. Also, with growing urban congestion, public transport operators are increasingly accepting load factors greater than one on some segments to accommodate rising demand on bus and rail routes with limited time and resources to expand capacity.

8-3-3 Urban Land Area Occupied by Public Transportation Systems

In dense urban areas, one of the primary motivations for developing and maintaining public transportation systems is the need to limit the total surface area occupied by the traveling public, especially at rush hour. In fact, one of the primary reasons that urban congestion occurs is that travelers in private vehicles with one or two passengers occupy not just the space around their physical bodies (on the order of 1 m^2 or less), but the entire space occupied by the vehicle. Public transportation reduces occupied area by consolidating passengers into vehicles that are larger than private vehicles but that can also transport much larger numbers of people. By this measure, the threshold for saving space may be as low as 6 or 8 passengers on a full-length public bus. This value is lower than the threshold for economic operation of the vehicle, where a significantly larger fraction of spaces must be occupied for fare revenues to cover all costs.

Since the goal of transportation is not to occupy space but to achieve throughput of passengers from origin to destination, a measure of occupied space should incorporate both time and area elements. One such measure is the *time-area factor*, $\overline{\text{TA}}$, which is calculated as follows:

$$\overline{\text{TA}} = \frac{WL}{q} = \frac{WL}{uk} \tag{8-7}$$

where W is the width of a single lane

L is the total length occupied by the vehicle

q is the flow rate, vehicles/hour

u is the speed, mph or km/h

k is the density, vehicles/km or vehicles/mile

It is desirable for achieving efficiency that the value of $\overline{\text{TA}}$ be as small as possible. From Eq. (8-7), the units for $\overline{TA}$ are thus m^2-hour/vehicle (or alternatively whatever unit is used to measure time, the units could instead be, for instance, m^2-second/vehicle). The length of the vehicle used in the calculation is the full length including both the physical length of the vehicle and the *shadow* required to maintain safe following distance between vehicles.

Equation (8-7) can be modified to measure time and space consumption in terms of passengers rather than vehicles by taking into account the average occupancy of vehicles α_{avg} to measure the *per-person time-area factor*, $\overline{\text{TA}}_p$:

$$\overline{\text{TA}}_p = \frac{WL}{\alpha_{avg} q} = \frac{WL}{\alpha_{avg} uk} \tag{8-8}$$

Urban Mode	Typical Maximum Throughput per Lane (1,000 passengers/hour)
Private vehicles	
Street traffic	1–2
Urban expressway	2–5
Public transportation	
Street transit (bus, streetcar)	3–15
BRT, LRT	5–20
Rapid transit	10–60

Source: Adapted from Vuchic (2007, p.192).

TABLE 8-1 Comparison of Typical Maximum Throughput of Passengers per Lane for Urban Transportation Modes

Note that since α_{avg} is measured in persons per vehicle, the units for $\overline{TA}_p$ are meter²-hour/person, or whatever units might reflect the chosen unit for time.

Typically, when comparing $\overline{TA}_p$ for private cars versus public transportation, the flow q for private cars may in many (though not all) cases be greater than that for public transportation, thanks to the former's ability to travel from point to point without intermediate stops to discharge and board passengers. However, due to the much greater values of α_{avg} possible in public transportation, the resulting value of $\overline{TA}_p$ will almost always be lower, unless the occupancy of the public transportation vehicle is very low. Again, a low value of $\overline{TA}_p$ is desirable from the point of view of conserving urban space.

Table 8-1 generalizes the capacity of passengers per hour per lane-mile for different urban transportation options. The values in the table express the maximum throughput that could be achieved per lane of road or single track for rail if all available spaces on the vehicle are occupied. Taking the two highest-performing urban modes as examples, the maximum number of spaces available on rapid transit, on the order of 40,000 per hour in some instances, is many times larger than that of expressways (5,000 per hour). An expressway system can compensate for the relatively low value per lane by providing many lanes of travel, and some urban expressways in the United States and elsewhere may provide 8 lanes or more for travel in each direction. The large width of such an expressway takes away space from other possible urban uses, creates stormwater management issues, and may have other detrimental effects. Example 8-3 provides an application of Eqs. (8-7) and (8-8) to a typical urban public transit situation.

Example 8-3S Urban passenger buses travel in an arterial street with lane width of 15 ft at an average speed of 14 mph including stops. The average occupancy of the buses is $\alpha = 18$ persons/vehicle, and the length is 40 ft. The buses maintain an average shadow of 33 ft. What is the time-area factor and per-person time-area factor?

Solution The total length occupied by each bus is 40 + 33 = 73 ft. Next, the speed and vehicle density are used to calculate flow q. Taking the given data:

$$k = \frac{5{,}280 \text{ ft/mile}}{73 \text{ ft/vehicle}} = 72.3 \text{ vehicles/mile}$$

$$q = uk = (14)(72.3) = 1013 \text{ vehicles/hour}$$

All quantities to calculate $\overline{\mathrm{TA}}$ are now known:

$$\overline{\mathrm{TA}} = \frac{WL}{q} = \frac{(15)(73)}{1{,}013} = \frac{1{,}095}{1{,}013} = 1.081 \text{ ft}^2 \cdot \text{hours/vehicle}$$

Next, we use the average occupancy value to calculate $\overline{\mathrm{TA}}_p$:

$$\overline{\mathrm{TA}}_p = \frac{WL}{\alpha q} = \frac{1{,}095}{(18)(1013)} = 0.0601 \text{ ft}^2 \cdot \text{hours/person}$$

Example 8-3M Urban passenger buses travel in an arterial street with lane width of 4.55 m at an average speed of 22.4 km/hour including stops. The average occupancy of the buses is $\alpha = 18$ persons/vehicle, and the length is 12 m. The buses maintain an average shadow of 10 m. What is the time-area factor and per-person time-area factor?

Solution The total length occupied by each bus is $12 + 10 = 22$ m. Next, the speed and vehicle density are used to calculate flow q. Taking the given data:

$$k = \frac{1000 \text{ m/km}}{22 \text{ m/vehicle}} = 45.45 \text{ vehicles/km}$$

$$q = uk = (22.4)(45.45) = 1{,}018 \text{ vehicles/hour}$$

All quantities to calculate $\overline{\mathrm{TA}}$ are now known:

$$\overline{\mathrm{TA}} = \frac{WL}{q} = \frac{(4.55)\,(22)}{1{,}018} = \frac{100.1}{1{,}018} = 0.0536 \text{ m}^2 \cdot \text{hours/vehicle}$$

Next, we use the average occupancy value to calculate $\overline{\mathrm{TA}}_p$:

$$\overline{\mathrm{TA}}_p = \frac{WL}{\alpha q} = \frac{100.1}{(18)(1018)} = 0.00298 \text{ m}^2 \cdot \text{hours/person}$$

Example 8-3 shows how public transportation can achieve relatively low values of $\overline{\mathrm{TA}}_p$. A stream of private vehicles traveling in the same arterial street lane as the bus in the example might achieve higher values of q, on the order of 1,700 to 2,000 vehicles/hour, thanks to faster speeds and greater density possible with shorter vehicles and shorter shadow lengths. However, due to smaller seating capacity and lower occupancy rates, the values of $\overline{\mathrm{TA}}_p$ will be higher, perhaps on the order of 0.2 to 0.4. (Calculation of $\overline{\mathrm{TA}}_p$ for a specific case is left as an end-of-chapter exercise.) Higher values of u for private cars do encourage the individual traveler to choose car over bus, since they may desire a shorter commute time, even if the bus can better achieve the societal aim of reducing space occupied by urban transportation.

A further space advantage accrues to the bus because of the parking requirements for private cars once they reach their workplace destination. A full-day accounting of space required for work-trip transportation incorporates both the time-area factor during commuting and parking space require, typically for 6 to 8 hours, between morning and evening work trips. Urban public transportation vehicles, whether bus or rail, are effectively always either in use during the work day (i.e., either stopped at a stop to board and discharge passengers, or in motion along the route) or else parked at a *depot* or *transportation center* which is almost always located away from the urban center on land that is not of as high value per square foot or square meter. Therefore, public transportation does not incur a parking time-area factor in the urban center, unlike private cars.

Lastly, one can use values from Table 8-1 to compare values of $\overline{TA}_p$ that might arise for urban freeways, heavy rail systems, and so on. Suppose a freeway moves vehicles at a rate of 2,100 vehicles/hour/lane with average occupancy of 1.2 passengers/vehicle. Each freeway lane is then moving approximately 2,500 passengers/hour. Increasing average occupancy to 2.4 passengers/vehicle would move on the order of 5,000 passengers/lane/hour, equivalent to the highest value shown in the table. In theory, if one could fill most or all of the average 5 seats available in each private vehicle, much larger values could be achieved, on the order of 9,000 to 12,000 passengers/lane/hour. Experience shows, however, that it is simply not practical to so closely coordinate passengers to ride together all in the same vehicle, given all different time and destination requirements. Similarly, in theory one could have higher maximum throughput per lane for rail rapid transit. As shown in Table 8-1, the observed capacity of rail rapid transit is already much higher than that of private cars. However, the physical capacity of rail transit could be even higher: 12-car subway trains might each carry 2,500 passengers at rush hour, operating at 90 seconds headways and resulting in a maximum flow of 100,000 passengers/track/hour. Here again, however, it would be impossible to coordinate the movement of so many passengers in and out of stations and on and off trains. Therefore the maximum value in the table is lower, although systems in New York and London, for example, do achieve values on the order of 60,000/hour with long trains and short headways

Total Door-to-Door Time for Public Transport Systems

Previously we have discussed the line capacity Q for a public transport system, which reflects both the vehicle capacity and the frequency of vehicles per hour. A related measure of interest to the traveler is the average travel speed of the vehicle, which is independent of the frequency and headway. A typical ROW C system in traffic, whether bus or streetcar, may achieve speeds of 11 to 15 mph, including stops at intersections and transit stops. Thus a *line-haul* distance of 4 miles might require 16 to 22 minutes using street transit.

To fully understand the traveler's choice of mode for an urban trip, however, it is necessary to consider not only the line haul travel time t_{line}, but all other elements of the total door-to-door time t_{OD}, including access time from origin to boarding stop t_a, waiting time at stop t_w, and egress time from alighting stop to final destination t_e:

$$t_{\text{OD}} = t_a + t_w + t_{\text{line}} + t_e \tag{8-9}$$

For public transport routes with short headways, it is reasonable to assume that, on average, the passenger arriving at the boarding stop will wait half of the headway time h before the arrival of the next service. Thus we can rewrite Eq. (8-9) as follows:

$$t_{\text{OD}} = t_a + h/2 + t_{\text{line}} + t_e \tag{8-10}$$

Empirical evidence suggests that when headways are approximately 15 minutes or longer, passengers will plan their arrival at the stop to coincide with the service they plan to use, rather than arriving at the stop without consulting a schedule, and waiting for the next available service. In this situation, the assumption of $t_w = h/2$ no longer applies. For short headways, however, the traveler typically does not coordinate their arrival at the stop with the arrival of the service.

The implication of Eqs. (8-9) and (8-10) is that public transportation should be assessed on the basis of t_{OD} and not solely on the basis of t_{line}. Suppose a public transport journey consists of a walk of 5 minutes to a stop, a 3-mile ride on a bus traveling 12 mph with a 15-minute headway, and a 5-minute walk to the destination. The line-haul time is 15 minutes, and the door-to-door time is 32.5 minutes, so that the former is 46% of the latter. For comparison, suppose travel by car were available to the traveler with the same 3-mile trip distance, 15 mph average speed in city traffic, and 2 minutes at each end to get from home to the parked car and from the car to inside the door of the destination. The total trip time would be 16 minutes, of which 12 minutes or 75% is line-haul time.

Based on this comparison, two recommendations for attractive public transportation emerge. First, for short trips, high frequency and short access distance are important to hold down t_a, $h/2$, and t_e. Therefore having a route network that thoroughly covers an area with frequent service is important. For longer trips, t_{line} dominates total trip time, so high average linehaul speed is important.

8-4 Methods to Enhance Public Transportation System Performance

To increase modal share and have a greater impact on the overall efficiency of urban transportation in a given region, public transportation systems must improve performance. Recent experience has already shown that individual agencies have been capable of launching outstanding new or revamped service, and that when such service is provided, travelers respond by shifting their travel choices from other modes. In the first part of this section we consider improvements to infrastructure, and in the second part operational improvements to service, once infrastructure is in place. In the last part, we provide a case study of LRT and BRT systems to highlight how these methods can be applied. Whereas the previous section considered the planning of individual routes, this section considers enhancements at all three levels (routes, networks, and network connectivity).

Innovation in public transportation systems occurs in all parts of the world and on every continent. It is not possible in the limited space of this chapter to provide an exhaustive list of all types of possible innovations, with international examples of each. The intention is instead to provide a limited and representative list of examples that illustrate the breadth of what is possible.

8-4-1 Infrastructure Enhancements

Since ROW type is one of the most fundamental characteristics of public transportation infrastructure, one of the most significant possible enhancements is to upgrade ROW. Thus a bus or streetcar operating on ROW C can achieve faster speed and improved reliability by upgrading to ROW B with a separate ROW adjacent to general traffic. In this way, street transit can become LRT or BRT. In the same way, an existing LRT or BRT can be enhanced by construction of a viaduct or tunnel to create full vertical separation on some segments as well as lateral separation on others. It is common practice for transit systems that operate either ROW C streetcars or ROW B LRT systems to create ROW A infrastructure in the most congested parts of the network, typically in or near the CBD. Examples include Boston, Mass.; Philadelphia, Pa.; and The Hague, The Netherlands.

One form of ROW C to ROW B conversion applicable to either bus or streetcar/LRT is the *transit mall*, in which an arterial street in the CBD is converted to transit use only, by excluding general private vehicle traffic (taxis and bicycles may or may not be allowed). Buses and rail vehicles can access the transit mall from connecting streets (ROW C) or dedicated bus or rail infrastructure (ROW B). If parallel streets can adequately accommodate general traffic, the transit mall can improve public transport without causing excess congestion elsewhere. Also, the ambiance of the transit mall can make it an attractive CBD destination, although impact on retail sales levels can be an issue, since in some cases retailing on a transit mall can be an extra attraction to shoppers, but in other cases removal of cars can depress retail sales levels if shoppers find it difficult to carry merchandise to their cars. Examples include Denver, Colo. and Portland, Oreg. as well as numerous cities in Europe.

Public transportation infrastructure can specifically be upgraded to consolidate ROW C routes that travel on parallel but separate routes from the hinterland to the CBD onto a single higher-performing artery. Figure 8-5 shows three different options in current use in the Philadelphia (U.S.) metropolitan region. In all three options, routes originate in the hinterland on the periphery of the urban area. Option A represents the most basic version, where buses travel on parallel routes before converging on the terminus at the CBD. In the enhanced Option B or Option C, the routes converge on A grade-separated (ROW A) length of infrastructure, called a *trunk section*, which provides faster travel to the passenger. In other applications, the trunk section could be ROW B instead of ROW A. Vehicles arriving from the hinterland travel the entire distance to the CBD without terminating, as in Option B. They may also terminate at the entry to the trunk section, in which case travelers transfer to a dedicated trunk section vehicle, as in the

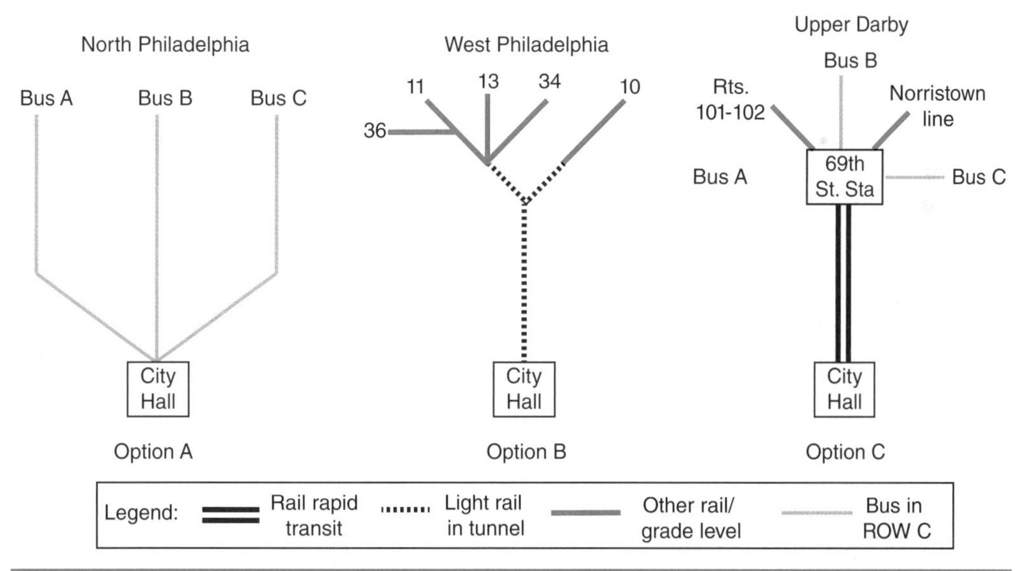

Figure 8-5 Examples from Philadelphia, Pennsylvania: (*a*) street transit, (*b*) LRT with routes in tunnel in CBD, and (*c*) feeder system for trunk line rail rapid transit system on full ROW A.

Notes: Diagrams are not to scale and do not have consistent compass orientation. "Bus A" etc. in Options A and C are not actual route numbers but represent an array of bus routes in street traffic converging on a network hub. See text for further explanation.

Source: Adapted from Bruun (2013, p. 65).

heavy rail line in Option C. Several variations on Options A to C are possible; for example, bus or BRT vehicles could be used in place of rail.

Each of the three options in Options A to C has advantages and disadvantages. Option A provides the most area coverage, since the three routes shown (additional routes exist in the actual North Philadelphia network) cover a wide geographic area for most of the distance approaching the CBD. The downside is slow travel speed, however, so many urban passenger transit networks have moved toward some variant of Options B or C, with peripheral bus routes feeding passengers to the trunk line, rather than traveling the entire distance to the CBD. Option B has the advantage over Option C of direct travel to the CBD without transfer, which empirical evidence suggests riders prefer to avoid if they can. Option C provides better balance of short headways on both the feeder and trunk lines, because both terminate at the transfer hub (the 69th St. terminal in Upper Darby in the example) and are therefore able to turn around and serve a shorter segment. In Option B, with many lines needing to share the trunk section, the service will tend toward very short headways on the trunk line but longer headways on the individual routes at the periphery, which may be unsatisfactory to some riders.

Enhanced Transfer between Routes and between Modes

Initially, the goal of any urban public transportation system at a network level is to provide area coverage, that is, accessible public transportation for as much of the surface area or population of an urban region, with sufficiently frequent service that passengers will choose public transportation over other options. In larger urban regions (especially with population 500,000 or more), upgrading parts of the network from ROW C to ROW B or A will further enhance the attractiveness of the network.

An equally important though historically overlooked aspect is *network connectivity* for passengers needing to transfer between vehicles, either between vehicles of the same mode, or between modes. Infrastructure enhancements can support this goal (operational aspects of network connectivity are also discussed in the section). For example, design of interchange points that minimize distance that passengers must travel between vehicles can save time and confusion. An interesting example from Hong Kong and elsewhere is the coordinated design of subway stations to facilitate "cross-platform" transfers from one line to another. In the simplest form of cross-platform transfer, subway trains traveling in the same direction arrive on either side of a platform, and passengers can change from one to the other line by walking across the platform between trains.

In the case of the Hong Kong network, however, consecutive stations are designed so that at one station passengers can walk across the platform to go in the opposite direction, and then at the next station passengers transfer across platforms to go in the same direction, as shown in Fig. 8-6 for the "Prince Edward" and "Mong Kok" stations. For example, a passenger arriving from the Po Lam direction can alight on the platform at Prince Edward and transfer across the platform to board trains in the opposite direction toward Tsuen Wan. Passengers with destinations on Hong Kong Island would stay on the arriving train for one more stop and alight at Mong Kok, where they would walk across the platform to board continuing trains. Note that the platforms in the stations are shown side by side in the figure because it is drawn in two dimensions; in practice they might be constructed vertically with one platform on a level above the other.

Efficient transfers between modes are important as well. At key network nodes, many cities have developed multimodal transportation centers designed to streamline the transfer of passengers between modes (e.g., from bus to rail networks) in a manner

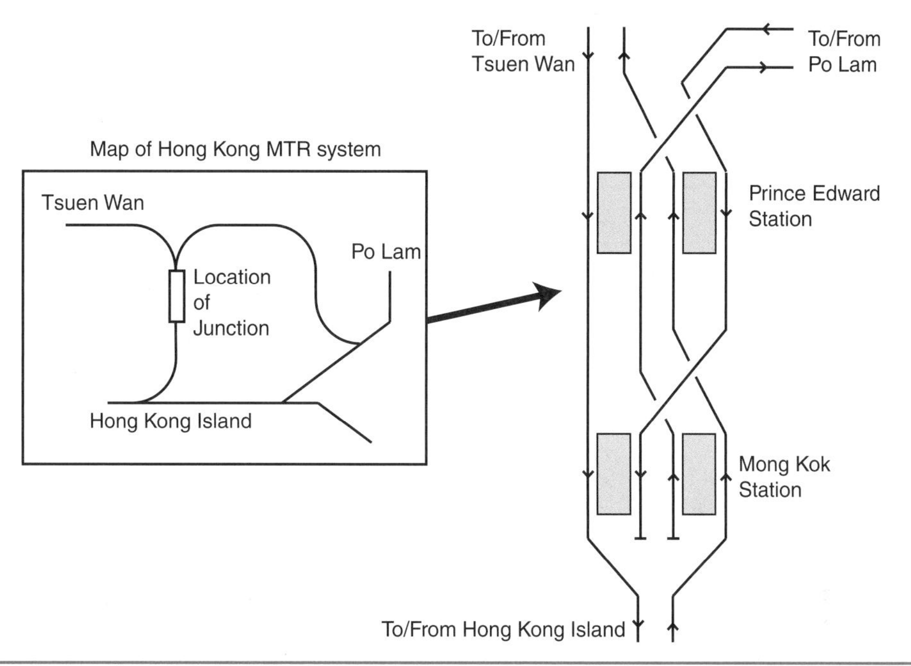

FIGURE 8-6 Coordinated station design at Prince Edward and Mong Kok metro stations in Hong Kong heavy rail network.

that is both convenient and easy to understand. In the metro Washington, D.C. area, for example, major transportation centers have grown up around metro (subway) stops on the perimeter of the District of Columbia, that are served by metro lines such as the Red Line or Orange Line, the bus network of WMATA (Washington Metro Area Transportation Authority), as well as the bus networks of adjacent counties in the states of Maryland or Virginia. Taxi and bikeshare services are also available.

As an example of a multimodal transportation center that incorporates both urban and intercity travel in a smaller-population urban region, the city of Syracuse, N.Y. (metro region population approximately 660,000), has created a multimodal center served by the metro bus network, intercity bus lines, and intercity rail (Amtrak) (Fig. 8-7). Thus travelers arriving by rail or bus into the Syracuse region can transfer to the bus network to reach the CBD. The center is also served with a taxi stand and is accessible by bicycle or on foot, in addition to providing parking for private vehicles.

Use of Information Technology

As mentioned previously, information technology can usefully be applied to transportation systems in the form of intelligent transportation systems (ITS) to improve the flow of transportation, manage disruptions in the network, and provide information to travelers in a clear, timely, and accurate format.

One of the simplest applications to implement for public transportation systems is an electronic passenger information system that provides schedule information at stops and terminals on screens or other electronic devices. Management can update information remotely, so that when schedules change, up-to-date information is provided, rather than outdated schedules as can happen when printed schedules are used at stops.

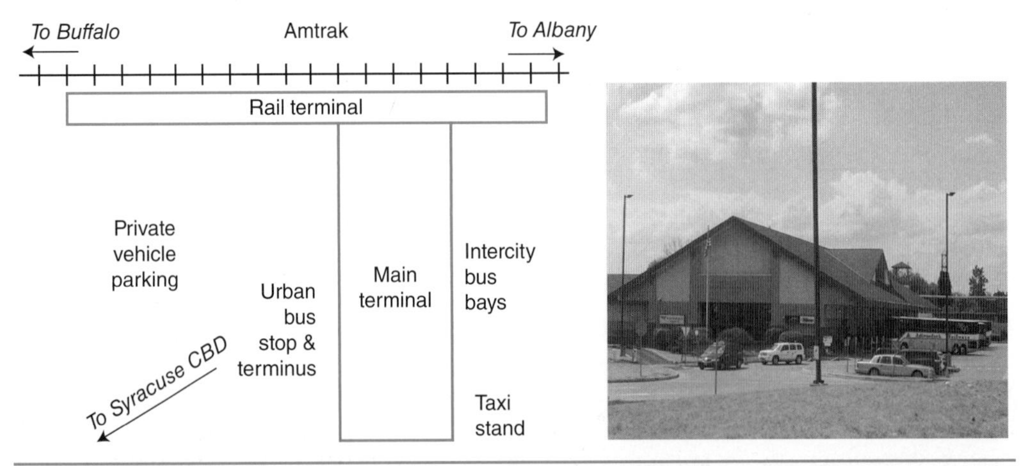

FIGURE 8-7 Multimodal transportation center in Syracuse, N.Y., including plan view schematic (left) and elevation view photograph (right).

A more sophisticated and very useful application is to provide not only schedule information but also real-time information about the actual location of arriving service. Thanks to GPS devices mounted on vehicles and wireless communications, it is possible for each vehicle to communicate its real-time position to the network, which can then provide information at each stop or station. Waiting passengers can then learn the expected number of minutes until the next arrival, and can be given information about any delays or service disruptions that may pertain. The same information can in principle be provided to handheld smartphone devices.

Intelligent transportation systems can also be used to improve system service. An ideal application of ITS to public transit is the *transit signal priority (TSP) system*, in which a public transportation vehicle (e.g., bus or LRT car) approaching a signalized intersection can communicate wirelessly with the traffic signals. If stops are located on the far side of the intersection, the instruction is to hold the green signal until the vehicle is able to pass through the intersection. If stops are on the near side, the driver can push a button when ready to depart, and the bus will receive an early green before the other lanes. The result is a reduction in travel time between stops, and shorter average journey times for passengers.

8-4-2 Service and Operational Enhancements

Once infrastructure is in place for public transportation system, there are many enhancements on the service and operational side that can improve performance. One recent trend has been toward the provision of greater capacity in vehicles to compensate for limited ability to increase physical capacity of infrastructure. Thus, railcar makers have developed double-deck cars for commuter rail systems, while bus and light-rail vehicle makers have expanded the manufacturing of single- or multiple-articulated buses or railcars. One potential advantage of rail vehicles over buses is that, since the rail system provides a precise guideway for the vehicle's travel, it is easier to couple multiple vehicles up to the maximum length attainable given length of station platforms or staging areas in maintenance yards. Where feasible, heavy rail or LRT operators can add railcars to each train so as to increase the capacity of each scheduled arrival.

Optimizing the scheduling of public transportation services so as to maximize the practicality of transferring between services is a complex and demanding challenge. Since there are naturally tradeoffs between the convenience of system use for travelers in one part of the network versus another, planners must carefully weigh the number of possible transfers and the likelihood of travelers making each possible transfer when deciding schedules. This type of optimization is an ideal application for analytical tools from the field of operations research, since the analyst must search among a very large number of possible combinations to identify a solution that is at or near the optimal value.

One useful tool especially relevant to bus networks is the *timed-transfer concept*, where bus services converge on a central node in the network on a regular basis (typically with frequency of one or two times per hour). When the service arrives, passengers have time to transfer to buses going in any direction, before the buses simultaneously depart. The routes are then designed so that all vehicles return to the timed-transfer node on a regular basis, either at the end of a single run, or else after visiting some other terminal point and then returning on the next run.

Figure 8-8 shows a timed-transfer network with a hub and three spokes in use in Ithaca, N.Y. A single hub is situated at the CBD, from which bus routes serve East Hill, West Hill, and South Hill. A *pulse* of the timed-transfer system occurs every 30 minutes; buses in the figure are shown in their relative positions at the beginning of a pulse. The number of buses on each route varies with route length and level of demand. For instance, the South Hill route is relatively short, so that the bus can

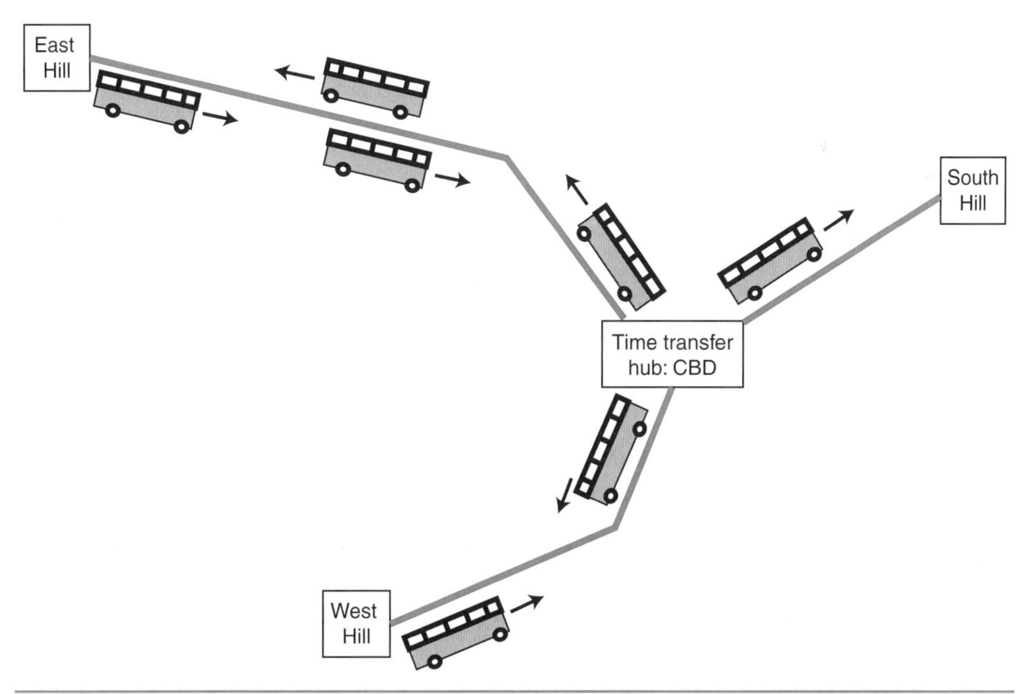

Figure 8-8 Timed-transfer network with hubs at CBD serving three routes of varying length and headway. See text for explanation.

return within the 30-minute window for the next pulse. The West Hill route is longer so two buses are needed; when one bus is leaving from the CBD, the other is leaving from the far terminus, and at the next pulse their positions are reversed. The East Hill route is of similar time duration (i.e., two pulses per round trip) but due to higher demand the headway is 15 minutes and four buses are deployed. Thus, two buses participate in the timed transfer and two do not, and riders wishing to transfer from the East Hill route to the other two with minimum waiting time must plan their choice of bus accordingly. In general, the routes should be designed so that they require most, but not all, of the time between pulses. For example, if the headway is 30 minutes as in this example, the route might be designed to take 25 minutes to complete under good traffic conditions, leaving 5 minutes of *dwell time* at the end points between runs. Efficient use of time between pulses also holds down total dwell time at the terminus and accompanying emissions from excessive idling of diesel engines. If travel times become unreliable, timed transfer systems also become unreliable, and measures need to be taken to restore reliability.

Approaches to Managing Peaking of Public Transport Demand

One of the key challenges to the public transport operator is to manage the peaking of passenger demand during the weekday morning and afternoon rush hour periods. (On weekends and holidays, there are generally not pronounced periods of peak and ebb demand, although demand in the early morning and late evening is usually low compared to demand in the middle of the day.) The operator wishes to avoid excessive peaking of equipment requirements, since, at the margins, certain vehicles or railcars will be purchased that generate very little revenue and are difficult to justify financially. At the same time, the rush hour traveler desires relatively short headways and adequate comfort during the commute to justify the choice of public transport over other possible modes. A tradeoff therefore arises between these competing goals.

A common solution is to shorten headways. For instance, a route that runs at 30-minute headways in general may adjust to 20- or 15-minute headways during the peak, to handle the increase in demand. Vehicles used in this way may see 4 hours of activity per peak period, or 8 hours per weekday; some amount of underutilization is inevitable. One strategy is to use the newest vehicles continuously (e.g., from approximately 6 a.m. to 10 p.m.), and then use older vehicles during the peak periods only. Uneven labor requirements are an issue as well: agencies may offer *split shifts* that provide drivers an 8-hour shift by allocating 4 hours during the morning peak, then time off, and then 4 more hours during the afternoon or evening peak. Even with shortened headways and added capacity measured in seats per hour, vehicle occupancy during the peak may be *standing room only*, that is, all seats are occupied, and passengers who board later must stand for all or part of their journey. However, as long as passengers are not prevented from boarding because the driver judges the bus to be full and refuses to allow more passengers on board, passengers will generally tolerate some loss of comfort during the rush hour, as other options (driving private car or bicycling in congested conditions) are also disadvantageous.

Given that public transport operators may need to increase service frequency at rush hour in any case, some of these services can effectively be operated as *express* services, which stop at only the most important or highest-demand stops so that travelers moving long distances along a route can make their journey more quickly. Operators may also designate certain services as being *semi-express*, meaning that they stop less frequently than ordinary service (also known as *local* service) but more frequently than express service.

A successful express or semi-express service must consider the needs of passengers who wish to travel longer distances with few stops, and remaining passengers who need access to short distance travel on local service. The operator ideally provides reasonably regular service to both types of riders. Coordination between the two types of services is also important. This feature is especially visible in metro and regional rail services, where express and local arrivals at key stations are synchronized so that passengers can exchange between the two types, as dictated by their origin and destination station. During off-hours on weekdays or on weekends, if the level of demand does not justify operation of an express service, operations may revert to local-only service, since in any case the operator is obligated to provide some amount of service to all stops in the system. Coordination of express and local service is also possible in bus, BRT, and LRT systems, though it is not as common. In general, such coordination requires station layout modifications if it is a modification to an existing system. Furthermore, express/local services are greatly simplified to design and operate if there are multiple tracks for rail or multiple lanes for road vehicles.

The fare structure in a public transport system can be used alongside operations planning and management to encourage some leveling of demand between peak and off-peak periods. Off-peak fares can be discounted: work commuters who must travel during the peak period pay the full fare out of necessity, but travelers making discretionary trips are encouraged to wait until the peak period ends to obtain the discounted fare. The exact fares to charge must be set in a way that provides a sufficient discount to move discretionary trips out of the peak period, while at the same time meeting the revenue goals of the system. As an example, a representative peak trip from the outskirts to the center in the Washington, D.C., subway system might cost $3.30, but the same trip costs $2.15 off-peak.[2] A variation on the off-peak fare structure in use in London, England, is an off-peak day pass, sometimes called a *day saver*, which for a single price allows unlimited travel starting from the end of the rush period at 9:30 or 10 for the remainder of the day. Although the day-saver passengers then mix with peak-hour travelers during the afternoon peak period, experience shows that the timing of the homeward work journey in London is more spread out than the morning, so that the peaking is not as great.

Use of peak and off-peak fares to shape public transport usage is not without its detractors. There is a second school of thought that says that increasing fares in the peak discourages use of public transport right when it most needs to be encouraged in order to reduce pollution and congestion. In practice, many systems have no spare capacity in the peak anyway so they charge what the market will bear, and use the additional revenue to support off-peak service.

8-4-3 Flexible Routing in Bus Service and Demand Response

One potential advantage of bus over rail service is that, since buses are not limited to a fixed guideway, they can easily deviate from a route as the situation dictates. One application is that during road reconstruction, buses can either change lanes to avoid a lane closure, or travel via a different street to avoid construction or temporary traffic congestion. The situation for ROW C rail vehicles (trolleys, streetcars) is more complex. Here street reconstruction requires careful coordination with street maintenance authorities. In some

[2]Price in January 2013.

cities where an adequate grid of streets with tracks exists, such as Prague, Czech Republic, or Melbourne, Australia, it is possible to close a section of track for repairs and detour the affected streetcar routes around the affected area on parallel streets. Many other streetcar networks are too sparse for such a contingency to be practical, however.

Fixed-Route Bus Service with Route Deviation

With the adaptability of bus routing comes the possibility of *flex-routing*, also known as *route deviation*, to better serve passengers in the near vicinity of a route. With route deviation, the vehicle travels entire length of the fixed route and visits all scheduled stops to serve fixed route passengers. Between scheduled stops, the vehicle can deviate from the route briefly to visit a particular address for passenger pickup or dropoff that has been scheduled in advance. The Potomac and Rappahannock Transportation Commission in northern Virginia outside of Washington, D.C., operates one of the early pilot projects of this type of system.

Figure 8-9 illustrates the function of route deviation. The bus first visits Stop 1, and then a representative rider is picked up between Fixed Stops 1 and 2. The vehicle then returns to the route at or close to the point where it deviated, and in any case before the next scheduled Stop 2. The rider is then dropped off between Stops 2 and 3. In anticipation of route deviation, some slack is built into the schedule, but delays are not permitted to the point of delivering poor service to passengers traveling between fixed stops.

Scheduling off-route pickups and dropoffs requires extra sophistication on the part of the transit agency. Passengers requiring a route-deviation pickup or drop-off call the dispatching office of the transit agency and then are scheduled into the run of the route, either in advance of the departure of the run or possibly while it is en route if there is sufficient time. The underlying decision support system evaluates the passenger's

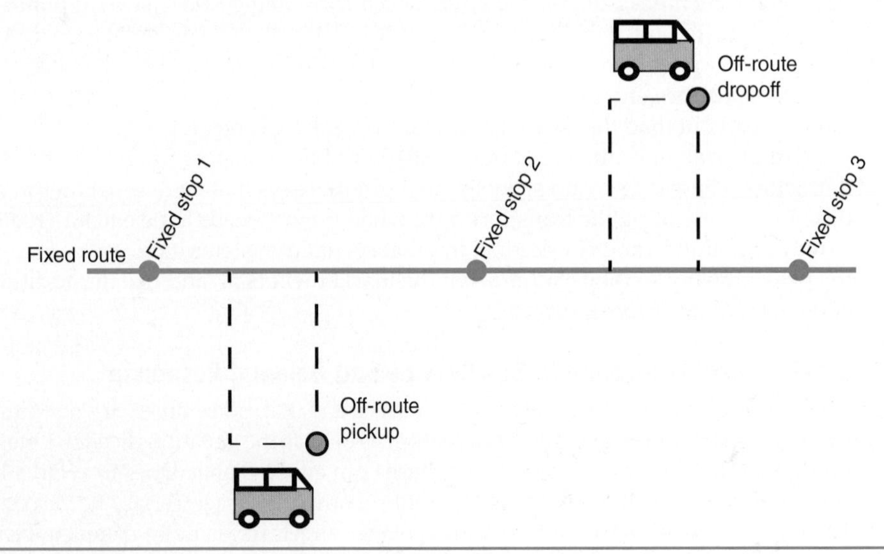

FIGURE 8-9 Example of route deviation from a fixed route to pick up and drop off passengers in vicinity of route.

Source: Adapted from Bruun and Marx (2006).

request and can approve it instantaneously over the phone, provided that its algorithm determines that the amount of deviation meets schedule adherence requirements. A route-management system on the bus with graphical interface and wireless communication with the dispatching office helps the driver to manage route deviation stops, directing the driver to the appropriate address and confirming pickup and dropoff.

The business argument for flex-routing is that in return for a relatively modest investment in ITS equipment (decision-support software, wireless communications, on-board graphical interface for the driver), the route can potentially generate significant additional ridership. Part of the ridership comes from passengers whose schedules do not allow access and egress time to and from fixed stops but are able to ride the bus if the service is provided door to door. Another part comes from *paratransit* ridership that is able to use the new service. Paratransit riders are those who are elderly or disabled and cannot travel to a fixed stop on their own, and for whom the 1990 Americans with Disabilities Act (ADA) requires the public transport operator to provide an alternate means of transportation, usually in minivans or minibuses that have facilities for, among other things, loading and unloading wheelchairs. One segment of the paratransit market consists of riders who are not able to walk ¼ to ½ mile to a fixed stop but are able to walk to the curb of the street outside their residence if the flex-routed bus can meet them there. Since from the agency's perspective full paratransit service is expensive to provide on a per-passenger basis, the potential to both increase ridership on scheduled routes and reduce paratransit cost is attractive.

Demand-Responsive Bus Service

As an alternative to route deviation, a public transit operator can provide a service that is entirely determined by requests from passengers for dropoff or pickup. Such a service might link a terminus at a focal destination such as the CBD of an urban area or a major transit center with a less densely populated hinterland. Runs of the demand-response service leave the terminus on a schedule, and then travel through the area within the demand-response catchment zone, either dropping off outbound passengers who have boarded at the beginning of the run or picking up inbound passengers intending to travel back to the terminus. As an example, Tompkins Consolidated Area Transit (TCAT) in Ithaca, N.Y., operates a demand-response route originating on the campus of Cornell University and servicing a mixed suburban and rural zone approximately 7 miles long by 2 miles wide (Fig. 8-10).

8-4-4 Potential Impact of Increasing Public Transport Modal Share on Energy Efficiency

Enhanced public transportation that uses the approaches and techniques described in this section to win ridership has the potential to improve overall urban passenger transportation energy efficiency. This outcome assumes that the service is well-utilized, with high space average load factors and low per-person time area factors. On the other hand, it takes time to grow public transportation service, since fixed infrastructure, vehicle fleets, and workforce must all be expanded in response to available investment. If an urban area is initially dominated by the use of private cars in terms of trip or passenger-kilometers modal share, there are limits on how much impact modal shifting can have on total energy consumption in the short to medium term.

Example 8-4 illustrates the prospects and limitations for energy savings in a growing urban transportation market.

FIGURE 8-10 TCAT Route 41 demand-response service in Ithaca, N.Y.

Example 8-4S A metropolitan region generates 4.7 billion passenger-miles (pmi) of transportation demand per year and is experiencing 2% per annum growth in total pmi, and a 0.5% annual decline in energy intensity of passenger-miles. Current energy intensity of automobile travel and public transportation is 3,300 Btu/pmi and 1,500 Btu/pmi, respectively, and share of passenger-miles is 90% and 10%, respectively, for those two modes. For the purposes of this example, ignore all other types of transportation. Suppose the government institutes a comprehensive transportation program that increases the passenger-miles share for public transportation by 5 percentage points over 10 years. Calculate (a) the baseline energy consumption in the present year, (b) the baseline energy consumption after 10 years, and (c) the reduction in energy consumption in the tenth year relative to the baseline in that year due to the policy, assuming that the increase in public transit modal share is achieved.

Solution

(a) For the current year, the total energy consumption is the sum of the consumption for the two modes, as follows:

$$\text{For car: } (4.7 \times 10^9 \text{ pmi})(0.9)\left(3{,}300 \frac{\text{Btu}}{\text{pmi}}\right) = 13.96 \text{ tril.Btu}$$

$$\text{For public trans.: } (4.7 \times 10^9 \text{ pmi})(0.1)\left(1{,}500 \frac{\text{Btu}}{\text{pmi}}\right) = 0.705 \text{ tril.Btu}$$

The total energy is therefore 13.96 + 0.705 = 14.66 tril. Btu.

(b) For the tenth year, we account for the growth in passenger-kilometers as follows:

$$(4.7 \times 10^9 \text{ pmi})(1 + 0.02)^{10} = 5.73 \times 10^9 \text{ pmi}$$

Modal energy intensities have also changed, and these are recalculated as

$$\text{For car: }\left(3{,}300\frac{\text{Btu}}{\text{pmi}}\right)(1-0.005)^{10} = 3{,}138\frac{\text{Btu}}{\text{pmi}}$$

$$\text{For public trans.: }\left(1{,}500\frac{\text{Btu}}{\text{pmi}}\right)(1-0.005)^{10} = 1{,}427\frac{\text{Btu}}{\text{pmi}}$$

Recalculating the energy consumption by repeating the calculation in part (a) with the increased passenger-kilometers and reduced energy intensity gives 16.18 tril. Btu for car, 0.82 tril. Btu for public transport, and a total value of 16.18 + 0.82 = 17.0 tril. Btu for the entire system.

(c) This alternative assumes that in the tenth year the modal split is 85% car / 15% public transport. We can therefore recalculate as in part (a) using the new values:

$$\text{For car: }(5.73\times10^9\text{ pmi})(0.85)\left(3{,}138\frac{\text{Btu}}{\text{pmi}}\right) = 15.28\text{ tril.Btu}$$

$$\text{For public trans: }(5.73\times10^9\text{ pmi})(0.15)\left(1{,}427\frac{\text{Btu}}{\text{pmi}}\right) = 1.23\text{ tril.Btu}$$

The combined total is 15.28 + 1.23 = 16.51 tril. Btu. Therefore, the energy reduction is 17.00 – 16.51 = 0.49 tril. Btu.

Example 8-4M A metropolitan region generates 7.5 billion passenger-km (pkm) of transportation demand per year and is experiencing 2% per annum growth in total pkm, and a 0.5% annual decline in energy intensity of pkm. Current energy intensity of automobile travel and public transportation is 2200 kJ/pkm and 1000 kJ/pkm, respectively, and share of pkm is 90% and 10%, respectively, for those two modes. For the purposes of this example, ignore all other types of transportation. Suppose the government institutes a comprehensive transportation program that increases the pkm share for public transportation by 5 percentage points over 10 years. Calculate (a) the baseline energy consumption in the present year, (b) the baseline energy consumption after 10 years, and (c) the reduction in energy consumption in the tenth year relative to the baseline in that year due to the policy, assuming that the increase in public transit modal share is achieved.

Solution

(a) For the current year, the total energy consumption is the sum of the consumption for the two modes, as follows:

$$\text{For car: }(7.5\times10^9\text{pkm})(0.9)\left(2200\frac{\text{kJ}}{\text{pkm}}\right)\left(10^{-12}\frac{\text{PJ}}{\text{kJ}}\right) = 14.85\text{ PJ}$$

$$\text{For public trans.: }(7.5\times10^9\text{pkm})(0.1)\left(1000\frac{\text{kJ}}{\text{pkm}}\right)\left(10^{-12}\frac{\text{PJ}}{\text{kJ}}\right) = 0.75\text{ PJ}$$

The total energy is therefore 14.85 PJ + 0.75 PJ = 15.6 PJ.

(b) For the tenth year, we account for the growth in passenger-kilometers as follows:

$$(7.5\times10^9\text{ pkm})(1+0.02)^{10} = 9.14\times10^9\text{ pkm}$$

Modal energy intensities have also changed, and these are recalculated as

$$\text{For car: }\left(2{,}200\frac{\text{kJ}}{\text{pkm}}\right)(1-0.005)^{10} = 2{,}092\frac{\text{kJ}}{\text{pkm}}$$

$$\text{For public trans.: }\left(1{,}000\frac{\text{kJ}}{\text{pkm}}\right)(1-0.005)^{10} = 951\frac{\text{kJ}}{\text{pkm}}$$

Recalculating the energy consumption by repeating the calculation in part (a) with the increased passenger-kilometers and reduced energy intensity gives 17.22 PJ for car, 0.87 PJ for public transport, and a total value of 17.22 + 0.87 = 18.09 PJ for the entire system.

(c) This alternative assumes that in the tenth year the modal split is 85% car/15% public transport. We can therefore recalculate as in part (a) using the new values:

$$\text{For car: } (9.14 \times 10^9 \text{ pkm})(0.85)\left(2092 \frac{\text{kJ}}{\text{pkm}}\right)\left(10^{-12} \frac{\text{PJ}}{\text{kJ}}\right) = 16.26 \text{ PJ}$$

$$\text{For public trans.: } (9.14 \times 10^9 \text{ pkm})(0.15)\left(951 \frac{\text{kJ}}{\text{pkm}}\right)\left(10^{-12} \frac{\text{PJ}}{\text{kJ}}\right) = 1.30 \text{ PJ}$$

The combined total is 17.56 PJ. Therefore, the energy reduction is 18.09 – 17.56 = 0.52 PJ.

The results of Example 8-4 show the challenge of curbing growth in transportation energy consumption in the face of continually expanding passenger-kilometers values. Despite a 50% increase in the amount of public transportation over 10 years, energy consumption is higher in year 10 than in the start year. Nevertheless, using metric units, public transportation has made a measurable reduction in energy use of 0.52 PJ, which is equivalent to 21% of the 2.49 PJ increase that occurs with no increase in public transport modal share.

One limitation of this type of analysis is that it assumes each unit of passenger-kilometers shifted to public transportation will achieve energy savings based on the difference between the average energy intensity of the two transportation options. For a more accurate estimate, the analyst might build a computer model of the transportation network that includes the modeling of residents' modal choices for different types of trips. The analyst would first verify that the model reproduces the baseline system in the real world within some degree of accuracy and then impose the new expanded public transportation system on the model to evaluate the new modal split and total energy consumption. Such a modeling exercise gives more reliable results but requires a far greater amount of effort.

8-5 Case Study of BRT and LRT: Los Angeles Orange Line

In this section, we present an extended case study of a public transportation application so as to synthesize themes from previous sections into a tangible project where the reader can better understand how the system design and operation unfolds.

The chosen location is the city of Los Angeles, which suffers from some of the worst congestion and urban air quality in the United States. Though not the sole medium by which city, state, and federal officials seek to address L.A.'s transportation challenges, developing public transportation is very much part of the mix, and since the 1980s the Los Angeles Metropolitan Transportation Authority (LAMTA) has been advancing ROW A and ROW B systems beyond the original network of bus routes, which had been in existence since the demise of L.A.'s urban passenger rail network in the early 1960s. With the opening of the Blue Line LRT system mentioned above in 1990 and the Red Line underground subway in 1993, L.A. entered the modern age of urban rapid transit. Indeed, the chosen subject of this case study, the Orange Line, is an existing ROW B system that entered service in 2005, with an extension on the western end of the line entering service in 2012. In comparison to the case of Manhattan described above in Fig. 8-3, however, given the relatively recent advent of rapid transit in L.A. and the much greater road capacity, the modal share of trips for the entire rapid transit system in L.A. is much lower than that of Manhattan.

8-5-1 General Characteristics of the Orange Line and Modal Options

The Orange Line rapid transit system takes advantage of an unused 14-mile long rail ROW that is available to the City of Los Angeles for building public transit. The ROW runs across the San Fernando Valley region from North Hollywood in the southeast to Chatsworth in the northwest.[3] Although the entire San Fernando region has a population of approximately 1.8 million (2010 data), the Orange Line is presumably most useful to those who live nearest to it, or those who can conveniently drive to it and then take a journey along its length, possibly connecting with the Red Line subway for downtown Los Angeles. At the western terminus in Chatsworth, passengers can transfer to MetroLink trains to points north or east.

The goals of the Orange Line project are the following:

1. Provide an alternative mode of transportation for some residents and certain destinations that is more convenient than driving a personal car.
2. Incrementally reduce overall demand on the road network by removing a certain number of daily trips from the San Fernando Valley road network.
3. Provide a rapid transit extension to the MTA Red Line from its terminus in North Hollywood across the San Fernando Valley.
4. Provide a rapid transit system at a lower total system cost than would be required for a ROW A underground or elevated rail rapid transit system.

For the purpose of economic analysis, a discount rate of 7% is used, along with an investment lifetime of 12 years for vehicles and 24 years for infrastructure. Other lifetime lengths around these values could be chosen, but it is convenient that the infrastructure life is exactly twice that of the vehicles, since it makes the calculations simpler.

Vehicles traveling along the route, whether buses or rail vehicles, frequently encounter grade crossings and cross traffic at intersections with fairly dense traffic volumes. This condition is a challenge for either BRT or LRT because of the limits created on shortening headways without causing unacceptable delays to other traffic at intersections. In other words, if BRT or LRT vehicles arrive too frequently at signalized intersections and must be given a turn to pass through these intersections, vehicles arriving to the intersection from streets will form long queues.

BRT Option

Analysis of the cost of the BRT system includes capital cost for the ROW and vehicles, and any additional annual operating cost. ROW costs are estimated at $23 million per mile, or

[3]Originally the actual Orange Line ran from North Hollywood to Pierce College, with the extension from Pierce College to Chatsworth opened later. For the purposes of this case study, the line is treated as a single project from North Hollywood to Chatsworth. Project cost per mile and average speed data for this case study are taken from Stanger (2007), who reviewed the actual Orange Line implementation by the LAMTA. Readers wishing to see further details beyond what can be covered in the limited case study presented here are referred to this extended Orange Line study, and also to Bruun (2005) for a general comparison of BRT and LRT cost.

$322 million for a 14-mile corridor. This figure includes stations, signage, access and egress, and intersection signals, but not maintenance facilities or vehicles. Although the BRT vehicles might use an existing maintenance facility for storage and repairs with relatively low-cost modification, $10 million is budgeted to construct a maintenance facility adjacent to the route for the particular type of bus used for BRT, giving a total of $332 million for infrastructure.

Total capital cost of buses depends on the number of buses required, which in turn depends on the headway and length of the route. During peak demand periods, a headway of 5 minutes is required, equivalent to a frequency of 12 buses per hour. Suppose that the average speed of the buses is 20 mph, as has been observed in actual Orange Line operations. The density k in buses per mile is therefore:

$$k = \frac{12 \text{ bus/hour}}{20 \text{ miles/hour}} = 0.6 \frac{\text{bus}}{\text{mile}}$$

Given the route length of 14 miles and the need to be able to maintain 4-minute headways in both directions, the total number of buses distributed around the route is

$$(0.6 \text{ bus/mile})(14 \text{ miles})2 =\sim 17 \text{ buses}$$

This calculation is consistent with observed typical BRT service speeds in the United States of 19 to 20 mph (Federal Transit Administration, 2011). If 3 spare buses are required beyond peak deployment, the total vehicle requirements is 20 buses.

For maximum capacity, double-articulated buses are chosen at a cost of $1.5 million per vehicle with a capacity of 130 riders, including both standing and sitting. Thus the total capital cost is $30 million.

The total annual cost generated per bus owned (i.e., operating cost), not including capital cost, but including wages, fuel, maintenance, and overhead, is $450,000. Therefore the total operating cost for 20 buses is $9 million.

Lastly, we annualize capital cost to calculate the total cost of the system per year. Using the format (A/P, i%, N) to indicate the required discounting factor,[4] the annualized cost of the infrastructure and vehicle purchases are, respectively:

$$(A/P,\ 7\%,\ 24) = 0.0872$$
$$A = (\$332M)(0.0872) = \$28.95M$$
$$(A/P,\ 7\%,\ 12) = 0.1259$$
$$A = (\$30M)(0.1259) = \$3.78M$$

The total cost per year is then as follows:

$$\begin{aligned}\text{TotCost} &= \text{InfraCost} + \text{VehCost} + \text{OpCost} \\ &= \$28.9M + \$3.8M + \$9.0M = \$41.7M\end{aligned}$$

or $41.7 million per year. For this investment, the system achieves the maximum space capacity per hour of

$$Q = mC_v f = (1)(130)(12) = 1{,}560 \frac{\text{spaces}}{\text{hour}}$$

[4]See the appendix for explanation of the calculation of engineering economic discounting factors.

LRT Option

Many elements of the LRT calculation are similar to those of BRT, although unit costs are of course different. As is typical of LRT, the infrastructure system including tracks and overhead catenary (i.e., electric line used to deliver power to the railcars) is more expensive than that of BRT, so cost per mile is $60 million. Thus total cost is $840 million for 14 miles. Since there are no previously existing LRT facilities in the area, an additional $30 million must be spent on a maintenance center, which includes a link to the main ROW, storage areas, and covered facilities for maintenance work. Total infrastructure costs are therefore $870 million.

LRT vehicles are more specialized and built in smaller numbers, so their cost of $4 million per unit is higher than that of BRT vehicles, although they also provide 175 spaces per vehicle, or 45 more than BRT. It is assumed based on typical U.S. operating experience that the LRT system will operate at a higher average speed of 29 mph. Therefore with the same headway of 5 minutes at peak demand and using the same steps as in the calculation for BRT vehicles above, the total number of LRT vehicles is 15, or 12 vehicles to operate the maximum frequency plus 3 extra. The total capital cost of the vehicles is thus $60 million. The same annual operating cost for wages, energy, maintenance, and overhead is assumed for LRT as for BRT of $450,000 per vehicle, resulting in $6.75 million in non-capital cost annually. Equal operating cost per vehicle could be justified on the basis that energy costs are likely cheaper for LRT (electricity being cheaper than diesel fuel), but maintenance costs might be higher, so that the two cancel each other out. Annualizing capital cost gives

$$(A/P,\ 7\%,\ 24) = 0.0872$$
$$A = (\$870M)(0.0872) = \$75.8M$$
$$(A/P,\ 7\%,\ 12) = 0.1259$$
$$A = (\$60M)(0.1259) = \$8.5M$$

The total cost per year is then as follows:

$$\begin{aligned}\text{TotCost} &= \text{InfraCost} + \text{VehCost} + \text{OpCost}\\ &= \$75.8M + \$8.5M + \$6.75M = \$91.1M\end{aligned}$$

or $91.1 million per year. This figure represents a $49.4 million increase over the cost of the BRT system. On the other hand, the maximum space capacity of the system at the peak hour increases to

$$Q = mC_v f = (1)(175)(12) = 2{,}100\,\frac{\text{spaces}}{\text{hour}}$$

8-5-2 Comparison and Discussion

Table 8-2 gives a comparison of options showing the total cost per year including both capital and operating costs, and the maximum spaces per hour provided with arrivals at 5 minute headways. From the table it is evident that there is a substantial jump in total annual cost in going from BRT to LRT, and at the same time LRT provides a higher peak capacity. Since frequency of 12 arrivals per hour is near the maximum possible value due to the presence of cross traffic at intersections with busy arterial streets and the need for the vehicles to be able to clear intersections to maintain schedules, a possible option is

Option	Infrastructure Cost (Million $)	Vehicle Cap Cost (Million $)	Operating Cost (Million $)	Total (Million $)	Peak Capacity (Spaces/hour)
BRT	28.95	3.78	9.00	41.73	1,560
LRT	75.85	8.50	6.75	91.10	2,100

TABLE 8-2 Comparison of Options for Orange Line Case Study

doubling the value of m from 1 to 2. In the case of LRT, two vehicles are coupled together, and the service can be operated with a single driver. In the case of BRT, two vehicles must travel together (called *platooning* in public transportation jargon), and two drivers are required. This change assumes that station platforms are long enough to accommodate the longer vehicle presence. Either increase would make the infrastructure investment more productive, since any additional investment in fixed assets to accommodate two vehicles arriving simultaneously at a stop would be small. At the same time, the advantage of LRT over BRT in terms of peak capacity would be accentuated, as BRT would increase to approximately 3,100 spaces/hour but that of LRT would increase to 4,200 spaces/hour.

Some concluding thoughts are in order regarding this case study. First, the comparison of BRT and LRT cost and capacity points to a logical evolution of this type of service. An agency may launch rapid transit service in the form of BRT on a ROW with maximum capacity in the range of approximately 1,000 to 1,500 spaces/hour to provide accelerated service to customers compared to ROW C while keeping overall cost lower than that of LRT. As demand grows, however, transformation to LRT may become attractive or even inevitable because of its ability to accommodate higher maximum demand in terms of required spaces per hour. The LAMTA may eventually decide to change the Orange Line over from BRT to LRT for these reasons.

A second, and related, observation is that because of limitations of the ROW B classification and the extent of cross traffic at major intersections, even LRT will not be able to accommodate increased demand beyond a point, because it is not possible to shorten headways without upgrading all or part of the infrastructure to ROW A. For example, if the agency wished to increase capacity by shortening headways, they might construct grade separation at selected intersections so that vehicles on the Orange Line would not encounter cross-traffic at those intersections. Alternatively, they might upgrade the entire Orange Line to ROW A by upgrading the entire existing ROW or building a new one parallel to it. This approach would be more expensive than selective upgrading but would give the best possible level of service.

Lastly, the potential capacity of the Orange Line to meet travel demand must be seen in the context of overall trip requirements for the region. In July 2008, the LAMTA observed 680,000 rides taken on the Orange Line; if this boarding rate were maintained year round, the annual figure would be just over 8 million rides. Yet the population of the San Fernando Valley is large enough that in a year's time it will generate many more trips than 8 million so that, by itself, the Orange Line will not dramatically reduce congestion. Still, it should not be concluded that investment in the project is not worthwhile. As part of a widespread and sustained effort to develop other rapid transit lines, increase connecting bus service, encourage other alternatives to the private automobile, and change land-use patterns so that less passenger miles are required, the Orange Line can make a measurable contribution.

8-6 Future Prospects for Public Transportation

In the short to medium term, there are several reasons why demand for public transportation will, at a minimum, remain at current levels or increase modestly. Whether it could grow dramatically so as to change the national modal share given in Fig. 8-3 is less certain. Both issues are discussed here.

One issue that will continue to put pressure on the demand for public transportation is the relatively high cost of oil and hence transportation fuels derived from oil such as gasoline and diesel. With the growing use of motor vehicles around the world, demand for oil is likely to continue to grow, which will keep prices at or above current levels. In the long term, both consumers and transportation providers can adapt. Some drivers may at first, when faced with sustained higher gasoline prices, continue to drive and pay a larger part of their income for fuel. However, after observing their sustained higher spending at the pump over time, some fraction of these drivers may turn to using public transport for some of their travel needs in place of their car, if an appropriate route is available. Similarly, public transportation operators may not be able to change their offerings in immediate response to higher prices, but when they observe sustained higher prices over time, they are willing to expand service, knowing that potential new customers are available among the ranks of financially pressured motorists. Rising oil prices also measurably affect the overall cost of providing public transportation modes that use diesel fuel, but cost per vehicle mile for public transportation is less sensitive to changes in oil prices than is cost per mile for private light-duty vehicles, so that the impact on modal share favors public transport.

A second issue that will continue to favor public transportation is the chronic issue of congestion. Major cities in the United States and elsewhere continue to suffer from high levels of congestion, as they have done for many years. Many trips and itineraries of individual travelers in large cities do not lend themselves to any mode other than private car. However, for trips that are "transit-viable," there is a strong incentive to continue to expand public transportation to moderately reduce roadway demand and also to provide relief to travelers able to ride transit to complete their origin-to-destination trips. This goal is reflected, for instance, in the overall plan for expanding public transportation in greater Los Angeles and the resulting development of the Orange Line.

These two considerations may continue to grow transit usage gradually, but in order to grow transit in the United States to 10% to 20% of all trips nationally in the space of two or three decades, other catalysts would need to be involved. One is the outcome of a national debate about multimodalism. In the United States at present the extent to which a multimodal approach to transportation is appropriate is being vigorously debated. Proponents of multimodalism argue that diversion of revenues obtained from private cars, such as the gas tax, into public transportation projects helps the entire traveling public as well as the transit rider by making the system less congested for everyone. They are also critical of payment of road costs in part from general government revenues rather than user fees, leading to access to roads in private vehicles at artificially low cost per vehicle-mile. Detractors of public transportation claim that this funding transfer reduces funding available for maintenance of the road network without significantly contributing to ending congestion. The outcome of this debate can either slow or accelerate the development of public transportation. Cities in peer countries in the European Union are typically further ahead on increasing modal share for public transportation, and these efforts may serve as an example for the United States and other countries with relatively low public transport modal share (EPOMM, 2013).[5]

A second potential catalyst is the possibility of changing land use patterns so that gradually the urban form becomes more accessible to transit. Taking the example of the San Fernando Valley again, most of the land use in this region was developed in the post-World War II era around use of the car for both work and non-work trips. A system that is retrofitted onto this land use pattern can meet some demand up to a point, but there are many origins and destinations that are not transit-accessible. In response, the transportation community has established *transit-oriented development* (TOD) as a model of urban planning where residence, employment, and amenities are all centered on transit. Widespread implementation of TOD over time could lead to much larger transit modal shares.

In summary, as seen in the case of the Orange Line case study, when ROW B or ROW A high quality public transportation systems can run into the tens of millions of dollars per route per year, these systems may appear expensive. Yet the value of public transportation comes in many forms: energy and emissions savings, congestion reduction, accessibility for the elderly and disabled, and economic opportunity for low-income members of the workforce who cannot afford to own a car. Its benefit must be assessed in terms of the sum of its contributions in all of these areas.

8-7 Summary

In this chapter, we have defined what constitutes a public transportation system and described major urban road and rail modes, including heavy rail, light rail, bus rapid transit, and street transit, drawing an important distinction between *rapid* and *ordinary* transit. The design of public transportation systems can be understood in terms of planning of routes, evaluation of capacity provided, and calculation of utilization factors using actual versus ideal ridership rates.

The *time-area factor* measures how efficiently urban transportation modes utilize scarce urban space; public transportation vehicles, by having a relatively small footprint per passenger moved, can achieve a relatively low time-area factor value. Lastly, a wide range of possible innovations are available to public transportation operators in terms of infrastructure design, route planning and operation, and fare structures. These innovations can help grow demand for and effectiveness of public transportation in the future.

References

Bruun, E. (2005). "Bus Rapid Transit and Light Rail: Comparing Cost with a Parametric Cost Model." *Transportation Research Record* 1927:11–25.

Bruun, E. (2013). *Better Public Transit Systems: Analyzing Investments and Performance, 2d ed.* Routledge, London.

[5]The European Platform on Mobility Management at *www.epomm.eu* provides data and examples of the effect of "integrated measures for achieving sustainable mobility."

Bruun, E. and E. Marx (2006). "Omnilink: Case Study of Successful Flex Route–Capable Intelligent Transportation System Implementation." *Transportation Research Record* 1971:91–98.

Cornell University (2008). *Transportation, Cornell, and the Community: Transportation Generic Environmental Impact Statement. Technical Report,* Cornell University, Ithaca, NY.

European Platform for Mobility Management (EPOMM) (2013). *Mobility Management in the CIVITAS Initiative.* Electronic resource, available at *www.epomm.eu*. Accessed Dec. 1, 2013.

Federal Transit Administration (2011). *Metro Orange Line BRT Project Evaluation.* FTA, Washington, DC.

Schiller, P., E. Bruun, and J. Kenworthy (2010). *An Introduction to Sustainable Transportation: Policy, Planning, and Implementation.* Earthscan, London.

Stanger, R. (2007). "An Evaluation of Los Angeles's Orange Line Busway." *Journal of Public Transportation* 10(1):103-119.

U.S. Department of Transportation (2011). *National Household Travel Survey.* USDOT, Washington, DC.

U.S. Department of Transportation (2012). *Summary of National Transportation Statistics.* USDOT, Washington, DC.

Vuchic, V. (2005). *Urban Transit: Operations, Planning and Economics.* Wiley, New York.

Vuchic, V. (2007). *Urban Transit: Systems and Technology.* Wiley, New York.

Further Readings

Allen, D. and G. Hufstedler (2006). "Bus-and-Rail and All-Bus Transit Systems: Experience in Dallas and Houston, Texas, 1985 to 2003." *Transportation Research Record* 1986:127–136.

Carruthers, R., M. Dick, and A. Saurkar (2005). *Affordability of Public Transport in Developing Countries.* Technical Report, The World Bank, Washington, DC.

Center for Urban Transportation Research (2002). *Journal of Public Transportation Special Issue on Bus Rapid Transport.* CUTR, University of South Florida, Tampa, FL.

Currie, G. and J. Phung (2008). "Understanding Links between Transit Ridership and Gasoline Prices: Evidence from the United States and Australia." *Transportation Research Record* 2063:133–142.

Eliasson, J. and M. Lundberg (2012). "Do Cost–Benefit Analyses Influence Transport Investment Decisions? Experiences from the Swedish Transport Investment Plan 2010–21." *Transport Reviews* 32(1):29–48.

Hensher, D. and J. Rose (2007). "Development of Commuter and Non-Commuter Mode Choice Models for the Assessment of New Public Transport Infrastructure Projects: A Case Study." *Transportation Research Part A: Policy and Practice* 41(5):428–443.

Hook, W. and L. Wright (2007). *BRT Planning Guide.* Institute for Transportation & Development Policy, New York.

Hook, W. (2013). "BRT Brings Cleveland Back to Health." *Sustainable Transport* (winter 2013):13–15.

Kim, S. and G. Ulfarsson (2012). "Commitment to Light Rail Transit Patronage: Case study for St. Louis Metrolink." *Journal of Urban Planning and Development* 138(3):227–234.

Lindau, L., D. Hidalgo, and D. Facchini (2010). "Bus Rapid Transit in Curitiba, Brazil: A Look at the Outcome after 35 Years of Bus-Oriented Development." *Transportation Research Record* 2193: 17–27.

Litman, T. (2012). "Bus Rapid Transit: Bus System Design Features That Significantly Improve Service Quality and Cost Efficiency." *TDM Encyclopedia*. On-line resource, available at *www.vtpi.org*. Accessed Jun. 27, 2013.

Morichi, S., S. R. Acharya, eds. (2013). *Transportation Development in Asian Megacities: A New Perspective*. Springer, Heidelberg.

Morlok, E. (1978). *Introduction to Transportation Engineering and Planning*. McGraw-Hill, New York.

Vasconcellos, E. A. (2001). *Urban Transport, Environment and Equity—The Case for Developing Countries*. Earthscan, London.

Vuchic, V. (1981). *Urban Public Transportation: Systems and Technology*. Prentice-Hall, Englewood Cliffs, NJ.

Vuchic, V. (1999). *Transportation for Livable Cities*. Center for Urban Policy Research, New Brunswick, NJ.

Exercises

8-1. An urban arterial street with width of 15 ft is observed to have vehicle velocity u as a function of density k in the form of $u = 55 - 0.381k$ for values of u between 0 and 45 mph. Private vehicle traffic travels at an average speed of 17 mph. The average occupancy of the vehicles is $\alpha = 1.2$ persons/vehicle, and the length is 16 ft. What is the average shadow per vehicle, the time-area factor, and the per-person time-area factor? Suppose 40′ buses with average $\alpha = 22$ persons/vehicle and the same shadow as the private vehicles travel on the same street. Ignoring time spent stopping and starting at bus stops, what is the time-area factor and per-person time-area factor for the bus?

8-2. An origin-destination (OD) matrix for a bus route with 8 stops where the distance between each stop is 0.5 miles is given below. (a) What is the boarding and alighting count for each stop 1 through 8? (b) What is the space-averaged load factor for this run? (c) What is the segment of the run with the highest number of passengers on board, and what is the number on board in that segment? If the maximum occupancy of the bus is 22 seated and 44 standing, is the maximum capacity exceeded?

Stops	S2	S3	S4	S5	S6	S7	S8
S1	1	2	2	4	3	1	9
S2		0	1	0	1	3	6
S3			1	1	0	2	4
S4				3	2	0	8
S5					1	3	6
S6						2	5
S7							6

8-3. A transit bus costs $300,000 and has an expected life of 12 years with $0 salvage value; it drives 25,000 miles/year. It gets 3.5 mpg diesel, and diesel costs the transit agency $3/gal. The bus completes a round trip route of 10 miles in 1 hour. The bus must recover $50/hour for wages to pay not only for the driver wages but office and support staff as well. The discount rate is 7%. (a) If the cost of the bus is amortized on an annual (as opposed to monthly) basis, what is the annualized cost of the bus? (b) Calculate the total cost per round trip for the bus, taking into account capital cost, fuel cost, and wage cost values given above; do not include any other costs. (c) Calculate the number of riders that must board at an average fare of $1.50 for the bus to break even. (d) Name one cost component that the calculation does not include (more than one possible right answer).

8-4. A light rail transit (LRT) system runs once every 15 minutes from 6 a.m. to 11 p.m. and carries a total of 15,000 riders in to the city per day (the same 15,000 riders then return via LRT to their point of origin). The line maintains a uniform schedule with departures at :00, :15, :30, and :45 past the hour, with the first departure leaving at 6 a.m. from each endpoint and the last departure leaving at 10 p.m. from each endpoint. The line runs for 20 miles in each direction. The LRT averages 17 mph when in operation from one endpoint to the other, and then waits for the next departure time before leaving to return in the other direction. Among the riders of the new service, 90% are previous car commuters who traveled in single occupant vehicles (SOVs) an average of 15 miles one-way to work, but now leave their cars at home. (a) To the nearest minute, how long does the LRT vehicle wait at the end of its run before starting its return run in the opposite direction? (b) How many LRT vehicles are required to maintain the schedule with 15-minute headways and each run ending and starting as described in part (a)? You can ignore the need for spare vehicles. (c) If both car and LRT vehicle miles traveled (vmt) are taken into account, how many net vmts per day are reduced by shifting the SOV commuters to LRT?

CHAPTER 9

Personal Mobility and Accessibility

9-1 Overview

This chapter looks at ways to improve transportation sustainability in the urban setting by providing a diverse range of options other than public transportation (discussed in Chap. 8). First, the urban traveler can move away from a "one-size-fits-all" solution where the same vehicle is used for all trips, to one where the vehicle choice is adjusted to the most suitable option, including different sizes of highway vehicles, different types of limited-use vehicles, or nonmotorized options. Second, the traveler can take advantage of carsharing or bikesharing systems to meet local transportation needs. Last, they can use telecommuting and other alternatives to physical travel to reduce their travel requirements.

9-2 Introduction

As discussed in the previous chapter, the objective of modal shifting relied on a substantial commitment by local and regional governments to provide public transportation service as an alternative to travel by car. Even without using public transportation, however, the individual traveler has a wide range of other choices, and these choices are the focus of this chapter.

In presenting personal options we maintain a distinction between *mobility* and *accessibility*. By mobility we mean a solution to obtaining an amenity that involves actual transportation of a traveler from one place to another. The traveler may use a motorized or nonmotorized mode, and they may use a means of locomotion that they own or one to which they have access through some sort of sharing system. In using the term accessibility, by contrast, we mean access to the amenity by some means that avoids transportation of the traveler.

Personal mobility and personal accessibility options can be divided into three main categories, as follows:

- *Adjustment to personal modal option:* An urban traveler may have many personal modal options besides their own private car, should they happen to own one. Travelers who live in households that own more than one vehicle may be able to optimize their vehicle choice so as to minimize environmental impact.

Travelers may also be able to choose a nonmotorized mode such as bicycling or walking. Lastly, travelers may be able to band together in a permanent arrangement called *carpooling* and *vanpooling*, or on a one-off case-by-case basis called *ridesharing*.

- *Access to a shared vehicle fleet:* Instead of using a vehicle that they already own, or that a vehicle pool member already owns, travelers can participate in a vehicle-sharing system, commonly known as *carsharing* or *bikesharing*. A separate agency then provides the service of making available vehicles, managing reservations, and providing maintenance on the vehicle fleet. Generally, use of these vehicles is limited to members who have been vetted as part of an application process, although some bikesharing systems allow one-time use of bicycles without formal membership.
- *Telecommuting and other substitutes for physical travel:* In an age of increasing cost and inconvenience associated with physical travel, a wide range of amenities can be obtained by exploiting substitutes for physical travel that achieve accessibility rather than mobility. These substitutes usually take advantage of information technology and electronic communication in some form. One application is *telecommuting*, in which the would-be traveler substitutes remote access to their workplace for physical travel to the workplace. Other activities that fall in this category are using electronic communication for educational or business activities, or on-line shopping where a delivery vehicle comes to the home rather than the individual traveling to a "brick-and-mortar" retail outlet.

These three options comprise the three main sections of this chapter.

9-3 Adjustment to Personal Modal Option

One of the simplest ways of reducing negative impact of transportation for travelers who drive to own different vehicles for different purposes, assuming they have the necessary financial means. For these travelers, especially those with higher incomes, it may be practical to own a smaller vehicle for single-occupant work or nonwork trips, and a larger vehicle such as a van, sport-utility vehicle (SUV), or truck for occasions where the motorist is either carrying a large amount of goods or has several passengers on board. The situation of the driver traveling by himself or herself alone is termed a *single-occupant vehicle* (SOV), while a vehicle with multiple passengers is a *high-occupancy vehicle* (HOV). Highway departments in urban areas specify the number of occupants required for an HOV; for example, HOV-2 for HOVs with two or more occupants, HOV-3 for three or more occupants, and so on.

From an energy-efficiency point of view, it is desirable to avoid larger vehicles, whether large passenger cars or light trucks, traveling as SOVs. For example, an SUV that delivers 15 mpg (6.34 km/L, or 15.8 L/100 km) during urban travel with a single occupant has an intensity of approximately 5,000 kJ/passenger-km, which is much higher than the average U.S. value across all transportation modes. There is anecdotal evidence that, with the increase in gasoline costs in the U.S. market since 2005, drivers who own both a compact car and a light truck increasingly are choosing the compact car for single-occupant travel in order to reduce their fuel expenditures. Example 9-1 illustrates the benefit of shifting travel in this way over an entire year.

Example 9-1S A household has access to both an SUV (15 mpg fuel economy) and a subcompact car with 30 mpg delivered fuel economy. Suppose one of the breadwinners in the household in the previous year drove 15,000 miles in the SUV. Then, in the subsequent year, they recognize that 10,000 miles could be driven in the subcompact, and the remaining 5,000 miles require the extra capacity of the SUV for carrying either large numbers of people or goods. If gas costs $4/gal, what are the number of gallons of fuel saved, the financial savings on reduced fuel costs, the energy savings, and the CO_2 savings?

Solution The SUV driving 15,000 miles/year will consume 1,000 gal of gasoline, costing $4,000/year. If gasoline has an energy content of 115,000 Btu/gal and emits 19.5 lb of CO_2 per gallon, the energy consumption and CO_2 emissions are respectively 115 million Btu and 19,500 lb/year, respectively.

In the new scenario, the SUV's figures are cut by two-thirds, leading to values of 333 gal, $1,333 cost, 38.3 million Btu, and 6,500 lb CO_2 per year, respectively. Since the subcompact is exactly twice as efficient and also travels twice as far, its figures are all the same as those for the SUV with twice as many miles traveled. Therefore, the new annual total is the sum of the two, or 667 gal, $2,667 cost, 76.7 million Btu, and 13,000 lb CO_2 per year, respectively. The savings are thus 333 gal fuel, $1,333 cost, 38.3 million Btu, and 6,500 lb CO_2 per year, respectively.

Example 9-1M A household has access to both an SUV (for simplicity fuel economy is rounded to 16 L/100 km fuel consumption) and a subcompact car with 8 L/100 km fuel consumption. Suppose one of the breadwinners in the household in the previous year drove 24,000 km in the SUV. Then, in the subsequent year, they recognize that 16,000 km could be driven in the subcompact, and the remaining 8,000 km require the extra capacity of the SUV for carrying either large numbers of people or goods. If gas costs $1/L, what are the number of gallons of fuel saved, the financial savings on reduced fuel costs, the energy savings, and the CO_2 savings?

Solution The SUV driving 24,000 km/year will consume 3,840 L of gasoline, costing $3,840/year. If gasoline has an energy content of 32.1 MJ/L and emits 2.34 kg of CO_2 per liter, the energy consumption and CO_2 emissions are respectively 123.3 GJ and 8,986 kg CO_2 per year, respectively.

In the new scenario, the SUV's figures are cut by two-thirds, leading to 1,280 L, $1,280 cost, 41.1 GJ, and 2,995 kg CO_2 per year, respectively. Since the subcompact is exactly twice as efficient and also travels twice as far, its figures are all the same as those for the SUV with twice as many miles traveled. Therefore, the new annual total is the sum of the two, or 2,560 L, $2,560 cost, 82.2 GJ, and 5,990 kg CO_2 per year, respectively. The savings are thus 1,280 L, $1,280 cost, 41.1 GJ, and 2,995 kg CO_2 per year, respectively.

The above example shows the large potential benefit from use of a more efficient vehicle. In particular, using standard units, the average U.S. CO_2 emissions per capita are on the order of 20 short tons per year. The savings of 6,500 lb or 3.25 short tons of CO_2 per year are therefore a measurable fraction of the total per-capita value.

9-3-1 Niche Motorized Vehicles

Outside of highway-capable light-duty vehicles (i.e., vehicles able to travel at the full range of speed limits, including expressway speeds), other options exist that can further allow the motorist to mode-shift toward the travel solution that meets the needs in the most efficient way possible:

- *Limited-use vehicles* (*LUVs*): Small battery-electric or gasoline vehicles with limited top speed and range that are useful for short trips in an urban setting. Some of these vehicles have necessary signaling and safety equipment to be road legal, such as the GEM Electric LUV (Fig. 9-1). Basic electric LUVs cost in the range of $9,000 to $12,000 for a two-seat model, with larger LUVs with either more seats

Figure 9-1 Limited-use vehicle (LUV) from the GEM (Global Electric Motors) division of Chrysler.

Notes: The turn signals, headlights, seatbelts, and license plate seen in the image allow for legal operation on public roadways. The letters "LU" in the New York State license plate signify that the vehicle is only legal to operate on roads with a speed limit of 35 mph (approx. 60 km/h) or less.

or more cargo capacity costing more. In other situations, a nonroad legal vehicle can be used on separate roadways, such as in golf communities where residents use golf carts within the community for golf, shopping, and other amenities.

- *Motorcycles, scooters, and other motorized two- and three-wheel vehicles*: When the motorist is traveling alone and does not have many goods to carry, these vehicles provide a very energy-efficient option for local travel. Vehicles may be powered electrically or with internal combustion engines. Some options like motorcycles have been in existence for about the same length of time as the automobile. Others, like the "Segway" motorized two-wheel platform (Fig. 9-2), are relatively new entrants. The Segway uses weight sensing technology to accelerate, decelerate, and change direction as the rider leans forward, to the sides, or pulls back on the handles; cost is in the $5,000 to $8,000 range, depending on options. Electric-assist bicycles are treated separately in the section on nonmotorized transportation (NMT).

Ownership of a niche motorized vehicle for non-recreational transportation is justified when the user can choose it over a conventional car frequently enough to merit the extra up-front cost. A motorcycle, moped, or scooter owner must be able to take trips in a condition where they are exposed to the elements, and may therefore need access to a conventional vehicle some of the time. An LUV or Segway owner is limited by the

FIGURE 9-2 The Segway personal transporter used by a tour guide, Paris, France, 2007.

Notes: One application of this technology is the "Segway tour," in which the guide leads a party of tourists each riding on their own Segway.

range and maximum speed of the vehicle, but if daily mileage requirement and point-to-point trip distances are short enough, either vehicle may be practical and also beneficial, since it is easier to park, and costs less for fuel cost (electricity in this case) and maintenance. A typical application of an LUV is a university or industrial campus, where vehicles can be charged on-site at night and then used during the day for many short, routine trips, thereby saving operating cost for the owner of the campus.

9-3-2 Methods to Encourage Diversified Use of Private Cars

The preceding section considered several niche vehicles that may appeal to short-distance travelers as alternatives to cars and light trucks. In addition, there are a number of organizational and economic mechanisms that a population center or region can make available to travelers to encourage use of cars and light trucks in a more efficient way. These mechanisms include systems to organize travel behavior, infrastructure systems, and economic incentives and disincentives.

Carpooling and Ridesharing Opportunities

From a transportation policy perspective, a *carpool* is a permanent or semi-permanent arrangement in which commuters join together to travel to their workplace in a single vehicle rather than each person driving their own vehicle. In a typical arrangement, each member of the pool will take their turn driving their car one day of the week, or for one week at a time, before the next person in the pool has their turn; the driver will pick each member up from their home before completing the line haul trip to the workplace.

In a variation on carpooling, commuters may join a *vanpool* where members of the pool, or their employer, pay an outside organization to provide a van for the purpose of shared commuting to work. Usually one of the members of the pool becomes the permanent driver and is responsible for driving each day and keeping the van at their home on nights and weekends.

As opposed to carpooling, *ridesharing* is another means of increasing vehicle occupancy on a case-by-case basis, where a rider joins a driver for a single ride. An established form of organized ridesharing is the use of ridesharing *stations* on the perimeter of cities such as Washington, D.C., to match riders to drivers for the commute into the city. At these stations, riders (called *slugs* for the purpose of the ridesharing system) form a queue in the station and, as each driver pulls up to the queue, the desired number of riders board the vehicle. This system allows drivers to enter the HOV/HOT lane network who otherwise would not qualify.

A recent variation of ridesharing made possible by wireless communications and smart phone technology is the *dynamic ridesharing* concept, in which participants have the possibility of anywhere-to-anywhere transportation. Members of a dynamic ridesharing network are first vetted by the organizing company and pay an annual membership fee. They can then make themselves available to provide rides when they are driving, or request a ride if they are taking a trip and do not have their private vehicle. with them at the time. When a rider makes a request, the ridesharing software performs an analysis in real time of available drivers and attempts to make a match. The software also tracks rides given and received, and periodically charges members for their usage. Dynamic ridesharing can potentially reduce members' overall transportation costs by delivering the same total number of trips with fewer vehicle miles driven.

High-Occupancy Vehicle and High-Occupancy Toll Lanes

Participation in these shared vehicle arrangements may allow travelers to take advantage of *high-occupancy vehicle lanes* (HOV lanes) to which they would not have access if they were traveling by themselves. HOV lanes usually require a minimum of two or three passengers in each vehicle traveling in the lane. They generally avoid becoming congested at rush hour, even if the parallel general traffic lanes are experiencing congestion.

In an adaptation of the concept of HOV lanes, some metropolitan areas have started implementing *high-occupancy toll lanes* (HOT lanes), where drivers can qualify to enter the lane either by traveling in a high-occupancy vehicle or by electronically paying a toll, whose price varies according to the number of vehicles already in the HOT lane. The State Route 91 freeway in southern California exemplifies the use of HOT lanes. In principle, as the volume in the HOT lane increases, the price increases to the point where further vehicles are discouraged from entering, thus leading to an equilibrium. If the right balance is struck, HOT lanes provide more relief for congestion than HOV lanes since they could on average be expected to carry more vehicles, while at the same time improving traffic conditions in regular lanes and reducing fuel consumption and emissions.

Congestion Pricing Corridors and Zones

Congestion pricing is a general term for the charging of additional cost to travelers (usually motorists in private vehicles) at times of peak demand to discourage excessive travel at those times, reduce congestion, and improve average travel speeds. The actual value of the charge to be applied can be calculated by evaluating the marginal cost of adding a vehicle to those already traveling, and charging the user some premium beyond the costs that they are already incurring to make sure that they experience the full marginal social cost. Unlike the provision of an HOV lane that serves as an economic "carrot" to encourage a certain behavioral choice, congestion pricing is an economic "stick" to make sure that only travelers willing to pay the extra cost choose to travel at the time in question. However, the two approaches can work together, as in the HOT lane described above, where the traveler is given the choice of paying extra to travel individually or avoiding the cost by traveling in an HOV.

A number of variants on congestion pricing exist. The price can be fixed according to a timetable (for instance, during weekday business hours) at a set value that does not change over time, or it can vary in real time based on conditions observed. The latter more accurately reflects the true cost of congestion but is also more complicated and expensive to implement, since devices must continually measure conditions and convey prices to the traveling public. Congestion pricing can also be implemented in the form of a *ring* or *zone*, where all travelers incur the cost when they enter the zone regardless of route chosen, or in the form of a *corridor*, which may be a more desirable road that still allows an alternative less desirable but free-of-charge alternative to the destination. Notable world examples of congestion pricing are Singapore's Electronic Road Pricing (ERP), launched in 1998, and London's Congestion Charge Zone (CCZ), launched in 2003.

9-3-3 Self-Driving Cars: Emerging Concept for Automatic Control

Extending the concept of specialized vehicles to improve urban passenger transportation is the self-driving car, which provides a possible tool for reducing congestion, improving fuel economy, and relieving the shortage of urban parking spaces. A self-driving car is mechanically similar to a conventional one, with internal combustion engine, drivetrain, brakes, and so on. However, it uses a range of sensors, video equipment, and an onboard computer to drive the car in place of the driver. The self-drive system can both plan and execute the origin-destination route requested by the driver and respond to surrounding conditions, including precipitation, hazards, the presence of other vehicles/cycles/pedestrians, and so on.

One milestone in the evolution of self-driving cars was the passage of a law in the state of California in September 2012 allowing them to operate in the state, and setting out requirements for legally bringing them onto the road. The law still requires that a licensed driver sit behind the wheel in a position to take over driving of the car if for some reason the self-driving system is disabled.

Self-driving cars are appealing for several reasons. They can reduce traffic accidents by eliminating the role of human error, which arises among other reasons from driver fatigue or unsafe human choices (e.g., texting while driving, drinking unsafe amounts of alcohol prior to making a trip). Because self-driving cars can drive more accurately and efficiently, for instance by traveling with closer spacing or accelerating/decelerating more smoothly, they can also reduce congestion and increase capacity of roads. The latter objective should be approached cautiously, because the wide penetration of self-driving cars might allow more cars to travel in urban areas with less time lost due to congestion, but increase overall energy consumption and carbon emissions. Lastly, self-driving cars still require a driver at present. However, driverless operation in the future might eventually allow vehicles to operate on an *on-call* basis, where vehicles *float* in some holding location so that they do not need to find parking in congested neighborhoods when not in use. For instance, they could remain on stationary standby in an out-of-the way location, or slowly circulate through neighborhoods until summoned.

9-3-4 Nonmotorized Modes

Nonmotorized transportation (NMT) comprises any type of transportation not requiring a powered mechanism, thus relying on the travelers on power instead.

- Cycling mode: Any type of pedal-powered cycle. Although this mode includes primarily bicycles, the term *cycle* is used so that pedal-powered vehicles with three or four wheels can be included as well. The role of electric assistance to the pedal-powered vehicle is considered at the end of Sec. 9-3.
- Pedestrian mode: Walking as a means of origin-destination travel. Walking is somewhat unusual as a *mode* in that it is a form of transportation that does not involve the use of a vehicle *per se*. Nevertheless, it is possible from surveys to gather data on length and type of walking trips, and estimate total passenger-miles or passenger-kilometers for a population. For example, the European statistical agency Eurostat publishes separate data on both walking and cycling. The U.K. government publishes pedestrian and cycling passenger-mile data in the annual volume *Transport Statistics Great Britain*, finding, for example, that in 2006, of 1,037 trips taken by the average citizen, 249 were on foot and 16 were by cycle (U.K. Department for Transport, 2008).
- Other types of nonmotorized conveyance: These include in-line skates, skateboards, longboards, and so on. At present these are generally viewed as forms of recreation as opposed to modes of transportation, so there are no data sets on volumes of demand for nonrecreational transport, but in principle they could in the future evolve into nonmotorized modes along with bicycling and walking. One possible advantage of these modes is that compared to walking they offer the additional speed of a wheeled conveyance, but the means of conveyance is more portable by car or public transportation than a bicycle or other cycle. They are also more compact and therefore easier to store.

Nonmotorized modes have several advantages for sustainable transportation. They can relieve pressure on the urban road system and are by themselves quite compact in terms of land requirement. They require no energy source at the end-use stage. Looking at the situation holistically, the traveler who uses NMT will on average require somewhat more nutrition, and the infrastructure and where applicable vehicles for NMT have some embodied energy, but even taking these factors into account, life-cycle energy consumption per passenger-mile is low compared to motorized modes. These modes also have the benefit of supporting cardiovascular health, although it is important that the air quality of the region not be detrimental to the cyclist or pedestrian at the same time that they are getting exercise.

Although NMT is not the dominant mode of travel in industrialized countries, whether measured by modal share of passenger-miles or trips, the efforts of some urban areas to promote NMT has led to higher share levels. Taking the case of cycling, Fig. 9-3 shows modal share of trips for several cities around the world that have supported cycling. Both very large cities (Amsterdam, Copenhagen, Osaka) and smaller cities (Ferrara, Malmo) have been able to achieve high fractions.

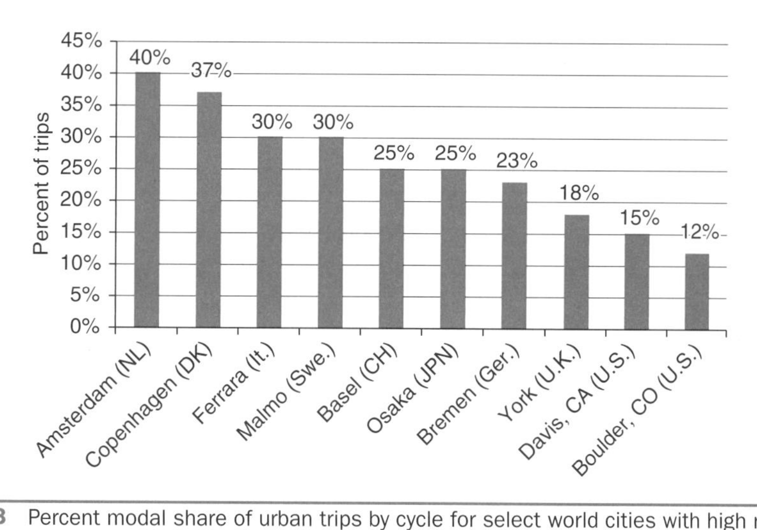

Figure 9-3 Percent modal share of urban trips by cycle for select world cities with high rates of cycling.

Source: Glaser et al. (2009).

Supporting Infrastructure for Nonmotorized Transport

Often it is not the lack of a vehicle but rather the lack of a cycle- and pedestrian-friendly infrastructure that prevents greater use of these modes. In response, many urban regions have been expanding and improving the infrastructure for NMT, based on the following categories, listed in increasing order of investment:

- *Shared roadways:* Streets or roads marked with roadside signage or other indications to motor vehicle traffic reminding them of the presence of NMT.
- *Safe passage at intersections and across roadways*: Measures include marked crosswalks or cycle lanes across intersections, marked or signalized crosswalks across major arterial streets, and separate signals for pedestrians and/or cycles.
- *NMT connector trails to provide traffic-free access:* A safe, low motorized traffic volume route for NMT can sometimes be provided by opening NMT-only pathways in key locations to link networks of roadways. An example is shown in Fig. 9-4. In the map of the region shown, networks of residential streets (labeled "Beckett Woods," "Highgate Circle," and "Winthrop Heights") surround the shopping district on several sides. Each network by design is laid out to prevent through motor vehicle traffic from reaching the shopping district, so that traffic volumes will be low and quality of life for residents on those streets enhanced. By adding a short NMT link between the neighborhoods and the shopping district, residents are given walking or cycling access, and pedestrians and cyclists from further distances can also access the shopping district via a safer, quieter route. Without the NMT links, routes might be much more circuitous. For example, since cyclists and pedestrians from Beckett Woods and Highgate Circle cannot access the Route 13 expressway, they would be forced to detour around to Oak Crest Rd to access the shopping district.
- *Cycle lanes:* Separate marked traffic lane for cycles, typically with a width of approximately 5 ft (~1.5 m), to keep cycles separate from motor vehicle traffic.

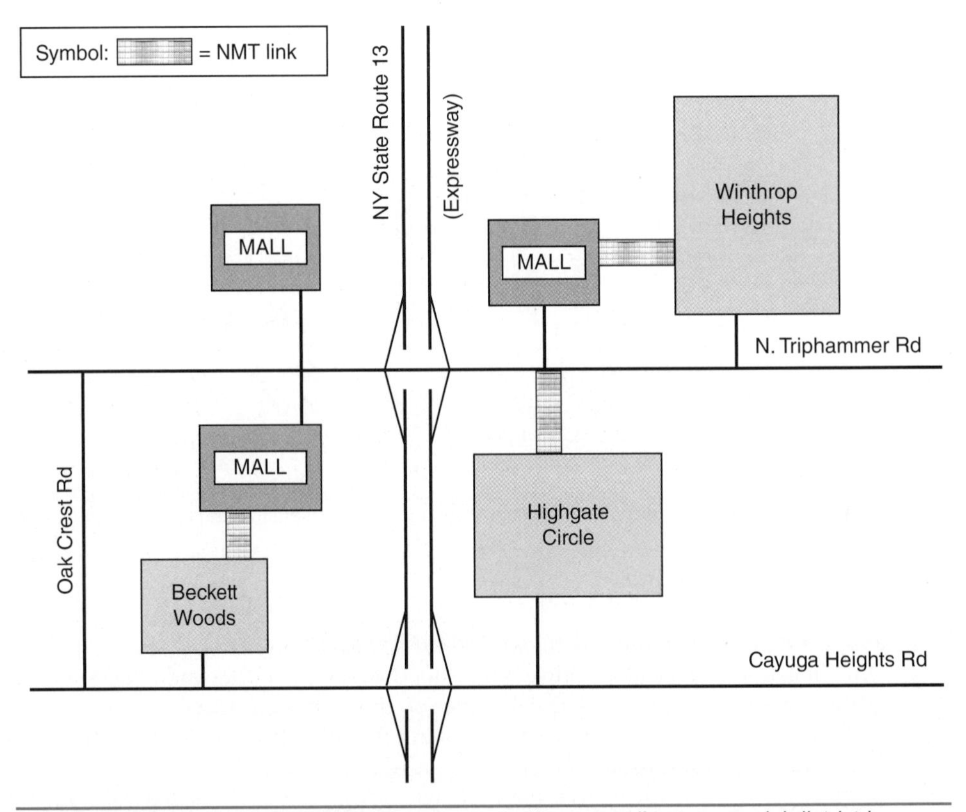

FIGURE 9-4 NMT existing and proposed connector trails surrounding commercial district in Ithaca, N.Y.

Source: Adapted from Vanek and Spindler (1999).

Unlike a cycle path, a cycle lane is part of the same contiguous paved surface as the adjacent lane(s) for motorized traffic. A cycle lane may travel in the same direction as motorized traffic, possibly located between the travel lane and the lane dedicated to on-street parking. *Counterflow* cycle lanes are another possibility, in which the cycle lane allows cyclists to travel in the opposite direction on a street that is marked for one-way travel for motorized traffic.

- *Multiuse trails*: Laterally separate trails, either adjacent to a roadway or on its own alignment (e.g., through an urban park or adjacent to a waterway), which allow mixed use by cyclists, pedestrians, and other NMT. Multiuse trails commonly mix recreational and nonrecreational travel: for some users the trail provides a safe location for aerobic exercise, while for others it may serve as a traffic-free link between origin and destination.
- *Separated cycle and pedestrian paths:* Separate cycle and pedestrian paths adjacent to an urban thoroughfare. This infrastructure requires a higher level of investment than a multiuse trail, but is also safer because the faster cycle traffic is separate from the slower pedestrians. A typical configuration found in each

direction of travel in northern European countries (such as the Netherlands or Denmark) is a pedestrian path at the edge of the roadway, motor vehicle lanes in the center, and a cycle path between the motor vehicle lanes and pedestrian path.

- *Grade-separated passage for NMT:* Includes overpasses and tunnels over and under major arterials. Although a significantly higher level of investment than a signalized crosswalk or cycle crossing, grade separation may be the only practical, safe solution for roadways that are especially wide or have very high volumes of traffic.
- *Elevation climbing assistance for cycles:* In some urban areas, hills create a barrier that reduces the use of cycling for point-to-point cycling. In response, urban areas can provide means of conveyance that allow cyclists to climb the elevation with their cycle but without needing to exert themselves. In Trondheim, Norway, the experimental "Trampe" system pushes cyclists to the top of the hill while they sit on their cycle without pedaling. Another option is to equip public buses with collapsible cycle racks on the front of the bus to carry passengers' cycles (See Fig.8-1*e*).

The challenge for supporting infrastructure for NMT is to make the financial case for investment in improving safety, since especially the later options on the above list require significant investment, not only for construction but also for setting aside the needed land area. In some instances, national government funding is available, such as the Chester Creek Rail Trail project near Philadelphia, Pa., which in 2007 received funding from the U.S. Environmental Protection Agency's *Congestion Mitigation and Air Quality* (CMAQ) program.

Electric-Assisted Cycles

An electric-assisted cycle is a human-powered cycle with an optional electric assistance system that the rider can engage or disengage. The electric-assist function allows the rider to either travel faster or respond to a high load condition, such as going up a grade, with less physical strain. The rider also has the option of turning off the electric-assist system, in which case the vehicle acts like an ordinary human-powered cycle. Electricity is provided by a battery that is recharged at night; some electric-assist cycles also have regenerative braking that returns charge to the battery when decelerating by turning the motor into a generator and creating drag on the cycle.

Although an electric-assist cycle could also be viewed as a niche-powered vehicle (see Sec. 9-3-1), it is covered under NMT for purposes of this book because this type of cycle requires a significant amount of pedaling from the rider and because traffic codes treat them the same as ordinary cycles. They are required to keep to the side of the road when not traveling at or near posted speeds. They can be operated without a driver's license and without registering the cycle, unlike motorcycles. They also have the same access to NMT infrastructure as nonelectrics, including cycle lanes, cycle paths, and multiuse trails.

Figure 9-5 shows an example *retrofitted* electric-assist cycle, where the owner has purchased a conventional cycle (semi-recumbent in this case) and then adapted it with a conversion kit that has been purchased separately. Although the example is a retrofit, cycles built from the ground up with electric-assist are also available in the marketplace. In Fig. 9-5(*a*), the conventional front wheel of the original cycle has been replaced

(*a*)

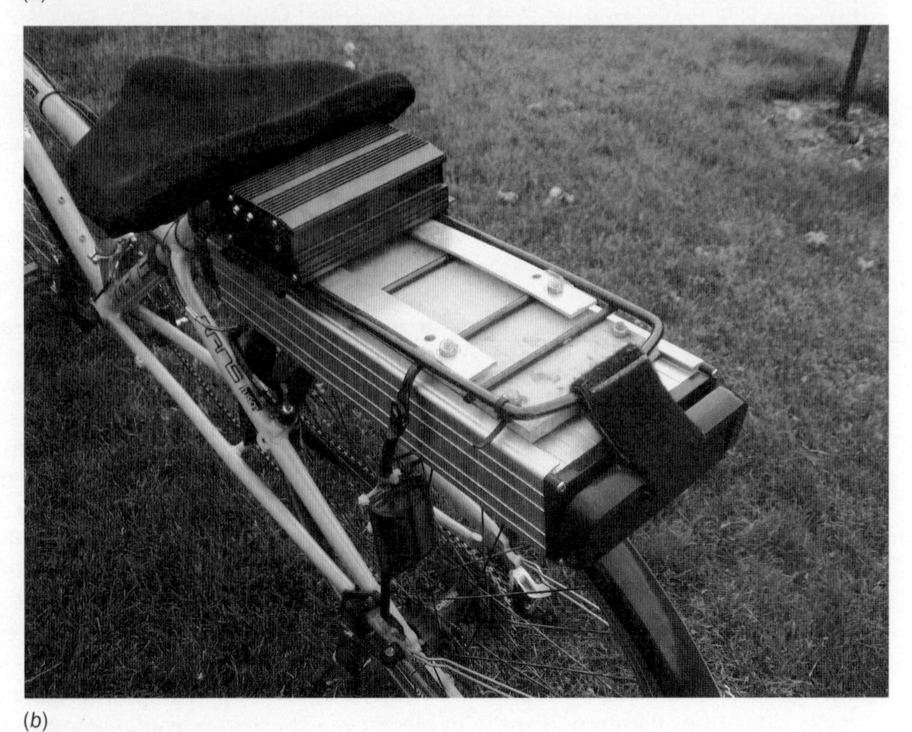

(*b*)

FIGURE 9-5 Rans "crank-forward" (semi-recumbent) bicycle with electric-assist: (*a*) full-length view with electric-assist hub in front; (*b*) close-up of rear-mounted battery and controller.

with a motorized wheel, along with a hand grip on the handlebar which the rider uses to activate and deactivate the motor. The original rear chain-based drivetrain of the cycle remains intact. Thus the rider can decide how much to use the electric drive and how much to use the power of their own legs. The less they rely on electric power, the longer the charge lasts. Even when the cycle is traveling on level ground at urban street speeds of 25 to 30 mph (speeds that most conventional cycles cannot maintain), the cycle can continue to be driven partly by human power, since the overall power available from the mix of human and electric input determines the speed.

Figure 9-5(*b*) shows the battery and charge or motor controller for the electric-assist. The battery storage consists of a single unit attached directly below the rear luggage rack of the cycle. Above it is the controller, with on-off switch visible, which responds to the signal from the rider's handgrip by increasing or decreasing current to the front wheel. For reasons of both safety and comfort, it is important that the motor torque be changed precisely, hence the need for a controller separate from the battery.

To reduce weight, electric-assist cycles typically use advanced battery compounds with lithium or nickel rather than lead-acid. Battery weight, size, and range configurations vary widely, but as an example, the company BionX markets a retrofit kit with a battery capacity of 9.6 amp-hours at 26 V. Under favorable conditions (low level of electrical assist demanded by the rider, good road conditions, flat to moderately hilly terrain) this system has a range of up to 65 km (40 miles). Range will be shorter if any of these conditions do not apply, and in any case, actual delivered range varies so much with the rider's behavior that it is difficult to provide definitive numbers. These uncertainties notwithstanding, electric-assist cycles can potentially increase cycle commuting and hence reduce energy consumed by cars and light trucks, because they enable some commuters to travel by cycle who otherwise would be prevented by long distances, hilly terrain, or other factors.

9-4 Vehicle Sharing: Carsharing and Bikesharing

Carsharing and bikesharing are membership plans for giving members access to a fleet of automobiles or cycles so that the members can use the cars or cycles only as needed and return them when not in use. These organizations are generally run by a professional staff and charge members for usage and then administer reservation systems, vehicle maintenance, and other issues. Because carsharing uses essentially the same vehicle technology as personal car ownership, the focus in this section is on the economic signals that might spur a consumer to shift from car ownership to carsharing. Congestion and ecological benefits are then presumed to follow when carsharing is implemented effectively. Most of the remainder of this section focuses on carsharing as opposed to bikesharing; a subsection on bikesharing at the end highlights unique features of using cycles in a vehicle sharing system.

In its essential form, a carsharing system consists of a group of members all having access to a fleet of cars that each has its own fixed parking space (see Fig. 9-6). Members have access to cars through a reservation system. When a vehicle has been reserved the member generally accesses the vehicle using some sort of log-in system such as a fob or swipe-card. An electronic recognition system on-board the vehicle recognizes the member and unlocks the vehicle when the log-in device is presented. The carsharing organization has the capacity to rent automobiles for short periods of time and for short

FIGURE 9-6 Ithaca Carshare in Ithaca, N.Y.; a carsharing example.

Notes: Vehicles are clearly marked to facilitate finding the vehicle in its parking place when the carshare member first approaches. Each vehicle has a dedicated 24-hour parking space, often in business districts where parking spaces are difficult to find, to which it is returned any time it is not in use.

distances (longer times and distances are also possible), and makes them available in distributed locations around the urban area so that they are easy to access from one's home address, workplace, or any location where access may be useful. In this way, carsharing is different from the rental car industry, which is geared more toward rental of cars for periods of 24 hours or longer, and for longer distances.

9-4-1 Motivations for Carsharing

There are several reasons to provide carsharing, both from the perspective of the individual user and the community in which it is based:

- Lower total cost for access to a car than personal ownership: For drivers on the lower end of the range of annual miles driven per person per year, and who do not need a car to commute to a full-time, "9 to 5" job, carsharing can be a cheaper option than owning a car outright. Carsharing achieves this advantage because compared to car ownership the fixed annual cost is much lower, although variable cost is higher.
- Inconvenience of parking and maintenance of own vehicle: Carsharing benefits from dedicated parking spaces for the vehicle, which allows the user to retrieve

the vehicle from a guaranteed location and return the vehicle to the same location without the nuisance of needing to look for scarce parking. Alternatively, where one-way carsharing is available, the user may simply park and lock the car near their final destination, and then the next user rents and drives the car from that location. The carsharing organization also takes responsibility for maintenance.

- Encourages diverse mix of modal choices: Members of carsharing organizations may join in addition to owning their own car, but for those who join as a substitute for car ownership, carsharing encourages the individual to diversify choice of transportation modes, since he or she no longer has the sunk cost of owning a vehicle. Members see the full cost of car usage in time and mileage cost each time they opt for carsharing, so they may be more likely to choose public transportation, NMT, or telecommuting.
- Access to optimal vehicle: In Sec. 9-3, the situation of a driver choosing a subcompact for single-occupant trips and an SUV for trips requiring high capacity was discussed as an example of how a traveler might reduce fuel cost and ecological impact. While the appeal of having different options for different types of trips is clear, many travelers find it impossible to create these opportunities for themselves, both for reasons of the additional capital cost of owning different vehicles and the practical requirement for space to store and maintain multiple vehicles in and around one's residence. The need to occasionally carry a large amount of goods or a large number of passengers often dictates that the individual who can only afford one vehicle purchase a large one, and then travel at most times with *excess capacity* in terms of passenger seats or volume of cargo space. The excess capacity in turn translates into additional weight that must be moved around when the vehicle is in use, which increases both personal financial expense and energy consumption. Carsharing gives members access to specific types of vehicles at the times when they need them (e.g., compact cars for solo trips and minivans or light trucks for moving large amounts of goods or large numbers of people) without the need to own and maintain the vehicle on one's own premises. Matching the size of the vehicle to the needs of the trip in this way helps to reduce energy consumption.

Worked examples in the remainder of this section quantify some of the benefits of the four points above.

Brief History of Carsharing

Carsharing first launched in Europe in Switzerland in 1987, and arrived in North America in the 1990s, to Quebec in Canada in 1994 and to the United States in Portland, Oregon, in 1998. As is typical of many technological innovations, initially there were many small entrants into the field and single-city, small, independent carshare operations were emblematic. As carsharing took root and membership entered a rapid growth phase (see Fig. 9-7), operators consolidated the business so that as of 2012 a limited number of large organizations have emerged: Zipcar, Inc., the world's largest carsharing company by number of members or vehicles; the Carsharing Association, which represents 15 independent carsharing organizations in Canada and the United States, and several other large players.

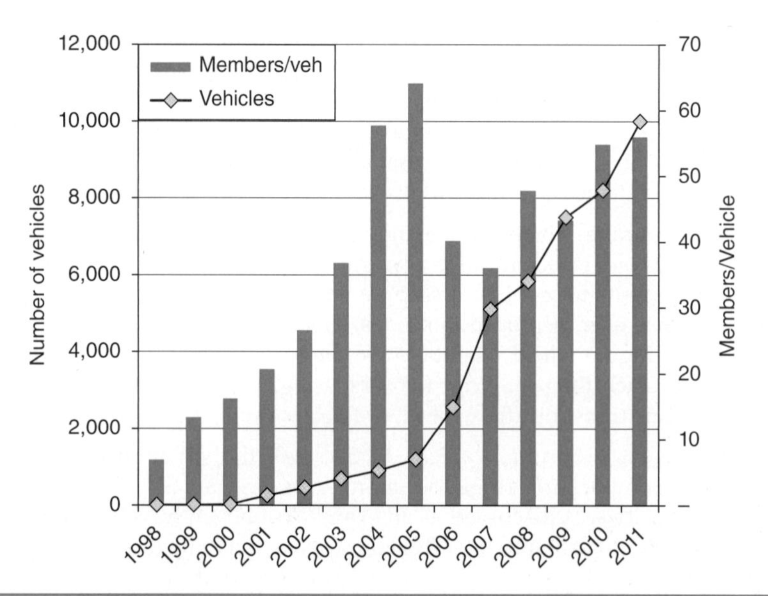

Figure 9-7 Accelerating growth of carsharing membership and vehicle fleets in Canada and the United States, 1998 to 2008.

Source: Shaheen (2011).

Contrasting Carsharing with Car Ownership and Traditional Car Rental

The advantage of carshare membership over individual car ownership is the avoided large cost of car ownership. Other than a modest annual membership fee, all costs to the carshare user are incurred at the time of use of vehicle in the form of charges for time of use and miles driven. Thus there is almost no sunk cost and the total cost of driving a vehicle is transparent to the user. Uncertainties in the cost of driving, such as unexpected large repair cost, are borne by the carshare company, not by the individual. Also, some individuals who own a personal car pay a large fraction of their overall cost of ownership in the form of large fixed costs, such as monthly car payments, cost of a location to store the vehicle, or to some extent insurance cost, that are incurred even if the vehicle is not driven or driven only a very short distance. Depending on the total distance driven, the total amount paid per year for driving may be less for carshare than for comparable use of a personal car, especially if the total miles driven is below some breakeven threshold (see Example 9-3).

Carsharing is also different from traditional car rental companies in a number of ways (see Table 9-1). (Note that divisions between carsharing and car rental companies are blurring, as discussed in the following section.) Car rental companies keep a large fleet of vehicles in a central location; the goal of carsharing companies is to distribute the parking places of the vehicles around the community so that ideally each member has several cars to choose from within a short distance from their starting location. Also, car rental companies are open to any customer, so the customer must be vetted each time they rent, which takes time. Carsharing companies vet the member once when they apply for membership, and after that, the reservation system is completely automatic, allowing the member to reserve a car and begin using it as quickly as possible when they arrive at its location. Lastly, car-rental companies are organized around longer periods of ownership and longer distances driven. Typically, cars are rented on a

Organizational Aspect	Traditional Car Rental	Carsharing System
Location of vehicles	Centralized in company office to make different models available in single location.	Distributed to permanently reserved parking spaces throughout community to be close to members.
Customer base	Open to any new customer with appropriate access to credit card and driving record.	Open to members only (some carsharing organizations allow members from other cities to reserve).
Time required to take out a vehicle	Slow: occasional customer must travel to rental office, fill out paperwork, etc. Regular customers have access to streamlined process.	Fast: customer is already a member and previously approved, so reservation is made in seconds. Vehicles ideally located near customer so access and egress rapid as well.
Time increments	Long; rented on a 24-hour day basis or longer. Customer typically gets unlimited mileage to facilitate driving long distances.	Short; typically charge for each additional 15 minutes. No free mileage: time and mileage charge encourages many short rentals instead of small number of long ones.
One-way rentals possible?	Generally yes, although rental companies may charge high additional fees if they need to discourage an imbalance in the flow of vehicles between two locations.	Generally no, due to the complexities and cost of using staff time to reposition vehicles when necessary. Since approximately 2010, however, some companies have been developing this service.

TABLE 9-1 Differences between Traditional Car Rental and Carsharing

24-hour day basis with unlimited mileage, with discounts that apply for periods of one week or more. The goal of carsharing is to allow several members to use a single vehicle in a given 24-hour period, so cars can be reserved in increments of as short as 15 minutes, and charges accrue for time of rental in addition to miles driven. Carshare systems may provide incentives for members who reserve a car for longer periods; for example, providing a vehicle to a member for a full 24-hour day at no extra charge if they reserve for the first 10 hours. Most reservation periods are, however, for shorter periods, on the order of 3 hours or less per reservation.

9-4-2 Recent Innovations in Carsharing: Rental Company, One-Way, and Peer-to-Peer Carsharing

The success of *traditional* carsharing has spawned a number of new developments and innovations. One development is the entry of major car rental companies into the carsharing market, for example, by offering hourly car rentals or unattended access to vehicles. To offer these services, car rental companies are beginning to equip some of their vehicles with wireless communications technologies. Car rental companies have also entered the market by acquisition of carsharing companies. Vehicle makers such as Daimler and BMW are entering the carsharing market as well, by factory-equipping the vehicles with communications technologies and by offering membership carsharing services using the vehicles that they manufacture.

Another significant development is one-way carsharing, in which the carsharing company allows the user to rent the vehicle one-way for a trip to a particular destination, and then discharge the vehicle near that location. Subsequent users then look online at a real-time location map to find a vehicle close to their origin, which they then use to make another one-way trip, and the cycle continues. As of 2012, one-way carsharing had reached seven countries on two continents.[1] As an example, Communauto from Quebec City, Canada, which launched in 2011, combines one-way carsharing and alternative fuels by using all-electric Nissan Leaf vehicles made available for one-way trips. At the end of each business day, Communauto operators replace the most deeply discharged vehicles with two freshly charged ones from their offices, and then recharge the former until the next day.

Lastly, the growth of peer-to-peer (P2P) carsharing has allowed the spread of the service from vehicles owned exclusively by the company to vehicles owned by private parties. The goal of this approach is to reduce the overall cost to the organization by eliminating the capital investment needed in the vehicle, although some investment in overhead or enabling technologies (websites, wireless communications, etc.) is still required. P2P carsharing can exist on several different levels. A traditional carsharing company can offer P2P as an add-on service, thus expanding their available fleet to become a mixture of company-owned and privately-owned cars. A P2P service may also deal exclusively in privately-owned vehicles. They may offer a high level of service in return for higher fees, including online location and reservation information, insurance, dispute resolution in case of mishaps, and installation of communications equipment in a privately owned vehicle for unattended access. At the other extreme, the service provider may act as a simple broker that for a lower fee provides a forum to bring lender and renter together, but stays out of negotiations about price, location, or loss and damage claims arising.

9-4-3 Cost of Carshare Usage to Members and Carsharing Organizations

The business argument for carsharing revolves around the potential savings available to a member compared to owning their own car, and at the same time the potential net revenue stream available to a carsharing organization from income from a well-used carshare car, after paying costs incurred to keep the vehicle on the road and available to members. In particular, the prospective member must consider how many miles they drive per year to decide whether or not to join. A member who does not require a car for the daily work commute can potentially spend less by substituting carsharing for a personal car. Also, for a two-car household where the second car is not in continuous use, owning a single car plus having each driver with a carsharing membership may be financially preferable to owning two cars.

The carsharing business can be divided into two components: the costs incurred by the vehicles, and the overhead costs of the organization. The vehicle cost has the following components:

---

[1]Shaheen and Cohen (2012) report one-way carsharing available in Austria, France, Germany, the Netherlands, and the United Kingdom in Europe, and in the United States and Canada in North America.

- *Capital cost of vehicles:* This includes primarily the cost of the vehicles, but also a small additional cost for equipment needed to share the vehicle (member identification system, wireless communications with reservation center). Carshare vehicles are sold on the secondary market after 4 to 5 years of use, so capital cost calculations should include salvage value.
- *Insurance cost:* Because the carshare vehicle is used by a range of different drivers, insurance per vehicle costs more than it would for an individual policy for a private owner. This cost disadvantage can be offset by defraying the annual premium value over a larger number of vehicle miles traveled (VMT), since a carshare car can expect to see higher utilization than a private car.
- *Fuel cost:* The make and model of car chosen for use as a carshare vehicle affects the fuel economy and hence annual fuel expenditure, so this item is broken out of the total cost analysis. Unlike traditional car rental, the price of gas is included in the reservation cost. In practice, when the tank on a carshare car runs low, the member takes it to a filling station but the charge for refueling is made directly by electronic transfer from the carsharing organization to the fuel vendor, so that the member is not involved in payment.
- *Maintenance:* Carshare customers put a premium on vehicle reliability, so carshare organizations typically purchase new vehicles for the fleet and sell them before they reach an age where frequent repairs become an issue. Therefore, this component is relatively small compared to other components.

In terms of revenues, the need to estimate both mileage and time charges complicates the prediction of total revenue from a vehicle. A useful shortcut is to estimate how much reservation time is associated with each unit of distance driven, as a way to convert time charges into equivalent mileage charges. The total earnings can then be estimated based on projected miles driven per year. Carsharing organizations often have a formula for estimating how many miles will be driven for each hour of rented time; Ithaca Carshare, for example, uses a figure of 8 miles driven for each hour, or 0.125 hours/mile, when projecting the cost of a reservation for members on their website. Also, different organizations and regions have developed different approaches in regard to whether to emphasize charging by unit of distance or time. For instance, Canadian carsharing tends to emphasize per-mile cost, whereas U.S. carsharing emphasizes per-hour cost.[2] Example 9-2 shows how cost, gross revenue, and net revenue can be calculated for a representative carshare vehicle

Example 9-2S A carsharing organization uses a subcompact hatchback car as a fleet vehicle, with the following characteristics: list price new, $16,000; resale value in good condition after 5 years, $11,000;[3] fuel economy, 28 mpg; average maintenance cost, $400 per year. The price of gas is $4.00/gal, the insurance cost is $2,500/year, and the discount rate is 10%. The carsharing organization earns on average $0.20/mile and $6.50/hour of rental, and recent records show that each mile of driving corresponds to 0.2 hours of rented time. The total miles driven by all customers using the vehicle over

---

[2]As an observation, in a study of 14 U.S. and 11 Canadian carsharing organizations, Shaheen et al. (2006) found that, for a representative rental of 50 miles of driving, Canadian and U.S. total cost increased by 37% and 106%, respectively, when the duration of the rental increased from 2 to 8 hours.

[3]Car values based on age, number of miles, and condition for actual makes and models are available at car information websites such as Kelley Blue Book (*www.kbb.com*) or *www.cars.com*.

the course of 1 year is projected to be 10,000 miles. What is (a) the total cost per year for the vehicle, and the breakdown of cost by category, and (b) the total revenue earned per year?

Solution

(a) To calculate annual cost, it is necessary to annualize both the purchase and resale costs. Assuming the car will be sold after 5 years and that therefore this is the investment lifetime:[4]

$$\begin{aligned} A &= P(A/P, i, N) \\ &= P\frac{i(1+i)^N}{(1+i)^N - 1} = (16{,}000)\frac{0.1(1+0.1)^5}{(1+0.1)^5 - 1} = (16{,}000)(0.2638) = \$4{,}221 \\ A &= F(A/F, i, N) \\ &= F\frac{i}{(1+i)^N - 1} = (11{,}000)\frac{0.1}{(1+0.1)^5 - 1} = (11{,}000)(0.1638) = \$1{,}802 \\ \text{Net} &= \$4{,}221 - \$1{,}802 = \$2{,}419 / \text{year} \end{aligned}$$

Next, the annual fuel cost is computed based on fuel economy and price per gallon for gasoline:

$$\text{AnnFuelCost} = \left(10{,}000\ \frac{\text{miles}}{\text{year}}\right)\left(\frac{1\ \text{gal}}{28\ \text{miles}}\right)(\$4/\text{gal}) = \$1{,}429$$

Values for all costs including fixed insurance cost and maintenance cost are combined to give the total cost per year:

$$\begin{aligned} \text{TotCost} &= \text{CapCost} + \text{Fuel} + \text{Insurance} + \text{Mtc} \\ &= \$2{,}419 + \$1{,}429 + \$2{,}500 + \$400 = \$6{,}748 / \text{year} \end{aligned}$$

The total cost is thus \$6,748/year, or \$0.67/mile.

(b) The total revenue earned per year is based on 10,000 miles driven and the corresponding number of hours reserved. Since it is expected that each mile driven requires 0.2 hours, the hour charge for the year is

$$\begin{aligned} \text{HourChg} &= \left(10{,}000\ \frac{\text{miles}}{\text{year}}\right)\left(0.2\frac{\text{hour}}{\text{miles}}\right)(\$6.50/\text{hour}) \\ &= \left(2{,}000\ \frac{\text{hour}}{\text{year}}\right)(\$6.50/\text{hour}) = \$13{,}000/\text{year} \end{aligned}$$

The mileage charge is simply 10,000 miles times \$0.20/mile, or \$2,000. Therefore, total revenue per year for the vehicle is \$15,000.

(c) The net revenue is therefore \$15,000 – \$6,748 = \$8,252.

Example 9-2M A carsharing organization uses a subcompact hatchback car as a fleet vehicle, with the following characteristics: list price new, \$16,000; resale value in good condition after 5 years, \$11,000;[5] fuel economy, 8.45 L/100 km; average maintenance cost, \$400/year. The price of gas is \$1.057/L, the insurance cost is \$2,500/year, and the discount rate is 10%. The carsharing organization earns on average \$0.125/km and \$6.50/hour of rental, and recent records show that each kilometer of driving corresponds to 0.125 hours of rented time. The total miles driven by all customers using the vehicle over the course of 1 year is projected to be 16,000 km. What is (a) the total cost per year for the vehicle, and the breakdown of cost by category, and (b) the total revenue earned per year?

---

[4]See appendix for background on discounted cash flow used in this example.

[5]Car values based on age, number of miles, and condition for actual makes and models are available at car information websites such as Kelley Blue Book *(www.kbb.com)* or *www.cars.com.*

Solution

(a) Annual cost is the same as Example 9-2S, in other words \$2,419/year. The annual fuel cost is computed based on fuel economy and price per liter for gasoline:

$$\text{AnnFuelCost} = \left(16{,}000\frac{\text{km}}{\text{year}}\right)\left(\frac{8.45\text{ L}}{100\text{ km}}\right)(\$1.057/\text{L}) = \$1429/\text{year}$$

Values for all costs including fixed insurance cost and maintenance cost are combined to give the total cost per year:

$$\begin{aligned}\text{TotCost} &= \text{CapCost} + \text{Fuel} + \text{Insurance} + \text{Mtc}\\ &= \$2{,}419 + \$14 + \$2{,}500 + \$400 = \$6748/\text{year}\end{aligned}$$

The total cost is thus \$6,748/year, or \$0.422/km.

(b) The total revenue earned per year is based on 16,000 km driven and the corresponding number of hours reserved. Since it is expected that each kilometer driven requires 0.125 hours, the hour charge for the year is

$$\begin{aligned}\text{HourChg} &= \left(16{,}000\ \frac{\text{km}}{\text{year}}\right)\left(0.125\ \frac{\text{hours}}{\text{km}}\right)(\$6.50/\text{hour})\\ &= \left(2{,}000\frac{\text{hour}}{\text{year}}\right)(\$6.50/\text{hour}) = \$13{,}000/\text{year}\end{aligned}$$

The mileage charge is 16,000 km times \$0.125/km, or \$2,000. Therefore, total revenue per year for the vehicle is \$15,000.

(c) The net revenue is therefore \$15,000 – \$6,748 = \$8,252.

The calculations in Example 9-2 show how capital and insurance cost tend to dominate the overall cost of operating the vehicle (Fig. 9-8). Together, these two components constitute 73% of the total cost. On a per-mile basis, the total cost of \$0.67 per mile for the vehicle is broken down among insurance, capital, fuel, and maintenance as follows: \$0.25,

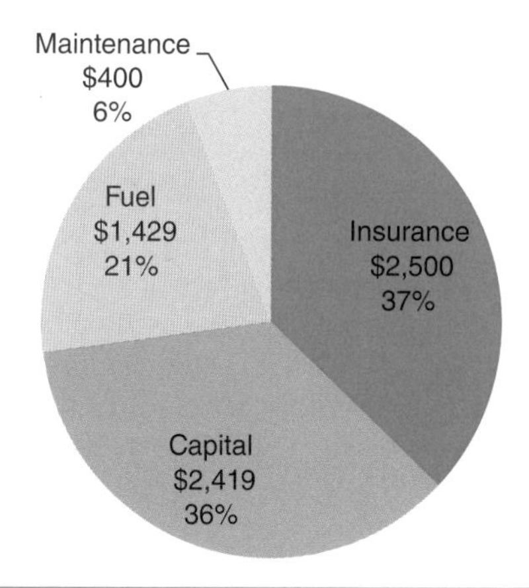

FIGURE 9-8 Breakdown of annual cost for 10,000 miles of driving for a representative carshare car, as calculated in Example 9-2.

Note: Total cost is \$6,748/year, or \$0.67/mile (\$0.42/km).

$0.24, $0.14, and $0.04, respectively. The overall cost per mile is similar to the U.S. national average, since the insurance cost is somewhat higher than the average but the capital, fuel, and maintenance costs are lower for a smaller, more fuel-efficient vehicle.

Sensitivity Analysis of Example: Cost and Revenue Calculation

Many of the cost, lifetime, and utilization figures used in the example could vary between carsharing applications, so a sensitivity analysis is in order. First, the discount rate of 10% is conservative when compared to many commercial car loan rates in recent years. At a rate of 3.9% and keeping all other conditions fixed, the annual capital cost of the vehicle would decrease to $3,584.

On the other hand, if the carshare company billed fewer hours per mile driven or sold the car on the second-hand market for a lower price, the net revenue figure might be less generous. Suppose the company bills 0.125 hours/mile driven (equal to the value used by Ithaca Carshare) and earns $6,000 on the sale of the vehicle after 5 years, with the price reduced due to relatively high wear in everyday carshare use. Holding all other figures fixed, the gross earnings are reduced to $10,125, and the cost increases to $7,567, so that the net proceeds from the car are now $2,558. In this new scenario the vehicle still operates in the black and contributes to the overhead cost of the business, but not as much as before.

Business Considerations for Net Earnings from Vehicles

Assuming the numbers used in Example 9-2, since the car costs $6,748 to operate but earns the business $15,000, the net proceeds of $8,252 are available to run the carsharing business, including overhead costs such as wages of employees, maintenance of office space, or upgrades to information technology. As long as vehicles see sufficient usage, the net proceeds should cover all costs with some funds left for other projects, such as investment in new vehicles to grow the fleet. Vehicles that do not generate sufficient usage, however, will have difficulty contributing to revenues because capital and insurance costs are essentially fixed: fuel and maintenance costs may decline with fewer VMT per year, but earnings from miles and hours will decline substantially. Fortunately, the carsharing organization has the possibility to move such a vehicle to another location with the goal of generating more usage. Relocating vehicles can be done relatively easily and with little cost; steps include research to find a better location, working with authorities to get the necessary permission to allocate space for carshare parking, and moving signage and markings.

Another observation is that the amount of time the vehicle is in circulation and generating revenue for the organization is relatively high compared to the typical usage pattern of a privately owned car. From Example 9-2, the vehicle is projected to be in use (i.e., charged out to a member and accruing per-hour revenues) for 5.5 hours or 23% of a 24-hour day and drive 27 miles daily. No one member is expected to drive a large fraction of the 10,000 VMT generated per year, but there are many users, so the total mileage and time in use is relatively high. Also, the cost per mile is increased by relatively high insurance cost, but drivers are primarily concerned with total cost per year and not cost for each individual mile. Since members drive fewer miles per year than drivers who own cars and use them for all their transportation needs, total cost per year is lower.

As a final observation, Example 9-2 shows that most earnings from the vehicle come from hours reserved, and not from miles driven. The typical usage profile of a carshare vehicle is relatively short reservation periods (3 hours or less) spent either in slow, urban traffic, or making stops to run errands where time charges are being accumulated but mileage charges are not. A carshare member might also use the vehicle for longer-distance trips; for example, for a day-long business trip. Suppose under the price regime from Example 9-2, a member leaves at 6 a.m. for a 250-mile business trip to another city, meets from 10 a.m. to 2 p.m., and then returns home by 6 p.m. In this situation, the driver would incur $100 in mileage fees and $78 in time fees. Experience shows that this type of usage accounts for a minority of total earnings for carsharing. This may be in part due to the competition between carsharing and conventional car rental for this market: a user might opt instead for a car rental with a fixed 24-hour cost, free unlimited mileage, and the only additional charge is for fuel ($70 for a 500-mile trip assuming $0.14/mile for fuel cost, based on the given fuel economy).

9-4-4 Breakeven Distance between Car Ownership and Carsharing

By implication, the discussion in Example 9-2 suggests that carsharing membership is of particular interest to individuals who drive less, and that therefore a breakeven distance exists below which carsharing is more cost-effective than car ownership. Carsharing has very low annual fixed cost—essentially only the annual membership cost. Personal vehicle owners must invest their own capital, take out a car loan, or lease a vehicle, and this commitment must be upheld regardless of number of miles driven. The same is true of insurance, for which premiums must be paid on a regular basis even if the vehicle is idle.

Formulaically, breakeven is calculated by comparing fixed cost *FC* and variable cost *VC*. Suppose options 1 and 2—for example, personal ownership and carsharing—have fixed and variable costs FC_1 and FC_2, and VC_1 and VC_2, respectively. Here variable cost may be per unit of miles driven, time elapsed, or some other variable; in this case the variable of interest is distance *D* in miles. Since at breakeven distance $D_{\text{breakeven}}$ the combination of fixed and variable cost must be equal, the following equation holds:

$$FC_1 + VC_1 D_{\text{breakeven}} = FC_2 + VC_2 D_{\text{breakeven}} \tag{9-1}$$

Rearranging Eq. (9-1) to solve for the value of $D_{\text{breakeven}}$ gives

$$D_{\text{breakeven}} = \frac{FC_1 - FC_2}{VC_2 - VC_1} \tag{9-2}$$

Example 9-3 illustrates the calculation using values given in Sec. 9-4-3.

Example 9-3S A prospective carshare member is deciding between buying a personal vehicle and joining carsharing. For carsharing, membership costs $200/year, and since they expect to be using the carshare heavily, they qualify for an hourly rate of $5/hour, lower than the $6.50/hour used in Example 9-2. Mileage charge remains at $0.20/mile. The other option is to lease the identical vehicle for $200/month and $1,500/year for comprehensive insurance with low-deductible collision (required by the terms of the lease). All maintenance costs are covered by the lease, so the only additional cost is fuel at 14.3 cents/mile. What is the breakeven distance below which carshare is preferred, and above which car ownership is preferred?

Solution For the personal car option, since the monthly charge is $200, the annual capital cost is $2,400. On this basis FC_1 and VC_1 can be calculated:

$$FC_1 = \$2{,}400 + \$1{,}500 = \$3{,}900$$
$$VC_1 = \$0.143$$

For carsharing, it is again necessary to translate time charges into mileage. Since each mile driven is projected to require 0.2 hours, the time-based umbrella charge per mile is

$$\left(0.2\frac{\text{hours}}{\text{mile}}\right)(\$5/\text{hour}) = \$1.00/\text{mile}$$

Since the distance-based mileage charge is \$0.20/mile, FC_2 and VC_2 can now be presented:

$$FC_2 = \$200$$
$$VC_2 = \text{TimeChg} + \text{MileChg} = \$1.00 + \$0.20 = \$1.20$$

Breakeven distance is therefore:

$$D_{\text{breakeven}} = \frac{FC_1 - FC_2}{VC_2 - VC_1} = \frac{\$3{,}900 - \$200}{\$1.20 - \$0.143} = \frac{\$3{,}700}{\$1.057} = \sim 3{,}500 \text{ miles}$$

In other words, if the driver expects to drive less than 3,500 miles, they should choose to join carsharing, all other things equal; if more than 3,500 miles, leasing the car is preferred.

Example 9-3M A prospective carshare member is deciding between buying a personal vehicle and joining carsharing. For carsharing, membership costs \$200/year, and since they expect to be using the carshare heavily, they qualify for an hourly rate of \$5/hour, lower than the \$6.50/hour used in Example 9-2. Mileage charge remains at \$0.125/km. The other option is to lease the identical vehicle for \$200/month and \$1,500/year for comprehensive insurance with low-deductible collision (required by the terms of the lease). All maintenance costs are covered by the lease, so the only additional cost is fuel at 8.9 cents/km. What is the breakeven distance below which carshare is preferred, and above which car ownership is preferred?

Solution For the personal car option, since the monthly charge is \$200, the annual capital cost is \$2,400. On this basis FC_1 and VC_1 can be calculated:

$$FC_1 = \$2{,}400 + \$1{,}500 = \$3{,}900$$
$$VC_1 = \$0.089$$

For carsharing, it is again necessary to translate time charges into mileage. Since as above each mile driven is projected to require 0.2 hours, the time-based umbrella charge per mile is

$$\left(0.125\frac{\text{hours}}{\text{km}}\right)(\$5/\text{hour}) = \$0.625/\text{km}$$

Since the distance-based mileage charge is \$0.125/km, FC_2 and VC_2 can now be presented:

$$FC_2 = \$200$$
$$VC_2 = \text{TimeChg} + \text{MileChg} = \$0.625 + \$0.125 = \$0.75$$

Breakeven distance is therefore

$$D_{\text{breakeven}} = \frac{FC_1 - FC_2}{VC_2 - VC_1} = \frac{\$3{,}900 - \$200}{\$0.75 - \$0.089} = \frac{\$3{,}700}{\$0.661} = \sim 5{,}400 \text{ km}$$

In other words, if the driver expects to drive less than 5,400 km, they should choose to join carsharing, all other things equal; if more than 5,400 km, leasing the car is preferred.

A related application of breakeven distance calculation concerns the choice of membership tier for carsharing organizations that offer more than one tier. An organization may offer a low annual fee tier with higher cost per hour for members who use carshare vehicles only occasionally, and a higher fee/lower cost per hour option for more regular users. If so, a breakeven analysis can be performed to decide which option is preferable.

The outcome in Example 9-3 of $D_{\text{breakeven}}$ = 3,500 miles/year, which is illustrated graphically in Fig. 9-9, is just one possibility. There are many variables that affect breakeven distance. To begin with, the input variables are all subject to change, depending on individual circumstances. For carsharing, the cost per hour, the cost per mile, and the expected miles driven per hour all affect the breakeven distance. The terms of the purchased vehicle also have an impact, including whether the maintenance cost is included, or how the life-cycle capital cost of the vehicle is incorporated.

Another consideration is the size of vehicle that the driver might choose, if they choose the own car option. Calculations for both options in Example 9-3 are based on purchase and operating cost of a representative subcompact hatchback, since carsharing organizations commonly deploy this type of vehicle. However, if the driver in the example chooses the "own car" option, they may on occasion need more capacity than a hatchback provides, and may opt for a larger vehicle, leading to higher capital, insurance, fuel, and maintenance charges, and therefore a higher value for breakeven distance.

The analysis in Example 9-3 could also be made more sophisticated by considering the impact of choice of carshare or personal car on annual VMT. Carsharing tends to encourage choice of alternatives to driving such as public or nonmotorized transport, while at the same time encouraging more efficient use of automobiles as a transportation alternative, since all costs are paid out of pocket in the form of time and mileage fees. Also, the separation between residence and carshare vehicle location, and the need to place a reservation for each use, may discourage the tendency to "jump in the car" on very short notice. Therefore the use of a single value for both options could be replaced

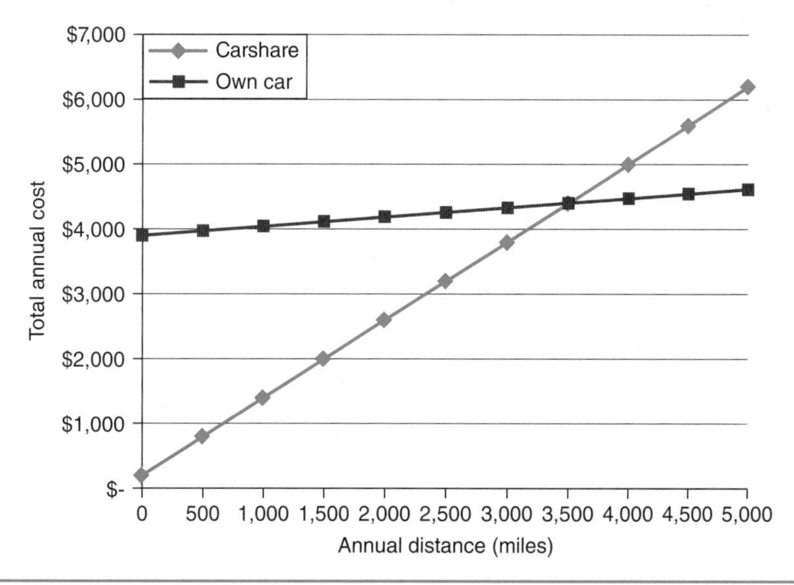

FIGURE 9-9 Total annual cost for "carshare" and "own car" options in Example 9-3.

Notes: Figure shows fixed cost at 0 miles driven per year, combined fixed and variable cost for distances between 0 and 5,000 miles, and break even at approximately 3,500 miles.

with a slightly higher value for personal car, to be compared with a slightly lower value for carshare.

Finally, the nature of anticipated vehicle use and not just the VMT per year affect the choice of carsharing versus personal car ownership. Take the case of a driver who commutes to their workplace by car and then leaves it parked from morning until late afternoon every weekday. Even if the round-trip distance to and from the workplace is short, the use of a carshare car would be prohibitively expensive because of the high hourly charges for leaving the vehicle parked in the workplace parking lot for such a long time. Nor would this use serve the purposes of carsharing well, since the goal is to serve the automobile needs of many users with each vehicle.

9-4-5 Carsharing as a Catalyst for Diversification of Modal Choices

In the preceding section it was suggested that substituting carsharing for car ownership diversifies urban passenger transportation modal choice, so in this section we take a closer look at this impact.

Experience shows that shifting from car ownership to carsharing generally encourages the carsharing member to rely less on the automobile (in this case, the vehicle belonging to the carsharing organization) to meet transportation needs and instead on a more diverse mix of carsharing, public transportation, and nonmotorized transportation. This shift depends on several factors. First, for carsharing to be successful, any individual who is expected to work regular full-day shifts at a fixed location must have regular access to public transport, NMT, or a mixture of both, due to the excessive cost of reserving a carshare vehicle for an entire work day (for which most of the time the vehicle would be parked and idle at the workplace, but accruing time charges). Second, for trips that require *trip chaining*—that is, trips that combine several destinations and purposes into a single trip from home back to home—that chaining must be feasible using carsharing, public transport, or NMT. For example, trip chaining might involve a shopping errand, a leisure activity, and a childcare-related stop all in the same trip. Lastly, the right mix of carshare/transit/NMT must be available close enough to the user's residence that they do not need to own a car to reach the other options; otherwise giving up car ownership will not work.

Decisions about carsharing versus car ownership have a large subjective component, and may come down to personal preference in a gray area where the two are competitive with each other. Example 9-4 illustrates choices a user might face in evaluating weekly transportation costs where car ownership, carsharing, transit, and bicycling are all options.

Example 9-4 Consider the transportation choices of an individual living in an urban area who can either own a car or be a member of carsharing, but not both. For simplicity, the individual's urban travel pattern is reduced to just three types: work commuting, leisure trips (linear out-and-back to a single location), and "running errands" (a circuitous route involving several stops and ending up back at home). Parameters for each type of trip are in the table below. (The abbreviation "r/t" is used for "round trip.")

The individual has the following modes at their disposal, with the costs shown in the table.

Personal car (Mode 1) means that the traveler has invested the sunk cost necessary to own a car; in this case $3,900/year, or $75/week. After that, the traveler makes trip-by-trip decisions based on the out-of-pocket cost. For Carshare (Mode 2), the individual pays the hourly cost for the entire duration of the trip, not just when the vehicle is moving; this is also the only mode that has a specific time component in the cost. Transit (Mode 3) in this example serves only the work trip, since the transit network near the individual's residence is too sparse to support other types of trips. Also, transit charges a flat fare in each direction, rather than by the mile. Lastly, Bicycle (Mode 4) appears as a zero-cost mode in the example. Although realistically a bicycle would have some life-cycle costs, these are ignored in this problem to make it simpler. Some caveats: for professional reasons, the work trip cannot be done by bicycle, and the errands trip must use car or carshare because of the items that must be dropped off or picked up. (a) What is the range of total weekly cost values for carsharing if the individual uses bicycling for 100%, 50%, or 0% of the leisure trips? (b) What is the answer to the same question if the individual owns a car? (c) What is the weekly cost for (b) if the weekly fixed cost is ignored and only the variable cost is considered?

Data for use in problem

Trip Type	Distance (r/t, Miles)	Time (Hours)	Frequency (Events/Week)
Work	20	10	4
Leisure	8	2	3
Errands	20	3	1

Travel Requirements

Mode	Sunk Cost ($)	Out-of-Pocket Cost ($/Miles)	($/Hours)
Personal car	$75/week	$0.15	$0
Carshare	$0	$0.20	$5.00
Transit	$0	$1.50/trip	$0
Bicycle	$0	$0	$0

Modal Cost Options

Solution

(a) Carsharing weekly cost: Because use of carshare would require 10 hours of rental time per day for use for work trips, and bicycle has been disallowed, only transit is feasible for this option. Transit costs $3/day, or $12/trip. The weekly errands trip costs $4 for mileage and $15 for 3 hours of use, or $19 total. Thus the basic price is $31/week if the member always travels on bicycle for leisure trips. At the other extreme, each leisure trip by carshare costs $1.60 for mileage and $10 for time, or $11.60 total, so 100% use of carshare for all three leisure trips per week would cost $34.80/week, with a combined total of $65.80/week. 50% bicycling then gives a value halfway between the two, or $17.40 for leisure trips per week and $48.40 total per week. Therefore, the range of values is $34.80 to $65.80 per week.

(b) In the case of car ownership, since the car trip or use of transit for work have the same out-of-pocket cost ($12/week), it is very likely the user will choose car for reasons of convenience and certainty. Therefore, the lowest cost with 100% use of bicycling for leisure trips is $75 for fixed cost plus 100 miles at $0.15/mile, or $15/week, for a total of $90/week. If either 50% or 0% bicycling is chosen instead, miles driven for leisure trips increases to 12 and 24 per week, and cost increases to $91.80 and $93.60, respectively. Therefore, the range of values is $90 to $93.60 per week.

(c) Out-of-pocket cost in the case of car ownership is simply the cost without the $75/week fixed cost. The result is a cost of $15 to $18.60 per week.

Discussion of Modal Choice Example: Implications and Alternative Scenarios

A number of inferences can be made from Example 9-4 about the effect of carsharing. First, the weekly cost calculations illustrate the observation made above that the difference between out of pocket and sunk cost is significant. For the car owner, having committed the large sum per week to own the vehicle, the additional cost per week is relatively small, so that individual decisions about whether or not to use the personal car have relatively little impact. For the carshare member, on the other hand, each trip costs between $11.60 and $19 (in other words, a similar magnitude to the entire week of out-of-pocket cost for the car owner), so they make decisions about modal choice more carefully.

The importance of having modal choices available for the work trip in the carshare case is made clear as well. Any commuter who is required to stay at a job location distant from their residence for the entire work day must have other options available. In this case, bicycle was ruled out due to work requirements, although for some members of the workforce, commuting by bicycle is of course feasible (depending in part on distance, terrain, physical fitness, etc.). If neither public transport nor NMT are feasible, the commuter will have no choice but to purchase a personal car, at which point the financial feasibility of carshare in addition to personal car ownership is questionable. (Some carshare organizations offer just-in-case memberships, in case a personal car breaks down and is not available on a given day, or for other similar reasons.)

In the example, transit and personal out-of-pocket cost for the work trip were arbitrarily set to be equal. It was assumed that workplace parking did not cost the commuter. In this situation, many commuters would choose to drive their personal car, as assumed in this particular solution. In some situations, the user might choose public transportation, for example, if the drive to work or finding parking is particularly stressful, if they prefer having the time while traveling on transit for other activities, or if they appreciate the ecological benefits of not driving. For many travelers, however, the convenience and predictability of traveling in their personal car along with the time penalty for access and egress to transit as well as the slower linehaul speed will favor choosing the personal car. Thus the user needs ideally both strong transit and strong carshare options to choose not to purchase a car.

The example also shows how the carshare member has a stronger financial incentive to use bicycling as a substitute for driving on the leisure trip, when weather conditions allow. The financial savings for each avoided driving trip are much larger for the carshare member, namely $11.60/trip versus $1.20 out of pocket for the car owner. Full use of bicycling on every occasion year round (i.e., 0% carshare usage from Example 9-4) may not be realistic due to inclement weather on a given day or the hot or cold season making bicycling too uncomfortable in general. Depending on the region, however, values of 50% or more bicycle usage may be achievable, and these figures translate into a substantial annual transportation cost savings.

The full cost per week for the carshare and personal-car options can be compared graphically as shown in Fig. 9-10. For the purposes of fixed cost for carsharing, a membership cost of $200/year is spread over 52 weeks, resulting in a cost of $3.85 added to projected total transportation cost. The scenarios are then divided up into a low-cost ("Lo") scenario, where all leisure trips are by bicycle, a high-cost ("Hi") scenario, where they are all by carshare, and a mid-cost ("Mid") scenario that is between the two. The dramatic effect of adding carshare trips is illustrated: with between 1.5 and 3 leisure trips per week added, the increase in weekly cost is quite visible. (Note that the figure of 1.5 trips per week is actually more realistically 3 trips every 2 weeks, or 2 trips in Week 1, and 1 trip in Week 2, etc.)

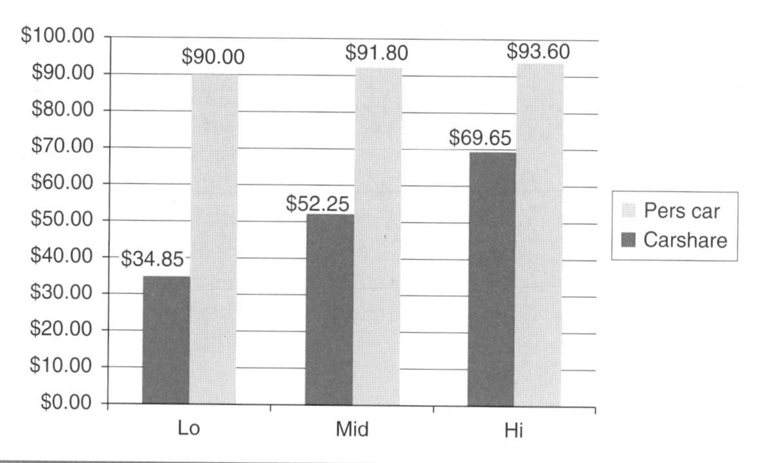

FIGURE 9-10 Comparison of weekly cost for carshare or personal car: Lo, Mid, and Hi cost scenarios.

The main implication of the example is that by trading in their own car for carsharing, the user ends up spending less for transportation, once sunk cost of owning a vehicle is taken into account. Looking from the opposite perspective, however, if the user cannot save very much money (depending on their perception of purchasing power) by trading in their car for carsharing, they may be inclined to keep the car for reasons of convenience and peace of mind. In the "Hi" scenario, the cost for carshare is on the order of $3,600/year, versus $4,700/year for the own car option. The user may or may not be interested in this level of savings, especially if the annual savings of $1,100 is small compared to their disposable income. All of this assumes that carshare vehicles are available in sufficient numbers near the user's starting location, that NMT and/or public transportation options are adequate, and that other transportation needs throughout the year, such as one-off intercity transportation requirements, are not significant factors in making the decision.

9-4-6 Policy Measures to Support Carsharing

Up to the present, carsharing has evolved primarily as a neighborhood-based alternative to individual car ownership. In the most common type of deployment, vehicles are distributed throughout the neighborhoods that surround the central business district (CBD) of an urban area, with sufficient density that any member has second or third choices close to their location in case their closest vehicle is already reserved. In the future, carsharing will have opportunities to expand into other niches as well. Carsharing fleets can be based at large businesses where employees need transportation to run errands in the middle of the day, on college campuses to provide an alternative to personal car ownership for students, and in low-income neighborhoods where car ownership is financially out of reach for residents.

There are several policy steps that governments and other stakeholders can support to accelerate the growth of carsharing. Some are already in widespread use, while others are at a more embryonic stage:

- *Loosen parking regulations for urban development:* Zoning requirements often require urban residential developments to provide a certain number of parking spaces per number of anticipated residents. These requirements can be both expensive and logistically difficult to meet, given constraints on urban space. Codes can be changed to allow developers to reduce the number of spaces required in return for guarantees of a certain number of carsharing vehicles provided, on the premise that for each new carshare vehicle a certain number of newly arrived residents will opt to join carsharing instead of bringing a personal vehicle. Going a step further, a residential developer might require new residents to agree not to bring a private car to their residence as a condition of living there, as a way of ascertaining that the parking burden will not be excessive. If it were effective, such a policy might be particularly effective in promoting carshare usage, although enforcement of the agreement could be challenging.

- *Provide financial support for carsharing membership:* Municipalities can provide parking spaces free of charge to carsharing organizations, as a way to support their growth. In places other than their home parking space, carshare vehicles can receive preferential parking; that is, parking spaces reserved for carpool vehicles or alternative fuel vehicles can also be made available to carshare vehicles. Tax policy can be used as well, for example, by exempting carsharing charges from sales or tourism taxes. Government, universities, and private businesses can subsidize or pay for carsharing memberships or ongoing carsharing costs for their employees, the same way they might provide free parking or transit passes. These policies must be approached carefully. In many of these instances, government foregoes revenues it might otherwise earn from metered parking, local taxes, and so on. In an age of austere budgets and complaints about preferential treatment of some citizens over others, these measures may be controversial, but where supported, carsharing can benefit. On the other hand, carsharing may be treated by some localities as a luxury like a hotel stay and taxed accordingly, thus overlooking the public good (e.g., lower energy consumption, reduced traffic) when travelers choose carsharing over other options.[6]
- *Create fleet partnerships to support carsharing:* Large stakeholders such as municipal governments, large universities, or private industry can merge part or all of their fleets with carsharing fleets. Both fleet users and carsharing members can benefit, since the average utilization of the vehicles will likely increase, given that the pool of users increases. Fleet owners may also be able to reduce costs, since the total number of vehicles they must purchase or lease decreases. Vehicles may be allocated according to a formula, such as giving fleet users exclusive access during 9 to 5 business hours, and carshare members access on nights and weekends.
- *Combine carsharing and public transportation access:* Transit operators can offer discounts to carshare members when they purchase passes or other types of tickets. Conversely, carsharing organizations can offer discounts for transit use. The goal is to incentivize shifting away from personal vehicle ownership by offering a carsharing-transit package that provides greater financial discounts than either option considered separately.

---

[6]For example, Bieszczat and Schwieterman (2011) found average taxation rates on carsharing to fall between 14% and 18%, depending on the length of the reservation, compared to average sales tax just over 8%.

- *Specific collaboration between carsharing and universities:* Universities and colleges have specific transportation needs that are well met using carsharing. They have a large student population that in general needs only occasional access to a vehicle, since campuses are usually organized so that students can walk or bicycle from dormitories to classes. Universities and colleges are often large enough that they can financially self-insure fleets of vehicles, rather than paying an outside provider for insurance services. Finding affordable insurance can be a barrier to launching carsharing, especially in the United States, so having the academic institution provide insurance may be a way to launch carsharing in a community where it might otherwise not be viable. Alternatively, these institutions can invite large carsharing chain organizations to come to a campus and launch a carsharing branch that is specifically targeted toward supporting students, faculty, and staff.

9-4-7 Future Prospects for Carsharing

Carsharing organizations currently are studying several interesting design challenges that, if they can be solved successfully, might allow the further expansion of carsharing. One challenge is the work trip question: if a carshare vehicle were used for commuting to/from work in the morning and evening, could the vehicle be put back into circulation for other members to use for short trips starting and ending at the workplace in the middle of the day? If so, the cost burden on the commuting member might be shared with others who use the vehicle to run short errands. Another challenge is the development of a vehicle repositioning system that would allow a user to reserve a carshare vehicle from one parking location and return it to another. It is possible that individual members' trips might allow the vehicle usage to balance out; for example, Member 1 drives from A to B, Member 2 from B to C, Member 3 from C back to A, and so on, although without exact balancing, the carshare workforce might at times need to reposition vehicles to rebalance the network, a potentially time-consuming and expensive task.

In conclusion, even without solving these challenges, carsharing is already on a solid footing in Europe and North America. Originally launched as an experimental concept whose place was not guaranteed next to personal vehicle ownership and traditional car rental, carsharing has grown into an international establishment with thousands of vehicles and members. In one U.S. city alone, Philly Carshare of Philadelphia, Pa., claims more than 50,000 members. There are several reasons for the ongoing strength of carsharing: (1) it takes advantage of wireless communication and is consistent with the skills of an IT-savvy generation, (2) it is bolstered by the rising cost of personal car ownership and the increasing inconvenience of parking and auto dependence in many locations, and (3) it is consistent with the growing embrace of multimodalism, in which automobiles are just one of several choices next to bicycling, walking, transit, and telecommuting.

The large number of members for each vehicle in mature carsharing deployments in cities large and small also suggests that the concept is in healthy shape. New carshare deployments in cities where they previously did not exist inevitably start with low numbers of members per vehicle, because members want to see how many vehicles are available before they commit to joining. The ratio of members to vehicles can be expected, however, to quickly rise to the levels seen in mature organizations, as long as the launch is successful.

9-4-8 Overview of Bikesharing

Similar to the function of carsharing, the goal of bikesharing is to reduce auto dependence by conveniently providing bicycles for certain trips where the user needs flexible mobility and bicycling provides an attractive alternative to driving. Many of the organizational aspects of bikesharing are similar to those of carsharing, so for brevity we will focus on distinguishing features here.

Figure 9-11 illustrates bikesharing deployments in several locations. The Paris bikesharing station (Fig. 9-11(*a*)) represents the commercial approach to bikesharing: residents can pay a fee to join the system, and then are charged for each time they rent a bicycle from a station. Electronic tracking of each bicycle rental allows the user to rent at one location and return at another, facilitating one-way travel. As another example, the Bixi system (Fig. 9-11(*b*) and 9-11(*c*)) first launched in Montreal and more recently in Ottawa and Toronto, Canada, allows tourists as well as resident users to swipe in to the system with a credit card and begin using bikesharing immediately, all with the goal of making travel without a private car in the urban core more attractive. A billboard next to the bikeshare station in the photo advertises the availability of the service.

The Cornell University bikesharing shown in Fig. 9-11(*d*) represents a university-based variant of bikesharing. Bicycles are provided to students, faculty, and staff free of charge with the goal of encouraging them not to bring private cars to the central campus, and the bikes are administered by the library system, with the procedure for borrowing a bicycle resembling that of borrowing a book. Since the customers for this

(*a*)

Figure 9-11 Bikesharing systems: (*a*) Velib system, Paris, France; (*b*) Bixi System, Montreal, Canada; (*c*) closeup of Bixi fare payment kiosk; (*d*) university-based bikesharing system at Cornell University.

(b)

(c)

Figure 9-11 (Continued)

(d)

FIGURE 9-11 (Continued)

system are already registered with the university library system, the facility is simpler than that of Paris or Montreal: riders borrow a key from inside the library to unlock the bicycle, so that no special equipment is needed at the rental station other than the rack itself.

Bikesharing systems use a variety of funding mechanisms to purchase and maintain equipment. The Bixi system in Canada uses a traditional not-for-profit format, like that of public transit agencies: user fees cover part of the cost, and then the municipal government covers the rest by making regular financial contributions to the operator. The Velib system, on the other hand, uses an advertising model, where advertisers pay for the right to promote products on signs at rental stations, with the proceeds used to fund the operation.

One challenge for bikesharing usually not associated with carsharing is vandalism, due in part to the relatively small size of each individual bicycle compared to that of a carshare car. Bikesharing organizations have needed to contend with loss and damage to bicycles, even as they continue to expand.

One operating aspect that is simpler for bikesharing than for carsharing is the repositioning of *vehicles* not currently in use, and this factor allows the bikesharing member or user to rent a bicycle at one location and return it at another. The bikesharing organization can initially build extra parking spaces into its network of bike rental stations so that usually when a user arrives at a destination station there will be open slots available. If a user arrives and all slots are full, they can be allowed extra rental time free of charge to be directed to another nearby station where one is available. Also, organizational staff can monitor the number of bikes parked at each station, and if necessary use a van or pickup truck to redistribute several bikes at once from one station to another to avoid running out of either bicycles or parking slots at any location. In the case of the Cornell University system, bikes are dispensed at three locations around the campus, and university maintenance staff members periodically redistribute them between the locations to rebalance the number available at each. This capability is potentially an advantage for bikesharing, since, unlike carsharing, users can plan on one-way trips between stations, without needing to return to the station of origin.

9-5 Telecommuting and Other Substitutes for Mobility

Solutions for urban passenger transportation up to this point have typically combined the objective of reducing dependence on automobiles with other objectives. For instance, public transit has the advantage of providing increased mobility to individuals who do not have access to a private car. NMT provides opportunities for improved physical health through cardiovascular exercise. Either of these approaches continues to provide mobility to the urban resident, albeit in an alternate way.

Another approach is to eliminate trips entirely through the use of telecommunications networks and other amenities—in other words, *substituting accessibility for mobility*. Mobility may be replaced entirely in cases where the resident stays in their residence, but it may also be replaced only partially where the resident travels to a nearby location, such as a coffee shop or library, and accesses the telecommunications network from there. Where the mobility concerns the would-be traveler's workplace, the activity is called *telecommuting*.

The dictionary defines telecommuting as "working at home by the use of an electronic linkup with a central office."[7] More broadly speaking, telecommuting is the use of information technology, electronic communications, and other enabling technology to perform in a remote location tasks that would otherwise be performed in an employee's default work location, thereby substituting networked communication for the physical need to travel. Telecommuting need not always take place at home; it may also take place in telecommuting satellite locations that are near the worker's residence compared to a relatively longer distance to the workplace. In this case, the benefit of telecommuting to the transportation system comes from the reduction in travel from the telecommuter not traveling the entire distance to the central office.

As such, not all electronic communication constitutes telecommuting. Certain electronic functions that are performed exclusively electronically without physical need to travel, such as online ordering of products and services, are not telecommuting. Nor is every type of remote work telecommuting. For instance, a college professor who takes papers or exams to a remote location away from their office to carry out evaluation and grading is working remotely, but is not telecommuting because they are not relying on an electronic linkup for this work.

Emergence and Current Status of Telecommuting

First seen on a small scale in the 1980s, telecommuting has advanced to various degrees in the industrial and industrializing countries of the world. In the United States

[7]Source: *www.meriam-webster.com.*

telecommuting is thought to have played a role in the employment patterns of about 8% of the workforce in 1996. By 2008, this number had increased to 22%.[8] The rapid advances in information technology, the common interest of both employers and employees in improving quality of life, and the burgeoning problems with the congested transportation network in urban areas all appear to be pushing a rapid expansion in telecommuting access.

9-5-1 Telecommuting Benefits and Challenges

Telecommuting is of interest in modern times because it has potential benefits for three important stakeholders, as follows:

- *Employees:* Employees can benefit from telecommuting because the time required to commute to and from the central office becomes available for other purposes. Telecommuting may facilitate completing certain errands that are carried out close to the home, while still allowing sufficient time to complete all daily tasks that are required professionally. Employees may also benefit from an improvement in quality of life because they can spend part of their working time at home, as opposed to spending it entirely at the central office.
- *Employers:* Employers can benefit from increased productivity of their employees, after paying for any startup costs for which they may be responsible, such as enhanced linkups or faster wireless communication. Care must be taken that the productivity does in fact increase, since an undisciplined employee may actually be distracted by their home environment and suffer from reduced productivity. Employers may also be able to avoid certain infrastructure costs, such as office or parking spaces, since by having a certain fraction of their work force always out on telecommute, the total amount of space needed at the central office declines.
- *General public:* Because electronic communication has a relatively compact physical footprint compared to transportation infrastructure, the public can benefit when a certain fraction of the population telecommutes rather than physically commuting. With fewer travelers, fewer roads need to be built, avoiding both commitment of land surface area and financial commitments to highway infrastructure. Telecommuting can also lead to less pollution and climate change with fewer travelers moving, with the proviso that the capacity created is not then absorbed by *latent demand* in the transportation network. Also, by keeping both employers and employees happier and with a better quality of life, the general public can reap the benefits of a healthier society.

At the same time, there are several challenges for quantifying costs and benefits of telecommuting and determining whether telecommuting is beneficial overall in a given situation. The following list is indicative:

- *Presence versus intensity of telecommuting:* There are limits to the accuracy of efforts to estimate the volume of telecommuting, since measures of *telecommuting*

[8]1996 figure is from Shafizadeh et al (2007). 2008 figure is based on an estimate of 33.7 million telecommuters according to the website *www.suitecommute.com,* out of a workforce total of 155 million according to the U.S. Department of Labor.

workers or *telecommuting households* capture only whether they are engaged in telecommuting or not, and not the intensity of telecommuting; for example, number of telecommuting days per work week or per year.

- *Value of commute time:* The time spent commuting to work could be valued at the employee's average earnings per hour, on the grounds that if they were not commuting, they could be earning income with the same amount of time. However, the commuter may also use time in the car in other productive ways, such as listening to the news on the radio, so that the hourly cost might be valued less than their earnings rate.
- *Value of avoided cost:* The avoided cost of driving a motor vehicle could be valued at the average cost per mile to operate the vehicle, but it also might be evaluated at the marginal cost of the miles driven, since some vehicle costs do not increase linearly with distance driven.
- *Relative productivity at home versus in workplace*: In some instances, it might reasonably be assumed that the home is a more productive environment for the employee, since the distractions of the workplace do not surround them. Both employee and employer might therefore have an interest in supporting telecommuting. Some employees may, however, be less productive at home.
- *Relative cost of utilities*: When the employee telecommutes, they may incur utility costs for climate control (heating or cooling of home space) or IT use that they would not otherwise incur had they left their home and gone to the office to work. On the other hand, some utility costs are ongoing in any case, or perhaps the employee makes no thermostat adjustment whether they are absent or present. From the employer's vantage point, having the employee on-site at the office may measurably increase their energy footprint.
- *Investment required for IT*: To telecommute effectively, the employee may require investments in hardware and software beyond the basic requirements for leisure internet connectivity (e.g., basic desktop or laptop computer, tablet, or smart phone). These costs are incurred in addition to utility costs for electricity to operate equipment. Costs are generally independent of the number of telecommuting days per year, and may be incurred on a one-time-only or annual basis.
- *Stakeholder bearing financial responsibility for investment:* To varying degrees, employers may pay for various parts of the employee's telecommuting package, or may require the employee to pay. Each must then evaluate total costs and benefits. It is possible to have a situation where telecommuting is net cost-effective for one but not the other.

9-5-2 Evaluation of Telecommuting: A Case Study

To explore some of the considerations raised in the previous section, a case study based on a Monte Carlo simulation model is presented here. The case study assumes the perspective of an employee who would like to telecommute, and wishes to know the range of possible outcomes for the ratio of benefits to costs. The employee puts a value per hour on time spent traveling, and also commutes in an SOV, so that reduced driving to work is a possible benefit of telecommuting. However, telecommuting will require an annual investment in additional equipment, and increase home utility costs. These four factors are randomized in the case study. Other factors such as number of telecommuting days per year are treated as constant for brevity. Calculations for the case study are provided in Example 9-5:[9]

Example 9-5 A commuter is considering a shift to telecommuting for part of their work week, and would like to know the benefit-cost ratio (B/C ratio) to help them make a decision. Many inputs into the B/C calculation are uncertain so they use a Monte Carlo simulation to study the range of possible outcomes. Random variables include travel time and cost per round trip for vehicle (including fuel, maintenance, etc.), utility costs for working from home, and annual setup costs for telecommuting. All random variables are assumed normally distributed with mean and standard deviation values in the following table. Assume that the number of telecommuting days per year is fixed at 50 (i.e., 1 per week for a year with 2 weeks of holiday), and that the value of avoided commuting time is \$11.40/hour. Number of telecommuting days is held fixed to limit the size of the example, but in principle this parameter could be treated as a random variable as well. For this simulation, (a) calculate the deterministic solution, (b) calculate the value of a single iterate if the standard normal generator returns values of –0.243, –0.242, 0.707, and 0.740 for the four required inputs, and (c) run a Monte Carlo simulation with 1,000 iterates and calculate the percentage of iterates for which $B/C < 1$.

Item	Mean	SD
Travel time (hours)	0.688	0.237
Travel cost	\$3.85	\$1.32
Utility cost	\$4.80	\$0.92
IT cost	\$250.00	\$50.00

Table of Values

Solution

(a) The deterministic solution is calculated using the expected value for each input to calculate a B/C ratio. Since the value of travel time is given in terms of expected number of hours per round trip to the central office, we use the given value of \$11.40/hour to convert to a monetary value:

$$\left(0.688 \frac{\text{hours}}{\text{roundtrip}}\right)(\$11.40/\text{hour}) = \$7.84$$

The B/C ratio is then calculated using the number of days per year of $N = 50$ and the various monetary values given. Note that because IT cost is incurred annually; it is not multiplied by the number of telecommuting days per year:

$$B/C = \frac{(\text{TimeValue} + \text{TravelCost})N}{(\text{UtilityCost})N + \text{ITCost}}$$
$$= \frac{(\$7.84 + \$3.85)50}{(\$4.80)50 + \$250} = 1.19$$

Since the calculated ratio of $B/C = 1.19$ is greater than unity, it is expected that adopting a pattern of telecommuting at this level would be attractive to an employee.

---

[9]This calculation is based on a longer case study in Shafizadeh et al. (2007), and the reader is referred to this source for more extensive results, including a model with more factors, presentation from the employer's perspective, and discussion of telecommuting benefits in general. Also, see Chap.3 for an introduction to Monte Carlo simulation.

(b) For this individual iterate, the returned values for each input must be calculated using the mean, standard deviation, and sampled value from the standard normal. For example, the standard normal value for the travel time (t_{travel}) returned is –0.243. The value of t_{travel} for this iterate is therefore

$$t_{travel} = \mu + z \cdot \sigma = 0.688 + (-0.243)(0.237) = 0.630 \text{ hours}$$

The monetary value is then (0.630 hours)($11.40/hour) = $7.18. Repeating this procedure for the other three inputs results in the following table:

Item	Std. Norm. Val.	Val. Returned
Travel time	–0.243	$7.18
Travel cost	–0.242	$3.53
Utility cost	0.707	$5.45
IT cost	0.740	$287.00

The values in the "Value Returned" column can now be used to calculate B/C for this iterate:

$$B/C = \frac{(\$7.18 + \$3.53)50}{(\$5.45)50 + \$287} = \frac{\$536}{\$560} = 0.96$$

The iterate therefore returns a value slightly unfavorable to telecommuting, because the benefit values in the numerator are lower than expected and the cost values in the denominator are higher than expected. With the uncertainties surrounding the inputs for this individual, it is realistic to expect that the shift to telecommuting might work out negatively in a minority of outcomes, even if the expected value is positive.

(c) The simulation is carried out by repeating the steps in part (b) in an automated way for 1,000 iterates, in this case using a spreadsheet model. No two runs of the Monte Carlo simulation will be identical, but the results for a particular run gave an average value of $B/C = 1.21$, compared to the expected value from (a) of $B/C = 1.19$. Also, 28.3% of iterates resulted in values of $B/C < 1$.

As illustrated in Example 9-5, if we look at the problem deterministically, the commuter should expect to be 19% better off according to the deterministic results; that is, for every $1 spent on telecommuting needs, $1.19 is returned in savings on travel time and vehicle costs. However, since there is considerable uncertainty in the input values, the output *B/C* values in the simulated solution are scattered across a wide range. In 1,000 iterates, a value as low as $B/C = 0.31$ and as high as $B/C = 3.35$ were returned for the particular simulation run in question. Also, when the results of the simulation are sorted into bins as shown in Fig. 9-12, some 167 iterates result in *B/C* values that are slightly below breakeven ($0.8 < B/C < 1$), and 116 iterates that are strongly below breakeven ($B/C < 0.8$). If the commuter's primary purpose is to save money by telecommuting and if they are risk averse, then they may not be willing to make the investment. In any case, the example shows the power of the Monte Carlo simulation: if we only calculate the deterministic solution, then we know the expected value of savings from telecommuting but we do not obtain any quantitative information about the risk that the investment will not break even.

Example 9-5 is also simplified in a number of ways. It is assumed that all regular commuting takes place in a single-occupant vehicle, and that when the worker

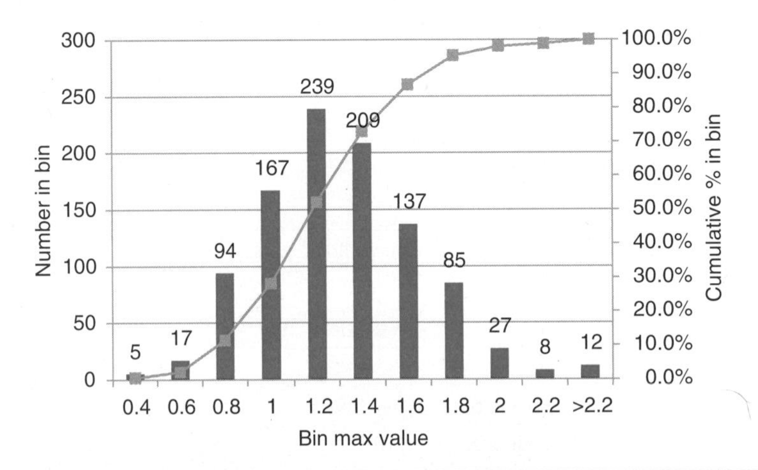

FIGURE 9-12 Distribution of 1,000 iterates into bins for representative Monte Carlo simulation run ($\mu = 1.21$, CV = 31%).

Note: In the leftmost bin, 5 out of 1,000 iterates returned values of B/C < 0.4; in the next, 17 returned values in the range 0.4 ≤ B/C < 0.6, etc.

telecommutes, the full value of avoided driving will accrue. However, a different commuter might use some mixture of SOV, public transport, NMT, or carpooling, in which case the value of telecommuting as a replacement for these other modal choices would be different. The worker may also have planned some errands on the drive to or from work, in which case the avoided work trip may lead to an added nonwork trip later to complete the errands. In this case, telecommuting would replace a combination work-errand trip with a shorter errand-only trip, so the benefit would need to be calculated more carefully.

9-5-3 Opportunities for Telecommuting and Future Prospects

Looking to the future, it is of interest to know about how people use telecommuting; that is, not only whether but also how frequently workers telecommute. There is also the parallel question of total penetration of telecommuting. As an example, Fig. 9-13 shows a representative penetration pathway going forward assuming effectively 0% penetration in the year 1980, figures given above for the years 1996 and 2008, and an eventual saturation at 30% of the workforce. A substitution model curve is then fit to the observed data to project values beyond 2008.[10] The curve suggests that by 2020 penetration might surpass 29%. The curve assumes that the remaining 70% to 71% of workforce positions are in fields such as retail sales that require a personal presence at the work site. However, as the tasks required in different lines of work change and evolve, it is possible that the maximum possible penetration of telecommuting may increase.

[10]See presentation of substitution curve model in Chap. 3. Best-fit parameters for curve in Fig. 9-13 that minimize RMSD are $F = 0.3$, $c_1 = -3.799$, and $c_2 = 0.1724$.

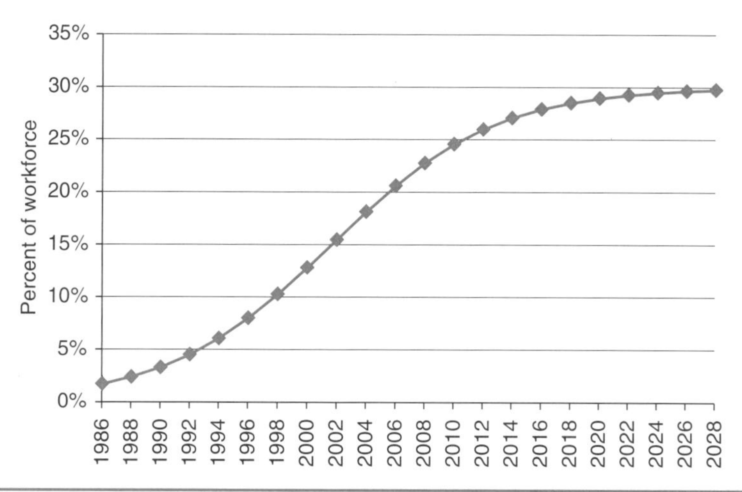

Figure 9-13 Possible pathway for telecommuting penetration of workforce 1986 to 2028 based on observed penetration levels up to 2008.

9-6 Discussion of Personal Mobility and Accessibility Options: Case of Portland, Oregon

The focus of this chapter has been on improving the sustainability of transportation at a personal level, including diversifying the types of vehicles available to individual users, formal vehicle sharing including carsharing and bikesharing, and telecommuting. We bring these policy steps together in this section using a case study or *exemplar* of a specific city, namely Portland, Oregon, United States.[11]

By way of background, Portland is a city with a population of 1.9 million living in the greater Portland metropolitan area and average income of $37,342, thus a fairly affluent city by U.S. standards. Portland is well known among U.S. cities for efforts to contain urban sprawl since the 1980s by encouraging development within an urban ring around the city center and discouraging development outside that ring. Focusing on public transportation, Portland has an LRT network that is well-utilized by U.S. standards, and recently introduced an urban streetcar in the city center; per-capita transit use increased by 26% from 1990 to 2005. In terms of transportation statistics, the average person in Portland drives 10,753 km (6,638 miles) per year and boards transit 58 times per year. The former number is relatively low, and the latter number relatively high, for a U.S. city.

Turning to personal mobility and accessibility, Portland has achieved or is developing the following transportation measures:

- *Infrastructure to encourage more optimal modal choice:* The metro Portland region has made a substantial effort to provide bicycling and walking infrastructure.

[11]Schiller et al. (2010) further develop the idea of exemplar cities as models for improving transportation sustainability, and present an international list including Portland as well as Boulder, Colo.; Freiburg, Germany; Surabaya, Indonesia; Seoul, South Korea; and Vancouver, Canada. The reader is referred to these examples for more details.

As an example, an urban freeway along the Willamette River was dismantled and replaced with the Tom McCall urban park. Modal share of trips for cycling increased from 1.1% in 1990 to 6.8% in 2011 (Pucher and Buehler, 2013).

- *Carsharing:* Portland was the first city in the United States in which carsharing was successfully launched. Today Zipcar operates a carsharing operation in Portland with several hundred vehicles available.
- *Bikesharing:* Portland is currently developing a bikesharing operation that is expected to begin operation in spring 2014.

Information on telecommuting specific to the Portland area is difficult to obtain, but it can be assumed that as a city with a modern and vibrant city center, Portland has all the amenities needed for successful telecommuting. The city possesses institutions typical of a high-tech research and knowledge sector, including research universities such as Portland State University and medical research facilities such as Providence Portland Medical Center. As an approximate indication of the availability of locations available for remote teleworking, a search for coffee shops of one of the national chains, Starbucks Coffee, found approximately 150 Starbucks in greater Portland, with at least 50 independent coffee houses in addition.

To summarize, the combination of policies that encourage more compact land use, discourage excessive use of personal cars, and emphasize the development of public transportation, especially LRT, create a foundation for transportation success. One result has been relatively fewer hours lost to congestion compared to other large U.S. cities: the Texas Transportation Institute estimates this figure at 37 hours/year in 2010, down from 42 hours in 2005, and much lower than the highest 2010 value in the United States, namely Los Angeles at 72 hours/year.

9-7 Summary

In this chapter, we have seen how adjustments to transportation at a more personal level can improve sustainability. Availability of the appropriate motorized vehicles, including limited-use, two-wheel vehicles, or eventually self-driving vehicles, can reduce impact. Changes to infrastructure and commuting practices including HOV/HOT lanes, congestion pricing, carpooling, and dynamic ridesharing can all make more efficient use of motorized vehicle-miles traveled. Use of NMT modes has the further advantage of providing an opportunity for physical exercise. Infrastructure for NMT and electric-assist cycles can both encourage their use. New modes are emerging, including other NMT modes besides walking and cycling, and motorized modes that have access to sidewalks or cycle paths, such as Segway transporters or electric-assist cycles. Emerging carsharing and bikesharing systems allow individuals to avoid the responsibility of personal ownership of vehicles, and at the same time ensure that each vehicle gets more utilization. Travelers may incorporate these modes for their travel needs, or they may avoid traveling entirely by telecommuting instead of physically commuting. The appeal of telecommuting to many workers is clear and its popularity has been growing, but because many of its costs and benefits are uncertain, it is a suitable subject for static (Monte Carlo) simulation modeling.

References

Bieszczat, A. and J. Schwieterman (2011). *Are Taxes on Carsharing Too High? A Review of the Public Benefits and Tax Burden of an Expanding Transportation Sector.* Report, Chaddick Institute for Metropolitan Development, DePaul University, Chicago, IL. Electronic resources, available at *www.las.depaul.edu*. Accessed Nov. 23, 2013.

Glaser, M., P. Madruga, and E. Griswold (2009). "The World's Most Bicycle Friendly Cities." Electronic resource, available at *www.copenhagenize.com*. Accessed Jun. 12, 2013.

Pucher, J. and M. Buehler (2013). "Cycling to the Future: Lessons from Cities across the Globe." Presentation at University of Washington, Jun. 18, 2013.

Schiller, P., E. Bruun, and J. Kenworthy (2010). *An Introduction to Sustainable Transportation: Policy, Planning, and Implementation*. Earthscan, London.

Shafizadeh, K. R., D. A. Niemeier, P. L. Mokhtarian, and I. Salomon (2007). "Costs and Benefits of Home-Based Telecommuting: A Monte Carlo Simulation Model Incorporating Telecommuter, Employer, and Public Sector Perspectives." *Journal of Infrastructure Systems* 13(1):12–26.

Shaheen, S., A. P. Cohen, and J. D. Roberts (2006). "Carsharing in North America: Market Growth, Current Developments, and Future Potential." *Transportation Research Record* 1986:116–124, Transportation Research Board, Washington, DC.

Shaheen, S. and A. Cohen (2012). *Innovative Mobility Carsharing Outlook: Carsharing Market Overview, Analysis, and Trends.* Report, Transportation Sustainability Research Center, University of California Berkeley. Fall 2012 ed, Calif.

U.K. Department for Transport. (2008) *Transport Statistics Great Britain: 2008 Edition.* Her Majesty's Stationery Office, London.

Further Readings

Fan, W., R. Machemehl, and N. Lownes (2008). "Carsharing: Dynamic Decision-Making Problem for Vehicle Allocation." *Transportation Research Record* 2063:97–104, Transportation Research Board, Washington, DC.

Portland Metropolitan Area Government (2013). "Urban Growth Boundary." Electronic resource, available at *www.oregonmetro.gov.* Accessed Jun. 26, 2013.

Pucher, J., R. Buehler, and M. Seinen (2011). "Bicycling Renaissance in North America? An Update and Reassessment of Cycling Trends and Policies," *Transportation Research A* 45 (6):451–475.

Shaheen, S, A. Cohen and M. Chung (2009). "Carsharing in North America: a Ten-Year Retrospective." *Transportation Research Record* 2110:35–44.

Shaheen, S. (2011). "Carsharing: A Strategy for Reducing Carbon Footprint & Parking Policy Approaches." Presentation, 2011 CCPA Conference, Oakland, CA, Nov. 3, 2011.

Shaheen, S., M. Mallery, and K. Kingsley (2012). "Personal Vehicle Sharing Services in North America." *Research in Transportation Business and Management.* 3:71–81.

Shaheen, S., S. Guzman, and H. Zhang (2010). "Bikesharing in Europe, the Americas, and Asia: Past, Present, and Future." *Transportation Research Record* 2143: 159–167.

Shaheen, S. and A. Cohen (2012). "Carsharing and Personal Vehicle Services: Worldwide Market Developments and Emerging Trends." *International Journal of Sustainable Transportation* 7(1):5–34.

Spielberg, F. and P. Shapiro (2000). "Mating Habits of Slugs: Dynamic Carpool Formation in the I-95/I-395 Corridor in Northern Virginia." *Transportation Research Record* 1711:31–38.

Tao, C. and C. Chen (2007). "Dynamic Ridesharing Matching Algorithm for the Taxipooling System Based on Intelligent Transportation Systems Technology." *International Conference on Management, Science, and Technology*, Harbin, China.

Vanek, F. and S. Spindler (1999). "Community Cycling Accessibility Initiative: Enhancing Cycling to Decrease Auto Dependency. *Journal of International Association of Traffic and Safety Sciences* 23(2):36–42.

Exercises

9-1. A carsharing organization offers two rate tiers, $50/year and $8/hour, or $200/year and $5/hour. For either tier, the mileage charge is $0.20/mile. The anticipated hours required for 1 mile of driving is 0.125 hours. A prospective carshare member anticipates reserving the car for 3 hours on each occasion. What is the minimum frequency at which they could reserve the vehicle and still prefer the cheaper hourly rate?

9-2. An urban resident is debating whether or not to either join a carsharing organization or purchase a new car. Joining carsharing costs $300/year, and after that, for simplicity of this problem the carsharing rate structure is based entirely on miles driven. The chosen vehicle is a compact hatchback, purchase cost $16,000, salvage value $8,000, life 5 years, insurance $2,500/year, fuel economy 28 mpg, and maintenance cost $400/year. Use a discount rate of 10% throughout. The resident is concerned on the vehicle purchasing side that she will not have enough space for the occasional long trip, large group of friends, or large load, so has chosen a midsize sedan as the vehicle she will purchase, to meet these occasional needs. (With carsharing, if these needs arise, she will address them some other way since in any case she is not tied to renting a carshare vehicle for every trip.) The sedan lists for $28,000 has a fuel economy of 24 mpg overall, costs $1,500/year for insurance, and incurs $0.04/mile for repairs and maintenance. It should be discounted over 10 years with 0 salvage value. Assume $4/gal for gas for either option. (a) Calculate the cost per mile for carsharing. (b) Calculate the fixed and variable cost components for buying the sedan. (c) If the carsharing agency charges twice the cost of driving per mile to cover all of their overhead costs, what is the breakeven distance in miles? (d) If the resident expects to drive 4,000 miles or 10,000 miles/year, what is the difference in cost between the less and more expensive option for meeting their driving needs?

9-3. A carsharing organization uses a subcompact hatchback car as a fleet vehicle, with the following characteristics: list price new, $18,000; resale value after 5 years, $8,000; fuel economy, 27 mpg; and average maintenance cost, $500/year. The price of gas is $3.80/gal, the insurance cost is $2,700/year, and the discount rate is 6%. Recent records show that each mile of driving corresponds to 0.125 hours of rented time. The organization desires to earn 25% of revenue from per-mile charges and the rest from per-hour charges. The total miles driven by all customers using the vehicle over the course of 1 year is projected to be 11,000 miles. (a) Calculate the average cost per mile to the nearest whole cent. (b) Calculate the percent of per-mile cost allocated to each of capital cost, insurance, maintenance, and fuel. (c) Suppose the carshare company needs to earn 1.5 times the annual cost of the vehicle to cover their overhead cost. What should they charge per mile and per hour for this vehicle? (d) What percent of hours in a 24-hour day must the vehicle be rented to achieve the hourly goals?

9-4. In this problem, you are to build a Monte Carlo simulation model with 1,000 iterations for the *B/C* ratio for telecommuting from the employee's perspective, taking into account the following components in the ratio: (1) driving cost for vehicle, (2) time cost of time spent driving, (3) home utility cost, and (4) annual IT cost to be able to telecommute. The following table gives five stochastic inputs; these should be "sampled" to calculate the *B/C* ratio for each run of the simulation. In addition, the following inputs are fixed ("deterministic") in each run of the simulation: fuel economy of 28 mpg, average travel speed of 30 mph, 46 work weeks per year, 1.5 days per week of telecommuting, driving cost other than fuel of $0.10/mile, and gas cost of $4/gal. (a) Before conducting the simulation, calculate the fixed-value output from the model. What is the expected value of B/C? (b) Suppose that RAND() returns for the four stochastic inputs driving distance, value of time, utility cost, and IT cost the following four values: 0.724, 0.812, 0.728, and 0.932, respectively. What is the resulting value of *B/C* in this case? (c) Run a simulation in the software package of your choosing and compute 1,000 runs for your simulation model. Deliver (1) a page of run results, with the input values and resulting *B/C* value for each run (one line per run is expected); (2) the overall average *B/C* ratio across the 1,000 runs; (3) the percent deviation of the simulation average *B/C* value from the fixed-value average from part (a). (d) Produce a histogram with bin sizes of your choosing for the 1,000 results from the simulation. (e) Short answer: What information do the simulation and histogram give you that is not already available from the fixed-value average *B/C* in part (a)? 2 to 3 sentences maximum.

Variables	Distance	Units	(μ, •)*	(*a*,*b*)
Average one-way distance	Normal	miles	(11.6,4)	n/a
Value of avoided time	Uniform	$/hour	n/a	(6,10.20)
Home utility costs	Normal	$/event	(2.75,0.5)	n/a
Annual IT costs	Normal	$/year	(250,40)	n/a

CHAPTER 10

Intercity Passenger Transportation

10-1 Overview

This chapter considers approaches to improving the sustainability of intercity passenger modes other than passenger cars and other light-duty vehicles, which have already been considered extensively in previous chapters. A special emphasis is given to the high-speed rail (HSR) and air modes, since at the present time these two are growing the most in total passenger-kilometers and ecological impact, and are receiving the most attention from governments and the general public. With respect to HSR, an overview section provides history and current technological function of HSR trains and systems, and then a formulaic treatment of force and power requirements as a function of speed allows assessment of potential energy efficiency, financial cost, and emissions impact. Next, current status and governing equations for contemporary aviation is discussed, leading to a comparison of aviation and HSR. Trends in other modes, including intercity bus and waterborne modes, are important as well, so the chapter rounds out with a discussion of these modes.

10-2 Introduction

This chapter focuses on intercity passenger travel (i.e., travel that is not "local," thus not associated with short-range daily trips in either urban or rural areas: trips for work, education, shopping, and so on), and in particular travel in modes other than privately owned, light-duty vehicles (LDVs). Options for LDVs have been considered in some depth in previous chapters, and many of the solutions for urban travel are transferrable to intercity travel; for example, systems for ridesharing that facilitate drivers and riders sharing vehicles to increase occupancy and reduce both cost and energy consumption per passenger, can be adapted for intercity travel as well. Thus the modes of interest in this chapter include HSR and air and, to a lesser extent, intercity public bus and waterborne modes.

Several recent trends are shaping global intercity passenger travel, as follows:

- *Growth in global passenger volume:* In both industrial (European Union, Japan, United States, etc.) and industrializing countries (China, India, etc.), demand for intercity travel, measured in either passenger-kilometers or passenger-miles, has grown as GDP per capita has grown in recent years. Some intercity travel is discretionary and therefore influenced by economic or political events. For example, in the United States, times of economic downturn or the aftermath of the September 2001 terrorist attacks have led to periods of reduced intercity travel. The overall trend has been upward for the last three decades and, because of the major influx of new consumers of intercity travel in industrializing countries such as China and India, the global growth has been significant.

- *Growth in high-speed transportation:* Growth of high-speed modes (aviation and HSR) has been particularly large because of the growing desire of passengers to travel long distances between countries and continents in a timely fashion. Aviation has captured the largest share of this market because it is the only one of these two that is practical for travel over water or over the longest distances (on the order of thousands of kilometers), and because HSR networks are still being built up. Year-on-year percentage growth in HSR network size and demand, however, has also been significant in the last several decades. Another contributor has been the development of low-cost carriers such as JetBlue in the United States or RyanAir in Europe that can serve similar origin-destination pairs as *legacy* carriers (Delta Airlines, British Airways, etc.) at lower cost and that have expanded the market for air travel by making trips by air possible that previously might have been made by highway or not at all.
- *Congestion of highway networks and airports, leading to increased interest in HSR:* Expansion of highway and airport infrastructure has generally not kept pace with growing demand for road and air travel, so that the total physical capacity of the infrastructure is strained by high demand, especially during peak periods. This situation has led several countries to plan and develop HSR networks that might take some of the traffic off highways and out of airports, so that congestion would decrease.

On balance, intercity passenger transportation poses a major challenge for sustainability, both from the strain on infrastructure (with its resulting social strain on travelers) and from the pressure on the natural environment to absorb pollution and other impacts, notably the effect on global climate from greenhouse gases (GHGs) such as CO_2. There are some mitigating factors as well: intercity travel is not as large an energy consumer or GHG emitter as nonintercity passenger travel, and modes such as HSR or conventional rail provide the opportunity for propulsion without CO_2 emissions in the end-use stage of the life cycle (i.e., ignoring embodied emissions in vehicles and infrastructure) if the electricity is derived from non-CO_2 emitting sources.

10-2-1 Comparison of Intercity Modal Split: Case of Japan and United States

Countries in the industrialized world have come to different solutions for meeting their intercity transportation demand, which makes international comparison possible. Figs.10-1 and 10-2 compare the total volume of passenger-kilometers (pkm) and modal share for intercity travel in Japan and the United States in 2008. The United States, with a population of 304 million, had an intercity travel volume of 7,209 pkm per capita, while Japan, with a population of 128 million, had a notably smaller volume of 3,318 pkm per capita. Modal share values are quite different as well. While intercity rail passenger-kilometers in the United States has only a very small share of the total

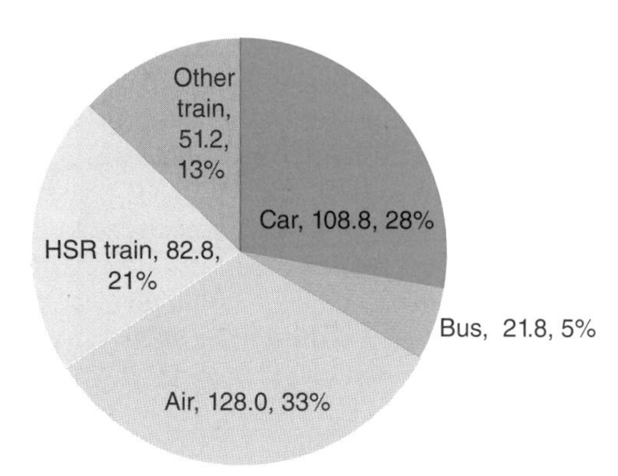

FIGURE 10-1 Modal share of intercity passenger travel in Japan, 2008 (Total = 393 billion pkm).
Sources: Adapted from Lipscy and Schipper (2012) and Public Transportation Library (2010).

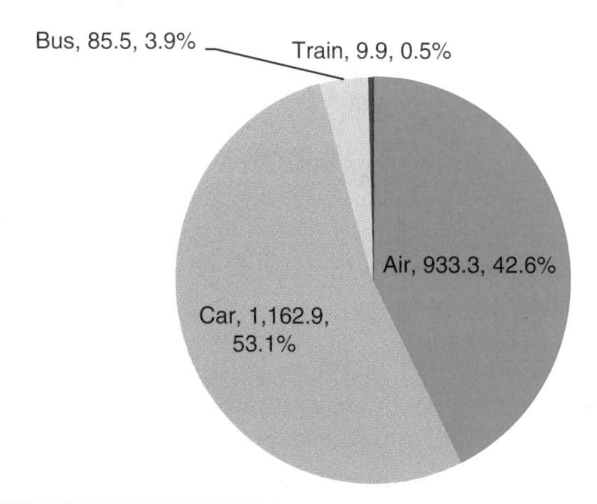

FIGURE 10-2 Modal split of 2008 intercity passenger-kilometers for the United States (Total = 2.19 trillion pkm).
Source: Adapted from U.S. Bureau of Transportation Statistics (2012), USDOT (2011).[1]

(9.9 billion pkm, or 0.5% of the total 2,191 billion pkm), rail in Japan has captured 34% of total pkm between HSR and conventional intercity rail (134 billion out of 393 billion pkm). Air is a significant mode in both countries, with a share of 33% and 43%, respectively, for Japan and the United States, as air travel has found a niche especially at the longest travel distances. Note that the "Car" mode in the figures includes all types of LDVs, such as passenger cars, light trucks, minivans, and sport-utility vehicles (SUVs).

Both geographic and policy factors explain these differences. The United States has approximately 25 times the land area of Japan (9.37 million versus 378,000 km^2) and

[1]Some fraction of reported U.S. train passenger-kilometers qualify as HSR, but the breakdown was not available from the underlying source, so all passenger-kilometers are reported together.

as a result a much lower population density (32 vs. 339 persons/km^2). All other things equal, higher population density favors HSR development since it leads to shorter, more densely populated corridors where HSR can be deployed effectively. The most heavily traveled origin-destination (OD) pair in Japan, Tokyo to Osaka, has a distance of 380 km, whereas a significant fraction of U.S. passenger-kilometers is incurred between the east and west coasts, with distances in the thousands of kilometers.

National policy in the two countries has also been markedly different. Both countries upgraded rail systems in the 1960s, with Japan opening the first Tokyo-Osaka leg in their HSR network for the 1964 Tokyo Summer Olympics and the United States upgrading the electrified northeast corridor main line between Washington, D.C., and New York to operate in places at HSR speeds. Thereafter, the paths of the two countries diverged. Japan had created a stand-alone HSR technology that could then be extended to all other parts of the country and eventually exported (e.g., to Taiwan), while the United States discontinued further improvements to the national passenger rail system, instead prioritizing highway and air modes for intercity passenger travel. Instead of developing passenger service, major private U.S. railroads (Burlington-Northern, Union Pacific, Norfolk Southern, CSX, and so on) put their efforts into innovations entirely on the freight transportation side. These private rail developments accelerated after the deregulation of U.S. railroads in 1980.

10-2-2 Modal Comparison of Delivered Energy Intensity

The potential for reducing energy consumption and emissions is one of the motivations that is currently driving interest in alternative intercity modes. To evaluate potential savings, we must evaluate all factors involved. Total energy consumption depends on delivered energy intensity in kilojoules/passenger-kilometer, which in turn depends both on the inherent technical efficiency of the vehicle and the load factors of the service provided (i.e., percent of available seats filled).

Figure 10-3 shows energy intensity for three intercity passenger modes in the United States, namely, bus, air, and rail. Although the rail mode is generally very efficient when

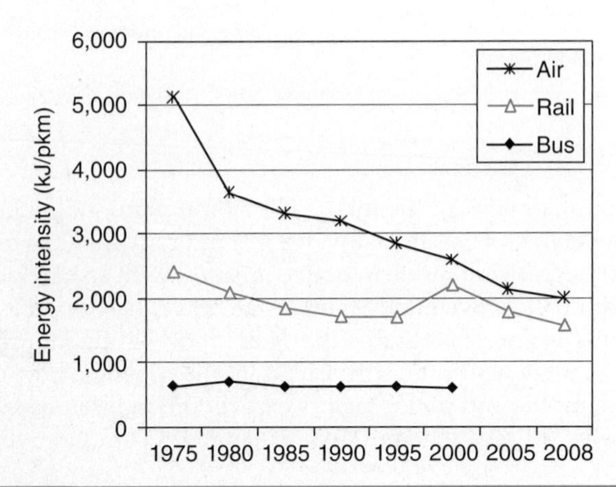

FIGURE 10-3 Intercity passenger transportation energy per passenger-kilometer, 1975 to 2008.

Note: Intercity bus data were not available after 2000.

Source: U.S. Department of Energy.

load factors are high, intercity rail in the United States suffers from low occupancy, so that energy intensity per passenger-kilometer is only slightly lower than that of airline travel (1,600 vs. 2,000 kJ/pkm in 2008). By contrast, the bus mode is consistently more efficient than the other two up to the last year of available data in 2000. The air mode has been able to reduce energy intensity by 61% over the time period shown, due to both improved aircraft and engine technology and the use of yield management to maximize seat occupancy on flights. Data on LDV passenger energy intensity divided between intercity and urban geographic scope were not available, so intercity car travel does not appear in the figure. However, for comparison, combined urban and intercity car and light truck intensity values in 2008 were 2,266 and 2,401 kJ/pkm, respectively, which are both higher than values for rail or air.

Figure 10-4 shows the trend in load factors for air and rail, the two U.S. modes for which data were available through 2008. Load factors are calculated slightly differently for each mode due to data limitations. For commercial aircraft, load factors are given in terms of percent of seats filled on average for the years 1975 to 2008. However, for rail, these data are not available directly, so instead a measure of passenger-kilometers delivered per traincar-kilometers moved is used. Data are also available on train-kilometers of movement, or distance traveled by entire trains, but traincar-kilometers are used as a basis for this figure since they more closely reflect the capacity provided. Because two different measures of load factors are used for the two modes, the values are presented in relative terms indexed to a value of 1975 = 1. The graph shows that load factors have steadily increased for air by 24 percentage points from approximately 55% in 1975 to 79% in 2008, while for rail the number of passenger-kilometers delivered per traincar-kilometer of movement rose 53% over the same period (from 14.8 to 22.7), although with more fluctuation in the intervening years. Note that for the rail mode the use of passenger·km/traincar·kilometers is not entirely precise because the average number of seats per traincar may change over time.

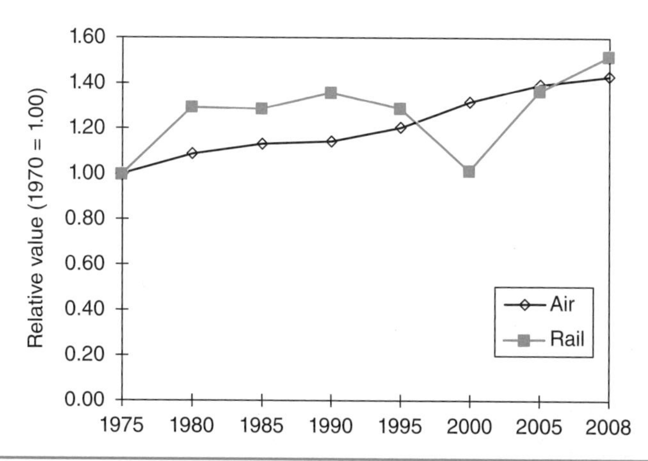

FIGURE 10-4 Relative usage of available capacity for U.S. commercial air and passenger rail modes, 1975 to 2008, indexed to 1975 = 1.00.

Notes: In 1975, for air, 54.9% of seats were occupied, and for rail, each traincar-kilometer delivered 14.8 passenger-kilometer of service.

Source: U.S. Department of Energy.

The United States is unusual among the industrial nations for having a particularly low share of passenger-kilometers and passenger energy consumption for bus and rail. Most other industrial nations have a higher share for these two modes (as reflected in the comparison with Japan above), and also lower energy intensity for rail, since load factors are higher. The importance of efficient technology and efficient systems for maximizing productivity should be emphasized, however. A mode such as rail that at first glance has a high potential for efficient operation must be utilized effectively (i.e., high occupancy rate), and the technology must be continuously pushed toward cutting edge propulsion systems, aerodynamics, and so on, so as to truly create an advantage compared to competing modes.

10-3 High-Speed Rail

High-speed rail is typically defined as intercity rail service that travels at or above a threshold speed of approximately 200 km/h (125 mph) during line haul movements. HSR rights of way are generally used exclusively for passenger transportation, but there are instances where HSR trains are used for express package service, either mixed with passenger-carrying service or in dedicated unit trains. HSR can be either *new-build* or *incremental*, meaning that the right of way can be developed from the ground up on an all-new alignment (e.g., the approach first implemented for the HSR system in Japan in the 1960s), or existing conventional railway tracks can be upgraded to accommodate the higher speeds (e.g., in the United States, the Chicago to St. Louis upgrade project that started construction in 2011).

10-3-1 Trainset and Right of Way Design Considerations

High-speed rail propulsion is for the most part electric with electricity supplied from overhead *catenary* or wires via a pantograph to electric motors that drive the wheels. This system allows for high power transmission enabling sustained acceleration to speeds as high as 350 km/h (220 mph) and continued propulsion at these speeds without needing to incur the weight penalty of carrying the electricity generation system on board the train, since the electricity is generated off-site at a central power plant. As a counterexample, British Railways has developed the "Intercity 125" high-speed train (HST) capable of diesel-propelled high-speed travel at 125 mph, using a *power car* at either end of the trainset, which carries a cylinder-and-piston diesel engine that in turn drives an electric generator to provide power to the wheels.[2] The HST thereby gives the operators flexibility to run trains on tracks either with or without catenary, although noise and local emissions are higher.

Figure 10-5 illustrates a number of the features of an HSR right of way: full separation from the surrounding landscape using excavated embankments or fences; no grade crossings, with cross-vehicle traffic passing the right of way with either bridges over or tunnels under; and long turning radius of curves to allow the trains to pass at high speeds without causing passenger discomfort. As an example of the latter point,

[2] A "trainset" is a term for the entire HSR train including power car with engineer's cabin at either end of the train and all passenger or café-restaurant cars in between, in total usually comprising between 8 and 12 individual cars including the power cars. Regarding the use of the diesel and electric combination, due to the large power requirements and the complexity of providing a purely mechanical transmission from a rotating diesel engine to the wheels of a locomotive, virtually all diesel locomotives and propulsion units are in fact diesel-electric.

Figure 10-5 Representative HSR line with passing trainset (French TGV northbound from Lyon to Paris).

Notes: The photograph illustrates several principles of HSR system design: full separation of right of way with walled, fenced track; long-radius, gentle curves and elevation changes; and energy supply from overhead electrical catenary.

the Japanese HSR track is designed for full-speed operation with a 4-km turn radius, meaning that a 180° turn would require the train to go around a circle with a diameter of 8 km (5 miles). Some HSR (and fast conventional) trains (such as trains in Sweden, Italy, or Spain) use *tilt technology* that actively tilt the train beyond the angle of tilt built into the track so that passengers can maintain comfort in turns at a higher maximum speed.

10-3-2 Historical Development and Current Status of HSR

As shown in Table 10-1, the first true HSR service was launched in 1964 in Japan. Known as the "Shinkansen" (in English, "new trunk line"), the Japanese HSR operates on its own network of tracks separate from conventional railways and has a different gauge (width between the two rails of 1,435 mm for the Shinkansen vs. 1,067 mm for conventional rail), so Japanese HSR trains cannot exit the HSR network to travel on Japan's network of conventional railway lines.[3] (Note: This capability is called a *closed* as opposed to *open* network in the table.) HSR arrived in Europe in 1976 with the opening in France of the *Train a Grand Vitesse* (TGV, in English *high-speed train*) between Paris and Lyon.

[3]The Spanish railways have developed the Talgo passenger train technology that can adjust the width of the wheels to be able to travel on different gauges in different countries, but this technology has not been widely adopted.

Country	Year Opened	Length, km (2012)	Open Network?[1]	Selected Cities Served
Belgium/France/U.K.[2]	1993	738	No	Paris, London, Brussels
China	2008	9356	No	Beijing, Shanghai, Wuhan, Guangzhou
France	1976	968	Yes	Paris, Lyon, Strasbourg
Germany	1988	686	Yes	Munich, Hamburg, Cologne
Italy	1978	848	Yes	Rome, Florence, Milan
Japan	1964	3482	No	Tokyo, Osaka
Korea	2004	409	No	Seoul, Pusan
Spain	1992	1411	No	Madrid, Barcelona, Seville
Taiwan	2007	345	No	Taipei, Kaoshung
Turkey	2008	533	No	Ankara, Eskisehir
United States[3]	1968	365	Yes	Washington, New York

Notes:

1. *Open Network:* Whether or not trains operating on HSR tracks can also operate on conventional rail tracks, thus extending the potential size of the network to serve more destinations.
2. Entry for Belgium/France/U.K. is the Eurostar train that connects Brussels, Paris, and London via the Channel Tunnel under the English Channel.
3. The U.S. distance shown is for the northeast rail corridor between Washington and New York, for which portions were upgraded to be able to operate at 125 mph starting in 1968. HSR trainsets were introduced only in 2001 with the launch of the "Acela" service.

Sources: Adapted from Albalate and Bel (2012), Public Transit Library (2010).

TABLE 10-1 High-Speed Rail Development in Select Countries, 1964 to Present

The French system allowed for interconnectivity, so international HSR service eventually followed. For example, HSR service between Geneva, Switzerland, and Paris consists of a direct train that leaves Geneva and travels on conventional track until it reaches the HSR line between Lyon and Paris, and then travels the rest of the distance at maximum speed to Paris.

After Japan and France, HSR spread to numerous countries in Europe (including Belgium, Germany, Italy, Spain, Turkey, and the United Kingdom) and Asia (China, South Korea, and Taiwan). In North America, passenger train service along the northeast corridor, where 125-mph travel had been possible since the 1960s, was upgraded in 2001 with the introduction of HSR trainsets similar to those used by the TGV system and extension of electric catenary from New Haven, Connecticut, to Boston, which allowed travelers on the new trainsets to travel continuously from Washington to Boston without needing to change trains.

Maglev as an Alternative to Steel Wheel / Steel Rail Technology

One challenge for HSR is the need to maintain a smooth ride quality at high speeds so that passengers do not experience excessive vibration or other negative sensations. The concept of *magnetic levitation* or "maglev" has been under development since the 1960s as an alternative to conventional HSR steel wheel/steel rail technology to address this concern, especially at the highest speeds. With maglev, a powerful magnetic field creates a gap between the vehicle and the railbed, allowing the vehicle to travel on the railbed without encountering rolling resistance. As with conventional HSR, the maglev unit must still overcome aerodynamic resistance, and a continuous power supply is required to maintain the magnetic field. The potential advantages of maglev compared to steel wheel/steel rail include smoother operation at the highest speeds (350 km/h or more) and a higher maximum service speed. The term *high-speed ground transport* (HSGT) is sometimes used in place of HSR to indicate that high-speed surface transportation includes both rail and nonrail options.

Building on experience with test track maglev systems in Germany and Japan, the world's first commercial maglev was opened in Shanghai in 2004, connecting Shanghai Pudong International Airport with one of the urban metro stations on Shanghai's metro system over a 30-km (18-mile) track, where trains reach a maximum speed of 430 km/h (267 mph) in regular service. A longer maglev system for the Berlin-Hamburg corridor in Germany (length of 290 km or 180 miles) was planned in the 1990s but never built. The decision to postpone indefinitely the building of the Berlin-Hamburg maglev reflected concerns about the high capital cost and high ongoing energy cost for maglev compared to HSR. A fatal maglev accident at Lathen in western Germany on September 22, 2006, further dampened enthusiasm for developing the technology. A conventional HSR line was eventually built instead on the Berlin-Hamburg corridor.

10-3-3 HSR in Comparison to Other Modes

The favorable market niche for HSR service is for trips with a line haul distance of between approximately 200 and 1,000 km. The dominant modes at <200 km and >1,000 km are road and air, respectively. High-speed rail rights of way are expensive to develop on a per-distance basis, so the network of HSR tracks is sparse compared to the network of conventional rail or limited access highways. Therefore, unless a traveler going a shorter distance happens to live along an HSR corridor, it is likely that travel via HSR will be too circuitous due to the distance required for access to and egress from the HSR network, and these trips will instead be made by private car, intercity bus, or conventional rail. At distances longer than 1,000 kilometers, air travel becomes dominant, since the fixed time to access airports (check-in, security clearance, time for aircraft to take off and land, etc.) is offset by faster line-haul speeds. Aircraft at cruising altitude travel at speeds around 850 to 900 km/h (528 to 559 mph), compared to speeds of 350 km/h (217 mph) or under for HSR, so the advantage is substantial.

Potential Advantages of HSR

HSR has the potential to contribute to transportation planning and ecological protection goals of many regional and national governments around the world, as indicated in the following list:

- *Contribution to a more compact built environment:* Compared to a limited-access highway with two or more lanes in each direction, an HSR line with two tracks requires less lateral space and therefore reduces total land consumption for an intercity surface transportation artery. Flow on HSR lines may be increased by allowing HSR trains making all station stops to pull off the through track at some stations to allow limited-stop HSR trains to pass, as is done with the Japanese Shinkansen system. HSR trains generally travel from city center to city center, requiring less space for stations than is required for airports at the perimeter of urban areas. In addition, for the world's largest urban areas that occupy the greatest geographic expanse, where HSR lines make more than one stop in the urban area, a secondary use of HSR is to rapidly connect different parts of the urban area. For instance, for the greater Tokyo metropolitan area in Japan, whose 26 million people occupy most of the Kantoh Plain, smaller HSR stops on the perimeter of the region can provide a rapid nonstop connection into downtown Tokyo.

- *Relief from congestion on highways and in airports:* HSR has the potential to encourage modal shifting away from highway and air travel, thus reducing congestion on the most congested intercity freeways or in the busiest airports. This possibility must be interpreted with caution, since the launch of HSR may shift some travelers from air and rail, thereby freeing up capacity for new trips on highways and in airports, so that congestion levels remain unchanged.
- *Improved energy efficiency and reduced emissions:* For O-D pairs within its ideal range of 200 to 1000 km, HSR is often one of the most energy-efficient travel options, assuming that occupancy rates of available seats are sufficiently high. Electrified HSR lines also contribute to reduction of petroleum dependency and harmful emissions, since the electricity used is typically generated from a diverse range of sources rather than petroleum, and the use of non-CO_2 emitting primary sources in electricity generation can reduce overall GHG emissions per passenger-kilometer compared to automobile or air. HSR on new-build track has the potential to reduce end use energy consumption per passenger-kilometer thanks to reduced circuity, that is, because of the high speeds required, the tracks travel in a straighter alignment from origin to destination, thus reducing total distance traveled for a given O-D pair compared to conventional railways, many of which were surveyed and built decades ago.
- *Improved safety in intercity travel:* HSR has the potential for extremely safe intercity travel, with a particularly low risk of injury or death in an accident compared to air travel and especially automobile travel. For example, of the three oldest new-build HSR systems, the Japanese and French systems have never had a fatal accident, and the German system has had only one fatal accident since inauguration in 1988, namely, the Eschede accident of June 1998. Since air travel is also very safe (very few fatalities per millions of passenger-kilometers), this advantage for HSR is in comparison to driving, not flying.

HSR systems also have other advantages that accrue specifically to individual travelers or to railroad companies. For passengers traveling between city centers, there is a potential for reduced door-to-door travel time compared to air travel because of the avoided extra time to travel to the airport on the periphery of the urban area and to pass through check-in and security. Some airports also suffer from congestion to the point that boarding and takeoff can be delayed compared to scheduled times, so that HSR travel serves as a more reliable mode of transit in terms of arriving at the destination within a limited number of minutes of the published arrival time. For railway operators and national rail systems (such as those of Japan, France, or Germany), HSR has the benefit of providing access to high value-added, high-speed travel that can improve the revenue situation of the rail system as a whole. A comparison of prices charged by HSR or air operators compared to those for bus or conventional rail indicate that travelers are willing to pay a premium for fast travel, and HSR development allows rail operators to access that higher-value market. For example, although the National Passenger Railroad Corporation in the United States (the nationwide intercity train operator commonly known by its commercial name "Amtrak") requires an annual operating subsidy from the federal government because its revenue does not cover all its costs, the "Acela" HSR service between Washington and Boston earned a profit of $200 million in 2011. Acela's success on the corridor emanates from the combination of high occupancy rates and the possibility of charging high fares, thanks to comparatively rapid full-journey times on specific city pairs on the corridor (e.g., Boston-New York, New York-Washington, Philadelphia-Washington), coupled with congestion on major alternatives (airports, superhighways).

Disadvantages and Critiques of HSR

Along with the above advantages, HSR systems are subject to a number of disadvantages based on their physical nature as a technology. Experience has also shown that their deployment in various countries is subject to the distorting effect of political influence and can result in counterproductive outcomes. The following points are representative:

- *Inability to compete economically with air travel over long distances:* For HSR, infrastructure cost increases linearly with distance, so that a hypothetical long distance link on the order of thousands of kilometers between two cities (such as Chicago to Los Angeles) is much more expensive than a link on the order of hundreds of kilometers (Chicago to St. Louis). At the same time, infrastructure cost for air travel comes from the cost of constructing and maintaining airports, so the distance between two destinations is not a factor. Therefore, for long-distance travel, not only is HSR markedly slower than air travel but infrastructure costs are much higher.
- *Slowing effect of intermediate stations and transfers on overall travel time:* In addition to the previous point, HSR travel compared to air suffers from the slowing effect of intermediate stops. For example, a trip from Geneva, Switzerland, to Edinburgh, Scotland (a straight-line distance of 1,257 km or 781 miles) could be made almost entirely by HSR by using the TGV service to Paris, changing stations, taking the Eurostar to London, changing stations again, and completing the trip on the HST. The total time required would be on the order of 10 to 12 hours depending on schedules, compared to 2.5 hours in a direct flight. The much longer trip duration for HSR occurs despite cruising speeds[4] of between 200 and 300 km/h on line-haul portions of the trip.
- *Embodied energy and emissions in infrastructure development:* Although end-use energy consumption for HSR operations (i.e., energy used to propel the trainsets) may have low emissions per unit of energy consumed, the construction of an HSR alignment, with its numerous bridges and tunnels, is itself an energy and greenhouse gas-intensive process.

[4]Definition: *cruising speed* is the typical sustained maximum speeds at which HSR trainsets operate under ideal conditions: straight track, level terrain, not approaching or leaving a station stop.

- *Potential isolation of intermediate cities compared to conventional rail routes:* When HSR connections are built parallel to conventional rail tracks, smaller cities and towns along the route may lose service entirely or have less service once HSR displaces conventional rail. HSR may thereby have the effect of concentrating economic activity in the endpoint cities (e.g., Paris and Lyon for the Southeast TGV line in France) at the expense of intermediate cities.
- *Expensive infrastructure that cannot be shared:* In the case of new-build HSR tracks, a substantial investment must be made in tracks, right of way, tunnels, and bridges, which once complete can be used only by HSR trainsets. Other slower-moving or heavier trains (such as freight trains) cannot share the tracks because of the high speeds that must be maintained. Regarding the use of HSR infrastructure for freight, for many types of goods, the requirement for higher maximum weight on the rails would entail a stronger, more expensive grade of infrastructure that railway companies are not willing to finance. A survey of HSR routes opened since 2006 in France, Germany, Italy, and Spain found cost values between $21 and $138 million per mile, with cost values at the upper end driven by mountainous terrain requiring extensive tunneling or bridges to allow the straight alignments that HSR requires. The limited access to HSR tracks could be compared with a limited access highway, which although limited to much slower speeds, can mix privately owned automobiles, public intercity buses, and freight vehicles, all on the same infrastructure.

In summary, the decision to finance and build HSR is often driven by a mixture of the practical need to provide additional transportation capacity or relief to other modal options on the one hand, and the opportunity to (1) enhance national prestige by overcoming some perceived technological challenge, or (2) connect underserved regions to the larger HSR network to advance social or political aims. The latter factors can tip the decision between build and no-build in the direction of making the commitment to build, in a situation where a project might be rejected on purely economic and commercial grounds. The completion of the Channel Tunnel exemplifies motivation 1, while the Joetsu Shinkansen line in Japan (which links the more isolated Sea of Japan coast on the western perimeter of the country to the more populated eastern seaboard) represents motivation 2.

Examining the case of the Channel Tunnel more closely, the history of the launch of the Eurostar HSR service via the tunnel exemplifies the tension between these two motivations.[5] Greater London and greater Paris, and more recently Brussels as the seat of the European Union administration, are clearly important destination cities that generate heavy volumes of passenger traffic, and the Eurostar service has been successful in winning large market shares in these markets since inauguration in 1994. At the same time, both construction cost projections and initial revenue forecasts for the tunnel proved overly optimistic once it was finally opened—in other words the final cost figure of approximately 12 billion British pounds (GBP) was more than twice the pre-project estimate of 4.9 billion GBP, and revenues from passengers and freight did not meet expectations at first. In 1998, under pressure from very large debt payments,

[5]Note that the Channel Tunnel is actually a combination HSR and shuttle infrastructure link, since it also provides a car and heavy truck shuttle, services previously provided by ferryboats across the English Channel.

the Channel Tunnel group completed a debt restructuring to provide financial relief to their operations. As of the end of 2011, there remained a long-term debt of 3.23 billion GBP. This experience has been repeated with a number of other HSR projects around the world. High infrastructure cost of the type observed with the Channel Tunnel project will likely remain a challenge for HSR projects for the foreseeable future, especially when tunneling or mountainous terrain is involved.

Confrontation and Collaboration between HSR and Air Modes

HSR and commercial air operators have at times had a confrontational relationship in regard to service development, but there are also instances of collaboration between the two modes. As an example of confrontation, a private consortium won a franchise from the government of the state of Texas in the early 1990s to build an HSR network linking the major cities of Dallas, Houston, and San Antonio, but a campaign led by Southwest Airlines (a dominant air carrier in the state) succeeded in blocking the project from moving forward. More recently, airlines and HSR have collaborated in certain corridors to facilitate intermodal transfers between air departures and arrivals and HSR. The Frankfurt Airport in Germany, a major hub for European and intercontinental flights, provides streamlined connections with the German HSR network. Passengers traveling through U.S. airports along the Northeast Corridor such as Newark Liberty International Airport outside New York City are provided timetable information to onward rail connections and arrival times in destination stations in other cities along the corridor.

10-3-4 HSR Case Studies

This section considers two specific case studies of HSR, representing the incremental and new-build approaches. With a large number of lines already built in Europe and financial problems since the economic downturn of 2009 slowing further development there, the chosen case studies are instead in North America (state of Illinois in the United States) and Asia (network linking eastern Chinese cities).

Case Study 1: Chicago to St. Louis HSR Upgrade

This project consists of the upgrading of the existing passenger rail service on a double-track line between Chicago and St. Louis, a distance of 284 miles (455 km) to 110-mph speeds (180 km/h). The existing service operates at or under the national rail speed limit of 79 mph, and has only a small share of the market, with five services per day in each direction. Most passenger traffic is carried either by air on direct flights between the two cities or in private cars on Interstate 55, a limited-access highway that parallels the rail route.

The required changes include improvements to the tracks, changes to the signaling system to allow safe operation at higher speeds, and new rights of way in select locations where the existing right of way cannot be upgraded for HSR operations. Upgrades are being carried out in phases between 2011 and 2014. In October 2012, the project passed a major milestone when 110-mph operation was successfully tested on the first section of track fully upgraded near Pontiac, Ill. Steps in the latter phases of the project include upgrades to the approach to the terminal stations in Chicago and St. Louis to make access and egress within the metropolitan area faster. Because of the built-up urban area, this phase of the project is the most complex and expensive per mile, and is therefore scheduled at the end. Total cost for upgrading the infrastructure is estimated at $1.3 billion ($2.9 million per mile or $4.6 million per km), with a portion funded from the Recovery Act of 2009.

The eventual goal of the project by 2017 is to have 110-mph service on the entire length of the route, using newly purchased diesel-propelled trainsets to maintain higher speeds with greater passenger comfort. A further goal is to use more frequent, higher-quality service to win a larger percent market share of the Chicago to St. Louis market, taking market share away from road and air. The project, in its current form, does not meet a number of HSR expectations, including minimum speed of 125 mph, electrification of the line, and full separation of the right of way (some number of level grade crossings will continue to exist along the line). However, the goal is for a positive experience on this corridor to lead to other HSR projects around the region, with new-build HSR elsewhere achieving speeds of up to 220 mph (350 km/h). The Chicago to St. Louis link may prove to be the first leg in an eventual Midwest HSR network serving the Great Lakes-Midwest Megaregion, with Chicago as a hub and links to other cities including Minneapolis/St. Paul, Cincinnati, Detroit, and Cleveland. Such a regional HSR network would then take its place among several others around the country, including ones in Florida, Texas, California, and the Pacific Northwest.

Case Study 2: China HSR Network Program

In contrast with the incremental approach of the HSR program in the United States, the HSR program in China is designed to meet more pressing needs and is therefore markedly more rapid in its deployment. In the United States, the already built-out national network of limited access highways and airports means that the role of HSR is to curb additional congestion. Therefore the focus is on several disconnected regional networks rather than a single national HSR grid, to be constructed over an open-ended time frame. China faces a much larger population, more rapid economic growth, and much faster growth in travel demand than is currently occurring in the United States, due to the large number of Chinese citizens entering the middle class as their incomes grow. The solution in China is to rapidly build a full network of HSR in the eastern half of the country, covering the distance from the cities of Harbin in the north to Guangzhou in the south, and many coastal and inland cities in between. Starting from a project launch in the year 2000, as of the end of 2012 a network with a length of approximately 9,400 km (5,900 miles) had been completed. The overall project aims to complete 18,000 km of HSR by the year 2020, with an anticipated total investment of $300 billion.

The Chinese railway industry, as overseen by the Ministry of Railways of the Chinese national government, has developed domestic manufacturing of the trainsets by working with global manufacturers such as Bombardier of France, Siemens of Germany, and Kawasaki Heavy Industries of Japan. The resulting trainsets combine technology transfer with domestically developed component designs, creating a domestic HSR manufacturing industry. Start-to-finish infrastructure project times have been rapid by international standards. For example, the 968-km (601-mile) Wuhan-Guangzhou line, connecting the interior with the south coast, took just 9 years from concept in 2000 to the first passengers riding in 2009.

The rapid pace of HSR deployment has led to some program missteps in the 2010 to 2012 period. A fatal collision between two trains due to a signaling error at Wenzhou, Zhejiang Province, on July 23, 2011, and instances of poorly constructed civil engineering works failing during heavy rains, led to a slowdown in the pace of construction. Nevertheless, demand for passenger travel by HSR continued to grow, and the year 2012 saw ridership reach 486 million passengers, the highest figure to date.

10-3-5 Energy, Cost, and Emissions Analysis of HSR

Two of the primary drivers for exploiting HSR technology are its potential energy security and ecological benefits, since electrified HSR in particular uses energy from a variety of primary sources (coal, natural gas, renewable, nuclear) and has no direct emissions from the train itself. To estimate energy requirements and resulting emissions for HSR as a tool for comparing with other modal options, we require a quantitative understanding of the relationship between the travel speed of the HSR trainset, resistance forces acting against it, and the power necessary to overcome those forces.

Several equations are presented in this section that estimate force or power as a function of speed. Note that here we focus on trainsets cruising (i.e., operating at steady-state, linehaul speeds) on flat terrain, and ignore starting and stopping, the effects of curves in the line, impact of changing elevation, and ancillary loads.

Force and power equations are similar to those discussed in Chap. 5 for light-duty road vehicles. Because trainset designs, internal components, lengths of trains, and operating conditions for HSR systems are different both within national networks and in comparing one country to another, there is no one theoretical equation that can be used to evaluate all HSR trains. Instead, empirical equations for individual classes of HSR trainsets are used. In this case, we use a French TGV as an example.[6] Let D = drag on TGV train in kilonewtons (kN) and V = velocity in m/s. Then drag as a function of velocity is

$$D(V) = a + bV + cV^2 \tag{10-1}$$

where a, b, c are parameters used to fit a curve to observed values of drag as a function of speed, in this case having the values $a = 3.82$, $b = 0.1404$, and $c = 0.006532$. Alternatively, if V is given in kilometers per hour, the parameters are $a = 3.82$, $b = 0.039$, and $c = 0.000504$, or if given in miles per hour and D in kilopounds force, the values are $a = 0.9587$, $b = 0.01403$, and $c = 0.00029$.

Regardless of units, the drag on the trainset consists of three components: (1) static, (2) linear, and (3) nonlinear. Hereafter we use V in units of miles per second (m/s), since it is convenient for converting drag to power. The power requirement P in units of kilowatts to maintain a given speed V is

$$\begin{aligned} P &= DV \\ &= aV + bV^2 + cV^3 \end{aligned} \tag{10-2}$$

Using drag values from Eq. (10-1) and velocity increasing from 0 to 100 m/s, we can calculate total power requirement for different speeds, including the contribution of static, linear, and nonlinear components to the total value, as shown in Fig. 10-6. For comparison, the maximum speed shown of 100 m/s is equivalent to 360 km/h or 225 mph. These maximum speed values represent the highest speeds reached by the most advanced HSR trainsets and infrastructure at present. As shown in the figure, by the time speed reaches 40 m/s (144 km/h or 90 mph), which is well below HSR cruising speeds, the nonlinear component of the power requirement dominates, reflecting the importance of aerodynamic drag in determining propulsion power required.

---

[6]The source for the TGV equation and also the estimated 75% efficiency of electrical power arriving at the pantograph that is delivered to the wheels of the trainset is Hopkins et al. (1999).

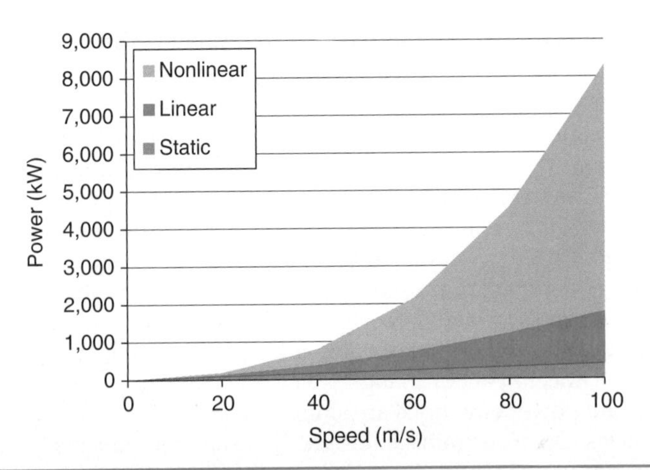

FIGURE 10-6 Propulsion power required as a function of operating speed in miles/second, divided between static, linear, and nonlinear components, as computed using Eq. (10-2).

Next, the energy requirement in kWh to travel for a given amount of time t, measured in hours, and taking into account the efficiency of the HSR train η_{HSR} and delivery efficiency of the grid η_{grid} from generating source to the train network is

$$E = \frac{Pt}{\eta_{HSR}\eta_{grid}} \tag{10-3}$$

The term η_{HSR} takes into account all losses from the point of purchase of the electricity from the grid to the train network and includes local transmission losses and conversion losses in the electric motors that drive the train. A figure of η_{HSR} = 75% to 80% is typical. Losses in the term η_{grid} include all transmission losses from point of generating electricity (gas- or coal-fired power plant, wind farm, solar PV array) to the point of sale, including losses in stepping voltages up and down; η_{grid} = 88% to 90% is representative. Note that conversion losses, for example, thermodynamic losses when combusting fuels to spin a turbine in a thermal powerplant, are not included in η_{grid}.

Life-cycle comparisons of primary energy sources between electrified HSR and other transportation alternatives must be conducted with care. Take the example of an electric HSR trainset using electricity primarily from some mix of fossil fuels (coal, oil, natural gas) compared to a diesel HSR like the British HST. In the electrified life cycle, the biggest portion of losses occur in the conversion of thermal energy in the fuel to mechanical and then electrical energy in the power plant. For the HST, this large conversion loss occurs onboard the trainset as diesel fuel is combusted and converted to electricity to drive the train. Thus Eq. (10-3) may make the electrified HSR appear very efficient compared to the energy content of the diesel used by an HST for an equivalent trip. A full accounting of life cycle energy consumption, however, requires the analysis to look upstream at energy content of fuels used to make the required electricity.

Estimation of electricity consumption leads next to the calculation of total energy cost per HSR trip or per passenger. Electricity costs vary widely between regions depending on the mix and efficiency of generating sources used, regulatory policies that affect the price of electricity, the cost of distribution from electricity generating station to train line, and other factors. As a general rule, the order of magnitude of this cost is \$0.08 to \$0.12/kWh including both generation and transmission. Example 10-1 illustrates the calculation of drag, power, energy, and energy cost for a representative HSR journey.

Example 10-1 An HSR trainset with 384 passenger capacity travels 1,000 km at a speed of 250 km/h, or 69.44 m/s. The trainset efficiency is 75%, the grid efficiency is 90%, and electricity costs \$0.10/kWh. Assuming the train has 100% occupancy and ignoring starting and stopping, and changes in speed due to terrain, calculate (a) the instantaneous drag, instantaneous power requirement, and electrical energy generation required for the trip; (b) total cost of electricity, and cost per passenger for electricity; and (c) supposing the same trip is made by car in a vehicle with 5 occupants that has a fuel economy of 13 km/L (30.8 mpg) on the highway, if gasoline costs \$1/L (\$3.79/gal), the energy cost per passenger for the trip by car.

Solution

(a) Since the train travels 1,000 km at 250 km/h, the time requirement is 4 hours. Accordingly, using Eqs. (10-1) to (10-3) with speed in units of miles per second to calculate *D*, *P*, and *E*:

$$D(V) = a + bV + cV^2$$

$$D(69.44\ \text{m/s}) = 3.82 + (0.1404)(69.44) + (0.006532)(69.44)^2 = 45.07\ \text{kN}$$

$$P(V) = DV$$

$$P(69.44) = (45.07)(69.44) = 3130\ \text{kW}$$

$$E(V) = \frac{Pt}{\eta_{\text{HSR}}\eta_{\text{grid}}}$$

$$E(69.44) = \frac{(3130)(4)}{(0.75)(0.9)} = 18{,}547\ \text{kWh}$$

(b) Assuming the electricity costs \$0.10/kWh and that there are 384 passengers on the train, the total cost and cost per passenger values are

$$(18{,}547\ \text{kWh})(\$0.10/\text{kWh}) = \$1{,}855$$

$$\frac{\$1855}{384\ \text{person}} = \$4.83/\text{person}$$

The resulting value of \$4.83 per person for energy cost for a 4-hour journey gives some indication of the energy cost-effectiveness of a well-utilized HSR train.

(c) In the case of the trip made by car, the total fuel consumption and cost of fuel are

$$(1000\ \text{km})\left(\frac{1\ \text{km}}{13\ \text{L}}\right) = 76.9\ \text{L}$$

$$(76.9\ \text{L})(\$1/\text{L}) = \$76.90$$

Dividing this amount among 5 passengers gives \$15.38 per passenger. Thus the trip by car costs more than three times as much per person even though the cruising speed is on the order of 100 to 120 km/h rather than 250 km/h.

Comparison of French TGV and Spanish AVE Power Requirement Models

Along with Eqs. (10-1) and (10-2), observation of rated power output from either locomotives (for conventional rail) or trainsets (for HSR) provides another means of understanding the effect of velocity on power requirements for HSR. We make a distinction between locomotives and trainsets because, in general, a conventional train is pulled by a detachable locomotive or string of locomotives that may be moved between trains,

V (km/h)	P required (kW)	P rated (kW)
0	—	—
50	98	728
100	354	1,796
150	875	3,045
200	1,766	4,429
250	3,130	5,922
300	5,073	7,509
350	7,701	9,178

Sources: Hopkins et al. (1999) for power required, Alvarez (2010) for rated power observed.

TABLE 10-2 Predicted Power Requirement for Propulsion and Rated Power Observed in Trainsets or Locomotives as a Function of Speed in Kilometers per Hour

whereas in an HSR trainset the *power car* is an integral part of the trainset that cannot be reallocated. An assessment of maximum speeds achieved by conventional trains and "Alta Velocidad Espanola" or AVE HSR trains in the Spanish rail network revealed the following empirical relationship between V_{max} in kilometers per hour and the rated maximum power of either the locomotive or the trainset:[7]

$$P_{rated} = 4.468 \cdot V_{max}^{1.3021} \tag{10-4}$$

For example, the fastest AVE HSR attains speeds of 350 km/h, so the maximum power of which the trainset is capable is predicted as follows:

$$P_{rated} = 4.468 \cdot 350^{1.3021} = 9178 \text{ kW}$$

Similarly, a conventional train capable of speeds of 150 km/h would have a rated power value of 3,045 kW according to Eq. (10-4), and so on.

Table 10-2 compares the predicted power requirement from Eqs. (10-1) and (10-2) and rated power predicted from either the locomotive or the trainset for speeds between 0 and 350 km/h. Although the equations are for HSR technology from different countries (France and Spain) and also different time periods (the underlying train data from Spain were gathered more recently), there is reasonable agreement between the two sources. For instance, taking the 250 km/h HSR train that was the subject of Example 10-1, the table gives a value of 3,130-kW instantaneous power requirement for steady-state operation (maintaining constant speed on level ground). The predicted rated power for the trainset is 5,922 kW, so the additional power is available for acceleration or to take into account losses in conversion of electrical power to forward motion of the trainset.

[7]The empirical formula in Eq. (10-4) is due to Alvarez (2010)."Alta Velocidad Espanola" translates to "Spanish High Speed."

Emissions from HSR Operations Based on Electricity Mix

The preceding discussion of energy consumption in HSR alluded to the challenge of estimating true life-cycle energy consumption to support the operation of HSR, due to a lack of knowledge of the energy losses in extracting fossil fuels and converting them to electricity. The study of emissions, and of CO_2 emissions as contributors to climate change in particular, allows for a more accurate assessment of the true impact of electric-powered HSR because most of the emissions occur at the plant (as opposed to upstream from it), and because average emissions from different types of plants are documented. A relatively small amount of residual emissions occur upstream from the plant, for example, for fossil-fuel-powered equipment and vehicles that extract and transport fuels. A more complete assessment would include these stages as well, but it is a reasonable simplification to exclude them.

Compared to other transportation alternatives, HSR offers the potential for reduced CO_2 per pkm of travel because of the possibility of deriving some or all of the electricity from zero- or reduced-CO_2 sources. A diversified portfolio of sources can include either renewable or nuclear electricity, which emit no CO_2 at the generating site, or high-efficiency natural gas, which emits relatively little CO_2 per kWh generated. The recent electricity mix in New York State in the United States provides an example of this situation. A relatively small fraction comes from coal-fired electricity plants, which emit on average 0.850 kg CO_2 per kWh (NYSERDA, 2009). However, most of the electricity consumed by the state came from either non-CO_2 emitting sources such as hydro, nuclear, and wind, or from natural gas (which averaged 0.325 kg CO_2 per kWh), so that the overall state average was 0.256 kg CO_2 per kWh. Some countries with a substantial HSR infrastructure such as Japan and France have relatively low CO_2 emissions from their electricity sector, due to the lack of coal generation in the mix. Example 10-2 illustrates how electricity source impacts emissions from HSR.

Example 10-2 Suppose the same HSR trainset in Example 10-1 (i.e., 250 km/h, 384 passengers, 1,000-km trip) derives its electricity supply either entirely from coal or from a typical mixed grid supply. If the New York State values are used, what are the emissions from the trip for the trainset as a whole and for each individual passenger?

Solution The energy consumption for the trip is the same as in Example 10-1, or 18,547 kWh, and the emissions rates were introduced above at 0.850 and 0.256 kg CO_2 per kWh. Total emissions from the trip are therefore

$$\text{Coal: } (18547 \text{ kWh})(0.85 \text{ kg CO}_2/\text{kWh}) = 15{,}758 \text{ kg} = 18.8 \text{ tonne}$$
$$\text{Mixed: } (18547 \text{ kWh})(0.256 \text{ kg CO}_2/\text{kWh}) = 4{,}748 \text{ kg} = 4.7 \text{ tonne}$$

Dividing by 384 passengers on board gives the emissions per person:

$$\text{Coal: } \frac{15{,}758 \text{ kg}}{384 \text{ person}} = 41.1 \text{ kg CO}_2/\text{person}$$
$$\text{Mixed: } \frac{4{,}748 \text{ kg}}{384 \text{ person}} = 12.4 \text{ kg CO}_2/\text{person}$$

The values calculated in Example 10-2 can be compared to an equivalent trip in a passenger car. Suppose the car seats 5 persons and has a fuel economy of 30 mpg, or 12.7 km/L, on the highway. At a tailpipe emissions rate of 2.35 kg CO_2 per liter of fuel consumed, the 1,000-km trip would emit 185.6 kg CO_2, or 37.1 kg CO_2 per passenger assuming 100% occupancy. Although the emissions per passenger for HSR powered by 100% coal-derived electricity is similar in magnitude, the mixed-electricity HSR is substantially less at 12.4 kg CO_2 per passenger, a figure that can be further reduced by displacing coal and natural gas with non-CO_2 emitting alternatives.

10-4 Aviation

Aviation can generally be divided into *general* and *commercial* aviation, respectively meaning air travel in usually small, privately owned aircraft, and for-hire carriage of passengers in larger craft. The great majority of air travel passenger-kilometers take place in commercial aircraft, so these are the focus of the remainder of this section. Historically, piston-and-cylinder driven propeller engines were used in commercial aircraft, but this technology has given way, in almost all cases, to jet engines in larger craft or a choice between either jet or turboprop engines in smaller craft. Either type of propulsion system uses aviation fuel, a high-octane liquid fuel derived from crude oil.

10-4-1 Aviation Efficiency: Recent Advances and Current Challenges

Commercial pressure on the passenger aviation industry has catalyzed significant innovation to reduce the amount of aviation fuel per passenger. The impact of these changes on energy per passenger-kilometer was already shown in Fig. 10-3. Improvements to aviation technology or best practices fall into one of several categories, as follows:

- *More efficient engines:* Advances in fluid mechanics and materials science have led to new generations of jet engines that provide more thrust per unit of thermal energy available in the fuel. For instance, more precise, higher-temperature ceramic components in jet turbine blades allow engines to withstand higher maximum temperatures, which improve maximum potential efficiency in line with the laws of thermodynamics.
- *Improvements to aerodynamic shape of aircraft:* Improvement in the understanding of how aircraft create aerodynamic drag as they fly led to innovations in the design of aircraft fuselage and wings to reduce drag. One of the examples that is most visible to the traveling public are the winglets of various designs at the end of airplane wings of both short-haul and long-haul wide body jet aircraft. Where these wings in the past ended in a simple flat tip, aircraft today have a winglet that protrudes at some angle to the main winging, breaking up the tendency of the wingtip to produce a circular eddy or *vortex* and thus reducing drag.
- *Use of lightweight materials:* Shifting toward lighter materials for an aircraft of a given size and strength reduces end-use energy consumption, all other things equal. Around the year 1950 commercial aircraft had shifted to using aluminum for the fuselage and wings as a material with a high strength-to-weight ratio, as well as corrosion resistance. Increasingly in recent years, major makers are replacing aluminum with composite materials, such as carbon fibers, to further reduce weight.

- *"Yield management" to maximize occupancy:* Yield management, also known as perishable asset review management (PARM), is the practice widely used by airlines of mixing advance- and short-notice seat sales to maximize both revenue and occupancy of aircraft (i.e., revenue passenger-kilometer delivered per available seat-kilometer flown). Airlines use historical sales data to sell a fraction of the seats on a given flight well in advance by discounting the price, and then retain the remaining seats to be sold closer to the flight date at a higher price. Airlines may also overbook flights—in other words, sell more seats than are actually on the flight, knowing that some small fraction of passengers will not take the flight on the actual day of departure. This practice has the ecological benefit of increasing average occupancy rates, although it is unpopular among the traveling public, who are fearful of not getting the seating that they expect.

The result of the above efforts in combination has been a major reduction in average energy required per passenger flown. For the United States, the figure has declined from over 5,000 kJ/pkm in 1975 to around 2,000 in 2008.

Remaining Challenges for Sustainable Aviation

The above achievements notwithstanding, the lack of diversity of fuel sources remains one of the major challenges for aviation. There are at present no major CO_2-free energy sources, although a small amount of biofuel-based aviation fuel is in use today, whose carbon content comes from crops and other biological sources rather than fossil fuels. The first published use of a biofuel in a commercial flight took place on February 24, 2008, when a Virgin Atlantic Boeing 747 flew with an 80/20 mixture of jet- and bio-fuel in one of four engines on the London-Amsterdam leg of a long-haul flight. The following year, the Boeing Company announced that it had developed a biofuel that equaled or exceeded the performance specifications for petroleum-derived jet fuel. Thereafter various airlines inaugurated occasional use of biofuel from 2009 to 2012. While use of biofuels has continued to grow slowly, one of the main limitations is the lack of a *sustainable supply chain*, in other words, a source of biologically derived raw material that could be grown in large enough quantities to displace most or all of the conventional aviation fuel currently in use. Use of hydrogen as an alternative to jet fuel is also being considered but is still in the concept stage.

According to the International Panel on Climate Change (IPCC), the aviation sector contributed 3% of worldwide CO_2 emissions in 2008. One consideration with aviation is the phenomenon of *uplift*, in which CO_2 emitted from aircraft at cruising altitude has a proportionally larger impact on climate change per ton of emissions because it enters the atmosphere already several kilometers above the earth's surface.

10-4-2 Commercial Aircraft Performance, Energy Requirements, and Emissions

The performance of commercial jet aircraft can be evaluated based on physical dimensions, weight, maximum range, and other factors. Table 10-3 gives specifications for several models of aircraft from the two largest global suppliers of aircraft, the European-based Airbus and the U.S.-based Boeing. As shown in the table, each maker offers models to compete in the various air travel markets. For example, the Airbus A319 and Boeing 737 compete in the short-haul market, the A330 and 787 compete in the long-haul market with seating requirements in the range of 250-300 persons,

Make	Model	Length (m)	Width (m)	Seats*	Max Wt* (tonnes)	Range (km)
Airbus	A319	33.84	34.1	124	64.4	6,850
	A330	63.69	60.3	295	230.9	11,900
	A380	72.72	79.75	525	562	15,700
Boeing	737	33.6	35.8	126	70.1	6,370
	747	64.4	70.6	416	363	13,450
	787	57	60	250	227.9	15,200

*Values shown are for a representative variant of the given model. Each model is offered in the form of several different variants, each with slightly different specifications, for example, seating, maximum weight, range, etc. Seating value shown is for a typical multiclass (either 2- or 3- passenger classes) configuration; a single economy-class configuration will have higher total number of seats. Maximum weight is the weight before takeoff with full fuel tanks, weight on landing or with no fuel on board will be less than the value shown due to fuel consumption in flight.

Sources: Airbus Corporation; Boeing Corporation.

TABLE 10-3 Basic Specifications for Select Commercial Aircraft from Airbus and Boeing Companies in 2013

and the A380 and 747 compete in the long-haul market with the largest seating requirements.

In terms of instantaneous power and energy requirements of aircraft, the equations for estimating aircraft performance measures are similar to those for road vehicle or HSR presented earlier, but with important differences as well. Aircraft encounter no road or rail friction, so the only drag is aerodynamic drag, which decreases at cruising altitudes thanks to lower air density (on the order of 0.4 kg/m^3 compared to ~1.0 kg/m^3 at sea level). However, aircraft must provide aerodynamic *lift* to stay aloft, which adds to the power requirements. The mass and dimensional values (length of aircraft, wing-span width) from Table 10-3 are incorporated in both drag and lift elements.

In equation form, total power required P_{total} is a combination of drag and lift, each of which can be calculated based on cruising speed and various other parameters:

$$\begin{aligned} P_{\text{total}} &= P_{\text{drag}} + P_{\text{lift}} \\ &= 0.5\rho C_d A_p V^3 + 0.5\frac{(mg)^2}{\rho V A_s} \end{aligned} \qquad (10\text{-}4)$$

where ρ is the density of air

C_d is the aerodynamic drag coefficient specific to the particular aircraft

A_p is the frontal area of the aircraft

V is the speed

m is the mass of the aircraft

g is the gravitational constant

A_s is the effective area of the aircraft supported by aerodynamic lift (which is the product of fuselage length and wingspan width).

Since V increases P_{drag} but decreases P_{lift}, there is a tradeoff in choosing the design cruising speed of the aircraft. From this calculation, an optimal speed that minimizes P_{total} emerges, so an energy-saving operating strategy arises for commercial flights of climbing to cruising altitude and then cruising at or near the optimal speed. Example 10-3 illustrates the computation of power and energy requirements for a specific aircraft, namely a Boeing 747.

Example 10-3 A Boeing 747 has the values given in Table 10-3 and an aerodynamic drag coefficient of 0.03, a frontal area of 180 m^2, and jet engines with an efficiency of 33%.[8] At cruising altitude with air density of 0.4 kg/m^3 it travels at a speed of 254 m/s (914 km/h). (a) What is the instantaneous power requirement for the 747 to maintain the given speed and altitude? (b) If the 747 travels a distance of 1,000 km under these conditions, what is the energy content of the required amount of aviation fuel? (c) If jet fuel has an energy content of 35,900 kJ/L and costs $0.75/L, how does the energy cost per passenger per trip compare to that of HSR or driving from Example 10-1? Note that although jet fuel has higher energy content per liter than gasoline, the paid price is lower because the motorist is paying fuel taxes for highway maintenance in the cost per liter above that are not paid when purchasing jet fuel.

Solution

(a) Supported area A_s can be calculated from length and width of the 747 from the table:

$$A_s = L \cdot W = (70.6)(64.4) = 4547\ \text{m}^2$$

All other values are given so P_{total} can be calculated using Eq. (10-4):

$$\begin{aligned} P_{total} &= 0.5\rho C_d A_p V^3 + 0.5\frac{(mg)^2}{\rho V A_s} \\ &= 0.5(0.4)(0.03)(180)(254)^3 + 0.5\frac{(363{,}000\ \text{kg}\cdot 9.8\ \text{m/s}^2)^2}{(0.4)(254)(4547)} \\ &= 1.76\times 10^7\ \text{W} + 1.37\times 10^7\ \text{W} = 3.13\times 10^7\ \text{W} = 31.3\ \text{MW} \end{aligned}$$

(b) Since the distance traveled is 1,000 km and the speed is equivalent to 914 km/h, the time required is

$$\frac{1{,}000\ \text{km}}{914\ \text{km/h}} = 1.09\ \text{h}$$

The output energy E_{out} required to maintain the needed power for 1.1 hours and the input energy E_{in} based on the efficiency of the jet engines are calculated as follows:

$$\begin{aligned} E_{out} &= P_{total}t = (31.3\ \text{MW})(1.09\ \text{h}) = 34.3\ \text{MWh} \\ E_{in} &= \frac{E_{out}}{\eta_{engines}} = \frac{34.3}{0.33} = 104\ \text{MWh} \end{aligned}$$

(c) To calculate fuel cost, E_{in} is first converted from megawatt-hours to gigajoules:

$$(104\ \text{MWh})\frac{3.6\ \text{GJ}}{1\ \text{MWh}} = 374.4\ \text{GJ}$$

Since the fuel energy content is 35,900 kJ/L, total fuel consumption is

$$\frac{374.4\ \text{GJ}}{0.0359\ \text{GJ/L}} =\sim 10{,}400\ \text{L}$$

[8]Values for drag coefficient and jet engine efficiency are taken from Mackay (2009).

The total expenditure and expenditure per passenger per trip are then

$$(10,400\text{ L})(\$0.75/\text{L}) = \$7,800$$

$$\frac{\$7,800}{416} = \$18.75/\text{passenger}$$

This cost is substantially more than that of HSR (\$4.83), and modestly more than car travel (\$15.38). Line-haul travel time is on the other hand faster than HSR and much faster than driving.

Note that in Example 10-3 both aerodynamic drag and lift requirements contribute substantially to the overall power consumption. In the specific example of the 747 cruising at 254 m/s, the contribution of each component is 56% and 44%, respectively. Using an electronic solver and keeping all inputs into Example 10-3 fixed, it can be found that the optimal speed that minimizes energy required for the aircraft is 238 m/s (857 km/h), and that the energy required in this case is E_{out} = 103.2 MWh. Thus the speed used in the example leads to an energy requirement that is about 1% higher than the optimal. Note that the speed increases by about 7%. This comparison suggests that aircraft operators can adjust cruising speed within some range around the energy-optimal value, for instance to make up for lost time in case of delays, to meet their scheduling or operational needs without significant increase in overall energy consumption. Figure 10-7 shows how energy consumption as speed increases from 218 to 258 m/s.

CO_2 Emissions from Aircraft at Cruising Speed and Altitude

End-use CO_2 emissions from aircraft are directly proportional to the fuel consumed in liters, since the combustion of 1 L of aviation fuel results in 2.528 kg of CO_2 emitted from the jet engine. In addition to the tailpipe emissions from jet engines in flight, there are additional upstream emissions from the extraction, processing, and transportation of crude oil and jet fuel, but in terms of life-cycle emissions, these may represent just 10% to 12%, with the remaining 88% to 90% emitted from the jet engines during end use. If the aviation fuel comes from a nonconventional petroleum source such as tar sands or shale oil, however, the upstream fraction may be substantially higher.

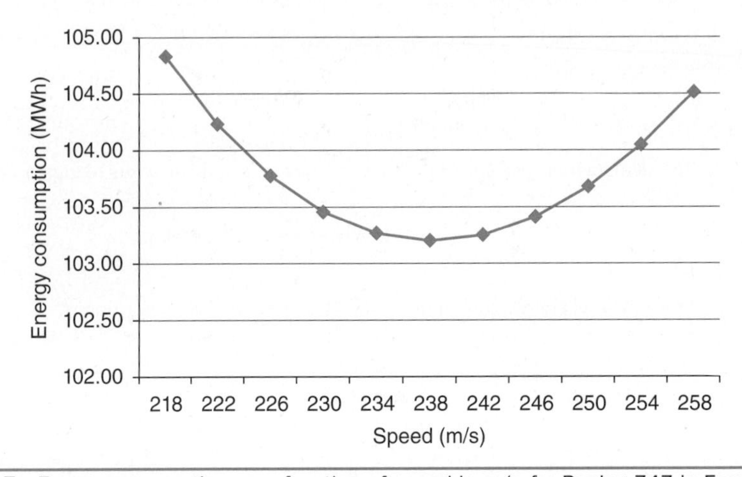

FIGURE 10-7 Energy consumption as a function of speed in m/s for Boeing 747 in Example 10-3.

Example 10-4 extends the previous example of a Boeing 747 by calculating associated inflight CO_2 emissions.

Example 10-4 (a) What are the total CO_2 emissions for the 1,000-km flight in Example 10-3? (b) If the aircraft is full, what are the emissions per passenger for the trip? How do the emissions compare to the HSR service in Example 10-2? (c) If emissions for gasoline are 2.31 kg/liter, how do the emissions for travel by car presented in Example 10-1 compare to aviation?

Solution

(a) From the previous example, fuel consumption is 10,400 L. Since emissions are 2.528 kg CO_2 per liter, total emissions are

$$(10,400\ \text{L})(2.528\ \text{kgCO}_2/\text{L}) = \sim 26,400\ \text{kgCO}_2$$

(b) From Table 10-3, the seating capacity of the 747 is 416 passengers, assuming a standard cabin configuration with economy, business, and first-class compartments. Emissions per person for a full flight are therefore

$$\frac{26,400}{416} = 63.43\ \text{kgCO}_2/\text{passenger}$$

The figure of 63.4 kg CO_2 per passenger can be compared to a range of emissions from 12.4 to 41.0 kg CO_2 per passenger for HSR. Note that in this case emissions per passenger for HSR vary across a wide range because some mixtures of electricity sources are much more CO_2-intensive per kWh than others. Aviation fuel by contrast is more or less constant in terms of CO_2 per unit consumed.

(c) From the previous example, the car consumes 76.9 L to travel 1,000 km. Emissions are therefore

$$(76.9\ \text{L})(2.31\ \text{kgCO}_2/\text{L}) = 177.6\ \text{kgCO}_2$$

$$\frac{177.6\ \text{kg}}{5} = 35.5\ \text{kgCO}_2/\text{person/trip}$$

Therefore CO_2 emissions are nearly twice as large for flying as they are for driving.

Comparison of Examples 10-2 and 10-4 suggests that HSR might provide a substantial emissions reduction on linehaul distances where both modes are available for a given origin-destination pair. In particular, HSR has the potential to exploit non-CO_2 emitting energy sources, whereas aviation is largely limited to petroleum-derived aviation fuel, although some biofuels are beginning to make inroads. Also, comparing Examples 10-3 and 10-4, travel by air has a relatively small energy cost disadvantage and a larger cost disadvantage, because aviation is paying substantially less per unit of energy in liquid fuel than a motorist pays—jet fuel is both cheaper and more energy-dense per liter.

Caveats: Limitations on Comparisons between Air and HSR Modes

Because the previous example calculations for air and HSR omit several important details, the following caveats apply:

- *Relative occupancy of air and HSR:* The preceding discussion of CO_2 emissions per passenger-kilometer assumed that aircraft, trainsets, or private automobiles all travel with full occupancy. This assumption removes the need to determine an appropriate average occupancy value less than 100%, which in real-world conditions is of course the norm. We typically observe higher occupancy in aviation compared to rail including HSR, which would tend to narrow the gap between the two in terms of emissions per trip. Driven in part by high energy cost, air operators have been forced to market seats aggressively to fill aircraft (see section "Aviation Best Practices"). HSR also has the disadvantage that a scheduled service of an HSR trainset makes many intermediate stops during its trip where passengers board and alight. Relatively few passengers travel the entire distance of the service from endpoint to endpoint, and when seats become available partway through a trip, it is more difficult to fill them than the case of an airline flight. In the case of aviation the aircraft is typically emptied at the end of each flight to be filled for an onward flight. Many passengers make trips requiring connections at hub airports, but usually they disembark and transfer to a different aircraft, which usually starts out empty, giving the airline full latitude to fill as many seats as possible using yield management.

- *Nonlinearity of aviation energy consumption:* The preceding examples did not consider the disproportionate amount of energy required to lift an aircraft from ground level to cruising altitude. Depending on the overall length of a flight, the impact of this omission on total energy and emissions will be more or less substantial. Consider a representative aircraft climbing from sea level to 11,000 m (approximately 36,000 ft). Although actual climb patterns are complicated by different climb rate and airspeed requirements at different stages as the aircraft gains altitude, suppose that our sample aircraft climbs at a rate of 600 vertical m/minute, with a horizontal speed of 130 m/second (all typical average numbers). The entire climb takes 18.3 minutes, during which time the aircraft travels a horizontal distance of approximately 143 km. For the longest intercontinental flights (e.g., Hong Kong, Singapore, or Sydney to either Europe or North America), the distance involved is a small fraction of the total. For a short-haul flight in a small jet such as the A319 or 737 in the table, on the other hand, it is significant.
- *Advantage of speed of aviation and prohibitive time requirement for long-distance travel in HSR:* One of the explanations for the higher emissions per traveler for aviation is the much shorter travel time of 1.1 hours cruising versus 4 hours in the HSR. Leaving aside the impact of time required to climb and land for aviation and time for intermediate stops in HSR, for many travelers it may be impractical to take the time to travel by the slower mode, even if they desire to choose HSR to reduce emissions. Furthermore, many important origin-destination pairs are separated by distances much longer than 1,000 km, and in these cases the time requirement becomes even more challenging.
- *Inability of HSR to travel over water:* There are a number of important shorter-distance travel markets, such as to and from the Caribbean islands, between countries of southeast and east Asia, and between the British Isles and mainland Europe, where distances are shorter and air is the only mode that is physically possible. Undersea tunnels may be constructed for short distances, but the existing Channel Tunnel (50.4 km) and Seikan Tunnel (Honshu to Hokkaido Islands in Japan, 53.8 km) represent the longest distances for which tunneling is economically feasible. Ferry services that carry entire railway trains are another possibility that was more prevalent in the past, but the additional time required to load and unload trains mean that this option is seldom used today (see discussion of potential niches for intercity travel by ship under "Other Intercity Alternatives").

10-4-3 Aviation Best Practice: Maximizing Utilization through Yield Management

Commercial airlines must combine efficient technology with operational techniques to assure that operating costs are minimized, so it is fitting that they have adopted yield management techniques to encourage maximum possible sales of available seats on flights. In this case, there are synergies between business motives for reducing operating cost and increasing the ecological benefits of providing a given amount of transportation volume (measured in revenue passenger-kilometers) with the lowest possible energy consumption and emissions from flights.

The modern practice of yield management is an outcome of advances in the ability to use information technology both for accurate digital record-keeping and to run sophisticated computer models that can quickly determine how prices should be adjusted. The models make decisions about pricing and then updated information is provided to prospective travelers via the internet, so that seat price and availability information is easily obtainable.

As a business practice, yield management works by meeting the basic revenue requirements of airlines with a stream of business travelers that pay relatively high prices, and then topping up revenue with additional leisure travelers that purchase tickets in advance at lower prices. Yield management also benefits the environment because it results in planes that fly mostly full and at lower average energy consumption per passenger per flight. Prior to the development of yield management, many flights carried mostly business travelers (or other wealthy travelers) at high prices, but flew with many more empty seats. An alternative would be to sell all seats at a fixed discounted price, which would also result in full flights but with inadequate revenue for airlines to stay in business. Even low-cost carriers that have entered the market in recent years practice some form of yield management, such as rewarding passengers for purchasing in advance or flying on flights at inconvenient times with lower ticket prices.

Price and Demand Assumptions

Illustration of the practice of yield management depends on an understanding of the relationship between changes in price charged per ticket and the resulting demand for seats purchased by the traveling public. Many factors enter into the traveler's response to prices offered, including the distance traveled, the time of day at which the flight is offered, the time of year and whether or not the flight takes place during peak travel season or not, and so on. For the purposes of this section, we will focus on an arbitrarily chosen origin-destination pair of cities and leave out the impact of time of day or time of year.

The difference between two travel markets, business (subscript *B*) and leisure (subscript *L*), is of particular interest for yield management, as shown in Eq. (10-5):

$$Q_B = A_B - B_B P_B$$
$$Q_L = A_L - B_L P_L \qquad (10\text{-}5)$$

Here Q is the number of seats demanded, P is the price per ticket, and A and B are parameters used to fit the demand curve to observed demand in the marketplace as ticket sales vary in response to changes in prices charged by the airlines.

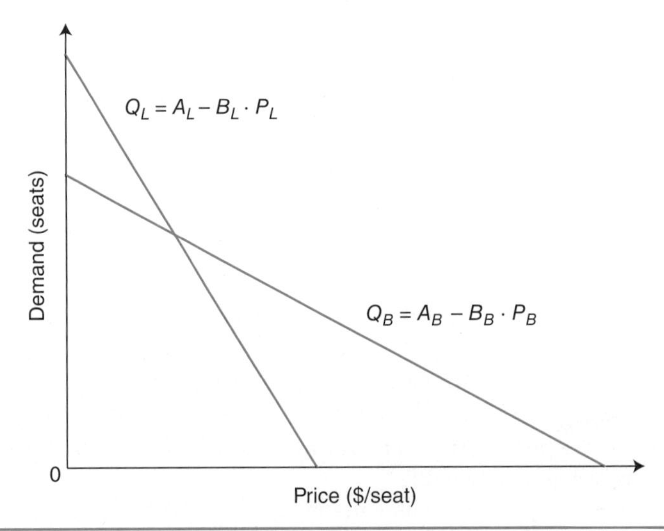

FIGURE 10-8 Number of seats of demand on a flight as a function of ticket price P_B or P_L for business travelers (Q_B) and leisure travelers (Q_L), respectively.

The reader should not confuse travel markets with the distinction between first, business, and economy or coach classes in the physical compartment of the aircraft. The type of class is concerned with the quality and space allotted to seating in physically separate parts of the aircraft, while in the aircraft represented by Eq. (10-5), the same physical airline seats are being divided between two different markets.

The two demand curves from Eq. (10-5) are shown in Fig. 10-8 in a generic form on a single set of price-demand axes. In practical terms, the linear curve is fit to observed price and demand values where both price and demand are nonzero, and then extrapolated to the x- and y-intercepts. Of the two demand curves in the figure, the curve for Q_B will always be the one with the slower rate of decline that has nonzero values at higher ticket prices, because business travelers who often travel alone and on short notice are the ones who can afford to purchase tickets at these prices. By contrast, the curve for Q_L falls off steeply to zero as price increases, reflecting the tendency of leisure travelers to change either dates or destinations for a particular trip rather than accept the higher prices of the business travel market.

Optimal Solution of a Two-Tier Yield Management Scheme

We next consider the optimal solution of a simple two-tier yield management scheme (business and leisure travelers), which in principle could be applied to situations with more than two tiers. Additionally, we assume that the capacity of the aircraft is fixed at some number of seats, C, that is, we exclude the possibility of overbooking to increase actual occupancy on the day of the flight.

From the airline's perspective, total revenue R earned from the flight can be written as

$$R = Q_B P_B + Q_L P_L \tag{10-6}$$

Substituting Eq. (10-5) above, R can be rewritten solely in terms of P_B and P_L:

$$R = (A_B - B_B P_B)P_B + (A_L - B_L P_L)P_L \tag{10-7}$$

Similarly, the equation for C can be rewritten in terms of P_B and P_L:

$$\begin{aligned} C &= Q_B + Q_L \\ &= (A_B - B_B P_B) + (A_L - B_L P_L) \end{aligned} \tag{10-8}$$

The values of P_B and P_L that maximize R can now be found. One approach is to rewrite the problem as an optimization constrained by the capacity of the aircraft C, which is the general approach for more complex variants of the yield management problem with multiple time periods, days in advance of purchase date, and so on:

$$\begin{aligned} &\text{Max } Z = (A_B - B_B P_B)P_B + (A_L - B_L P_L)P_L \\ &s.t. \\ &(A_B - B_B P_B) + (A_L - B_L P_L) \le C P_B, P_L \ge 0 \end{aligned} \tag{10-9}$$

Alternatively, Eqs. (10-7) and (10-8) can be rewritten as a single equation in either P_B or P_L and solved for the price value that maximizes R using differentiation. This approach is illustrated in Example 10-5.

Example 10-5 An airline flies an aircraft that seats 100 passengers on a route that has observed business and leisure demand curves with parameters $A_B = 100$, $B_B = 0.15$, $A_L = 140$, and $B_L = 0.5$. Calculate (a) the revenue-maximizing prices for business and leisure tickets, (b) the corresponding number of seats sold to each type of travelers, (c) the total revenue earned, and (d) the fraction of total revenue earned from each type of traveler.

Solution From the description of the aircraft, we know that $C = 100$. The demand functions for business and leisure travelers are written as follows:

$$\begin{aligned} Q_B &= 100 - 0.15\,P_B \\ Q_L &= 140 - 0.5\,P_L \end{aligned}$$

(a) To solve for optimal prices, solve for P_B in terms of P_L:

$$\begin{aligned} &Q_B + Q_L = 100 \\ &(100 - 0.15\,P_B) + (140 - 0.5\,P_L) = 100 \\ &0.15\,P_B = 140 - 0.5\,P_L \\ &P_B = 933 - 3.33\,P_L \end{aligned}$$

Next, substitute known values into Eq. (10-7) to reduce to an equation with a single unknown P_L:

$$\begin{aligned} R &= (A_B - B_B P_B)P_B + (A_L - B_L P_L)P_L \\ &= (100 - 0.15P_B)P_B + (140 - 0.5P_L)P_L \\ &= [100 - 0.15(933 - 3.33P_L)](933 - 3.33P_L) + (140 - 0.5P_L)P_L \\ &= -2.135P_L^2 + 3734P_L - 37273 \end{aligned}$$

To maximize R, take the derivative of the equation for R and set equal to zero to solve for P_L and P_B:

$$\frac{dR}{dP_L} = -4.27\,P_L + 734$$

$$-4.27\,P_L + 734 = 0$$

$$P_L = \frac{734}{4.27} =\sim \$172$$

$$P_B = 933 - 3.33(\$172) =\sim \$360$$

Thus the optimal business and leisure fares are \$360 and \$172, respectively.

(b) The number of seats sold of each type is a function of the chosen prices, as follows:

$$Q_B = 100 - 0.15(360) =\sim 46$$

$$Q_L = 140 - 0.5(172) =\sim 54$$

(c) The total revenue is calculated by substituting prices and demand into Eq. (10-6):

$$R = Q_B P_B + Q_L P_L = (46)(\$360) + (54)(\$172) = \$25{,}848$$

(d) From (c), the business travel portion of sales contributes \$16,560 or 64% of the revenue, whereas the leisure traveler contributes \$9,288 or 36%.

Example 10-5 illustrates how the business travelers account for 45% of the seats but generate 64% of the revenue. Also, solutions with ticket prices close to P_L = \$171 have total revenue values close to the maximum observed value of \$25,785, as shown in Fig. 10-9. If the ticket price is decreased to \$160 or increased to \$184, with corresponding

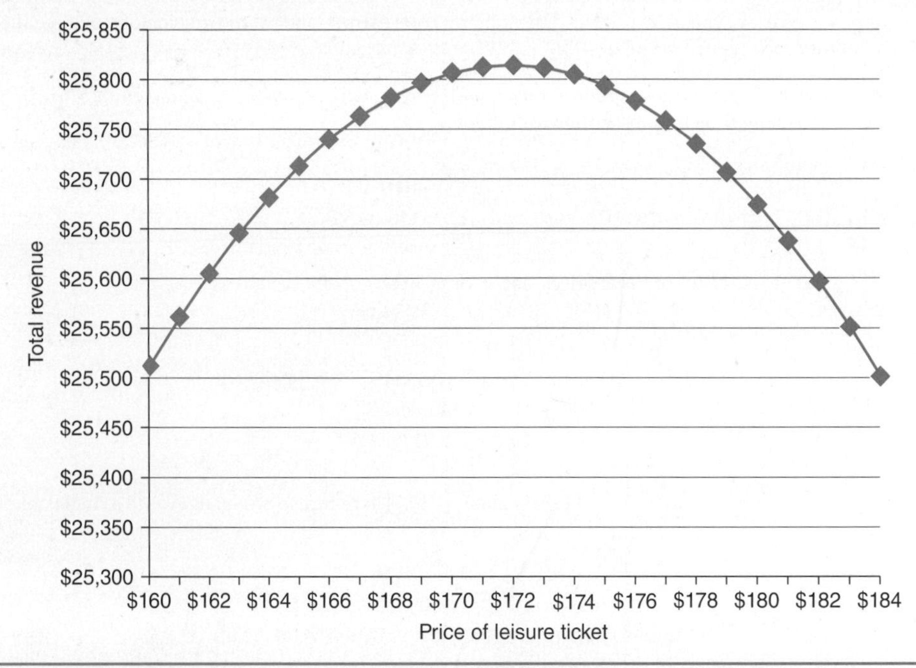

FIGURE 10-9 Revenue earned by full aircraft as a function of price charged for leisure ticket P_L.

change in P_B to create the right amount of demand to fill the plane, revenue falls to around $25,500 in either direction. Thus the revenue earned is not very sensitive to the exact fare charged in the vicinity of the optimal. (Note that the revenue figures in the figure are slightly different from the solution to Example 10-5 due to rounding.)

Practical Considerations for Implementing Yield Management

The actual implementation of yield management requires several steps beyond those presented in Example 10-5. First, the fare structure may have more than two tiers; for example, leisure travel might be divided between very budget conscious travelers who can only travel at the lowest fare and more affluent travelers who can afford to spend more and therefore purchase a more expensive ticket closer to the departure time. Also, the price and quantity suggested by the calculation using the method of Example 10-5 is in fact a target. In practice, the airline will monitor sales as the departure date approaches and discontinue sales at price P_L once a threshold value is reached. Using the 100-seat aircraft in the example as an illustration, the actual value of this threshold may be above or below 54 seats, depending on circumstances. If the sales trend, compared to historical records, suggests that they might be able to sell more seats at price P_B than the target, then the number of seats available at price P_L may be reduced to increase overall revenues, and vice versa.

Although the aviation industry has been a leader in implementing yield management, the intercity bus and rail industries use this system as well. Rail providers in both Europe and North America have adopted the practice of rewarding travelers who purchase early with low prices. The railway makes available a limited number of seats at the lowest possible price tier, and when these seats sell out the traveler must either pay for a more expensive seat or change their travel date.[9] The seats on trains that travel at the most desirable time of day, such as arriving at a destination just before the 9 a.m. to 5 p.m. workday starts or leaving just after it ends, may also fill up the fastest so that some leisure travelers may be pushed into less full trains that travel at less convenient times.

10-5 Other Intercity Alternatives

Besides HSR and aviation, other commercial alternatives for intercity travel (i.e., excluding travel in private cars) include public buses, conventional rail, and various marine modes. Demand for intercity bus transport has grown in recent years as it provides a low-cost alternative to rail or air travel, and also has been increasingly competitive with travel in private car, as a result of rising motor fuel costs. Although time series data for bus energy intensity from the U.S. Energy Information Administration are available only up to the year 2000, data up to that point indicate that bus travel was consistently more energy efficient than either train or air, with intensity values on the order of 700 to 800 kJ/pkm over many years.

The recent growth of low-cost bus service such as Megabus or Boltbus in the United States has further expanded opportunities for bus travel. These carriers use exclusively Web-based reservation systems and curbside pickup and drop-off to avoid the costs of

[9]As an anecdotal example, on Friday Jan. 4, 2013, Amtrak's lowest offered fare one-way New York to Washington was $49 for a trip on Sunday Jan. 13 at an off-peak time, but by Sunday Jan. 7 that fare for the same date was no longer offered, and the lowest offered fare on the same train was $93.

brick-and-mortar terminals and ticket outlets, holding down costs. As an example of the market for low-cost bus service, the Megabus network comprised 120 cities in Canada and the United States in 2012, and Megabus recorded 22 million boardings in the period 2006 to 2012. A potential concern for low-cost bus service is safety: a number of fatal crashes involving either driver error or mechanical failure such as a crash near Syracuse, N.Y., in 2010 or near Litchfield, Ill., in 2012 have drawn criticism in the media. On the other hand, U.S. Department of Transportation records indicate that low-cost carrier buses do not fail safety inspections at a higher rate than other carriers. It remains an open question as to whether the difference between the safety standards of low-cost carriers and their conventional cost structure competitors is statistically significant or not over time.

Turning to conventional rail service, although much of the growth in rail service has been in the area of HSR, conventional rail has grown in some markets as well, in some cases benefitting from the growth in fuel costs that has affected intercity travel in private cars. In the case of Amtrak in the United States, ridership increased by 55% between 1997 and 2012, reaching 31 million boardings total. Price volatility and overall rising prices of gasoline characterized the latter part of this period from 2005 onward, convincing some travelers to use rail instead of driving. Other types of conventional train service have lost ground recently. For instance, in Europe, the growth of low-cost air travel led by discount airlines such as EasyJet and RyanAir made long-distance overnight trains uncompetitive in many markets, so that many of these services have been discontinued. In some Latin American countries such as Mexico and Nicaragua, passenger rail has been unable to compete with intercity bus service and the entire system has been discontinued.

Finally, passenger travel on ships and ferries continues to fill an important niche, although it is limited by geography. Low line-haul speeds of marine vessels dictate that, where road, rail, and/or air infrastructure are well-developed, travelers will opt for these other modes rather than take the time to travel by ship. This situation limits the use of marine passenger travel to situations where travel over water provides a geographic shortcut (e.g., crossing the North Sea between Britain and continental Europe rather than going around by the Channel Tunnel) or situations where travelers are transporting a private car for onward travel once they disembark. *Fast ferry* technologies that use advanced hull designs to allow ferries for either passengers-only or passengers and highway vehicles to travel at higher linehaul speeds have helped to keep ship travel competitive in some markets.

It is possible that travel options involving marine travel may reduce CO_2 emissions if efforts are made to maintain high efficiency or use low-CO_2 energy sources. A case in point is travel to and from Ireland via the Holyhead (Wales) to Dublin ferry across the Irish Sea. Since many private vehicle owners on either side require transportation of their vehicles for use in onward travel after disembarking, some sort of car ferry service will be essential in this market for the foreseeable future. In 1999, the Irish Ferries Company inaugurated a high-speed ferry that cut travel time from 3½ to 1½ hours. This ferry service combined with express train service to and from major cities in central and northern England (e.g., Birmingham, Manchester) provide a service that can potentially compete with flying on a combination out-of-pocket cost and travel time basis. Travel time is longer via the train-boat combination than via a direct flight, but some travelers may opt for the former if they can travel at lower cost. With use of low-CO_2 fuels such as biofuel blends in diesel locomotives or on the ferry, plus CO_2-free electricity for electrified rail, CO_2 emissions per traveler on this corridor might be reduced as well compared to flying.

10-6 Discussion: Directions for Sustainable Intercity Travel

From the discussion of different intercity modes in this chapter, we can make several observations about the future of intercity travel.

First, the need to reduce CO_2 emissions and conserve built-up space indicates that the regional development of HSR is a primary focus. Electrified HSR can operate on a diverse range of primary sources of electricity and, as the fraction of electricity from renewable sources continues to grow, average CO_2 per kWh will decline. A well-utilized HSR line with frequent train service and high occupancy in each trainset is also an efficient use of space. Conventional rail and intercity bus service can play a role as well.

With HSR, conventional rail, and bus serving regional travel, aviation can be reoriented toward long-distance and intercontinental travel. At present, many existing airports have limited available additional capacity and few opportunities to expand either the number of aircraft departures or capacity for landside access to/from airport terminals. It therefore makes sense to prioritize long-distance flights for allocating this capacity, since HSR is not practical at long distances, as discussed before.

The growth of HSR has positive synergies with another important ongoing trend, namely, the growth of *megaregions*, or agglomerations of regional metropolitan areas into coordinated administrative and economic blocks. A megaregion need not be continuously urbanized throughout; instead, it uses urbanized areas within the region as the focus of population density and economic activity, and intersperses agricultural and forest lands in between. Just as the twentieth century featured the transition of cities into *metropolitan areas*, in the twenty-first century, metropolitan areas in different parts of the world will join together to form megaregions. For instance, in the United States, the northeast corridor from Washington, D.C., to Boston qualifies as an existing megaregion, while another megaregion will potentially form in the lower Great Lakes incorporating the areas around Chicago, Detroit, Toledo, and Cleveland. Because the typical size of a megaregion is similar to the ideal distance for HSR travel (200 to 1000 km, or 150 to 600 miles), HSR can potentially serve as the main provider of intercity travel within the megaregion. This phenomenon has already been observed with the opening of the Taiwanese HSR system that runs for 345 km between the capital city Taipei in the north and Kaohsiung, the largest city in the south. The entire trip takes 90 minutes including stops, similar to the amount of time required to cross some of the world's largest metropolitan areas by conventional surface transportation. Therefore, a significant effect of HSR has been to bind the entire region together into a single unit in a way that it was not previously.

Lastly, transportation providers will continue to make inroads in reducing CO_2 emissions from aviation. Aircraft makers are developing new more efficient models such as the Boeing 787 Dreamliner and, as new aircraft replace older ones, emissions will decline. As with other end uses of energy, fossil fuels must not only be conserved but eventually replaced with nonfossil alternatives to eliminate CO_2 emissions entirely. Efforts to develop and phase in biofuels or hydrogen are ongoing (see Chap. 16 for discussion of biofuels in particular).

It is sensible, however, to work on replacing fossil fuels with non-fossil alternatives in other transportation applications as well as stationary energy uses first, due to the time required to fully phase out CO_2 emissions from commercial aviation. Because of the difficult operating environment of aviation (cold temperatures at high altitude, absolute criticality of a reliable fuel supply for safety reasons), biofuels or hydrogen for use in aircraft are not the low-hanging fruit as far as reducing CO_2 is concerned. At the same time, even when taking into account the problem of uplift discussed earlier and the relatively rapid growth of CO_2 from aviation, this sector still represents only a small portion of total anthropogenic CO_2. Political leaders, businesses, and individuals will all have numerous other ways to eliminate CO_2 as a permanent solution for CO_2-free aviation develops gradually over time.

All of these developments will take place in the context of a market economy where consumers choose the most cost-effective transportation options that meet their needs. The government can facilitate the choice of green transportation options through mandates for low emissions, tax breaks and other fiscal incentives for sustainable transportation projects, and support for R&D.

10-7 Summary

Intercity passenger transportation by commercial carriers (i.e., other than travelers using their own private vehicle for intercity trips) can generally be divided between air, high-speed rail (HSR), conventional rail, intercity bus, and long-distance passenger ferries (either with or without the capability of carrying motor vehicles). Among these, air and HSR are especially critical for sustainable transportation because they are the modes that can best compete with driving to reduce highway congestion, thanks to their rapid linehaul speeds of 200 km/h (125 mph) or more. Between the two, HSR has benefit of easier access to non-CO_2 emitting fuel in the form of electricity generated from non-fossil sources, so it can be used as a tool to reduce CO_2 from intercity passenger transportation. For aviation, reducing CO_2 per passenger-kilometers is more challenging because of the difficulties in deploying plentiful CO_2-free substitutes for petroleum derived aviation fuels. On the other hand, much can be done to reduce the CO_2 burden from aviation by developing the most efficient technology possible and by maximizing aircraft occupancy through techniques such as *yield management*, which enable air carriers to fill more of the available seats on each flight. A possible strategy for the future of commercial intercity passenger transportation that addresses both congestion and climate change entails (1) developing HSR capacity for medium-distance travel and encouraging modal shifting from private cars and short-haul aviation to HSR, (2) refocusing aviation on long-haul flights where other modes are not viable and eventually transitioning to system-wide use of nonfossil fuels, and (3) using conventional rail, intercity bus services, and ferryboat connections in appropriate niches not served by HSR or air.

References

Albalate, D. and G. Bel (2012). "High-Speed Rail: Lessons for Policy Makers from Experiences Abroad." *Public Administration Review* 72(3):336–349.

Hopkins, T., J. Silva, and B. Marder (1999). "Maglift Monorail: A High-Speed, Low-Cost, and Low Risk Solution for High-Speed Ground Transportation."*High-Speed Ground Transportation Annual Conference*, Seattle, WA.

Lipscy, P. and L. Schipper (2012). *Energy Efficiency in the Japanese Transport Sector.* Technical report. Electronic resource, available at: *http://www.stanford.edu/~plipscy/JapanTransport2012-2-22.pdf.* Accessed Nov. 2, 2013.

MacKay, D. (2009). *Sustainable Energy: Without the Hot Air.* UIT Press, Cambridge, U.K.

NYSERDA (2009). *Patterns and Trends.* New York State Energy Research & Development Authority, Albany, NY.

Public Transit Library (2010). *Japan High-Speed Rail Passenger Traffic Statistics.* Electronic Resource, available at *http://publictransit.us/ptlibrary/trafficdensity/JapanHSRTrafficDensity2010.pdf.* Accessed Nov. 2, 2013.

USBTS (2012). *National Transportation Statistics.* U.S. Bureau of Transportation Statistics., Washington, DC.

USDOT (2011). *2009 National Household Travel Survey: Summary of Travel Trends.* U.S. Department of Transportation, Washington, DC.

Further Readings

National Committee for America 2050 (2010). "America 2050: A Prospectus" America 2050, New York. Electronic resources, available at *www.America2050.org.* Accessed Feb. 15, 2013.

Albalate, D. and G. Bel (2012). *The Economics and Politics of High-Speed Rail: Lessons from Experiences Abroad.* Lexington Books, Lanham, MD.

Alvarez, A. (2010). "Energy Consumption and Emissions of High-Speed Trains." *Transportation Research Record* 2159:27–35.

Behrens C. and E. Pels (2012). "Intermodal Competition in the London-Paris Passenger Market: High-Speed Rail and Air Transport." *Journal of Urban Economics* 71(3): 278–288.

Brookings Institute (2013). *A New Alignment: Strengthening America's Commitment to Passenger Rail.* Technical report, Brookings Institute, Washington, DC. Electronic resource, available at *www.brookings.edu.* Accessed Mar. 21, 2013.

CAHSRA (2012). *California High-Speed Rail Program Revised 2012 Business Plan.* California High-Speed Rail Authority, electronic resource, available at *www.staging.cahighspeedrail.co.gov.* Accessed Feb. 5, 2013.

Chang, L. and P. Chen (2001). "Build-Operate-Transfer Model: Taiwan High Speed Rail Case."*Journal of Construction Engineering and Management* (May/June 2001): 214–222.

Chen, X. and M. Zhang (2010). "High-Speed Rail Project Development Processes in the United States and China."*Transportation Research Record* 2159:9–17.

Daggett, D., O. Hadaller, R. Hendricks, and R. Walther (2006). *Alternative Fuels and Their Potential Impact on Aviation.* Tech. Cleveland: National Aeronautics and Space Administration-Glenn Research Center.

Davies, J., M. Grant, J. Venezia, and J. Aamidor. (2006) "Greenhouse Gas Emissions of the U.S. Transportation Sector: Trends, Uncertainties, and Methodological Improvements." *Transportation Research Record.* 2017:44–51.

Givoni, M. (2007) "Environmental Benefits from Mode Substitution: Comparison of the Environmental Impact from Aircraft and High-Speed Train Operations."*International Journal of Sustainable Transportation.* 1(4):209–230.

Lee, J.J., S. Lukachko, P. Waitz (2001). "Historical and Future Trends in Aircraft Performance, Cost and Emissions," *Annual Review of Energy and the Environment.* 26:167–200.

National Passenger Rail Corporation (2009). *A Vision for High-Speed Rail in the Northeast Corridor.* Technical Report, Washington, DC. Electronic resource, available at *www.amtrak.com.* Accessed Feb. 4, 2013.

Pohl, H. W. and V. V. Malychev (1997). "Hydrogen In Future Civil Aviation."*International Journal of Hydrogen Energy*. 22:1061–1069.

Penner, J. E., D. Lister, and D. J. Griggs (1999). *Aviation and the Global Atmosphere*, Special Report of the Intergovernmental Panel on Climate Change. Cambridge University Press, New York.

Qiao, H. (2013). "China's High-Speed Programme Back on Track." *International Railway Journal.* January 13, 2013.

Saynor, B., A. Bauen, and M. Leach (2003). *The Potential for Renewable Energy Sources in Aviation*, Technical report. Imperial College Center for Energy Policy and Technology. Electronic resource, available at *http://www3.imperial.ac.uk/pls/portallive/docs/1/7294712.pdf*. Accessed Feb. 5, 2013.

Schafer, A., J. Jacoby, and J. Heywood (2009). "The Other Climate Threat: Transportation." *American Scientist* 97:476–483.

Tanaka, Y. and M. Monji (2010). "Application of Postassessment of Kyushyu Shinkansen Network to Proposed U.S. High-Speed Railway Project." *Transportation Research Record* 2159:1–8.

Tomer, A. (2012). "More Lessons to Learn: Continued Knowledge Gaps in American High-Speed Rail." *Public Administration Review* 72(3):349–350.

USDOT (1997). *High-Speed Ground Transportation for America.* Technical Report, U.S. Department of Transportation, Washington, DC.

Exercises

10-1. A high-speed rail trainset with 10 passenger cars and a power car at either end has a capacity of 460 passengers. It operates on a line between a large city and a rural "hinterland," which is broken up into 6 segments, each approximately 100-miles long. (In other words, there are 7 stations on the line. You could picture a situation like the HSR that operates in Great Britain between London and the north of England and Scotland, where clearly the major population center is at the southern end, and the market is more sparsely populated at the northern end.) On a given service, the train leaves the major city on segment 1, and carries the numbers of passengers shown in the table below on each of the six segments. It then turns around and comes back carrying the numbers of passengers for the return journey (note that you would read up to know the order of carrying passengers, since the train travels segment 6, then 5, 4, etc.). (a) Calculate seat-miles, passenger-miles, and utilization. (b) Why might it be difficult for this route to achieve utilization at or near 100%? Explain in one or two sentences.

Segment	Outward Journey	Return Journey
1	404	422
2	399	344
3	289	279
4	224	160
5	154	171
6	65	60

10-2 You are to evaluate the total operating cost and profit or loss of a high-speed rail (HSR system, based on four cost components: (1) energy cost (electricity), (2) capital repayment cost, (3) wages of the work force, and (4) overhead (which includes the usual business overhead plus all maintenance cost and any other cost associated with the business). Energy costs are based on an average operating speed of 150 mph (240 km/h). You can assume the HSR trains are 75% efficient and the electricity delivery is 90% efficient. The operation owns 160 HSR trainsets which each travel 648,000 miles/year. For calculating electricity consumption, assume a cost of 9 cents/kWh and treat the trains as if they travel constantly at the operating speed—for simplicity, ignore start-up and stopping, or slowing to go around turns, etc. Capital repayment cost is $10 million/trainset/year. For wages, assume that each trainset requires 50 employees in the company workforce for all functions (train crew, repair shop, ticketing, etc.) and that the average cost to the company of an employee is $100,000. Lastly, overhead is 25% of the sum of the first three cost components. For revenues, the system carries 45 million passengers/year, and the average fare is $85.00. (a) Calculate total revenue and total operating cost for the operation. What is its annual profit or loss? (b) What percent of the total operating cost does the energy cost constitute? (c) Suppose the HSR trainsets in this example operate at 250 mph instead of 150 mph. Compare the power consumption of the two: for a 67% increase in speed, by what percent does power increase?

10-3. An Airbus A319 has the specification values given in Table 10-3 and an aerodynamic drag coefficient of 0.03, a frontal area of 57 m^2, and jet engines with an efficiency of 33%. At cruising altitude with air density of 0.4 kg/m^3 it travels at a speed of 230 m/s (828 km/h). Assume it is operating at maximum weight. Use an energy content and CO_2 emissions rate of 35.9 MJ/liter and 2.528 kg CO_2 per liter, respectively. (a) What is the instantaneous power requirement for the A319 to maintain the given speed while maintaining constant altitude? (b) What is its jet fuel consumption per kilometer? (c) If the A319 is filled to capacity, what are the CO_2 emissions per passenger-km under these conditions, ignoring the effects of takeoff and landing?

10-5. Solve Example 10-5 using optimization. First write out the objective function and constraints in mathematical form. Then use the software alternative and coding approach of your choice, and show that the resulting answer agrees with that of Example 10-5.

10-6. An airline offers service between Ithaca and Philadelphia that attracts business travelers and nonbusiness (leisure) travelers. The service is provided by a 50-seat regional jet, and the demand functions for the two types of travelers are $Q_B = 60 - 0.2\ P_B$ and $Q_L = 70 - 0.55\ P_L$, respectively, where the one-way segment prices (P_B and P_L) are in dollars. (a) If the airline seeks to maximize revenue from its flight each morning, what prices should be charged to the two market segments? (b) What are the resulting numbers of business and leisure travelers on each flight? (c) What is the revenue obtained for the flight? (d) What percent of the revenue is contributed by business and leisure travelers?

UNIT 4

Freight Transportation

CHAPTER 11

Overview of Freight Transportation

11-1 Overview

This chapter provides a framework for sustainable freight in three parts. The first part is a review of the growth of freight transportation activity by different modes, freight energy consumption, and a survey of environmental impacts stemming from freight. The second part introduces the concept of total logistics cost and economic order quantity as important determinants of freight decision-making, including the size and frequency of shipments. The third part contrasts two different perspectives on freight energy consumption, namely, mode-based and commodity-based, and provides illustrations of each.

11-2 Introduction

Freight transportation systems make a critical contribution to the efficient functioning of modern society. They make possible access to goods used by households and businesses. They also enable manufacturers to move raw materials and components in modern supply chains toward assembly into finished products. They give regional industrial producers access to national and global markets, spurring economic growth. They also facilitate movement of foodstuffs from region to region, so that when a part of the world suffers drought or other causes of low food productivity, dietary needs can be met with imports from other producing regions; conversely, when a region enjoys higher than average productivity, food products can be exported, rather than being left unharvested for lack of demand.

The pursuit of sustainable freight transportation systems encompasses several directions that are discussed in this and the subsequent two chapters:

- *Current status and trends in freight transportation:* (current chapter) We review recent growth in demand for freight transportation from both a modal and commodity perspective, including comparisons between the United States and United Kingdom, as two representative industrialized countries. We also introduce logistical pressures on the freight transportation that encourage more efficient use of resources in some ways but encourage greater freight intensity and hence energy consumption and emissions in others.

- *Intramodal, intermodal, and supply chain solutions:* (Chap. 12) We present several approaches to improving sustainability currently available. First, each freight mode can minimize its impact by adopting efficient technology and best practices. Some modes are inherently more energy-efficient than others, so shifting freight to more efficient modes is another possible strategy. Lastly, having the shippers and carriers work together at a supply chain level can further boost efficiency.
- *Geographic and spatial distribution aspects of sustainable freight:* (Chap.13) We explore how on the one hand the longer distances of freight transportation and the more complex supply chains in a global economy have increased the total volume of freight. We also explore how opportunities for substituting shorter distance freight might be identified and encouraged with the goal of reducing energy consumption and emissions as well as local impacts.

11-2-1 Overall Growth in Freight Activity and Change in Modal Share

Rapid growth in average length of haul for shipments, increasing complexity of supply chains, and the increasing importance of imports and exports in the total mix of goods being moved have all contributed to rapid growth of freight activity in the United States since 1970 (Fig. 11-1). One of the key developments is that shippers of goods have increased their service expectations, so that a long-term modal shift toward faster and more energy-intensive modes, namely, truck and airfreight, has taken place, especially for the more valuable finished products.

Rail and water modes continue to move large volumes of bulk goods, such as energy products (coal, petroleum products) or bulk agricultural products (grains, feeds). However, the loss of market share of total ton-miles of higher-value finished products for railroads over this period has been significant. For example, in 1977, railroads carried 45% of all food product ton-miles, which include any value-added foods that have been canned, packaged, or prepared in some other way. (Grains and other unprocessed agricultural output are classified as *agricultural products* and are therefore not included in this figure.) By 1997, rail's share of food product ton-miles had fallen to 23%.[1] This type of modal shift has been seen in many European countries and Japan as well. In Fig. 11-1, rail shows overall growth in ton-miles—on the order of 130% over 38 years—but much of this growth has been in the form of bulk commodities.

Comparison of U.S. and U.K. Freight Trends

For comparison purposes, freight tonne-kilometer (tonne-km) trends for the United Kingdom are shown in Fig. 11-2. Due to the longer time series data available from the U.K. government, figures are available starting in 1955, near the beginning of the long-term rise of freight activity led by the truck mode. (Other freight modes are not shown in the figure: marine freight is not as significant as truck although it has also grown gradually during 1955 to 2005, and pipeline and domestic airfreight are very small.) In that year, truck and rail tonne-kilometers were nearly equivalent at 40 and 35 billion, respectively (note different axes in the figure). During the ensuing 50 years, truck activity increased threefold to 160 billion tonne-km, while rail actually declined by 37% to 22 billion tonne-km, although it has been rising since 1995. Thus some of the trends for

---

[1]Data for 1977 and 1997 obtained from the U.S. Commodity Flow Survey.

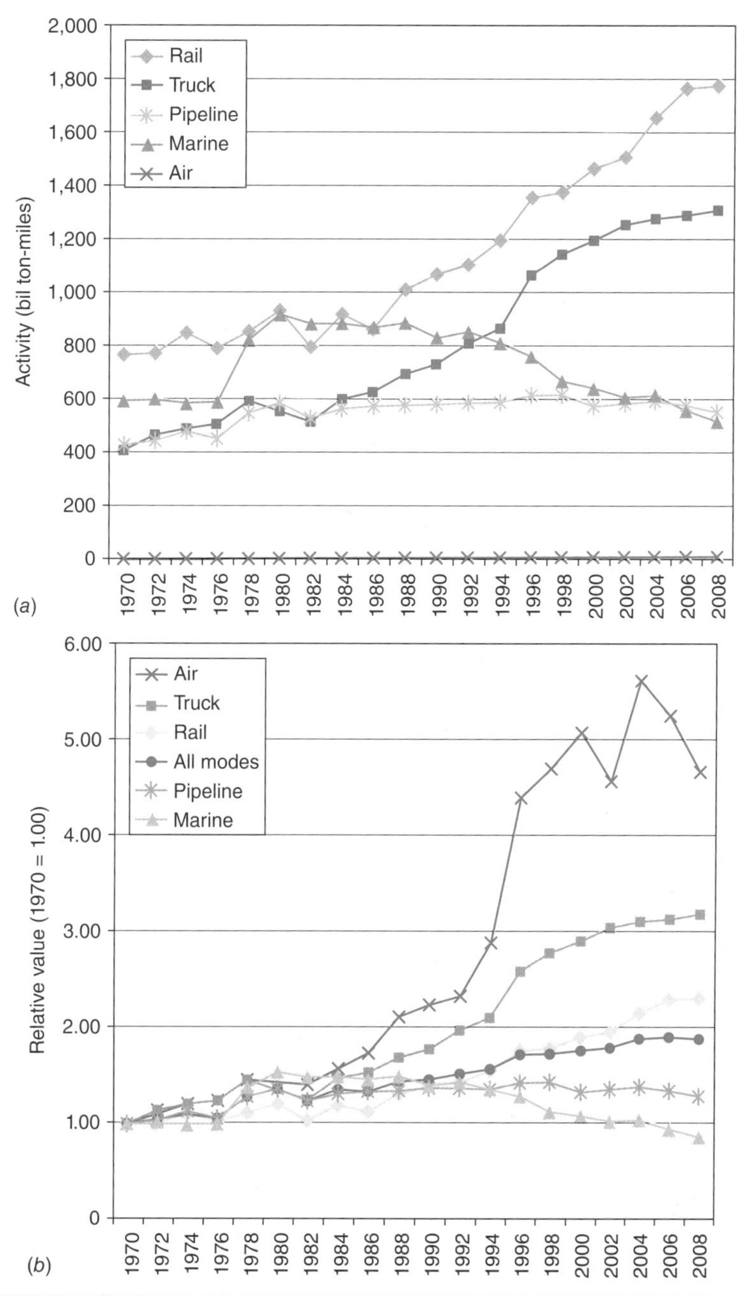

FIGURE 11-1 Growth in freight modal activity 1970 to 2008 (*a*) in absolute amounts and (*b*) indexed to 1970 = 1.00.

Notes: (1) Rail ton-miles are for Class I railroads only, and overlook a small number of additional non Class I ton-miles; (2) 2008 truck and pipeline ton-miles were not published directly in BTS data sources, and are therefore estimated using linear extrapolation of trend from preceding years.

Source: Vanek & Morlok (2000) for data 1970 to 1994; U.S. Bureau of Transportation Statistics for data 1996 to 2008.

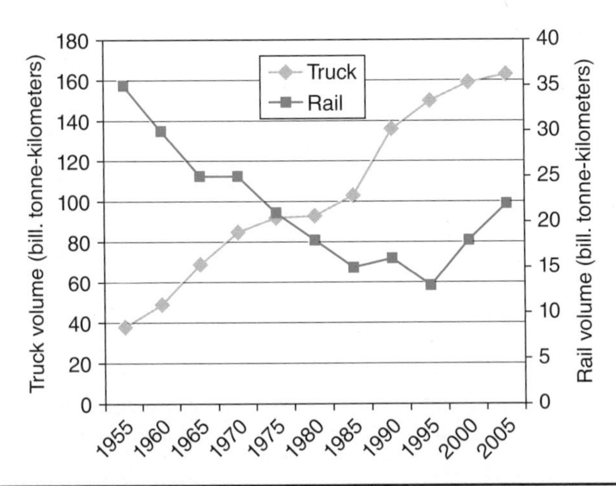

FIGURE 11-2 U.K. truck and rail freight volume in billion tonne-kilometers, 1955 to 2005.

Notes: Rail volume shown on separate axis to make annual values more visible. Conversion: 1.45 tonne-km = 1 ton-mile.

Source: Transport Statistics Great Britain 2008.

the United Kingdom are the same as the United States, while others are different. Notably, overall freight activity has expanded with the growing economy as in the U.S. case, but U.K. rail freight volume has dwindled, different from the United States. One factor contributing to the small modal share for rail is that the U.K. rail network, like that of many peer countries in Europe and elsewhere, and prioritizes rail passenger volumes over goods movement, for reasons discussed in Chap. 10.

Comparing the U.K. figure of 2005 to the U.S. figure, the combination of truck and rail amounts to 185 billion tonne-kilometers, or 128 billion ton-miles. U.S. combined truck and rail in that year amounts to 2,830 billion tonne-kilometers. Since the 2005 populations for the two countries are 60.2 and 296 million persons, respectively, the per-capita figures are approximately 2,100 and 9,600 ton-miles/person/year. The much lower figure for the United Kingdom can be explained in part by the less materially intensive lifestyles (as reflected, e.g., in smaller average home size). It can also be explained in part by geography and population density: the respective land areas of the two countries are 95,000 and 3.8 million square miles, so the population densities are accordingly 636 and 78 persons/square miles.

Impact of Import and Export Traffic on U.S. Freight Demand

Growth in import and export traffic in a globalizing economy is another important driver of freight growth. Both intercity truck movements of import/export goods and shipments of containers to and from ports strongly influence freight demand. Because neither container ton-miles nor weight of freight in containers are easily measured, activity is instead measured in twenty-foot equivalent units (TEUs) of containerized freight passing through a given port. In Fig. 11-3, U.S. container freight activity trends are compared to GDP, total truck ton-miles, and overall freight ton-miles. Due to limitations on container shipment data, the period shown is from 1996 to 2008. Since 1996 GDP has been growing faster than truck ton-miles, but the fastest growth overall has

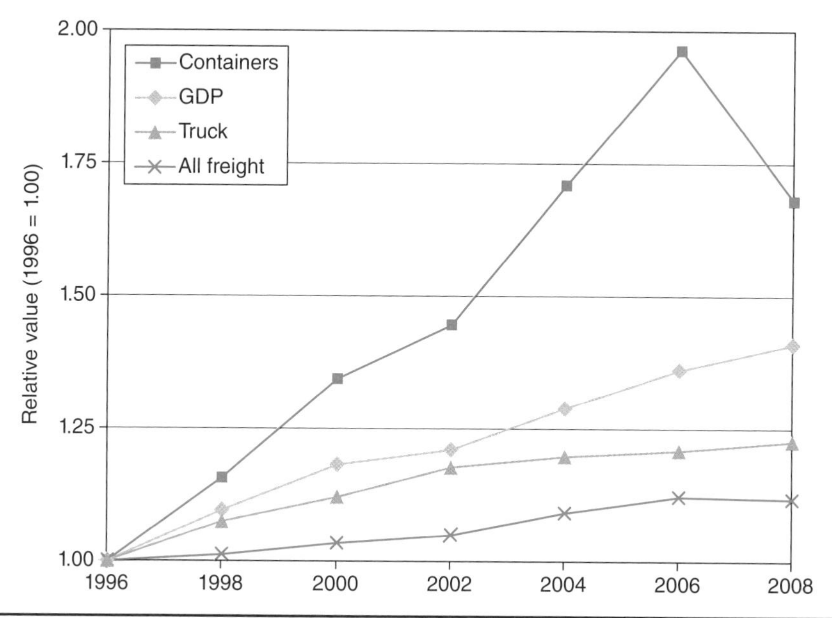

FIGURE 11-3 Comparison of indexed time-series data for GDP in constant $2,000, overall freight ton-miles, truck ton-miles, and container shipping TEUs, 1996 to 2008.

Notes: In 1996, container shipping was 22.6 million TEUs, GDP was $8.3 trillion, truck volume was 1.07 trillion ton-miles, and all freight volume was 3.8 trillion ton-miles.

been container freight TEUs either imported or exported through U.S. ports, which outpaced even the growth of GDP, albeit unevenly, as reflected in the drop from 2006 to 2008.

11-2-2 Commodity Perspective on Freight Activity: Ton-Miles versus Economic Value

Along with dividing freight activity among different modes, freight activity can be divided based on the commodity moved into three general types: low-value bulk, ordinary finished products, and high-tech finished products, as shown in Fig. 11-4. (This allocation is discussed more extensively later in the chapter as part of the section on freight energy.) As shown, the low-value bulk commodities are a substantial amount of the total ton-miles but only a small fraction of the value. These commodities are typically shipped by the rail and marine modes. Bulk commodities may sometimes move by truck, but only over relatively short distances where a rail or marine link is not available. At the other extreme, high-value goods such as electronics, pharmaceuticals, and the like, account for a small fraction of ton-miles and a large amount of value. These goods rely on truck and airfreight the most, as they are the most difficult to move by other modes due to their time-sensitivity and high value. The remaining *mid-value* products lie somewhere in the middle: food products and beverages, paper and wood products, basic metal products, ordinary consumer goods, and the like. These products make extensive use of the truck and air modes but also to some extent are shipped by rail and occasionally marine. Note that the value of shipped goods shown in Fig.11-4(*b*)

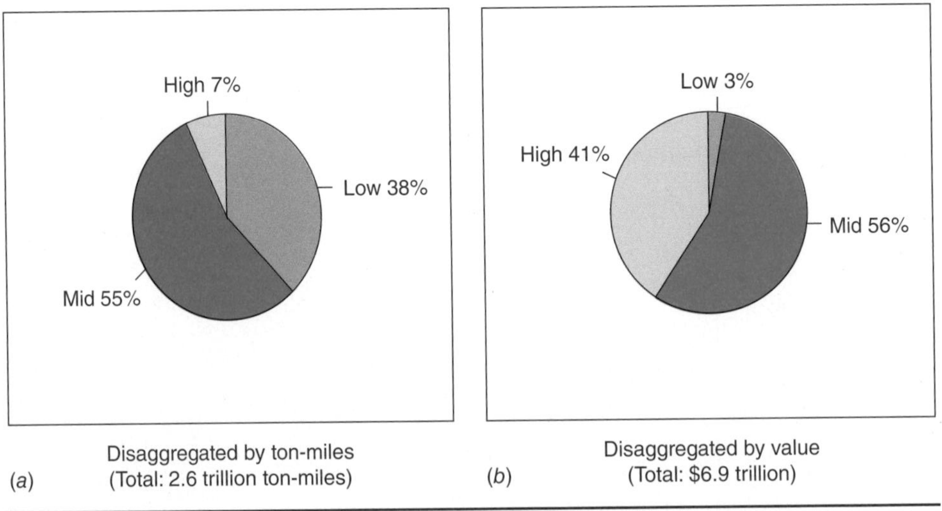

FIGURE 11-4 Comparison of three types of freight in the United States in 1997: low-, mid-, and high-value.

Source: 1997 U.S. Commodity Flow Survey.

is different from the total value of expenditure on freight transportation services in 1997, which is significantly less than the $6.9 trillion total value shown.

From a sustainability perspective, the mid-value products are the area where the most gains can be made in reducing the impact of freight shipments on infrastructure, congestion, and the environment, through changes to logistical practices and modal shifting. Unlike the high-value products, they represent a large fraction of the total ton-miles and by extension total energy consumption and ecological burden. At the same time, due to their higher value than the low-value products, they are more likely to be shipped in an energy-intensive way. Prioritizing the mid-value products could be likened to a *triage* process: of the three groups, one prioritizes the area where improvement is both possible and available in relatively large quantity.

11-2-3 Overview of Energy Consumption and Environmental Impact

With the prominence of freight in the United States, the United Kingdom, and other industrialized countries, this sector inevitably contributes measurably to energy consumption and environmental problems. The combustion of fossil fuels in internal combustion engines for freight applications leads to air pollution, heavy reliance on the extraction and processing of petroleum, and emissions of greenhouse gases.

In the United States, the majority of intercity freight energy use is for domestic movements, such as long-distance movements of trucks and trains, energy use in dedicated freight aircraft such as parcel freighters used by FedEx or UPS, energy consumed in barges and coastal shipping, and so on. Lesser amounts of energy are used in the portion of energy consumed in international shipping that can be attributed to the overall U.S. energy budget, whether transoceanic or to and from neighboring countries of Canada and Mexico. All told, energy consumption for freight of 8.3 quads in the United States in 2008 was on a par with energy consumption *for all end uses* for some peer

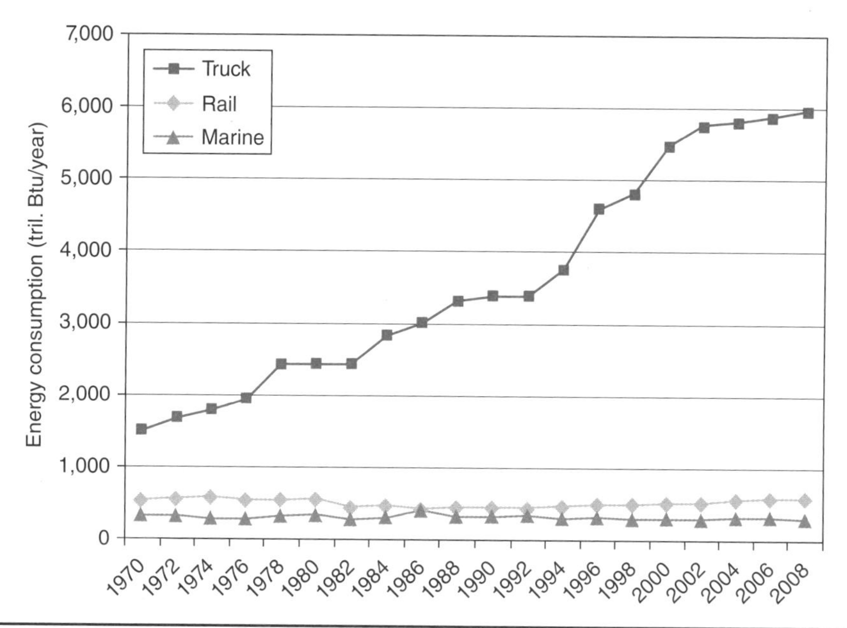

Figure 11-5 Total U.S. freight transportation energy consumption by mode in trillion Btu, 1970 to 2008.

Notes: Airfreight and pipeline energy consumption not included due to inadequate year-on-year data. For comparison, 2006 figures for energy consumption were an estimated 1.0 quad for pipeline and 0.45 quad for air freight.

Sources: Vanek & Morlok (2000), for data 1970 to 1994; Davis & McFarlin (2009) for data 1996 to 2008.[2]

countries in Europe such as Italy (8.1 quads). Of the five major freight modes, truck, rail, and ship (marine) time series energy consumption data are available for the period from 1970 onward and are shown in Fig. 11-5.

The increase in U.S. freight energy consumption is driven by the road mode, which grew from 1.5 quads in 1970 to 6.0 quads in 2008. Improvements in truck technology have in general favored saving energy, and growing demand and operational choices (e. g., by what mode to ship, how quickly, in what shipment size) have in general been putting upward pressure on total energy consumption. Freight vehicles have benefited from improvements in engine efficiency, use of lightweight materials, improved aerodynamics, and other advances. Nevertheless, total truck energy consumption grew nearly fourfold, as shown. Other modes including water and rail held more or less

[2]Truck energy consumption 1996 to 2008 is based on published annual ton-miles and estimated energy consumption based on published energy intensity in Btu/vehicle-mile and estimate of Btu/ton-mile using 1994 value of 4,318 Btu/ton-mile and increasing or decreasing per ton-mile intensity according to per vehicle-mile intensity. Example: Btu/vehicle-mile is 22,193 and 22,109 in 1994 and 1996, respectively, so per ton-mile intensity in 1996 = $(4318) \times (22109)/(22193)$ = 4,302 Btu/ton-mile. Most recent data did not provide energy intensity for 2008, so 2008 total energy consumption estimated by multiplying published 2008 ton-miles by 2006 energy intensity values. 2004 data was smoothed by averaging 2002 and 2006 data, due to problems with the published values for 2004 from USDOE sources.

constant due to modest increases in total freight tonne-kilometers combined with gradually increasing energy efficiency. Year-on-year changes in air freight energy consumption are not shown in Fig. 11-5 for lack of adequate time series data. Available data do, however, suggest that improvements in energy efficiency largely offset the nearly five-fold growth in activity, so that total energy consumption grew only moderately.

Types of Environmental Impact from Freight

Truck, marine, and rail freight modes are almost entirely dependent on diesel engines of different sizes for propulsion, so virtually all of their energy comes from diesel fuel. On this basis, energy consumption values for 2008 for the three modes represent CO_2 emissions of 522, 52, and 26 million tons, respectively.[3] A small number of freight vehicles use natural gas rather than diesel as a fuel, which is more than 95% methane. Methane is the second most important greenhouse gas after CO_2 in terms of its anthropogenic global warming impact. Small amounts of methane may leak in the supply chain that delivers natural gas to freight vehicles, but their contribution to climate change is negligible compared to the direct emissions of CO_2 from truck, marine, and rail modes because the CO_2 volume is so much larger. Note also that, to the extent that peer countries rely on electric rather than diesel locomotives for freight rail and derive electricity from nonfossil sources, CO_2 and methane emissions are avoided.

Freight also contributes to air quality problems especially in the form of NO*x* and particulate emissions because of heavy dependence on diesel as opposed to gasoline engines in trucks, locomotives, and ships. State-of-the-art diesel engine technology in conjunction with ultra-low sulphur diesel (ULSD) fuel greatly reduces especially particulate emissions. Problems remain with older diesel engines that are still in use and have high emissions, especially in less economically advanced countries that cannot yet afford to upgrade to the new technology.

Freight transportation also generates a variety of other environmental impacts and nuisances.[4] Surface freight vehicles as well as aircraft operating in densely populated urban areas create noise and vibration nuisances, as well as the negative effect of having a large vehicle navigate a congested area. Logistics facilities such as regional distribution centers (RDCs) similarly generate concentrated volumes of vehicle traffic, noise, and in some cases light pollution. An example of a *nonlinear* impact is the contribution of freight transportation to the spread of invasive species between different parts of the world. Unlike energy use or CO_2 emissions, the impact of freight on invasives cannot be estimated by multiplying impact per vehicle-mile or ton-mile by total levels of activity. Instead, most freight movements result in no invasives impact at all, but occasionally a single "low-probability high-consequence" event may have profound effects on both the environment and the regional economy. For example, in the period from 2000 to 2013, the Emerald Ash Borer, an invasive insect from Asia, migrated to the United States and gradually worked its way to the east coast, consuming many ash trees as its population spread. Not only does an invasive such as this deplete the biodiversity of U.S. forests by decimating the native population, but the regional economy suffers the loss of a valuable resource.

---

[3]Based on conversions of 128,700 Btu/ gallon for diesel (the lower heating value) and 22.4 pounds of CO_2 emitted per gallon.

[4]Schiller et al. (2010) Chap. 5 on "Moving Freight, Logistics, and Supply Chains in a More Sustainable Direction" provides a survey of these impacts.

11-2-4 Intercity versus Urban Freight

The majority of freight activity, energy consumption, and emissions occur in intercity movements. In the reverse pattern from passenger transportation, roughly 80% of all ton-miles in a typical industrialized country are intercity rather than urban, so the focus of this chapter is mostly on intercity freight. Some brief comments on the special characteristics of urban freight are offered here.

The dominant vehicle for urban freight is the small delivery truck, of the type operated by express package services such as Federal Express or UPS. Freight typically arrives into an urban area by air, rail, or by full-length articulated truck at some sort of terminal, and eventually is transferred to a smaller vehicle for final delivery. The process works in reverse as well, as small package pickups are made throughout the urban area to be consolidated into a full truck or container load at some sort of local transfer point. Alternatively, freight such as food and beverage products may arrive from an out-of-town RDC to be delivered all the way to their destination retail outlet (supermarket or independent smaller grocery store) in a full-length vehicle. In either case, the most common urban mode is the truck or road mode, but other concepts exist. From 1906 to 1959, a network of small underground railcars operated underneath the downtown area of Chicago, making deliveries and at the same time reducing traffic at the street level. As urban traffic congestion remains a chronic problem, other cities have considered variants of this system.

The dominance of regional and national decision-making in determining where and how freight moves in an urban region leads to a challenge for urban transportation planners working at the level of a single metropolitan area. Local representatives of the freight system such as store managers or drivers may have little input into decisions, so the planner does not have easy access to a large organization such as a retail chain or express package-delivery company. The situation is different from that of urban passenger transportation, where the planner can communicate with either local employers who generate work trips or individual citizens about the behavior of local commuters.

11-3 Total Logistics Cost and Economic Order Quantity Model

In the preceding section, we have seen that low, middle, and high-value commodities vary greatly in their share of ton-miles compared to share of value of products shipped. The ratio of dollar value to ton-miles from the data in Fig. 11-4 indicates the level of variability. For the low-value commodities, the value of this ratio is 0.21 (e. g., $207 billion divided by 988 billion ton-miles) versus 2.7 and 15.5 for mid- and high-value products, respectively.

The value of a commodity relative to the number of ton-miles it generates affects how decisions are made regarding its shipment—modal choice, speed of delivery, and so on. Firms that carry or ship inventory are subject to the time value of money, meaning that committing financial assets to one purpose or another costs the firm money over time because of interest paid on borrowed working capital or because of the opportunity cost from foregone alternatives for which assets could have been used instead. Therefore, the value of the product dictates its *inventory holding cost*, or amount that the owner of the unit of commodity will incur to keep the product in inventory. If a firm manufactures a product, they are responsible for its inventory cost until it is sold. If they purchase the product from a supplier, then they incur inventory cost from the time of purchase to the time of sale.

For many types of products, the shipper is charged by the carrier based on a fixed cost for an entire vehicle (e.g., railcar in the case of rail, or trailer in the case of truck), regardless of how full the firm fills it.[5] A tradeoff therefore emerges between filling each shipment as full as possible to minimize the cost per unit of product shipped, and shipping as frequently as possible to minimize the time product is kept in inventory and minimize inventory cost. The components of shipping cost per container and inventory cost are combined into the *Total Logistics Cost* (TLC) for the product and the answer that optimizes TLC is called the *Economic Order Quantity* (EOQ). The resulting cost model is given the name *EOQ Model* hereafter although it is alternatively called the Total Logistics Cost Model in some sources.

11-3-1 Components of the EOQ Model

We introduce the EOQ model starting with the average time that a product is expected to be in inventory. Let Q be the lane volume in units per unit of time (e. g., kitchen appliances per year) and V be the shipment size (units per shipment). Suppose that, as a simplifying assumption, all units that arrive in the initial shipment are sold before the next shipment arrives, and that since the units are sold at a steady rate, the last unit sells out just before the next arrival. If a load of product arrives into inventory in a location, individual units might be sold soon after arrival or just before the next arrival, but an average product is expected to stay in inventory for half of the time between one arrival and the next. Therefore time in inventory is

$$\textit{Avge time in inventory} = \frac{1}{2}\textit{ time between shipments} = \frac{V}{2Q} \qquad (11\text{-}1)$$

Example 11-1 provides an illustration.

Example 11-1 Suppose the volume in a given distribution channel is 100 units/year, and the shipment size has been specified at 25 units/shipment. What is the expected value of the average time in inventory?

Solution The term V/Q gives 0.25 years per shipment, or that an interval of ¼ year, or roughly 91 days, will pass between the arrival of one shipment and the next. Therefore, the average time in inventory will be half of this amount, or 45.5 days.

Time in inventory can be converted to economic cost. Let R be the cost of inventory given in units of percent of value per unit of time (percent per day, week, year, etc.). Let P be the value per unit of product, then PR is the cost of inventory in units of \$/time/unit. Therefore, the total cost of holding an item in inventory is the amount of time multiplied by the cost per unit of time, or $(PRV)/(2Q)$.

We can now state and define the three components of the EOQ model, which quantifies the total logistics cost for a given shipment size V. Let C = total logistics cost per unit, then C has three main cost components:

$$C = \textit{Holding Inventory Cost} + \textit{TransitInventory Cost} + \textit{Shipment Cost} \qquad (11\text{-}2)$$

[5]Recall that a *shipper* is an entity that requires the shipment of a product, and a *carrier* is an entity that provides the service of moving products on a commercial basis for shippers. A shipper may also provide the service for itself using its own private fleet.

Define T as the length of time required for shipment, S as the buffer length of time to cover variability in T, F the cost of shipment per container (i.e., per railcar, truck trailer, shipping container, etc.), and D the charge for placing the order per container. Note that it is common practice to charge separately for the line haul of moving the shipment from origin to destination, and for the process of placing the order, hence the distinction between F and D. We can now rewrite the equation for C in terms of known quantities:

$$C = \frac{PRV}{2Q} + PR(T+S) + \frac{F+D}{V} \tag{11-3}$$

The explanation is the following. The product of inventory cost per unit of time and average time in inventory gives the holding inventory cost (first term). The second term for transit inventory is similar, namely, the product of cost per unit of time and time in transit. Units must be consistent for the first two terms: inventory cost is usually given in cost per year, whereas average time in inventory or transit is usually in days. Lastly, the total cost per shipment, $F + D$, is divided by the number of units per order to give the cost of shipment per unit.

Next we can simplify the equation for C by defining $T' = T + S$ and $F' = F + D$. Substituting gives

$$C = \frac{PRV}{2Q} + PRT' + \frac{F'}{V} \tag{11-4}$$

The equation for C is now used to find an *optimal* shipment size V^* that minimizes total logistics cost, taking into account the competing interests of reducing inventory cost and shipment cost per unit. Taking Eq. (11-4) as a starting point, the partial differential of C with respect to V gives

$$\frac{\partial C}{\partial V} = \frac{PR}{2Q} + 0 - \frac{F'}{V^2}$$

Rearranging and simplifying

$$\frac{PR}{2Q} - \frac{F'}{V^2} = 0$$

$$\frac{V^2}{F'} = \frac{2Q}{PR}$$

$$V^2 = \frac{2F'Q}{PR}$$

The solution to the above equation is thus the optimal solution V^*:

$$V^* = \sqrt{\frac{2F'Q}{PR}} \tag{11-5}$$

The value V^* can then be substituted back into the equation for C to give total logistics cost per unit of product. Such a calculation assumes that $V^* \leq V_{max}$, the maximum number of units per vehicle; where $V^* > V_{max}$, the analyst can calculate the difference in TLC between V^* and V_{max}, and as long as they are not substantially different, it may be acceptable to designate V_{max} as the number of units per shipment. Note also that transit time T' does not appear in the equation for V^*: intuitively, inventory cost in transit does not affect TLC per unit because each unit must incur the same inventory cost while being moved, regardless of shipment size. Further implications of the calculation of TLC and EOQ are illustrated in Example 11-2.

Example 11-2 Washing machines with a value of \$500 per unit are shipped by truckload freight over a shipping distance requiring 1 day (the need for a buffer period can be ignored). The maximum number of machines per truckload is 50, and each shipment costs \$750 for the line haul plus \$50 to order. Demand at the destination is 5 units/week. Inventory cost is 25% per year.

What order size should the shipper adopt, and what is the TLC at that shipment size?

Solution To set up the solution to V^*, we first calculate $F' = F + D = 750 + 50 = 800$. We also need the demand Q in units per year; since demand is 5 units per week, assuming 52 weeks in the business year gives 260 units per year. Optimal shipment size is then

$$V^* = \sqrt{\frac{2F'Q}{PR}} = \sqrt{\frac{2(800)260}{500(0.25)}} =\sim 58 \text{ units/shipment}$$

Thus, 58 units to the nearest whole number. Note that in this situation $V^* > V_{max}$. If this were not the case, we would simply adopt V^* and calculate TLC, but here we calculate TLC for both V^* and V_{max} and make a comparison. Accordingly:

$$C(V^*) = \frac{PRV^*}{2Q} + PRT' + \frac{F'}{V^*} = \frac{125(58)}{2(260)} + 125\left(\frac{1}{365}\right) + \frac{800}{58} = 13.94 + 0.34 + 13.79$$
$$= \$28.08/\text{unit}$$

$$C(V_{max}) = \frac{PRV_{max}}{2Q} + PRT' + \frac{F'}{V_{max}} = \frac{125(50)}{2(260)} + 125\left(\frac{1}{365}\right) + \frac{800}{50} = 12.02 + 0.34 + 16.00$$
$$= \$28.36/\text{unit}$$

Based on this calculation, the shipper can conveniently ship full trailers with only a slight increase in TLC per unit.

Variants on Interpretation of Total Logistics Cost in EOQ Model

The reader should be aware of two possible variants in the EOQ Model. First, holding inventory cost could be interpreted as the time spent by the product awaiting shipment at the origin plus the time spent awaiting sale at the destination. According to this variant, total time in inventory is twice the value given above in Eq. (11-1), or V/Q instead of $V/(2Q)$. Therefore the first term in Eq. (11-3) becomes PRV/Q by this interpretation. Second, total logistics cost can be defined as total cost incurred by all units of the product in a given unit of time rather than on a per-unit basis, in which case the right-hand side of Eq. (11-3) is multiplied by Q throughout, giving

$$C = PRV + PR(T + S)Q + \frac{(F + D)Q}{V} \tag{11-6}$$

These alternate assumptions may be encountered elsewhere in the literature, although the assumptions of Eq. (11-3) are used throughout this chapter.

11-3-2 Implications of Total Logistics Cost for Sustainability

Two observations can be made regarding the effect of shipment size on total logistics cost, both in Example 11-2 and in general. First, as shown in Fig. 11-6 when looking at the

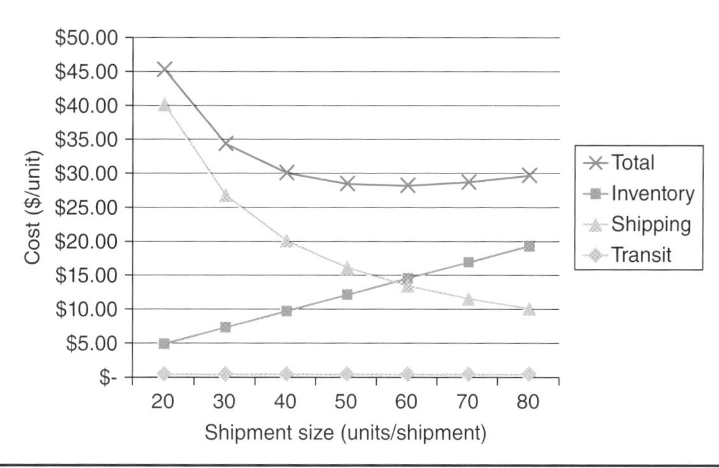

FIGURE 11-6 Total logistics cost (TLC) as a function of shipment size for washing machine example. *Notes:* "Inventory" = holding cost while not in transit, "Transit" = holding cost while being moved, "Shipping" = cost to each unit for its part of fee charge by transportation provider.

"Total" curve, on either side of the optimal shipment size, the value of TLC is not very sensitive to adjustments in chosen shipment size up to a point. Thus shippers can use V^* as a starting point for considering choice of shipment size, and then make adjustments as needed to take into account practical considerations. This was illustrated in Example 11-2 with the increase of $0.28/unit in TLC to ship at V_{max} rather than V^* (if V^* had been physically possible). Indeed, in Fig. 11-6, the shipper could choose values of shipment size anywhere between 40 and 50 units and in no case exceed a value of $30 per unit for TLC, whereas the optimal value is $28.08. On the other hand, some products are sold to customers with low profit margins by the vendor, in which case shipping at or as close as possible to the EOQ value may be important for the profitability of the business.

Secondly, in recent times freight transportation has generally become more efficient and hence costs on average less per vehicle load shipped for a given distance. At the same time, the price of keeping inventory has risen. Therefore, many different types of products have moved toward smaller optimal shipment sizes. This change may help the bottom line but it also has implications for the total volume of freight movement and for the resulting energy consumption and ecological impact. On balance, smaller shipment size can encourage lower average lading per vehicle and more trips per unit of time, so that vehicle activity, energy consumption, and emissions increase. We continue to explore these topics in the following section.

11-4 Disaggregation of Freight Energy Consumption

In the face of the rapid growth in freight energy use illustrated in Fig. 11-5, it is important for the United States and other countries (both industrialized and industrializing, such as the BRIC countries)[6] to develop strategies that can meet long-term energy

[6]BRIC = Brazil, Russia, India, China.

security and carbon emissions goals. Strategies might start with a *fair-share* approach to energy use and carbon reduction targets, that is, percentage reductions in the freight sector should be in proportion to the percentage reduced across all sectors. For example, if the goal of 80% reduction in CO_2 from all sectors between 2005 and 2050 is adopted, then the freight sector should contribute its fair share by reducing total emissions by 80% compared to the estimated 2005 value. As in other sectors, such a deep reduction poses a major challenge for freight, since it involves not only becoming more efficient but also developing alternative energy resources to fossil fuels. Disaggregation of freight energy consumption into different categories provides a tool for understanding how that consumption takes place.

11-4-1 Mode- and Commodity-Based Approaches to Understanding Freight Energy Use

A conventional framework for estimating energy use in freight that is widely used in the literature is the *mode-based approach*, which focuses on energy requirements for the movement of a given volume of freight activity (see Fig. 11-5). If modal activity (ton-miles or tonne-km) and total modal energy consumption are given, average modal energy intensity can be estimated. Alternatively, if activity and intensity are given, total energy requirement for an amount of modal activity can be calculated. This framework looks only at the end-use energy required directly for the movement of freight; it does not take into account life-cycle energy consumption, including in the refining and distribution of petroleum products to freight vehicles, in constructing vehicles and infrastructure, or in indirect energy uses such as electricity consumption in maintenance facilities. Viewed generally, the main factors can be combined in Eq. (11-7):

$$E_{\text{tot}} = \sum_k \mu_k \cdot s_k \cdot \text{XD} \tag{11-7}$$

where E_{tot} is total energy consumption
μ_k is energy intensity [Btu/ton-mile or kJ/tonne-km]
s_k is percent modal share for mode k
XD is total ton-miles or tonne-km moved, and
$\sum_k s_k$ is 1

According to this model, energy use for a given demand level for freight XD is a function of the modal share s_k multiplied by the energy intensity of that mode μ_k. Based on this framework, four directions for reducing energy use emerge:

1. *Improve technological efficiency*: Improve engine (or electric motor where applicable) and vehicle technology and deploy the technology in as much of the fleet as possible to move a given number of vehicle-miles with less energy. This approach is particularly relevant for the truck mode, which has the highest energy intensity per ton-mile among surface modes.
2. *Improve operational efficiency*: Move a given quantity of freight by a given mode using less energy by increasing the average load per vehicle, by reducing the number of empty vehicle moves, or by other means. This approach assumes the technology remains fixed (unlike point #1) and that μ_k is reduced by making changes in operations.

3. *Shift to more efficient modes*: Increase the modal share s_k of more energy efficient modes, such as intermodal truck-rail instead of truck.
4. *Reduce total volume of freight activity*: Reduce total ton-miles by *rationalizing* freight transportation networks (meet demand for goods with less freight activity) or by *dematerializing* goods (products that weigh less but deliver the same functions demanded by consumers generate fewer ton-miles).

In an alternative view of the freight energy problem, the overall freight distribution network generating total levels of freight activity is broken down into its component commodities (e. g., total truck ton-miles split into food, machinery, paper, etc., shipped by truck, the same for rail, marine, etc.), a methodology known as the *commodity-based approach*. The transformation of the analysis results in the following modification of Eq. (11-7):

$$E_{\text{tot}} = \sum_{c}\sum_{k} \mu_{c,k} \cdot s_{c,k} \cdot \text{XD}_c \qquad (11\text{-}8)$$

where E_{tot} is total energy consumption
μ is energy intensity [Btu/ton-mile or kJ/tonne - km]
s is percent modal share
XD_c is tonne - km or ton-miles of commodity c
c is subscript for commodity
k is subscript for mode
$\sum_{k} s_{c,k}$ *is* $1, \forall c$

An additional subscript c is introduced to represent the various commodity groups; freight activity, modal shares, and energy efficiencies can now be differentiated by commodity as well as mode.

From the commodity-based perspective, the total volume of freight is divided among many commodity sectors and is therefore more closely connected to the shippers and carriers of that commodity.[7] This approach sheds light on the role of different commodities in generating consumption of freight energy, including the intensity of freight energy requirements, the trend over time for the product in question, and the contribution of the different freight transportation modes to the total freight energy for the product. This information can then be used to carry out more targeted improvement of freight energy efficiency. For example, it may be of particular interest to work on improving energy consumption with a sector of the economy that (1) uses a large fraction of the total freight energy budget, or (2) is energy intensive relative to the value or weight of goods moved, or (3) is increasing its freight energy consumption rapidly.

11-4-2 Assessment of Freight Energy Use at a Modal Level

Comparison of Modal Energy Intensities

At present, the truck mode is on average much more energy intensive than the water or rail modes. In the year 2006, the average energy intensity values for these three modes

---

[7]The comparison of mode-based and commodity-based freight energy analysis are developed further in Vanek and Morlok (2000).

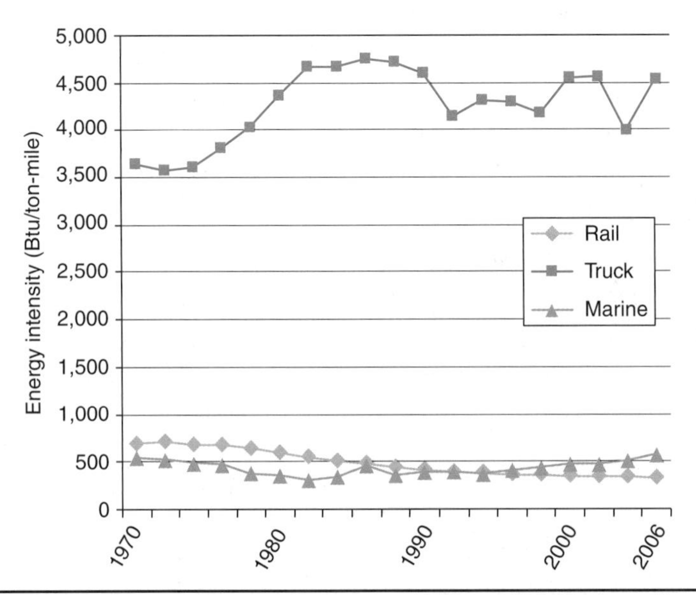

FIGURE 11-7 Energy intensity of U.S. surface modes 1970 to 2006, Btu/ton-mile.
Note: 2008 values not included due to inadequate data.
Source: Davis et al (2013).[8]

were 4,542, 571, and 330 Btu/ton-mile, respectively (Fig. 11-7). For the truck mode, the largest user of U.S. freight energy, energy intensity increased from 3,600 Btu/ton-mile in 1970 to 4,800 Btu/ton-mile in 1986, and then declined slightly to 2006 as shown. Note that this measure masks the impact of bulk goods on energy efficiency of water and rail, since they tend to be densely packed and move slowly, allowing these modes to achieve efficiency values that are not possible for more high value, time sensitive goods. It is still more efficient to move high-value goods by rail than by truck, but the efficiency gain is not as great as might be implied by the approximate 10:1 advantage based on the overall average modal energy intensities observed in Fig. 11-7 alone.

Truck Operational Efficiency Measured in Energy and Ton-Miles per Vehicle-Mile

Operational efficiency of trucking can be measured in a number of ways, including the percentage of distance moved loaded (commonly called the *load factor*, and measured by dividing loaded distance by distance moved either loaded or empty), the average weight of freight loaded per movement, the average percentage of capacity used, the average energy per vehicle-mile, or the average ton-miles delivered per vehicle-mile moved. For brevity, we study only the latter two measures for the United States.

The truck energy intensity trend per vehicle-mile has been similar to the per ton-mile trend: in 1970, trucks averaged around 25,000 Btu per vehicle-mile in the United States; they then rose to 25,100 Btu/vehicle-mile in 1981, but by 2006 declined to around

[8]The decline in truck energy intensity in 2004 compared to 2002 or 2006, which were approximately the same, may be due to data gathering anomalies rather than actual improvement in delivered intensity.

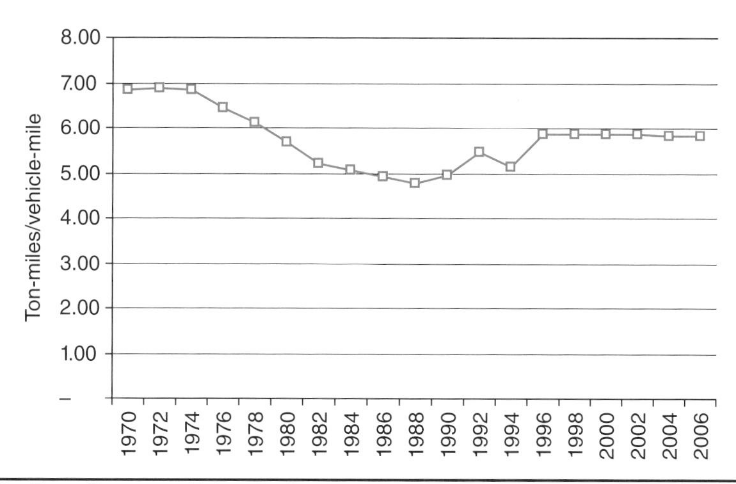

FIGURE 11-8 Ton-miles per vehicle mile for U.S. trucks, 1970 to 2006.

Source: Own calculation based on ton-mile data from Bureau of Transportation Statistics and vehicle-mile data from Eno Transportation Foundation (1970 to 1994) and U.S. Federal Highway Administration (1996 to 2006).

23,300 Btu/vehicle-mile. Given the number of advances in engine technology, materials, and so on, this rather mild reduction in energy consumption reflects the counteracting influences of operational pressures to move smaller loads faster and over longer distances, as well as the application of improved technology to increasing overall power rather than improving efficiency.

Ideally, an operationally efficient system should deliver the maximum possible number of ton-miles per vehicle-mile (tmi/vmi), which would reflect high average loading of trucks with tons of freight. As shown in Fig. 11-8, it appears that during the period from 1970 to 1994 the tmi/vmi ratio first rose slightly and then in general declined to 1994, after which the measure entered a plateau up to 2006 with values between 5 and 6 ton-miles per vehicle mile. Fundamental logistical trends such as just-in-time delivery and smaller consignment size exerted downward pressure on the operational efficiency of trucking in general, while improving technology and advances in vehicle routing and scheduling (VRS) aided efficiency.[9] Over the 1970 to 2006 time period taken as a whole, positive pressures on operational efficiency were not as strong as negative ones, and tmi/vmi declined.

11-4-3 Assessment of Freight Energy Use at a Commodity Level

Focusing on specific *commodities* as well as modes can provide another basis for making freight more energy-efficient. Estimation of energy intensity by both commodity and mode, that is, $\mu_{c,k}$ as opposed to the original μ_k, allows the analyst to take into account characteristics specific to individual products. Availability of $\mu_{c,k}$ values combined with ton-miles by commodity and mode enables the analysis of overall energy consumption for the movement of different commodities.

[9]The acronym VRS represents a range of software products and logistical specialists that use computerized decision support systems for planning freight vehicle movements.

Mode-Commodity Energy Intensity

The total energy use for each commodity group is based on the modal volumes for the five major modes. Not every mode applies to every commodity group; for example, movement of low value density (i.e., dollar value per ton of product) bulk commodities by air freight, or movement of high value manufactured goods by pipeline. Truck and rail energy intensity values by commodity are available for a wide range of commodities, whereas for air, marine, and pipeline modes, the energy intensity values are less well understood.

Since truck and rail are the most significant movers of ton-miles, and consumers of energy, across a wide range of both bulk and finished commodities, Table 11-1 provides estimated commodity-specific truck and rail energy intensities for 13 commodities representing much of the freight moved in the United States (a number of small-volume commodities are not shown). Taking the 2007 values, truck intensities are significantly higher than those of rail across the board, ranging from 3,705 Btu/ton-mile for petroleum and coal products to 5,395 Btu/ton-mile for lumber and wood products. Rail intensities are lower, but they also have wider variability—coal consumes on average just 247 Btu/ton-mile, whereas transportation equipment consumes 1,147 Btu, for a ratio of more than 4 to 1. An average value for the five raw material commodities and

	Truck	Rail
Overall modal average	4628	366
Fabricated metal products	4542	791
Transport equipment	4792	1147
Coal	4345	247
Pulp/paper	4592	501
Primary metal products	4044	393
Farm products	4736	472
Petroleum/coal products	3705	427
Non-metallic minerals	4070	291
Lumber/wood products	5395	597
Chemicals	4232	380
Food/kindred products	4915	530
Textiles & apparel	5316	1031
Machinery	4247	1098
Raw material average	4218	363
Finished commodity average	4730	761

Notes: Values calculated using data from 2007 U.S. Commodity Flow Survey based on methodology from Vanek & Morlok (1998).

The raw material average is the arithmetic average of coal, petroleum products, farm products, chemicals and nonmetallic minerals; the finished commodity average includes the remaining eight commodities in the table. Conversion: To obtain value in kilojoules/tonne-kilometers, multiply value shown by 0.725.

Table 11-1 U.S. Truck and Rail Energy Intensities by Commodity, 2007 (Btu/ton-mile)

for the remaining eight finished commodities is given as well at the bottom of the table. Mode-commodity energy intensity estimates form a basis for estimating overall commodity energy intensity μ_c, which is a weighted average of values of $\mu_{c,k}$ taking into account ton-miles of commodity c moved by each mode k, i. e.,

$$\mu_c = \sum_k \mu_{c,k} \cdot s_{c,k}, \forall c \tag{11-9}$$

Overall Freight Energy Use Disaggregated by Commodity

A comparison of commodity-based analysis of intercity freight energy use in the United States and United Kingdom is given in Figs.11-9(*a*) and 11-9(*b*). For the U.K. figure, 13 commodity groups are shown plus a miscellaneous shipments category ("Misc. Products"). The data did not support disaggregation of rail and water modes in the United Kingdom by commodity. Therefore, the contribution of these modes is not included in the commodity disaggregate values, and instead is shown separately in aggregate form on the right side with white colored bars. The remaining values are then just for road energy consumption. For the U.S. figures, 14 commodity groups are disaggregated, including energy consumption for all modes (e. g., road, rail, water) for each group. Both figures exclude pipeline energy consumption, energy use for urban movement of freight, and energy consumption in outbound international airfreight movements.

A common point between the two figures is that in both countries, shipping of food products is a large consumer of energy relative to other commodities. The agricultural products (U.K.) or farm products (U.S.) groups are also fairly large consumers of transportation energy, so the total energy balance for the entire delivery of food items from crops on the farm to the final consumer, as a fraction of total freight energy, is even larger: roughly 25% of the total energy consumption covered in each of the three figures.

Comparison of Freight Energy Growth by Commodity Group

When freight energy consumption values by commodity are known in different years, it is possible to compare commodity groups over time to determine which groups are contributing relatively more or less to the overall growth in freight energy consumption. Fig. 11-10 shows percent growth in energy consumption for 10 groups for which comparable data were available for the period 1972 to 2002. For each commodity group, energy consumption in the start and end year is determined from modal share of ton-miles and mode-commodity energy intensity, and then the percent change in total energy is calculated. Note that some of the groups from Fig. 11-9 are not included because certain commodities were disaggregated differently in the respective versions of the commodity flow survey, so that a direct comparison could not be made. From Fig. 11-10, fabricated metal products and petroleum or coal products made only a modest contribution to overall growth, whereas coal, chemicals, machinery, and food products made a larger contribution. From Fig. 11-5, overall energy growth for combined truck, rail, or marine from 1972 to 2002 is 128%, while individual commodity growth ranges from a low of 8% to a high of 180%. Time series calculations with the commodity flow survey should be treated with some caution, since changes in the survey methodology or definitions of commodity groups may affect the comparability between years, even if the commodity group bears the same name.

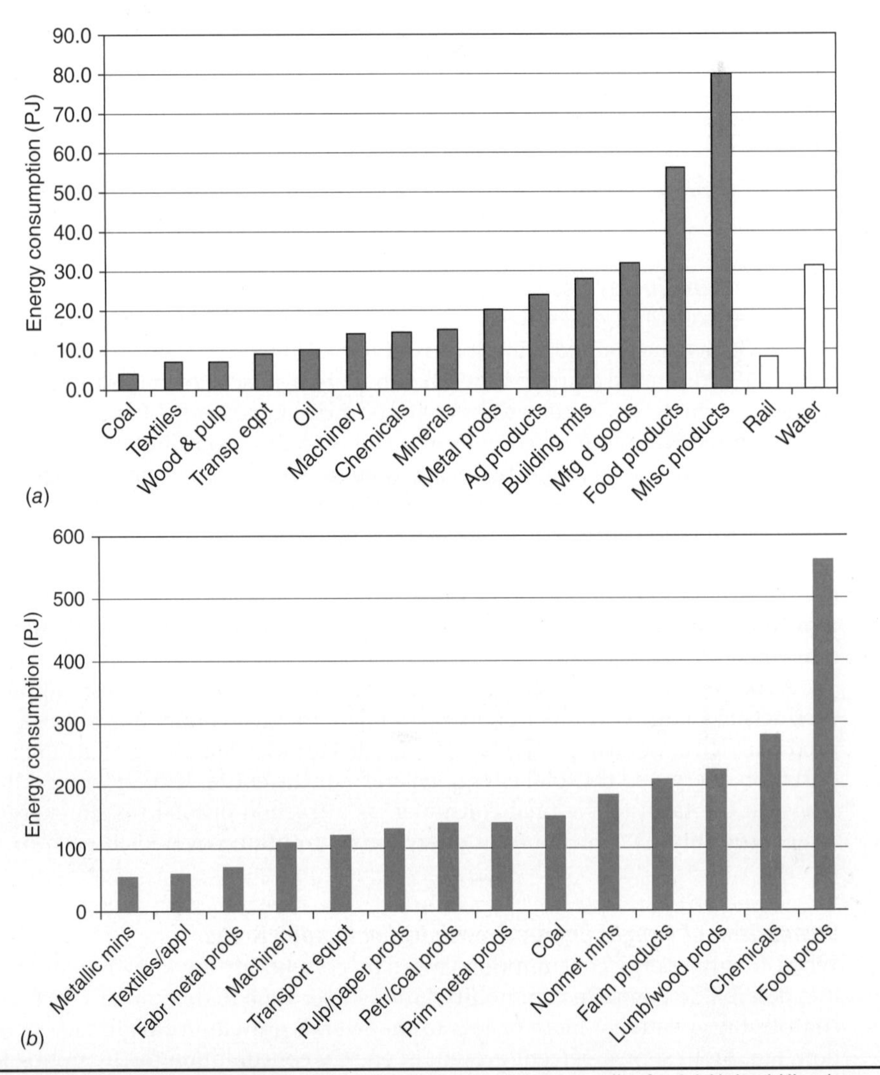

FIGURE 11-9 Freight energy consumption disaggregated by commodity for (*a*) United Kingdom, 1995 and (*b*) United States, 2002.

Sources: For (*a*) Vanek & Campbell (1999); for (*b*) own calculation based on data from 2002 U.S. Commodity Flow Survey.

Conversions: For trillion Btu, multiply values shown by 0.948.[10] 1 PJ = 1×10^{15} joule.

Comparison of Freight Transportation and Production Energy Use

At a macroscopic level, we can compare the estimates of transport energy use by commodity to estimates of energy use in the production of those commodities to assess the

[10]For reference, the full names of the categories in Fig.11-9*c* are, from left, "Metallic minerals," "Textiles and apparel," "Fabricated metal products," "Machinery," "Transportation equipment," "Pulp and paper products," "Primary metal products," "Coal," "Nonmetallic minerals," "Farm products," "Lumber and wood products," "Chemicals," "Food products."

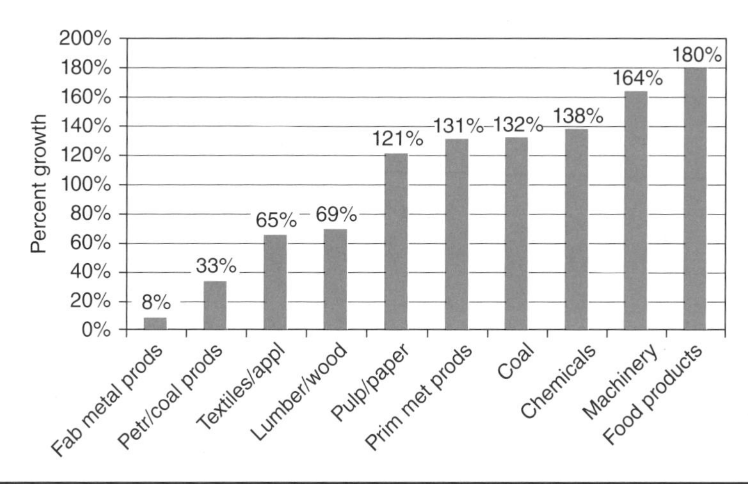

FIGURE 11-10 Percentage growth in freight energy consumption by commodity group, 1972 to 2002. *Source:* Own calculations based on U.S. Commodity Flow Survey (USBTS, 2004).

relative importance of these two stages in the life cycle energy consumption for that commodity (Table 11-2). Here consideration of the end-use and disposal stages of the life cycle is left out. However, for the groups included, end-use is significant only for transport equipment and machinery because these are the only two commodities where the product consumes energy (e. g., motor fuel or electricity) during its lifetime of use. Also, the disposal stage is expected to have relatively low energy use for any of the groups, as it mainly involves a single (and usually short) move from end-use either to a landfill or to a recycling center. In addition, once at the landfill or recycling center, ongoing energy use per unit of commodity is thought to be relatively small. Note that for six of the commodities at the bottom of the table, direct comparison between transportation and production was not possible because some commodities are classified differently in the Standard Industrial Code used in manufacturing. (For the remaining 10 commodities comparison was possible.)

For most of the 10 commodity groups where comparison was possible, energy in production was found to be the dominant user of energy. However, certain groups such as foods and kindred products, lumber and wood products, and apparel had a large component of transportation energy use—up to 70% of the value for production in the case of apparel. These results suggest two key ideas:

1. Energy use in transportation is significant, and thus it is essential to include the transportation energy consumed in any estimate of a product's life cycle energy use.
2. Producers and others may uncover significant opportunities for improving their product's life-cycle energy use in the transport stage, which, if implemented, can make a measurable contribution to saving energy from an overall life-cycle perspective.

More globally, for any commodity group, improving production and transportation are by no means mutually exclusive, and a combination of manufacturing and

Directly Comparable	Production	Transportation	Ratio
Fabricated metal products	307	77	4.0
Transport equip.	323	111	2.9
Pulp/paper	2,506	177	14.1
Primary metal products	2,467	163	15.1
Petroleum/coal products*	2,198	216	10.2
Lumber/wood products	4,51	292	1.5
Chemicals*	1,861	255	7.3
Food/kindred products	9,56	577	1.7
Textile mill products	2,74	35	7.8
Apparel & textile prods	44	30	1.4
Total (comparable only)	11,388	1,934	5.9
Not directly comparable			
Coal	—	151	—
Farm Products	—	265	—
Nonmetallic minerals	—	227	—
Industrial machinery & equipment	250	—	—
Electric/electronic equip.	224	—	—
Machinery & computers	—	51	—

Notes: Blank cell means comparable data was not available for this category.

*Production energy shown is USDOE figure minus 63% for feedstocks. Before reduction figures are 6,289 and 5,324 trillion Btu, respectively.

Source: U.S. Department of Energy (1994), for production energy use; Vanek and Morlok (1998), for transportation energy use. Note that production data for 1991 was used, because data for 1993 were not available.

TABLE 11-2 Comparison of U.S. Production and Transportation Energy Use by Commodity Group (Trillion Btu)

transportation improvements can lead to the most efficient outcome of all. Especially in cases where a commodity has relatively high energy intensity per ton-mile or tonne-kilometers, improving transportation may make a valuable contribution.

Translating predictions at the macro level about potential energy savings into results at the micro level (individual product) will inevitably point up some of the inaccuracy inherent in using aggregate data to make generalizations about a wide range of products. Individual products within a given sector may vary widely in both production and transportation energy use. Also, much of the production and transportation of a number of commodity groups (e.g., apparel) occurs outside of the United States and beyond the range of the data gathering which underlies the figures presented here.

Nevertheless, the group-wide life-cycle analysis need not predict the energy use of each individual product in order to motivate commodity sectors to seek out overall gains in efficiency. Sectors with high transport intensity will be motivated to become more efficient, thereby improving not only the transportation energy consumption of individual products but also their overall life-cycle energy use.

11-5 Discussion: Toward Greater Sustainability in Freight Transportation

Earlier in this chapter, it was established that total demand for freight activity is currently at a high level (approximately 14,000 ton-miles per capita in the United States in 2008),[11] leading to high levels of energy consumption and CO_2 emissions. Many solutions are at hand to address these challenges, and these are discussed in the remaining two chapters in the freight unit.

Chapter 12 surveys near-term solutions for both freight technology and operations. Technological solutions can reduce energy consumption, vehicle-miles, or both, and are to varying degrees cost-effective in terms of recouping the initial investment by reducing ongoing expenditures. Beyond changes that can be made within each mode, transfer of freight from less to more efficient modes can further reduce energy consumption and emissions. In the spirit of the commodity-base as opposed to mode-based approach to addressing freight energy consumption, when shippers that control supply chains for specific products take the initiative to find more efficient ways to ship and warehouse those products, they can seek out the best possible combination of advanced technology, best operational practices, and modal shifting.

Chapter 13 addresses the longer-term consideration of spatial patterns in freight transportation and supply chains, which have generally been spreading but which could also be returned to a smaller geographic expanse in the future to assist with reducing ecological impact. In the first instance, it is useful for the practitioner to understand how both more distant sources and markets, and more complex supply chains within a more agile multimodal transportation system, have led to spatial spreading since the dawn of the logistics revolution in the 1950s. It is also valuable to note that greater congestion in the transportation system and higher fuel costs at present, as well as opportunities to produce, consume, and recycle some products on a more regional basis, might provide opportunities for the opposite of spatial spreading, namely spatial redistribution.

11-5-1 Greater Sustainability through Increased Stakeholder Involvement

The freight system relies on many stakeholders to operate, so involving as many of these stakeholders as possible in the solution of sustainability-related problems (such as congestion, energy security, and climate change) will allow for a better outcome. The list of stakeholders includes the following:

- *Shippers:* Generators of shipments of products and raw materials who wish to benefit from a system that functions well and also to contribute to improved efficiency and reduced impact.
- *Carriers:* Providers of transportation services to shippers, who can implement new technologies or best practices
- *Government agencies:* Mixture of elected leaders and career public servants who implement laws, inspect the function of private shippers and carriers, and oversee the provision of infrastructure, vehicles, and fuel.

---

[11]In other words, approximately 4.2 trillion ton-miles divided by population of 300 million.

- *Vehicle manufacturers:* Private manufacturers who incorporate requirements for cleaner or more efficient vehicles into their products.
- *Infrastructure providers:* Private companies that build and maintain roads, railroads, and other infrastructure on behalf of the government or private companies such as private railroads, who can help improve operations.
- *Fuel providers:* Refiners and distributors of motor fuels who can provide greener fuel options, either proactively or in response to new requirements from government.
- *General public:* Individual members of the public who in addition to their role in electing government officials can also respond to positive or negative developments with the sustainability of freight and petition government or other stakeholders to address local or global sustainability concerns.

Freight sustainability efforts have in the past focused on two players, namely, the government through its various agencies and vehicle manufacturers who develop improved technologies or fit emissions control equipment. Larger improvements are possible through the growing involvement of other stakeholders. For example, shippers (both manufacturers and retailers) and carriers together control product transportation, and therefore make the decisions affecting number of legs in the supply chain, total distance traveled, and choice of mode. Looking specifically at the case of energy efficiency, new opportunities within this relationship will encourage cooperation. These opportunities can be seen in the following examples:

1. *Cooperation in analyzing energy efficiency*: Working independently, a shipper may be unable to analyze energy use in specific transportation moves because it has little information about modal choice and operational characteristics of the carrier. However, if a shipper works with a carrier, the two together can evaluate all viable alternatives for distribution and then the carrier can evaluate the energy use of each. With each player providing the information about the part of the decision it knows best, an alternative which is both practical and uses the minimum energy can be found.
2. *Cooperation in investing in energy-efficient transportation*: Improving vehicle technology and shifting modes remain viable alternatives for saving energy. At present, a carrier may be reluctant to invest in programs such as more efficient vehicles or improved infrastructure for intermodal truck-rail if it feels that its customers are unwilling to bear these costs in higher shipment prices. However, if the shipper is itself interested in reducing total freight energy in its operation by improving energy use in trucks or increasing the use of rail, it may readily agree to a program of improvements worked out with the carrier.

In conclusion, a broader coalition of stakeholders can do more to advance the sustainability of freight than any one stakeholder can do individually.

11-6 Summary

In the latter half of the twentieth century and beginning of the twenty-first, freight transportation demand measured in ton-miles or tonne-kilometers has grown rapidly in industrialized countries. In particular, truck freight has grown by a factor of at least three in the United States and the United Kingdom during this period. Freight demand can also be viewed from the perspective of the value rather than weight of product shipped, in which case high-value products make up a much larger share of the demand, and low-value products much smaller. Whether measured in physical activity or product value, growing freight demand supports a modern industrialized way of life but also leads to increased energy consumption and a range of harmful impacts on the environment.

Total logistics cost (TLC) considers the complete range of costs added to a product once it is finished to move and store it until it is sold to a customer. TLC includes both the cost per unit to transport the item and inventory cost that accumulates over time as the product awaits sale. Economic order quantity (EOQ) is the shipment size that minimizes TLC. Since smaller shipment sizes and faster transportation modes can reduce inventory cost, sometimes the pursuit of lower TLC can inadvertently increase energy consumption or emissions per product handled in the supply chain.

Energy consumption in freight transportation can be disaggregated from either a mode-based or commodity-based perspective. A mode-based approach strives to make each mode as efficient as possible in the aggregate, and shift freight to more environmentally benign modes where practical. A commodity-based approach highlights differences in energy intensity for a given mode depending on what type of commodity is being moved. It also illuminates the relative contribution of different commodities to overall freight energy consumption and growth over time, as well as the proportion of a product's life-cycle energy consumption incurred by freight. Viewing energy consumption from both modal and commodity perspectives can assist in moving freight in a more sustainable direction.

References

Davis, S., and D. McFarlin (2009). *Transportation Energy Data Book: Ed. 28.* Oak Ridge National Laboratories, Oak Ridge, TN.

Davis, S., S. Diegel, and R. Boundy (2013). *Transportation Energy Data Book: Ed. 32.* Oak Ridge National Laboratories, Oak Ridge, TN.

Schiller, P., E. Bruun, and J. Kenworthy (2010). *An Introduction to Sustainable Transportation: Policy, Planning, and Implementation.* Earthscan, London.

USBTS (2004). *U.S. Commodity Flow Survey*, U.S. Bureau of Transportation Statistics, Washington, DC.[12]

U.S. Dept. of Energy (1994). *Manufacturing Consumption of Energy 1991.* Energy Information Administration, Washington, DC.

Vanek, F. and E. Morlok (1998). "Freight Energy Use Disaggregated by Commodity: Comparisons and Discussion. *Transportation Research Record* 1641:3–8.

Vanek, F. and J. Campbell (1999). "UK Road Freight Energy Use by Product: Trends and Analysis from 1985 to 1995," *Transport Policy* 6:237–246.

Vanek, F. and E. Morlok (2000). "Reducing US Freight Energy Use through Commodity Based Analysis: Justification and Implementation." *Transportation Research Part D* 5(1):11–29.

[12]The U.S. Commodity Flow Survey is jointly undertaken with the U.S. Department of Commerce and the U.S. Bureau of the Census. Publications from previous USCFS editions (2002, 1997, etc.) are not listed separately but are published by BTS.

Further Readings

Abacus Technology Corporation (1991). *Rail vs. Truck Fuel Efficiency*. U.S. Department of Transportation, Springfield, VA.

American Association of Railroads (2011). *Railroad Facts*. AAR, Washington, DC.

Bernardini, O. and R. Galli. (1993) "Dematerialization: Long-Term Trends in the Intensity of Use of Materials and Energy. *Futures* 25:431–448.

Davies, J., M. Grant, and J. Venezia (2007). "Greenhouse Gas Emissions of the U.S. Transportation Sector: Trends, Uncertainties, and Methodological Improvements." *Transportation Research Record* 2017:41-46.

Fitch, J. (ed.) (1994). *Motor Truck Engineering Handbook, 4th ed.* Society of Automotive Engineers, Warrendale, PA.

Gerondeau, C. (1996). "Freight Transport in Western Europe: The Case for Using New Units of Measurement." *Transportation Quarterly* 50(3):51–58.

Greene, D. and Y. Fan (1994). "Transportation Energy Intensity Trends: 1972–1992." *Transportation Research Record* 1475:10–19.

Greene, D. (1996). *Transportation and Energy*. Eno Foundation, Washington, DC.

Institute of Energy Economics, Tokyo (1993). *Modal Shift and Energy Efficiency ("Modaru Shifuto to Enerugi Koritsu")*. I.E.E., Tokyo.

Korpela, J., K. Kylaheiko, and A. Lehmusvaara (2001). "The Effect of Ecological Factors on Distribution Network Evaluation." *International Journal of Logistics* 4:2.

Mckinnon, A., S. Cullinane, and A. Whiteing, (eds.) (2009). *Green Logistics: Improving the Environmental Sustainability of Logistics*. Kogan Page, London.

McKinnon, A. and M. Piecyk (2011). *Measuring and Managing CO_2 Emissions of European Chemical Transport*. European Chemical Industry Council. Electronic resource, available at *www.cefic.org*. Accessed Aug. 11, 2013.

Moffat, B. (2011). "The Chicago Freight Tunnels." *MAS Context* 9(Spring):72–83. Electronic resource, available at *www.mascontext.com*. Accessed Aug. 11, 2013.

New York State Department of Transportation (2001). *Data Needs in the Changing World of Logistics and Freight Transportation: Conference Synthesis*. NYSDOT, Albany, NY.

Schafer, A., H. Jacoby, and J. Heywood (2009). "The Other Climate Threat: Transportation." *American Scientist* 79:476–482.

Schipper, L., L. Scholl, and L. Price (1997). "Energy Use and Carbon Emissions from Freight in 10 Industrialized Countries: An Analysis of Trends from 1973 to 1992." *Transportation Research Part D* 2(1):57–75.

Sheffi, Y., B. Eskandari, and H. Koutsopoulos (1988). "Transportation Mode Choice based on Total Logistics Cost." *Journal of Business Logistics* 9(2):137–153.

Winebrake, J., J. Corbett, and A. Falzarano (2008). "Assessing Energy, Environmental, and Economic Tradeoffs in Intermodal Freight Transportation." *Journal of the Air and Waste Management Association* 58(8):1004–1013.

Exercises

11-1. Compare two options for shipping a product, either by truck or by intermodal rail (IM). Data for the two modes are the following: for truck: V_{max} = 38,000 lb and shipping time = 2 days; for IM, V_{max} = 43,000 lb and shipping time = 4 days. The product has the following characteristics: the unit cost is $4.03 per pound, and the inventory rate is 18% per year. Demand is given in terms of full truckloads at 12 truckloads per year, regardless of which mode is chosen. The shipping distance in both cases is 734 miles. For trucking, there is a $100 flat charge and then $1.25/mile additional, and also an order fee of $35 per order. For IM, there is a $400 flat charge and then $0.70/mile additional, and also an order fee of $35 per order. (a) Calculate the optimal shipment size and total logistics cost per pound for the product, for both truck and IM rail. (b) Indicate which modal option minimizes TLC. (c) If the sales cost is the sum of value of product and total logistics cost (TLC), what is the sales cost per pound and the percent contribution of TLC to sales cost in each of the truck and IM cases? (*Hint:* Throughout the problem, the differences between truck and rail are small, so carry enough decimal places to show the difference.)

11-2. Ton-mile and energy consumption data for U.S. trucks and railroads for the period 1980 to 2000 are given in the following table. Use Divisia decomposition to create a table and a graph for the period 1980 to 2000, showing four curves: (1) actual fuel consumption, (2) trended fuel consumption, and the contribution of (3) energy intensity, and (4) structural changes to the difference between actual and trended fuel consumption.

	Activity		Energy	
	Truck	Rail	Truck	Rail
Year	Bil. tkm	Bil. tkm	EJ	EJ
1980	836	1342	1.89	0.583
1985	925	1342	1.96	0.485
1990	1045	1565	2.17	0.478
1995	1194	1733	2.4	0.469
2000	1524	2168	3.24	0.555

11-3. Below are the data in standard units for U.S. rail and marine for the period 1970 to 2005. Use Divisia decomposition to create a table and a graph for the period 1970 to 2005, showing four curves: (1) actual fuel consumption, (2) trended fuel consumption, and the contribution of (3) energy intensity, and (4) structural changes to the difference between actual and trended fuel consumption.

	Rail		Marine	
Mode	Bil. tkm	Tril. Btu	Bil. tkm	Tril. Btu
1970	765	533	596	325
1975	754	560	566	280
1980	919	556	922	330
1985	877	560	893	356
1990	1,034	450	834	328
1995	1,306	483	808	308
2000	1,466	517	646	306
2005	1,696	575	591	319

CHAPTER 12

Modal and Supply Chain Management Approaches

12-1 Overview

In this chapter, we survey technologies and practices for achieving sustainable freight and logistics, divided into three approaches. First, an *intramodal approach* considers changes that can be made within each mode to make it more sustainable. Since truck is the freight mode that has the largest impact on the human-built environment and causes the largest amount of ecological impact, this section focuses primarily on truck freight. Second, an *intermodal approach* considers how shifting freight from more intensive modes such as truck to less intensive modes such as rail or marine can reduce impact. Particular attention is paid to intermodal truck-rail, since this is the mode that most closely resembles trucking in its level of service and is therefore most likely to be competitive with truck. Third, a *supply-chain approach* focuses on the role of the shippers and carriers whose collective decisions on both a strategic- and operational-level generate freight activity. Given their central role in the freight transportation system, these stakeholders are capable of making freight more efficient while reducing the direct financial cost they incur for transportation and logistics services. The discussion of the role of freight carriers providing shipment services in single or multiple modes as well as supply chain managers representing the shipper also provides background for exploration of spatial redistribution of freight patterns and increased local and regional sourcing in Chap. 13.

12-2 Introduction

As discussed in Chap. 11, the growth of freight transportation activity both internally in industrialized countries and internationally in the global economy has led to many positive economic benefits, but also an increasing burden on the natural environment and on quality of life in many communities. Current levels and year-on-year trends in overall levels of freight activity can be disaggregated into the role of different modes, with growth in truck and air-freight activity being particularly large because of their high level of service. Overall freight activity can also be disaggregated by the contribution of different commodities in terms of the freight volume that different products generate, depending on their mass, economic value, and distance from source to market.

In this chapter, we turn from describing the current situation to outlining strategies for making freight transportation and logistics more efficient and less environmentally harmful. The overall goal of improving sustainability is divided into three major approaches:

1. *Intramodal approach:* This approach looks at each mode in isolation, takes current levels of technology and best practices, and considers how sustainable transportation can be pursued by increasing levels of penetration. For example, a mode may have a current level of activity in ton-miles and energy use in petajoules (PJ) or trillion Btu; the penetration of a certain energy-efficient technology or practice may initially be at X%, and by increasing penetration, energy consumption can be reduced, other factors held constant.
2. *Intermodal approach:* This approach evaluates the relative impact of various modes in areas such as congestion or energy consumption, as well as the modal share of different modes. Where modal share may be shifting in a more energy- or impact-intensive direction, the intermodal approach seeks to understand what forces may be driving such a trend. It then seeks opportunities to shift freight volumes from more to less harmful modes.
3. *Supply chain approach:* This approach recognizes that the way businesses manage their supply chains has a strong impact on modal activity for different modes and the resulting impact on sustainability goals. At the same time, supply-chain managers have a compelling interest in improving efficiency of supply chains, for example, by reducing unnecessary vehicle movements or avoiding loss of products due to damage or spoilage. Therefore, the supply chain approach harnesses inherent motivations to improve all aspects of the triple bottom line: financial, ecological, and social.

Note that these three approaches are not mutually exclusive. Intramodal and intermodal approaches are, as the names suggest, primarily mode-based approaches to improving sustainability performance. On the other hand, the nature of specific commodities figures strongly in the supply-chain management approach, since commodity characteristics dictate required shipment speed, vehicle choice, handling requirements, and so on. It therefore follows that a shipper who is following a supply chain approach to improving performance will take an interest in both intramodal and intermodal improvements.

12-3 Intramodal Approach

The objective of the intramodal approach is to make improvements internally among each of the major modes. Among these modes, trucking stands out as having a major impact. Along with light-duty vehicles and aviation, trucking is one of the three major modes that contributes to CO_2 emissions from the transportation sector (as presented in Chap. 2, emissions from rail, marine, and pipeline are not as significant). Trucking CO_2 emissions have also been growing rapidly, at a similar rate to those of light-duty vehicles (LDVs) and aviation, in recent decades in both the United States and peer countries. Thus the discussion of the intramodal approach highlights trucking. In principle, many of the same concepts could be applied to other freight modes.

Economic sustainability concerns as well as ecological and social ones are driving intramodal changes in the truck mode at this time. The rising cost of fuel has been a major concern. Fuel cost has risen or been volatile in recent years, and when these costs rise they drive up total operating cost for trucking. As a result of these types of changes in the price regime, new technologies for saving fuel that previously were not cost-effective become viable at the margins, meaning that vehicle operators with the highest fuel costs currently reach a breakeven point where the additional up-front cost of the technology is paid back by fuel-cost savings within their target investment lifetime.

Another concern has been the increase in freight volume in general, which has translated into major growth for the truck mode. Fig. 12-1 shows growth from 1995 to 2000 for containerized freight passing through U.S. ports: in the most extreme example, the port of Los Angeles grew by 75% and that of Long Beach by 50% in just these 5 years. More broadly, the 10 busiest ports in the United States expanded their combined throughput by 45% and the combination of all major container ports by 34%. This period marked a time of rapid globalization and expansion of imported and exported products. Looking over the longer term, overall U.S. container traffic (combined imports and exports) grew from 8 million TEUs (twenty-foot equivalent units) in 1980 to 43 million TEUs in 2012. Taking the case of imports, as shipments passed through port cities inbound to inland markets, many containers are transferred to trucks for transportation to their final destination. These additional shipments added to congestion, wear and tear on the highway infrastructure, energy consumption, air pollution, and climate change. Some possible changes may address all of these concerns, such as increased efficiency in packing trucks. With the same amount of goods moved with fewer vehicle movements, all five concerns are addressed. On the other hand, some countermeasures affect only a subset of the concerns. Changes to the truck engine or body do not affect the truck's number of vehicle-miles traveled (VMT) per year, but they do reduce energy consumption and emissions per unit of VMT incurred.

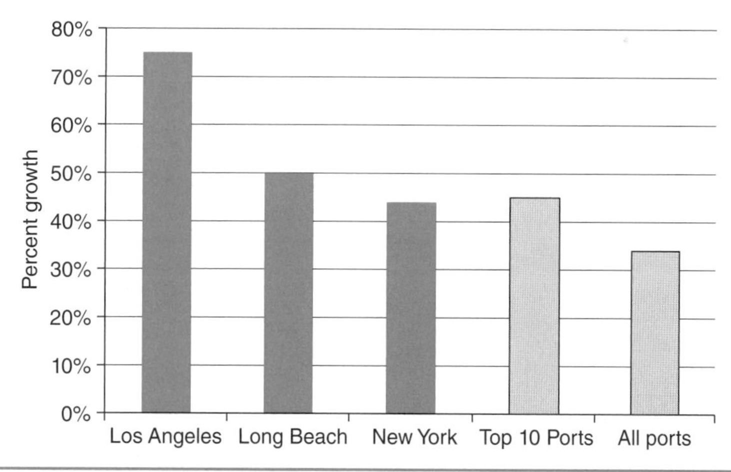

FIGURE 12-1 Growth in container shipment TEUs[1] during rapid growth phase for U.S. Ports, 1995 to 2000.

Source: U.S. Bureau of Transportation Statistics.

[1]A 20-foot container is 1 TEU, a 40-ft container 2 TEUs, etc.

12-3-1 Description of Approaches to Reducing Truck Impact

Broadly speaking, reductions in truck impact come from three areas: (1) reduction in aerodynamic or friction resistance, (2) reductions in weight, or (3) changes to operating practices (see Table 12-1). The technologies in this section focus on reducing energy and emissions, but some of the techniques for reducing impact also improve other aspects such as reducing truck noise and nuisance. Also, in many major industrial countries (e.g., Australia, Canada, European Union, and United States) the use of tractor-trailers or semi-articulated trucks, as opposed to single-body trucks, has become the primary type of truck in terms of total ton-miles of freight moved. Therefore, the discussion focuses primarily on tractor-trailers; some of the techniques also apply to single-body (also known as "rigid-body") trucks.[2]

Name of Technique	Description	Advantages and Disadvantages
Cab fairing	Modifying shape of truck cab to reduce aerodynamic drag	Little impact on truck driver experience. Already in widespread use.
Trailer fairing	Modifying shape of trailer, such as underbody in front of wheels, to reduce aerodynamic drag	Large potential for adoption.
Single-wide tires	Replacing standard double-tires on tractor and trailer with single, wider tire to reduce rolling resistance	Large potential for adoption. Operators have concerns about vehicle stability and safety in case of blowout.
Advanced lubricants	Replacing engine oil and transmission fluid with advanced compounds that reduce friction losses in drivetrain	Large potential for adoption. Operators have concerns about possible negative impact on drivetrain durability.
Vehicle weight reduction	Reducing tare weight of vehicle by replacing heavier steel with lighter substitutes (aluminum, composites)	Large potential for adoption. Up-front cost is high relative to value of fuel savings, except in cases with ancillary benefits (e.g., for chilled foods, where aluminum avoids corrosion that might harm foods).
Anti-idle strategies	Restricting extended idling of vehicles when not in motion, and enforcing restrictions and/or provide alternatives, e.g., auxiliary power units (APUs)	Operators reluctant to turn off engines when stationary. Enforcement often inadequate. Large opportunities for APU stations that provide alternative power during extended stops, e.g., at rest areas on major truck routes.
Speed reduction	Reducing total energy consumption per trip by lowering cruising speed	Additional benefits, e.g., improved highway safety. Lower speed may conflict with pressure to meet delivery deadlines or make greatest productive use of drivers' hourly wage.

Source: Adapted from Ang-Olsson and Schroeer (2002).

TABLE 12-1 Techniques for Reducing Truck Impact

[2]The reader is referred to Ang-Olson and Schroeer (2002) for a more extensive survey of energy-efficient technologies and practices.

One of the most widespread techniques currently in use is the aerodynamic truck tractor, in which the tractor is equipped with fairings and airfoils that allow air currents to pass more easily as the tractor travels at highway speeds, reducing aerodynamic drag. Because these additions do not substantially change the manufacturing cost of a tractor and because they have the potential to measurably reduce fuel consumption, over the past two decades a new generation of aerodynamic tractors has penetrated the fleet as older tractors have been retired, where other usage requirements did not prevent this transition. This transition to aerodynamic tractors is largely complete. In some instances, drivers opt for a retro look tractor that does not have aerodynamic features for aesthetic reasons, but given the high cost of fuel, these instances are and will likely remain limited.[3]

Truck trailers, on the other hand, represent a relatively untapped opportunity for reducing aerodynamic resistance and improving fuel economy. A typical trailer has an open space underneath the trailer through which air can pass when it is in motion. A system of fairings in this space can reduce drag. Similarly, the tail end of the trailer can be reshaped to reduce eddy currents generated by the typical square end of a box-shaped *van* trailer. Ultimately, the industry may evolve toward the tractor and trailer having integrated aerodynamic design so that air currents pass seamlessly from the front to the rear of the entire vehicle. Such a program would require coordination between tractor and trailer designers, since tractors typically pick up, transport, and drop off a mixture of trailers over time, rather than being continuously paired with the same trailer. (Historically, this capability was in fact one of the motivations for developing tractor-trailers in the first place.)

Along with aerodynamic resistance, truck operators can also reduce rolling resistance by converting drive wheels of tractors and trailer wheels to wide-base tire technology (Fig. 12-2). A single wide-base tire carries the same weight as two conventional truck tires, but reduces rolling resistance because its total width of tire tread touching the asphalt surface is less. Wide-based tires may also reduce lifetime maintenance cost because the overall tire system is simpler; for example, there is only one rim compared to two rims in a conventional tire system. Penetration into the U.S. market up to approximately 2011 was modest, but appears to be accelerating as of 2013. Some operators avoided shifting to wide-base tires in the past because of a perception that they are unsafe in the event of a tire blowout. Take the case of a conventionally equipped trailer with eight tires at the rear: it is reasonable to expect that only one tire will fail at a time, and in the event of a tire failure, the remaining seven tires can easily support the weight of the trailer until the next opportunity to replace the failed tire. With only four tires, the failure of one was of greater concern.

Advanced lubricants used in truck engines and transmissions provide another opportunity for fuel savings by reducing friction. As with wide-base tire technology, there is interest in the potential financial savings but concern about the reliability of these lubricants and about possible premature component may fail if operators switch from conventional versions.

---

[3]The impact of fuel cost on trucking industry decision-making can be better understood by comparing it to the experience of a private car driver. A car driving 10,000 miles/year at 20 mpg (miles per gallon) will consume 500 gal of gasoline, which at $4.00/gal would cost $2,000/year. A truck driving 96,000 miles/year at 6 mpg will consume 16,000 gal of diesel, which at $4.50/gal would cost $72,000/year.

(*a*) Conventional tire

(*b*) Wide-base

FIGURE 12-2 Tire configurations on truck trailers: (*a*) conventional configuration with four tires, (*b*) wide-base tire configuration with two tires.

Reducing the tare weight of the tractor and empty trailer can also reduce operating cost. Take the case of a trailer that is fully loaded to the legal gross weight limit, also known as *weighing out*, that is, the weight of loaded freight is the limiting factor that prevents the loading of additional cargo. In such a case, a lighter weight trailer (reduced tare weight) would allow the truck to carry a larger amount of cargo without exceeding the limit. This change would in turn allow the same total weight of goods to be moved over an extended period of time such as a month or a year with fewer vehicle movements. Two factors limit its greater adoption. First, the additional capital cost can be substantial. A van trailer (that is, a rectangular fully enclosed trailer typically on the order of 10 ft or 3 m high, the length of the trailer, as opposed to a flatbed or tanker trailer) built of aluminum rather than steel increases the gross weight that can be carried, but costs enough more than steel that operating cost savings will have difficulty repaying the initial investment. Here a government incentive program to assist with upgrading to aluminum, or a carbon tax that favors energy-efficient alternatives, can make a difference. The second factor is the tendency of some shipments to *cube out* before they weigh out, that is, before the maximum weight is reached the volume of the product fully occupies the space available so that no further cargo can be added. Low-density items such as certain foodstuffs or textile products are subject to cubing out. In such a situation, the lighter-weight trailer will not have much effect on energy efficiency.

Best Practices to Reduce Impact

Anti-idling programs represent one type of best-practice solution for reducing truck impact. Heavy-duty vehicles such as trucks are at times required to idle for a short time, for example, while stopped in traffic, but once they have reached a stationary point, for example, when loading or unloading or while the driver rests, turning the engine off can save energy and address local noise and air quality concerns. At the same time, the driver may wish to continue to idle the engine to maintain power supply for ancillary loads such as electronics or cab climate control, or to avoid difficulties when restarting. Both a carrot and stick approach to reducing excessive idling are in use: provision of off-board alternative power stations at rest areas and other locations frequented by long-distance truckers, and local ordinances that ban idling beyond a short time limit such as 5 minutes. For the latter, enforcement remains an issue; truck drivers and other heavy-duty vehicle operators ignore ordinances if they perceive that there is no consequence for noncompliance.

Last among the techniques in Table 12-1, *speed reduction* entails lowering the maximum highway speeds attained to conserve fuel. Although the truck will then spend more time en route, the extra time during which fuel is consumed is more than offset by the lower fuel consumption rate, leading to fuel savings. Note that this relationship does not hold at lower speeds such as urban stop-and-go traffic or congested highway conditions, where reduced speed can actually lead to both longer duration of the trip and increased fuel consumption. Speed reduction by 5 or 10 mph (8 to 16 km/h) does require greater planning and may conflict with logistical requirements for delivery deadlines or the desire of the firm to hold down wages, if slower travel speeds lead to more hours of work required of drivers. Optimization of gear ratios can also help. For each gear in the truck's gearbox, there is an optimal driving speed at which efficiency is maximized. Gear ratios should be chosen so that regardless of the chosen cruising speed (e.g., 65 or 70 mph), there is an available gear ratio that is optimal for that speed.

Example 12-1 illustrates the effect of speed reduction on travel time and energy consumption.

Example 12-1S Suppose a combination truck consumes energy embodied in diesel fuel at the rate of 1.365 million Btu/hour at a speed of 70 mph. Diesel fuel has an energy content of 128,700 Btu/gal. Reducing speed to 65 mph reduced energy consumption rate by 13%. For a 70-mile trip, and assuming the vehicle maintains constant speed, what is the quantity of energy saved in kWh? How many gallons of fuel are saved?

Solution Since the trip is 70 miles and the original speed is 70 mph, the trip will take exactly 1 hour, and the energy consumption will be 1.365 million Btu. The reduction in energy consumption rate @ 65 mph is

$$1.365 \times 10^6 \text{ Btu/h} \cdot 0.87 = 1.187 \times 10^6 \text{ Btu/h}$$

The time required for the journey at 65 mph is

$$\text{Time} = \frac{70 \text{ mi}}{65 \text{ mph}} = 1.08 \text{ h}$$

The energy consumption at the new speed is

$$\text{Energy} = (1.187 \times 10^6 \text{ Btu/h})(1.08 \text{ h}) = 1.282 \times 10^6 \text{ Btu}$$

Therefore, the reduced speed reduces energy consumption by 82,400 Btu or 6.3% and increases travel time by 8%. This shows how increased travel time undercuts the 13% reduction of instantaneous power required.

Based on energy content of 128,700 Btu/gal, the truck in the base case uses 10.61 gal and in the improved case 9.96 gal. Therefore the savings constitute 0.64 gal of diesel. Note that the fuel economy improves from 6.6 to 7.0 mpg.

Example 12-1M Suppose a combination truck consumes energy embodied in diesel fuel at the rate of 400 kW at a speed of 112 km/h. Diesel fuel has an energy content of 35.87 MJ/L or 9.96 kWh/L. Reducing speed to 104 km/h reduced energy consumption rate by 13%. For a 112-km trip, and assuming the vehicle maintains constant speed, what is the quantity of energy saved in kWh? What is the fuel savings in liters?

Solution Since the trip is 112 km and the original speed is 112 km/h, the trip will take exactly 1 hour, and the energy consumption will be 400 kWh. The reduction in energy consumption rate at 104 km/h is calculated as shown:

$$\text{Rate} = (400 \text{ kW})(0.87) = 348 \text{ kW}$$

The time required for the journey at 104 km/h is

$$\text{Time} = \frac{112 \text{ km}}{104 \text{ km/h}} = 1.08 \text{ h}$$

The energy consumption at the new speed is

$$\text{Energy} = (348 \text{ kW})(1.08 \text{ hour}) = 375 \text{ kWh}$$

Therefore, the reduced speed reduces energy consumption by 25 kWh or 6.3% and increases travel time by 8%. This shows how increased travel time undercuts the 13% reduction of instantaneous power required.

Based on an energy content of 9.96 kWh/L, the truck in the base case uses 40.14 L and in the improved case 37.63 L. Therefore the savings constitute 2.51 L of diesel. Note that the fuel economy improves from 35.84 to 33.60 L per 100 km.

Education as a Means of Increasing Effectiveness

The professional driver is central to efficient truck operation, so not only best practice but also technological approaches benefit from driver education. Anti-idle and speed-reduction education can help the driver to operate the truck without needing to idle for extended periods of time, or to accelerate, cruise, and decelerate in the most efficient way possible, and without exceeding the desired speed limit of the shipper or carrier. Education about adaptations to equipment can also help. For instance, awareness about advanced lubricants can keep the driver alert to any changes in maintenance practices that might be required. Training around energy savings devices can make sure that these are kept in good working order.

Extension to Other Modes besides Truck

The focus in this section has been on trucks because they are such important freight energy consumers. In the case of the United States, trucks consumed 6 quads of energy in 2008 compared to 0.6 quads and 0.3 quads for rail and marine, respectively. However, many of the same approaches can in principle be applied to the two other modes. Since these other two modes have in common with trucks the diesel engine and at least partially mechanical drivetrains (diesel rail locomotives use electricity for a portion of the transmission from engine block to wheels, hence the name *diesel-electric locomotive*), advanced lubricants can improve fuel economy. Reducing speed where shippers' demands allow can also reduce fuel consumption. One limitation is that rail and marine already have, on average, relatively low-energy consumption per average ton of goods moved, so it may be difficult to justify further reductions in fuel consumption on economic grounds.

In the case of the marine mode, there are some exceptions to the rule of inherently low energy intensity per ton-mile or tonne-kilometer. One example is the "roll-on roll-off" ferry (also called RoRo) with mixed passenger car and truck traffic. Since it is in the commercial interest of ferry operators to travel at higher speeds of 13 knots or more to attract car traffic, trucks using this service end up with higher average energy intensity than general bulk or container traffic on dedicated freight ships.[4] Another example is the "fast ship" concept for providing container shipping on either coastal or trans oceanic routes, which would fill the market between conventional freight ships and airfreight by using advanced hulls to travel at speeds up to 37 knots. First generating interest in the 1990s, fast ship exists today in concept only at this point, due in large part to continued high fuel costs and concern about economic competitiveness. However, if commercialized, fast ships would lead to higher marine freight energy intensity, although lower than airfreight for a given origin-destination pair.

12-3-2 Implementation of Intramodal Improvements: Penetration Issues

From the discussion in the preceding section, it is observed that a technique such as aerodynamics or weight reduction might yield a certain percent reduction in VMT or energy use. This information is useful at the level of the individual vehicle or firm operation. From a public policy perspective, however, we are also interested in the impact on total emissions over time. Specifically, three aspects are of interest:

- *Maximum penetration possible:* Since most techniques are limited to a subset of the entire truck fleet on the road and will not be relevant to the remainder, an estimate of the percent of the total fleet is useful for calculating maximum possible savings. For example, APUs are not relevant to trucks that engage in short-distance haulage where they will have no need to stop at a rest area along an interstate.

---

[4]Conversion: 1 knot = 1.15 mph = 1.84 km/h.

- *Maximum rate of penetration:* Along with the eventual maximum penetration of the technology, the pathway of the technological penetration and year-on-year changes in impact are important. Among other things, the predicted rate of change must be realistic, or, to quote thc classical economist Alfred Marshall, "nature does not make jumps," and by extension, neither can economic activity. For example, a technology cannot be substituted into the fleet faster than the industry that manufactures it can supply that technology when working at maximum output.
- *Background changes in total fleet activity:* The number of vehicles in the fleet and their average amount of activity per vehicle (e.g., VMT/year) will change over time, and this will affect the background amount of impact against which to measure progress thanks to techniques introduced. One concern is that growth in total fleet activity will wash out improvements in level of impact per unit of activity, so that no net progress is made toward overall reduction in impact. Even in such a situation, there is some benefit to the penetration of the technology, since the outcome is still better than the "no-change" case where the total activity measured in VMT increases and impact per unit of activity stays constant. However, the expectation of net progress will not be met.

Case Study of Implementation: Auxiliary Power Units

As an illustration, in this section we study the possible penetration pathway of an impact reduction technology, namely, APUs, against the background of the United States national truck fleet. This technology is not currently in frequent use but has strong potential for penetration because of the benefits to both the public and the individual truck operator. The technique to be used is the technological substitution curve based on the logistics curve (see Chap. 3 for background). Recall that this curve has the shape of form:

$$f = \frac{F \cdot e^{(c_1 + c_2 t)}}{1 + e^{(c_1 + c_2 t)}}$$

where t is the number of years elapsed since the start
f is the percent penetration in year t
F is the ultimate penetration
c_1 and c_2 are curve-fitting parameters.

Example 12-2 shows the calculations.

Example 12-2 According to the Oak Ridge National Laboratories, there were approximately 2.5 million combination trucks in use on U.S. roads in 2010. Of these, 1% were making regular use of off-board auxiliary power units in place of engine power for ancillary power requirements (climate control, in-cab electricity, etc.), because at that point the technology was quite new. Suppose the levels for years 2011 to 2013 are 1.8%, 2.4%, and 4%, respectively. The ultimate penetration thought to be possible for all vehicles is 38%. (a) Using the penetration achieved in years 2010 to 2013 with year 2009 as $t = 0$, and the ultimate penetration of 38%, project the penetration of the strategy into the future and identify the year that it has captured 95% of the ultimate penetration value. (b) In what year will the strategy achieve the maximum rate of increase in penetration, and what is the increase in penetration in that year, measured in number of vehicles?

Solution

(a) Since there are 2.5 million trucks total, it is expected that 38% or 950,000 vehicles will eventually make use of APUs. The desired threshold is 95% of this figure, or 902,500 vehicles.

The given data for years 2010 to 2013 and for the ultimate level of penetration must be fit to the substitution curve introduced previously. The best-fit values of parameters c_1 and c_2 are found using a computer solver, whose use is demonstrated here.

An initial table is created using values $c_1 = -4$ and $c_2 = 0.5$ to create estimated values as a starting point for calculating the optimal parameter values. A table is created with one row for each of the years 2010 to 2013:

Year	t-value	Observed	Modeled	(Error)2
2010	1	0.01	0.01114	1.297E-06
2011	2	0.018	0.01802	4.766E-10
2012	3	0.024	0.02883	2.329E-05
2013	4	0.04	0.04530	2.806E-05

The column for "Modeled" values uses the substitution curve equation to find the appropriate value. For example, for the year 2010 when $t = 1$, the following value is returned:

$$f = \frac{0.38 \cdot e^{(-4.0+0.5\cdot 1)}}{1+e^{(-4.0+0.5\cdot 1)}} = 0.0111 = 1.11\%$$

The root mean square deviation (RMSD) based on the values in the right column is 0.003628. The electronic solver takes the RMSD as its objective and seeks values of c_1 and c_2 that minimize RMSD. Running the solver gives values of $c_1 = -4.0449$ and $c_2 = 0.4727$, which reduces RMSD to 0.001187. The new values based on improved parameter values are given in the following table:

Year	t-value	Observed	Modeled	(Error)2
2010	1	0.01	0.01038	1.474E-07
2011	2	0.018	0.01639	2.597E-06
2012	3	0.024	0.02562	2.640E-06
2013	4	0.04	0.03950	2.495E-07

Extrapolating the modeled curve outward shows that in the year 2023 there are 877,300 vehicles using the APUs, and in 2024 the figure has grown to 903,100 vehicles. Thus 2024 is the first year in which the threshold is exceeded.

(b) Number of vehicles penetrating the market is equivalent to the difference between the number of vehicles in a year with the number in the year before. For example, between 2023 and 2024, the difference is 903,100 − 877,300 = 25,800 vehicles. The maximum is found to be 110,000 vehicles between years 2018 and 2017, using the same approach.

Extrapolating the fit curve out to the future by which time most of the penetration would have taken place gives the curve shown in Fig. 12-3. The curve shows the typical penetration curve of "Slow start" in the early years of 2010 to 2014 in this case, "Rapid uptake" in the middle years from approximately 2014 to 2021, and asymptotic growth at the end from 2021 onward. Note also that the period of maximum increase from (b) in Example 12-2 is occurring in the period 2017 to 2019 when penetration grows from 16% to 24% and is growing at the rate of 4% per year.

One question that the penetration curve can help answer is whether or not the maximum penetration per year is realistic. As an extreme example, it is not realistic to add enough APUs in a single year to support all 950,000 vehicles eventually expected to use them. Instead, a ramping up of the installation process is more realistic. A further investigation of

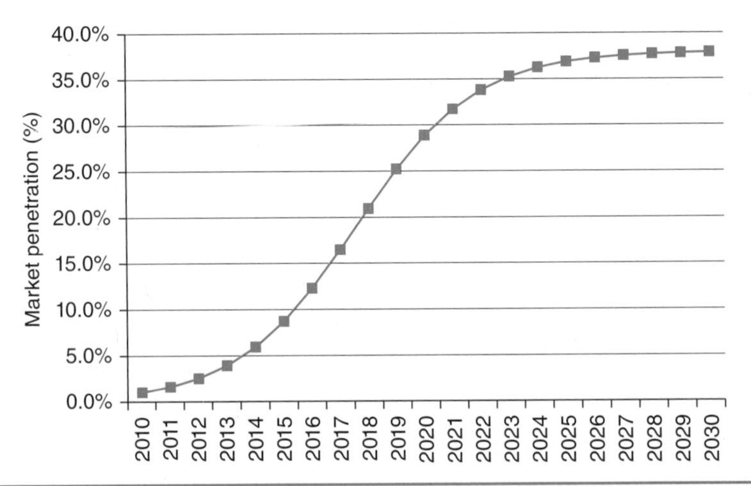

FIGURE 12-3 Projection of market penetration of APUs for period 2010 to 2030 from Example 12-3.

the manufacturers that provide the APUs would reveal whether or not they had the capacity to add APUs for 110,000 additional trucks, as calculated in Example 12-2 for the period 2017 to 2018 at the point of maximum growth.

Lastly, some of the limitations of Example 12-2 should be noted. Only the number of vehicles affected, and not their impact on overall energy consumption, were calculated. Length of truck idling time is a nonuniform characteristic so the amount of idling avoided is unknown; furthermore, the types of trucks beginning to use APUs may change over the period of penetration, so the average benefit changes as well. Also, the number of vehicles was assumed fixed; a more complete example would take into account the changing number of vehicles on the highways. As an indication of the extent of change possible in the U.S. vehicle fleet, the number of combination trucks grew from 2.1 to 2.5 million during the period 2000 to 2010.

12-4 Intermodal Approach

The discussion in the preceding section assumed that each mode would minimize energy consumption in isolation, and that there would not be shifting of freight among the different modes. Making intermodal changes can, however, bring further gains. As a rough indicator, consider that the average energy consumption for truck, rail, and water modes in the United States in 2006 was 4,542 Btu, 330 Btu, and 571 Btu per ton-mile, respectively. These values must be interpreted carefully, since they reflect not only relative technical efficiencies of different modes but also the type of products carried; truck, in particular, is penalized because it must carry the bulk of the high-value, time-sensitive products. Nevertheless, these values give some indication that significant savings could be achieved by shifting away from truck to other modes.

The most important concept for diversifying modal choice today is the concept of *intermodalism*, or the creation of a seamless freight transportation service that delivers the shipment from origin to destination with a high level of service (LOS) while using different modes for different segments of the journey where each mode has a comparative advantage. The freight industry has come to recognize that, over long distances, modes such as rail (and in certain situations water) have attractive advantages in terms not only of energy efficiency but also reduced cost and labor requirement. At the same time, most shipments today must at some point move by truck, since trucks provide the most convenient access to origins and destinations. Some locations, such as large manufacturing plants or mining facilities, have direct access to the rail and/or water network. However, in today's dynamic economy, locations change rapidly, and it is much more practical to link a new facility to the road network than to the rail or water network. Also, for small- and medium-sized facilities, it is often more financially attractive to both the business and the transportation operators to send shipments via truck to the nearest transit point to the rail or water modes, rather than incur the high cost of building a dedicated rail or water link.

A number of systems exist to transfer shipments from the road to other modes at the intermodal transit point, as follows:

- *Roll-on roll-off:* The single-body truck or tractor-trailer enters and leaves the railcar or marine vessel as a complete unit.
- *Loading of truck trailers:* The trailer is separated from the truck tractor and loaded onto the railcar (also known as "piggyback" rail service). Separate loading of trailers onto marine vessels is less common, though possible.
- *Loading of shipping containers:* When arriving by road to a rail or marine intermodal facility, the chassis (wheeled underbody that allows a shipping container to move over the road) is removed and then the container is loaded onto the railcar or marine vessel. Direct loading or unloading between rail and marine is also possible. Modern intermodal railcars allow *double-stacking* of containers on railcars to maximize the productivity of each train.

On the basis of these different systems, a wide variety of applications are possible. For example, between Salerno, Italy, and Valencia, Spain, the European Union has been supporting the development of a roll-on roll-off service aimed at trucks that would allow them to avoid driving along the perimeter of the Mediterranean Sea via France between the two countries. Although a motorway exists along the land route, it is circuitous and passes through a number of highly populated areas, so it is desirable to transfer some truck traffic to the seas. Elsewhere, the Swiss government has developed a network of roll-on roll-off trains for trucks to allow them to transit the Alpine region without driving. The main driver of this program is the reduction of air pollution in an environmentally sensitive region, but there are also CO_2-reduction benefits, since the Swiss railways are almost entirely electrified and Swiss electricity comes from hydro- and nuclear power.

One of the largest intermodal freight shipping operations in the world today is the shipping of containers and truck trailers in continental North America, along the rail networks of Canada, the United States, and Mexico (See Fig. 12-4). In this market, some freight shipments move very long distances (1,200 to 2,000 miles or more) to and from population centers in the interior to shipping ports along the Atlantic and Pacific. Over these long distances, it makes financial sense to bundle a large number of shipments onto a single train so as to save energy and labor costs, as a single double-stack train with crew of three in the locomotive can replace 300 or more trucks each carrying one container.

FIGURE 12-4 Double-stack intermodal train carrying shipping containers.
Source: BNSF Railroad. Reprinted with permission.

12-4-1 Reasons for Improved Performance from Environment-Friendly Modes

Reductions in environmental impact from eco-friendly modes come from moving a given shipment via rail or water instead of over the road, reducing congestion, road wear and tear, noise and vibrations, energy consumption, air pollution, and contribution to climate change on each tonne-kilometer of movement. Intermodal shipments sometimes require truck movements between the origin and destination and the intermodal terminal that are not necessary when the shipment moves by truck directly from origin to destination using the intercity highway network. Nevertheless, for most long-distance movements where intermodal transportation is financially competitive with direct shipment by truck, the energy savings can be substantial.

The rail and water modes achieve these gains in several ways:

1. *Lower resistance to motion*: For either rail or water, at typical line haul operating speeds there is reduced friction against the surface. For rail, the steel wheel on steel rail encounters less rolling resistance than a rubber tire on asphalt or concrete, saving energy. Similarly, a boat passing through water encounters less friction than a tire. This is especially true for large ships; as the size of the ship increases, the energy consumption per ton-mile or tonne-kilometer decreases. Also, ship energy savings comes from travel at relatively slow speeds; high-speed ships such as those used by the military consume much more energy per gross ton-mile or tonne-kilometer moved. The advantage of the truck tire on pavement is that it can climb steep grades, turn much tighter radius corners, and steer on any part of the paved surface (the lack of a fixed guideway allows the driver to steer around congestion, bypass vehicles that have broken down in the travel lane, and so on). This flexibility is achieved, however, at the expense of greater frictional losses.

2. *More level trajectory over intercity distances*: Rail and water avoid large grades for which energy must be expended climbing and then other energy lost to friction when descending. This is especially true of inland waterway travel: for example, from its source, the Mississippi River in the United States descends just 1,475 vertical ft over 2,530 miles to the Gulf of Mexico, and for the final 1,183 miles from St. Louis, just 465 ft. Rail also climbs and descends only relatively mild grades (unless it is equipped with a cog railway system, which is rarely used for freight) and when descending is capable of using *regenerative braking* to generate electrical energy which can then be reused, especially in the case of hybrid or all-electric locomotives. One negative consequence of the relatively small changes in grade for rail and water is the extra *circuity* involved in routing via rail or water from origin to destination compared to truck.[5] Even after factoring in circuity, however, significant savings are possible.
3. *No interference from cross traffic:* Although trains or ships may occasionally encounter congestion in the line where they are forced to idle, thus wasting fuel, the amount of unexpected traffic encountered en route is small relative to truck because the rail or water corridor is completely controlled by the operator. Entry and exit into a rail mainline is controlled by a central dispatcher, so once it is underway, a train will not encounter another train unexpectedly unless there has been a breakdown in the system, the way a truck driver might unexpectedly encounter congestion due to a traffic accident that would lead to both a delayed arrival and wasted fuel. Marine operations on inland rivers and canals are analogous, and obviously not subject to constraints on large open bodies of water such as lakes and seas.

Downside of Eco-friendly Modes: Complexity of Coordination

The advantages of the eco-friendly modes of rail and marine (larger volume of freight per train or per ship, less resistance to motion, less congested line haul pathway, and so on) come at the expense of greater complexity in consolidating freight into a single shipment at the beginning of a line haul and then deconsolidating it at the end. The requirement for additional complexity is inherent in the fundamental nature of these modes; it is not possible to have the efficiency gains without the complexity increase.

Lack of complexity also explains some of the appeal of the truck mode. If it were only a matter of efficiency and direct cost, rail and marine should be able to dominate truck, since they require both lower energy and labor costs. However, truck is also relatively quick and reliable: trucks can reach any shipper and any destination (consignee), and many truck shipments travel directly door-to-door. Truck LOS can be compared with that of rail or water, which may involve *drayage* to bring freight by truck to and from rail or water terminals, as well as consolidation during the course of the journey. Rail in particular may require that a container or railcar be de-coupled at an intermediate *classification yard* to be transferred from one train to another before it reaches its final destination in the network. This complexity leads to slower average delivery time and also unpredictability in the delivery time, both of which hurt LOS compared to truck.

---

[5]*Circuity* is quantified as the ratio of actual distance traveled to straight-line distance. For example, if two locations are separated by 100 miles in a straight line but the rail route between them is 150 miles long, the circuity factor is 1.5.

One opportunity for rail and marine modes to become more competitive with truck therefore is to improve their operations so that they become faster and more reliable. Since line-haul cruising speeds for both ships and rail locomotives are typically limited by the technology, terminal operations are an especially good opportunity to make improvements. If shipments can be moved quickly through terminals without loss or damage, overall improvements in LOS can be achieved.

Example of Calculation of Power Requirements for Rail Mode

To further explore potential benefits of alternative modes to truck, in this section we take the case of the rail mode and present formulaically the power requirements based on mass and speed. These power requirements can in turn be used to calculate energy consumption over a line haul of a certain distance, fuel consumption, and emissions.

In this presentation, we build on examples from Chap. 5 regarding the impact of aerodynamic and rolling resistance on vehicles generally to consider the specific case of freight railroads. In this case, resistance of both rail cars and locomotives are evaluated in U.S. customary units and are a function of car weight and resistance. Let R = resistance in pounds, T = weight in tons, V = speed in mph, N = number of axles on locomotives (does not apply to railcars). The resistance of a railcar on level surface traveling at constant speed is then

$$R_{\text{CAR}} = aT + b + cTV + dV^2 \tag{12-1}$$

For a representative railcars, values a = 1.5, b = 72.5, c = 0.015, and d = 0.055 are appropriate. Next, for the locomotive, the following equation applies

$$R_{\text{LOCOMOTIVE}} = aT + bN + cTV + dV^2 \tag{12-2}$$

For a generic locomotive, representative values are a = 1.3, b = 29, c = 0.03, and d = 0.0312. Lastly, the maximum tractive effort (TE) of the locomotive is measured in pounds and is a function of the maximum horsepower P of the locomotive and V; as the locomotives speed increases, the maximum tractive effort decreases according to

$$TE = aP/V \tag{12-3}$$

where the value a = 308 is representative. Note that Eq. (12-3) is not intended for calculations at very low values of V, since as V approaches 0, the equation suggests that TE approaches infinity, when in fact it remains roughly constant at a high value in this range of V values. However, since we are generally interested in steady-state energy consumption at line haul speeds, this limitation is not a concern here.

From a railroad operations perspective, the key planning criterion is to provide sufficient locomotives capable of delivering sufficient value TE at desired speed V to overcome the total resistance calculated using Eqs. (12-1) and (12-2). The process entails adding locomotives until the total value TE exceeds the required amount of force. Excess *TE* beyond what is required is not a concern and is actually useful since climbing grades or acceleration increases power requirements. Indeed, Eqs. (12-1) and (12-2) could be extended by including the effect of instantaneous grade (ascending or descending) and acceleration (positive or negative). Example 12-3 illustrates these calculations for a representative freight train.

Example 12-3 A freight trains consists of 50 freight cars, each weighing 30 tons empty and carrying a load of 90 tons (typical of, e.g., a fully load hopper car for carrying coal), at 50 mph on a level track.

The train is to be pulled by a string of locomotives each with maximum power of 2, 500 hp, weight of 120 tons, and 4 axles. (a) What is the total resistance of just the pulled load of the train, not including the locomotives, in pounds? (b) What is the minimum number of locomotives required to pull the load at the desired speed, and the resulting total resistance of the train in pounds? Assume all parameter values for Eqs. (12-1) to (12-3) as given in the passage.

Solution Equations are shown with parameter values substituted.

(a) The resistance per car is calculated using Eq. (12-1) based on $V = 50$ mph and $T = 120$ tons total for the combined car and freight:

$$R_{\text{CAR}} = 1.5\,T + 72.5 + 0.015\,TV + 0.055\,V^2$$
$$= 1.5(120) + 72.5 + 0.015(120 \cdot 50) + 0.055(50)^2 = 480 \text{ lb}$$

Therefore, the resistance for all 50 cars is (50)(480) = 24,000 lb.

(b) The resistance per locomotive is calculated using Eq. (12-2):

$$R_{\text{LOCOMOTIVE}} = 1.3\,T + 29\,N + 0.03\,TV + 0.312\,V^2$$
$$= 1.3(120) + 29(4) + 0.03(120)(50) + 0.312(50)^2 = 1232 \text{ lb}$$

The tractive effort at $V = 50$ mph per locomotive is calculated as

$$\text{TE} = 308(2500)/50 = 15,400 \text{ lb}$$

From further calculations, a train with one locomotive has a resistance of 24,000 lb + 1,232 lb = 25,232 lb and maximum tractive effort of 15,400 lb, and will therefore not be able to travel at 50 mph since the resistance exceeds the available tractive effort. A train with two locomotives, however, has a resistance of 26,464 lb and TE = 30,800 lb, so that it will be able to attain the required speed. Therefore, the number of locomotives required is 2, and $R_{\text{TOT}} = 26{,}464$ lb.

Equations (12-1) to (12-3) can be used to calculate resistance in steady-state operations and also to compare energy requirements for rail to other modes. The ratio of maximum power to total weight pulled is called *horsepower per trailing ton* and is used as a measure to indicate the likely maximum speed or ability of a train to accelerate rapidly once it has been dispatched for the line haul. For instance, in Example 12-3, the train has 5,000 hp total available and is pulling 6,000 tons, and therefore has 0.833 horsepower per trailing ton.[6]

Comparison of Rail and Truck Power Requirements

Example 12-3 can next be extended by comparing the results for rail freight to equivalent movement by truck. In this case, we divide the overall load on the train into individual loads suitable for carrying by either rail or truck, and then compare power requirements per load. The necessary calculations are presented in Example 12-4, where inputs and results from Example 12-3 have been converted to metric for convenience of using the tractive power formula for road vehicles.

Example 12-4 Compare the tractive power requirement on a per-load basis for the diesel train in Example 12-3 with an articulated truck (tractor-trailer) with the following parameters: weight empty 15 tonnes, 9.2 m² cross-sectional area, drag coefficient of 0.4, and rolling resistance coefficient of 0.006. Use the tractive power formula from Chap. 5. Assume air density of 1.1 kg/m³. For converting quantities from standard to metric, the 50 mph speed of the train is equivalent to 22.2 m/s, and using

[6]For more information about railroad operations and powering of freight trains, interested readers seeking a full-length work on the topic are referred to Armstrong (1993).

a conversion of 4.448 N per pound of force, the force required to maintain the speed is 117,712 N. Assume that each load of 90 tons per traincar is divided into 4 loads of 22.5 short tons or 20.45 tonnes. (a) What is the tractive power required per load to move the freight by rail, in kW? (b) What is the same power requirement by truck, in kW?

Solution

(a) Since there are 50 cars and 4 loads per car, there are a total of 200 loads on the train. Using the relationship power equals force times velocity:

$$P_{\text{PerLoad}} = \frac{(117{,}712\ \text{N})(22.2\ \text{m/second})}{200} = 13.1\ \text{kW}$$

Therefore each unit of load on the train requires 13.1 kW power to maintain 22.2 m/s.

(b) The gross weight of the truck is 15 + 20.45 = 35.45 tonnes. Other needed quantities are given. Tractive power is calculated using the appropriate equation:

$$\begin{aligned} P_{\text{TR}} &= 0.5\rho A_F C_D V^3 + mgVC_o \\ &= 0.5(1.1)(9.2)(0.4)(22.2)^3 + (35{,}450)(9.8)(22.2)(0.006) \\ &= 68.5\ \text{kW} \end{aligned}$$

Therefore a truck carrying one unit of the load at 22.2 m/s requires 68.5 kW of power. Thus the truck is slightly more than 5 times as power-intensive as the train.

This example supports the observation first introduced in Chap. 11 that trains can significantly reduce energy consumption per unit of freight moved. The comparison is between just one of many possible combinations of vehicle weights, freight densities, and travel speeds, however, so a broader investigation would be needed to predict the average efficiency advantage of rail over truck across the freight system as a whole. In this instance the circumstances favored the train, since most of the weight moved (82 out of 109 tonnes in each railcar) is net as opposed to tare weight, whereas for the truck the ratio is not as favorable. The large mass of freight carried per railcar suggests a dense bulk commodity for which rail is especially well suited. In the case of low-density, high-value consumer goods the weight per load might be on the order of 5 to 10 tonnes, but the tare weight for the train might be largely unchanged. In such a situation, the efficiency of the train would not be as high compared to the truck.

As a simplification, Example 12-4 also looks only at the tractive power required from the vehicle to maintain the desired speed. To calculate energy content required in fuel consumed needs information about the efficiency of the drivetrain from diesel fuel on board the locomotive or truck tractor through the internal combustion engine (ICE) and drivetrain to power delivered to move the vehicle down the road or track. For both rail and truck modes, the losses from chemical energy in unburned fuel to kinetic energy of the moving vehicle are substantial, especially due to the thermodynamic losses in the ICE.

Note also that the calculation of tractive power per load for rail deliberately incorporates the impact of locomotives on resistance force by apportioning equally the force incurred by the total number of locomotives among the loads carried. This approach is more accurate than simply looking at the resistance force per car or per carload, because just as the truck tractor is an integral part of the vehicle in the case of road freight, the rail load cannot move without the needed locomotive(s), so they are essential in the calculation. Furthermore, a train with high LOS requirements may require more horsepower per trailing ton, which will add to the resistance force per load and increase tractive power required, all other things equal.

12-4-2 System-Wide Impact of Modal Shifting: Example of Intermodal Rail

We now shift from the power requirements of the individual freight train in the preceding section to the potential impact of shifting large volumes of freight to the rail mode as a whole. In this case, we focus on shifting from truck to intermodal rail (IM), as it is the mode that most closely resembles truck in terms of its LOS.

The popularity of IM in recent years is observed in Fig. 12-5. Between 1990 and 2012, carload freight (i.e., where the freight is loaded directly into the railcar, as opposed to IM where either the container or trailer filled with freight is loaded on or off a specialized IM car) grew by 35%, but IM grew by 105%. The number of IM loadings in the figure is the sum of trailers and containers (where each type counts for one loading) but increasingly in the North American market the unit of choice is the shipping container, rather than the complete trailer with wheels attached. When transferred off the IM train and onto the road, shipping containers require an undercarriage called a *chassis* to be transported by a truck tractor. Since the freight industry has developed experience with this technology, an increasing number of trailers for domestic as well as international shipping have a removable chassis, allowing them to be shipped in dedicated railcars.

Several important innovations have accompanied the growth of IM, some of which are illustrated in Fig. 12-6. The *well car* allows the double-stacking of containers, since the lower container in the stack rides as low to the rails as possible so that a second container can be stacked on top without interfering with overhead bridges or other structures

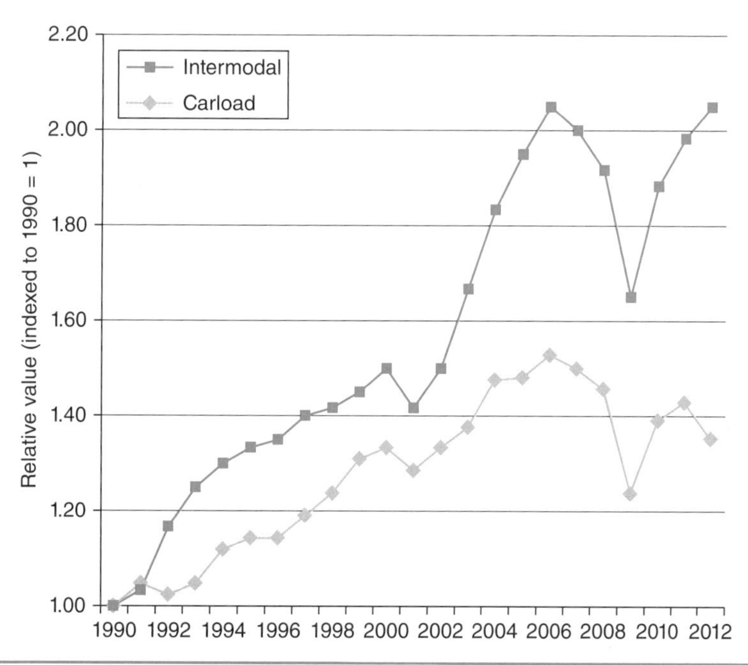

FIGURE 12-5 Indexed growth in carloads originated and intermodal trailers and containers shipped, 1990 to 2012, indexed to 1990 = 1.00.

Note: In 1990, intermodal value was 6 million containers and trailers, and carload value was 21 million loads.

Source: American Association of Railroads.

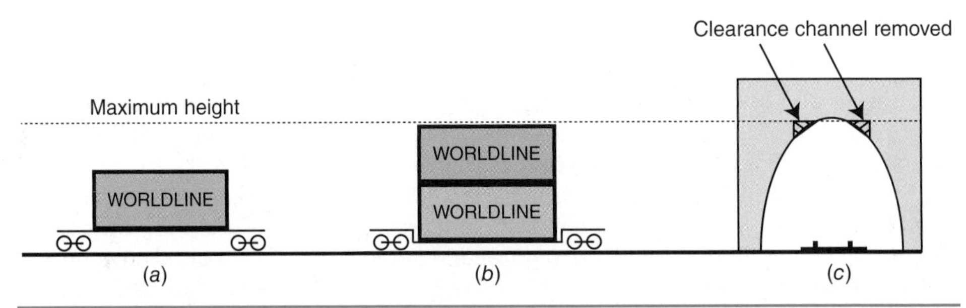

FIGURE 12-6 Aspects of intermodal container-on-rail service: (*a*) single-stack conventional flatcar, (*b*) double-stack "well" car, (*c*) tunnel modification for double-stack clearance.

along the rail route. Where necessary, *doublestack clearance projects* have been undertaken to widen or raise any obstacles that might interfere with a doublestack train that uses well cars, since in some cases even the well car does not allow the train to pass obstacles in their original form. For example, a tunnel may not have high enough roof clearance to allow a doublestack railcar to pass, so channels can be cut in the roof of the tunnel along its entire length to prevent contact, as shown in the figure. As an additional adaptation to provide an attractive LOS, multiple cars are coupled into a single articulated unit to reduce the mechanical shocks that would otherwise pass through the train as it accelerates and decelerates. A train of articulated well-car units with approximately 30 containers visible in the photograph is being pulled by the set of locomotives in Fig. 12-4.

Example 12-4 illustrates the potential energy savings available when a measurable fraction of loads are carried by IM instead of by truck. Note that in this example, three possible intensity values are used for IM, namely low-, mid-, and high-intensity. Since truck energy intensity is fixed in the example and the IM low-intensity option has the lowest energy intensity per tonne-kilometers or ton-mile, the low-intensity option results in the highest energy savings, and conversely the high-intensity option results in the least savings.

Example 12-4S: In the year 2000, U.S. trucks averaged 3,337 Btu of energy per ton-mile moved. Published estimates of average energy consumption across all intermodal shipments were not directly available, but a plausible range for this intensity value is 1,200 to 2,305 Btu/ton-mile. Suppose in that year approximately 42 billion ton-miles that would have moved by long-distance truck instead move by intermodal. What is the energy savings?

Solution Taking the case of the average IM intensity value of 1752 Btu/ton-mile, the difference in intensity between the two modes is 3,337 – 1,752 = 1585 Btu/ton-mile. Then the approximate energy savings is the ton-miles multiplied by intensity per ton-mile:

$$(4.2 \times 10^{10} \text{ ton-mile})\left(1585 \frac{\text{Btu}}{\text{ton-mile}}\right) = 6.7 \times 10^{13} \text{ Btu}$$

Thus 67 trillion Btu are saved. Repeating the calculations for the low- and high-intensity scenarios gives savings between 43 and 90 trillion Btu. Note that the low-intensity scenario has the largest savings, since the intermodal option consumes the least energy per ton-mile.

Example 12-4M In the year 2000, U.S. trucks averaged 2,420 kJ/tonne-km. Published estimates of average energy consumption across all intermodal shipments were not directly available, but a plausible range for this intensity value is 870 to 1672 kJ/tonne-km. Suppose in that year approximately 61 billion tonne-km that would have moved by long-distance truck instead move by intermodal. What is the energy savings?

Solution Taking the case of the average IM intensity value of 1271 kJ/tonne-km, the difference in intensity between the two modes is 2420 - 1271 = 1149 kJ/tonne-km. Then the approximate energy savings is the tonne-kilometers multiplied by intensity per tonne-kilometer:

$$(6.1 \times 10^{10} \text{ tonne-km})\left(1149 \frac{\text{kJ}}{\text{tonne-km}}\right) = 7.02 \times 10^{13} \text{ kJ} = 70.2 \text{ PJ}$$

Thus approximately 70 PJ are saved. Repeating the calculations for the low- and high-intensity scenarios gives savings between 46 and 95 PJ. Note that the low-intensity scenario has the largest savings, since the intermodal option consumes the least energy per ton-mile.

Using the same pattern of calculation as presented in Example 12-4, it is possible to estimate the growth in energy consumption avoided due to U.S. containers and trailers moved by rail rather than over the road since the inception of modern container-on-rail service around 1980. The range of intensity estimates for the year 2000 was used in other years to translate the number of intermodal shipments from 1980 to 2005 into an estimated range of energy savings. As shown in Fig. 12-7, in 1980 the energy savings ranged between 5 and 22 PJ, and by 2005, the savings had risen to approximately 40 to 120 PJ or 0.7% to 2.1% of energy used by intercity trucking in that year. From Fig. 12-4 it is clear that the economic slowdown of 2008 to 2010 decreased the potential energy savings from growing use of IM, but at the same time the reduction in economic activity decreased demand for truck freight as well.

Future Prospects for the Intermodal Approach

The expansion of intermodalism requires financial investment in the necessary equipment, including not only vehicles and rights of way (rail lines or waterways), but also

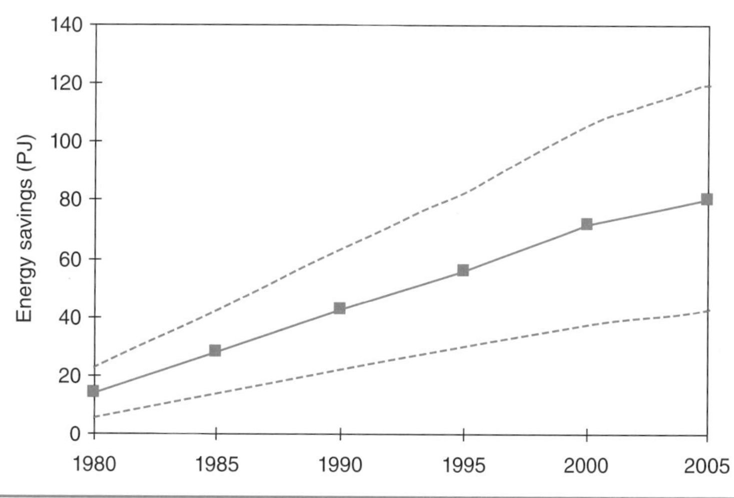

FIGURE 12-7 Savings attributable to using intermodal rail in the United States, 1980 to 2005, under lo-, mid-, and hi-savings scenarios.

Note: Conversion: 1 PJ = 955 billion Btu.

Mode	1993 (mil. tons)	1993 (bil. ton-miles)	2007 (mil. tons)	2007 (bil. ton-miles)
Truck	6,477	872	8,779	1,342
IM	38	43	226	197
Percent	0.6%	4.9%	2.6%	14.7%

Sources: Vanek and Morlok (2000), for 1993 data; 2007 U.S. Commodity Flow Survey, for 2007 data.

TABLE 12-2 Comparison of Truck and Intermodal Truck-Rail Volumes, 1993 and 2007.

intermodal transfer facilities that make possible the smooth transition from one mode to another. If either transfer points or long-distance corridors do not function correctly, then the entire system becomes unattractive to shippers, and the potential to move freight off the highways is lost. On the other hand, successful intermodal freight is a win-win situation for governments, shippers, rail and marine operators, and even for trucking firms, who can profit from transferring segments of truck freight movements to the rail or water modes. For all of these entities, it brings the benefits of taking pressure off the overburdened and overcongested road network. From an energy perspective, it also brings the benefit of moving freight at lower energy intensity.

In the short run, one of the main limitations on intermodalism is lack of terminal capacity to move shipments onto and off of the network. Table 12-2 compares national statistics for truck and intermodal tons and ton-miles, including the ratio of IM to truck values (row labeled "Percent" in the table). The figures show how ton-miles and in particular tons are small for intermodal compared to truck. The difference between tons and ton-miles suggests that relatively few tons of product are being loaded onto the intermodal network, and those that are loaded tend to move long distances.

Table 12-2 also shows the growth in IM tons and ton-miles during the period from 1993 to 2007. Although a larger share of ton-miles are moving by intermodal, with 197 billion ton-miles for IM in 2007 compared to 1.3 trillion for truck, the relative volume of tons loaded remains low. We can infer that most intermodal shipments are moving long distances based on the high ton-mile to ton ratio (872 ton-miles/ton in 2007), and that there are a large volume of short- and medium-haul shipments moving mostly by truck, on origin-destination pairs where intermodal service is not competing in the market. Furthermore, the capacity of the terminals that load and unload intermodal shipments is sure to reflect the current demand level for transshipments, in the sense that terminals are built to handle some average volume of demand plus X% for peak demand situations. They could not easily, however, absorb a fivefold or tenfold increase in throughput without major capital investments in expanding capacity. Thus a change in the value of tons loaded from 2.6% in 2007 to on the order of 20% to 30%, that might truly alter the environmental and livability impact of freight, could not happen without such a major investment

The result of the disparity between IM and truck activity appears in volume density of the national rail and highway networks shown in Fig. 12-8. In these figures, the lines show routes on which major volumes of IM or truck freight are traveling, and then the thickness indicates the relative flow volume. (Note the different units for each map,

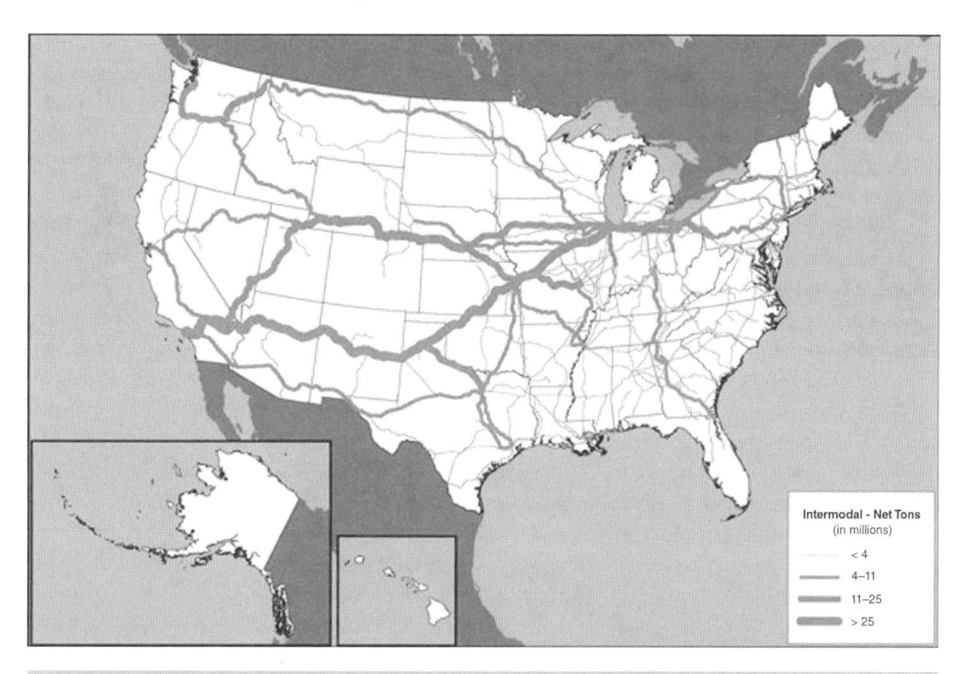

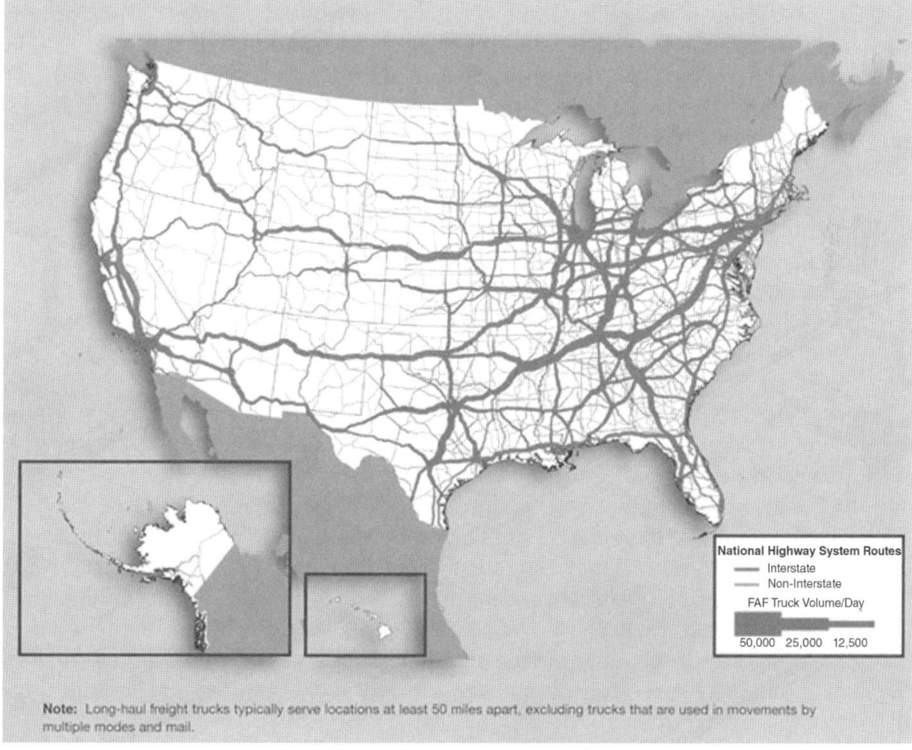

FIGURE 12-8 Comparison of intermodal and truck volumes on major links in the U.S. national network.

Source: U.S. Bureau of Transportation Statistics. Public domain.

millions of tons per year for the IM map, and thousands of trucks per day for the trucking map. A set of maps where both truck and IM flows are given in common units, e.g., truckload equivalents or tons, was not available.) The IM network is relatively sparse compared to the highway network that handles major volumes of freight, so that the opportunities to serve as much freight as trucking will not exist, and the volume served by IM will inevitably be smaller. The more complete highway network is in fact used for access and egress of IM shipments to and from the rail network. Also, the busiest links on the truck network handle 50,000 trucks per day. If each truck carried on average 6 tons, the tonnage per year would be more than 100 million tons per year, or four times as large as the highest bracket in the intermodal rail figure (>25 million tons).

Although line-haul trucking has dominated the shipment of higher-value goods over short to medium distances over the last two or three decades, innovators continue to expand IM into shorter distance markets by using advances in IM technology as well as the fuel cost advantage of rail to win market share. For example, the Modalohr system uses a flatcar with rotating platform to simultaneously load and unload articulated trailers without carrying the truck tractor on board and without use of cranes.[7] By lowering capital cost of intermodal terminals and accelerating trailer throughput, innovations such as these make IM increasingly competitive.

12-5 Supply Chain Management Approach

The preceding sections on intramodal and intermodal approaches describe possible changes but do not stipulate which stakeholder(s) would take leadership responsibility for executing them. The logical choice for leadership is collaboration between shippers and carriers, overseen and assisted by national and regional government as the representative of the general public.

Changes in government policy can encourage both intra- and intermodal changes, for example, in the form of tax policy that encourages investment in intermodal connections at key nodes. Alternatively, a shipper or carrier might undertake these changes internally independent of changes in policy, for example, when a new energy-saving technology emerges at a competitive price. Changes might also happen in the freight marketplace as a whole: a carrier in one mode might win a shipper's business from a carrier in another mode, leading to modal shifting.

At present, interest in improving the energy efficiency of freight transportation reflects both environmental pressure from the public and the competitive pressures of business. Concern has been raised about the relatively rapid growth in environmental impact, and in particular energy use, occurring in the freight transport sector. At the same time, the potential cost savings from reduced energy use have only grown as energy costs have increased. Firms have therefore come under mounting pressure to improve the fuel efficiency of their freight transportation operations. This impetus leads to the supply chain approach, in which the firms that control the supply chain take leadership of reducing its overall impact. This section of the chapter gives an overview of the supply chain approach and then presents a case study involving *benchmarking*, or making a survey of shipment practices and energy use patterns at multiple firms to compare performance.

---

[7]See *www.modalohr.com* for a downloadable video illustrating the loading and unloading function of the system.

12-5-1 Definition of Supply Chain and Supply Chain Management

A *supply chain* is a set of relationships between material and product sources, intermediate holding points such as warehouses and distribution centers, transportation linkages, and final consumers of products that define the process by which materials and products are used to meet a customer need.

If the supply chain is the web of relationships, then *supply chain management* is the process of actively managing those relationships from a big-picture, holistic perspective. For instance, the Council of Supply Chain Management Professionals defines the term *supply-chain management* as follows:

> Supply chain management encompasses the planning and management of all activities involved in sourcing and procurement, conversion, and all logistics management activities. Importantly, it also includes coordination and collaboration with channel partners, which can be suppliers, intermediaries, third-party service providers, and customers. In essence, supply chain management integrates supply and demand management within and across companies. – CSCMP[8]

Supply-chain management therefore seeks to optimize the number, location, and function of intermediate distribution centers, the connections between distribution centers and retail outlets, and the identity of suppliers bringing materials and goods into the network. Logistics managers speak of the *supply-chain management revolution* to describe the growth since the end of World War II of the field from simple management of shipping activities to optimizing entire supply chains, thanks in large part to the power of modern information technology and electronic communication.

Figure 12-9 shows a hypothetical supply chain, with one or more bubbles for manufacturing or warehousing activities at each stage, and arrows showing freight movements between locations. In general, the locations represented by the bubbles are geographically separate from each-other thus requiring that materials or products are shipped by some

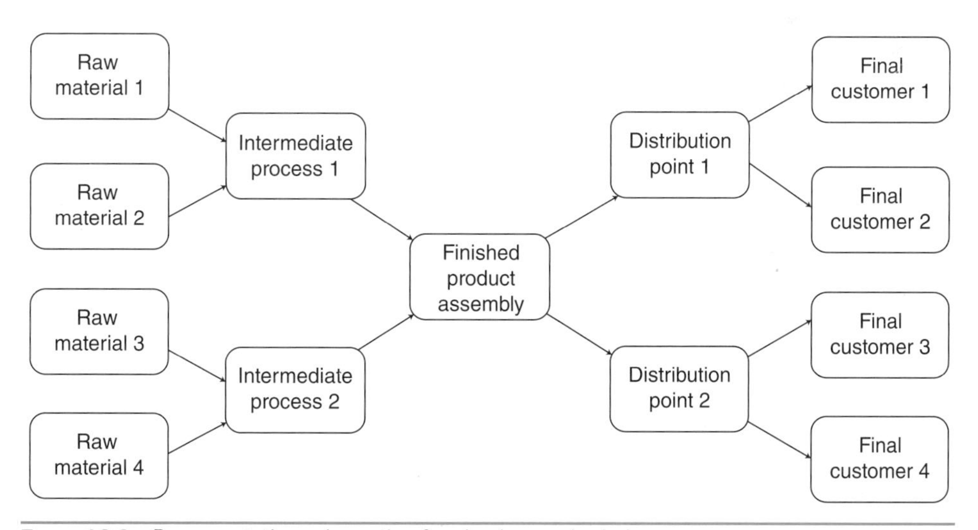

FIGURE 12-9 Representative schematic of a simple supply chain.

[8]As published at website *www.cscmp.org*. Accessed July 23, 2012.

freight mode. In the supply chain shown, raw materials enter from four distinct suppliers, are transformed into parts or semi-finished raw materials at two intermediate manufacturing centers, and then reach a single factory for final assembly into the finished product. Since for most products it is not convenient to sell directly to the retail customer at the factory, on leaving the product passes to one of two distribution centers, also known as *regional distribution centers* (RDCs), where it is held before onward shipment to retail outlets. In the diagram, the final stage in the supply chain is the four bubbles labeled *final customer,* which represent the retail outlets where the customer purchases the product and the product exits the supply chain. It is also possible to view the customer's home as the final stage in the chain and explicitly include the link between home and retail outlet in the supply chain mapping, although this was not done here.

Complexity of Actual Supply Chains

In the modern economy with its vast array of products for sale, there is much heterogeneity in supply chains that cannot be captured visually in the simple diagram in Fig. 12-9. This heterogeneity takes several forms:

- *Multiple nodes at each layer:* Depending on its complexity, a product may have many more than two or four nodes for sourcing materials or converting them into intermediate products.
- *Multiple points of final manufacture:* A firm may also manufacture the same product in more than one location. Thus the *finished product assembly* bubble in the diagram may be replicated in other locations, and multiple locations may or may not share incoming supplies from intermediate processing points and destination distribution points for finished product leaving the plant.
- *Multiple layers of intermediate processing or distribution:* The diagram shows just one layer between raw material sources and final product assembly, and one layer between the latter and the final customer. Depending on the complexity of the supply chain, however, additional layers may appear either upstream or downstream of final assembly.
- *Option for bypassing layers of supply chain:* For some products, additional channels beyond those shown in the diagram may exist, where the product bypasses one or more layers on its way to the final customer. For example, a facility that assembles laptop computers may ship to a distribution center, which in turn ships to a network of retail outlets. However, the customer may also be able to custom-order a laptop with specific components, which is then assembled at the factory and shipped directly to the customer's address, bypassing the distribution and retail outlet layers.
- *Coexistence of multiple supply chains for multiple products:* Although the diagram shows a supply chain for a single product, most or all of the facilities represented by the various bubbles are actually participants in multiple supply chains, each of which could be represented by a separate diagram. Therefore a *raw material* supplier represented by any one of the bubbles sells raw material for multiple products, the *finished product assembly* facility assembles multiple products (e.g., food products or vehicles), and so on.
- *Distinction between vendor, manufacturer, and retailer ownership:* Not shown in the diagram is the ownership of the facility represented by a given bubble.

For many products, the standard business model is to separate the retailing and manufacturing functions, allowing the manufacturer to sell to multiple retailers, and retailers to purchase from multiple manufacturers. Manufacturers in turn must decide whether to retain possession of upstream functions in the supply chain, such as raw materials extraction or intermediate processing, or *outsourcing* those functions to separate vendors.

Role of Supply Chain Management in Supporting Intramodal and Intermodal Approaches

The supply chain approach has a symbiotic relationship with the intra- or intermodal approaches. Rather than the one precluding the other, the firm using supply-chain approach to reducing impact may conclude that they need to improve their management within a given mode or pursue modal shifting to achieve supply chain goals. Thus a firm that depends on trucks for its freight transportation needs may discover that it is not meeting its fuel-cost goals and respond by adopting measures from the intramodal list in Table 12-1, such as increased aerodynamics, wide-based tires, or advanced lubricants to cut fuel costs.

Not all supply-chain approach tools fall under the heading of intra- or intermodal approach, however. A firm may, in analyzing its supply-chain activities, discover that its logistics management practices are inefficient. A typical response would be to tighten controls so that smaller volumes of stocks are kept on hand, vehicles are on average more fully loaded, and so on. These changes do not involve introducing new technologies or best practices, nor do they shift modes, but the resulting changes nevertheless reduce truck vehicle-miles, energy use, and emissions.

12-5-2 Implementation of the Supply Chain Approach to Improving Sustainability

A business may implement a supply-chain approach for reasons of needing to reduce cost, comply with regulation, or derive benefit from proactively becoming greener or more socially acceptable than the minimum requirement, so as to attain a business advantage. Implementation entails calculating baseline values for performance measures such as total expenditures on logistics, total transportation volume in terms of VMT of the firm's truck fleet, total energy consumption in transportation or warehousing operations, and total emissions. The firm can then set goals for moving performance measures in a positive direction that reduces cost or negative impact on the community or the environment. These measures could be evaluated in absolute terms or per unit of gross revenue, since growth in activity of the firm may lead to upward pressure on all supply chain activities.

In recent years, both the statement of environmental goals and progress towards them have found an outlet in *corporate environmental reporting* as a growing number of major businesses have begun publishing an annual environmental report. Corporate environmental reports apply to many of the firms involved in supply-chain management. Major retailers that generate a large volume of freight transportation between manufacturers, distribution centers, and retail outlets have an interest in truck travel volume, energy consumption, and emissions, as do carriers and third-party logistics providers. In some cases, environmental progress is reported as part of overall triple bottom line efforts: the Sainsbury supermarket chain in the United Kingdom includes an extensive section on environmental report in their overall *Corporate Responsibility Report* that also includes sections on ethical sourcing of food products, support for healthy diets for customers, and good corporate citizenship in the communities in which they work and sell.[9] The culture surrounding corporate environmental reporting has evolved significantly compared to the secretive posture that many businesses took several decades ago regarding their ecological impact. It has become acceptable in corporate environmental reports to publish shortfalls against goals as well as successes, perhaps because the public and other specific stakeholders are more concerned that businesses have a plan and are pursuing it, rather than being concerned that firms execute their environmental plan perfectly every year. At the same time, internal funding of efforts aimed at corporate environmental responsibility remain vulnerable to contraction in times of economic hardship because they may not yield an immediate return. Economic recession since 2008 has slowed the expansion of environmental reporting and other steps for this reason.

The firm may carry out activities to improve the sustainability of supply-chain management activities independently, as is often the case with results found in corporate environmental reports. It may also conduct them jointly in concert with other members of a coalition, as described in the case study in the following section.

12-5-3 Case Study: Benchmarking Study of Food and Beverage Sector in the United Kingdom

In the United Kingdom as in other industrialized countries, the food and beverage transportation sector depends heavily on trucking for freight movement as well as refrigeration to keep foodstuffs cool or frozen in transit, and is therefore a large transportation energy user. Responding to this situation, three stakeholders came together in the 1990s to form a coalition to study best practices in this sector, namely, the U.K. Environment Ministry, the Chilled Storage and Distribution Federation (or CSDF, the trade organization representing the major movers of temperature-controlled food and beverages in the United Kingdom), and the Logistics Research Centre at Heriot-Watt University in Edinburgh, Scotland.[10] Members were drawn both by the opportunity to save energy and protect the environment, and also by the potential to reduce operating cost and learn from one another.

Study participants agreed at the beginning of the study to a benchmarking format, where individual identities are kept confidential but individual performance can be compared to the worst, average, and best, which are published. After the initial year of the project, it evolved into an ongoing program where the results from one study informed the design and execution of the next. By benchmarking we mean a type of study in which all participants submit their supply-chain management data with the agreement that their data will not be revealed to their competitors. The data analysts for the study (the Logistics Research Centre in this case) then calculate performance measures (also called key performance indicators, or KPIs) for previously agreed categories of interest and report best, worst, and average values, along with the values for each participant with numbers used instead of the firms' identities to protect anonymity.

---

[9]Based on publicly available report from Sainsbury plc (2012). For a general discussion of corporate environmental reporting, see, e.g., Welford (1998).

[10]For further details see, e.g., McKinnon and Ge (2004).

Benchmarking allows participants to consider the benefits from a number of possible scenarios, including:

1. *Performance relative to technical optimal:* How well did a participant perform on average compared to the best value that might be technically possible, and what would be the benefit (e.g., for traffic, emissions, etc.) of the participant moving up to this level?
2. *Performance relative to benchmarking study best-in-class:* Since it is possible that no firm achieved the maximum possible technical performance for a given measure, it may be more realistic to ask how well a participant performed on average compared to the best value that was achieved among the participants, and what would the benefit be from this improvement.
3. *Benefit from having all lower-performing participants improve their performance:* The system-wide benefit of having all those participants who did not reach a certain performance improve to some target. For instance, the threshold might be the mean value of performance, and the target might be to have all lower performance improve to the mean. Example 12-6 illustrates this application.

The study was conducted on a *time-synchronized* basis, meaning that the participants all chose the same 48-hour period to gather the data needed for benchmarking, rather than each gathering the data at their convenience. The period was chosen to be close to *average* for year-round conditions, namely, two consecutive weekdays in autumn, and the time-synchronized nature was chosen so that each participant would be seeing largely the same conditions in terms of traffic, weather, level of demand, and so on.

In practice, the execution of the study meant having each participant record in a database the activities of each vehicle in service, including information such as type of activity, quantity of goods on board, distance traveled, and adherence to schedule. For example, in a given generation of the study, 28 companies participated, representing 3,650 vehicles that traveled 1.45 million km total over the 48 hours (i.e., on average about 200 km per vehicle per 24-hour day). Note that the pool of vehicles used was not a statistical sample of the average U.K. food movement fleet, since the participants self-selected to join the study or not. For the purposes of the government agencies and participants involved, this method for assembling the survey fleet did not detract from the value of the study. Also, inspection of the average characteristics for the surveyed vehicles showed that they resembled those of vehicles surveyed in the Continuing Survey of Road Goods Transport (CSRGT) that is conducted annually in the United Kingdom.

Development of Metrics and Sample Calculation of a Metric Value

The stakeholders in the study chose performance measures based on tractability of calculation and value of resulting data for assisting participating fleets with evaluating operational efficiency and making fleet or operational decisions. The following list of five metrics resulted:

1. *Average loading of vehicles—fraction of maximum weight or volume occupied.* Measure that gives a sense of how well the vehicles are utilized
2. *Extent of empty running:* Total kilometers and fraction of total kilometers driven during which the vehicle is empty

3. *Allocation of each hour of 48-hour period by category of activity:* Each hour allocated to one of several possible activities (e.g., driving over the road, loading, unloading, etc.) and then the total number of hours in each category tallied
4. *Average energy intensity:* Amount of energy required per unit of useful service delivered
5. *Adherence to schedule and extent of delays:* Percent of trips subject to delays and extent of delays in minutes (Note that this is different from the measure of congestion, since vehicles may encounter traffic congestion that is expected and that is therefore built into the scheduled delivery targets.)

In this section, we focus on the fourth metric, namely, energy intensity, as an example of how a performance measure is calculated and compared. In discussion with the participants, the research team developed the intensity measure of *fuel per pallet-kilometer* that would be measured in the benchmarking study. This measure is similar to the familiar energy per tonne-kilometer or ton-mile used in national statistical databases, but has more meaning to the survey participants since their operations measure the delivery of product in pallets, and not in tons. The measure incorporates energy consumption in both moving the vehicle over the road and in running refrigeration equipment on board the vehicle, which typically each draw from separate fuel supplies that are refueled at separate times. It also measures productive output from freight activity in pallet-kilometers by taking into account the number of pallets on board the vehicle during different legs of the journey made by a food and beverage truck during the course of a typical workday.

The measure is calculated as shown:

$$\eta_{\text{p-km}} = \frac{(D_{\text{tot}}/F + E_{\text{refrig}})}{(D_{\text{laden}}N_{\text{pallets}})} \cdot \left(\frac{1000\text{ mL}}{1\text{ L}}\right) \quad (12\text{–}4)$$

where D_{tot} is the total distance traveled by vehicle including both loaded and empty miles

F is the average fuel economy, km/L

E_{refrig} is total fuel consumed by refrigeration, L

D_{laden} is total distance traveled laden with freight, km

$N_{pallets}$ is weighted average number of pallets across all laden trips

η_{pkm} is fuel intensity, milliliters of diesel fuel per pallet-kilometer.

Note that $N_{pallets}$ includes only laden trips. Example 12-5 illustrates the application of the equation.

Example 12-5 A vehicle participates in an audit for the purposes of calculating average fuel intensity and sharing with a benchmarking study. The fleet uses average fuel economy from the previous fiscal year, during which time 12 vehicles in the fleet traveled 180,000 km and consumed 62,000 L of fuel. For a particular entry in the survey, the vehicle starts out from a regional distribution center with 30 pallets loaded, travels 130 km to deliver the first 8 pallets, travels another 70 km to deliver the remaining 22 pallets, and drives empty 50 km more to return to the RDC. During the journey, the refrigeration unit on the vehicle consumes 11.6 L of fuel. What is the fuel consumption intensity in milliliters fuel per pallet-kilometer?

Solution First, the average fuel economy per liter of fuel is calculated from the given data. Accordingly, the figures give

$$F = \frac{180{,}000}{62{,}000} = 2.9 \text{ km/L}$$

Thus $F = 2.9$ km/L for purposes of the vehicle in question (equivalent to 6.9 mpg in standard units). Next, the laden portion of the trip must be divided into percentages to calculate N_{pallets}. Since the first portion of the trip is 130 km and the second 70 km, the percentages are 65% and 35%, respectively. Therefore, N_{pallets} is calculated as

$$N_{\text{pallets}} = (0.65)30 + (0.35)22 = 27.2$$

Using this information, we calculate intensity using Eq. (12-4):

$$\eta_{\text{p-km}} = \frac{(250/2.9 + 11.6)}{(200)(27.2)} \cdot \left(\frac{1000 \text{ mL}}{1 \text{ L}} \right) = 5.28$$

Thus the overall intensity is 5.28 mL fuel per pallet-km.

Note that in Example 12-5, the fuel consumption is based on annual average fuel economy multiplied by distance driven. Ideally, one might have observed fuel consumption data for each leg of each journey, so that the actual amount could be entered into Eq. (12-4). However, if the truck is not equipped with a device for logging fuel consumption on each journey, consumption can be measured only occasionally when the vehicle refuels.

Also, in Example 12-5 the value of $\eta_{p\text{-km}}$ is fairly low, and the trip is hence fairly efficient, because during much of the trip the vehicle is full or nearly full. The value of $\eta_{p\text{-km}}$ can be minimized by maintaining utilization of available space in the vehicles at high levels. It can also be minimized by finding inbound loads to the distribution center, either reverse logistics (i.e., transport of packaging materials for recycling) or product from suppliers, and letting the inbound loads replace the outbound products as they are distributed. For example, for the last 50 km the truck is empty, so it might visit a manufacturer near the second retail outlet and load the truck with manufactured food products destined for the RDC, if the schedule can be coordinated.

Reporting of Performance Measures and Potential Energy Savings

Once the benchmarking survey period has ended, the survey administrator analyzes the data and produces for each participant and for each performance measure an average value. Figure 12-10 shows the measures of $\eta_{p\text{-km}}$ estimated for a subset of 15 fleets from among those covered in one generation of the benchmarking study.[11] The bars in the figure are divided between *primary distribution* (transport from manufacturers to RDCs) and *secondary distribution* (from RDCs to retail outlets). Fleets are shown in the figure with a reference number only to maintain anonymity; participants are told their number in the figure, and then all participants are shown the range of intensity values so that they can see how they compare. Energy intensity values are reported for each fleet, although an individual retailer or third-party logistics provider may operate more than one participating fleet and may therefore appear more than once in the figure.

[11]Reported in McKinnon & Ge (2004).

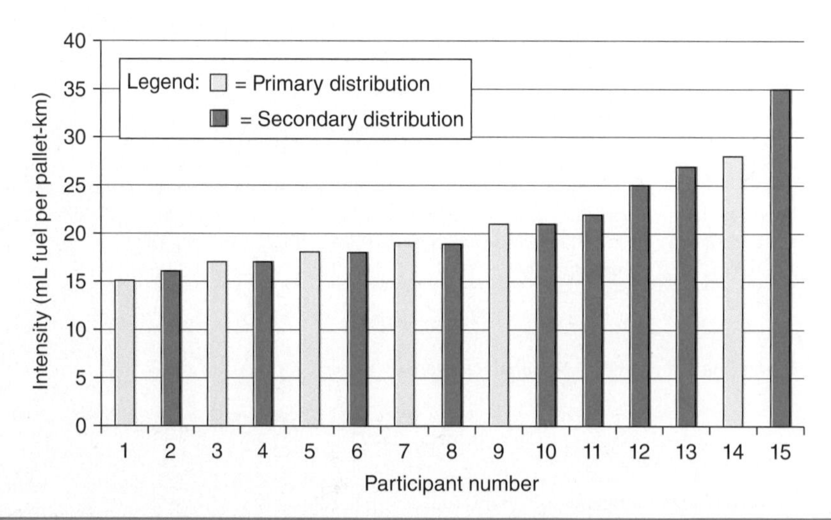

Figure 12-10 Comparison of average energy consumption per pallet-kilometer for benchmarking study participants.

Notes: Lower value is preferable.

Source: Adapted from McKinnon and Ge (2004).

Figure 12-10 suggests that energy intensity is somewhat explained by type of distribution: primary distribution tends to involve larger volumes per shipment of product from factories to RDCs and is therefore relatively efficient compared to secondary distribution, where deliveries may be smaller. Nevertheless, there is also considerable variability within each type of distribution, so an inefficient participant might recognize that they may have room for improvement (if they cannot explain the differences due to factors beyond their control, such as geography), and take steps accordingly. Example 12-6 shows how improvements in several fleets in the survey might improve overall average performance.

Example 12-6 Consider the six primary distribution fleets in the study, and suppose that in a given period of time, each generates 10,000 pallet-kilometers of volume. (a) What is the total amount of fuel consumed in liters? (b) If all fleets with intensity values higher than the average improve their efficiency so that they achieve the average, how many liters of fuel does each affected fleet save, and how many liters are saved on the whole?

Solution

(a) For the six fleets included (Numbers 1, 3, 5, 7, 9, and 14), the intensity values are 15, 17, 18, 19, 21, and 28 mL/p-km (diesel), respectively. The arithmetic average of these six values is 19.67 mL/p-km. Multiplying each fleet intensity by 10,000 pallet-km and summing up the resulting liters of fuel consumption gives 1,180 L of diesel.

(b) The fleets with higher than 19.67 mL/pallet-km have values of 21 and 28, respectively. Therefore, the savings for each fleet is

$$(10{,}000 \text{ pallet-km})\left(\frac{21 - 19.67 \text{ mL}}{\text{pallet-km}}\right) = 13 \text{ L}$$

$$(10{,}000 \text{ pallet-km})\left(\frac{28 - 19.67 \text{ mL}}{\text{pallet-km}}\right) = 83 \text{ L}$$

Initially, these two fleets consume 210 L and 280 L fuel, so that the savings are 6% and 30%, respectively, of the original value. The total fuel saved is 96 L, or 8% of the original 1,180 L.

Overall Benefits of Benchmarking

To conclude, the benchmarking approach gives supply-chain managers actionable information that they can use to motivate improvements, or if they are already near the top in terms of performance, make sure they stay there in subsequent versions of the benchmarking study. It also gives the government agency a snapshot of how well the sector is performing, and indicates trends if repeated over time. The benchmarking approach is somewhat crude since participants may change over time, with some companies leaving and others joining. Also, the content of the survey may change, both to reflect learning from previous surveys, and because the group is committed to meeting whatever needs are most important to the participants when the new survey is drawn up. It is therefore hard to control for mitigating factors and isolate the impact of benchmarking on overall performance. Nevertheless, the real-world experience gained from the benchmarking program reviewed in this case study inspires confidence in its value because of anecdotal evidence that participants improved performance, and because by and large participants were retained from one generation of the study to the next.

Benchmarking is also relevant to the discussion of spatial analysis of freight patterns in Chap. 13. In the above example, the focus of benchmarking is on each participant in the study attempting to maximize the efficiency of their operations by reducing fuel consumption, minimizing empty running of vehicles, and other measures. The next chapters consider the overall spatial patterns of different industries, including the major locations that generate the product of that industry and the locations from which the major markets draw the product. Spatial analysis of production patterns can deliver the same overall volume of product (measured in either total economic value or tonnage), thus satisfying market demand, while finding sources closer to the market so that fewer ton-miles are required, for example, by increasing local and regional production. One way to achieve this outcome is to apply benchmarking within a consortium of shippers to understand the nature of the overall freight pattern. Then arrangements can be found that benefit benchmarking study members by reducing total transportation cost and also reduce overall ton-miles generated.

12-6 Summary

This chapter has shown how freight transportation performance can be improved by changes within individual modes, shifts between modes, or by changes to how supply chains are managed. Among freight modes, trucks have the largest negative impact, however a wide range of improvements from efficient trailer bodies, tires, and power supply systems to best practices are available to address these concerns. For advanced technologies or best practices that are just beginning to penetrate a given freight mode or market, the rate of penetration and eventual saturation level can be used to more accurately predict their impact on sustainability goals (e.g., energy consumption per unit of freight activity) over a multiyear time horizon. In terms of modal shifting, features such as improved railcar technology or double-stacking capability for shipping containers make intermodal truck-rail the most promising substitute for long-haul trucking over the road network. The absolute value of freight volume moved by intermodal truck-rail is still small compared to trucking generally, even in a freight market such as that of the United States where railroads are generally quite successful and a model for other countries. Intermodal volume is nevertheless growing and can continue to grow in the future. The supply chain approach gives the shipper of products and manager of supply chains the role of agent for spurring improvements not only in terms of making intra- and intermodal improvements, but promoting more efficient logistical practices generally to assure improvement. Shippers can independently make supply chain changes to improve sustainability, or they can collaborate with other firms in their sector through sector-wide projects such as benchmarking studies.

References

Ang-Olson, J. and W. Schroeer (2002). "Energy Efficiency Strategies for Freight Trucking Potential Impact on Fuel Use and Greenhouse Gas Emissions." *Transportation Research Record* 1815:11–18.

Armstrong, J. (1993) *The Railroad: What It Is, What It Does. The Introduction to Railroading.* Simmons-Boardman, Omaha, NB.

McKinnon, A. and Y. Ge (2004). "Use of a Synchronised Vehicle Audit to Determine Opportunities for Improving Transport Efficiency in a Supply Chain." *International Journal of Logistics: Research and Applications* 7(3):219–238.

Sainsbury plc. (2012). *2011 Corporate Responsibility Report.* J. Sainsbury plc, London. Electronic resource, available at *www.j-sainsbury.co.uk*. Accessed Jul. 16, 2013.

Vanek, F. and E. Morlok (2000). "Reducing US Freight Energy Use Through Commodity Based Analysis: Justification and Implementation." *Transportation Research Part D*5(1):11–29.

Welford, R. (1998). *Corporate Environmental Management*. Earthscan, London.

Further Readings

American Association of Railroads (2011). *Railroad Facts.* AAR, Washington, DC.

Campbell, J., A. McKinnon, and F. Vanek (2000). "Refrigerator on Wheels: Key Performance Indicators of Logistical Efficiency and Energy Use in the U.K. Temperature Controlled Supply Chain." Paper presented at 2000 Transportation Research Board Conference, Washington, DC.

Christopher, M. (2011). *Logistics and Supply Chain Management. (4th ed.).* Prentice Hall Financial Times Edition, London.

Coyle, J., J. Langley, and R. Novack (2012). *Supply Chain Management: A Logistics Perspective. 9th Edition.* Southwest College Publishing, Nashville, TN.

Greene, D. and Y. Fan (1995). "Freight Transportation Intensity Trends: 1972–1992." *Transportation Research Record* 1475:10–19.

McKinnon, A. (1989). *Physical Distribution Systems.* Routledge Kegan & Paul, London.

Schewel, L. and L. Schipper (2012). "Shop til We Drop: A Historical and Policy Analysis of Retail Goods Movement in the U.S." *Environmental Science & Technology* 46(18):9813–9821.

Schiller, P., E. Bruun, and J. Kenworthy (2010). *An Introduction to Sustainable Transportation: Policy, Planning, and Implementation*. Earthscan, London.

Vanek, F. and E. Morlok (1998). "Freight Energy Use Disaggregated by Commodity: Comparisons and Discussion." *Transportation Research Record* 1641, 3–8.
Vanek, J. and F. Vanek (1999). "Systems, Location, Ecology, and Society: Theoretical and Empirical Analysis." *Economic Analysis* 2(3):209–222.

Exercises

12-1. A tractor-trailer truck weighs 34,500 kg, has a cross-sectional area of 8.8 m^2, a drag coefficient of 0.4, and a coefficient of rolling resistance of 0.0056. Assume an air density of 1.1 kg/m^3. The truck travels a distance of 400 km at a constant speed of 112 km/h without stopping, and at that speed, the tank to wheel (TTW) efficiency is 26%. (a) How many liters of diesel fuel does the truck consume? (b) Suppose that the truck now travels the same distance at a reduced speed of 104 km/h to reduce CO_2 emissions. The TTW efficiency is again 26%. How many kilograms of CO_2 emissions are saved over the same 400-km distance compared to driving at 112 km/h?

12-2. Studies show that converting trucks from conventional to wide-based tires can reduce fuel consumption by 2.6% per participating vehicle. It is believed that the maximum potential penetration for this technology into the market is 100%. The technology in the year 2010 has penetrated the market to the level of 5%. For years 2011 to 2013, the penetration values are 5.4%, 5.7%, and 6.6%, respectively. In 2010, there are 1.78 million combination trucks (ignore single-body trucks for this problem), each driving on average 98,000 miles/year. In the base case, trucks that do not have wide-based tires achieve a fuel economy of 5.6 mpg diesel. For the purposes of this problem, you can assume that total number of trucks, miles per truck per year, and fuel economy stay constant for the duration. (a) Using only the penetration achieved in years 2010 to 2013 and the ultimate penetration value, use the technological substitution "*s*-curve" to estimate the percent market penetration in the year 2030 to the nearest 0.1%. Include in your answer the values of the curve-fitting parameters c_1 and c_2 that you deduce. (b) Create a graph of year-by-year penetration of the technology from 2010 to 2030, including both observed and modeled curves. (c) Using the data given, how many gallons of fuel are saved in the year 2030 compared to a situation where no wide-based tires have been installed?

12-3. A freight trains consists of 80 freight cars, each weighing 30 tons empty and carrying a load of 40 tons (typical of consumer goods), at 79 mph (the U.S. national rail speed limit) on a level track. The train is to be pulled by a string of locomotives each with maximum power of 3,700 hp, weight of 142 tons, and 6 axles. (a) What is the total resistance of just the pulled load of the train, not including the locomotives, in pounds? (b) What is the minimum number of locomotives required to pull the load at the desired speed, and the resulting total resistance of the train in pounds?

12-4. For a given freight market, the cost per shipping container is fixed price \$125 and \$1.75/mile for truck, or fixed price \$600 plus \$0.80/mile for IM rail. (a) What is the breakeven distance below which truck is cheaper and above which IM rail is cheaper? (b) Suppose a truck with a load of one container consumes on average 0.167 gal of diesel per mile, and an intermodal train on average 0.05 gallons per container per mile. Each gallon of diesel combusted results in 22.4 lb of CO_2 produced. Suppose a carbon tax of \$100/ton CO_2 is introduced. What is the new breakeven distance?

12-5. A trucking company whose tractor-trailers average 98,000 miles/year is considering an investment in an energy saving technology that costs \$20,000 per truck. The device must pay for itself in a maximum of 5 years, simple payback. Fuel costs \$4.75/gal, the baseline fuel economy is 6 mpg, and the technology increase fuel economy by 5%. Does the technology meet the maximum allowable payback time criterion? Calculate the actual payback time required to support your answer.

12-6. Compute energy intensity per pallet-kilometer for a vehicle run around the itinerary presented in the following table, in which the distances are from the previous stop, and the number of pallets is the delivery size at the current location (RDC = regional distribution center). The truck performing the run is part of a fleet of 350 trucks that in the preceding 12 months consumed 28 million liters of diesel and drove 8.48 million km. The capacity of the truck is 30 pallets. Diesel consumption for refrigeration is on average 29-mL diesel for each loaded kilometer; once the truck is completely unloaded the refrigeration is turned off. (a) What is the energy intensity of the journey, in L/pallet-km? (*Hint*: for this problem, calculate N_{pallet} using a weighted average, taking into account the length of each leg of the journey.) (b) What is the weighted average percent utilization of the truck's pallet capacity, including both laden and empty segments?

Location	Km Traveled	Pallets Delivered
RDC	0	0
1	20	4
2	70	8
3	30	7
4	30	5
5	25	6
RDC	50	0

CHAPTER 13

Spatial and Geographic Aspects of Freight Transportation

13-1 Overview

This chapter considers spatial patterns and geographic aspects of freight transportation, including average distance per shipment, flow patterns observed in origin-destination flow matrices, and trends observed as geographic patterns of freight change over time. The first part of the chapter surveys spatial patterns of freight: why they exhibit the patterns they do and what types of goods are more or less affected. The second part presents a possible strategy of rearranging spatial patterns to reduce distances and hence total volume of movement on a commodity by commodity basis. To make the strategy more tangible, a prototype analysis of the pulp and paper sector is developed to illustrate how spatial patterns might be modified, and then the extension of the strategy to products from other sectors of the economy is discussed.

13-2 Introduction

Over the last three decades, freight systems in industrialized countries such as the United States have witnessed the spatial spreading of freight patterns. In 1960, in the United States, the average distance traveled by a shipment was 366 miles, but by 2007 this figure increased to 619 miles. Similarly, from 1993 to 2007, the total tons of freight shipped grew by 27% from 9.9 to 12.5 billion, but the total ton-miles increased 38% from 2.4 to 3.4 trillion, so that average ton-miles per ton increased from 242 to 272.[1] All of these figures show the growing geographic expanse and complexity of freight activity. Whenever the average distance a shipment travels increases or the number of ton-miles generated for each ton of freight moved increases the system becomes more spatially intensive, whether by each shipment moving farther from origin to destination or individual goods requiring a larger number of steps over their shipment lifetimes.

[1]Note that average miles per shipment and average ton-miles per ton are slightly different metrics. Average miles per shipment takes all shipments captured in a survey and calculates their arithmetic average whether the shipment size is large or small. Ton-miles per ton is the sum of all ton-miles divided by the sum of all tons, and therefore tends to weigh large shipments more heavily.

The total volume of freight ton-miles in recent years also translates into a high volume per capita. If one divides total ton-miles by total U.S. population, the ton-miles per capita figure stood at 9,387 in 1993. By 2007, the figure had grown to 11,085. Figures like these indicate that a wealthy and materially intensive society like that of the United States can also be freight transportation intensive.

Although longer average shipment distance and increased ton-mile per ton may put upward pressure on total impact from freight (e.g., air pollutant and greenhouse gas emissions, infrastructure wear and tear, contribution to congestion, etc.), there are a number of advantages that drive the system in this direction. Concentrated production in larger facilities can reduce the average cost per unit for production, even after taking into account increased freight cost. Also, *globalization* has encouraged transition to global suppliers of raw materials, components, and finished goods, especially in countries with low manufacturing costs. In some instances, manufacturing in either factories or countries that are highly efficient in a specialized area may reduce ecological burden, since these facilities may produce each unit at the lowest possible impact level per unit. Centralized manufacturing may have the lowest possible emissions and ecological burden even when both production and transportation are taken into account, if the manufacturing process is efficient enough. On the other hand, if the facilities are inefficient, (e.g., if they use electricity that is produced with a high CO_2 emissions per kWh) then long-distance shipping is making a bad situation worse, since the ecological impact from both manufacturing and freight transportation is worse than the case of a smaller, low-impact factory producing for a more regional market.

13-3 Background on the Study of Freight Spatial Patterns

The concept of reducing environmental or infrastructural impact through spatial redistribution is akin to the financial optimization of freight flows within the individual firm, which dates back to the birth of linear programming in the 1940s and the use of information technology to optimally solve data-intensive transportation problems that are not easily solved by hand. In this early work, the models minimized shipment cost rather than ton-miles, energy use, or emissions. With the tremendous progress made over the subsequent decades in data gathering and computerized problem solving, it can be assumed that today a shipper or carrier will use software to optimally trade off production cost, delivery timing, product protection, and distance and routing, so that at this level there will be little unnecessary movement of goods or extraneous ton-miles/vehicle-miles.

Nevertheless, there are many factors at work such that total ton-miles of freight, or vehicle-miles traveled by freight vehicles, have risen substantially over the years, even beyond what might be predicted by growing population or purchasing power of an increasingly wealthy population. First, as the economy has become more global, far for near substitution has occurred in the procurement of goods from the world market, and at the same time regional manufacturers have reached out to national and international markets to find outlets for their products. Also, a desire for rapid delivery leads to vehicles traveling less full. Lastly, more complex products and supply chains may lead to more steps, and hence longer overall distances, as raw materials are converted to components and eventually to completed finished products.

A useful metric for understanding the growth in complexity of supply chains is the measure of *lane-miles* required for the complete assembly of a product. Each mile of movement of any component or raw material used in the product as it is transported from one location to the next in the supply chain counts as a lane-mile, regardless of the size or weight of the component. A sum of the total number of lane-miles (e.g., in a cell phone, copy machine, or automobile) gives some indication of the complexity of the supply chain.

The underlying assumption is that one can use total ton-miles (or vehicle-miles) as a general proxy for the wide range of both global and local negative impacts that come from freight activity. Although this relationship is lumpy and nonlinear, it can reasonably be assumed to be positive; therefore, a reduction of ton-miles, or at least a decrease in their rate of growth, is a useful public policy goal.

Despite the advantages of better understanding this phenomenon, research on spatial spreading of freight is at a preliminary level, for example, compared to the amount of research that has been done on improving spatial decisions within the individual firm, such as vehicle routing and scheduling (VRS). This can be explained perhaps because there is an immediate return to the individual business from improving vehicle routing, since it can reduce operating costs and in some cases improve delivery reliability and the condition of the delivered good, thus directly contributing to the profitability of the firm. The problem of spatial spreading at a regional level is more complex.

Table 13-1 provides some representative recent studies from Europe and Japan that seek to understand freight spatial patterns at a system-wide level. In one example, the Royal Commission on Environmental Pollution (RCEP), a standing body in the British

Name of Study and Focus Region	Description
Royal Commission on Environmental Pollution, 1994, United Kingdom	Projected future growth of total tonne-kilometers of freight for the United Kingdom based on current trends, and advocated reducing growth in tonne-kilometers compared to the *business as usual* (BAU) baseline by reconsidering the spread of geographic patterns of freight movement.
Boge, 1995, Germany	Studied the supply chain for the total production of a "well-traveled yogurt pot," i.e., raw materials and final ingredients in the production of packaged, flavored yogurts, and advocated a rethinking of the supply chain to reduce freight requirements.
Vanek, 2000 and 2001, United States	Studied relationship between spatial patterns and total tonne-kilometers in the United States pulp and paper sector using gravity model (2000) and linear programming (2001).
Nijkamp et al., 2004, European Union	Used neural networks and discrete choice models to model the spreading of interregional trade patterns and overall growth in demand for freight tonne-kilometers due to European unification and the removal of barriers to cross-border movement.
Wisetijndawat et al., 2006, Japan	Incorporated spatial interaction between alternative choices available to a customer and resulting freight movements, with focus on case study of freight flows to and from Tokyo market.

Table 13-1 A Sample of Geographically Focused Studies of Freight Transportation

government, considered the role of growing shipment distance and its ecological implications as part of a larger treatment of the impact of freight transportation. They advocated steps to curb the trajectory of tonne-kilometers growth in future years even as the economy continues to grow. Boge's study traces back materials used in the manufacture and distribution of a single-serving container of yogurt, finding that the elements of this apparently simple product came from a surprising distance. The study's title is "the well-travelled yogurt pot," reflecting the way in which even a seemingly mundane product can generate a large amount of transportation activity. Vanek's study of the U.S. pulp and paper sector explores opportunities to reduce transportation volume on a sector-wide basis. (These studies are the basis of the case study later in this chapter.) Nijkamp et al. studied the impact of removal of border controls in the European Union, which led to rapid growth in freight activity and average shipment distance as businesses shifted from domestic to pan-European sourcing. Wisetijndawat et al.'s study looks at Japan, which due to high population and manufacturing density naturally generates a large amount of freight activity, the geographic patterns of which are of concern for societal reasons.

Underlying the challenge of developing decision-making tools for tackling the spatial distribution issue is a fundamental question of whether or not measures which seek to curb the growth of freight volume should even be contemplated. There are two schools of thought. The first considers the ease of movement of freight to be paramount; hence changes to spatial patterns of freight movement are not a legitimate means to reduce freight impact. Instead, growth of ton-miles should be taken as given and any damage from that growth should be minimized by improving technology and shifting modes. The second school of thought argues that given the severe pressure created on other policy measures by rapid spatial spreading, direct consideration of ton-mile growth should be rightfully considered alongside other policy countermeasures. The reader should in any case be aware that while an understanding of why average shipment distance or ton-miles per ton are growing may be of interest, there is by no means consensus that these trends are on the whole detrimental and should be reversed.

13-3-1 Availability of Data and Creation of Metrics

Commodity flow data are gathered and published by national governments in a number of countries. Government agencies usually gather the underlying primary data by surveying a subset of all freight transportation movements and inferring characteristics about the entire population of movements based on this sample. For instance, in the United States, the federal government carriers out the *U.S. Commodity Flow Survey (USCFS)* approximately every 5 years (in years ending in -2 and -7). In the United Kingdom, the *Continuing Survey of Road Goods Transport* (CSRGT) is conducted annually.

One data challenge is that certain data types such as origin-destination (OD) flows are expensive to produce, even after the raw data has been gathered by the surveying agency. In 1993, the USCFS committed the resources to produce OD tables, so we are able to demonstrate their use in the following examples. Subsequent versions of USCFS provide summary tables of total ton-miles and tons, but these do not support the creation of commodity-specific OD tables. OD flow data are also available from private companies, but these data sets are expensive and generally affordable only to larger agencies and businesses rather than individual professionals and students.

Commodity data are usually broken down into at least 10 or 15 major commodities, with small amounts of unusual commodities combined into *another* or *miscellaneous* category. For each commodity, most or all of the measures below will be available:

1. *Total weight of commodity moved* (metric or U.S. customary tons)
2. *Total activity attributable to product* (ton-miles or tonne-kilometers)
3. *Average shipment distance* (based on distance for each shipment captured in the survey)
4. *Value of commodity being transported* (of commodity only, not considering the cost of transportation)

The publishing agency aims to provide each measure for each commodity; however, in some instances a value for a specific commodity may not be published if either the data are of inadequate quality or if providing the value would reveal sensitive commercial information about the activities of specific firms. For example, if it is known that one firm dominates the movement of a given commodity, publishing its value of ton-miles would be equivalent to telling the public, including competitors, about the economic inner workings of that firm.

To maintain distinction between highly aggregated commodity groups and disaggregated, specific commodities and products, survey compilers typically use a system of digits in which an increasing number of digits indicates increasing level of detail. For example, in the system used in the 1993 USCFS, the Standard Transportation Classification Code (STCC), the two-digit classification "26" indicates the entire pulp and paper commodity group, the three-digit code "264" indicates miscellaneous paper products, and the four-digit code "2643" indicates paper bags. From 1997 onward, USCFS has used the Standard Classification of Transported Goods (SCTG), which is similar to STCC but with some updates to improve usability in analyzing commodity flows. It is important that any transportation classification system correspond to some degree with systems used on the industrial side, such as the Standard Industrial Classification (SIC) and North American Industrial Classification System (NAICS).

From the list of four published data sets above, several additional metrics can be calculated to characterize the movement of various commodities. U.S. customary units are used here, but equivalent metric units could be used throughout. The first of these is *value density*, defined as

$$\text{Value Density} = \frac{\text{Total Economic Value}}{\text{Total Weight Moved}} \quad (\$/\text{ton}) \qquad (13\text{-}1)$$

Thus if one divides the total economic value of the products in a given commodity group by the total weight of those products, the result is an approximate measure of the value of the product. Note that in this case the same product may appear more than once in the survey: first as part of a movement from locations A to B, and later from locations B to C. However, because both value and weight are measured twice, the double-counting does not distort the overall measure of value density. A second metric that can be derived is the ton-miles/ton measure, simply

$$\text{Ton-miles per Ton} = \frac{\text{Total Ton-Miles Moved}}{\text{Total Tons Moved}} \quad (\text{ton-miles/ton}) \qquad (13\text{-}2)$$

Here total ton-miles for a given commodity instead of total value is divided by total weight. Ton-miles/ton gives a similar sense of how far each product is being moved each time it is shipped the average shipment distance, but in this case, because the metric is on a per-ton rather than per-shipment basis, heavier shipments will influence the value of the metric more than lightweight ones.

Lastly, one difficulty with any effort at surveying commodity movements, alluded to in the discussion of value density, is that a physical product may be measured more than once as part of the survey. The term *tons loaded* (sometimes called *tons lifted*) is used to indicate that total values obtained in a survey[2] are a measure of the total weight moved in the system, and not the total *tons produced*. It is likely that a given unit produced will incur more than one movement from point of production to final destination, so that the tons loaded measure will be larger than that of tons produced.

If data on tons produced are available, it is possible to calculate an additional metric called a *handling factor:*

$$\text{Handling Factor} = \frac{\text{Total Weight Moved}}{\text{Total Weight Produced}} \quad \text{(tons loaded/ton produced)} \qquad (13\text{-}3)$$

While tons loaded are usually readily available when a commodity flow survey such as the USCFS or CSRGT is conducted by a national government or international agency, it is often more difficult to estimate the tons of commodity produced in a publishable form, so it may not always be possible to calculate the handling factor. Where the metric can be obtained, it indicates how the complexity of supply chains (i.e., number of times a product is loaded and reloaded as it passes through the supply chain network) contributes to overall freight transportation intensity.

13-3-2 The Role of Commodity Type and Value in Spatial Patterns

In Chap. 11, it was mentioned that different commodities could be divided into three major categories, namely, (1) high-value (i.e., expensive, high-technology goods), (2) mid-value (i.e., general finished goods), and (3) low-value goods (i.e., bulk raw materials and semi-finished products). Table 13-2 gives examples of each category. In all, there are 15 commodity groups in the table, which represent some 85% of all U.S. freight ton-miles.

Of the three categories, high-value products tend to be the most transportation-intensive because the additional cost of expensive or long-distance movement has little impact on the final sale price of the product—for instance, a laptop computer custom assembled in Asia and then shipped by airfreight to a customer in North America or Europe. At the other extreme, low-value products are greatly affected by shipping cost; hence, producers attempt to keep these costs to a minimum, either by using low-cost modes such as rail or marine, or by keeping shipping distances short if using trucking. For example, nonmetallic minerals such as gravel may move by truck, but will be sourced near their point of end use to keep cost down. This leaves the mid-value products, which constitute a substantial share of both total ton-miles and total value because they are valuable enough to justify shipping over longer distances but inexpensive enough that they are purchased in large quantities measured in mass or volume.

---

[2]For instance, using published values from the 1993 USCFS, 1.13 billion tons loaded of coal or 545 million tons loaded of chemicals.

Commodity Category	Example Commodity Groups
High-value	Textiles and apparel Machinery (including electronics) Transportation equipment Pharmaceutical products
Mid-value	Food products Primary metal products Fabricated metal products Pulp and paper products Lumber and wood products
Low-value	Metallic minerals Nonmetallic minerals Coal* Petroleum and coal products* Farm products Chemicals

*The STCC and SCTG make a distinction between *coal* in an unprocessed form and *coal products* which have undergone some value-added process.

Table 13-2 Examples of High-, Mid-, and Low-Value Products

As another example, food products are rarely as valuable per pound as consumer electronics or pharmaceuticals. However, foodstuffs are of course necessary every day for nutrition, and middle- and upper-class consumers are willing to spend a substantial fraction of their disposable income to eat a diverse diet, including some foods that are shipped from other regions or countries. The cost of the diverse food, including the extra transportation cost from shipping long distances instead of procuring the item from the closest possible source, is absorbed by the consumer because of the higher quality of life the consumer obtains from increased utility or personal satisfaction that comes from a diverse diet. (There are also instances where local foods are favored over those obtained from a long distance—this issue is revisited at the end of this chapter.)

The effect of a product's value on shipping distance can be further explored by comparing three scatter charts of representative products, namely, machinery, paper, and nonmetallic minerals, as shown in Fig. 13-1. These data are taken from the state-to-state flows between the U. S. contiguous states (excluding Alaska and Hawaii) published in the 1993 USCFS. (Note that units are metric rather than standard.) The annual tonnes are the total shipped per year for a given OD pair, and the length of haul is the total tonne-kilometers for that OD pair divided by total tonnes. The complete 48-by-48 state OD table has over 2,300 data points, so for clarity in the scatter charts only the largest OD flows are present, while many OD pairs with zero or very small amounts of flow are omitted.

As a first observation, note that the volumes in tons represented on the y axis are much higher for nonmetallic minerals than for pulp and paper and especially for machinery. Not surprisingly, nonmetallic minerals (gravel, sand, rocks of various types, etc.) are consumed in large tonnages, especially compared to machinery, where each unit of product may cost thousands to tens of thousands of dollars or more. Thus the

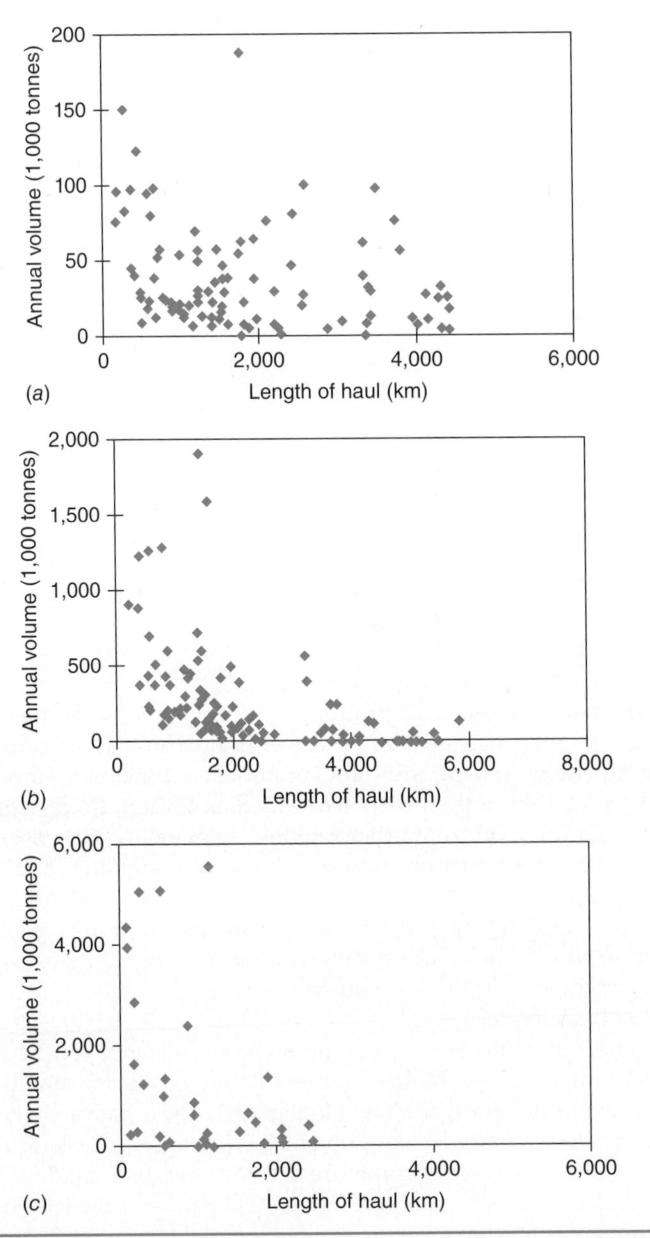

FIGURE 13-1 Comparison of scatter of tonnage as a function of distance for (*a*) high value machinery, (*b*) mid value pulp and paper products, and (c) low value non-metallic minerals, for select U.S. origin-destination flows.

Note: Value density for commodities in parts (*a*), (*b*), and (c) are \$13,000/ton (\$14,300/tonne) for machinery, \$900/ton (\$1,000/tonne) for pulp and paper, and \$12/ton (\$13/tonne) for nonmetallic minerals, respectively.

Source: 1993 U.S. Commodity Flow Survey.

maximum value on the y axis for the minerals is 6 million tonnes/year, whereas for machinery it is only 200,000 tonnes/year. When value density is taken into account, however, the positions are reversed: 200,000 tonnes of machinery at a value density of $14,300/tonne, has a value of some $296 billion, whereas 6 million tonnes of nonmetallic minerals at $13/tonne are worth just $79 million. With such high value densities for machinery, the economy will only support the sale and resulting transportation of a relatively small number of tonnes/year, compared to a low-value product like nonmetallic minerals.

Value density also impacts the trend in the scatter chart as we move to the right toward increasing shipment distances. In the case of nonmetallic minerals, flows as large as 5.5 million tonnes/year are seen at distances of 1,200 km (750 miles), but above this distance, the largest flows drop off dramatically to no more than 1.5 million at 1,900 km (1,200 miles) and 500,000 at 2,500 km (1,500 miles). Above 2,500 km no substantial flows are observed at all. Paper flows in Fig. 13-1(*b*) are less sensitive to distance, with flows of 1.8 million tonnes seen at distances around 2,000 km and 600,000 at 3,000 km. These flows tend to favor regional producers of paper, but also allow for long movements across the continental United States as purchasers access distant sources: the 600,000 tonnes mentioned are the total flows from Arkansas in the central United States to California on the west coast. With its high value density, machinery is predictably the least affected by distance: the largest flow of 180,000 tons/year is observed at 1,900 km, but a flow of 100,000 tons is observed at 3,800 km.

Thus the underlying intuition is confirmed: purchasers are unwilling to incur the costs of long distance shipment for low-value products, but for high value products, the added cost of long distance shipment is not an obstacle. For instance, if shipment by truck costs on average $0.20/ton/mile, and the purchaser is willing to add up to 10% to the cost of a machinery product with a value density of $13,000/ton, a distance of 6,500 miles/truck (i.e., costing a total of $1,300) would be acceptable for this purchase—longer than the maximum distance across the continental United States (!).

Limitations on the Ability to Constrain Spatial Spreading: Competing Factors

While it is clear that spatial spreading contributes to the overall growth of the amount of freight activity measured in ton-miles, and growth in ton-miles in turn puts upward pressure on energy consumption, CO_2 emissions, and other ecological and social concerns, it is far from clear that there is an easy way to reverse the trend. Realistically, the competing factors that push in the other direction are strong, and the ability of transportation policy to reverse that direction may be limited. Nevertheless, given the importance of addressing energy security and climate change, it is important to consider the competing factors and tradeoffs, with a view towards reducing spatial spreading where possible.

All other things being equal, spatial spreading adds vehicle-miles to a product's life and therefore adds to its overall cost. There are, however, at least three other important factors that may make this extra cost acceptable in the larger picture, as follows:

1. *Complexity of products:* As many manufactured products, ranging from microelectronics (such as cellular telephones or personal computers) to transportation equipment grow more complex, their manufacture requires more steps. Inevitably, some of these steps are carried out at geographically dispersed locations, leading to an increase in the total number of *lane-miles* (i.e., distance

travelled by any component in a product regardless of weight or size) required for the product.

2. *Benefit of centralized production:* It is often more efficient, in terms of material resource, energy, or financial cost, to manufacture a given volume of products per year at a single, large facility rather than at many smaller ones. For example, the large facility may use a single, maximum-size processing machine that costs less per unit produced to operate than a smaller machine that serves the same purpose.
3. *Ability to use lower cost labor:* Many manufacturing processes that require unskilled or semi-skilled labor can be conducted in countries with lower cost of living and hence lower wages than those of OECD countries in North America or Europe. Industrializing countries such as China and India have been especially successful at attracting these sorts of industrial jobs. While the loss of manufacturing jobs may be highly detrimental to certain communities, for instance in the United States, when these relatively well-paying jobs leave for overseas locations, from an economic point of view, they do help to hold down the overall cost of the product.

Any of these three factors may take economic priority over increased transportation cost. As alluded to in the case of transportation equipment above, the value density of many high-value, manufactured products is such that an increase in transportation cost can be more than offset by a decrease in other product life-cycle costs.

In addition to economic advantages, there may at times be ecological advantages to centralized production with higher transportation requirements, or at least reasons why such changes are neutral rather than negative. In the first instance, if the centralized facility is more energy efficient, or if it has access to cleaner energy sources, it may offset the ecological burden from added transportation. For many products, manufacturing is a much larger fraction of product life-cycle energy use than transportation, so it makes sense that an efficient, centralized production facility could offset increases in transportation energy consumption (See Chap. 11). In the second instance, considering the entire supply chain from raw material to delivery of final product is important as well. Suppose the product is dependent on a unique raw material source, or a limited number of sources, all of which are distant from a given market. If one were to reduce transportation requirements for delivery of the final product by locating the manufacturing plant adjacent to the market, total transportation requirements might not change: there would be less ton-miles dedicated to finished product, but more dedicated to raw materials.

At the other extreme, there are also reasons why spatial spreading and movement of manufacturing to offshore locations in a globalizing economy is negative from a manufacturing as well as transportation point of view. Typically, low-wage manufacturing countries also have energy sectors that are not as well developed as those in the wealthy countries, and, for instance, release more pounds of CO_2 per kWh of electricity produced, as is the case in comparing the United States to China (which has a higher fraction of electricity plants powered by coal and a lower fraction by other sources such as renewables or natural gas). Furthermore, manufacturers may move production out of wealthy, high-wage countries precisely because emissions standards are rigorous and therefore expensive to meet. In this case, green production loses twice: once due to increased emissions from manufacturing, and again due to rising ton-miles.

To conclude, there are some cases where increasing spatial spreading is a win for both the economics of the product and for the environment. In general, though, it is difficult to reconcile environmental and economic aims in the supply chain. Decisions that have a large ecological cost, such as the choice of a long, complicated, and convoluted supply chain for a modern product, have only modest impact on the overall cost of the product.

13-4 Spatial Spreading and Spatial Redistribution: Paper Sector Case Study

In the first part of this chapter, we considered spatial patterns of freight at a general level, considering how the different value density classes of commodities behave differently in terms of their spatial characteristics. The phenomenon of *spatial spreading* was observed at the aggregate level and applies to many products from consumer electronics to food products. As the freight transportation system has evolved, long-distance and international shipment of products has become more prevalent, allowing consumers to access products from the global market, and manufacturers to develop long, complex supply chains to assemble products. In this section, we introduce the new concept of *spatial redistribution*, which seeks to counteract the effect of spatial spreading by encouraging more compact supply chains.

To look more closely at spatial patterns, it is helpful to focus on one commodity in particular, so for the purposes of a more detailed analysis, the pulp and paper commodity group (hereafter "pulp and paper") at the two-digit level (i.e., the broadest possible level of commodity) was chosen as a case study subject.[3] The case study will show how spatial spreading has already occurred in the case of pulp and paper, with major paper-producing regions selling to the national as opposed to regional market. The potential for spatial redistribution to a more regionally focused supply system is then discussed at a preliminary level. Although spatial redistribution as a tool for the overall sustainability of freight transportation is in an embryonic state compared to some of the measures discussed in Chaps. 11 and 12, there is anecdotal evidence that the concept is being discussed in sectors from food products to electronics manufacturing, and it has holistic advantages that are discussed at the end of this chapter.

13-4-1 General Background on Pulp and Paper Sector

Pulp and paper as a group comprises *pulp* or wood that has been processed into a material used in the production of paper, and various *paper products* ranging from boxes to paperboard. Some paper products are sold to consumers for end-use consumption, while others, such as printing paper, are sold to various publishers and transformed into another product, *printed matter*, which has its own separate two-digit group that can be tracked and analyzed in a commodity flow survey.

The data for the case study are taken from the 1993 USCFS, which as discussed above is the most recent version of the survey that can support this type of study. There is anecdotal evidence that changes in the industry since 1993 have led to shifts in shipping patterns. Therefore the practical value of the example for the industry is reduced, even if its value as an example of a methodology is maintained. Production and distribution of

---

[3]The case study presented in this section is based on the content of Vanek (2001).

paper is somewhat dictated by where forests can be grown for use in making pulp, and the four main regions of the country with these resources, namely, northeast, southeast, north central, and west coast, continue to possess large areas of forest land. However, other trends between 1990 and 2010, such as electronic publishing or the recession of 2008 to 2009, will certainly have affected both total demand for paper and the patterns of distribution in the national network.

Data on pulp and paper shipment volumes from the USCFS and on energy consumption for transportation from other sources reveal that this commodity group was the seventh largest energy consumer of the 32 groups in the survey, consuming 5.4% of the total energy use for freight across all commodities while accounting for only 2.9% of the ton-miles. Compared to the average across all commodities, pulp and paper had higher average ton-miles/ton (465 versus 250) but lower average shipment distance (186 versus 424 miles), thanks to a relatively large number of small shipments traveling short distances while long-distance shipments driving the ton-mile/ton figure were few in number.

Table 13-3 gives the values of published and derived metrics for both pulp and paper, and the total across all commodities, for both the 1993 and 2007 USCFS. Tons, ton-miles, and value of pulp and paper grew more slowly over this 14-year interval than the total value across all commodities, perhaps due to the slowdown in paper consumption as many paper-based applications moved onto the internet. Similarly, the average value density of pulp and paper products, based on the reported total value and tons shipped, increased, but not as rapidly as value density across all shipments. Nevertheless, pulp and paper products remain more valuable per ton than the average ton of product. Lastly, the ton-mile/ton metric exhibited a lower percentage increase than the average product (5.9% vs. 7%).

	Pulp and Paper[1]		All Commodities	
Category	1993	2007	1993	2007
Tons, million	217	227	9,690	12,500
Ton-miles, billion	101	112	2,420	3,340
Value, $ billion	195	245	5,850	11,700
Average distance, mi[2]	186	n/a	424	619
Value density,$/ton	$899	$1,078	$604	$936
Ton-miles/ton	465	493	250	267

Notes:

1. 2007 pulp and paper values shown in the table are the sum of two published commodity groups, "Pulp, newsprint, paper, and paperboard" (SCTG Group 27) and "Paper or paperboard articles" (SCTG Group 28). The combination of these two groups is equivalent to the STCC "Pulp and paper" group from 1993.
2. Because the 2007 pulp and paper value is a combination of two published values, it is not possible to report an average shipment distance across the two published values. For the two published categories (see previous note), the average shipment distance was 297 and 512 miles, respectively.

Sources: USDOC (1996) for 1993 data; USDOC (2009) for 2007 data.

TABLE 13-3 Activity and Dollar Value Measures for Pulp/Paper and All Commodities from U.S. Commodity Flow Survey, 1993 and 2007

Pulp and paper was chosen because of an interest in exploring the potential for more regionalized production, consumption, and recycling of paper products with the growth of the recycled paper industry in the United States in recent years. This commodity group was also of interest because of the potential for reducing its total transportation burden through replacement of paper with electronic media.

Definition of Target Audience for Spatial Analysis of Freight

Government authorities responsible for infrastructure, private industry, nongovernmental organizations that pursue environmental protection, and the general public all have an interest in the spatial analysis of freight transportation and logistics. Although both government and private industry are interested in a transportation network that allows efficient and economical access to nodes both near and far, they also have an interest in not overburdening that network, because of both congestion and excessive wear over the long term. Also, reducing environmental impact and burden on infrastructure is of direct interest to both transportation and environmental regulators within the government. It will also be of increasing interest to major players in various sectors of the economy as environmental management becomes a core component of doing business. Rapid and long-distance transportation can create economic opportunities for products, but rationalized, shorter-distance shipment patterns can reduce the transportation cost component of a product and hold down overall costs.

On this basis, the target decision-maker for the methodology is either the government analyst, the representatives of a given industrial sector, third-party organizations, or some combination of the three. Analysis by a group of firms rather than an individual firm is seen to be advantageous in that a group can collectively evaluate current demand for freight ton-miles, potential for improvement, and the feasible contribution of each member of such an association to achieve an improved pattern.

Extraction and Investigation of Geographic Paper Flow Data from CFS

As a starting point for the case study, an OD matrix of paper between the 48 contiguous states (excluding Alaska and Hawaii) was extracted from the CFS data. An OD matrix consists of all geographic points in a network (48 individual states in this case) appearing in both the rows and the columns of the matrix, so that flow volumes between any two points can be presented. In the case of the 48 states, it is also possible to include *intrastate* flows on the diagonal of the matrix, which represent flow volumes within the state. Volumes can be presented in either tons or ton-miles. A subset of the largest flows measured in tons/year from this matrix were presented graphically above in Fig. 13-1*b*.

An example OD matrix extracted from the complete 48×48 matrix (i.e., including the 48 contiguous U.S. states) for pulp and paper products, including 9 states, is shown in Table 13-4. States appear in alphabetical order. Since a commodity flow survey captures a large number of short-distance shipments, many such shipments stay within the geographic boundaries of the state, so that the intrastate volumes tend to be large relative to interstate volumes. Flows to adjacent states are often relatively large as well; for example, as shown in the table, Alabama > Georgia, Georgia > Florida, Arizona > California. Interstate flows between small states or states at a great distance (Delaware > Connecticut, Colorado > Florida) are often zero or negligible.

Due to limits on statistical reliability, not all ton-miles of freight for a given commodity reported at an aggregate level are published in the cells of the OD matrix. As shown in Fig. 13-2, 18% of the 101 billion ton-miles of pulp and paper estimated to have

Origins	AL	AZ	AR	CA	CO	CT	DE	FL	GA
Destinations									
AL	1674	0	131	0	0	0	0	201	517
AZ	0	436	17	305	9	0	0	0	2
AR	0	0	1,232	0	0	0	0	69	221
CA	270	292	624	11,873	22	0	0	71	134
CO	0	0	35	41	1032	0	0	0	33
CT	0	0	37	0	0	555	0	45	0
DE	0	0	0	0	0	6	35	0	0
FL	657	0	109	0	0	0	0	3089	1384
GA	2,133	0	217	54	0	11	0	452	4493

Notes: Abbreviations are AL = Alabama, AZ = Arizona, AR = Arkansas, CA = California, CO = Colorado, CT = Connecticut, DE = Delaware, FL = Florida, GA = Georgia. Alaska not included. Source states are in columns and destination states are in rows, e.g., right-most column, first row shows that 517,000 tons of pulp and paper were moved from Georgia to Alabama. Cells on diagonal represent intrastate flows, e.g., 11.9 million tons of product shipped were intrastate within California, where both the origin and destination were within that state.

TABLE 13-4 Example of OD Matrix for Pulp and Paper Movements for the United States A to G in 1,000 Tons/Year, 1993

been generated in the United States in that year were not published in the OD matrix because of these reliability limits. The remaining ton-miles that are published in the OD matrix are primarily interstate flows of paper and paper products (64%), with additional interstate flows of pulp products (9%) and intrastate movements (8%) of all types, including pulp products, paper, and paper products.

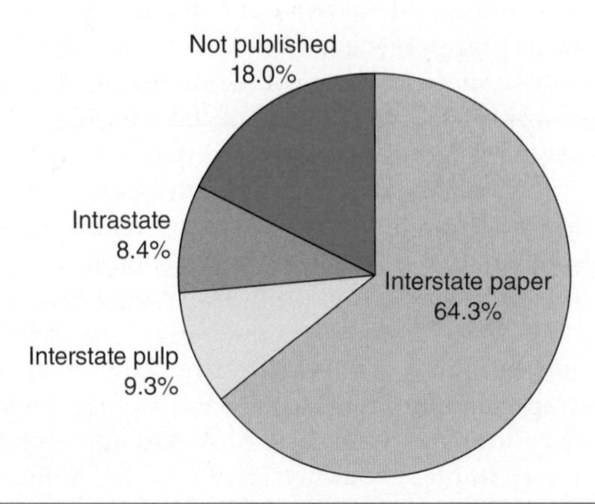

FIGURE 13-2 Breakdown of pulp and paper ton-miles among component segments (Total = 101 billion ton-miles/year).

Note: "Not published" means the ton-miles appear in the total ton-miles for pulp and paper at the aggregate level (See Table 13-2), but are not assigned to any cell in the national OD matrix.

Source: 1993 U.S. Commodity Flow Survey.

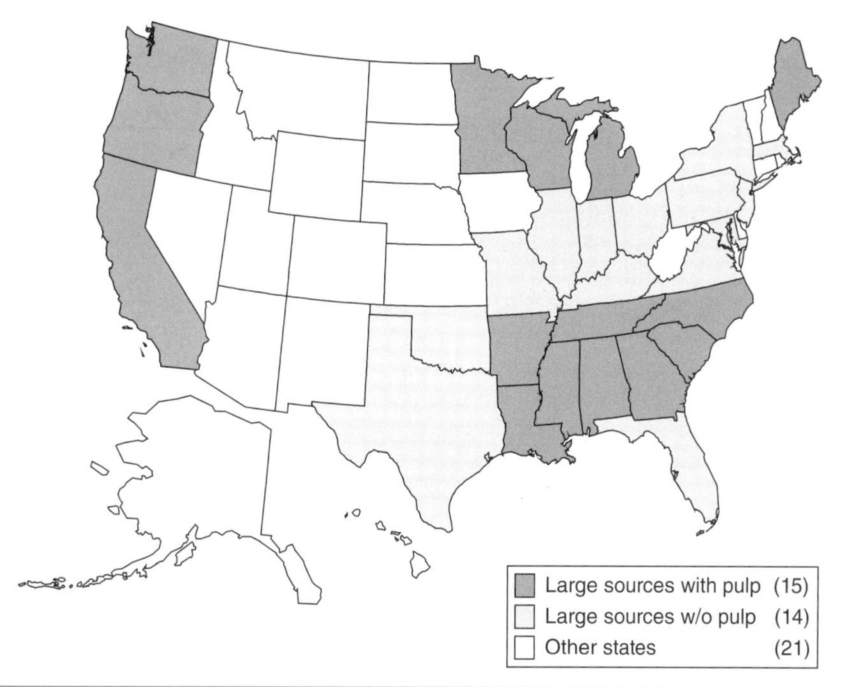

FIGURE 13-3 Major origin states for paper products with pulp resources and without.
Source: 1993 U.S. Commodity Flow Survey.

The interstate shipments can be further broken down among three categories of states, based on the division of the approximately 75 billion ton-miles of interstate flows of pulp and paper products from Fig. 13-2 (i.e., 73.6% of 101 billion ton-miles). Logically, a state that produces a large amount of paper pulp is also likely to have an accompanying paper products industry (e.g., industrial plants to convert woody material grown in-state into paper pulp). In addition, some states without a pulp industry may nevertheless be large producers of paper products (for example, if they are adjacent to pulp-producing states, or if they generally have a large manufacturing base). Therefore, in Fig. 13-3, states are divided into three categories: (1) large shipment origins with pulp resources (>1 million tons/year pulp and paper shipments, and >0.1 million tons/year pulp products), (2) large sources without pulp (same but <0.1 million tons/year pulp products), and (3) small sources (<1 million tons/year pulp and paper). On the origin side of the matrix, these three categories account for 56%, 34%, and 10%, respectively, of pulp and paper tons shipped interstate from all states. On the destination side, they account for 33%, 58%, and 9% of the interstate tons delivered.

Estimation of OD Pair Distance and Observation of Patterns

Along with total tons of flow per year, travel distances between OD pairs pose another consideration for modeling commodity flows. In reality, flows within and between states occur between many individual origins and destinations scattered throughout each state. For simplicity, these paths are reduced to a single path between any OD pair (or within the state for intrastate flows) that can be visualized as occurring between single point "centroids" located somewhere in the approximate center of transportation activity for the state.

Distances between centroids are estimated using USCFS data on tons and ton-miles. As a part of the data publication, USCFS has estimated ton-miles for each OD pair using a national multimodal transportation network model developed by the Center for Transportation Analysis (CTA) at Oak Ridge National Laboratory, Tennessee. Not every origin-destination postal code has a node in the CTA network, so shipments are first routed between postal codes and the nearest node in the network, and then routed through the network using an all-or-nothing assignment which minimizes impedance; in other words, providing the most convenient path in the network. Real-world flows might travel by any one of several alternative paths depending on circumstances at the time of shipment, but this simplification is thought to be acceptable without loss of accuracy. For all references to distance on OD pairs in this case study, distances in the network were calculated by dividing total ton-miles by the total tons moved according to USCFS. In some cases, there were 0 tons of pulp and paper moved on a given OD pair, or such a negligible amount that it was recorded as zero in USCFS (see Table 13-4, which includes many such instances). Since in this situation neither ton nor ton-mile values are available to compute an estimated distance value, the total ton and ton-mile values across all commodity groups are used as a substitute for calculating an approximate distance (these were found to be always non-zero for any pair of states in the 1993 USCFS). Also, total supply of products from each state is set to be the total tons originated by that state, and demand is set at the total tons arriving into the state.

The resulting pattern of OD flow movements, from observing the entire matrix for the 48 contiguous states, shows two main patterns:

1. *Trend 1:* Pulp and paper shipments are largely dictated by proximity to the nearest of the four main regions of pulp and paper production: in order of productive size, southeast, far west, north central, and northeast. Thus many paper products arrive at their destination either from a nearby pulp-producing region, or a paper producer that is in turn near the pulp producing region. For example, Pennsylvania, a large consumer of paper products, receives significant tonnage from Tennessee and Georgia in the southeast region, and significant additional tonnage from Ohio, which is close to both the southeast and north central regions.
2. *Trend 2:* There are also measurable quantities of "cross-trade" flows of paper products from pulp-producing regions to non-pulp producing regions that are distant from the region of origin. Thus, the pulp-producing states of the southeast ship significant tonnage to the west coast states like California or Washington, while measurable volumes flow from the far west into eastern states like Pennsylvania or New York.

If the distribution of pulp and paper movements were to move partially or entirely away from the second trend and toward the first, some fraction of total ton-miles and attendant energy consumption, greenhouse gas (GHG) emissions, and so on, might be avoided. The optimization model and application in the next section is intended for this purpose.

13-4-2 Creation and Solution of National Optimization Model to Quantify Potential Transportation Reduction

In the previous treatment of the national volume and spatial pattern of pulp and paper movements, the following data have been gathered: (1) observed OD matrix flows between all of the 48 contiguous states; (2) approximate supply and demand for product in each state, which is estimated by summing total flows in and out of these states; and (3) a matrix of distances between any two states for the average paper shipment. On one level, these data provide a picture of the spatial patterns as they exist currently. As introduced at a simple level in Chap. 3, it is also possible to create an optimization model for generating a minimum-distance distribution pattern, which can then be used as a lower bound for reducing freight measured in ton-miles compared to the existing pattern. If the existing pattern is close to the lower bound (as might be expected in the case of a low-value per ton commodity like nonmetallic minerals in Fig. 13-1(*c*), then there may be little motivation to look for ways to reduce average distance per shipment or ton-miles/ton. If, on the other hand, there is a wide difference, then a spatially improved pattern may be able to make a significant improvement to the overall efficiency of distribution of the product.

The optimization model that minimizes total transportation ton-miles or cost is called the transportation problem. Here we consider a version that minimizes total ton-miles. Let Z be the value of total ton-miles (in a different version this could be total dollar cost instead), D_{ij} the distance between origin i and destination j, X_{ij} the volume of flow in tons assigned to the OD pair ij, A_i the capacity to provide product at source i, and B_j the demand for product at market j. The problem is then formulated:

$$\text{Minimize } Z = \sum_{i,j} D_{i,j} \cdot X_{c,i,j} \tag{13-4}$$

$$\text{Subject to } \sum_{j} X_{i,j} \le A_i, \forall i \tag{13-5}$$

$$\sum_{i} X_{i,j} = B_j, \forall j \tag{13-6}$$

$$X_{i,j} \ge 0 \text{ , } \forall i,j \tag{13-7}$$

Equation (13-5) prevents product demanded from i from exceeding the capacity at node i, Eq. (13-6) ascertains that demand at j is met, and Eq. (13-7) makes sure that all flow volumes assigned by the model are nonnegative.

Next, we extend the model from the single-commodity version in Eqs. (13-4) to (13-7) to a multicommodity version. Recall that like many commodity groups, paper is divided into several products (bulk paper rolls, paperboard, paper products such as boxes, etc.), so the problem is reformulated to take the presence of multiple commodities into account, with new or modified variables and parameters defined as well:

$$\text{Minimize } Z = \sum_{c,i,j} D_{i,j} \cdot X_{c,i,j} \tag{13-8}$$

$$\text{Subject to } \sum_{j} X_{c,i,j} \ge (1 - \eta_{c,i}^{\text{orig}}) \cdot A_{c,i}, \forall c,i \tag{13-9}$$

$$\sum_{c,j} X_{c,i,j} \leq A_i, \forall i \quad (13\text{-}10)$$

$$\sum_{i} X_{c,i,j} = B_{c,j}, \forall c, j \quad (13\text{-}11)$$

$$X_{c,i,j} \geq (1 - \eta^{arc}_{c,i,j}) \cdot \tilde{X}_{c,i,j} \quad (13\text{-}12)$$

$$X_{c,i,j} \geq 0 \quad (13\text{-}13)$$

where $X_{c,i,j}$ is tons of commodity c shipped between i and j
$A_{c,i}$ is original level of c at i in base case
$\eta^{\text{orig}}_{c,i}$ is maximum percent reduction allowed in output of c at origin i
$B_{c,j}$ is demand for c at destination j
$\eta^{\text{arc}}_{c,i,j}$ is maximum percent reduction in flow of c over arc ij
$\tilde{X}_{c,i,j}$ is original flow of c over arc ij
c is index for commodity type.

Like the single-commodity model, this formulation seeks to minimize ton-miles of transport in Eq. (13-8). Constraints on overall supply (Eq. 13-10) and demand (Eq. 13-11) are similar to the single-commodity case. New constraint equations are the following. Each origin is now allowed to produce a mix of commodities c. Eq. (13-9) protects each commodity from going below some fraction of its original output level; as the value of $\eta^{\text{orig}}_{c,i}$ is increased a greater fraction of $A_{c,i}$ will be available for conversion to a different product. Equation (13-12) protects the reduction in flow of a given commodity over a specific arc from going below a certain lower bound percentage of the base case flow $\tilde{X}_{c,i,j}$.

In this model, Eqs. (13-9) and (13-12) are used to illustrate the effect of different levels of product aggregation, and also the tradeoff between achieving the maximum possible reduction in freight ton-miles and minimizing the disruption of changing output mix and pairing of origins and destinations. According to Eq. (13-9), outputs of certain commodities can exceed the original level $A_{c,i}$, but Eq. (13-10) prevents the sum of all outputs at i from exceeding the overall level A_i. In principle, the model could be generalized to allow an increase in capacity above the given A_i to test the effect of such an increase on potential savings. For example, the right-hand side in Eq. (13-10) could appear as $(1 + e)A_i$ to indicate that the total might be increased by some amount e to reflect unused production capacity or a deliberate attempt to increase supply capacity in strategic locations. On the distribution side, Eq. (13-12) protects the shipper-consignee relationship for a given OD pair ij and commodity c. As the maximum allowable change increases, the model becomes more and more likely to recommend a flow pattern in which a particular commodity c must be sourced from a new location i, while the shipper at i must shift to a closer destination for his/her product. Lastly, one can suppress the subscript c resulting in a single commodity model.

Small-Matrix Demonstration Model to Illustrate the Concept in Practice

To illustrate the concept, we have created a demonstration network using flow data for the Pulp & Paper group, choosing four U.S. states representing the four major paper producing regions in the United States (Oregon = OR, Georgia = GA, Maine = ME, and Wisconsin = WI), along with five major destination states for paper products (California = CA, New York = NY, Florida = FL, Illinois = IL, and New Jersey = NJ). Tons shipped between OD pairs and average distances between OD pairs are obtained from USCFS. In this network, capacity at an origin is taken to be the sum of outgoing shipments across all destinations, and demand at a destination is set as the sum of all arriving shipments. The base case is shown in Table 13-5; the average length of haul in this base case is 810 miles, and the total flow value of 6.86 billion ton-miles represents 6.6% of the total pulp and paper ton-miles published in USCFS.

(*a*) Flows in 1000 Tons Per Year

	From				
To	**OR**	**WI**	**GA**	**ME**	**Demand**
CA	2,094	267	134	61	2,556
NY	74	283	466	479	1,302
FL	0	87	1,384	61	1,532
IL	44	1353	369	444	2,210
NJ	0	94	322	125	541
Supply	2,212	2083	2,675	1,170	

(*b*) Ton-Miles Generated for Each OD Pair from Flow Volume in Part (*a*) in Million Ton-Miles Per Year

	From:				
To	**OR**	**WI**	**GA**	**ME**	**Total**
CA	1868	629	361	203	3061
NY	232	298	519	245	1294
FL	0	123	474	102	699
IL	106	330	330	580	1346
NJ	0	95	299	61	456
Total	2206	1475	1983	1192	

(*c*) Distance for Each OD Pair, in Miles

[miles]	**OR**	**WI**	**GA**	**ME**
CA	892	2352	2687	3361
NY	3149	1053	1114	511
FL	3411	1414	343	1689
IL	2409	244	894	1306
NJ	3235	1021	929	488

Note: Mileage values in this table are the ton-mile per ton values based on the published ton and ton-mile values in tables (*a*) and (*b*), i.e., for a given OD pair, dividing ton-miles by tons gives the distance shown. Note that individual flows may not add up to totals due to rounding.

Table 13-5 Observed Flows, Ton-Miles, and Distance in Test Network Extracted from USCFS

Turning to energy consumption, movement of paper by truck is estimated to have required 2,844 Btu/ton-mile while movement by rail requires 533 Btu/ton-mile at the time of the 1993 USCFS. Truck and rail together account for most of the pulp and paper ton-miles recorded in USCFS, but the remaining combination of all other modes of pulp and paper movements have an estimated average energy intensity of 1904 Btu/ton-mile. The weighted average of these three individual intensity values gives an overall average of 1886 Btu/ton-mile. Example 13-1 presents the application of the models in Eqs. (13-4) to (13-13) to the observed base case OD flow pattern in Table 13-5.

Example 13-1 Test for the potential for reduced ton-miles and energy consumption based on the flow pattern observed in Table 13-5. Use linear programming to compare the base case to an improved configuration where the flows have been rearranged to minimize ton-miles. Calculate the energy savings assuming constant energy consumption per ton-mile for each OD pair. Assume that demand in the base case remains the same and supply at origins cannot be increased.

Solution The solution is equivalent to the programming problem in Eqs. (13-4) to (13-7) above, or alternatively Eqs. (13-8) to (13-13) with $\eta^{orig}_{c,i} = 1$ and $\eta^{arc}_{c,i,j} = 1$, so that the multicommodity problem becomes equivalent to a single commodity problem. The problem formulation is constrained to focus only on improvements achievable through changing transport patterns and the mix of output in an origin state, that is, the demands shown in Table 13-5 must still be met, and output in each origin is not allowed to increase. However, the presence of cross-trading in the base case, especially in instances where sources are linked to destinations at opposite ends of the country (e.g., OR-NY, ME-CA), provide substantial opportunities for reducing total ton-miles.

The actual linear programming problem consists of an objective function, four-supply constraint equations (one for each of four sources), five demand constraint equations (one for each of five markets served), and 20 nonnegativity constraints (one for each OD pair). The problem can be solved using dedicated math programming software, or in this case using a solver built in to a spreadsheet.

Solving the programming problem redistributes the pattern so as to eliminate the longest flows, decreasing the total ton-miles in the network by 22.8% to 5.29 billion ton-miles (Table 13-6), and resulting in an average length of haul of 626 miles.

Since the average energy intensity is given above as 1,886 Btu/ton-mile, the total energy savings is

$$(1886)(6.86 \times 10^9 - 5.29 \times 10^9) = (1886)(1.57 \times 10^9) = 2.95 \times 10^{12} \text{ Btu}$$

Thus the original 12.9 trillion Btu energy consumption is reduced by 2.9 trillion to 10 trillion Btu.

To/From	OR	WI	GA	ME	Demand
CA	2,212	0	344	0	2,556
NY	0	0	132	1170	1,302
FL	0	0	1,532	0	1,532
IL	0	2083	127	0	2,210
NJ	0	0	541	0	541
Supply	2,212	2,083	2,675	1,170	

Note: Individual flows may not add up to totals due to rounding.

Table 13-6 OD Matrix for Allocation of Flows in Improved Scenario

Solution of the Full Model at the National Level

Having illustrated in the preceding 4 × 5 node *demonstration model* in Table 13-5 the principles of redistributing flows to reduce transportation demand (*spatial redistribution*), we now apply the modeling methodology to the full body of pulp and paper transportation volume as presented in USCFS. The goal is to see whether the savings suggested by the demonstration model are repeated when the methodology is applied on a much larger scale.

The model is run initially with $\eta^{orig}_{c,i} = 1$ and $\eta^{arc}_{c,i,j} = 1$, thus relaxing constraint Eqs. (13-9) and (13-12) so that all paper products are treated as being completely homogeneous, so that the transportation problem becomes that of Eqs. (13-4) to (13-7). (Hereafter this is referred to as the *homogeneous flow scenario.* See Scenario 1 in Table 13-8.) This solution gives a preliminary estimate at the two-digit level of potential reductions in ton-miles from an optimal pairing of origin and destination states. The optimized pattern reduces ton-miles from 82.8 to 51.0 billion, a reduction of 38.4% compared to the CFS pattern. (Note that only the published OD-flow values which total 82.8 billion ton-miles are used as a baseline, rather than the full 101 billion.) Also, the average ton-miles\ton is reduced from 661 to 431. Since the model assumes that total tonnage of paper produced in any given location can be divided among tons allocated to whatever assortment of individual paper commodities (e.g., paperboard, paper boxes, etc.) will minimize total ton-miles of shipment, this optimal solution relies on extensive flexibility in adapting the commodity mix. However, it does preserve some of the tendencies of the CFS base case, for example, substantial flows from the large paper producers in the south-east and north-central states to the industrial states of the lower Midwest (Illinois, Indiana, Ohio) and northeast (Pennsylvania, New York) are maintained.

Next, changes in the value of $\eta^{orig}_{c,i}$ and $\eta^{arc}_{c,i,j}$ away from a value of unity are used to explore how the percentage mix of commodities in each state's output and each OD flow affects the potential for reducing freight volume. For example, a particular region might specialize in paperboard, requiring it to rely on other suppliers at a considerable distance for other paper products such as paper or paper boxes. If the region in question adopted a policy of diversifying its output, it might reduce the total freight generated by the delivery of paper products to that region. In this section, the concept is illustrated by adjusting the flexibility parameters in the optimization of the national model. The real-world repercussion of this type of analysis is to show the tradeoff between achieving reductions in total demand for freight ton-miles on the one hand and in preventing disruption to existing patterns of production and distribution on the other. As background, the breakdown of pulp and paper tons loaded and ton-miles demanded is shown in Table 13-7.

The output and arc constraints are at this point applied to the solution of the national linear programming problem with the goal of observing the effect of general constraints on the total potential reduction rather than tailoring specific equations to incorporate the characteristics of individual states. Optimization of the national model with $\eta^{orig}_{c,i} = 0$ for all origin states i and commodities c dissociates the problem into five separate problems, one for each commodity (Scenario 3). The potential savings were reduced only moderately, from 38.4% to 32%. This result suggests that for the paper commodity group, most of the potential savings under a spatial redistribution strategy would come from the elimination of long lengths of haul within three-digit commodities, rather than a changing of output and shipment mix between these commodities. Whether adjustments within

STCC Number	26	262	263	264	265	266
Tons, millions	196.1	31.8%	29.4%	13.8%	21.0%	4.2%
Ton-miles, bill.	91.4	36.5%	37.1%	13.6%	9.5%	3.3%
Ton-miles/ton	466	535	588	459	211	366

Source: U.S. Dept. of Commerce (1996). Totals in (26) do not include amounts for pulp mill products (261). Remaining codes are: 262 = unprocessed paper, 263 = paperboard, 264 = paper products, 265 = paper boxes, and 266 = building paper.

TABLE 13-7 Division of Pulp and Paper Commodity Group into Commodities, 1993

the three-digit commodity would require changing commodity mix at the four-digit level is an open question that cannot be answered with the available data. The run with $\eta^{orig}_{c,i} = 0.25$ (Scenario 2) shows that much of the gain from origin flexibility would come from the first 25% change in output mix.

The effect of protecting existing commodity mix versus allowing it to vary can be applied to the percentage breakdown of product by OD pair as well as by origin (Table 13-8). Here this concept is explored by first setting $\eta^{orig}_{c,i} = 0$ for all states, and then setting $\eta^{arc}_{c,i,j} = 1$ for the 10 largest origins by tons originated (GA, AL, LA, WI, PA, AR, OH, SC, VA, OR) and $\eta^{arc}_{c,i,j} = 0.5$ for all other states, on the grounds that the states with the biggest paper industries are best able to accommodate changes in trading patterns (Scenario 4). In this scenario, the model still achieves a 16.8% reduction in ton-miles compared to the base case, which represents over half of the savings from Scenario 3.

Spatial redistribution of this type involves consideration of two related issues, namely, product mix and modal share. First, producers in each origin must be able to tailor the range of products within the commodity group to the demands of the destinations nearest to them. Modern manufacturing technology and practice make this type of flexible output increasingly viable; for instance, the increasing range of paper products made from recycled sources suggests that this increase in production flexibility is

Scenario	$\eta^{orig}_{c,i}$	$\eta^{arc}_{c,i,j}$	Bill. ton-miles	Percent Reduction*
Base case	= 0	= 0	82.8	0.0%
1. Homogeneous flow	= 1	= 1	51.0	38.4%
2. Partly flexible sources	= 0.25	= 1	52.2	37.0%
3. All sources inflexible	= 0	= 1	56.3	32.0%
4. Sources/arcs inflexible	= 0	$\eta^{arc}_{c,lg,j} = 1$, $\eta^{arc}_{c,sm,j} = 0.5$	68.9	16.8%

*Note that savings are compared to a baseline of 82.8 billion ton-miles, rather than the full 101 billion ton-miles, due to the 18% of OD flows not published for lack of statistical reliability (see Fig. 13-2). Baseline ton-miles do include 17.9 billion ton-miles of intrastate and pulp movements, which remain fixed in order to provide the raw material and final delivery for paper products in each scenario.

TABLE 13-8 Summary of Major Scenarios in National Analysis

already occurring. Secondly, the effect of shortening average length of haul on modal share should be closely monitored, as the modal share for truck tends to increase for the shortest lengths of haul. Where necessary, some adjustments to service by other modes (carload rail and water in the case of paper) may be required to maintain constant modal splits in the redistributed scenario to achieve the full energy savings potential, as explored in the following section.

13-4-3 Alternative Model to Minimize Energy Usage

Since movement of paper by truck is much more energy intensive than by rail, erosion of modal share for rail under a spatial redistribution scenario could undercut the benefits of redistribution in the absence of efforts to maintain the original modal share. Based on the modal breakdown of shipments by length of shipment from USCFS (Fig. 13-4), rail's share of tons of paper shipments nationwide was roughly 40% at distances of 500 miles and up, with truck capturing 50% of the tons. For shorter distances, rail's share falls off to less than 5% in the 0 to 50 mile range, with truck taking over 90%. (The remaining tons in both cases are carried by a combination of other modes.) By implication, if the overall spatial pattern were to move toward more shorter-distance and fewer longer-distance shipments, and modal split stayed the same as shown in the figure, total ton-miles would decrease, but energy consumption might stay the same or rise.

Data from Fig. 13-4 can be used to estimate an energy intensity by distance block μ_d, where the modal share for each of the nine distance blocks (i.e., 0–50 miles, 50–100 miles, etc.) are used with the modal energy intensities μ_k from above to calculate a weighted average. As shown in Fig. 13-5, the highest values for μ_d occur when the truck mode is

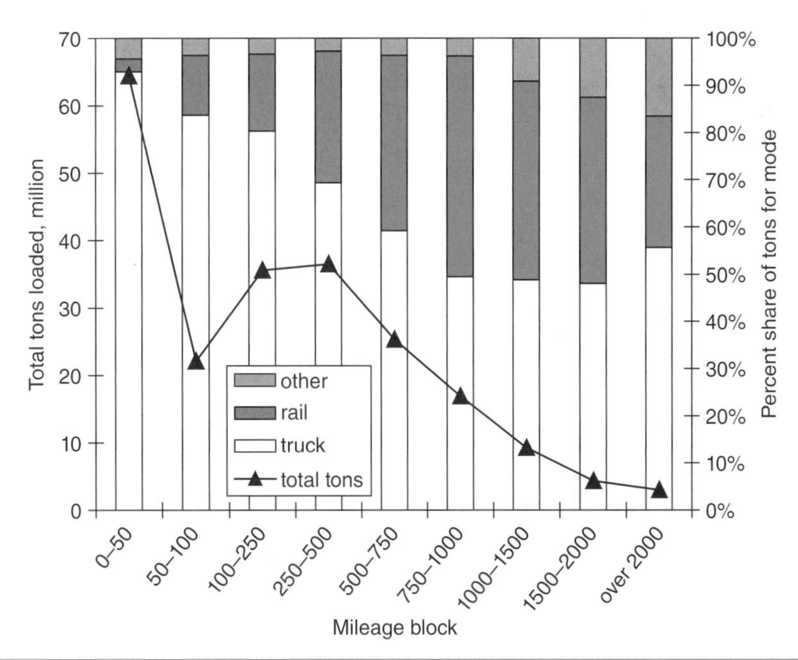

FIGURE 13-4 Pulp and paper total tons loaded and modal share by mileage block, 1993.
Source: U.S. Commodity Flow Survey.

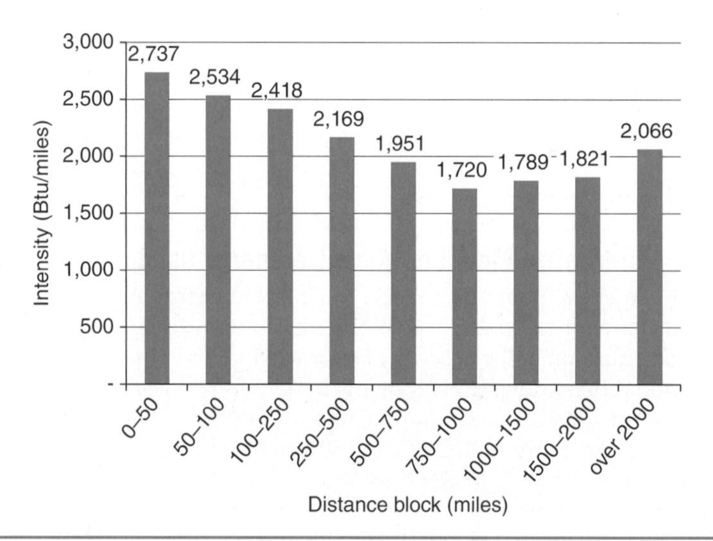

FIGURE 13-5 Energy intensity by distance block for average U.S. pulp and paper shipments.

dominant at short distances, and the lowest occur where rail is dominant in the 500- to 2,000-mile range. At the very highest distances (>2,000 miles), energy intensity actually increases slightly because the truck and "other" modes actually increase modal split at the expense of rail. This observation is counterintuitive, but because there is such a small number of tons shipped at these distances (see Fig. 13-4), it is possible that a limited number of shipments exist at these distances whose characteristics cannot be discerned from USCFS data alone but that nevertheless have some specialized needs that lead to unusual modal split requirements.

Solution of Demonstration Model for Energy Savings

In principle, one can explore the potential energy savings from redistributing flows while at the same time making maximal use of energy efficient modes by modifying the models above to focus on energy rather than ton-miles. In this case, we consider the single-commodity rather than multicommodity version of the transportation problem for reasons of brevity. Rewriting Eqs. (13-4) to (13-7) as an energy minimization problem with modal choice, and assigning tons by mode k as well as by OD pair ij, the problem is formulated as follows:

$$\text{Minimize } E = \sum_{i,j,k} \mu_{i,j,k} D_{i,j,k} Y_{i,j,k} \tag{13-14}$$

$$\text{Subject to } \sum_{j,k} Y_{i,j,k} \le A_i, \forall i \tag{13-15}$$

$$Y_{i,j,k} \le Y_{i,j,k}^{\max}, \forall i,j,k \tag{13-16}$$

$$\sum_{i,k} Y_{i,j,k} = B_j, \forall j \tag{13-17}$$

$$Y_{i,j,k} \ge 0 \tag{13-18}$$

Here the factor $\mu_{i,j,k}$ is the energy intensity per ton-mile of mode k modified to reflect the geographical or operational features for the OD pair i,j; $Y_{i,j,k}$ is the tons shipped between i and j by mode k; and $Y^{max}_{i,j,k}$ is the maximum tons allowed for the mode on the OD pair. Where information about these features proves insufficient, a uniform value μ_k for each mode could be used.

One difficulty arises out of the need to determine the modal capacity for each OD pair. This is particularly true for the rail mode, as its large energy efficiency advantage over truck means that the model will allocate shipments to rail wherever possible, so that the solution will be quite sensitive to the maximum volume that can be moved by rail on a given OD pair per year, that is, $Y^{max}_{i,j,k}$ where k = rail. Estimating maximum rail capacity for a large volume of rail OD pairs is beyond the scope of this case study, so as an approximation a modified version of the energy minimization model is run, in which the modal split for each OD pair is held constant, and energy intensity values from Fig. 13-5 are used to calculate total energy consumption.

The new model then takes the following form, where the modal choice variable k is collapsed into a single mode with the observed mix of truck, rail, or other for each OD pair and resulting average intensity $\mu_{i,j}$:

$$\text{Minimize } E = \sum_{i,j} \mu_{i,j} D_{i,j} Y_{i,j} \tag{13-19}$$

$$\text{Subject to } \sum_{j} Y_{i,j} \leq A_i, \forall i \tag{13-20}$$

$$\sum_{i} Y_{i,j} = B_j, \forall j \tag{13-21}$$

$$Y_{i,j} \geq 0 \tag{13-22}$$

Note that the OD pair capacity constraint Eq. (13-16) has been removed in the modified model. Example 13-2 repeats the solution of Example 13-1, but this time solving for minimum energy rather than minimum ton-miles.

Example 13-2 Consider the demonstration network shown in Table 13-5 and used in Example 13-1, with sources in Oregon, Wisconsin, Georgia, and Maine supplying markets in California, New York, New Jersey, Florida, and Illinois. Find the pattern of shipment that minimizes transportation energy consumption and compare the resulting energy level to the base case, taking into account the impact of shipping distance on energy intensity. Compare the shipment pattern to the solution from Example 13-1.

Solution Table 13-5(*c*) gives shipping distance by OD pair, and Fig. 13-5 gives energy intensity by distance block. Thus each OD pair can be assigned an energy intensity; for example, OR → CA is a distance of 892 miles, therefore it falls in the 750- to 1,000-mile block and has an intensity of 1,720 Btu/ton-mile. This process is repeated to construct an OD matrix of intensity values:

States	OR	WI	GA	ME
CA	1,720	2,066	2,066	2,066
NY	2,066	1,789	1,789	1,951
FL	2,066	1,789	2,169	1,821
IL	2,066	2,418	1,720	1,789
NJ	2,066	1,789	1,720	2,169

Next, for each OD pair, the value from the above table and the distance for the pair are multiplied to create a value of Btu/ton shipped over that pair. Using the OR → CA example again, the value is

$$(892 \text{ mi})(1720 \text{ Btu/tmi}) = 1.534 \times 10^6 \text{ Btu/ton}$$

Repeating this process for the entire table gives a new table in units of 1,000 Btu/ton shipped:

States	OR	WI	GA	ME
CA	1,534	4,858	5,550	6,942
NY	6,504	1,884	1,993	997
FL	7,046	2,529	744	3,075
IL	4,976	590	1,538	2,336
NJ	6,682	1,826	1,598	1,058

The linear programming problem to solve for the minimum-energy solution is very similar to the minimum ton-mile problem with the same number of variables and constraints. In this case, because the problem is small it actually results in the same optimal pattern of flows as seen in Table 13-6. However, because the figures for estimating energy consumption are different, the total energy before and after changing the pattern is slightly different. Instead of using the intensity figure of 1,886 Btu/ton-mile throughout, intensity varies between OD pairs, so now the energy figures are 13.0 trillion Btu before and 10.2 trillion Btu after making the change.

Also, the optimal pattern is unchanged for this small problem, but it need not always be so. For instance, in the larger full national model discussed next, the minimum energy pattern is different from the minimum ton-mile one.

Solution of Full National Model for Minimized Energy

The full national model can be explicitly solved for the pattern that minimizes energy, as opposed to solving for the pattern that minimizes ton-miles and then calculating the corresponding amount of energy consumed in that pattern. Solution of the energy-minimization problem in the homogeneous flow scenario (i.e., single-commodity problem) gives a reduction of 34.7% from 158 to 103 trillion Btu. Since much of the shifting of flows occurs between OD pairs above 500 miles where mode split is fairly stable, reductions in ton-miles translate directly into energy savings. Comparing this energy-savings estimate with the ton-mile reduction of 38.4% shown to be possible for the homogeneous flow scenario shows that although the constant mode split constraint somewhat diminishes the potential energy savings of spatial redistribution, the energy savings are still substantial.

For comparison, the energy use in the national-level minimum ton-mile problem can be calculated using the distance-specific energy intensity values μ_{ij} as opposed to the constant $\mu = 1886$ Btu/ton-mile. Let E_0 be the total energy consumption in this case, then

$$E_0 = \sum_{c,i,j} \mu_{i,j} D_{i,j} X_{c,i,j}$$

where $X_{c,i,j}$ represents the values of the optimal flows from solving the model in Eqs. (13-8) to (13-13). In the case of the homogeneous flow scenario (Scenario 1 in Table 13-8), the calculation of E_0 gives a 34.2% savings compared to the base case, so that in this instance, although the minimum ton-mile and minimum energy solutions are quite similar in their resulting predicted energy consumption, differences in assigned OD flows lead to slightly different energy consumption values. It is consistent with the

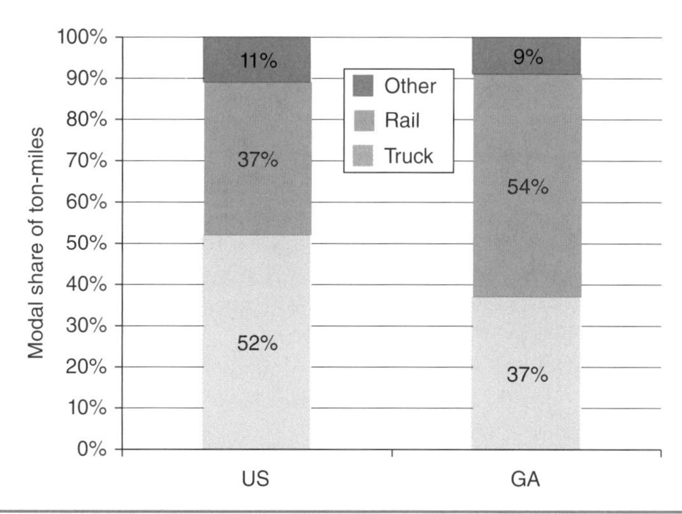

Figure 13-6 Comparison of modal share of ton-miles for U.S. overall average and state of Georgia. *Source:* U.S. Commodity Flow Survey.

expectation that the minimum energy solution has a greater savings (34.7% vs. 34.2%) since the objective of minimizing energy is inside the optimization problem. The two solutions happened to be close in their predicted energy savings in the case of the pulp and paper sector. Application of this technique to another commodity group might show a wider divergence between the minimization of ton-miles and energy use.

Impact of Origin State on Modal Split and Energy Intensity

Along with shipping distance, it is also possible to consider the effect of origin state on modal split and energy intensity. Differences in infrastructure and connectivity with road, rail, and waterway networks can affect the ability of shippers and carriers of goods to move products originating in a given state by a given mode. Figure 13-6 compares the U.S. national modal split for pulp and paper products with that of pulp and paper shipments originating in the state of Georgia, based on data from USCFS. Since ton-miles of pulp and paper shipments originating in Georgia have a higher modal share for rail than the national average (54% vs. 37%), their average energy intensity per ton-mile may be lower than the national average. The 1993 USCFS did not provide modal split for each of the 50 individual states by commodity and distance block, so the calculation of energy intensity in Example 13-2 and in the national minimum-energy model is calculated by distance block only, without taking origin state into account. However, with investment of additional resources in data analysis, this information could be extracted from the raw CFS data and then published to make possible a more accurate estimation of modal split and energy intensity.

13-5 Discussion: Prospects for Changes in Freight Spatial Patterns

As illustrated in the above case study from the pulp and paper sector, the first purpose of spatial analysis is to understand the spatial pattern for a given commodity as it currently exists. Is the pattern spatially intensive or compact? Is there some discernible logic to it, such as a limited set of dominant sources or markets that determine the overall pattern? Since the negative impacts from freight transportation are an important and growing concern, it is prudent for leading businesses in each commodity sector to understand their collective impact on the network.

Two factors are currently shaping spatial patterns of freight transportation. The first is the continued impact of spatial spreading as forces that have been at work for the past several decades continue to influence the geography of freight. Producers will continue to seek new markets to grow their business, and one way to do this is to look further afield toward new markets, possibly in more distant countries. The second factor is evolving consumer demand: consumers will seek out new products to meet changing tastes and/or demands, and again it is the more distant markets that are less likely to have been accessed before. In some cases, whole countries that have been politically isolated from the world community become able to participate at some point, contributing to further spatial spreading.

Counteracting the influence of spreading, and perhaps increasingly so, is the impact of *saturation* of the freight network. Both higher energy costs and increased congestion or other problems in the freight network have the tendency to curb growth in freight, measured in ton-miles generated per capita. In industrializing countries where infrastructure is still being built out and incomes are growing rapidly, the effect of saturation may be small. In industrialized countries, however, the transition to a global market economy is largely complete, so that going forward primarily population growth will increase total ton-miles of freight. In short, saturation is a third intermediate factor between spatial spreading on the one hand, which increases ton-miles/ton, and spatial redistribution on the other, which decreases them.

13-5-1 Spatial Redistribution Applied to Several Possible Sectors

The goal of spatial redistribution is to go beyond understanding the current freight spatial pattern and actively reduce total ton-miles, or at least curb their growth. In the pulp and paper case study above, the potential savings were substantial. A preliminary survey of length of haul and distribution of major sources for a wide range of commodity groups suggests that this methodology could find application for many types of products. For some of these groups, especially those with high value densities (e.g., electronics, fashion apparel), trends towards more distant sources of materials and increasingly complex supply chains will continue to put upward pressure on total ton-miles. These groups, however, cover only a minority of total freight ton-miles; mid-value products represent a much larger volume of ton-miles. The following examples show numerous ways in which the transportation impact of mid-value products can be reduced via spatial redistribution:

- *Food and beverage products:* Many regions of the industrialized world including Europe and North America have agricultural activity based on sufficient rainfall and adequate soil to grow a wide range of crops. These regions typically support some level of regional food production, although consumers also purchase from national and international sources to meet some of their food needs. Local and regional growers of fresh fruits and vegetables as well as regional food products made from local crops and produce can compete for sales on the basis of freshness and the "locavore" movement, which promotes local foods. Such a shift has the potential to reduce long-distance food ton-miles.

- *Paper, metal and glass products*: Measurable fractions of used paper, metal, and glass are recycled into raw materials for fabrication into new products in many industrialized countries. If regional production and distribution of products with recycled content can be developed, long-distance shipping could be reduced. Recycling programs are not 100% effective, so in each new generation of products a fraction of the raw material input is from virgin sources. These sources may be distant from the production location and therefore require shipment over long distances, and possibly globally. To the extent that the input is recycled, however, an opportunity for a more regional supply chain arises.
- *Wood products:* Wood products such as furniture or building materials do not generally lend themselves to recycling, although some can be salvaged and reused. However, as was shown in the pulp and paper case study, forest resources are dispersed to various parts of the continental United States. In Europe, they exist in the British Isles, Scandinavia, and Eastern Europe. To the extent that wood products can be sourced regionally and then have their resources recovered at the end of the life cycle (for example, through waste-to-energy conversion) the impact of transportation can be reduced.
- *Energy products:* A measurable percentage of total freight ton-miles is incurred in moving coal, petroleum, and various coal and petroleum products, through a mixture of barge, rail, and pipeline transportation. Energy product extraction tends to be concentrated in specific geographic locations; therefore distribution distances can be long. Many regions possess a mixture of solar and wind resources that could make them less dependent on long-distance imports of energy products. Also, primary energy (whether renewable, fossil, or nuclear) can be converted to energy carriers such as electricity or possibly in the future hydrogen, which could reduce ton-miles of freight carried on the surface transportation network.

Example of a Public-Private Consortium Applying Spatial Redistribution

A shipper of goods might apply the concept of spatial redistribution to their operations independently. They might also work as a member of a consortium with some or all major members of their particular economic sector, similar to the benchmarking coalition discussed in Chap. 12. In the case of a redistribution coalition, the target decision-makers include representatives of a number of producers in the sector (participants) and government statistical analysts responsible for transportation statistics (analysts). The consortium is motivated by an interest in reducing the environmental impact of freight distribution by reducing total ton-miles, or at least curbing the growth in total ton-miles, over a fixed time horizon. The steps they take include (1) analyzing the current distribution pattern and estimating the physical potential for reducing ton-miles, (2) setting targets based on the estimated physical lower bound, and (3) then carrying out a policy of attaining those targets over the lifetime of the project. Also, although the coalition in this case focuses on reducing ton-miles, an equivalent project could be undertaken to reduce some aspect of environmental impact, such as total energy use or carbon emissions.

According to the project's framework agreement, the consortium's ability to achieve the target will depend in part on demand growth over the project lifetime. If demand growth is strong, targets may need to be adjusted accordingly. Alternatively, the consortium might opt for a target value measured in ton-miles/ton rather than total ton-miles. Such a target could be achieved independent of changes in demand, but may not be agreeable to the analysts, who may feel compelled to pursue an absolute reduction in freight ton-miles to best serve the public interest.

After the target has been agreed upon, the analysts provide each member with a confidential firm-level profile of existing and optimal flows, and agree to provide analytical support to the consortium for the multiyear timetable of the project, while taking care not to divulge information sensitive to one participant to any of the others. This respect for confidentiality is one of the central characteristics of commodity flow data gathering, so methods for protecting sensitive information are well-established. At different points in time during the project as well as at the end, the analysts provide a general report to all participants evaluating how well the targets are being achieved based on flows published in the commodity flow survey at that time; they also provide a confidential, individual evaluation to each firm. After the first time horizon of the project is complete, the consortium can opt to continue the project with either more or less ambitious targets based on the growth in freight volume in the preceding time frame, the level of public pressure on the industry to reduce negative effects from freight, and the perceived feasibility of further spatial redistribution of flows.

Possible Extensions to Spatial Redistribution Strategy

As a sustainability strategy, spatial redistribution is a broad concept at an early stage of development, so many extensions are possible:

- The strategy presented in this chapter has focused primarily on substituting "near for far" sources, but reducing the circuity of movement of goods can also play a role.[4] Introducing a vehicle routing component into the modeling of freight movements can address circuity.
- Greater geographic detail would allow analysis of local and regional issues such as emissions of NO*x* or particulates and noise; it is possible that a spatial redistribution effort might create negative impact "hot spots" even as it reduces total ton-miles and energy use.
- The connection between changes in output mix by origin and total energy use and emissions on the production side could be included. An implicit assumption of the pulp and paper case study in this chapter is that impacts on the production side are equal, so that a gain on the freight side is a net gain overall. In a more complete model, however, improvements on the transportation side may be offset by deterioration on the production side, and vice versa.
- Limits of available data bound the geographic scope of any spatial redistribution study. In the case of the USCFS, flows over international borders are

[4]Whitelegg (1995, p. 40) has proposed the term "near for far," which is roughly synonymous with spatial redistribution, and has also advocated studying route circuity alongside overall distance between origin and destination.

not included. Although not available in USCFS, inclusion of the flows to and from Canada, Mexico, and other countries would support a more complete model. Similarly, in Europe, the agency Eurostat has been better able to measure flows within the European Union than in and out of it. The goal of making spatial pattern studies more international and global in scope is quite challenging, because government statistics offices have limited jurisdiction and therefore limited ability to survey commodity flows across international borders.

13-5-2 Limitations on Ability to Redistribute Freight Spatial Patterns

Spatial spreading of freight is an underlying driver of growth in negative societal and environmental effects, and spatial redistribution as a potential countermeasure is an elegant solution in that it can address many negative consequences at once: energy consumption, emissions, contribution to traffic congestion, and localized negative consequences adjacent to roads, railroads, and freight transshipment facilities. At the same time, it is more difficult to implement than other options because it works in the opposite direction of the trend in recent decades toward longer supply chains in a globalized marketplace.

Comparisons can be made to strategies discussed in Chap. 12. Intramodal improvements to individual modes, such as improvements to truck technology, are a natural extension of the use of technology to reduce operating costs. Improvements to truck-rail intermodal that win traffic from trucking are a similar proven use of technology to achieve energy savings and pollution reduction in a competitive marketplace. Growth of intermodal truck-rail also addresses some of the chronic problems of long-distance trucking, including driver shortages, highway congestion especially around urban areas, and deteriorating road infrastructure quality.

The most important limitation on spatial redistribution is the influence of economic forces that drive spatial spreading in the first place. Long-distance shipping and long supply chains with many steps are driven by opportunities to reduce total cost (including both production and logistics cost) and to earn profits by selling products into new markets that might not have been accessible before the advent of the logistics revolution. To some extent, saturation happening currently may slow or even stop the growth of total ton-miles of freight in the system in industrialized countries, but it is unlikely to substantially decrease them. The modern economy has come to depend on its current level of freight intensity.

Two possible forces might accelerate the application of spatial redistribution. First, a top-down intervention in the freight system from governments might push the system in the direction of fewer ton-miles per ton of product moved or per unit of GDP, as part of larger efforts to respond to climate change or peak oil. There is little indication of interest from political leaders, especially when these aims might be more easily achieved by shifting vehicles to nonpetroleum fuels. Second, consumer preference might give a marketing advantage to local or regional products as opposed to national or international products. Manufacturers might incur added cost by producing at a more regional level but be able to pass this cost along to some segment of consumers in the name of reducing the transportation burden as well as supporting the local economy. The rise of sales of fresh seasonal fruits and vegetables at farmer's markets in many parts of the United States provides an example.

13-6 Summary

This chapter considers spatial, or geographic, patterns of freight distribution and their contribution to freight volume, which has been growing in the last several decades, as reflected in the average length of shipment and the number of ton-miles generated by each ton of freight moved. It was observed that the increase in the volume of freight is not just a matter of more goods moving, but also of each unit of goods moving on average further; this phenomenon is known as *spatial spreading*. The trend may be toward growth in freight miles, but the type of commodity being moved also affects spatial patterns: low-value bulk commodities are less likely to move long distances, but generally finished products and especially high-value, high-technology products are likely to move long distances.

Analysis of geographic patterns can be used not only to understand current directions in the growth of freight, but also to uncover opportunities to reduce freight, and its resulting social and ecological burden, by changing patterns using a technique called *spatial redistribution*. This latter concept was illustrated using a case study from the pulp and paper sector. In the case study, levels of output at origins and demand in markets are held fixed, and then an optimization model creates scenarios where total freight ton-mile requirements are reduced while still delivering the desired number of tons needed in each market and not exceeding production capacities. For many commodities such as pulp and paper, modal split varies at different shipment distances, so the effect of modal split is considered in a modified version of the model that looks at minimizing energy consumption as opposed to ton-miles. Spatial redistribution is conceptually elegant because it addresses total ton-mile growth at the source with potential benefits for reducing energy consumption, emissions, impact on climate change, infrastructure wear-and-tear, and levels of local nuisance from freight. However, it is also difficult to implement compared to some more pragmatic solutions such as changing vehicle technology, so the chapter concludes with a discussion of its merits and disadvantages.

References

Boge, S. (1995). "The Well-Travelled Yogurt Pot: Lessons for New Freight Transport Policies and Regional Production. *World Transport Policy & Practice* 1(1): 7–11.

Nijkamp, P., A. Reggiani, and W. F. Tsang (2004). "Comparative Modelling of Interregional Transport Flows: Applications to Multimodal European Freight Transport." *European Journal of Operational Research* 155(3):584–602.

Royal Commission on Environmental Pollution (RCEP) (1994). *Eighteenth Report: Transport and the Environment*. HMSO, London.

USDOC (U.S. Bureau of the Census) (1996). *Commodity Flow Survey 1993*. U.S. Dept. of Commerce, Washington, DC.

USDOC (U.S. Bureau of the Census) (2009). *Commodity Flow Survey 2007: United States Summary*. U.S. Dept. of Commerce, Washington, DC.

Vanek, F. (2000). "The Transportation-Production Tradeoff in Regional Environmental Impact of Industrial Systems: A Case Study in the Paper Sector." *Environment and Planning A* 32:817–832.

Vanek, F. (2001). "Analysis of the Potential for Spatial Redistribution of Freight Using Mathematical Programming." *European Journal of Operational Research* 131(1):62–77.

Whitelegg, J. (1995). *Freight Transport, Logistics, and Sustainable Development.* World Wildlife Fund, Godalming, Surrey, UK.

Wisetjindawat, W., S. Kazushi, and M. Shoji (2006). "Commodity Distribution Model Incorporating Spatial Interactions for Urban Freight Movement." *Transportation Research Record* 1966:41–50.

Further Readings

Bleijenberg, A. (1996). *Freight Transport in Europe: in Search of a Sustainable Course*. Centrum voorEnergiebesparing en SchoneTechnologie ("Center for Energy Conservation and Clean Technology"), Delft, Netherlands.

Campisi, D. and M. Gastaldi (1996). "Environmental Protection, Economic Efficiency, and Intermodal Competition in Freight Transport." *Transportation Research C* 4/6:391–406.

Chisholm, M. and P. O'Sullivan (1973). *Freight Flows and Spatial Aspects of the British Economy*. Cambridge University Press, Cambridge, UK.

Dantzig, G. (1963). *Linear Programming and Extensions*. Princeton University Press, Princeton, NJ.

Hitchcock, F. (1941). "The Distribution of a Product from Several Sources to Numerous Localities." *Journal of Mathematics and Physics* 20:224–230.

Holzapfel, H. (1995) "Potential Focus of Regional Economic Cooperation to Reduce Goods Transport." *World Transport Policy and Practice* 1/2:34–39.

McKinnon, A. (1995). "Logistics and the Environment." *Logistics Europe (Journal of the European Logistics Association)* 3/3:16–22.

Morlok, E. and S. Riddle (1999). "Estimating the Capacity of Freight Transportation Systems: A Model and Its Application in Transport Planning and Logistics." *Transportation Research Record* 1653:1–8.

Payne, T. (2002). *U.S. Farmers Markets 2000: A Study of Emerging Trends*. Report to the U.S. Department of Agriculture, Washington, DC.

Schipper, L., L. Scholl, and L. Price (1997). "Energy Use and Carbon Emissions from Freight in 10 Industrialized Countries: An Analysis of Trends from 1973 to 1992." *Transportation Research Part D* 2/1:57–75.

Tavasszy, L. (1996). *Modelling European Freight Transport Flows*. TRAIL, Delft, Netherlands.

Vanek, F. and E. Morlok (1998). "Freight Energy Use Disaggregated by Commodity: Comparisons and Discussion." *Transportation Research Record 1641.*

Vanek, F. and E. Morlok (2000). "Improving the Energy Efficiency of Freight in the United States through Commodity-Based Analysis: Justification and Implementation." *Transportation Research Part D* 5:11–29.

Vanek, F. and Y. Sun (2008). "Transportation versus Perishability in Life-Cycle Energy Consumption: A Case Study of the Temperature-Controlled Food Product Supply Chain." *Transportation Research Part D*, 13:383–391.

Exercises

13-1. Reconstruct the optimization of Example 13-1 to verify the solution.

13-2. Reconstruct the calculation of energy intensity by distance block and the optimization of Example 13-2 to verify the solution. Use the following modal split percentages by distance block that are presented graphically in Fig. 13-4, and energy intensity values of 2,844, 533, and 1,904 Btu/trillion mile for truck, rail, and other, respectively.

Dist. (miles)	Other	Truck	Rail
0–50	4%	93%	3%
50–100	4%	84%	13%
100–250	3%	80%	16%
250–500	3%	69%	28%
500–750	4%	59%	37%
750–1000	4%	49%	47%
1000–1500	9%	49%	42%
1500–2000	13%	48%	39%
over 2000	17%	56%	28%

13-3. A retail firm operates a *decentralized* distribution system (factory to warehouse to shop), in which the system uses 6 smaller warehouses distributed around a region to receive a product from manufacturer (called *primary distribution*) and then sends product to retail outlets (called *secondary distribution*). The firm is offered the opportunity to shift to a *centralized* system, in which each unit of product will still undergo primary and secondary distribution, but now there will only be one warehouse in the middle of the region.

The transportation of the product incurs financial cost and energy consumption per vehicle-kilometer (vkm) of movement. In the case of the warehouse costs, inventory costs are incurred by virtue of needing to keep stock on hand in the warehouse between the time that the stock is purchased and the time it is sold, and energy is consumed to operate the warehouse. Transportation costs \$1.50 and consumes 19 MJ of energy per vehicle-kilometer. Assume that all other costs and rates of energy consumption are the same for either option, and therefore are not included in the calculation since they do not affect the outcome. Note also that the cost of energy is included in the warehouse and transportation cost values given.

a. Based on the data given, determine whether the decentralized or centralized system is preferred.
b. What is the environmental dilemma underlying this decision? Discuss in a short answer, up to 1 paragraph long.

Warehouse Costs and Energy Use per Warehouse

Alternative	Inventory Cost (\$1000/year)	Energy (GJ/year)
Decentralized	190	155
Centralized	710	730

Transportation Volume Generated per Year per Warehouse

1000 vkm/year	Primary	Secondary
Decentralized	35	82
Centralized	181	645

13-4. A food products company ships foods from three production plants to four markets as follows. The capacity of plants at Boise, Dubuque, and Charleston is 2,200 tonnes, 3,000 tonnes, and 2,000 tonnes, respectively. The demand at San Francisco, New York, Miami, and St. Louis is 2,002 tonnes, 1,784 tonnes, 1,355 tonnes, and 1,972 tonnes, respectively. The mode of shipment is by truck, and the energy intensity is 2,200 kJ/tonne-km. A table of distances in kilometers between cities is given below. (a) What is the allocation of shipments from plants to markets that meets all demands, does not exceed supplies available at any plant, and minimizes energy consumption? What is the value of energy consumption in this case? (b) Suppose that for all routes with distance of 1,500 km or more, an intermodal rail service is made available with energy intensity of 1400 kJ/tonne-km. Recalculate the allocation that minimizes energy. What is the new value of energy consumption? Does the shipment pattern from part (a) change?

City	Boise	Dubuque	Charleston
San Fran	1,427	3,763	4,299
New York	5,038	1,685	1,782
Miami	5,458	2,262	548
St Louis	3,855	390	1,431

13-5. Use the OD flow data and distance matrix below for 1993 annual pulp and paper movements in the United States to answer the following questions. (a) What is the total ton-miles of freight generated in the given pattern (hereafter called the *base case*), based on the tons shipped for each OD pair? (b) What is the estimated energy consumption, considering ton-miles and distance-based average energy intensity? (*Hint*: to estimate energy consumption by OD pair, you may wish to write a short macro or use conditional formatting in a spreadsheet.) (c) Assuming that the tons shipped from each origin represents the total amount available at that origin, and the tons arriving in the market represents the total tons consumed in that destination, create a table of capacity for origin states and market size for destination states, with tons/year for each one. (d) Use the transportation model in linear programming to find the OD pattern that minimizes total ton-miles. Report the OD flows in the optimal pattern in a new OD matrix, and give the new ton-mile total in this case, and the percent reduction in ton-miles compared to the base case. (e) Pick any origin state and show a map of the paper flows from that state for both the base case and optimal pattern. The map can be either hand drawn or computer generated. (f) If the freight flows according to the optimal pattern from part (c) what is the total energy consumption in the optimal case, and the amount of energy saved, in trillion Btus, by shifting to the optimal pattern? (g) Now revise the LP so that it finds the pattern that minimizes total energy consumption by directly adding up the energy consumed in each OD pair. What is the new energy consumption in this case? Is it the same optimal pattern and energy value as in the solution from part (d), or is it different? (h) Although the optimal pattern might be advantageous from a transportation intensity point of view, there are other factors that lead to a different, more transportation-intensive spatial pattern of freight represented by the base case pattern. Explain one such factor. (Short answer, 1 to 2 sentences max)

Shipment Volumes

1000 Tons	GA	LA	WI	AL	WA	MI	ME	OR	MN
PA	660	164	278	270	0	252	523	0	88
IL	369	585	1,353	497	87	586	444	44	481
OH	465	203	421	0	0	559	117	0	0
TN	398	133	213	543	0	117	127	0	55
NC	761	170	87	0	0	62	34	0	18
FL	1,384	259	87	657	0	55	61	0	41
TX	545	1,410	258	802	82	58	60	36	95
NY	466	122	283	142	150	152	479	74	58
NJ	322	184	93	200	0	80	125	0	53
CA	134	436	267	270	1,744	192	61	2094	161

Distances

Miles	GA	LA	WI	AL	WA	MI	ME	OR	MN
PA	914	1,415	878	1,059	2,703	730	730	2,762	1,182
IL	894	884	244	799	2,218	319	1,306	2,409	574
OH	763	1,025	570	392	2,405	356	1,017	2,421	766
TN	362	481	718	359	2,402	709	1504	2,355	1,055
NC	356	976	1,023	570	2,810	742	1,118	2,764	1,278
FL	342	749	1,414	559	2,993	1,327	1,689	2,947	1,780
TX	1,215	493	1,236	864	2,305	1,328	2,167	2,583	1,432
NY	1,114	1,525	1,053	1,317	3,640	579	511	3,149	1,190
NJ	929	1,462	1,022	1,150	2,842	825	488	2,902	1,472
CA	2,687	2,037	2,352	2,289	976	2,411	3,361	892	2,230

UNIT 5

Energy and Environment

CHAPTER 14

Overview of Alternative Fuels and Platforms

14-1 Overview

This chapter gives an overview of the development of alternative energy sources and platforms for transportation, and especially for light-duty road vehicles as the single largest user of transportation energy. The various vehicle propulsion options can be categorized among a limited number of generic *endpoint technologies*, and the strengths and weaknesses of each are considered. Thereafter we discuss hybrid-electric vehicles and advanced internal combustion engine vehicles as incumbent technologies against which other emerging vehicle options that use alternative energy sources can be compared. Well-to-wheel analysis and transition pathways as new technologies penetrate the market are also discussed.

14-2 Introduction

In the modern world of motorized cities, long-distance travel by jet or limited-access highway, and global trading of manufactured goods and commodities, it comes as no surprise that the transportation sector has become an enormous end user of energy. This sector is the single largest consumer of petroleum resources in the world today, and the second largest consumer of nonrenewable fossil fuels next to electric power conversion. Worldwide, the transportation sector accounted for approximately one-fourth of the total end-use energy consumption value of 480 quads (506 EJ) in 2009. In the United States alone, in that year the transportation sector accounted for 27.1 quads (28.6 EJ), which constitutes 28.6% of the total U.S. energy end-use budget of 94.5 quads (99.7EJ), or 5.6% of the world total. To put these values in context, the U.S. transportation energy consumption rate is equivalent to 9 billion 100-watt light bulbs burning continuously 24 hours a day, 7 days a week, all year long, or 29 light bulbs for every one of the U.S.'s approximate population of 307 million people in that year. The world transportation energy consumption figure is equivalent to 40 billion lightbulbs, or 6 lightbulbs per person.

Transportation energy consumption originates in the propulsion system of the vehicle, whether it is an internal combustion engine (ICE), a jet engine, or the electric motor of a railway locomotive. It is further influenced by other design choices of the vehicle, such as materials choices, which affect its overall weight, or styling, which affects its aerodynamic drag. Engineers in an R&D setting of a laboratory or design facility appreciate the benefit of designing a vehicle to be efficient, since reduced manufacturing cost or operating cost will make the product more appealing to the management of the company and to the prospective customer, respectively. However, the pursuit of efficiency must be weighed against other priorities, such as power or performance, and often in the pursuit of product sales it is the latter two criteria that are favored. The way in which the service of moving people and goods in the real world is delivered also affects total energy use, so that an identical vehicle may achieve different levels of energy efficiency in different situations, depending on the circumstances. Land use planning (i.e., the geographic location of amenities in a region), availability of transportation infrastructure, and extent of congestion also play a role.

14-2-1 Responses to Improved Efficiency: Causal Loop Diagram

As an example of how a broader perspective can help us understand more accurately the impact of improved energy efficiency of transportation technology, we can consider the *rebound effect*, as shown in the causal loop diagram in Fig. 14-1. A common response to rising energy use in the transportation sector is to introduce policies aimed at improving the energy efficiency of vehicles. These policies cause manufacturers to develop more efficient engines and other drivetrain components, so that the drivetrain is able to move a vehicle the same distance with less fuel consumption. One part of the effect of this change is to reduce energy consumption per unit of distance, thereby influencing total energy use in a downward direction. However, there are two other effects. One effect is that the more efficient drivetrain makes larger vehicles more affordable, increasing average vehicle size. The second effect is that, according to the laws of economics, if we make an activity such as driving cheaper, demand for that activity will increase. These latter two effects both influence the amount of energy consumption in an upward direction.

The causal linkages in Fig. 14-1 show us three possible pathways from the step of increasing the level of *policies to improve energy efficiency* back to *level of transportation energy use*, but they do not tell us whether, in the end, a net improvement in the amount of energy use will result. The outcome depends in part on the circumstances in each situation where such a policy is tried. In many cases, a government that enacts such a policy may reduce overall energy consumption compared to a baseline do-nothing scenario, even after taking losses due to the rebound effect into account. However, some erosion of energy-efficiency gains is almost inevitable, unless other policies specifically aimed at curbing growth in vehicle-kilometers or vehicle size are instituted at the same time. Also, since the overall long-term baseline in most industrialized and emerging countries is a steady increase in transportation energy consumption year after year, energy-efficiency policies may make reductions against the baseline but not be able to reduce overall energy consumption compared to its level at the beginning of the policy implementation.

14-2-2 Transition to Alternative Energy as Transcendental Opportunity

Up to the present, the impact of mechanized transportation, and the automobile in particular, has been profound, both positively and negatively. On the environmental front, the impact has been substantial, as can be measured in simply the quantities of air pollutants and greenhouse gases (GHGs) that have been emitted in the latter half of the

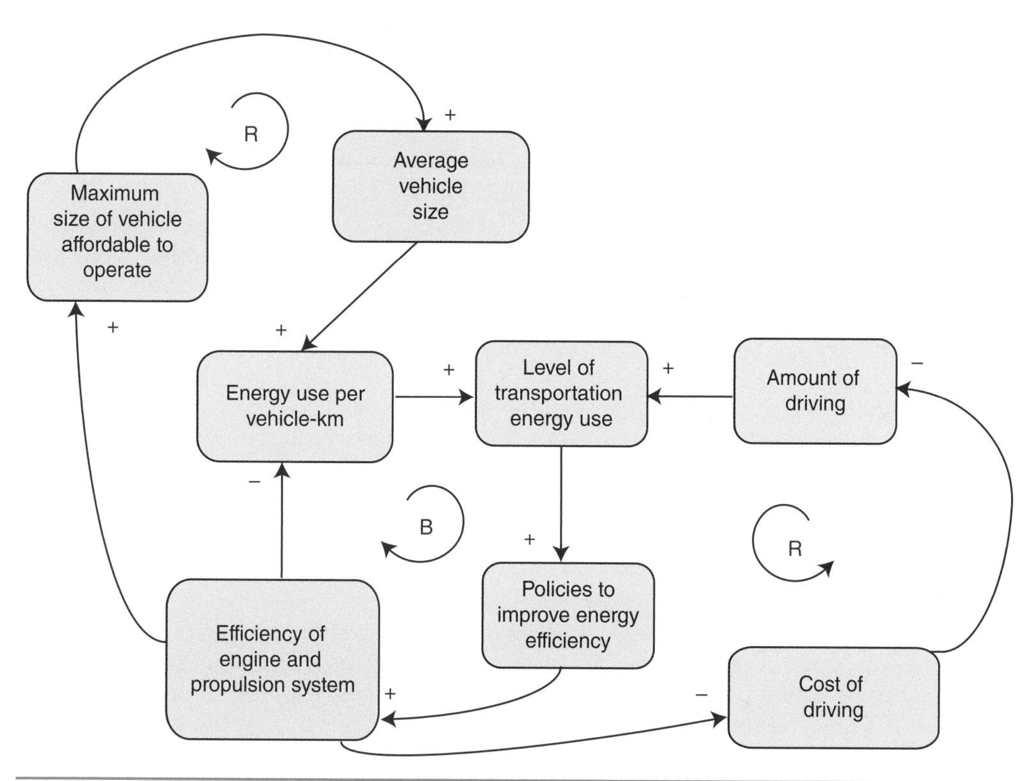

FIGURE 14-1 Causal loop diagram of the relationship between energy-efficiency policies, energy use per vehicle-kilometer or vehicle-mile traveled, average vehicle size, and demand for driving.

twentieth and early twenty-first century. Environmentalists have written at great length criticizing the impact of the system. The following quote from an ecological philosopher is representative: "Automobiles are now inefficient in the cities, poisonous to the air, deadly to the forests, subversive of neighborhood communities, and prohibitive of other modes of travel, such as walking or bicycling."[1]

For an extended period of time, this type of critique was met by efforts from the car industry's side to comply with regulations aimed at reducing environmental impact while at the same time working to slow their implementation. The relationship between the industry on the one hand and the government and environmental organizations on the other was often adversarial. The industry on occasion lobbied the government to slow the implementation of tighter emissions regulations and higher corporate average fuel economy (CAFÉ) standards, while eventually complying with regulations that emerged and cooperating with the U.S. Environmental Protection Agency on emissions testing and fuel economy ratings of different vehicle models.

By the turn of the century, however, the depth of the ecological and social challenge from the automobile and other forms of mechanized transportation created an opportunity to move beyond the status quo to a new paradigm. Emerging car markets in industrializing countries such as China and India accelerated the growth in the total

[1]*Source:* Thomas Berry, *The Dream of the Earth*, Sierra Club Press (San Francisco, 1988).

size of the world market. This growth in turn put pressure on world energy sources to provide sufficient energy for world demand. At the same time, growth in not only air pollutants but GHG emissions made the ecological situation more urgent. Auto makers in Europe, North America, and Asia began to look past the internal combustion engine burning petroleum products to alternative ways of powering vehicles that could "take the automobile out of the environmental equation."

Although stating this goal was only a first step and the path to its fulfillment remains long and arduous, the mere articulation of such a bold objective shows the seriousness of the impact of the car which all sides in the debate observe, as well as the opportunities for transforming its relationship with society in a truly profound way Properly conceived, alternative fuels might provide just such a way out. Vehicles could be powered by energy carriers such as electricity or hydrogen from carbon-free sources: renewable, nuclear, or fossil with sequestration. Alternatively, vehicles might still combust hydrocarbons in ICEs, but threats to the environment or long-term energy security might be eliminated in other ways.

14-3 Overview of Alternative Energy Endpoints

One of the defining characteristics of mechanized vehicles, and especially cars and light trucks that operate on road networks, is their capacity to store energy in a concentrated form on board the vehicle. These vehicles store a large quantity of energy per unit of weight and volume displaced in the vehicle, so that the weight or volume of the fuel does not limit passenger or cargo capacity. Some vehicles bypass the energy-storage problem by using the electric grid as a source of energy, through the use of electric catenary (e.g., electric trains with overhead wires, subways with third rails, trolleybuses, etc.), while others are nonmotorized (e.g., bicycles, cycleshaws, etc.). However, of the total global vehicle distance traveled each year, only a small fraction fall into these two categories. The great majority is traveled by vehicles that are (a) mechanized and therefore not relying on human power and (b) carry their own power source (sometimes referred to as *free-ranging*). It is therefore important to describe requirements that alternative fuel free-ranging vehicles must meet before presenting alternatives.

Most of the energy used by free-ranging vehicles comes from petroleum in either a gasoline or diesel form. Gasoline or diesel fuels have a number of characteristics that make them well-suited for use in motor vehicles: they do not require pressurization, and they are liquids, so they are relatively easily dispensed into the vehicle's fuel storage tank and combusted in the internal combustion engine. They also provide high specific weight and volume; that is, for the amount of space and payload taken in the vehicle, they provide a large amount of energy storage.

Many such alternative technologies exist on a small scale or in concept, including natural gas vehicles, fuels derived from coal, non-conventional crude oil sources such as oil shale, conversion of renewable energy sources into a form transferable to vehicles, and the like. However, in order to succeed, such technologies must be technically robust and financially viable. Under pressure from declining petroleum supplies, people may be willing to pay more for an alternative fuel source than they currently pay for gasoline, but they will refuse to support such a fuel if its price is exorbitant.

Furthermore, the energy source must be developed in parallel with the vehicles that use it and the infrastructure to distribute it, and all these things must fall into place in a way that keeps pace with the decline in availability of gasoline and diesel from the market. Mature technologies that use petroleum more efficiently, such as hybrid drivetrains, are already expanding in the marketplace, but they are not a permanent solution unless the energy source is changed from conventional petroleum. Cost of any new distribution infrastructure system is also a concern, although given the high value of the transportation fuels market—at an average cost of $4.00/gal, including taxes, the approximately 200 billion gal of gasoline, diesel, and jet fuel purchased in the United States in 2010 would have a retail value of some $800 billion dollars—it is clear that companies that deliver transportation energy products should be able to recoup a significant investment in new infrastructure through continued sales.

Along with the challenge of finding a clean and abundant source of energy from which to make a future energy source for transportation, the ability to transfer that energy source onto the vehicle poses a second daunting challenge. For example, nuclear energy and large-scale wind turbines are two proven technologies for generating electricity without CO_2 emissions that have a similar cost per kWh to fossil fuel energy generation. However, use of this energy for transportation would require an infrastructure to manufacture fuel cell or battery technologies on a scale to take nuclear or wind energy on board the vehicle as a substitute for gasoline (though in the case of electricity, home recharging might provide a partial solution). We also do not have a large-scale network of refueling stations available to the public to distribute and dispense electricity or hydrogen. These obstacles may favor instead the development of a petroleum substitute, such as a biofuel, that behaves like petroleum so that we can use our existing distribution and dispensing infrastructure, but that does not depend on nonrenewable resources or contribute to climate change.

To summarize, the development of a clean, abundant, and economical substitute for the petroleum-based transportation energy system is one of the major challenges facing the nations of the world today. Society would likely experience significant disruption from passing the peak oil point and not having a carefully prepared alternative waiting in the wings. Also, if the alternative were to be nonconventional petroleum sources, we might prolong for a time the worldwide transportation system based on liquefied fossil fuels but greatly aggravate the climate change problem if no suitable system for mitigating CO_2 emissions is in place. This challenge is arguably one of the most difficult technological and systems problems that we face in the pursuit of both sustainable transportation and sustainable energy.

14-3-1 Definition of Terms

The transportation literature refers to vehicles that run on a fuel other than gasoline or diesel as an *alternative fuel vehicle* (AFV); in some cases use of diesel in passenger cars is considered to be an alternative fuel, since, in markets such as that of the United States, diesel-powered cars have claimed only a small fraction of the light-duty vehicle market. Some alternative options for transportation energy, such as hybrid vehicles, do not use an alternative fuel but instead use an alternative propulsion *platform* to reduce the requirement for gasoline. Some sources refer to alternatives such as hybrids as advanced propulsion systems. It is also possible to have options that are a mixture of both; for example, a hybrid vehicle that runs on an alternative fuel. In this chapter, we use the umbrella term *alternative propulsion platforms and fuels* to describe the complete range of alternatives. We also interpret the term *fuel* broadly, so as to include electricity, even though electricity does not have the characteristics usually associated with the term fuel, such as combustibility.

Engines that use gasoline ignite the fuel using a spark, and are therefore called spark-ignition (SI) engines. Diesel-burning engines rely on the compression of the fuel alone to ignite the fuel, without the use of a spark, and are called compression-ignition (CI) engines. Although exceptions may occasionally arise, in general, when applied to conventionally-fueled vehicles, the use of the term SI engine implies gasoline combustion, and the use of the term CI engine implies diesel combustion. The performance of SI and CI engines can be modeled theoretically using temperature-entropy diagrams. These are not presented here but are widely available in thermodynamics texts (for example, Wark, 1983, or Cengel and Boles, 2010).

14-3-2 List of Available Energy Technology Endpoints

Transportation energy technologies that replace the use of petroleum for transportation as currently practiced must (1) be based on a more abundant supply of energy and (2) avoid permanently increasing the concentration of carbon in the atmosphere to avert climate change. Note that the second requirement does not preclude a carbon-based energy source. It also does not preclude emitting CO_2 from the vehicle tailpipe to the atmosphere at some point, as long as carbon emitted to the atmosphere is later removed so that the overall atmospheric concentration does not permanently increase.

Although there might appear to be a wide range of technologies competing to play this role, each can in its essence be reduced to one of the five *endpoint technologies*[2] that meets the objectives of abundant supply and protecting the climate. The five endpoint technologies are shown in Table 14-1. Some appear more promising at this time than others. For example, much research effort is going into developing hydrogen and biofuels as energy sources, into developing improved battery technology for vehicular use, and into sequestering carbon from the use of fossil fuels. However, given that we are at a very early stage of this transition and that it will take many years to complete, other less well-emphasized options are included in the table for thoroughness.

All five endpoint technologies share common characteristics:

1. They all present substantial technical, organizational, and financial challenges.
2. Whether or not the endpoint technologies require the introduction of new vehicle technologies or use existing ones, they all require large new infrastructure systems to generate energy and distribute it to vehicles. Some also require infrastructure to transform existing energy sources into a form that can be stored on a vehicle (e.g., electric vehicles, or EVs), and others require infrastructure to process CO_2 already in the atmosphere (e.g., internal combustion engine vehicles, ICEVs, that consume petroleum products), or to sequester CO_2 that is a byproduct of conversion to the energy currency used on board the vehicle (e.g., EVs that use electricity made from fossil fuels).
3. The endpoint technologies are not mutually exclusive. It is possible that one will eventually become dominant, and it is also possible that multiple ones will each claim some niche in meeting transportation energy demand. An analogy

[2]The use of the term *endpoint technology* in this chapter is similar to the term *backstop technology*. The latter was not adopted because the term in this chapter has a specific meaning in regard to the limited number of options, and because it was felt that the latter term is not well defined in the literature and may mean different things to different readers.

	Name	Description
Endpoint 1	Battery-electric	Zero-carbon electricity,* distributed to vehicles, which run on electricity between recharges.
Endpoint 2	Hydrogen	Zero-carbon hydrogen distributed to vehicles, which run on hydrogen between refills.†
Endpoint 3	Sustainable hydrocarbons	Hydrocarbon fuels resembling petroleum products made using biological processes,‡ or CO_2 emissions to atmosphere removed by separate process.
Endpoint 4	Alternative on-board energy storage	Alternative systems such as compressed air or flywheels used to power vehicle between recharges. Primary energy for systems from zero-carbon sources.
Endpoint 5	All mechanized vehicles attached to electric catenary	All mechanized vehicles are connected to grid via catenary. Grid electricity from zero-carbon sources.

**Zero-carbon electricity* means that the electricity is generated and delivered to the vehicle without increasing CO_2 concentration in the atmosphere. Thus electricity generated from biofuels could be used in this endpoint technology.

†Although it is likely that this endpoint technology would combine use of hydrogen with fuel-cell technology in order to achieve high efficiency, other conversion technologies (e.g., combustion in piston engine) are also possible.

‡Biological processes include crops, plant matter, or microbes. For sustainability, energy inputs should be renewable rather than from fossil fuel resources. Alternatively, fossil-fuel based energy source could be used if the resulting CO_2 is captured and sequestered.

Table 14-1 Alternative Endpoint Technologies for Petroleum-Free, Carbon-Free Transportation

could be made with today's situation, where most surface transportation (road, rail, and ship) is propelled by petroleum-fueled internal combustion engines, but a minority of rail service uses electricity supplied from outside the vehicle.

Endpoints 1 to 3: Primary Focus of Transportation Energy R&D

We first turn to the technologies that are the primary focus of current R&D efforts, namely, endpoints 1 to 3. These technologies have in common the use of some carrier that is stored on-board the vehicle, either electricity, hydrogen, or some type of hydrocarbon. To date, these three are in fact the only feasible terrestrial carriers for transportation energy, based on unsuccessful attempts to identify others.[3] All three have substantial technical hurdles, but also show greater potential than Endpoints 4 and 5, so we consider them in greater depth here.

Endpoint technologies 1 and 2 both have a relatively well-developed energy generation component, whereas the distribution, dispensing, and on-board technology are less developed. Large-scale production of carbon-free electricity (and by extension, hydrogen through electrolysis) from nonfossil sources already exists. In some circumstances, the electricity from these sources costs more than electricity from fossil fuels, but a concerted effort to build a transportation energy system around them would likely result in a reduction in cost so that any increase in cost per unit of energy equivalent delivered to the vehicle, compared to the current system, could be absorbed by

[3]For comparison, rocket motors for launching spacecraft can combust other fuels, such as combinations of hydrogen and nitrogen.

consumers as part of the transition. Alternatively, with widespread and cost-effective carbon sequestration, the electricity or hydrogen for these technologies could come from fossil-fuel combustion. In the case of endpoint technology 1, electrically powered vehicles might recharge at night when both power plant and grid usage are at a low point, so that the recharging function might be added to the tasks of the electric grid without requiring major expansion of generating and transmission infrastructure.[4]

Endpoints 4 and 5: Niche Endpoints

Endpoint technologies 4 and 5 are less promising as dominant means of providing transportation energy, but they are included here because of their potential in niche roles. Endpoint 4 bypasses the need for a fuel or currency altogether by storing and releasing energy in some way that does not involve combustion or the flow of electric current. Two forms currently available in an experimental form are (1) compressed air, which is pumped into the vehicle and then released through an engine that turns a driveshaft, and (2) onboard flywheel systems, which are set spinning by an external force prior to operation and then power the vehicle's drivetrain when it is in motion. For example, Moteur Development International of France is developing vehicles that use compressed air, with the expectation of eventually creating a sustained market for microcar versions of this technology in Europe. Endpoint 4 is in fact an umbrella category in that along with compressed air and flywheels, other options for unconventional onboard storage might be developed in the future, that we do not currently anticipate.

Turning to endpoint technology 5, relying on externally powered vehicles, the goal is to enhance or expand existing electrical catenary powered transportation systems. Taken to its logical conclusion, if 100% of free-ranging mechanized vehicles were to be discontinued in the name of eliminating emissions or petroleum reliance, transportation outside of the catenary network would require human or animal power. Therefore, because of the complexities of providing catenary (or third rails, etc.), the network of routes on which mechanized electrical vehicles operated would necessarily be sparser than the network currently accessible to motor vehicles. These vehicles would operate on main arterials in urban areas and on certain feeder streets, but side streets would have no powered vehicles. Similarly, rural areas might be served by some mixture of electrified rail vehicles (standard or light rail) and trolley buses, although many remote areas would be difficult to serve.

From a societal point of view, complete elimination of mechanized free-ranging vehicles would be very challenging because it would involve such a large change in social and economic patterns, including land use, business practices, and so on. The resulting urban landscape might resemble that of the late nineteenth century in certain cities in Europe and North America before the advent of the motor car, where electric tramways, horse-drawn vehicles, bicyclists, and pedestrians dominated the streets. Expanding the availability of catenary can, however, make an important contribution to sustainable energy for transportation.

Some amount of extension of electric catenary is underway in the world at present, such as recent projects to electrify railway lines between New Haven and Boston in the United States, or the electrification of the East Coast Main Line in Great Britain, as well as additions and conversions of transit lines to electric trolleys or streetcars in various

---

[4]For example, Georgia Power sells electricity at a discount from 11 p.m. to 7 a.m., but at a premium from 2 to 5 p.m.

cities around the world. Efforts aimed at developing *Personal Rapid Transit* (PRT) started in the 1970s and ongoing today might provide additional electrified transportation. Experience to date shows that it is prohibitively expensive to provide a comprehensive network of electrically powered guideways for vehicles throughout an urban area given that volumes (and hence economic return) on any leg in the network are low.[5] Nevertheless, development of PRT continues, as reflected in deployment of an automated system at Heathrow Airport in London or in the model Masdar City in the United Arab Emirates. To the extent that electrified rail or PRT systems currently use zero-carbon electricity, or can do so in the future, they can contribute to reducing the CO_2 burden from transportation.

We do not consider the underlying technologies required for endpoints 4 or 5 further in Chaps. 14 to 16. The electrical catenary component of endpoint 5 is, however, presented elsewhere in this book under the headings of public transportation (Chap. 8) and intercity transportation (Chap. 10).

14-3-3 Gaps in the Provision of a Complete Technological Solution

For endpoints 1 to 3, some technologies in the chain from energy source to vehicle propulsion are well-developed, but others are less so. In the case of endpoints 1 and 2, for electricity and hydrogen, the list of technologies in progress includes the following:

- *Adaptations to the electric grid*: Night recharging (as described in the Sec. 14-3-2) notwithstanding, the ability to recharge vehicles during the day would require a network of new charging stations, as well as possibly expansion of the transmission and distribution grid in certain locations, to deliver electricity to vehicles on a very large scale. Many municipalities and institutions such as hospitals and colleges are currently at work installing public charging stations to facilitate the adoption of electric vehicles.
- *Capacity for short-term storage*: In the case of endpoint 1, especially where intermittent energy sources such as solar or wind are used, if the electricity cannot be dispensed to the vehicle as it is generated, some means of short-term storage would be necessary to retain the energy content until it could be dispensed. For example, electricity could be converted to hydrogen for short-term storage.
- *Long-distance hydrogen infrastructure*: In the case of endpoint 2, it is envisioned that hydrogen would be produced in large central facilities to benefit from economies of scale. These facilities would require a new distribution grid for the product to reach end users.
- *Infrastructure for long-lasting and cost-effective batteries or fuel cells*: These technologies must be perfected for their respective pathways to succeed and, once perfected, a new manufacturing infrastructure to produce, distribute, and recycle batteries or fuel cells would be necessary.
- *On-board storage technology:* For endpoint 1, battery storage systems are available for electric vehicles already on the market but cost per unit of energy storage should be reduced if the technology is to penetrate the market more widely.

[5]For example, Vuchic (2007, p. 474) states that "the combination of the two features—small vehicles and complicated guideway—is paradoxical and makes the PRT mode impractical under all conditions."

For endpoint 2, the hydrogen fuel cell must be coupled with a means of storing sufficient hydrogen on board the vehicle in order for it to have sufficient range, while meeting cost and safety requirements.

Depending on which exact solution eventually takes shape, not all of the technologies on the list would be required, but some of them would be, and each requires substantial R&D to be made commercially viable.

By contrast, endpoint 3 takes advantage of the existing distribution and onboard conversion of the transportation energy resources more or less unchanged, and instead requires the development of a new energy-generation technology that creates from renewable resources a fuel that closely resembles, or is chemically identical to, today's gasoline or diesel. In many of the available variants on this endpoint, the original source of energy is the sun, with conversion of sunlight taking place in agricultural fields, in water-based resource *farms*, or in controlled facilities that use microorganisms to generate the raw materials for liquid fuels. Coal might also be used as an energy source instead of the sun, by converting the carbon content of the coal into a synthetic liquid fuel, and combusting this fuel in the vehicle. For any fossil fuel, the resulting CO_2 would be captured from the atmosphere.[6] The conversion of coal to synthetic fuel has already been used in the past in Germany and South Africa during times of war or political isolation. The attraction of endpoint 3 is that, once the hurdle of making the fuel or capturing CO_2 is overcome, companies that currently refine, distribute, and dispense motor fuels, or manufacture vehicles, could continue to use familiar technologies with only minor adaptations.

It should also be noted that combinations of endpoints 1 and 3 are possible. For example, a plug-in hybrid electric vehicle (see Chap. 15) might recharge using electricity from renewable resources from the grid (endpoint 1), and then refuel using a biofuel (endpoint 3).

14-3-4 Interactions between Emerging and Incumbent Technologies

In considering the transition to alternative transportation technologies, we must recognize the influence of the starting point of today and the existing worldwide fleet of motor vehicles, aircraft, and other consumers of transportation energy from petroleum, and the infrastructure that supplies this energy. While alternative endpoints are desirable in the long run, today they must be introduced in the context of a mature petroleum-based system that is the incumbent technology, and the expectations in terms of price, reliability, performance, and the like, which this system has created in consumers. The situation is therefore different from the early days of the automobile at the beginning of the twentieth century. At that time, use of horses had fallen out of favor for urban transportation, due to problems with fouling of streets. When "horseless carriage" designs emerged that used either the internal combustion engine, electricity, or steam for propulsion, each new technology had to compete with the other two for dominance of the new market, but horse propulsion was no longer a strong incumbent technology that could resist the rise of the automobile. Once the internal combustion engine had outvied electricity and steam, the market for both cars and gasoline could grow at a pace determined by the vehicles' success alone.

[6]Coal is used as an example here, but a similar system using nonconventional oil or natural gas might also be developed.

Historically, since the early success of the automobile there was a time during World War II when citizens accepted an alternative transportation energy option that has diminished performance to contribute to other objectives. Many of the countries involved experienced great changes in this area, such as the discontinuation of passenger car production in the United States between 1942 and 1944, or the dramatic rise in use of public transportation in many countries around the world. Although governments required citizens to participate through laws, rationing, and so on, these programs were in the end successful in nations on both sides of the conflict because, by and large, individuals recognized the contribution that the savings of energy and raw materials could make to the war effort.

In the same way, it is within the realm of possibility that the public might shift en masse from the current technology to a different one, the way that they have recently been shifting from land lines to cell phones, if the danger from environmental damage or resource scarcity was clear enough and the anticipated benefit of radically changing vehicle technology was convincing enough. Indeed, on an individual level, a small fraction of the population in various countries has already made such a switch for environmental reasons—for example, taking up regular commuting to work by bicycle rather than by car. However, a massive shift like this throughout the population presents a great challenge in terms of making social adjustments, and furthermore there are at present strong candidates for vehicles that use energy sustainably without a great loss of performance. Therefore, we focus in Chaps. 14 to 16 exclusively on options that can "oust the incumbent" technology (petroleum-driven ICEVs) by delivering equivalent performance while still achieving the desired reductions in CO_2 emissions and other environmental gains.

Recent Backsliding on Efficiency Gains in the U.S. Market

Before discussing emerging technologies, it should be noted that the incumbent gasoline technology does not stand still, and paradoxically has recently evolved in a negative direction despite the need for improved delivered efficiency. An automaker may introduce a model with one set of performance and fuel economy characteristics, and perceiving that the market may be pushing toward a larger, more powerful vehicle, engage in *size creep* to modify the model so that it is faster and more comfortable but also consumes more fuel. Table 14-2 gives the example of the Honda Accord: introduced as a fuel-efficient compact car in the 1970s, its weight increased by 78% and

Characteristic	Original	Recent
Model year	1976	2008
Weight, lb	2000	3567
Power, hp	68	268
Engine size, L	1.6	3.5
mpg (city/hwy)	32/46	19/29

Source: Sperling and Gordon (2009).

TABLE 14-2 Comparison of the First-Generation and Recent Honda Accord Models.

maximum horsepower by 294% as Honda sought more buyers in the midsize vehicle range. This trend has been widespread among many makes and models, although in certain models it has also gone in the other direction. For instance, the Chevrolet Malibu was a large car in the mid-1970s, but by the 2012 model year had been downsized to a midsized car with a 2.4 liter 4-cylinder engine and 22/33 mpg fuel economy rating.

Exceptions such as the Malibu notwithstanding, the general trend has been toward using increasing power available in engines not to drive the same size cars with less energy but to use the same amount of energy to drive larger cars. Figure 14-2 compares average fuel economy measured in miles per gallon and average vehicle efficiency measured in average ton-miles of vehicle moved per gallon of fuel consumed in the U.S. market between 1975 and 2004. At first, both measures improved in parallel, with fuel economy improving from 14 to 21 mpg and vehicle efficiency from 24 to 34 ton-miles/ between 1975 and 1985. After 1985, however, the trends diverged: vehicle efficiency continued to improve to 42 ton-miles/gal by 2004, but fuel economy remained constant at 21 mpg. The consequences for total fuel consumption in the United States were that as overall vehicle miles traveled (VMT) increased, so did total fuel consumption measured in gallons.

This evolution illustrates one of the challenges for efforts to reign in total transportation energy consumption. A strong economy generating disposable income, motorists with increasing wealth to spend on vehicles and fuel, low fuel costs up until the summer of 2005, and lack of strong policies to favor high-mpg vehicles resulted in the trends seen in Table 14-2 and Fig. 14-3. Larger vehicles and continued low fuel costs led to the *reinforcing* interactions illustrated in the causal loop diagram of Fig. 14-1. Not all industrialized countries went in the same direction as the United States, however; specifically, in Europe and Japan, deliberate policies to keep fuel taxes high as well as the advantages of smaller vehicles in narrower, more crowded streets favored small vehicles. As a result, average fuel economy in these countries continued to improve throughout the 1990s and 2000s.

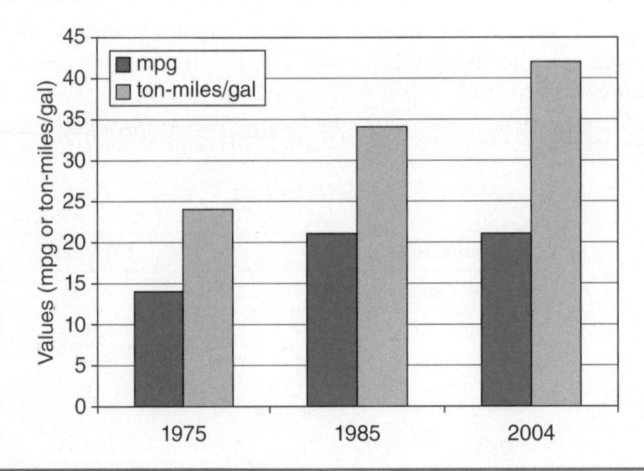

FIGURE 14-2 Comparison of fuel economy, mpg, and vehicle efficiency, ton-miles/gal, 1975–2004.

Note: If a vehicle achieves 30 ton-miles/gal and weighs 2 tons, it will travel 15 miles per gallon of fuel consumed; that is, 15 mpg fuel economy.

Source: Lutsey and Sperling (2005).

Divisia Analysis of Passenger Car and Light Truck Market in the United States since 1970

As was seen in Fig. 7-4, not only has growing light-duty vehicle energy demand driven up overall passenger transportation energy consumption in the United States in recent decades, but within this category it is the submode of light trucks [including pickup trucks, vans, and sport utility vehicles (SUVs)] that has been the driver of growth. Since 1970, while passenger car energy consumption remained more or less constant, light truck consumption grew substantially.

Divisia analysis provides insight into factors that contributed to this trend by considering the relative roles of these two main classes of light-duty vehicles. (The underlying technique for Divisia analysis is explained in Chap. 3.) In Table 14-3, activity levels are given in billions of vehicle-miles or vehicle-kilometers, and fuel consumption is given in billions of gallons or liters of fuel consumed; the subsequent Divisia analysis is carried out in metric units. While both cars and light trucks in the United States may have reduced energy intensity over the period from 1970 to 2000, in the case of cars, improving efficiency has kept pace with growing vehicle-kilometers, while in the case of light trucks it has not. As shown, fuel consumption for cars rises only slightly from

	Cars					
Year	Billion Miles	Billion Gal.	mpg	Billion.km	Billion.L	km/L
1970	917	68	13.5	1,467	257	5.7
1975	1034	74	14.0	1,654	280	5.9
1980	1112	70	15.9	1,779	265	6.7
1985	1247	72	17.4	1,995	271	7.4
1990	1408	69	20.3	2,253	263	8.6
1995	1438	68	21.1	2,301	258	8.9
2000	1600	73	21.9	2,560	277	9.3
	Light Trucks					
	Billion Miles	Billion Gal.	mpg	Billion.Km	Billion.Ls	km/L
1970	123	12	9.9	197	47	4.2
1975	201	20	10.2	322	75	4.3
1980	291	24	12.2	466	90	5.2
1985	391	27	14.2	626	104	6
1990	575	36	16.1	920	135	6.8
1995	790	46	17.3	1,264	173	7.3
2000	923	53	17.5	1,477	200	7.4

Note: Figures include the sum of gasoline and diesel fuel liters or gallons.
Source: Davis and Diegel (2013).

TABLE 14-3 Distance Traveled, Volume of Fuel Consumption, and Fuel Economy of U.S. Cars and Light Trucks 1970 to 2000

257 to 277 billion liters per year, while fuel consumption for light trucks increased approximately fourfold, from 47 to 200 billion liters. A significant shift in purchasing habits by American drivers pushed this trend: as Americans bought fewer passenger cars and more pickups, minivans, and SUVs, the total number of vehicle-kilometers traveled by light trucks increased more than sevenfold, while the increase in vehicle-kilometers for passenger cars was less than twofold.

To carry out the Divisia analysis, we need to know the kilometers, fuel consumption, and fuel economy of the combined fleet of cars and light trucks in the base year of 1970. These are obtained by adding together the respective values given in Table 14-3 and shown in Table 14-4; values from 2000 are also included in the table, for comparison. As shown, the overall fuel economy increases from 5.5 km/L to 8.5 km/L. Based on the increase in activity (kilometers of vehicle travel), the trended fuel consumption with no change in fuel economy would have been 737 billion liters:

$$\frac{4.037 \times 10^{12}\ \text{km}}{5.5\ \text{km/L}} = 7.37 \times 10^{11}\ \text{L}$$

The Divisia analysis is completed by calculating the effect of overall efficiency changes (the combined energy intensity of cars and light trucks) and structural changes (the relative share of the two vehicle types). The results are shown in Fig. 14-3. Of the two types of changes, efficiency changes have the stronger effect, amounting to a change of -301 billion liters in 2000. The increasing share of vehicle-kilometers for light trucks, from 12% to 37% of the total between 1970 and 2000, has an upward effect on fuel consumption equivalent to 40 billion liters in 2000. The two effects together explain the difference between actual and trended fuel consumption in the graph when fuel consumption E is given in units of billions of liters of fuel. Thus the relationship:

$$E_{\text{actual}} = E_{\text{trended}} + \Delta E_{\text{efficiency}} + \Delta E_{\text{structure}}$$

is verified as follows:

$$E_{\text{actual}} = 477$$

$$E_{\text{trended}} + \Delta E_{\text{efficiency}} + \Delta E_{\text{structure}} = 737 - 300 + 40 = 477$$

$$\therefore E_{\text{actual}} = E_{\text{trended}} + \Delta E_{\text{efficiency}} + \Delta E_{\text{structure}}$$

Note that the results of the Divisia analysis do not explain the substantial rise in both activity and energy consumption over the period 1970 to 2000, nor do they tell us to what extent the upward pressure on fuel consumption due to rising vehicle-kilometers could have been offset through increases in efficiency. One can surmise that, since

Year	Billion.Km	Billion.L	km/L
1970	1664	304	5.5
2000	4037	477	8.5

TABLE 14-4 Vehicle-Kilometers Traveled, Liters of Fuel Consumption, and Fuel Economy of all U.S. Light-Duty Vehicles, 1970 and 2000.

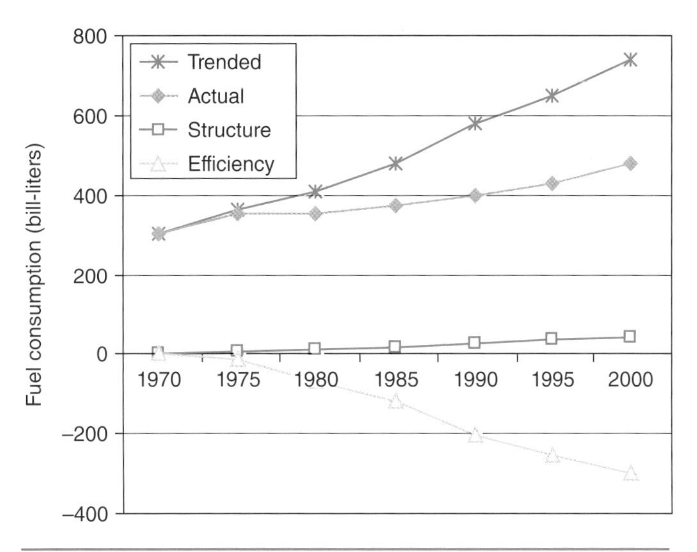

FIGURE 14-3 Divisia analysis of U.S. light-duty vehicle fuel consumption in billion liters, 1970 to 2000.

vehicle-kilometers increased by 143% over this period, it would have been difficult to prevent some amount of rise in absolute fuel consumption in any case—the average fuel economy would have needed to improve to 13.3 km/L to keep pace. The results do tell us that improvements in efficiency accounted for most of the improvement against the trended projection of where fuel consumption would have been in 2000. Although changes in the structure profoundly affected the nature of the light-duty vehicle market, with automakers selling many more light trucks, they only modestly increased fuel consumption relative to the trended value, as reflected in the value of $\Delta E_{structure}$ = 40 billion liters in Fig. 14-3.

Competition between Petroleum and Alternative Fuels: A Cautionary Tale from the 1980s AFV Programs

For an alternative fuel to take root, growth in energy outlets to supply the fuel must support the growth in the number of AFVs sold. As a cautionary tale about what can go wrong when support for AFVs is inadequate or ill-conceived, consider the efforts in the United States to introduce flex-fuel vehicles (i.e., able to use mixtures of gasoline and biofuels, or diesel and compressed natural gas, in different percent combinations) in the 1980s (Kreith et al., 2002). Through a system of mandates and financial incentives, several hundred thousand flex-fuel vehicles were sold, so that the potential existed for these vehicles to use biofuels and thereby reduce gasoline or diesel consumption.

Although the individual vehicles were technically proficient, the technology failed to take root because a business case was not established for building the infrastructure to deliver the alternative fuel. Many vehicles were driven in areas where they had no access to gasoline-biofuel mixtures, and never once during their driving life took advantage of the flex-fuel capability. The public judged the flex-fuel capability not to be useful, and it became unmarketable as soon as government incentives and mandates were removed. Fleet owners of flex-fuel vehicles found that if they had paid a premium for this capability when the vehicle was new, they could not pass this premium on in the used vehicle market because the buyers saw no added value in it.

Changing performance of gasoline technology in terms of air quality played a role as well. A major driver of introducing AFVs at the beginning of the program was the perception that they produced less air pollutants than gasoline vehicles, and that their introduction would help to improve urban air quality. However, over the lifetime of the AFV program, vehicle manufacturers were able to greatly reduce emissions per mile of many important pollutants, so that the air quality motivation no longer existed by the end of the program.

In the final analysis, the 1980s AFV program may have had some benefit in terms of forcing the emissions control technology for gasoline engines to improve or providing a technological platform on which biofuel programs that have been expanding since the year 2000 could build. Nevertheless, it failed as an attempt to create and sustain a permanent presence of flex-fuel vehicles using an alternative fuel-dispensing network. This experience can be contrasted with that of Brazil, where an alternative fuel industry based on the production of ethanol from sugar cane has for many years provided a substantial fraction of the motor fuel consumed (see Chap. 16).

14-4 Alternatives to ICEVs Today: Alternative Fuels and Propulsion Platforms

Alternative fuel and platform vehicles can be broadly divided into near-term alternatives, which are the focus of the rest of this chapter, and EVs, hydrogen fuel cell vehicles (HFCVs), and vehicles using biofuels, which are covered in Chaps. 15 and 16. The near-term alternatives include hybrid electric vehicles (HEVs), which take advantage of a fundamentally different propulsion platform to greatly reduce energy requirements, regardless of the energy source. They also include advanced ICEVs, such as high-efficiency diesel engines in light-duty vehicles that burn ultra-clean diesel fuel. All of the major makers in North America, Europe, and Asia are actively researching one or more of these short- or long-term technologies.

An alternative vehicle must deliver acceptable performance to the customer similar to that of the ICEV to succeed in the market, while reducing energy consumption and GHG emissions to succeed in environmental goals. Not only must it deliver sufficient power and driving range between refueling or recharging stops, it must also do so without incurring a large weight or volume penalty, since otherwise the handling of the vehicle or its capacity to carry cargo may be compromised. This standard applies to the alternative vehicles considered throughout this section.

14-4-1 Hybrid-Electric Vehicles

A recent trend that can benefit cars and light trucks alike is the emergence of the hybrid drivetrain, which uses a combination of internal combustion engine and electric motor to optimize energy use for propulsion. These drivetrains have been introduced primarily into passenger cars to date, with additional use of hybrid drivetrains in midsize delivery vehicles or urban passenger buses that operate in stop-and-go conditions. The most popular HEV in the U.S. market in terms of overall sales, the Toyota Prius, has a rated fuel economy value of 20.8 km/L (49.7 mpg) for combined city/highway fuel economy, versus 11 to 17 km/L (25–40 mpg) for typical conventional drivetrain vehicles. The Prius is not as efficient as the best-performing car in the European market, the Audi A2 diesel (fuel economy of 40 km/L diesel, or 95.3 mpg; taking into account the 11% greater energy content per gallon in diesel than gasoline, this is equivalent to 36 km/L or 85.9 or miles per gallon of gasoline equivalent), but the Prius also has a larger interior.

HEVs currently constitute a small part of the world's new passenger car market; however, concern about the environment and rising fuel costs have sparked interest, and sales have risen steadily, as shown in Fig. 14-4 for the sales of hybrids in the United States for the period 1999 to 2009. At the same time, with each passing year, automakers introduce new hybrid models in different market niches (sedans, minivans, SUVs, pickups, etc.), further increasing the opportunity to grow sales.

The HEV uses a combination of ICE and battery-electric technology to capture the advantages while avoiding some of the drawbacks of each. There are three basic energy flow concepts for the hybrid drivetrain:

1. *Series hybrid:* All mechanical power from the ICE is first converted to electrical energy, then to either a motor for driving the wheels, or to a battery system for storage.
2. *Parallel hybrid:* Mechanical power from the ICE goes to a transmission, where it is divided between driving the wheels and driving a generator for storing energy, depending on operating conditions.
3. *Series-parallel hybrid:* Hybrid system that combines aspects of both series and parallel systems.

In other words, the parallel drivetrain makes possible a direct mechanical connection between ICE and wheels, the same as an ICEV, and the series drivetrain does not, as shown in Fig. 14-5. In practice, however, a combination of these concepts may be implemented, as shown in Fig. 14-5(c), in the form of the series-parallel hybrid, which

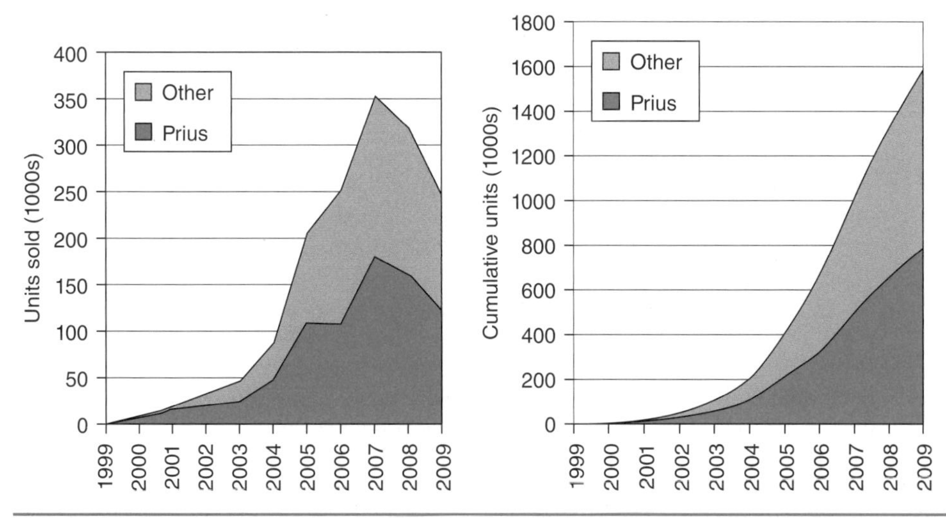

Figure 14-4 Growth in U.S. annual and cumulative hybrid vehicle sales 1999 to 2009.

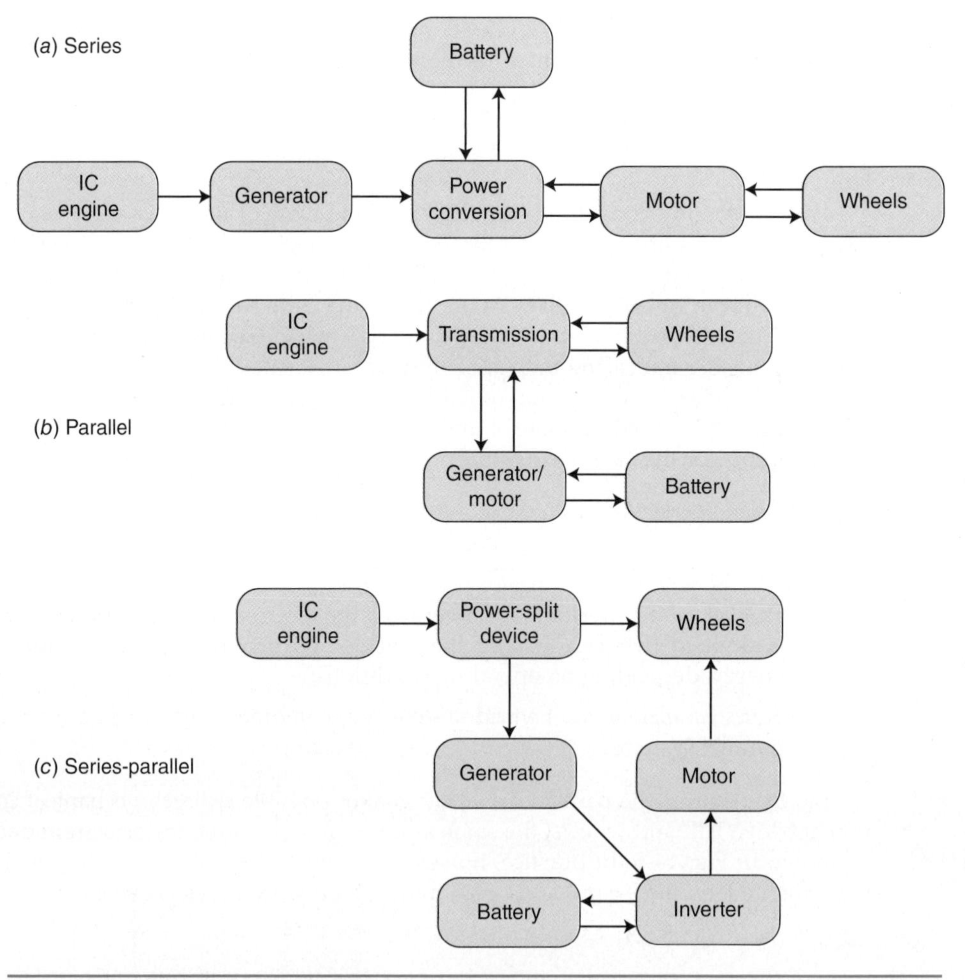

FIGURE 14-5 Schematic of basic energy-flow concepts for hybrid vehicle drivetrain: (*a*) series, (*b*) parallel, and (*c*) series-parallel.

provides both a series and parallel path from ICE to wheels. The series-parallel is presently the dominant hybrid drivetrain in the light-duty vehicle market, in the form of the Toyota full hybrid system used in the Prius and many other hybrid models (including versions adopted by Ford and Nissan). Example vehicles are shown in Fig. 14-6.

In most hybrid drive trains, some fraction of the power from the engine is used to move the vehicle when the load is low, and the remainder is used to charge the battery system. Mid-weight or heavy-duty vehicles such as trucks may use some other storage device (such as a hydraulic storage system) instead of batteries. When the load is too great for the power originating from the engine alone, the drivetrain withdraws energy from the battery system so that the combination of ICE and electrical power together can meet the load. Note that all hybrids use a *secondary battery system* to store and dispense electric charge to the drivetrain, which is separate from the primary battery system used in both HEVs and conventional ICEVs for auxiliary functions (lights, sound system, etc.).

A number of variations on the hybrid drivetrain are possible. A *through-the-road* hybrid (e.g., Toyota Highlander SUV, Dodge Durango SUV) uses the ICE to drive one set of wheels (front or rear) and the electric motor to drive the other. A *mild hybrid system* (e.g., 2007 Chevrolet Silverado truck) does not allow the vehicle to run entirely on electric power, but instead adds electric power as needed to smooth out load on the ICE. For heavy-duty vehicle applications where stop-and-go driving is common, such as in delivery vehicles or municipal waste haulage, companies such as Eaton and Parker-Hannifin provide hydraulic hybrid systems that store energy hydraulically on level ground or when braking, and release it in times of increased load. Diesel-electric hybrid drives are also in use in delivery vehicles belonging to companies such as FedEx and UPS.

Fuel Economy Advantage of the Hybrid System

The hybrid drivetrain has the three following advantages:

1. *Independence from external generating source*: All of the electric charge in the batteries comes originally from the energy in the fuel (gasoline or diesel), so it is not necessary to spend time recharging the vehicle from an off-board source of electricity while it is stationary, as in an EV. Also, since the electric charge is usually put to use in propulsion relatively soon after being generated (either from the ICE or from the drivetrain), the battery storage requirements are greatly reduced in terms of space, weight, and cost compared to an EV.
2. *Optimal ICE energy efficiency*: Since the ICE in a hybrid can depend on the battery-motor system for additional power under peak load, it need not be optimized for maximum power output at the expense of efficiency, as occurs in many ICEVs. To this end, some hybrids (e.g., Toyota Prius) use a thermodynamic cycle known as the *Atkinson* cycle, which is different from the *Otto* cycle in

(*a*)

FIGURE 14-6 Two approaches to marketing a hybrid passenger car: (*a*) Toyota Prius, developed "from the ground up" as a hybrid, and (*b*) Honda Civic Hybrid, adapted from the conventional Honda Civic ICEV.

(b)

Figure 14-6 (Continued)

conventional ICEVs. The Atkinson cycle emphasizes higher efficiency under average load at the expense of reduced peak power. Such a tradeoff would be unacceptable to most consumers in an ICEV, because they would not have a satisfactory amount of power available for rapid acceleration, climbing steep grades at speed, and other high-demand driving regimes. However, this arrangement works well in a hybrid, where the availability of combined engine and electric motor power under the heaviest loads allows it to improve fuel efficiency while still satisfying performance requirements.

3. *Regenerative braking*: The HEV can use regenerative braking to recover kinetic energy from the wheels back into the secondary battery when decelerating. Measured in round-trip energy efficiency, up to 40% of the energy originally available as kinetic energy in the rotation of the wheels is returned to the wheels after conversion from mechanical to electrical and back to mechanical energy.

Although point 3 is the feature that is more familiar to most drivers, it is point 2 that contributes the most to the improvement in fuel economy in many hybrids. In fact, the growth of the hybrid market has opened up new opportunities for the development of the Atkinson engine. The Atkinson cycle was originally invented by James Atkinson in 1882 as a way of circumventing the patent on the Otto cycle of Nicholas Otto. Although it did not come into widespread use at the time of its invention, the Atkinson cycle does have a combustion efficiency advantage, so that it was appropriate for use in the development of the hybrid drivetrain in the 1990s.[7]

In some models of HEVs, the hybrid drivetrain has been used to take larger models of cars that are fairly powerful and add additional power while at the same time modestly improving fuel economy. For example, for several years Honda produced a

[7]A variation on the Atkinson cycle is the *Miller cycle*, which was developed in the 1940s by adding supercharging (forced intake of air into the cylinder) to the Atkinson cycle. Both are referred to sometimes as "Atkinson-Miller cycle engines."

hybrid model of the Accord sedan with 6-cylinder engine has a maximum output (combined ICE + motor) of 190 kW (255 hp) and a combined city or highway average fuel economy of 11.7 km/L (28 mpg), versus 179 kW (240 hp) and 9.8 km/L (23.5 mpg) for the ICEV version with 6-cylinder engine. This model shows how the hybrid platform gives the automakers flexibility to either pursue maximum fuel efficiency or maximum performance with no loss in fuel economy. From an energy conservation or climate change point of view, the former option is desirable; however, the latter option may encourage some consumers, who might not consider the compact hybrids that achieve the best fuel economy, to purchase a hybrid and save some amount of fuel.

14-4-2 Hybrids in Comparison to Other Propulsion Alternatives

The potential of the hybrid drivetrain to reduce energy consumption across a wide range of vehicle applications can be exemplified by considering several current and recent passenger vehicle models (Table 14-5). Along with ICEVs, HEVs, and EVs, the table includes plug-in hybrid electric vehicles (PHEVs). (EVs and PHEVs are primarily discussed in Chap. 15.) Most vehicles are currently available in the U.S. market, although the Audi A2 is available only in Europe, and the Accord Hybrid and EV-1 have been discontinued. The Ford Escape is a small SUV; otherwise, all vehicles are 5-passenger coupes, sedans, or hatchbacks. The Toyota Echo is an ICEV that most closely resembles the 2001 Prius among Toyota's line of ICEVs, and is therefore included in the table for comparison. The VW Passat and Audi A2 are diesel vehicles, and the others are gasoline-powered.

A comparison of HEV and ICEV vehicles in the table shows that, in each case, the HEV vehicle delivers substantially improved fuel economy over the vehicle closest to it in size and performance. This is achieved despite a weight penalty for the additional drivetrain components and battery system. For example, the 2001 Prius weighs over 300 kg more than the Echo but is approximately 9 km/L better in terms of fuel economy.

In the case of the HEV and PHEV models (Volt, Civic, and Prius), the ICE power per unit of engine volume is low compared to other models in the table, consistent with the discussion above of the use of the Atkinson cycle. For the five vehicles shown, the power per unit displacement range is 28.0 to 54.8 kW/L, versus a range of 46.2 to 64.8 for the ICEVs, not including the 1976 Accord. (The Accord is included for historical reasons, to show how weight, power, and fuel economy have evolved since the first generation of the sale of this vehicle in the United States.) The HEV and PHEV models compensate for this lack of specific power with an electric motor that provides additional torque and power when needed.

Some of the advantages of HEVs over EVs are borne out in the table as well. For example, the 2011 Prius carries more passengers than the EV-1 (which is a two-passenger coupe), but weighs less and achieves similar fuel economy once upstream losses from delivering the respective energy source to the vehicle are taken into account. Battery requirements are a major contributor to the difference in curb weight; an EV-1 carries 395 kg of batteries, versus 27 kg for the secondary battery in the Prius. Also, the Nissan Leaf outweighs the Prius by 267 kg; while range is a limitation for the EV-1, the range of the Prius is actually better than many ICEVs in its class. A 2011 Prius has a nominal range of 946 km (591 miles), based on the tank size of 45 L (11.9 gal) and the overall fuel economy shown. Whether designed from the ground up (e.g., Prius) or adapted from an existing ICEV design (e.g., Honda Civic Hybrid), HEVs have similar engine compartment space requirements (Fig. 14-7).

Make and Model	Eng Size[a] (cc)	Max Power[b] (kW)	(hp)	Power/L[c] (kW/L)	Curb Wt (kg)	Overall Fuel Economy[d] (km/L)	(mpg)
2007 Audi A2	1190.0	55.0	73.8	46.2	930	35.7	85.5
2011 Chevrolet Volt	1400.0	111.9	150.0	42.6	1264	27.2	65.0
2005 Ford Escape (man)	2300.0	149.1	200.0	64.8	1575	9.9	23.7
2005 Ford Escape Hybrid	2260.0	99.2	133.0	n/a	1719	12.2	29.1
1976 Honda Accord (auto)	1600.0	50.7	68.0	31.7	909	16.0	38.3
2005 Honda Accord (man)	3000.0	179.0	240.0	59.7	1522	9.8	23.5
2005 Honda Accord Hybrid	3000.0	190.2	255.0	n/a	1591	11.7	28.0
2012 Honda Civic (man)	1798.0	104.4	140.0	58.1	1185	13.2	31.6
2012 Honda Civic Hybrid	1497.0	99.2	133.0	54.8	1297	18.4	44.0
2011 Nissan Leaf	n/a	80.0	107.3	n/a	1521	41.4	99.0
2001 Saturn EV-1	n/a	116.0	155.6	n/a	1350	25.1	60.0
2011 Tesla Roadster	n/a	215.0	288.0	n/a	1238	56.4	135.0
2013 Tesla 60 kWh Model S[e]	n/a	310.0	415.3	n/a	2112	39.7	94.0
2001 Toyota Echo (man)	1500.0	78.3	105.0	52.2	927	13.6	32.6
2001 Toyota Prius	1500.0	72.0	96.6	28	1259	17.4	41.6
2011 Toyota Prius	1800.0	133.0	178.4	40.6	1383	20.8	49.7
2013 Toyota Prius PHEV[f]	1800.0	133.0	178.4	40.6	1439	40.2	95.0
2008 VW Passat TDI	2000.0	126.8	170.0	63.4	1527	13.6	32.6

[a]Engine size values are displacement of internal combustion engine, where applicable. Size of electric motor not considered.

[b]Maximum power value for hybrids includes power from electric motor (specs not shown) in addition to power from ICE. Due to differences in the way that ICEs and electric motors function, each achieves maximum power output in different driving conditions, limiting the comparability of maximum power ratings between ICEVs, HEVs, PHEVs and EVs in this table. The calculation of maximum power for the hybrid models shown is the arithmetic sum of ICE and electric motor power.

[c]Power per liter of ICE. Values for 2001 and 2011 Toyota Prius, Chevy Volt, and Honda Civic Hybrid are based on 42 kW, 73 kW, 60 kW, and 82 kW, respectively, for the ICE only. Values for Ford Escape Hybrid not given due to ICE/motor breakdown not being available.

[d]Fuel economy value shown is the 55% city and 45% highway weighted average of economy values from USEPA, where available. USEPA values are post-2008 adjustment to reflect more realistic driving cycle. Exceptions: for EV-1, electricity consumption has been converted to gallons of gasoline equivalent and then fuel economy calculated. Fuel economy values for 1976 Honda and Tesla Roadster are from Sperling and Gordon (2009) and Randolph and Masters (2008), respectively. Chevrolet Volt fuel economy is average of USEPA electric and ICE fuel economy. Fuel economy values for Audi and VW diesel vehicles have been adjusted downward to reflect lower energy content per gallon of gasoline (assuming 89% of energy in 1 gal diesel).

[e]Tesla Model S sedan with 60 kWh of battery capacity.

[f]Fuel economy for PHEV Toyota Prius is USEPA rated value while charge is available in battery. Once charge is depleted, fuel economy is same as non-PHEV Prius.

Conversions and abbreviations: 1000 cc = 1 L. "man" = manual transmission, "auto" = automatic transmission.

Table 14-5 Performance Characteristics of Representative EV, HEV, PHEV, Gasoline ICEV, and Diesel ICEV Models

(*a*)

(*b*)

(*c*)

FIGURE 14-7 Top-down view of three engine compartments. (*a*) Conventional 2004 Honda Civic, (*b*) 2004 Honda Civic Hybrid, and (*c*) 2005 Toyota Prius.

Notes: In both the middle and lower images, the additional components of the hybrid drivetrain do not require any significant increase in the volume of the engine compartment.

Source: © Philip Glaser, 2007, philglaserphotography.com.

Quantification of the fuel economy figures in the rightmost two columns in Table 14-5 requires a detailed understanding of the drive cycles used by USEPA to test vehicles (or peer organizations in other countries or the European Union), and is therefore beyond the scope of this text. As a first approximation, however, we can assume a representative constant speed for city and highway conditions and compare the ratio of fuel economy over time to tractive power requirement to illustrate differences between different drivetrain platforms. Example 14-1 develops this comparison for a representative ICEV and HEV, namely, the 2001 Toyota Echo and 2001 Toyota Prius.

Example 14-1 From the USEPA website, the city or highway fuel economy values for the 2001 Echo and Prius are 29/37 and 42/41 mpg (12.26/15.64 and 17.75/17.33 km/L), respectively. The Echo has the following specifications in addition to those from Table 14-5: drag coefficient, 0.29; rolling resistance, 0.01; cross-sectional area, 2.51 m². For the Prius, the figures are drag coefficient, 0.29; rolling resistance, 0.01; cross-sectional area, 2.48 m². Assume air density of 1.15 kg/m³ and gasoline with energy content of 115,400 Btu/gal or 32.2 MJ/L. Use 25 mph (40 km/h) and 55 mph (88 km/h) as representative speeds for city and highway conditions. If the vehicles travel at constant speed under city and highway conditions for 1 hour and deliver the fuel economy values shown, what is the ratio of fuel consumed to tractive energy required to maintain these conditions?

Solution For convenience, we will use metric units in this example for comparability of fuel energy content to tractive energy required. First, comparing the two vehicles under city driving conditions, the speed of 40 km/h converts to 11.1 m/s. Plugging in known values to the equation for P_{TR} for the vehicles gives

$$\begin{aligned}P_{\text{TR-Echo}} &= 0.5\rho A_F C_D V^3 + \text{mgV}C_o\\ &= 0.5(1.15)(2.51)(0.29)(11.1)^3 + (927)(9.8)(11.1)(0.01) = 1{,}583\ W\\ P_{\text{TR-Prius}} &= 0.5\rho A_F C_D V^3 + \text{mgV}C_o\\ &= 0.5(1.15)(2.48)(0.29)(11.1)^3 + (1259)(9.8)(11.1)(0.01) = 1{,}938\ W\end{aligned}$$

The instantaneous power requirements can be converted to energy consumed in 1 hour in MJ as follows:

$$E_{\text{Echo}} = (1.58\text{ kW})(1\text{ hour})\left(\frac{3.6\text{ MJ}}{\text{kWh}}\right) = 5.70\text{ MJ}$$

$$E_{\text{Prius}} = (1.94\text{ kW})(1\text{ hour})\left(\frac{3.6\text{ MJ}}{\text{kWh}}\right) = 6.98\text{ MJ}$$

If the respective vehicles traveled for 1 hour averaging 40 km/h and delivered the rated fuel economy, their respective energy requirement values using 32.2 MJ/L fuel would be:

$$\text{Echo: }(40\text{ km})\left(\frac{1\text{L}}{12.26\text{ km}}\right)(32.2\text{ MJ/L}) = 105\text{ MJ}$$

$$\text{Prius: }(40\text{ km})\left(\frac{1\text{L}}{17.75\text{ km}}\right)(32.2\text{ MJ/L}) = 72.5\text{ MJ}$$

Thus the ratio value for the Echo is (105)/(5.7) = 18.4, and for the Prius is (72.5)/(6.98) = 10.4. Repeating this calculation for highway travel of 88 km using the same approach gives a tractive power requirement of 8.33 kW and 9.06 kW and a ratio value of 6.04 and 5.01 for the Echo and Prius, respectively.

Example 14-1 further highlights the ability of the HEV platform to deliver higher fuel economy while requiring more tractive power due to higher weight. For both the Echo and Prius, the actual energy requirement of stop-and-go city driving is more energy intensive than the tractive power requirement at 11.1 m/s constant speed suggests. Nevertheless the Prius ratio of 10.4 is much closer than that of the Echo, which consumes more energy in city driving despite weighing less. Since the two vehicles have the same drag coefficient and almost the same cross-sectional area (2.51 m^2 versus 2.48 m^2), differences in aerodynamic drag are very small. At highway speeds of 24.4 m/s the two vehicles are closer together in terms of both tractive power requirement and observed fuel economy. Still, the Prius has a closer ratio under highway conditions as well.

14-4-3 Advanced ICEV Technology: High-Efficiency Diesel Engine Platform

The primary means of advancing the efficiency of the ICEV platform available at present is to replace the gasoline engine with a high-efficiency diesel engine. Regarding this technology, conditions in the European versus North American markets have diverged, with European makers such as Volkswagen or Audi offering a wide range of diesel-powered vehicles, whereas the diesel offerings in the United States are quite limited. European governments laid the foundation for this development by requiring the refining of much cleaner diesel fuels (for example, with much lower sulfur content) for both heavy- and light-duty vehicles. European models such as the Audi A2 (Fig. 14-8) require clean diesel fuel for their engines to function correctly, and since these fuels only became available recently in the United States, their sale in the U.S. market was not possible in the meantime.

Table 14-5 shows the potential range of fuel economy values for diesel vehicles. The VW Passat is a midsize car emphasizing higher performance and comfort with moderately better fuel economy on a gallon per gasoline equivalent. Although the Passat compares well against gasoline ICEVs such as the Honda Accord or Civic in terms of engine size, curb weight, and overall fuel economy, it does not nearly achieve the very high fuel economy of the Audi A2, which has the highest fuel economy of any nonelectrified vehicle in the table despite not using a hybrid drivetrain.

FIGURE 14-8 Audi A2 5-door diesel subcompact.

Source: Rosemarie Vanek. Reprinted with permission.

The position of the Audi A2 and latest generation Toyota Prius suggests that the diesel ICE could be combined with a hybrid drivetrain to achieve the best possible fuel economy. The U.S. automakers adopted this approach in the 1990s in a government-sponsored R&D program known as the Partnership for a New Generation of Vehicles (PNGV), and developed concept cars that achieved 70 to 80 mpg using diesel hybrid drives on lightweight platforms. Peugeot-Citroen, GM, and Volkswagen, among others, have developed high-efficiency diesel hybrid prototypes, although they have not yet launched production because the sale price premium for the technology is prohibitive at this point.

Note that projections about the energy-saving benefits of diesel hybrids must be made carefully, since the diesel engine gains much of its efficiency advantage over gasoline in steady-state, highway driving, where HEVs generally have less of an efficiency advantage over ICEVs. From Table 14-5, if an efficient ICEV passenger car such as the Honda Civic averages 13 km/L (32 mpg) and the comparable gasoline Civic Hybrid achieves 18 km/L (44 mpg), one might assume that HEVs give a 40% increase in fuel economy across the board. However, it is unlikely that *hybridizing* of an efficient diesel ICEV such as the Audi A2 would result in a 40% increase in fuel economy for this vehicle on top of its already excellent rating.

14-4-4 Well-to-Wheel Analysis as a Means of Comparing Alternatives

It is clear from the variety of alternative fuel and propulsion platform options reviewed in the preceding section that it is essential to have some means of making an "apples-to-apples" comparison of quantitative effectiveness of each option in converting energy resources into vehicle propulsion while minimizing CO_2 emissions. The "*well-to-wheel*" (WTW) approach to energy efficiency is just such an objective measure of the overall energy efficiency of a transportation energy technology. WTW analysis includes both performance onboard the vehicle and also energy consumed upstream in extracting, transforming, and delivering an energy product to the vehicle. Letting η_{WTW} equal WTW efficiency, η_{WTT} equal well-to-tank (WTT) process efficiency and η_{TTW} equal tank-to-wheel (TTW) process efficiency, η_{WTW} can be written as

$$\eta_{WTW} = \eta_{WTT} \cdot \eta_{TTW} \tag{14-1}$$

The WTT process includes any energy expended to extract an energy product, transport it to a processing center, transform it into a fuel, and transporting the fuel onward to the dispensing point. The TTW process includes transforming the fuel into mechanical energy inside the vehicle and transmitting it to the wheels.

WTW values for three representative cases, namely ICEV, HEV, EV, and HFCV are given in Table 14-6. HFCVs are included even though they are not generally available for sale to the public since they are being sold by makers such as Toyota and GM in limited numbers, for example, to fleet operators.[8] The ICEV and HEV alternatives use petroleum refined into gasoline as basis for calculating WTW efficiency, while the EV uses electricity generated from the combustion of fuel oil derived from petroleum in a power plant to charge its onboard batteries. While fuel oil produces only a small fraction of the world's electricity, it is the closest to the gasoline used in the other two

[8]As of mid-2013 the fleet of the Port Authority of New York and New Jersey included 10 Toyota SUV HFCVs, refueled from a third-party fueling station on PANYNJ property.

vehicle examples. Electrical generation includes not only losses in electrical plants (where overall system efficiencies are on the order of 40% to 45%), but also energy expenditure in extracting and transporting natural gas to the plant and line losses in distribution of electricity from the power plant to the point of recharging the vehicle.

The WTW comparison illustrates a number of useful points about comparing alternative transportation technologies for the future. First, since the ICEV and HEV have the same WTT efficiency, the effect of changing to the hybrid drivetrain, shown in the increase in TTW efficiency from 16% to 28%, translates directly into an increase in overall WTW from 14% to 25%. The conventional ICEV is limited by a low TTW value. This drivetrain has already been the subject of many decades of research, and in recent years progress on increasing TTW efficiency for this platform has been slow in many mass-marketed vehicles. The high efficiency of the Audi A2 points to a possible way forward for TTW efficiency of ICEVs. High-efficiency diesels notwithstanding, achieving a major improvement while still keeping the gasoline engine as a starting point requires a more radical transformation, as exemplified by the conversion to HEV technology. For example, a 30% value of η_{TTW} is typical of HEVs today, and Toyota has set a goal of an HEV with WTW efficiency of 42%, or triple that of the ICEV, all by improving the TTW efficiency.

Secondly, the result for the EV shows the importance of looking at the whole energy cycle from the source. Looking at TTW efficiency alone, it appears that the EV is the most efficient approach, with 43% efficiency. Taking into account power plant and other distribution losses makes the comparison more realistic, showing that the representative EV is in fact less efficient than the HEV and only slightly better than the ICEV. Note that the numbers used in the table are intended to make general observations about the comparison between ICEVs, HEVs, and EVs; values for a specific vehicle and energy supply chain will vary somewhat from the values shown.

WTW analysis is the most useful when comparing mature alternatives over the short- to medium term; for example, where all the alternatives derive their energy from a single source, like fossil fuels in this case (a WTW comparison of solar energy converted to either electricity for EVs or biofuels for ICEVs/HEVs is also possible). However, if some of the sources do not emit GHGs and others do, then the relative WTW efficiency may be less important. For instance, if the EV in Table 14-6 were to derive its electricity from wind power instead of fuel oil, a low WTW efficiency in the table (due to the relatively low efficiency of wind turbines to convert the kinetic energy of the wind into electricity, compared to the efficiency of a combined cycle power plant to convert the chemical energy in methane into electrical energy) might not matter, because the end result of greatly reducing CO_2 emissions would have been achieved.

Availability of η_{WTT} values for different propulsion technologies allows for a more realistic comparison of the true fuel economy of a vehicle, reflecting not only published

Option	Well-to-Tank	Tank-to-Wheel	Well-to-Wheel
Conventional ICEV	88%	16%	14%
Gasoline HEV (Hi-Efficiency)	88%	30%	26%
EV w/electricity from gas	38%	43%	16%
HFCV w/hydrogen from gas	55%	45%	25%

TABLE 14-6 Well-to-Wheel Comparison of Propulsion Technologies for Representative Light-Duty Vehicles.

values but upstream energy losses incurred in delivering energy to the vehicle. Suppose we differentiate the nominated fuel economy mpg_{nom} from the effective fuel economy mpg_{eff}. The two values are related according to Eq. (14-2) because the well-to-tank efficiency captures the upstream losses:

$$mpg_{eff} = \eta_{TTW} mpg_{nom} \tag{14-2}$$

Example 14-2 develops this comparison for two vehicles both using natural gas as a primary energy source.

Example 14-2S The all-electric Nissan Leaf EV is to be compared with the natural-gas-powered Honda Civic GX, an ICEV that has been adapted by Honda to run solely on natural gas instead of gasoline. From Table 14-5, the Leaf has a nominal fuel economy of 99 mpg, and from USEPA the Civic GX achieves 28 mpg when measured in gallons of gasoline equivalent. For extraction of gas, transmission to a power plant, conversion to electricity, and transmission and charging of the Leaf, well-to-tank efficiency is 38%. Well-to-tank efficiency for the Civic GX including energy consumed in compressing the gas for use in the vehicle is 88%. What is the effective fuel economy for each vehicle?

Solution Using Eq. (14-2), the Nissan Leaf value is calculated as shown:

$$mpg_{eff} = \eta_{TTW} mpg_{nom} = (0.38)(99 \text{ mpg}) = 37.6 \text{ mpg}$$

Similarly, for the Civic GX:

$$mpg_{eff} = \eta_{TTW} mpg_{nom} = (0.88)(28 \text{ mpg}) = 24.6 \text{ mpg}$$

Thus the ratio of Leaf to Civic GX fuel economy, which is on the order of 3.5 to 1 when considering only nominal fuel economy, is reduced to 1.5 to 1 using effective fuel economy.

Example 14-2M The all-electric Nissan Leaf EV is to be compared with the natural-gas-powered Honda Civic GX, an ICEV that has been adapted by Honda to run solely on natural gas instead of gasoline. From Table 14-5, the Leaf has a nominal metric fuel economy of 41.4 km/L, and from USEPA the Civic GX achieves 11.7 km/L when measured in gasoline equivalent. For extraction of gas, transmission to a power plant, conversion to electricity, and transmission and charging of the Leaf, well-to-tank efficiency is 38%. Well-to-tank efficiency for the Civic GX including energy consumed in compressing the gas for use in the vehicle is 88%. What is the effective fuel economy for each vehicle?

Solution Using Eq. (14-2), the Nissan Leaf value is calculated as shown. Although the fuel economy is in metric units the variables mpg_{eff} and mpg_{nom} are used.

$$mpg_{eff} = \eta_{TTW} mpg_{nom} = (0.38)(41.4 \text{ km/L}) = 15.7 \text{ km/L}$$

Similarly, for the Civic GX:

$$mpg_{eff} = \eta_{TTW} mpg_{nom} = (0.88)(11.7 \text{ km/L}) = 10.3 \text{ km/L}$$

Thus the ratio of Leaf to Civic GX fuel economy, which is on the order of 3.5 to 1 when considering only nominal fuel economy, is reduced to 1.5 to 1 using effective fuel economy.

Example 14-2 shows that when the use of a certain primary energy source (natural gas in this case) involves combustion to eventually provide mechanical energy in the drivetrain, there are no easy ways to avoid significant losses in the chemical to mechanical energy conversion, using WTW analysis. Because of the upstream losses primarily in the power plant that converts gas to electricity, effective fuel economy of the Nissan Leaf is reduced to levels equivalent to that of an efficient gasoline ICEV (37.6 mpg in the example). The Leaf maintains its advantage over the Civic GX, but compared to a late model Toyota Prius, the Leaf has lower mpg_{eff} even after well-to-tank losses are taken into account (with $\eta_{WTT} = 88\%$, $mpg_{eff} = 43.7$ mpg for the 2011 Prius.) On the other hand, nominal fuel economy is still important for the consumer because it dictates energy cost per mile to drive different types of vehicles. In approximate numbers, whereas a typical U.S. ICEV in 2013 might cost \$4.00 in fuel to drive 25 miles, the same figure for the Civic GX is on the order of \$2.00, and for the Leaf on the order of \$1.00.[9]

14-5 Understanding Transition Issues

Comparison of new vehicle technologies such as HEVs in the preceding section suggests the type of energy savings that might be possible at a macroscopic level from large-scale transitions to those technologies. Many analyses of potential energy savings from changes to the transportation system are carried out on a *comparative static* basis, meaning that a comparison is made between two static solutions, one before some change is made and one after. While this approach is useful for quickly obtaining a preliminary estimate of potential benefits of changes, and may be the only option where data of adequate quality do not exist to support a more sophisticated model, it also ignores the *transition effects* of going from one state to another.

As discussed in Chap. 3 when a market transitions to a new technology, best practice, or other attribute, the shape of the penetration curve is often that of a logistics or triangle curve, with distinctive early, middle, and late stages. Figure 14-9 compares the historical market penetration curve of the automobile to that of three other major technological innovations, namely the landline telephone, the wireless radio, and connection of customers to the electrical grid. On the horizontal axis time is measured in years elapsed since the first introduction of the technology, and then the vertical axis gives the percent of market penetrated, that is, fraction of eligible consumers that possess the technology.

Examination of the individual curves is helpful for understanding previous experience with transitions. For each of the four technologies, the *early adopter* stage includes a period of germination where little if any progress is being made on penetrating the market. The length of this germination period varies: for electricity and telephone, it is on the order of 20 years, whereas for the auto and radio, it is as much as 30 years. Thereafter the length of the *rapid penetration* phase varies: electrification surpasses 90% penetration within 30 years, whereas the phone achieves this threshold in 80 years, and the car has achieved 70% penetration after 70 years. Both the phone and the auto experience a reversal of growth around the time of the Great Depression and World War II, although without major outside factors such as these, continuous nondecreasing growth is the norm. All the technologies except radio are seen to enter the late or *asymptotic growth* phase in the last years of their respective

---

[9]Cost per 25 mile figures are based on representative national figures for 2013 of \$0.12/kWh for electricity and \$4 per 1,000 ft^3 for gas.

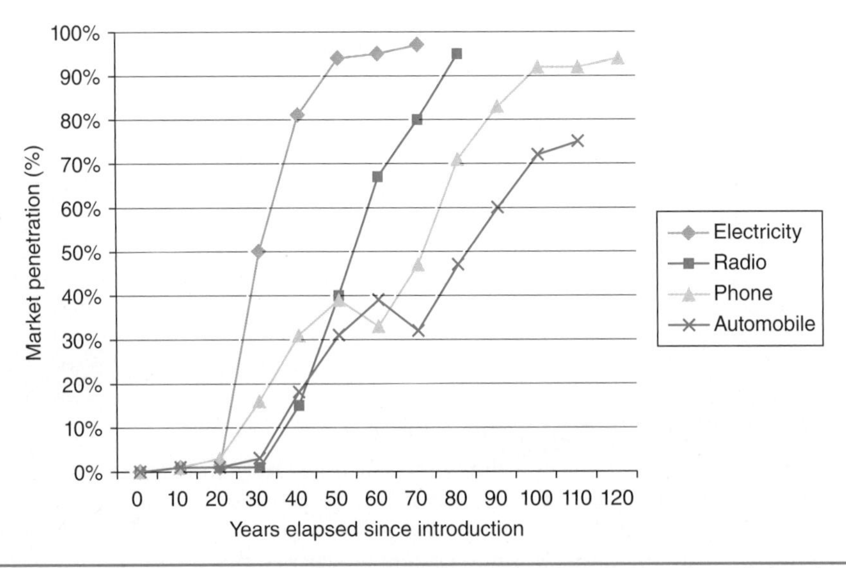

FIGURE 14-9 Market penetration in the U.S. market as a function of years elapsed since launch for electric grid connection, radio, landline telephone, and automobile.
Source: International Council on Systems Engineering (2006, p.2-7).

curves. The auto achieves 75% penetration 110 years after inauguration, and it is plausible given limits imposed by limited income, old age, or physical disability, that it might be difficult for the industry to develop products that are viable for all of the remaining 25%.

Although not shown in the figure, recent more compact technologies such as handheld wireless devices (mobile phones, smart phones, etc.) have penetrated the market more quickly, thanks in part to the greater agility of modern industry to produce and distribute products in a global marketplace. As was seen in Fig. 14-4, for a larger, higher capital cost asset such as an HEV, penetration progress into the market is likely to be slower. Therefore consideration of the penetration path and not just the starting and ending point are important.

14-5-1 Limitations of Comparative Static Analysis

Some of the limitations of comparative static analysis that ignores transition effects include:

- *Changes to baseline conditions during transition period are overlooked:* Transitions involving energy technology for transportation systems inevitably require long time horizons. During these time spans, all the underlying factors in a static analysis are subject to change, including total demand for energy, the efficiency of the incumbent technology, and the amount of GHG emissions. Furthermore, in addressing situations such as climate change or potential petroleum shortages, the timing of when the energy savings and CO_2 reductions are achieved is important. A comparative static analysis does not shed light on these issues.

- *The nature of the transition itself can shape the eventual outcome*: Certain factors that act on the system during the transition may act as an obstacle to its completion in the way that is expected in the static analysis. Projections that do not consider the transition as a possible barrier may overstate the benefits of the change.
- *Where transitions depend on government policy for support, the transition may fail to take root permanently:* While superior technological performance drives some transition (e.g., from paper-and-pencil to IT systems in transportation operations planning and management), others require the intervention of government through tax policy, subsidies, or regulations. Some transitions may require the permanent intervention of government in order for the technology to attract customers, which may or may not be financially or politically sustainable. Others may require an intervention long enough for the transition to take root, and will fail if it is too short. Where transitions fail for these reasons, the effect on energy consumption will not be the one that the static analysis predicts.

Fortunately, the research community has enough historical experience with technological transitions (e.g., the influx of a new, superior technology into a market, displacing an incumbent technology, or the government-led phasing out of an undesirable product) that we understand to some degree the shape that these transitions take, the forces at work during the transition, and even the future shape that a transition will take based on its early path. Where analysts can obtain the necessary data and have the time and resources to carry out a more careful, dynamic study of the transition, more accurate results are possible. Tools such as the logistics function or triangle curve are applied to mapping the transition pathway of new products and systems over time, leading to a more realistic understanding of when and to what extent energy savings will occur. Example 14-3 illustrates this process.

Example 14-3 Using the growth in hybrid sales in the United States shown in Fig. 14-4 as a starting point, consider the transition pathway to hybrid penetration into the fleet, in the following way. Suppose that the number of hybrid electric vehicles (HEVs) in the fleet, which starts in the year 2000 with 9,350 vehicles added, eventually tapers off to 13,000,000 units, and that each hybrid averages 15 km/L fuel efficiency, versus 8.5 km/L for the internal combustion engine vehicle (ICEV) alternative. Consider the year 2000 to be the year $t = 0$. Sales from 2001 to 2009, in order, number 20,287; 35,000; 47,500; 88,000; 207,000; 253,000; 355,000; 320,000; and 245,000. Note that the decline in the last few years is due to the severe recession of 2009. Assume that for each new sale another car is scrapped, so that the overall size of this segment of the fleet does not change, and that each car drives 16,000 km/year. (a) Use the logistics function to calculate in which year (call it year N) the number of hybrids in the fleet surpasses 99% of the 13 million target. (b) Calculate the cumulative fuel savings in liters for the shift to HEVs from years 0 to N, if the new cars are assumed to all be available on day 1 of each new year. For the years 2000 to 2009, use the sales estimated from the logistics curve model, rather than the actual sales, to calculate the number of HEVs in the fleet. (c) Compare the savings to the situation where a fleet of 13 million ICEVs is instantaneously transformed into HEVs at the beginning of year 0, and then driven to the end of year N.

Solution

(a) We begin by converting sales numbers into cumulative numbers in the fleet. The following table provides this information through year 2009:

Year	Number of Vehicles	
	Annual	Cumulative
2000	9,350	9,350
2001	20,287	29,637
2002	35,000	64,637
2003	47,500	112,137
2004	88,000	200,137
2005	207,000	407,137
2006	253,000	660,137
2007	355,000	1,015,137
2008	320,000	1,335,137
2009	245,000	1,580,137

We obtain all necessary parameters for the logistics function as follows. Based on full penetration of the 13 million market, we calculate fractional values of relative penetration, such as for 2000, $(9{,}350)/(1.3 \times 10^7) = 0.000719$, for 2001, $(29{,}637)/(1.3 \times 10^7) = 0.00228$, and so on. The value of c_1 and c_2 is solved in a spreadsheet by calculating the error between observed and model values for years 2000 to 2009, and then minimizing the square of the error terms, resulting in $c_1 = -5.26$ and $c_2 = 0.374$. For example, for year $t = 2$:

$$f(t) = \frac{e^{(c_1 + c_2 \cdot t)}}{1 + e^{(c_1 + c_2 \cdot t)}}$$

$$f(2) = \frac{e^{(-5.26 + 0.374 \cdot 2)}}{1 + e^{(-5.26 + 0.374 \cdot 2)}} = 0.0109$$

$$\text{HEV Total}_{\text{yr}=2002} = 0.0109 \cdot 1.3 \times 10^7 = 141{,}621$$

This estimate can be compared to the observed value of 64,637.

Plotting observed versus modeled penetration through 2009 gives the curve shown in Fig. 14-10.

In year $N = 27$, the penetration surpasses 99%:

$$p(26) = \frac{\exp(-5.26 + 0.374 \cdot 26)}{1 + \exp(-5.26 + 0.374 \cdot 26)} = 0.989$$

$$p(27) = \frac{\exp(-5.26 + 0.374 \cdot 27)}{1 + \exp(-5.26 + 0.374 \cdot 27)} = 0.992$$

(b) Since the total fleet size is constant at 13M, savings accrue from replacing some fraction of the fleet with hybrids. In each year, the baseline fuel consumption with no HEV penetration is

$$\frac{(1.3 \times 10^7 \text{ vehicles})(16{,}000 \text{ km/year})}{8.5 \text{ km/L}} = 2.45 \times 10^{10} \text{ L/year}$$

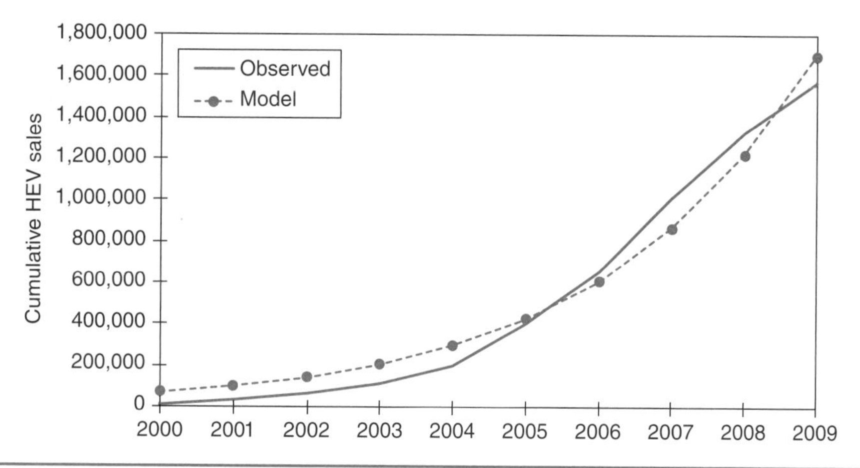

FIGURE 14-10 Actual versus modeled number of HEVs in fleet, 2000 to 2009.

The savings in each year is then the baseline minus the actual consumption, based on the mix of HEVs and ICEVs in the fleet. Taking year 2 as an example again, the HEV and ICEV fuel consumption values are respectively:

$$\frac{(3.992\times10^5\ \text{vehicles})(16{,}000\ \text{km/year})}{15\ \text{km/L}} = 4.26\times10^7\ \text{L/year}$$

$$\frac{(1.3\times10^7 - 3.992\times10^5\,\text{vehicles})(16{,}000\ \text{km/year})}{8.5\ \text{km/L}} = 2.44\times10^{10}\ \text{L/year}$$

The total fuel consumption is therefore little changed from the baseline, at 2.444×10^{10} L.

Repeating the calculation of annual savings for each year and summing to quantify cumulative savings gives the graph of fuel saved from 2000 through 2027 shown in Fig. 14-11. Based on these calculations, at the end of year 27, the savings has reached 1.2×10^{11} L, as shown.

(c) An instantaneous transition to 13 million HEVs gives the following energy consumption per annum:

$$\frac{(1.3\times10^7\ \text{vehicles})(16{,}000\ \text{km/year})}{15\ \text{km/L}} = 1.39\times10^{10}\ \text{L/year}$$

On this basis, the annual savings is $(2.45 \times 10^{10}) - (1.39 \times 10^{10}) = 1.06 \times 10^{10}$ L. Including the savings in year 0, the project has a 28 year time span. Therefore the cumulative savings is (1.06×10^{10}) (28 years) $= 2.97 \times 10^{11}$ liters of fuel.

Comparing the results of parts (b) and (c) in Example 14-3 shows that during the time of the project, considering the transition period results in a 58% reduction in the projected savings. Thus the benefits of reducing the consumption of petroleum or emissions of CO_2 would accrue much more slowly when one takes into account a realistic amount of time for the new vehicle technology to penetrate the market.

More realism could be added to this example by taking into account other real-world factors. Both HEV and ICEV fuel economy are likely to change over time. Also, stratifying the fleet by vehicle type, and then tracking sales and fuel economy of each vehicle type would make the projections of fuel savings more realistic. Furthermore, over the 27-year

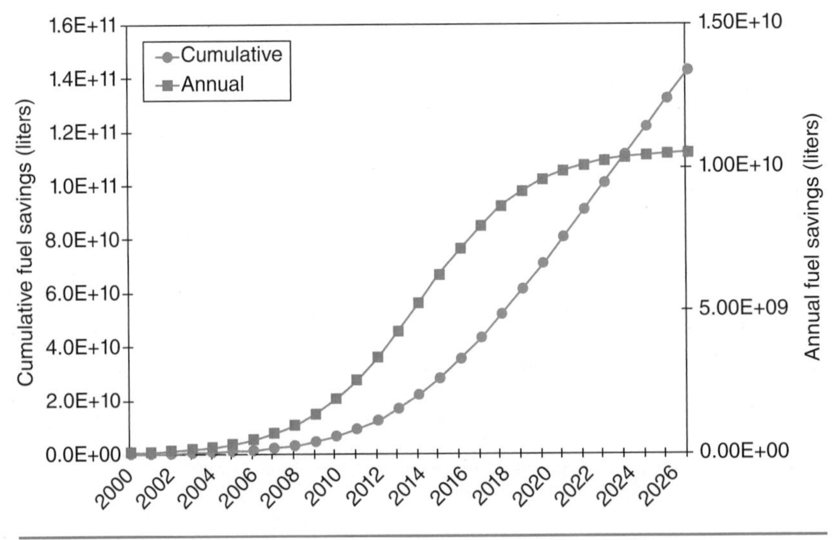

FIGURE 14-11 Annual and cumulative fuel savings, 2000 to 2027.

time span, most vehicles purchased in the first 10 to 15 years would have likely been retired, so an accounting of the turnover of both types of vehicles as part of the model would add accuracy.

14-5-2 Transition to Other Advanced Fuels and Drivetrain Platforms

The previous discussion and calculation in Example 14-3 considers the most prominent example in the near term of an advanced drivetrain platform that can replace the ICEV, namely the HEV. The reasoning is generally applicable to other prominent longer-term options, namely the electric vehicle or EV (and by extension the plug-in hybrid electric vehicle or PHEV); the hydrogen fuel cell vehicle or HFCV; and biofuel-powered variants of any of these where biofuels can be used, including ICEVs, HEVs, and PHEVs. Although none are as far along the transition curve shown in Fig. 14-9 as HEV technology, the transition of each can be modeled along its penetration pathway to estimate realistic energy and emissions savings compared to a business-as-usual baseline with no transition. Longer-term energy sources are the subject of the next two chapters: electricity and hydrogen in Chap. 15, and biofuels in Chap. 16.

14-6 Summary

In this chapter, we initially identify five possible endpoints for transportation energy technologies that could power vehicles, which are further reduced to three: (1) electricity, (2) hydrogen, and (3) sustainable hydrocarbons. The first two are the focus of Chap. 15 and the third is considered in Chap. 16. Several challenges face the adoption of these technologies as substitutes for internal combustion engine vehicles powered by petroleum products and emitting GHGs to the atmosphere. These obstacles include gaps in the complete chain of technologies needed for a successful replacement system, and stagnation in recent years in progress on improving vehicle efficiency.

New technologies for the short- to medium-term that improve efficiency while a more complete solution emerges include hybrid-electric vehicles, which combine an internal combustion engine with an electric motor and battery system to improve overall fuel economy. They also include advanced high-efficiency diesels that achieve high fuel economy without requiring a hybrid drivetrain. Overall efficiency of any alternative fuel or drivetrain platform can be measure in well-to-wheel efficiency, which considers both losses from the primary energy source to the vehicle and from the point of entry into the vehicle to the wheels moving down the road. Evaluation of the energy savings benefit of any new technology requires not only the endpoint efficiency of the technology once adopted in the future but the transition effects along its technological substitution pathway.

References

Cengel, Y. and M. Boles (2010). *Thermodynamics: An Engineering Approach, 7th ed.* McGraw Hill, Boston, MA.

Davis, S., S. Diegel, and R. Boundy (2013). *Transportation Energy Data Book, 32d ed.* Oak Ridge National Labs, Oak Ridge, TN.

INCOSE (2006). *Systems Engineering Handbook: A Guide for System Life Cycle Processes and Activities.* International Council on Systems Engineering, Seattle, WA.

Kreith, F., R.E. West, and B. Isler (2002). "Legislative and Technical Perspectives for Advanced Ground Transportation Systems." *Transportation Quarterly* 56(1/Winter 2002):51–73.

Lutsey, N. and D. Sperling (2005). "Energy Efficiency, Fuel Economy, and Policy Implications." *Transportation Research Record* 1941:8–25.

Randolph, J. and G. Masters (2008). *Energy for Sustainability: Technology, Planning, Policy.* Island Press, Washington, DC.

Sperling, D. and D. Gordon (2009). *Two Billion Cars: Driving Toward Sustainability.* Oxford University Press, Oxford.

Vuchic, V. (2007). *Urban Transit: Systems and Technology.* Wiley, New York.

Wark, K. (1984). *Thermodynamics, 4th ed.* McGraw Hill, New York.

Further Readings

An, F. and D. Santini (2003). "Assessing Tank-to-Wheel Efficiencies of Advanced Technology Vehicles."*SAE Technical Paper Series Report* 2003-01-0412. Society of Automotive Engineers, Warrendale, PA.

Braess, H. and U. Seiffert, eds. (2004). *Handbook of Automotive Engineering.* Society of Automotive Engineers, Warrendale, PA.

Commission on Engineering and Technical Systems (2000). *Review of the Research Program of the Partnership for a New Generation of Vehicles: Sixth Report.* National Academies Press, Washington, DC.

Fairley, P. (2004)."Hybrid's Rising Sun." *MIT Technology Review* (Apr.):34–42.

Fay, J. and D. Golomb (2002). *Energy and the Environment.* Oxford University Press, New York.

General Motors Corp. (2001). *Well-to-Wheel Energy Use and Greenhouse Gas Emissions of Advanced Fuel/Vehicle Systems—North American Analysis.* Argonne National

Laboratories, Argonne, IL. Available at *http://www.ipd.anl.gov/anlpubs/2001/08/40409.pdf.* Accessed Sep. 14, 2007.
Gillespie, T.D. (1999). *Fundamentals of Vehicle Dynamics.* Society of Automotive Engineers, Warrendale, PA.
Husain, I. (2010). *Electric and Hybrid Vehicles: Design Fundamentals, 2d ed.* CRC Press, Boca Raton, FL.
Jones, C., and D. Kammen (2014) "Spatial Distribution of U.S. Household Carbon Footprints Reveals
Suburbanization Undermines Greenhouse Gas Benefits of UrbanPopulation Density." *Environmental Science and Technology,* forthcoming.
Sager, J., J. Apte, and D. Kammen (2011). "Reduce Growth Rate of Light-Duty Vehicle Travel to Meet 2050 Global Climate Goals." *Environmental Research Letters* Vol. 6, pp. 1-6.
Stone, R. and J. Ball (2004). *Automotive Engineering Fundamentals.* SAE International, Warrendale, PA.
Tester, J., E. Drake, M. Driscoll, M. Golay, and W. Peters (2006). *Sustainable Energy: Choosing among Options.* MIT Press, Cambridge, MA.
Union of Concerned Scientists (2007). *HybridCenter.org: A Project of the UCS.* Web resource, available at *www.hybridcenter.org.* Accessed Sep. 7, 2007.
U. S. Dept of Energy, Energy Efficiency and Renewable Energy Office (2007). "Alternative Fuels Data Center." Informational Web site, available at *http://www.eere.energy.gov/afdc/altfuel/biodiesel.html.* Accessed Sep. 14, 2007.
Vanek, F., S. Galbraith, and I. Shapiro (2006). *Final Report: Alternative Fuel Vehicle Study for Suffolk County, New York.* Technical Report, New York State Energy Research and Development Authority (NYSERDA), Albany, NY.

Exercises

14-1. A compact passenger car that runs on natural gas emits 0.218 kg CO_2 per mile driven. For simplicity, treat natural gas as being pure methane (CH_4) having an energy content of 50 MJ/kg. (a) What is the mass and energy content of the fuel consumed per mile driven? (b) Suppose the vehicle is fueled with synthetic natural gas that is derived from coal. Of the original energy in the coal, 40% is used to convert the coal to the synthetic gas. Again, for simplicity, treat coal as pure carbon. What is the mass of coal consumed and new total CO_2 emissions per mile? Consider only the effect of using coal instead of gas as the original energy source, and the conversion loss; ignore all other factors.

14-2. Use the data in Table 14-3 to complete the Divisia analysis of car and light truck fuel consumption in the period 1970 to 2000 in metric units.

14-3. Use the data in Table 14-3 to complete the Divisia analysis of car and light truck fuel consumption in the period 1970 to 2000 in U.S. standard units.

14-4. Revisit Example 14-3, using the same fuel economy values for ICEVs and HEVs, but this time considering the entire U.S. passenger car fleet. Suppose that HEVs achieve 50% penetration of the national fleet by 2040. The fleet in 2000 consisted of 134 million vehicles; assume this number is fixed for the duration of the transition. Compare this transition to a scenario where there is no influx of HEVs. (a) Calculate the cumulative fuel savings for the period of 2000 to 2040 using a triangle function with peak rate of change in 2020. (b) Calculate the cumulative fuel savings for the period of 2000 to 2040 using a logistics function. Fit the logistics function to the data in years 2000 to2009 and 2040 only, with the assumed value in 2040 of 50% of the fleet; that is, 67 million vehicles. (c) Compare the results from calculations in parts (a) and (b). Discuss the differences.

14-5. Repeat Problem 14-4, but this time with changing overall fleet size and fuel efficiency. Assume that both the fleet size and average fuel economy grow linearly at the average annual rate from 1980 to 2000. Also, assume that HEV fuel economy improves by the same percent each year as the ICEV fleet for the period of 2000 to 2040. Obtain the necessary data to extrapolate rates of change from the internet or other source.

14-6. A car buyer is considering whether or not to spend extra for a hybrid in order to save money in the long run on gasoline expenditures. The buyer drives 25,000 km/year. The cost difference between the two vehicles is $4,000, and the fuel consumption is 7.9 L/100 km for the ICEV and 5.3 L/100 km for the HEV. The cost of fuel is $1.06/L (about $4/gal). (a) What is the simple payback for buying the HEV, in years? (b) If the buyer expects a 5% return on this investment, and either car will last for 10 years with negligible resale value at the end of that time, what is the NPV? (c) What is the NPV in (b) if the cost of fuel rises by 3% per year for the lifetime of the vehicle?

CHAPTER 15

Electricity and Hydrogen as Alternative Fuels

15-1 Overview

This chapter first presents electric vehicle technology, including both all-electric vehicles and plug-in hybrid electric vehicles, as a vehicular technology to reduce emissions and diversify primary energy sources for transportation at a vehicle level. Next, the benefits and challenges at a systems level of bringing electrified vehicles into the electric grid system and electricity markets are explored. This includes background on the function of the electricity grid, the role of fossil and nonfossil electricity sources, and the economic benefits of vehicle-to-grid systems. Hydrogen is discussed at the end of the chapter since many of the points regarding electrification also apply to hydrogen if it were adopted rather than electricity as an alternative to petroleum.

15-2 Introduction

The focus of this chapter is to extend the discussion of hybrid-electric vehicles (HEVs) of the Chap. 14 to vehicles that can draw power from the grid, namely, electric vehicles (EVs), plug-in hybrid electric vehicles (PHEVs), and hydrogen fuel cell vehicles (HFCVs). In the remainder of this introduction, we focus on EVs and PHEVs since they are more advanced in their evolution. However, hydrogen can potentially achieve many of the same benefits, for example, diversified primary energy sources, more stable energy prices, less ecological damage, and more reliable energy supply for the long term.

Electrification aids with sustainable transportation because it has the potential for achieving a higher overall well-to-wheel efficiency of conversion of a primary energy source into kinetic energy of a moving vehicle, compared to the internal combustion efficiency. Internal combustion engines must be able to deliver high levels of power when demanded by the driver for rapid acceleration or other heavy loads such as climbing a hill at speed. Optimizing internal combustion engines leads to sacrificing of efficiency at partial loads. The Atkinson-Miller cycle found in hybrid vehicles addresses this shortcoming by prioritizing high efficiency in steady-state operations rather than maximum power output at peak loading, and calling on the electric drive portion of the hybrid for additional power when needed. Electrification, on the other hand, takes advantage of much higher potential conversion efficiency values in centralized power plants, up to 55% or 60% in a state-of-the-art combined cycle plant. Even after taking into account losses in transmission and distribution, high conversion efficiency can be combined with high potential efficiency of the electric drivetrain to achieve higher overall well-to-wheel efficiency.

Electrification also decarbonizes transportation because it replaces gasoline or diesel derived from petroleum, whose carbon content comes entirely from crude oil, a fossil fuel. Even if a fraction of the electricity used by the electric vehicle is generated from fossil fuels, the presence of noncarbon (e.g., hydro, wind, nuclear, etc.) electricity in the mix can lead to reduced CO_2 emissions per mile compared to an internal-combustion engine vehicle (ICEV) using petroleum-derived liquid fuels.

15-3 Electric Vehicles

The electric vehicle (EV) uses electricity to operate a motor that then drives the wheels of the vehicle. Sometimes the abbreviation battery-electric vehicle (BEV) is used to distinguish EVs that carry their own stored energy in batteries from electric vehicles such as trolleybuses (also known as trackless trolleys) that also use an electric motor to propel the wheels but draw power from overhead wires.

15-3-1 Brief History of EV Development

The battery-electric vehicle has existed since the early days of the history of the automobile in a niche-market and prototype form. After automobile makers settled on the internal combustion engine (ICE) as the propulsion system of choice in the early 1900s, experimental work continued with the vehicles at a low level throughout the twentieth century, and by the 1970s prototypes existed with a range of 130 km (81 miles) per charge and acceleration of 0 to 96 km/h (0–60 mph) in 16 seconds. The World Solar Challenge between Aidelaide and Darwin, Australia, in 1987 was a catalyst for developing a new generation of more advanced EVs, and positioned the winning entrant, General Motors, to develop a prototype EV called the "Impact," launched in January 1990, that achieved a 145 km (~90 mile) range between charges, a 150 km/h (95 mph) top speed, and acceleration of 0 to 96 km/h (0–60 mph) in 8.6 seconds. Since it coincided with California's regulatory deliberations for tighter tailpipe standards, GM's introduction of the Impact concept EV had the unintended consequence of stimulating the Zero Emission Vehicle (ZEV) mandate.

In the late 1990s, a number of U.S. and Japanese makers entered EVs in the commercial passenger car and light truck market in California in the United States, and leased or sold several thousand vehicles, ranging from the Saturn EV-1 two-seat coupe (the production version of the Impact), to the Toyota RAV-4 small SUV, to the Ford Ranger pickup truck. In 2001, the automakers succeeded in weakening a mandate that would have required the expansion of EV sales in the California market, and by 2002 new EVs were no longer for sale from the major manufacturers. This unfolding of events was disputed for some time by the manufacturers on the one hand and individual EV owners and enthusiasts on the other. The former claimed that their efforts to sell the vehicle proved that an adequate market did not exist for a vehicle that lacked sufficient range for intercity driving, meaning that an owner would most likely need to own a second ICE-powered vehicle in order to meet all driving needs. The latter countered that there was in fact a sufficiently large market of drivers who could afford to own an EV as a second car, and that automakers had not made a good faith effort to promote the technology.

Electric Vehicles after the EV-1

Although the EV-1 and the generation of EVs that came with it were withdrawn after 2002, the experience proved that in terms of the technology EVs could be built so as to operate well on urban road networks otherwise dominated by ICEVs, even if the economics were questionable. It also provided a stepping stone toward the HEV technology that has been expanding steadily in the North American and Japanese markets since 1997, and at a faster rate since 2003 (see Chap. 14). Different forms of currently available EVs are shown in Fig. 15-1.

Mass-produced EVs have also reentered the passenger vehicle market both as full-performance EVs (i.e., capable of driving at highway speeds and possessing range per recharge of 80 miles or more) and niche electric vehicles. Nissan entered the U.S. market with the Leaf five-door hatchback in 2011, which sold approximately 10,000 units in the 2012 calendar year for a price around $30,000 (Figs.15-1(*a*) and (*b*). Other makers such as Ford and BMW were preparing to launch EV models in the U.S. market as of 2013. As an alternative to established ICEV makers that are launching EV models, the U.S.-based Tesla company represents an all-new EV-only manufacturer that has entered the market at the luxury end, starting with a limited-edition two-seat $100,000 sports coupe in 2010 and currently marketing a luxury sedan (the Model S) that costs between $70,000 and $110,000, depending on features and size of battery system.

On the niche EV side, manufacturers are marketing small EVs in Europe and Japan, good for short-distance travel and featuring ease of parking, such as the Italian 2-seat electric "Maranello 4" launched by Effidiin 2006. Limited-use vehicles (LUVs) powered by batteries that travel at a maximum speed of 50 to 60 km/h (30–35 mph) are available in the U.S. market, such as the GEM EV from Chrysler Corporation presented in Chap. 9 (see Fig. 9-1). While not legal for travel on highways, they are useful for niche applications such as national and state parks, corporate office parks, or academic campuses. LUVs are fitted with turn-signals, headlights, seatbelts, and license plates to allow for legal operation on public roadways with a speed limit of 35 mph (60 km/h) or less, which provides a platform that can be adapted for electric urban minibuses (Fig. 15-1*c*).

Retrofitting ICEVs to convert them to EVs, or using a "coaster" (i.e., ICEV sold by the car manufacturer without engine and drivetrain), is another method to market EVs. Siemens markets a complete electric drivetrain that can be retrofitted into an ICEV by removing the ICE drivetrain, and there are a number of small companies and individuals who provide retrofitting services to a few thousand vehicles each year, either using off-the-shelf EV drivetrains such as the Siemens product, or piecing together their own systems.

15-3-2 Electric Vehicle Drivetrain Design Considerations

Although early EV prototypes from past decades used DC current, contemporary EVs use AC motors due to superior performance. In such a vehicle, the drivetrain consists of a battery bank, a power controller that converts electrical energy coming from the battery system from DC to AC, a motor, and a transmission to the driven wheels, as shown in Fig. 15-2. Control over the vehicle is transmitted from the accelerator pedal electronically to the power converter, which then controls the rate of energy flow to the motor. Note that, although not shown in the figure, the EV drivetrain may also incorporate regenerative braking. Instead of dissipating kinetic energy as heat in the brake pads, as occurs in an ICEV, the EV can run the transmission in reverse, using the electric motor as a generator that creates drag on the wheels to slow the vehicle, while at the same time recharging the batteries.

(a)

(b)

(c)

Figure 15-1 Examples of electric vehicles: (*a*) Nissan Leaf full-performance hatchback; (*b*) Nissan Leaf with 110V home recharging cable attached to front port, (*c*) battery-powered electric minibus operated by Reseau de Transport de la Capitale (Capital Transport Region), Quebec City, Canada.

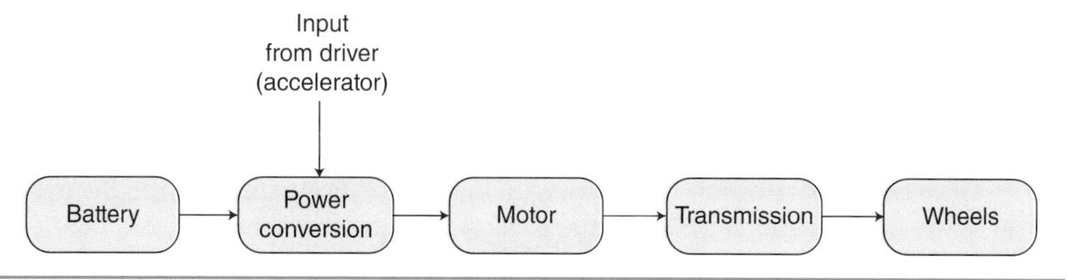

FIGURE 15-2 Schematic of components in EV drivetrain system.

Present-day EVs have settled on lithium-ion (Li-ion) batteries because they offer the most *charge density* per unit of weight in a large-scale battery system. Both the Leaf and the Tesla vehicles use Li-ion batteries. Li-ion charge density is estimated at 80 to 130 Wh/kg or 0.08 to 0.13 kWh/kg compared to 35 to 50 Wh/kg for lead-acid, the battery material of previous generations of EVs (Husain, 2010).

The desired selling price, allowable curb weight, and target driving range of the vehicle all influence the size of the battery system. The cost per kWh of storage capacity and total weight limit the range achievable for a given EV model. EV makers generally do not publicize their production cost per kWh for competitive reasons, but auto industry analysts are able to estimate approximate values. The price declined from approximately $500/kWh in 2010 to $400/kWh in 2013. The Leaf has a 24-kWh system while that of the Tesla Model S ranges between 40 and 85 kWh. On the basis of 2013 cost figures, the battery systems would cost approximately $10,000 for the Leaf and $34,000 for the largest Tesla Model S battery system.

One other limitation is the global supply of lithium, which is a relatively rare element compared to primary materials for some competing battery technologies, such as lead or nickel in nickel metal hydride batteries. In the short term, the number of EVs sold as a fraction of the world light-duty vehicle (LDV) market is small enough that lithium supply is not a concern. If in the future this quantity grows into a major fraction of the total, however, it remains to be seen whether a sustainable supply chain for extracting, distributing, and recycling lithium can be developed to keep up with demand.

Options for Recharging EVs: Household and Public Charging Devices

An important difference between EVs and ICEVs is the means of providing energy: EVs can be recharged in a wide variety of locations, but the downside is the long charge time required. For home recharging, either 120-V or 240-V systems are practical. In countries such as the United States where household electrical outlets at 120 V are the norm, recharging can be done at this voltage at little or no additional cost, since vehicles may come equipped with all necessary equipment to plug into a wall. The vehicle does, however, require many hours overnight to recharge, and may not recharge completely depending on the distance driven the previous day. Therefore, for reliable home charging on a regular basis, EV owners generally require a 240-V system, which may incur an additional $1,000 in total cost for materials and installation labor. Fast chargers of 480V installed in public places can restore partial charge sufficient to continue a daily commute in 20 to 40 minutes thanks to both higher voltage and current. Chargers of this size are impractical for home installation not only because of high cost but also due to safety and complexity issues that would arise with 480-V wiring in a residence.

15-3-3 Model of EV Range and Cost as Function of Battery Capacity

The charge density of the battery technology has a strong influence on the overall range of the vehicle, and is therefore the subject of an empirical model presented in this section. Charge density can be measured either in terms of energy per unit of weight or per unit of volume. The model presented uses density per unit of weight.

Suppose we take a simplified model of an EV in which charge availability is constant at all states of charge of the battery (i.e., whether full, half-full, etc.) down to the *maximum depth of discharge* DD_{max} (as a fraction of 100%). Assume further that the vehicle has constant average energy intensity in terms of the energy required per unit of mass of vehicle moved 1 km, defined as μ in units of Wh/kg-km. Let CD be the charge density, $W_{battery}$, the mass of batteries in kilograms, and $W_{vehicle}$, the mass of remainder of the vehicle in kilograms not including the battery system. The range R, in kilometers, can be estimated as a function of battery weight as follows:

$$R = \frac{CD \cdot DD_{max} \cdot W_{battery}}{\mu(W_{vehicle} + W_{battery})} \tag{15-1}$$

Eq. (15-1) can be rewritten to solve for $W_{battery}$ as a function of range R:

$$\begin{aligned} R\mu(W_{vehicle} + W_{battery}) &= CD \cdot DD_{max} \cdot W_{battery} \\ R\mu W_{vehicle} &= CD \cdot DD_{max} \cdot W_{battery} - R\mu W_{battery} \\ W_{battery} &= \frac{R \cdot \mu \cdot W_{vehicle}}{(CD \cdot DD_{max} - R \cdot \mu)} \end{aligned} \tag{15-2}$$

The overall size of the battery system has a direct impact on total vehicle cost as well as range. If we take the fixed cost of the vehicle to be C_{fixed} before adding the cost of the batteries, the total cost of the vehicle TC can be written as a function of range:

$$\begin{aligned} TC &= C_{fixed} + W_{battery} C_{battery} \\ &= C_{fixed} + \frac{R \cdot \mu \cdot W_{vehicle}}{(CD \cdot DD_{max} - R \cdot \mu)} C_{battery} \end{aligned} \tag{15-3}$$

where $C_{battery}$ is the unit cost of the batteries in dollars per kilogram. The use of these equations to estimate battery requirements and total cost as a function of range is illustrated in Example 15-1.

Example 15-1 A representative EV can be built with either lead-acid (Pb) or lithium-ion batteries. Regardless of battery type, the vehicle has the following values: $W_{vehicle}$ = 920 kg, μ = 0.128 Wh/kg-km, and C_{fixed} = $14,000. Lead acid batteries have values of CD = 55 Wh/kg and $C_{battery}$ = $125/kWh. Li-ion batteries have values of CD = 128 Wh/kg and $C_{battery}$ = $400/kWh. For both options, assume a maximum depth of discharge of 80%. What mass of batteries is required, and what is the total cost, of a vehicle that has a range of (a) 200 km and (b) 300 km?

Solution Mass of batteries for case (a), EV with 200-km range, from Eq. (15-1) the weight requirement for the lead-acid battery is

$$W_{\text{battery}} = \frac{R \cdot \mu \cdot W_{\text{vehicle}}}{(\text{CD} \cdot \text{DD}_{\max} - R \cdot \mu)} = \frac{200 \times 0.128 \times 920}{(55 \times 0.8 - 200 \times 0.128)} = 1280 \text{ kg}$$

Repeating for Li-ion batteries:

$$W_{\text{battery}} = \frac{R \cdot \mu \cdot W_{\text{vehicle}}}{(\text{CD} \cdot \text{DD}_{\max} - R \cdot \mu)} = \frac{200 \times 0.128 \times 920}{(128 \times 0.8 - 200 \times 0.128)} = 307 \text{ kg}$$

Mass of batteries for case (b): For a distance of 300 km, the above calculations are repeated but with $R = 300$ km, giving values of 6,309 kg for lead and 552 kg for Li-ion. Note that the value for lead-acid at 300-km range is unrealistic, as the battery system ways more than six times as much as the remainder of the vehicle. This result is discussed further below.

Turning to total cost of vehicle, it is necessary to compute C_{battery}, the cost per kg for batteries, from the given data. These values are computed for lead and Li-ion:

$$C_{\text{battery.Pb}} = (\$125/\text{kWh})\,(0.055 \text{ kWh/kg}) = \$6.88/\text{kg}$$
$$C_{\text{battery.Li}} = (\$400/\text{kWh})\,(0.128 \text{ kWh/kg}) = \$51.20/\text{kg}$$

We can solve for the total cost of the vehicle using Eq. (15-3) for the case of 200-km range:

$$\text{TC}_{\text{Pb}} = C_{\text{fixed}} + W_{\text{batt}} C_{\text{battery}} = \$14,000 + (1,280)\,(\$6.88) = \$22,800$$
$$\text{TC}_{\text{Li}} = C_{\text{fixed}} + W_{\text{batt}} C_{\text{battery}} = \$14,000 + (307)\,(\$51.20) = \$29,718$$

For case (b) with 300-km range, repeating the above calculations gives $57,374 for the EV with lead-acid batteries, and $42,262 for the EV with Li-ion batteries.

Example 15-1 shows how the higher charge density of the lithium-ion battery translates into lower additional mass to the vehicle, especially at longer ranges. This comparison is made visually in Fig. 15-3, where the range of the EV is plotted as a function of battery mass using Eq. (15-1) for values from 0 to 2,000 kg and other parameters taken from Example 15-1. At a range of 235 km, the batteries for the lead-acid system weigh 2,000 kg, which comprises more than two-thirds of the total mass of the vehicle in the case of lead-acid. Note also that this estimate is an optimistic simplification, since increasing vehicle weight by such a large amount will have a multiplier effect in terms of requiring additional strengthening of the vehicle body, a stronger electric motor, heavier brakes, and so on, to be able to support the extra batteries, further increasing weight. These limitations are considered below in the discussion of a more complete mass-range model.

Using cost, weight, and charge density figures from Example 15-1, it is also possible to compare total vehicle cost figures for the two battery technologies as a function of range (Fig. 15-4). At low-range values, there is a cost premium associated with the light weight of lithium-ion batteries, which is why applications such as golf carts—for which cost rather than range per charge is the primary concern—will continue to use lead-acid. However, at range values that begin to approach those of ICEVs (250 km and above), the cost premium narrows, and around 275 km the two technologies are equivalent in price. Above this range value, the cost for lead-acid increases rapidly, and the cost curve approaches a vertical asymptote around 344-km range, so that the vehicle cannot achieve this range for any cost. The problem of rising cost compounds the issue of excessive weight for lead-acid at significant range, as shown in Fig. 15-3. For instance, the Tesla Model S sedan with 85 kWh of storage capacity in theory adds 664 kg of battery mass to the vehicle, at a cost of $34,000, which compares favorably to lead-acid.

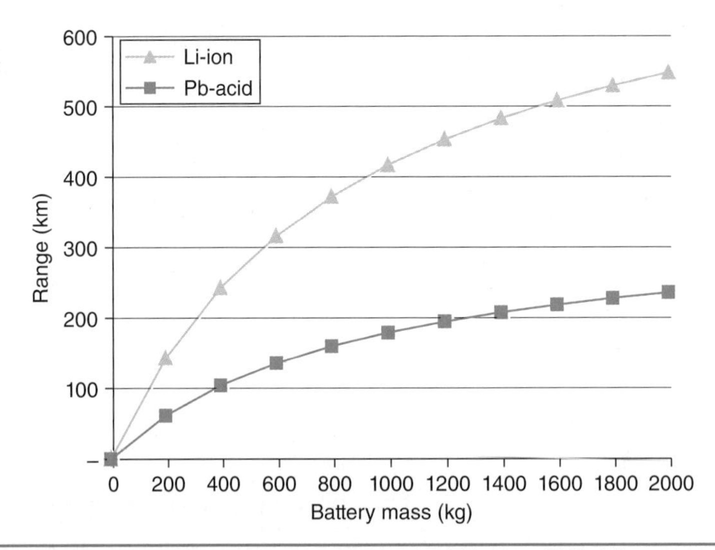

FIGURE 15-3 Maximum range for a representative lead-acid battery EV in kilometer on a single charge, as a function of mass of batteries installed. Parameters: Empty vehicle weighs 920 kg, intensity 0.128 Wh/kg-km.

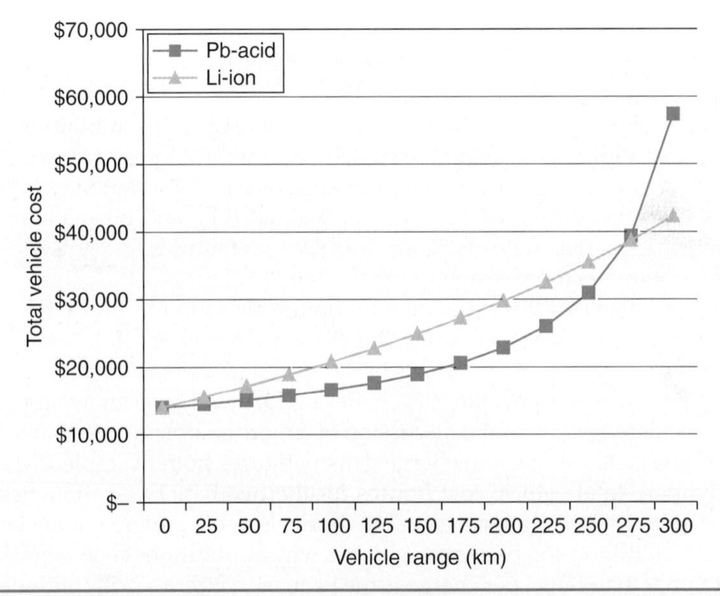

FIGURE 15-4 Vehicle cost as a function of range for EV. Parameters: Empty vehicle weighs 920 kg, intensity 0.128 Wh/kg-km.

Application of Range Model to Case of Tesla Model S Sedan

Since the Tesla Model S Sedan comes in different configurations in terms of the size of the battery system (measured in kWh), it provides a useful application for estimation of range based on battery system size and weight and for comparison to other published sources. Range estimates are obtained from three sources: (1) the model of Eq. (15-1),

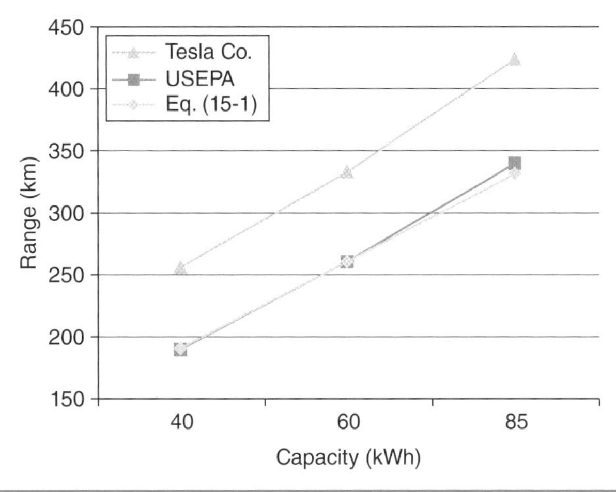

FIGURE 15-5 Prediction of average range per charge for three battery sizes of Tesla Model S from three alternative prediction sources. Note that the *y* axis does not start at 0-km range, to better show differences between curves.

Sources: Tesla Motors, for "Tesla Co." values; USEPA *fueleconomy.gov* Web site, for "USEPA" values.

(2) the USEPA at the Web site *fueleconomy.gov*, and (3) the manufacturers claimed range per charge. The Model S sedan is examined in 40 kWh, 60 kWh, or 85 kWh versions. Tesla decided to discontinue development of the 40 kWh model before it could be put into production, but the company did publish specifications and range estimates were made available, so those are used. The range model estimate uses a charge density value of 105 Wh/kg including the entire weight of the battery system and maximum depth of discharge of 95% (both of which are different from those of Example 15-1).[1] The intensity value μ is not known and must therefore be computed. Using above mentioned data, curb weight of 2,112 kg given by Tesla, and range given by USEPA of 261 km per charge gives a figure of $\mu = 0.104$ Wh/kg-km.

The results in Fig. 15-5 show that, having calibrated the range model using USEPA's estimated range for the 60-kWh version, the model's prediction for range are in good agreement with USEPA for the 40 kWh and 85 kWh versions. For the 40-kWh version the model predicts a range of 190.9 km compared to 190.0 km for USEPA, while for the 85-kWh version, the respective value are 331.7 and 340.0. Note that the claimed values in Fig. 15-5, which are higher than either the USEPA or model values, are not necessarily inaccurate; rather they reflect driving under conditions that may be more favorable to longer range per charge, such as a larger mix of highway driving.

Desired Characteristics of More Complete EV Range Model in the Future

It is desirable to have an enhanced analytical model that goes beyond the basic model of Eqs. (15-1) to (15-3) to predict EV range based on battery system characteristics, balance of vehicle characteristics, and operational conditions (terrain, driver

[1]Figures estimated based on information obtained from the Tesla Motors Web site, *www.teslamotors.com*.

behavior, and so on). Specifically, a range model might incorporate the following parameters:

- *Effect of increasing battery size on balance of vehicle mass:* As battery mass increases, at some point the various components of the vehicle will need to measurably increase in strength and weight to function correctly. The enhanced model could therefore include a multiplier factor or other means to adjust the mass of balance of vehicle ($W_{vehicle}$) upward as battery mass increases.
- *Effect of aerodynamic drag:* In the basic model, the impact of aerodynamic drag is subsumed into the intensity factor μ, since a vehicle with higher aerodynamic drag will require more energy to move 1 kg of its mass 1 km, other things equal. However, in an enhanced model, the aerodynamic drag coefficient could be broken out so that for fixed combined weight $W_{vehicle} + W_{battery}$ changes in drag would affect overall range per charge.
- *Effect of rolling resistance:* Similar to aerodynamic drag, the enhanced model could capture impact of changing rolling resistance on range for fixed total weight (see Chap. 5 for mathematical relationship between aerodynamic drag, rolling resistance, and tractive power requirement).
- *Effect of operating conditions* (e.g., driving speed, flat vs. rolling terrain): Incorporating operating conditions would allow the model to predict range based on predicted conditions. For instance, in hilly terrain, electric vehicle range will tend to be reduced as large amounts of charge are expended to climb elevation. Some of this charge is returned to the battery system using regenerative braking, but there are limits due to both inherent losses in the system and the maximum rate at which the motor working backwards can recharge the batteries. The driver's chosen highway cruising speed, aggressiveness of driving, use of air conditioning, and outside air temperature also have an effect.

At present, with the limited number of EVs on the road and lack of experience with the technology, it is premature to operationalize such a model for lack of real-world data. One solution in the interim is the use of a computerized driving cycle simulator, which takes a representative drive cycle or route between charges, measures power requirements at frequent intervals based on speed, acceleration, and grade, calculates instantaneous power requirement, and then adds up total energy consumption.[2] Although they ultimately represent the most accurate way of measuring range and other figures of merit for proposed EV designs, such models are time-consuming to implement and not accessible to a broad audience in the transportation community. There is therefore scope for a "middle ground" modeling option between the basic model already presented and the full simulator.

---

[2]For instance, Albertus et al. (2008) use a drive cycle simulator to model the movement of vehicles over a virtual test course with specific speed and grade requirements to estimate use and remaining availability of charge in the battery system.

15-3-4 Advantages and Disadvantages of EVs Compared to Alternatives

A summary list of benefits of EVs includes the following:

1. *Reduced emissions:* As already mentioned above, EVs are capable of reducing CO_2 emissions per mile either modestly or substantially, depending on whether the electricity is derived mostly from coal or mostly from a mixture of renewables, nuclear, or natural gas. Turning to criteria pollutants (e.g., NO*x*, VOCs, and so on) as opposed to greenhouse gases (GHGs), electrification entirely eliminates tailpipe emissions, and emissions of criteria pollutants can general be better controlled at centralized power plants. Furthermore, a large fraction of these plants are located away from urban population centers, so they contribute less to local air quality (AQ) problems, all other things equal. Thus emissions reductions are a strong argument in favor of electrification, on both the GHG and AQ fronts.
2. *Operating cost savings:* If one treats maintenance cost separately (see next point) operating cost can be interpreted as being equivalent to fuel or energy cost per mile of driving, ignoring any other operating cost. Based on the cost of electricity for recharging the EV, and the number of miles driven per full charge, the energy cost per mile compares favorably to that of the ICEV.
3. *Maintenance cost savings:* Electrification reduces year-on-year maintenance cost because on a whole-system basis, the EV drivetrain is simpler than that of the ICEV, needs less routine maintenance, and is subject to fewer mechanical breakdowns. For instance, the EV entirely avoids the need for an air intake and especially an exhaust system with pollution control equipment. EVs also do not need routine lubricant replacement, unlike the ICEV which usually requires this expenditure every 3,000 miles.
4. *Reduced noise and smooth ride:* Electrification reduces the contribution of light-duty vehicles to local noise and vibration problems. EVs make no noise when stationary (e.g., waiting for a red light) and relatively little noise when accelerating, cruising, and stopping in urban conditions. At highway speeds as well, EVs produce only tire noise (i.e., sound of tires rolling over pavement) and avoid the engine noise that is a source of local noise pollution from ICEVs. In addition, the smooth continuous transmission eliminates jerking or vibrations from gear shifts that are a concern in ICEVs.
5. *Eventual grid integration:* Electrification in the short to medium term lays a foundation for eventual grid integration of the demand for transportation energy as a major load alongside the various nonmobile loads that already draw electricity from the grid. With increasing penetration of renewable energy sources such as solar and wind into the grid, EV recharging could provide a logical load for electricity generation that has peaks and ebbs in its availability, since EVs like all LDVs spend much of their time at rest, either at the residence or work place of the owner.

Electrification also raises several disadvantages of EVs including management of the battery life cycle, possible capital cost of early replacement of the battery system if it does not last as long as the vehicle, limited driving range, and lengthy recharging time.

The latter two disadvantages pose the largest barriers to electrification and are therefore of particular concern.

Typical vehicle owners in the United States as well as other countries use their vehicles for urban driving on most days of the year. However, because they like to have the option of intercity travel in a private car—even if it is only a few days out of the year—many motorists are reluctant to switch from ICEVs to EVs, unless the EV is a second car that is used exclusively for urban motoring. Range figures from USEPA show that the concern is well founded: the estimated range per charge for the Nissan Leaf and the electric version of the Ford Focus is 75 and 76 miles per charge, respectively. The Tesla Model S with 60- to 85-kWh battery capacity achieves range values that approach those of typical ICEVs (according to Fig. 15-5, values in the range of 160 to 265 miles per charge, depending on battery system size and the source of the range estimate). As of 2013, the cost for the Tesla was prohibitive, however, ranging between $70,000 for the 60 kWh version to approximately $105,000 for the 85 kWh version, depending on options chosen.

Returning to supply-chain challenges with large-scale electrification of LDVs, a worldwide fleet numbering 1 to 2 billion in which a significant fraction are some mix of EVs (and PHEVs as well) would pose a strain on the environment without substantial battery recycling due to resource extraction, toxic byproduct handling, and waste disposal requirements. At present, the number of battery systems being deployed and then reclaimed as they reach the end of their life cycle is still small. As market penetration grows, the EV industry would be motivated to reuse and recycle materials to avoid the loss of raw materials. Government legislation to increase recycling can help as well. As an example of a mature battery technology, up to 95% of the material content of the battery in lead-acid primary batteries used in ICEVs can be recycled using current technology (Kannan et al., 2010). If the system that eventually emerges to recapture battery systems components is not effective in recovering a large fraction, however, the resulting waste stream could create another disadvantage for EVs.

15-3-5 Extending Vehicle Range with Plug-in Hybrid Electric Vehicles

The advantage of the EV is that it operates on electricity that can be made from a wide range of energy sources, including renewable or noncarbon ones, and that it does not create any tailpipe emissions. However, limited range per charge may be a concern for many drivers. Therefore, a promising alternative is the "plug-in" hybrid (PHEV), in which the secondary battery system has increased capacity that allow the vehicle to drive up to 50 miles on stored charge, either without using the ICE at all or using it at a reduced level.

Recent PHEV launches from the major automakers shown in Table 14-5 include the Chevrolet Volt (curb weight 1264 kg) and Toyota Prius PHEV (1439 kg). The Volt has the larger plug-in capacity, with a 17-kWh battery system that can drive the car on average 38 miles per charge. The comparable figures for the Prius PHEV are 4.4 kWh and 11 miles range per charge. Along with original equipment manufacturer (OEM) PHEVs, after-market companies offer plug-in retrofit kits that can convert an HEV to a PHEV and give the vehicle a range of 15 to 30 miles without using the ICE.

In some instances, the term "range-extended EV" may be used in place of "PHEV." The difference between the two terms is a question of emphasis. The term PHEV denotes a vehicle that is primarily an HEV, with a capacity to recharge overnight at a residence or another charge point to avoid consuming gasoline for the first number of miles. A range-extended EV is primarily an electric vehicle that has an onboard generator of electricity to recharge the batteries when the system drops below the maximum depth of discharge.

The PHEV system allows the user to charge at home at night in addition to refueling at the filling station, greatly increasing the "effective" gasoline fuel economy (i.e., total distance divided by total gasoline consumption) of the vehicle. Example 15-2 illustrates the difference between a PHEV and a comparable HEV with respect to fuel economy and energy cost per kilometer of driving.

Example 15-2S A PHEV has a fuel economy of 47.3 mpg when running on gasoline, and is able to travel 5 miles on 1 kWh of charge. Its range on a full charge is 50 miles, before it begins to use the ICE. Gasoline costs $4.00/gal and electricity costs $0.12/kWh.

Over a 2-week period, the vehicle accrues the following distance of travel each day:

Day	Miles	Day	Miles
1	38	8	114
2	17	9	90
3	66	10	21
4	14	11	24
5	39	12	44
6	45	13	16
7	267	14	33

Thus days 7 to 9 involve an extended intercity trip, and the other days involve driving in and around the vehicle's home base. Assuming that the vehicle has access to recharging each night, and that it only recharges the amount necessary to offset the kilometers traveled that day, calculate (a) the effective fuel economy, (b) the energy cost per mile (for combined electricity and gas), (c) the energy cost per km if it were only a hybrid with no plug-in capability.

Solution (a) For each day, we subtract off 50 miles of driving, and if there is a positive amount remaining, assign it to ICE propulsion. Thus on Day 1, there is 0-mile ICE use, and the vehicle uses (38)/(5 miles/kWh) = 7.6 kWh of electricity that must be recharged the following night; on Day 3, 16 miles are powered with the ICE, consuming 0.338 gal of gasoline, and so on.

On the basis of this type of daily calculation, in total 828 miles are driven in the 2 weeks, of which 337 miles require the use of the ICE. At the given fuel economy of 47.3 mpg, the fuel consumed is 7.1 gal. Thus the (effective) fuel economy is (828 miles)/(7.1 gal) = 117 mpg.

(b) At the given cost, 7.1 gal of fuel cost $28.48. The remaining 491 miles are driven using electric power, requiring 98.2 kWh of electricity, which costs $11.78. Thus dividing the total $40.26 of energy cost among the total distance driven gives $0.049/mile.

(c) Assuming that the gasoline-only HEV achieves the same 47.3 mpg economy over the entire distance traveled as does the PHEV when driving on gas, the fuel consumption is 17.5 gal, which costs $70.00. Dividing energy cost by distance driven gives ($70)/(828 miles) = $0.085/mile.

Example 15-2M A PHEV has a fuel economy of 20 km/L when running on gasoline, and is able to travel 8 km on 1 kWh of charge. Its range on a full charge is 80 km before it begins to use the ICE. Gasoline costs $1.06/L and electricity costs $0.12/kWh.

Over a 2-week period, the vehicle accrues the following distance of travel each day:

Day	km	Day	km
1	60.8	8	182.4
2	27.2	9	144
3	105.6	10	33.6
4	22.4	11	38.4
5	62.4	12	70.4
6	72	13	25.6
7	427.2	14	52.8

Thus days 7 to 9 involve an extended intercity trip, and the other days involve driving in and around the vehicle's home base. Assuming that the vehicle has access to recharging each night, and that it only recharges the amount necessary to offset the kilometers traveled that day, calculate (a) the effective fuel economy, (b) the energy cost per km (for combined electricity and gas), (c) the energy cost per kilometer if it were only a hybrid with no plug-in capability.

Solution (a) For each day, we subtract off 80 km of driving, and if there is a positive amount remaining, assign it to ICE propulsion. Thus on Day 1, there is 0 km ICE use, and the vehicle uses (60.8)/(8 km/kWh) = 7.6 kWh of electricity that must be recharged the following night; on Day 3, 25.6 km are powered with the ICE, consuming 1.3 L, and so on.

On the basis of this type of daily calculation, in total 1325 km (828 miles) are driven in the 2 weeks, of which 539 km require the use of the ICE. At the given fuel economy of 20 km/L, the fuel consumed is 27 L. Thus the (effective) fuel economy is (1325 km)/(27 L.) = 49.1 km/L.

(b) At the given cost, 27 L of fuel cost $28.48. The remaining 786 km (491 miles) are driven using electric power, requiring 98.2 kWh of electricity, which costs $11.78. Thus dividing the total $40.26 of energy cost among the total distance driven of 1,325 km gives $0.030/km.

(c) Assuming that the gasoline-only HEV achieves the same 20 km/L fuel economy over the entire distance traveled as does the PHEV when driving on gas, the fuel consumption is 66.2 L(17.5 gal), which costs $69.96. Dividing energy cost by distance driven gives ($69.96)/(1,325 km) = $0.053/km.

The treatment of fuel economy in Example 15-2 is simplistic because, depending on the design of the specific PHEV drivetrain, the drive controller might occasionally activate the ICE while there is still charge in the secondary battery in order to deliver sufficient total power for hard acceleration or climbing a grade. However, inclusion of this level of detail would likely not lead to a dramatic change in the results, so that the overall fuel economy would still exceed 100 mpg (42 km/L).

The focus on gasoline consumption as the figure of merit in this example stems from the concern about volatility of gas prices at the pump, and about energy security in countries such as the United States that depend heavily on imports. A more complete comparison of the HEV versus PHEV would take into account the total energy consumption for both gasoline and electricity, and the total CO_2 emissions. In some circumstances, a PHEV might save gasoline but increase energy and CO_2. The comparison in this case is left as an exercise at the end of the chapter.

Given that many passenger vehicle owners drive 30 miles or less on 70% to 80% of the days of the year, it is clear that a robust transition to PHEVs could significantly reduce petroleum consumption, as electricity from a wide range of sources (but very little from oil-fired electric power plants) would replace gasoline and diesel consumption in ICEs.

The PHEV could also have security benefits, since, in the case of a disruption in oil supplies, fleets of PHEVs in major urban centers could still drive a fixed distance each day on electric charge (which presumably is less vulnerable to crises in international relations), preserving some measure of mobility for the public.

For the PHEV technology to give a clear advantage over ICEVs in terms of CO_2 emissions and impact on air quality, it is important that any fossil-fuel generated electricity come from a source that has high thermal efficiency and up-to-date emissions controls to prevent the escape of pollutants into the atmosphere. In a scenario unfavorable to PHEVs, low efficiency or poor emissions controls might wash out any environmental advantage compared to the ICEV technology, even if the PHEV reduced consumption of petroleum. For either EV or PHEV technology, the composition and function of the grid is an important factor, so this topic is considered in the following section.

15-4 Background on Electric Grid Function

The electric grid consists of different generating assets linked by a transmission system (long distance, high voltage) and distribution (short distance, lower voltage) system to electric customers or "loads" large and small (Fig. 15-1(*a*). Different types of generating assets fill different niches. Type of fuel source, flexibility in stopping and starting, and fixed versus variable cost characteristics all influence the role of different generating assets. Initially we will focus strictly on fossil fuel options, and then introduce nonfossil options. For more detail the reader is referred to full-length works on the topic (for example, Grainger and Stevenson, 1994; Wood and Wollenberg, 1996).

Synergies between electrified LDV fleets and future advanced electric grids provide strong motivation for integrating the transportation energy supply with the grid. The grid has the potential to provide stable electricity to the transportation vehicles at predictable prices. The grid is already much more diverse than the transportation sector in terms of primary sources of energy. Whereas more than 98% of transportation energy comes from petroleum (in the form of gasoline, diesel, and jet fuel), the electricity sector draws on all three major types of fossil fuel (gas, coal, and a small amount of oil), nuclear energy, and a growing array of renewables.[3] Although underinvestment in the U.S. grid has in recent years led to reliability problems, assuming the grid is well maintained, diversity of sources should lead to greater reliability for the transportation energy supply: if any one primary energy source becomes expensive or unpredictable, the market can go to one of the others. The transportation market can help improve conditions for the electric grid as well. Electric vehicles can charge at off-peak times when other loads drop off (especially between 12 midnight and 6 a.m.), and if they are connected to the grid at appropriate times, they can also absorb peak periods of output from intermittent renewable sources such as solar and wind.

15-4-1 Composition of Generation Sources and Grid Components

Fossil fuels have dominated electricity generation in the United States for several decades, as they have in many peer countries (e.g., Canada, Germany, and the United Kingdom). Coal and natural gas are especially significant as primary sources of energy for generation, as the high cost of oil per barrel and high prices that can

[3]For transportation, a small percentage of primary energy comes either from biofuels or electrified rail transportation.

be obtained for transportation fuels make it desirable to preserve oil for conversion to gasoline, diesel, and aviation fuel. Not all industrialized countries rely on fossil fuels for electricity in this way; however, for instance France relied for 74% of its electricity on nuclear energy, with a large fraction of the remainder coming from hydropower.

Electricity market share values are also changing in many countries due to economic and environmental pressures. Figure 15-6 shows share of kWh generated in the United States for the years 2004 and 2011. Over this 7-year period, total demand stayed roughly constant around 4 trillion kWh/year, but gas share rose from 18% to 25%, while coal declined from 50% to 42%. Several issues contributed to this change. Between 2003 and 2010, the average price of anthracite and bituminous coal in the United States rose from \$35/ton to \$65/ton (not including transportation from the mine), while gas saw a drop between 2008 and 2011 from \$8 to \$4 per 1,000 ft^3 before delivery. Concern about air pollution and GHG emissions may have also encouraged greater investment in gas-generation capacity. Wind energy also grew between 2004 and 2011 from 0.4% to 2.9% of the total generation mix, thanks to an increase in installed capacity from 7 to 45 GW of capacity over this time period (at an average of 2 MW per turbine, 45 GW would represent 22,500 utility-scale wind turbines).

Along with generation capacity, the other part of the electric grid system is the transmission and distribution network, which may represent on the order of 40% of the total capital value of the entire system, with the generating stations representing the other 60%. As shown in Fig. 15-7, electric voltage values in the system are a tradeoff between the desire to minimize losses by increasing voltage versus reducing safety concerns by not exceeding some appropriate value for the part of the chain from generation to residential consumption in the case of the diagram. Electricity is generated at approxi-

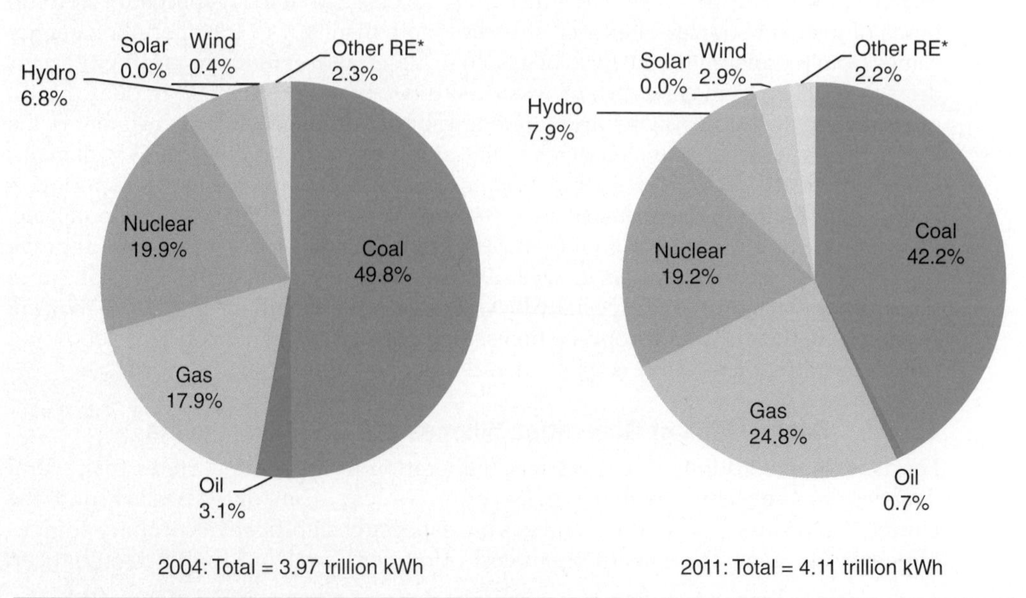

FIGURE 15-6 Share of U.S. electricity production, 2004 and 2011.
Source: U.S. Energy Information Administration (2012).

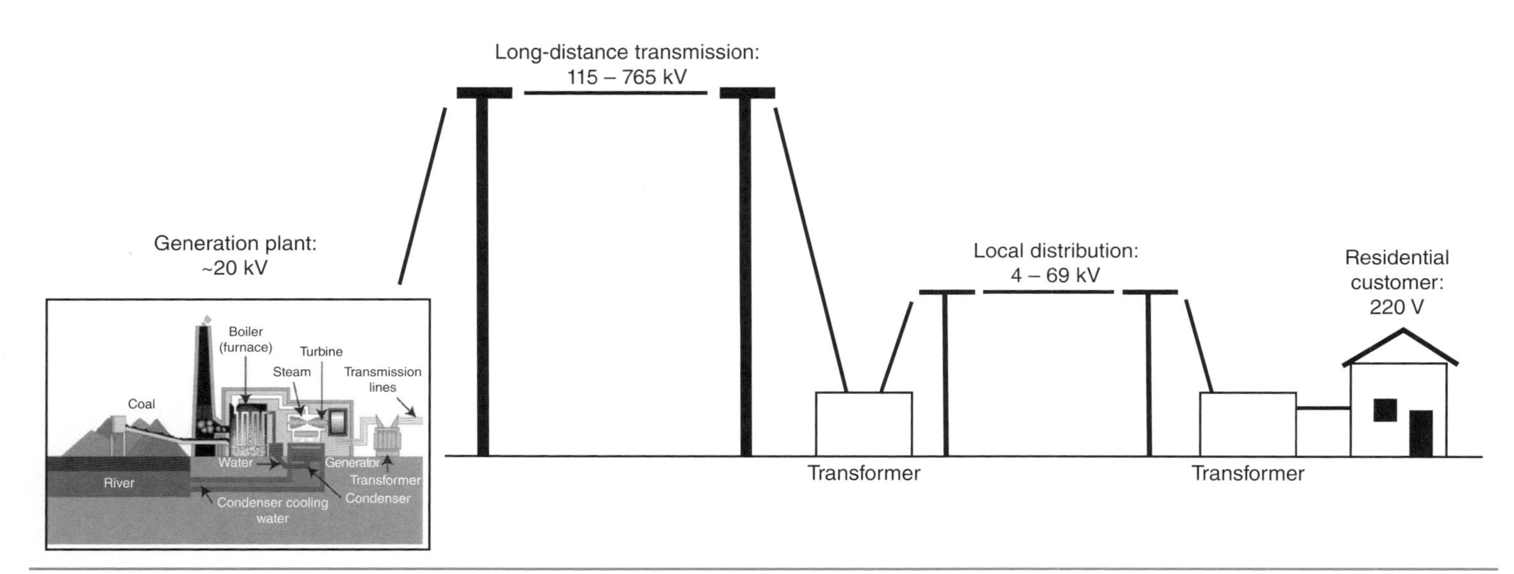

FIGURE 15-7 Generation, transmission, distribution, and load components of the grid system.
Source for power plant diagram: Tennessee Valley Authority, public domain image.

mately 20 kV in the power plant, and then stepped up to a value between 115 and 765 kV for long-distance transmission, with the highest voltages occurring on the lines with the heaviest loads. As the energy approaches its final destination for consumption (a residential neighborhood in this case), voltage values are stepped down to no more than 69 kV for local distribution, since distribution lines have less physical separation from roads, houses, and other components of the built environment. Finally, within the residence voltage is kept to no more than 220 V for safety reasons.

Types of Fossil Fuel Generating Assets

Fossil-fired generating asset types that provide electricity for the grid can be broadly divided into three types, as follows:

- *Baseline plants*: Typically these plants have high fixed capital cost but low variable cost due to low cost of fuel and high efficiency. Until recently the main fuel for these plants was coal, but gas may also be competitive for use in baseline plants going forward if recent low gas prices continue. As mentioned above and shown in Fig. 15-6, the share for gas has been increasing while that of coal has been declining. Since these plants operate for many or most of the hours of the year, the ability to start, stop, or vary output quickly is not as important.
- *Load-following plants*: Typically these plants have lower fixed capital cost but higher variable cost than baseline plants; on the other hand, capital cost is higher than peaking plants because additional investments in efficient operation are cost effective. As the name implies, the ability to vary output with load is emphasized, therefore short starting and stopping time is desirable. Gas is the fuel of choice among fossil alternatives.
- *Peaking plants*: These plants minimize fixed capital cost at the expense of high variable cost because they operate the fewest hours per year when demand is at or near its peak. Short start and stop time is emphasized. Fuel may be gas or oil.

The goal of reducing CO_2 emissions drives much of the interest in integrating transportation with the grid. Since in their current form none of the above plant types are carbon-free, carbon capture and sequestration (CCS) is proposed to remedy this problem. At present CCS happens only on a small scale in the United States, primarily at the Great Plains Coal Gasification facility in North Dakota, and does not happen commercially at any fossil-fired electric generation plants. Small-scale CCS operations at power plants have started operating in other countries, including Italy, Germany, and Poland (Table 15-1). In principle, however, any baseline or load-following plant that burns fossil fuels could be retrofitted so that the CO_2 stream would be diverted and sequestered (for peaking plants it might be prohibitively expensive given the limited generating hours per year).

Load Duration Curve

For planning purposes, all the stakeholders in the grid, including the electrical generators, grid owner, electrical supply companies, independent system operators (ISO), and public utilities commission of the government have an interest in ascertaining that there is adequate grid capacity to meet demand. To understand demand over the course of a year, analysts create a *load duration curve*, in which for the most recent year available the average demand in gigawatts or megawatts for each of the 8,760 hours of the year is

Plant Name	Country	Sequestration Rate	Target Year
		1,000 tonnes/year	
Brindisi Pilot project	Italy	8	Operating (2011)
Porto Tolle	Italy	<1,000	2015
Belchatow	Poland	100	2015
Vattenfall	Germany	600	2014
Luzhou*	China	60	Operating

*Carbon captured at Luzhou plant but used in industrial process rather than being sequestered. Inaugural year of operation not provided.

TABLE 15-1 Early Examples of Carbon Capture and Sequestration (CCS) Projects in Select Countries

tabulated, and then the hours of the year are arranged in order of decreasing load. A fictitious load-duration curve is shown for illustration in Fig. 15-8, in which peak loads between 11 and 18 GW occur for 2,000 hours/year, and the remaining 6,760 per year have lower loads between 6 and 11 GW. Figure 15-8(*a*) shows the curve with the original load data by hour, and in Fig. 15-8*b* the data has been smoothed so that it can be represented with piecewise linear functions. Note that the curve in Fig. 15-8(*a*) is continuously decreasing with a nonlinear path from highest to lowest hour of the year.

A load-duration curve of the size shown in Fig. 15-8 might represent the electricity market for an entire single- or multi-state region with millions of customers served by a single utility. The "spike" on the left side of the curve for the 500 or 1,000 highest load hours of the year is typical, and this is a significant concern for grid operation, since it means that a measurable fraction of the overall generating capacity will be used for relatively few hours per year. In any case, stakeholders can use the load duration curve to ascertain that the right mix of baseline, load-following, and peaking plants will be available to meet anticipated demand, with some extra capacity available for contingencies.

The future load-duration curve of course cannot be predicted with certainty, so grid operations are not scheduled for an entire year in advance as the load duration curve might suggest. Instead, the grid relies on a mix of day-in-advance and instantaneous (or "spot") markets in which plant owners offer to generate electricity with a certain cost and maximum capacity, and then the ISO chooses the mix of generators that meets anticipated demand at minimum cost. The ISO seeks to meet as much of the demand as possible with the day-in-advance market to lock in a favorable price, with the spot market used to fill in gaps in the moment, albeit at a higher average cost per kWh.

Once the available generators are known, thanks to the function of the electricity market, the ISO oversees the match between instantaneous load and electricity being fed into the grid to meet demand. On a continuing basis the ISO signals load-following plants to vary their output so that supply balances demand and voltage in the grid is maintained within an acceptable range. Instantaneous input from generators does not balance exactly with output to all types of loads (for instance, industrial users, institutions, households, and so on) down to the electron, but all electricity-using devices can

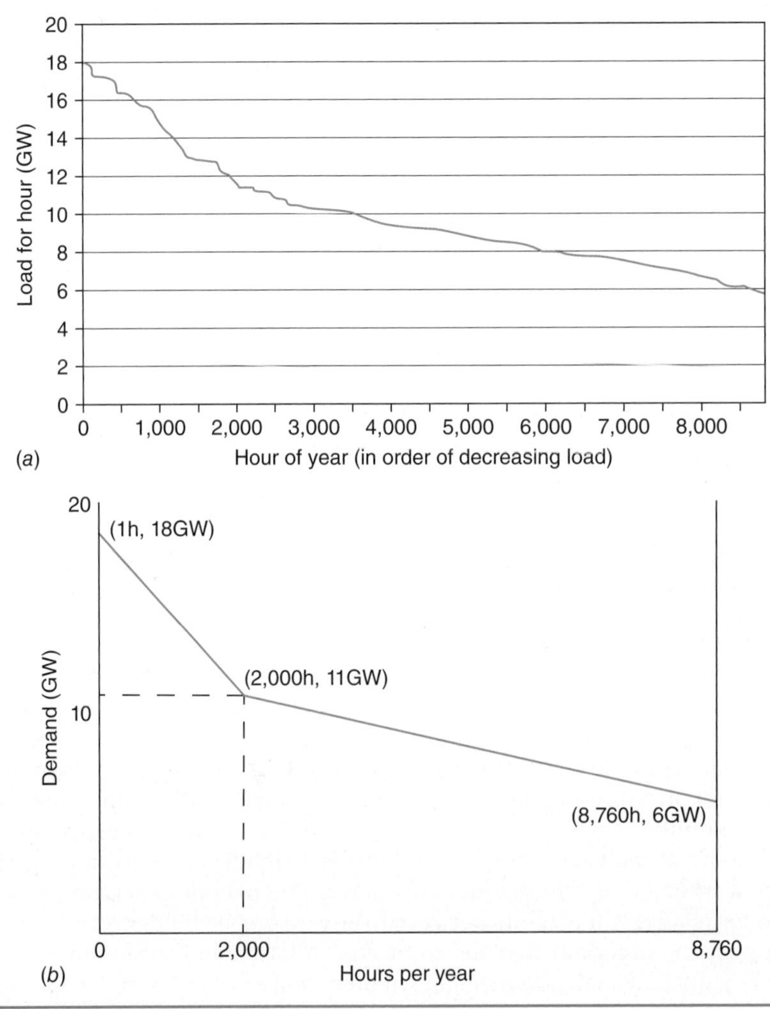

FIGURE 15-8 Load duration curves for a fictitious utility market having a minimum of 6 GW of demand for all 8,760 hours in the year, and a maximum of 18 GW of demand in the highest-demand hour of the year: (*a*) raw hourly values, (*b*) smoothed curve with demand as piecewise linear function of hour.

tolerate to some extent variations in input voltage. Therefore, the system can operate with a slight imbalance between total current flowing in from sources and total current flowing out to loads.

To further respond to ongoing rise and fall of total demand, the ISO can also call on electrical storage facilities, such as pumped storage or large stationary batteries. These assets are at present expensive per kWh of charge stored, and therefore used sparingly (see discussion of storage in Sec. 15-4-2). As a last resort, if the ISO cannot meet demand, for example, if a generating asset suddenly fails on a hot summer day and a dangerous drop in voltage is threatened, the grid will go down, incurring the consequences of customers suddenly losing supply rather than risking permanent, irreversible damage to many billions of dollars of electrically powered equipment. Consequences of large-scale grid failures can be dire: for instance, the Northeast Blackout of August 14, 2003 affected 55 million customers in Canada and the United States and is estimated to have had an economic cost of $6 billion (U.S. dollars). Grid failures are a factor in other parts of the world as well: a blackout in India on July 31, 2012, affected more than 500 million people. The ISO may be able to avoid a full-scale grid failure by putting in place advance agreements with select large customers that they will forego electrical supply in an emergency in return for reduced prices per kWh. A water-treatment plant in the Philadelphia area, for example, has such an agreement with the ISO, and has in an emergency agreed to rely on internal generation of electricity from waste-generated biogas and discontinuing nonessential loads to ride out any grid cut-off.

Example 15-3 illustrates calculations involving the load-duration curve and generation cost information to predict a mix of generating assets that can meet demand at minimized cost.

Example 15-3 A mix of generating plants with arbitrary sources of fossil fuel is used to meet the demand represented in the load-duration curve in Fig. 15-8. The table below gives fixed and variable cost characteristics for the plants. What mix of baseline, load-following, and peaking plant capacity in gigawatts could exactly meet the demand at minimal cost? Ignore the need for spare capacity to meet contingencies. Note also that for the variable cost the units should be interpreted as follows: for example, in the case of baseline plants," 1 kW generated for 1 hour incurs $0.025 of cost," and so on.

Plant Type	Fixed Cost	Variable Cost
	($/kW)	($/kWh)
Base	$180.00	$0.025
LF	$120.00	$0.040
Peak	$90.00	$0.065

Solution We can start with the cost information given in the above table to work out break-even hours of function for asset types in comparison to one another, independent of the information in the load duration curve. Once breakeven values are known, we can apply them to the particular demand pattern in question.

Taking the case of the baseline versus load-following plant as an example, at the breakeven number of hours per year (H), generating 1 kW of electricity will cost the same with either asset. Therefore the following equality holds:

$$\$180 + \$0.025\,\mathrm{H} = \$120 + \$0.04\,\mathrm{H}$$

$$\mathrm{H} = \frac{\$180 - \$120}{\$0.04 - \$0.025} = \frac{\$60}{\$0.015} = 4{,}000 \text{ hours}$$

Thus for more than 4,000 hours/year, it is cheaper to generate with the baseline plant, and for less than 4,000, it is cheaper using the load-following plant. Setting up a similar equation for the number of hours for comparing peaking and load-following plants:

$$\mathrm{H} = \frac{\$120 - \$90}{\$0.065 - \$0.04} = \frac{\$30}{\$0.025} = 1{,}200 \text{ hours}$$

Turning to the load-duration curve, we need to know the demand for the 1,200th and 4,000th hours of the year to calculate capacity. The two segments of the curve have the following equations:

$$\text{Demand}_{0-2,000} = (-0.0035)\ \text{hour} + 18$$
$$\text{Demand}_{2,000-87,60} = (-0.0007396)\ \text{hour} + 12.48$$

Plugging in the demand for the two breakeven points gives the following:

$$\text{Demand}(1,200\ \text{h}) = (-0.0035)(1,200) + 18 = 13.8\ \text{GW}$$
$$\text{Demand}(4,000\ \text{h}) = (-0.0007396)(4,000) + 12.48 = 9.52\ \text{GW}$$

From these figures, we infer that there will be for at least 4,000 hours/year the demand will be between 6 and 9.52 GW, and that this demand can be most economically served by the baseline plants because the fixed cost can be defrayed over a sufficiently large number of hours to keep overall cost to a minimum. For at least 1,200 hours/year the demand will be at or less than 13.8 GW. The demand above 9.52 GW and below 13.8 GW can most economically be met by the load-following plants. Peaking plants should meet remaining demand above 13.8 GW and up to 18 GW. Accordingly, the allocation of the total necessary 18 GW of capacity:

$$\text{Baseline}: 9.52\ \text{GW}$$
$$\text{Load Following}: 13.8 - 9.52 = 4.28\ \text{GW}$$
$$\text{Peaking}: 18 - 13.8 = 4.2\ \text{GW}$$
$$\text{Total}: 9.52 + 4.28 + 4.2 = 18\ \text{GW}$$

Two important simplifications are evident in Example 15-3. First, actual available capacity to meet demand going forward into the future would need to be larger than 18 GW, both because peak demand is variable and might exceed 18 GW in a subsequent year, and because the grid at all times needs a margin of reserve in case one or more generating asset fails unexpectedly. Suppose instantaneous demand were at 17.5 GW already, and a large 1 GW unit suddenly goes off-line. The grid would have only 500 MW in reserve, so the ISO would either need to quickly cut off a total of 500 MW of load, or else a blackout would result.

Secondly, the actual grid might resemble the allocation of capacity to baseline, load-following, and peaking assets, but with a buffer added. For example, 11.4 GW, 5.1 GW, and 5.0 GW, respectively, would represent the values calculated in the example with a 20% margin added to each. Actual values would vary slightly from these numbers, since plant sizes are discrete and might not add up exactly to 11.4 GW for baseline capacity, and so on. However, the function of the electricity markets and not the a priori planning exercise would determine allocation of supply. Suppose demand were exactly 13.8 GW: the above calculation implies that 69%, or 9.5 GW, would be generated by the baseline plants, and the remaining 4.3 GW or 31% by load-following plants. If the load-following plants had offered more favorable prices, their share might be larger than 31% under the assumption of up to 20% extra capacity, and vice versa.

15-4-2 Role of Nonfossil Generating Assets

The role of nonfossil (nuclear or renewable) generating assets depends very much on their specific characteristics. First, nuclear generating facilities are similar to large-scale gas- or coal-fired plants: they have high capital cost, low fuel cost (even lower than fossil due to high-energy-generating potential per dollar spent on fuel), and very long start and stop times. Existing nuclear plants are therefore ideal baseline plants. New nuclear plants are hindered at present in the United States by high capital costs and the repercussions from the Fukushima nuclear power station accident in Japan in 2011, although plants remain under construction in other countries such as China and South Korea.

Large-scale hydropower plants can vary output thanks to the water impounded upstream from the dam, so long as rain- or snowfall is sufficient and other needs such as irrigation have not excessively drained available water. Hydropower dams therefore function well as load-following plants and provide a CO_2-free alternative to gas-fired ones. Most hydropower dams are available as required by the grid, which is ideal for load-following, but some are "controlled release" on a once-per-day or once-per-week schedule and are therefore less flexible. One concern for hydro-power is the impact of climate change in producing either excessive rains at one extreme or drought at the other, either of which can hinder operations. In this sense, hydro-power can help to prevent climate change by avoiding the need to generate electricity from fossil fuels, but unfortunately may also become a victim of climate change in the coming years and decades. In any case, opportunities for new hydro-power installations are largely exhausted in the United States, with a U.S. Department of Energy study estimating the remaining potential at approximately 50 GW spread across 49 states, mostly in the form of smaller-scale and run-of-the-river hydro-power (USDOE, 2009).

Unlike large-scale hydropower, other large renewable resources such as wind and solar are intermittent and therefore their output cannot be scheduled in advance, so they do not fit in either the category of baseline or load-following. When intermittent renewable electricity is available, it is generally desirable to make full use of it as long as there is sufficient demand in the grid. Therefore, where sufficient quantities of centrally generated intermittent renewable energy exist (i.e., not distributed generation such as household PV systems), they can be conveniently treated on the "load" rather than the "supply" side of the balancing equation. The stakeholders in the grid can think of an "adjusted load" equivalent to the actual load minus the sum of all intermittent renewable electricity available. The adjusted load is the amount that must be met by the remaining combination of generating assets.

Other renewables such as biomass-fired or steam-driven surface geothermal power plants are *dispatchable* rather than being intermittent, so the same logic does not apply. Biomass or surface geothermal have a much smaller total potential available in nature compared to solar or wind biomass because it is limited to the amount that grasslands or forests can yield, and surface geothermal because it can only be produced in regions where underground steam can be accessed close to the surface. In coastal regions, tidal power, wave power, and ocean current power resources may be available as well, but obviously not in regions far from the coast. Regional solutions for meeting electricity demand remain important, however, and when those regions that have sufficient biomass or geothermal resources can meet their needs for transportation as well as stationary electricity loads with these resources, they depend less on other resources elsewhere.

15-4-3 Short-Term Storage Options for Extra Electricity Generated

Most of the electricity loaded into the grid is used in real time. This is fundamentally different from the transportation liquid fuels market, where petroleum can be refined into gasoline, diesel, or aviation fuels, and then stored in tankers, at filling stations, or in the reservoirs within the vehicles or aircraft themselves before being combusted in internal combustion engines or turbines to provide propulsion. There is a small and growing role for electrical storage, however. Electrical storage can be used to achieve price *arbitrage* (buying at night when rates are low and then selling during the day when rates are at their peak), to improve reliability in remote locations in the grid where blackouts are a concern, or as a buffer for intermittent renewable sources.

Options for storage with relevant examples include the following:

- *Pumped hydroelectric storage:* The principle of hydroelectricity can be used in reverse, with electrical energy expended to raise water into a reservoir, and then released again into the grid by allowing the water to run out of the reservoir through a typical hydroelectric turbine. New York Power Authority (NYPA) applies this principle at their 1,160 MW_e Blenheim-Gilboa pumped storage facility in the Catskill Mountains, where power stored at night can be used to relieve pressure on electricity rates for the greater New York City region during the day, especially during the peak summer cooling season.[4]
- *Compressed air energy storage:* In a similar concept to pumped hydro storage, air can be compressed in a natural or manufactured reservoir, and later allowed to expand through an air turbine to make electricity. Where available, natural reservoirs can greatly reduce system construction costs. A 150-MW facility is proposed for an abandoned salt mine underneath Watkins Glen, N.Y., in western New York State, where the underground cavity left by salt extraction provides a natural, leak-proof reservoir for storing compressed air.
- *Stationary battery storage:* New materials used for charge storage can gather electricity in large quantities at a fraction of the historical cost of consumer-sized or mobile batteries used in electronics and vehicles. The city of Presidio, Texas, is pioneering the use of a stationary battery storage system due to difficulty of providing reliable grid power to their remote location along the Rio Grande River. A 4-MW_e sodium-sulfur battery system provides temporary power for up to 8 hours when long-distance transmission is disrupted or must be taken off line for repair work.
- *Flywheels:* New materials and precision manufacturing have enabled the fabrication of flywheel systems that store large quantities of energy in a compact space, have minimal friction so as to reduce losses over time as the flywheel spins, and maintain environmental safety in case of any accident involving collision of other objects with the flywheel assembly. Beacon Power Corporation has installed a 20 MW_e flywheel storage station in Stephentown, N.Y., near the Albany capital region, to store and release electricity in a network of flywheels.

Regarding the transportation system, these storage devices might in general help the grid to function more reliably, so that the transportation sector could depend on electricity as an energy source option. The EVs and PHEVs themselves might also become another form of energy storage if they are connected to the grid at a charge point when not being driven, as discussed later in this chapter.

---

[4]The abbreviation MW_e stands for "MW of electrical output," as opposed to MW_t = "MW of thermal output."

15-5 Integrating Transportation Energy Demand and the Grid

Integration of transportation energy supply with a technologically updated *smart grid* can improve energy efficiency and system performance. Generally the focus of this effort will be through the growing use of PHEVs or EVs in the LDV fleet, because this is a large load consisting of many vehicles, each of which spends much of its time stationary over a 24-hour day and is therefore amenable to recharging through connection to the grid. Some of the findings in this section are also transferable to medium- and heavy-duty vehicles such as urban public buses and freight vehicles, as well as electrified rail systems, both urban and intercity. For brevity, we will focus mostly on LDVs, and focus the discussion on the case of EVs, with a discussion of special ramifications for PHEVs at the end.

Improvement opportunities exist both in the short and long run. In the short run, EVs might be recharged at night to take advantage of unused generating capacity in large generating assets, providing energy to vehicles at low cost. In the long run, grid operators might match growing capacity to store and use intermittent renewable energy with the capacity to store and use energy provided by these vehicles in the fleet.

In its fully developed form, the grid-to-vehicle system allows the ISO to coordinate the generation of electricity from both dispatchable and intermittent resources with the opportunistic charging of vehicles when they are connected at charge points, as shown in Fig. 15-9. For ideal operation in a system with high EV market penetration, charging is available not only at night at home but also during the day at the workplace. This arrangement makes some portion of the fleet available to receive charge for most of the hours of the day, so that the ISO has maximum flexibility to allocate charge to vehicles.

Both communication technology and price policy have a role in the envisioned charging system. Land-based or wireless communications networks would allow the ISO to know, at any given time, the number of vehicles available and the remaining uncharged capacity spread across those vehicles. An ability to forecast output from intermittent solar and wind resources up to several hours ahead would also facilitate the ISO's coordinating role. Differential pricing could be used to encourage vehicle owner participation; in other words, reduced cost per kWh to charge vehicles in return for availability when the vehicle is charged for long periods of time at home or at work.

As a further option, the ISO might oversee a *vehicle-to-grid* (V2G) system that not only optimizes the charging of the EV fleet but also allows the vehicles to sell electricity back to the grid to create a revenue stream for the owners, offsetting the incremental cost of including the battery system in an EV. The smart grid and EV fleet can function successfully without deployment of V2G, since the main requirement is that vehicles be connected through their chargers for extended periods of time so that the smart grid can provide charge at optimal times. In a non-V2G system, the two-directional flows between the grid and vehicles in Fig. 15-9 become single-direction arrows going to the vehicles only.

15-5-1 Near-Term Opportunities for Electrical Charging from the Grid

In the short run, EVs/PHEVs can take advantage of the uneven demand for electricity to charge at times of day when other sources of demand are low, even before deployment of large amounts of renewable generation or other CO_2-free electricity supply intended to reduce climate impact of transportation. Here we present hour-by-hour

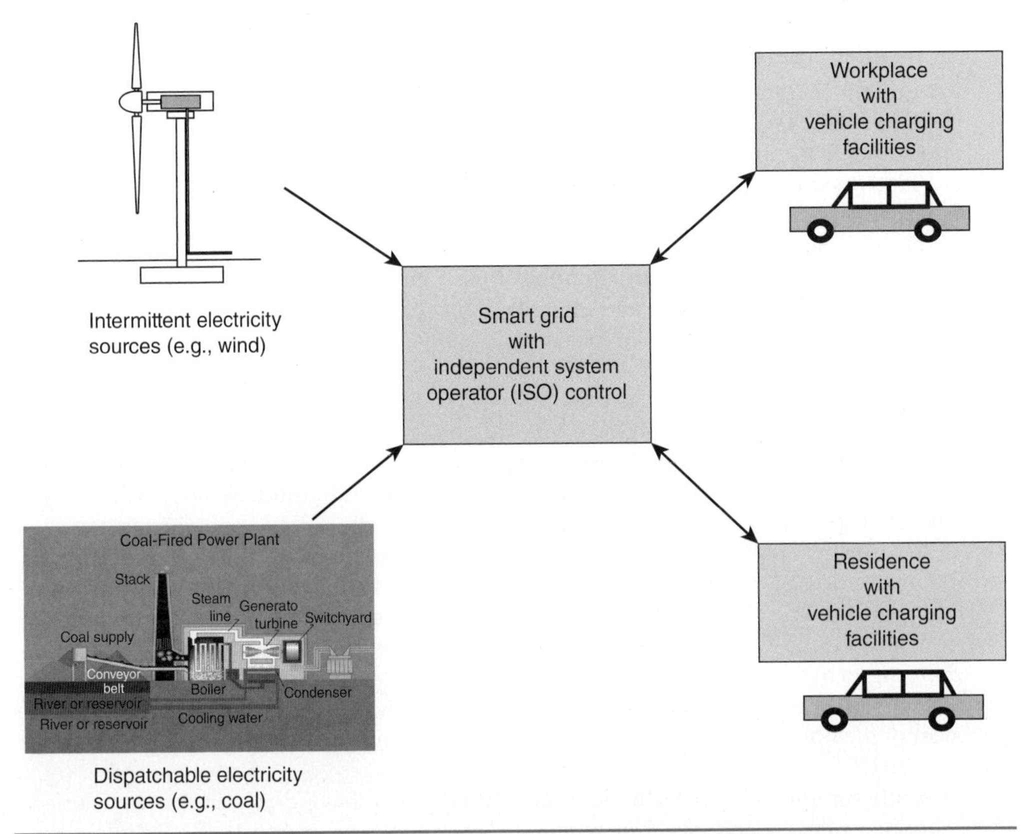

Figure 15-9 Components of smart grid based electricity generation and vehicle charging system, including vehicle-to-grid (V2G) option and home- or workplace-based charging.
Source: Powerplant public domain image courtesy of Tennessee Valley Authority.

variation in load from two regions, as an indication of general trends that might be observed across the United States and elsewhere. Figure 15-10 shows relative electrical power demand by time of day for all loads in the Tompkins County region surrounding Ithaca, N.Y. (Fig. 15-10(*a*) and household consumption only in El Paso, Texas (Fig. 15-10(*b*). In a pattern typical of most grid operations, a *trough* in demand exists between approximately 2200 and 0500 hours. The value of the decline is 29% in the case of Tompkins County (from 1800 to 0300 hours) and 52% for both weekdays and weekend days in the case of El Paso (between a peak at around 2000 hours and trough at 0300 hours on weekdays and 0500 hours on weekend days). Note that the inclusion of commercial and industrial loads in addition to residential in the case of El Paso might narrow the difference between the peak and the trough, although these data were not available from the source.

Low demand in the trough implies that plant capacity is available that could meet vehicle electrical charging requirements, so long as vehicles are controlled not to start charging before approximately 2,200 hours. This control policy could be achieved through pricing or designing charging hardware that only allows the vehicle to charge at certain times, with oversight by the ISO. Figure 15-10 also shows the inherent risk in

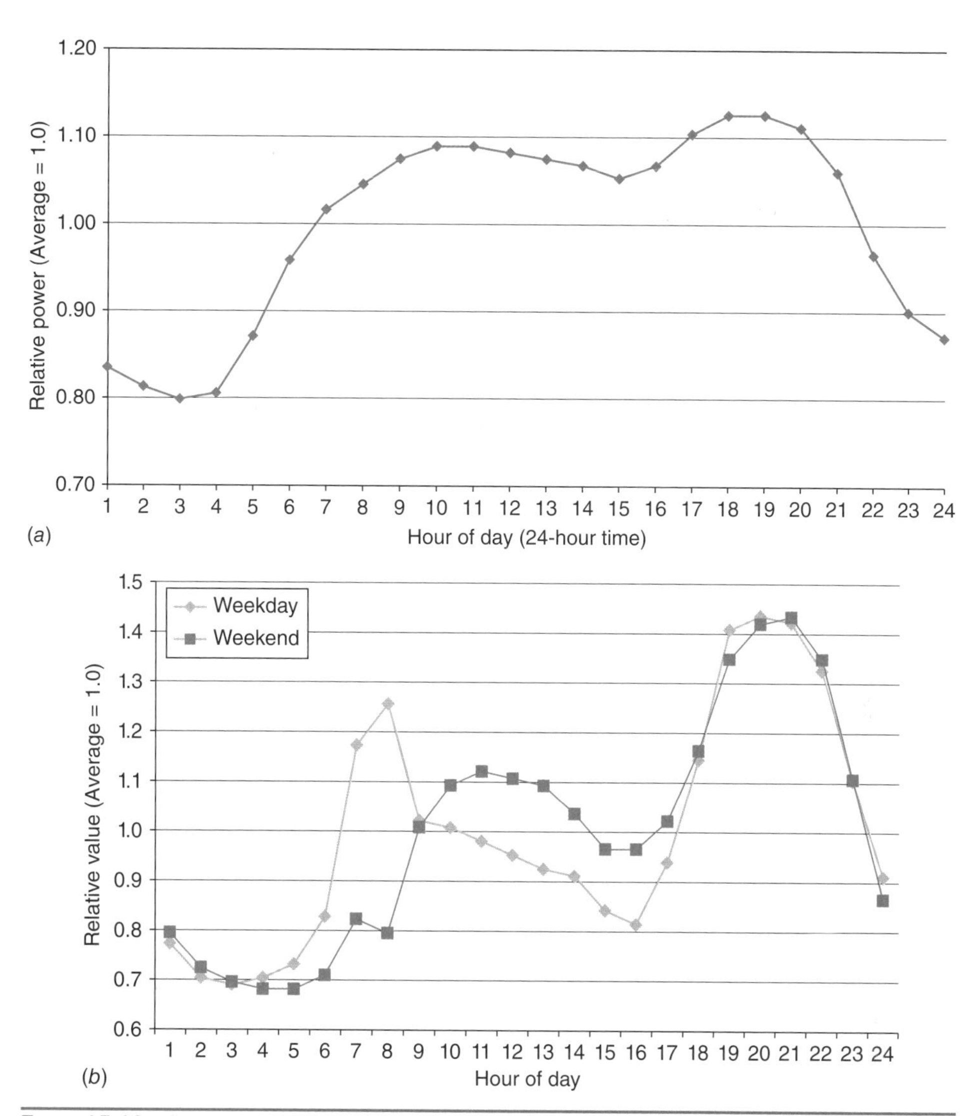

Figure 15-10 Average electricity demand over 24-hour period: (*a*) all loads in Tompkins County region in western New York state;[5] b) household loads only in El Paso, TX.
Source for part (b):
El Paso Electric Company.
Note: y-axis scale does not start at zero so as to better show variation in hourly load values.

[5]Attribution: Demand and wind energy data for Ithaca and Tompkins County quoted in this chapter originally compiled by Spring 2010 M. Eng project in Engineering Management, School of Civil & Environmental Engineering, Cornell University (Hoerig et al, 2010). Thanks to project team members Christina Hoerig, Dan Grew, Happiness Munedzimwe, Jun Wan, Karl Smolenski, Kim Campbell, Nicole Gumbs, Sandeep George, Tim Komsa, and Tyler Coatney.

widespread electrification of the fleet without such controls: in the worst-case scenario, vehicle owners might all return home around 1700 to 2100 hours and plug in their vehicles to begin charging just when household demand is at its peak, thus aggravating challenges with grid operations instead of filling in troughs in 24-hour demand. Lastly, this arrangement assumes that the vehicle charges fast enough to begin at 2200 hours and still be adequately charged by 0500 hours. EV owners may need 240V rather than 120V charging to achieve this goal.

Impact of Temporal Distribution of Travel Demand on Electrification Viability

The pattern of electricity demand can be compared to urban travel patterns by hour of day and day of week. Figure 15-11 shows distribution of trips by hour of day from midnight to midnight, and Fig. 15-12 shows distribution by day of week, both from the 2009 National Household Travel Survey (NHTS). The NHTS does not distinguish trips by both time of day and mode, so the figures shown are across all modes. However, because the large majority of trips are by car, it is reasonable to assume that the pattern for cars would look similar, namely a large peak during the day and a large trough at night from about 10 p.m. to 6 a.m. At the same time, the weekly pattern is fairly consistent from day to day, with the lowest figure on Sunday of about 12.5%, and the highest figure on Friday of about 15.5%. These statistics tell an important story for night recharging of electric vehicles. First, since the trough hours contain between 0.2% and 1.5% of the all the trips of the day, most vehicles will be idle during this time slot and available for recharging. Furthermore, because usage patterns by day of week are fairly consistent, the pattern of daytime usage and night recharging can be repeated daily without overly uneven usage of the necessary infrastructure.

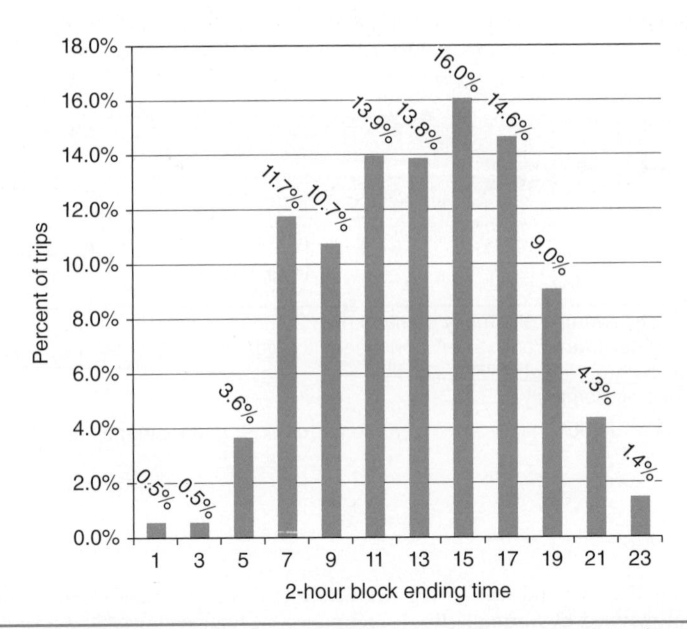

Figure 15-11 Distribution of U.S. household trips by time of day across all modes.

Source: National Household Travel Survey. Note: Block ending at 0100 hours includes trips in 2300–0100 period.

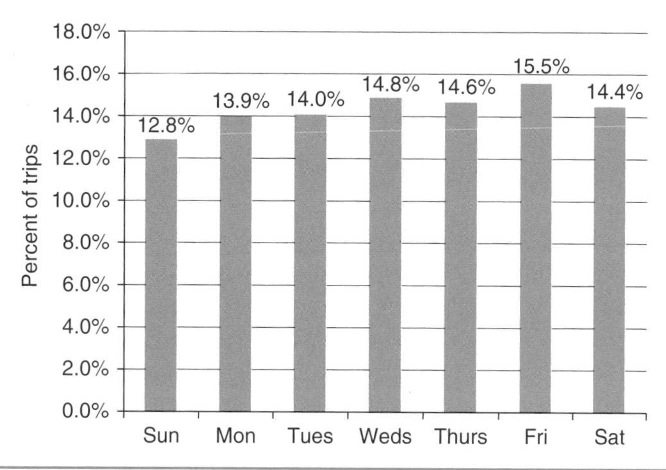

FIGURE 15-12 Distribution of U.S. household trips by day of week across all modes.
Source: National Passenger Transportation Survey.

Potential Impact of Electric Generation Mix on Emissions

Another concern is average CO_2 emissions per mile when vehicles switch from gasoline to grid electricity. Thanks to the presence of zero-carbon sources already in existence in the grid electricity mix, in many instances electrification reduces emissions. Table 15-2 shows the mix of different electricity sources for the United States in 2011, and the CO_2 emissions rate per kWh of electricity for each source. The table includes only CO_2 emissions at the generation stage of the life cycle (i.e., plant construction,

Source	Output	Total Emissions		Emissions Rate	
	Bil. kWh	Mil. tonnes	Mil. std. tons	kgCO2/kWh	lbCO2/kWh
Coal	1,734	1,723	1,895	0.994	2.186
Gas	1,017	409	450	0.402	0.885
Oil	28	27	30	0.959	2.109
Nuclear	790	—	—	—	—
Hydro	325	—	—	—	—
Solar	2	—	—	—	—
Wind	120	—	—	—	—
Other RE	90	—	—	—	—
Total	4,106	2,159	2,375	0.526	1.157

Sources: U.S. Energy Information Administration (2012), for generation data; U.S. Environmental Protection Agency, for CO_2 emissions data.

TABLE 15-2 2011 U.S. Electricity-Generation Mix and Average Emissions from Each Source equipment transportation, etc., are not included) so only fossil fuels have nonzero emissions rates. Summing total emissions from each source and dividing by the total kWh gives an overall figure of 0.526 kg CO_2 per kWh (1.157 lb CO_2), which is measurably lower than the 2004 national figure that could be derived from Fig. 15-6, due to the greater presence of renewables and gas, and the smaller amount of coal. Note that net imports from Canada are not included in the table. However, because these imports represent only a small fraction of the total 4.1 trillion kWh of demand, the emissions rates shown are reasonable estimates.

Example 15-4 illustrates some potential outcomes for a comparison of gasoline and electric propulsion when the electricity comes from a mixture of sources, and when it does not.

Example 15-4S Consider the CO_2 emissions performance of a plug-in hybrid of the type introduced in Example 15-2, with fuel economy of 47.3 mpg when running on gasoline and 5 miles/kWh when running on electricity. In one year, the vehicle travels 12,000 miles, of which 60% are powered by electricity and the remaining 40% by the ICE. Average emissions for electricity from the local grid are 1.157 lb CO_2 per kWh at the point of generation, with an average of 8% losses in transmission and distribution from the source of generation to the charge point for the vehicle. For ICE propulsion, CO_2 emissions are 19.49 lb/gal from the tailpipe plus 2.66 lb/gal upstream, or 22.15 lb/gal total. Suppose the vehicle were instead an HEV that traveled all 12,000 miles on ICE power at the given fuel economy. (a) What are the respective CO_2 emissions for these two alternatives, and which one has lower emissions? (b) Now suppose the PHEV instead used electricity generated entirely from coal with emissions of 2.186 lb CO_2 per kWh. How do the calculations in (a) change?

Solution

(a) We need to first adjust the emissions per kWh of electricity to reflect T & D losses, by factoring in the fraction of electricity delivered after losses are incurred:

$$\text{Emits}_{\text{Delivered}} = \frac{\text{Emits}_{\text{Plant}}}{1-\text{PctLoss}} = \frac{1.157}{1-0.08} = \frac{1.157}{0.92} = 1.257\ \text{lbCO}_2/\text{kWh}$$

Total emissions per year from EV operation of the PHEV are then

$$\text{TotEmits} = \left(7{,}200\frac{\text{mi}}{\text{year}}\right)\left(\frac{1\ \text{kWh}}{5\ \text{mi}}\right)\left(1.257\frac{\text{lbCO}_2}{\text{kWh}}\right) = 1{,}811\frac{\text{lbCO}_2}{\text{year}}$$

Additional total emissions per year from ICE operation of the PHEV are the following:

$$\text{TotEmits} = \left(4{,}800\frac{\text{mi}}{\text{year}}\right)\left(\frac{1\ \text{gal}}{47.3\ \text{mi}}\right)\left(22.15\frac{\text{lbCO}_2}{\text{gal}}\right) = 2{,}248\frac{\text{lbCO}_2}{\text{year}}$$

Total emissions from the PHEV are therefore the sum of the two sources, or 4,059 lb CO_2 per year. For the equivalent HEV, emissions are

$$\text{TotEmits} = \left(12{,}000\frac{\text{mi}}{\text{year}}\right)\left(\frac{1\ \text{gal}}{47.3\text{mi}}\right)\left(22.15\frac{\text{lbCO}_2}{\text{gal}}\right) = 5{,}619\frac{\text{lbCO}_2}{\text{year}}$$

Thus the reduction in emissions going from HEV to PHEV is 1,561 lb CO_2 per year, a decrease of 28%.

(b) As in part (a), we first adjust the emissions per kWh of electricity:

$$\text{Emits}_{\text{Delivered}} = \frac{\text{Emits}_{\text{Plant}}}{1-\text{PctLoss}} = \frac{2.186}{1-0.08} = \frac{2.186}{0.92} = 2.376\ \text{lbCO}_2/\text{kWh}$$

Repeating the calculation for the 7,200 miles driven in EV mode with an emissions factor of 2.376 lb CO_2 per kWh gives 3,421 lbCO_2 per year emitted. Emissions on the ICE side do not change, so total emissions are then:

$$3{,}421 + 2{,}248 = 5{,}669 \frac{\text{lbCO}_2}{\text{year}}$$

This amounts to a 1% increase compared to the HEV alternative from (a), or 49 lb CO_2 per year.

Example 15-4M Consider the CO_2 emissions performance of a plug-in hybrid of the type introduced in Example 15-2, with fuel economy of 20 km/L when running on gasoline and 8 km per kWh when running on electricity. In one year, the vehicle travels 19,200 km, of which 60% are powered by electricity and the remaining 40% by the ICE. Average emissions for electricity from the local grid are 0.526 kgCO_2 per kWh at the point of generation, with an average of 8% losses in transmission and distribution from the source of generation to the charge point for the vehicle. For ICE propulsion, CO_2 emissions are 2.34 kg/L from the tailpipe plus 0.32 kg/L upstream, or 2.66 kg CO_2 per liter total. Suppose the vehicle were instead an HEV that traveled all 19,200 kilometers on ICE power at the given fuel economy. (a) What are the respective CO_2 emissions for these two alternatives, and which one has lower emissions? (b) Now suppose the PHEV instead used electricity generated entirely from coal with emissions of 0.994 kg CO_2 per kWh. How do the calculations in part (a) change?

Solution

(a) We need to first adjust the emissions per kWh of electricity to reflect T & D losses, by factoring in the fraction of electricity delivered after losses are incurred:

$$\text{Emits}_{\text{Delivered}} = \frac{\text{Emits}_{\text{Plant}}}{1 - \text{PctLoss}} = \frac{0.526}{1 - 0.08} = \frac{0.526}{0.92} = 0.572 \text{ kgCO}_2/\text{kWh}$$

Total emissions per year from EV operation of the PHEV are then

$$\text{TotEmits} = \left(11{,}520 \frac{\text{km}}{\text{year}}\right)\left(\frac{1 \text{ kWh}}{8 \text{ km}}\right)\left(0.572 \frac{\text{kgCO}_2}{\text{kWh}}\right) = 823 \frac{\text{kgCO}_2}{\text{year}}$$

Additional total emissions per year from ICE operation of the PHEV are the following:

$$\text{TotEmits} = \left(7{,}680 \frac{\text{km}}{\text{year}}\right)\left(\frac{1 \text{ L}}{20 \text{ km}}\right)\left(2.66 \frac{\text{kgCO}_2}{\text{L}}\right) = 1{,}021 \frac{\text{kgCO}_2}{\text{year}}$$

Total emissions from the PHEV are therefore the sum of the two sources, or 1,844 kgCO_2 per year. For the equivalent HEV, emissions are

$$\text{TotEmits} = \left(19{,}200 \frac{\text{miles}}{\text{year}}\right)\left(\frac{1 \text{ L}}{20 \text{ km}}\right)\left(2.66 \frac{\text{kgCO}_2}{\text{L}}\right) = 2{,}554 \frac{\text{kgCO}_2}{\text{year}}$$

Thus the reduction in emissions going from HEV to PHEV are 709 kg/year, a decrease of 28%.

(b) As in part (a), we first adjust the emissions per kWh of electricity:

$$\text{Emits}_{\text{Delivered}} = \frac{\text{Emits}_{\text{Plant}}}{1 - \text{PctLoss}} = \frac{0.994}{1 - 0.08} = \frac{0.994}{0.92} = 1.08 \text{ kgCO}_2/\text{kWh}$$

Repeating the calculation for the 11,520 km driven in EV mode with an emissions factor of 1.08 kg CO_2 per kWh gives 1,555 kgCO_2 per year emitted. Emissions on the ICE side do not change, so total emissions are then:

$$1{,}555 + 1{,}021 = 2{,}576 \frac{\text{kgCO}_2}{\text{year}}$$

This amounts to an increase of 22 kg CO_2 per year or ~1% compared to the HEV alternative from part (a).

As illustrated in Example 15-4, whether the PHEV platform decreases CO_2 emissions compared to the HEV depends on many factors, including the exact fuel economy delivered by the vehicles compared, the extent of line losses in T & D, the emissions per kWh, and other factors. Some generalizations can be offered. First, for electrical grids with large amounts of gas or non-fossil energy generation, it is almost certain that PHEV will reduce emissions compared to HEV, irrespective of other factors. This is true because the primary energy source of the HEV is entirely petroleum, which is CO_2-emitting, giving PHEV an insurmountable advantage in reducing CO_2 as long as the share of carbon-free energy is large enough. Secondly, even in the case where coal dominates electricity production and PHEV has the potential to increase emissions, there may be motivation to introduce PHEVs. The grid provides an inroad to deliver carbon-free electricity in the future from renewables, nuclear, or fossil with CCS, so by the time PHEVs could significantly penetrate the light duty vehicle market, emissions of CO_2 may have been partially or mostly eliminated from the grid, leading to net CO_2 reductions.

Battery Swapping as an Alternative to Stationary Charging

An alternative to using the grid to recharge EVs when they are connected at a charging point is to recharge batteries off-board of vehicles when electricity is available, and then swap charged batteries for discharged ones when vehicles stop at a *swapping station*. Swapping stations require compatibility between the station mechanism, the battery, and the vehicles, so that the station can reliably remove the old battery and insert the new one. One potential advantage of swapping over recharging is that the swapping experience for the driver is similar to that of visiting a service station with an ICEV, where the entire operation is finished in a matter of minutes. EVs that are capable of using a swapping station can also be designed to connect to either household or public chargers, to give the owner the option of recharging as opposed to swapping.

The swapping station approach was being developed by the Better Place Company starting in 2005, and leading to the first demonstration swapping stations in Japan in April 2010 and Denmark in June 2011. Better Place collaborated with the Nissan/Renault conglomerate to modify the Nissan Leaf and Renault Fluence models so that their battery system would be compatible with the swapping system. Better Place encountered financial difficulties in 2013 which put in question the economic viability of swapping as opposed to ordinary recharging for EVs. However, even if swapping stations do not take root, the more general model of a business that acts as a broker for charging stations, green electricity, and tracking EV owners' electricity consumption for billing purposes may have more widespread appeal in the long run.

15-5-2 Integration of EVs with Renewables over the Longer Term

One of the premises of the short-term integration of EV/PHEVs is that conventional generating capacity, primarily from fossil fuels, is available to generate the extra electricity needed for the LDV fleet, especially at night when other loads are at their lowest. If, however, we move toward large-scale adoption of renewables over the longer term to eliminate CO_2 emissions or to move away from nonrenewable resources, the ability of renewable sources to meet transportation demand will become essential.

As discussed above, renewables can be broadly put in two groups. There are large-scale renewables that are also intermittent and relatively diffuse, and small-scale renewables that are dispatchable but have limited total potential capacity that could not match future U.S. passenger transportation demand of 15 to 20 quads plus another 5 to 10 quads for freight. The situation in many other major world regions (European Union, Russian Federation, China, and India) is similar: if renewables are to power as large a load as the transportation sector, a mix of solar and wind are required. We therefore focus on issues that arise from using a large but intermittent resource such as solar and wind in the remainder of this section.

Annual Average versus Seasonal and Diurnal Supply and Demand

First, taken as an aggregate resource, there is more than enough of either solar or wind energy to provide the entire electricity market for both conventional loads and transportation. Consider the following thought exercise: the entire U.S. electricity market consumes on the order of 40 quads of energy per year, and the LDV market another 20 quads. Thus the total is on the order of 60 quads, or roughly 60 EJ in metric. Converting this yearly amount into an average rate of energy flow gives roughly 2 TW. In the desert southwest, intensity of insolation may average 250 W/m^2, so if the productivity of solar devices were 10% of this amount, or 25 W/m^2, the area required would be 80,000 km^2, or less than one-third the land area of just one of the U.S. states, namely Arizona, at 295,000 km^2. Similarly, the Great Plains Wind Corridor in the states of North and South Dakota, Nebraska, and Kansas could meet all U.S. electricity needs on a year-round basis.

The above compares only the *total annual demand* from transportation output with the *total annual output* from solar and wind. Matching solar and wind to transportation demand must take into account both *seasonal* variability, or variability based on the month of the year, and *diurnal* variability, or variability based on the time of day. We begin with seasonal variability by considering relative productivity for representative solar and wind sites, as shown in Fig. 15-13. In Fig. 15-13(*a*), both curves are from sites near Ithaca in New York state in the northeastern United States; the solar curve is from an actual solar PV array for a representative year, and the wind curve is based on the average power available at the site of a proposed wind farm for a representative year. In this instance, the two sources are reasonably well-matched, with the peaking of wind output in the winter making up for the inevitable drop-off in solar output as the days become shorter. Figure 15-13(*b*) compares estimated average output per unit of capacity (i.e., electricity produced per month per unit of representative solar photovoltaic capacity) for four U.S. locations with varying amounts of average solar gain. Note that these values are modeled averages from the U.S. Department of Energy, which take into account the position of the sun, historical average weather data, and solar photovoltaic system component performance. Although there is a difference in productivity between locations (the least sunny, Binghamton, produces 74% of the most sunny, San Diego, in a year) and all locations experience a drop-off in output in the colder months, they all produce substantial amounts of power during any month of the year.

Depending on the part of the United States, solar will be more or less productive, with the southwest in particular having relatively high productivity in winter relative to summer. The northern tier of the country, on the other hand, from the northeast across the northern plains, will have a relatively wide spread between winter and summer, thanks to high latitudes and relatively cloudy weather in winter. It is more difficult to generalize about wind energy: many sites have the same pattern as the one shown in

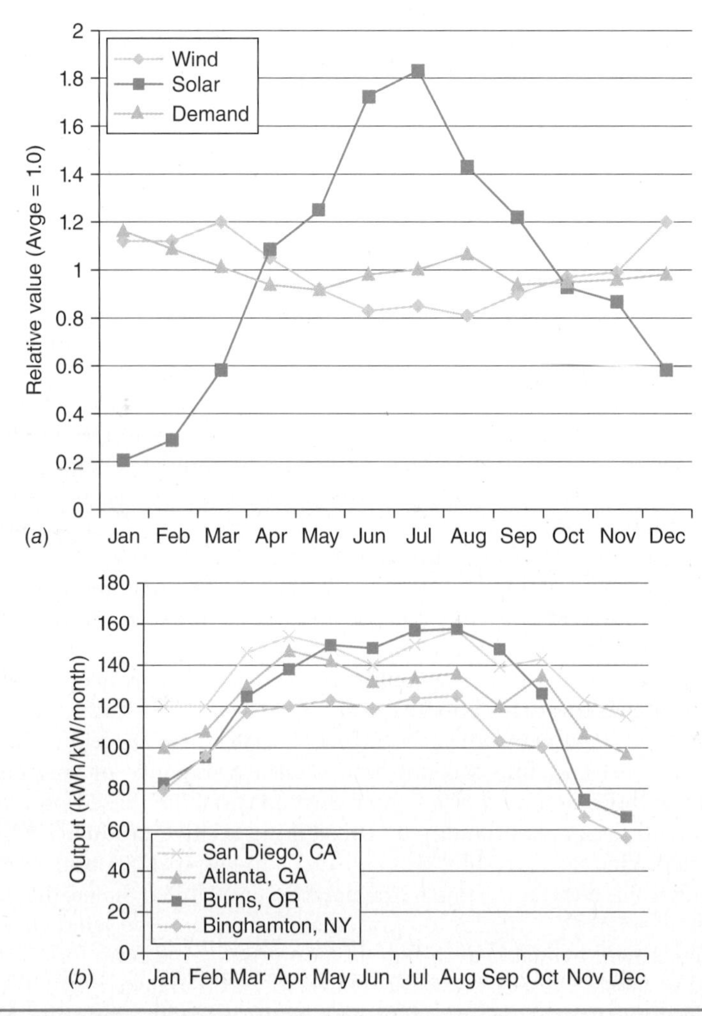

Figure 15-13 Variability in output by month: (*a*) for representative solar and wind sites and electricity demand level relative to annual average for Tompkins County in New York State; (*b*) solar PV output per kW of capacity for four U.S. locations: Binghamton, N.Y.; Burns, Oreg.; Atlanta, Ga.; San Diego, Calif.

Notes: Values in Fig. 15-13(*b*) are predicted electrical output per month in kWh per kW of installed capacity from U.S. Dept of Energy's "PVWatts" calculator (*www.nrel.gov/rredc/pvwatts/grid.html*). PV panels are assumed in fixed position facing south and raised to angle of local latitude, with 15% losses from PV cells to output from DC to AC inverter. Total annual output values for Binghamton, Burns, Atlanta, and San Diego are 1228, 1468, 1488, and 1656 kWh, respectively.

Source: National Renewable Energy Laboratory.

Fig. 15-13, where output peaks in the winter and drops in the summer, but not all do. Therefore, any reliance on a mixture of solar and wind to meet year-round transportation energy demand (or some large fraction thereof) assumes an upgraded grid transmission system to transmit renewable electricity long distances with losses that do not exceed some acceptable maximum.

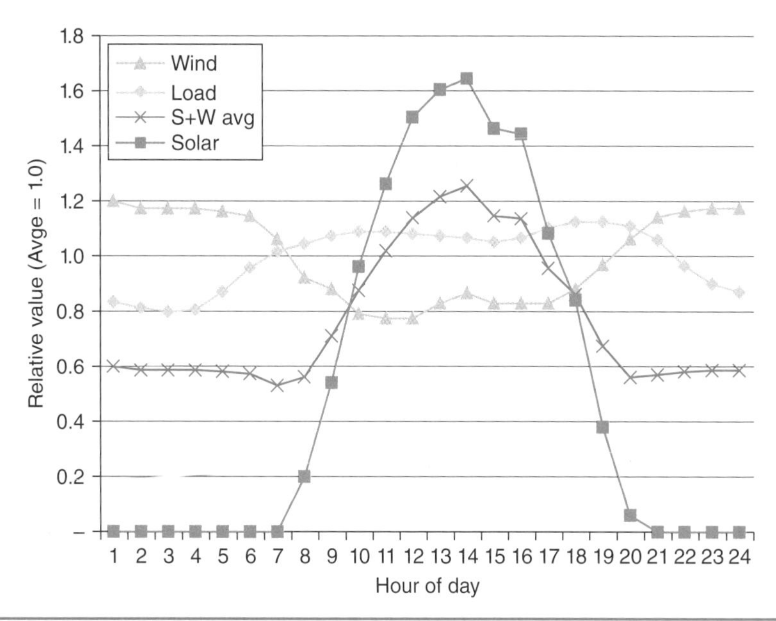

FIGURE 15-14 Relative solar energy output, wind energy output, average of solar and wind, and demand load for region around Ithaca, N.Y.

Notes: Relative solar values are compared to average of output between 7 a.m. and 8 p.m. only, since remaining hours of day had no output and are not included in the average. Example of wind and solar average curve: At 0100 hours, solar = 0, wind = 1.2, S + W avg = 0.6.

Diurnal variability is affected most pronouncedly in the case of solar by the absence of the sun at night, which varies in length between the winter and summer solstices but always leaves many hours of the day without the possibility of solar output, even on the longest days of the year. Figure 15-14 gives relative solar output for a representative summer day near the solstice along with average hourly wind speed for 365 days of the year. The data are for the same array and wind energy site as was presented in Fig. 15-13. The figure also shows average electric load requirements for the region around the solar and wind sites (Tompkins County, New York state), and for comparison to the load, the average of the wind and solar curves in the figure. Regarding solar energy, although the site is in a part of the United States that has a relatively modest amount of incoming solar energy per unit of area (approximately 3.8 kWh/m^2/day for the region, compared to 5.8 kWh/m^2/day for the region around San Diego, Calif.), the potential productivity of a solar installation with sufficient capacity is nevertheless significant. Also, the situation for solar is quite different from that of conventional power discussed above, the solar arrays are at rest at night and have no possibility of making output, so that solar cannot provide any extra capacity at night for charging vehicles.[6] Instead, solar arrays loosely follow the daily pattern of electricity demand shown in Fig. 15-10. The 24-hour

[6]Experimentation is under way with solar thermal generating system that would store sufficient thermal energy in a medium such as molten salt that steam could be generated to drive a turbine and make electricity at any hour of a 24-hour day. For purposes of this chapter, however, we focus on existing commercially available solar technology.

pattern for the local wind farm site under consideration is also shown; at the site, wind increases slightly at night, which could benefit night charging of vehicles. Diurnal patterns for wind, however, vary from site to site like the site seasonal pattern, so one cannot generalize this finding like that of the solar curve. Lastly, the curve showing the average of wind and solar is included for reference. This curve loosely follows the demand curve, with a rise from 0.6 at night to 1.25 in the early afternoon, showing that under favorable conditions the mix of solar and wind might be able to sustain the baseline load.

Both solar and wind curves must be interpreted with caution here: the curves shown represent average behavior that could be expected, but on any given day there may not be sufficient sun or wind to provide electricity at expected levels, so the system must be prepared to provide an alternative source of electricity in such an event.

Recommendations for Integrating Renewables and Transportation Energy Demand

The preceding information about renewables, and solar and wind in particular, can be generalized into several recommendations for integration with the transportation energy market:

- *Adequacy of solar and wind in general:* With sufficient capacity investment (i.e., enough generating capacity to meet demand plus some quantity *X*% to handle intermittency and other contingencies), wind and solar can on average provide sufficient electricity for both existing demand for electricity and transportation energy demand in the future.
- *Contingency needed for days when solar and wind resources are inadequate*: A system centered around solar and wind must provide some contingency for days when output is below average, since these are intermittent resources. If the goal were to power transportation demand strictly from renewables, this might provide an ideal role for niche renewables like hydro/biomass/geothermal that could be waiting in reserve and accessed to support solar/wind.
- *Importance of 24-hour access of vehicles to charge points:* Because solar is a resource that is never available at night, and not all wind farms have the behavior of increasing output at night as shown in Fig. 15-14, it will be highly beneficial if vehicles can charge during the day as well as at night. Since many vehicles have the usage pattern of rest overnight, movement during the morning and afternoon rush, and rest in the middle of the day, possibly at the place of employment, charge points should be widely deployed at places of employment to meet this goal.

Out-of-Pocket Cost of Renewable Electricity Compared to Alternatives

An analysis of the cost of supplying electricity from renewable sources to EVs under a grid system shows that once the buyer has made the commitment to the higher capital cost of a vehicle with batteries, the difference in cost between conventional and renewable electricity is relatively small. Therefore, they may be willing to pay extra for renewable electricity. The premium might go instead toward a surcharge for electricity from fossil with CCS or from newly-built nuclear, either of which has a premium cost per kWh compared to conventional electricity from existing sources thanks to additional up-front capital cost.

On an annual basis, the owner of a vehicle first sees the annualized capital cost of the vehicle; if the vehicle is an EV, then they also see the cost of the battery storage system, since this adds substantially to the total capital cost. Second, they see energy costs, in the form of gasoline cost for an ICEV or electricity cost for an EV. Third, they see road infrastructure maintenance cost, typically included in the cost per gasoline in the form of taxes; At present, EV users are able to power their vehicles without paying these taxes, however, for a fair comparison of the cost of EVs versus ICEVS one should assume that under a robust EV market scenario, a means for levying these costs on EV drivers would be established. Lastly, they see a balance of cost that includes insurance, maintenance, and any other miscellaneous costs such as parking. Of these costs, only fuel cost (with highway tax included) is paid for out of pocket, and only in the case of gasoline (an EV driver would only see the cost of electricity in their monthly electricity bill, unless they pay to use a public charge point), so it is more prominent in the mind of the owner. Example 15-5 illustrates the comparison.

Example 15-5 Compare three vehicle-fuel combinations as follows: (1) ICEV, (2) EV with conventional electricity, (3) EV with renewable (or carbon-free) electricity; all three options are based on a five-door compact hatchback model. The ICEV costs $16,000 and averages 20 mpg combined city and highway, with a fuel cost of $3.00/gal (excluding taxes, which are added later). The EV costs $23,000 without the battery system, a higher cost than the ICEV because it is manufactured in small numbers; the batteries add another $9,900, and energy consumption is 1 kWh for every 3 miles traveled. Conventional electricity costs $0.10/kWh delivered to the vehicle, and all-renewable electricity costs $0.18/kWh in the form of some seasonal mixture of solar and wind (typically more wind in the winter and more solar in the summer). Since any option must see some tax for highway usage, assume $500/year across all three options, which is equivalent to $1.00/gal for the gas option or the equivalent amount charged to the EV owner through their metering system (assume that the ISO makes the needed arrangements). Lastly, assume $1,500 for the balance of costs for each vehicle. If the vehicle drives 10,000 miles/year, the lifetime is 10 years, and the discount rate is 7%, (a) what is the total annual cost for each option, and (b) of this amount, how much is paid out of pocket for energy costs in each case?

Solution (a) Each vehicle is considered in turn. For each, the needed discounting factor is (A/P, 7%, 10y) = 0.1424. Starting with the ICEV, the annualized capital cost is

$$(P)(A/P, 7\%, 10) = (\$16,000)(0.1424) = \$2,278$$

Since the vehicle travels 10,000 miles/year and averages 20 mpg, the fuel consumption is 500 gal and the annual fuel cost at $3/gal is $1,500. Combining all annual cost components including provided values for tax and balance of cost gives

$$\begin{aligned} C_{\text{tot}} &= \text{Cap} + \text{Fuel} + \text{Tax} + \text{Other} \\ &= \$2,278 + \$1,500 + \$500 + \$1,500 \end{aligned}$$

Thus C_{tot} = $5,778. For both of the electric vehicle options, the annual capital cost is the combination of the vehicle and battery costs:

$$\begin{aligned} C_{\text{veh}} &= (\$23,000)(0.1424) = \$3,275 \\ C_{\text{bat}} &= (\$9,900)(0.1424) = \$1,367 \end{aligned}$$

Based on 10,000 annual miles and 3 miles/kWh, the annual electricity consumption is 3,333 kWh. For conventional and renewable electricity options, the annual cost is $333 or $600/year. Totalling up annual cost for the EV with conventional electricity gives

$$\begin{aligned} C_{\text{tot}} &= \text{Veh} + \text{Batt} + \text{Fuel} + \text{Tax} + \text{Other} \\ &= \$3,275 + \$1,367 + \$333 + \$500 + \$1,500 = \$6,975 \end{aligned}$$

The EV + RE option requires an increase of $267/year for the higher cost electricity, so the total cost increases to $7,242/year. The two options represent a 21% and 25% increase, respectively, in total annual cost compared to ICEV.

(b) The fuel costs for ICEV, EV, and EV+RE are $1,500, $333, and $600, respectively. Dividing each by the total annual cost gives the total cost of 26%, 5%, and 8%.

The annual cost values for the three options are presented graphically in Fig. 15-15. Although the total annual cost value for either EV is roughly one-fourth higher than that of the ICEV, the operating cost including all items except annual capital cost is lower by at least $900, depending on the option. The out-of-pocket cost for electricity is particularly low compared to that of gasoline. Experience shows that out-of-pocket costs (as opposed to sunk costs such as the commitment to the capital cost of the vehicle) factor more heavily in the financial decision-making of many consumers. Therefore, having avoided $1,500 in annual cost for gasoline (not including taxes) by purchasing an EV, the owner may opt to pay the $267 premium for electricity from renewable sources.

The preceding comparison of ICEVs and EVs raises a number of discussion points, some which work in favor of the viability of the EV and others against. For simplicity, it was assumed in the example that the battery system would last the full 100,000 miles or 10 years, but if it needed replacement or refurbishing earlier, the cost would increase. Also, a driving rate of 10,000 miles/year implies an average of slightly more than 27 miles per day over 365 days. Since EVs may have difficulty making long intercity

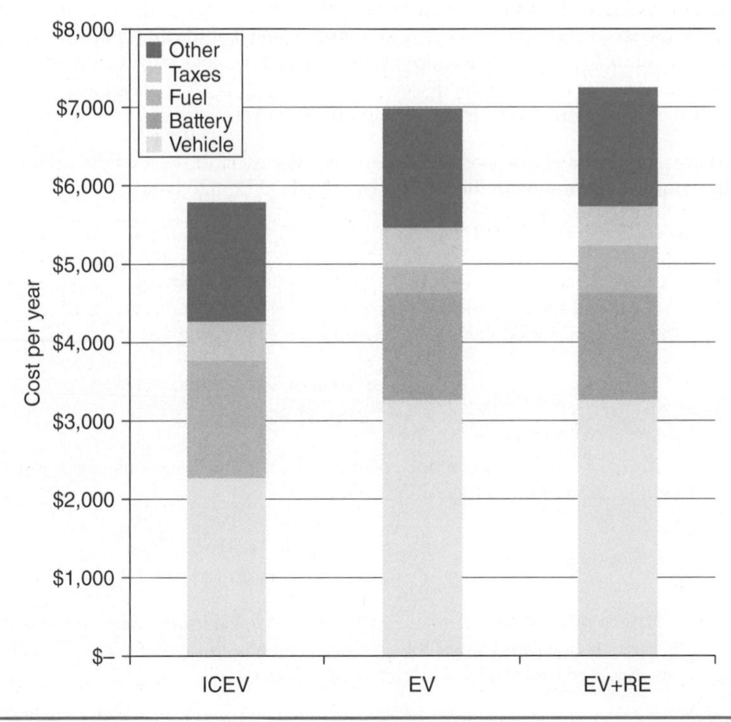

FIGURE 15-15 Comparison of cost components in annual cost for ICEV, EV, and EV with renewable electricity (EV + RE).

trips due to the lack of range, it may be difficult for an EV to drive this many miles in a year, which would change the economics. On the other hand, the cost of both the vehicle and the battery system for the EV are relatively high due to low volumes of production, and these values might decline in the future. Also, the drivetrain of an EV (battery system, motor, controller, and transmission) is simpler than that of an ICEV so an annual maintenance cost advantage for EVs could be expected, but this was not incorporated in the calculation.

15-5-3 Electric Vehicles with Vehicle-to-Grid Capability

Thus far we have only considered the benefits of one-way flow of energy from the grid to the electric vehicle to provide an alternative to petroleum that is greener and more reliable over the long term. V2G capability would allow electric vehicles to become fully integrated elements in the grid itself, providing storage and release of electricity while also providing propulsion to the vehicle (Kempton and Tomic, 2005a; 2005b).

Part of the motivation for expanding vehicle electrification from one-way provision of energy to vehicles to V2G is to cover the substantial incremental cost of battery storage systems compared to an ICEV that requires only a fuel tank for gasoline or diesel. The two-way arrows in Fig. 15-9 represent this potential flow both to and from the vehicles. For instance, the fleet of V2G-enabled vehicles might absorb a burst of electricity from renewable energy sources at low cost to the vehicle owners (since the incremental cost per kWh for generating electricity at solar or wind sites is low) and then sell it back to the grid later at a profit. This function may be especially valuable at times of peak demand for electricity, where each kWh that the vehicles can provide will be valuable. The vehicles can also assist with *grid regulation*; that is, providing small increments of electricity to help supply more precisely match demand, a service currently provided by a certain subset of the large central power plants.

Throughout this section, the application of V2G using EVs is assumed. The differences involved in using PHEVs instead of EVs are considered in the next section. Also, at present V2G is a concept at an embryonic stage that, while promising because of the numbers of vehicles that might eventually participate and the storage capacity that they could provide, nevertheless has many hurdles to overcome. Not only are there infrastructure questions around the numbers of vehicles, charge points, and control systems that would be required, but V2G raises issues of not excessively draining battery systems of charge that is needed by the driver, and also protecting both the individual and the grid operator from vulnerabilities due to the sensitivity of the controls.

Before discussing revenues and costs associated with V2G, we first consider physical requirements that must be met. Take the case of a residence with either 110- or 220-V lines available, that is connected to the grid via a local distribution system. The line from the battery system to the connection point between the vehicle and the residence, and the electric lines within the residence, must be able to support the current dispatched from the battery system within the vehicle. Typical maximum values for line currents might be on the order of 80 A, so the resulting maximum power would be 17.6-kWwith a 220 V line. Low speed charging of EVs using a 110-V outlet rated at 20 A (i.e., 2.2 kW) may be adequate in some instances in non-V2G situations when the user does not require rapid recharging. However, V2G participation may require higher maximum power, so some investment in upgrading the connection may be necessary to meet the requirements of a V2G contract.

Assuming physical requirements for connections are met, V2G brings revenue to the vehicle owner while at the same time providing a valuable service to the grid ISO by generating gross revenue from services in excess of the cost of participation. Some revenue is generated by selling power from the vehicle battery system when V2G calls for it, but because there are costs involved in storing and releasing power, the sale price must be higher than the purchase price for V2G to be attractive. Otherwise, the user relies on revenue from energy sales in tandem with standby charges earned. These charges are paid by the ISO on the basis of maximum power that could be drawn (in watts) and agreed-upon hours of availability per day or year, as the vehicle must be disconnected from the grid some of the time when it is in motion. The units of standby capacity are thus watt-hours, and the amount paid is measured in \$/Wh or \$/kWh.

On the cost side, the V2G participant incurs the capital cost of any equipment needed, such as a smart charging system that can communicate electronically with the ISO to either charge or discharge the battery. The capital cost of the portion of the battery system life that is used by discharging electricity to the grid for V2G rather than to the motor for propulsion is also included. Example 15-6 provides an illustration of V2G economics.

Example 15-6 An electric vehicle with characteristics like that of a Nissan Leaf is outfitted to participate in a V2G system. The battery system is considered part of the capital investment for participation; it has a capacity of 24 kWh, a maximum depth of discharge of 83% (i.e., 20 kWh available), an expected life of 130,000 1-kWh cycles, and a cost of \$10,000. The vehicle also requires a capital investment of \$2,000 for all equipment necessary to participate in V2G. Both pieces of capital equipment should be valued annually using a 10-year lifetime and a 10% discount rate. Electricity costs \$0.10/kWh whether being bought or sold by the vehicle, and the owner is paid for standby capacity at the rate of \$0.04/kWh for a maximum power flow of 15 kW and a time duration of 16 hours/day. Actual power dispatched to the grid averages 10 kWh/day. Ignore inverter losses. (a) Calculate expected annual net revenue. (b) Discuss any complicating considerations or limitations.

Solution

(a) First, the discounting factor is (A/P, 10%, 10) = \$0.1627. Therefore the two capital costs for battery and control system are

$$(\$10{,}000)(0.1627) = \$1{,}627$$
$$(\$2{,}000)(0.1627) = \$325$$

Assuming the battery lasts 10 years, on average 13,000 cycles could be expected per year, so the cost per cycle of wearing out the battery is (\$1,627)/(13,000) = \$0.13/kWh. The capital cost of the control system is incurred regardless of the number of kWh sold to V2G.

The earnings from standby capacity due to V2G participation are based on a maximum capacity of 15 kW and time of 16 hours per day:

$$(15\text{ kW})(16\text{ h/day})(365\text{ day/year})(\$0.04/\text{kWh}) = \$3{,}504$$

Earnings from electricity sold amount to

$$(10\text{ kWh/day})(365\text{ day/year})(\$0.10/\text{kWh}) = \$365$$

Thus the gross revenue per year is \$3,504 + \$365 = \$3,869. Total annual cost is based on 3,650 kWh cycled through the battery and sold to V2G:

$$(3{,}650\text{ kWh})(\$0.13/\text{kWh}) + \$325 = \$1{,}147$$

The difference is then the net revenue from V2G participation of \$3,869 – \$1,147 = \$2,722.

(b) Several consideration should be addressed. First, since the maximum energy flow is 15 kW, if we assume 220-V service available, the maximum amperage for the system is

$$(15{,}000\ \text{W})/(220\ \text{V}) = 68\ \text{A}$$

Thus in the case of residential parking the vehicle would require a stronger connection to the main household control panel than a typical individual circuit. This upgrade could be paid for as part of the $2,000 cost. Since 110-V service would require 134-A current levels, it is necessary to upgrade the connection to 220V and probably 80 A to provide an extra margin. Next, since the vehicle must be connected at the exact time the system calls on it and must have sufficient charge, the ISO has an interest in the actual energy drawn being small compared to the maximum potential per day. If the ratio is low, there will be sufficient redundancy built into the system, i.e., many participating vehicles potentially available so that in any instant a subset of them is very likely to be connected and having charge. The maximum potential if the vehicle were to be called at maximum power for the available time window is (15 kW)(16 hours) = 240 kWh. The value of this ratio is

$$10\ \text{kWh}/240\ \text{kWh} =\sim 4\%$$

This condition is therefore reasonably maintained. Lastly, the total number of cycles per year used by V2G should be small enough that the remainder is adequate for propelling the vehicle when it is moving. The average number of cycles expected is 13,000 kWh/year. Of these, 10 kWh/day or 3,650 kWh/year are expected to be used by V2G, so the remainder of 9,350 kWh (72% of the total) is available for propulsion.

A summary of the costs and revenues from the V2G system in Example 15-6 is shown in Table 15-3. As shown, V2G is profitable on an annual basis, although in a real system not 100% of the charge added to the battery is returned to the grid, so round-trip losses would increase annual cost somewhat. Setting aside these losses, the net $2,722/year

Capital Cost Side		Amount
Battery cost	Lifetime	$10,000
	Annual	$1,627
	Cycle cost	$0.125/kWh
V2G equipment cost	Lifetime	$2,000
	Annual	$325
Revenue side		
Total energy sold	kWh/year	3650
Capacity revenue	Annual	$3,504
Energy revenue	Annual	$365
Net earning		
Gross revenue	Annual	$3,869
Total cost	Annual	$1,147
Net revenue	Annual	$2,722

Table 15-3 Summary of Economics of Example V2G System with Cost, Revenue, and Net Revenue

could go toward repaying the higher initial cost of the vehicle. Notice that the net revenue is positive despite the energy costing the vehicle \$0.214/kWh and selling for \$0.10/kWh.[7] This is true because so much of the revenue comes from capacity rather than energy. To create this revenue stream, however, it is crucial that the vehicle be available at or near the 16 hours/day specified. Most private vehicles spend 90% to 95% of each 24-hour day idle and at rest, so it is reasonable to think that the vehicle would be parked either at home or at work, and could be accessed. However, the charge point must be available as well, and it must function properly and reliably.

The analysis in Example 15-6 could also be extended by considering the benefits of selling to the grid during periods of peak demand: if the vehicle could sell to the grid at prices higher than what was paid to purchase the electricity, additional profits would result. These opportunities are more difficult to quantify in a worked example because the number of hours where electricity can be sold at a peak price, and the price per kWh paid, are both unknowns. Furthermore, the ISO must see the need to pay the vehicle owner for the capacity service, in lieu of other options. The price used of \$0.04/kWh is in line with what might be paid by the electric utility industry for regulation services from other sources, but reliability of V2G might be an issue. Lastly, the vehicle owner can preset with the ISO the minimum amount of charge that must be kept on reserve in the vehicle so that the owner can drive some minimum distance before needing to recharge. In some cases, the average amount from the example of 10 kWh discharged per day to V2G might be more than the driver will allow.

15-6 Hydrogen Fuel Cell Systems and Vehicles

Having considered both the potential and limitations of electric vehicles, we next discuss the use of hydrogen as a transportation fuel. Although the primary focus of this section is on the aspects of hydrogen use that are unique, there are commonalities with electrification as well: if used with a hydrogen fuel cell (HFC), hydrogen must be converted to electricity and conducted through a motor to drive the wheels of the vehicle, so this stage of the drivetrain is similar to that of the EV or PHEV. At the same time, hydrogen has potential advantages in terms of energy density of storage that make it more suitable for heavy vehicles (e.g., full-length urban public buses) or long distances between refueling stops that might allow it to occupy a separate niche in the domain of transportation fuels in the future, if the technology can be developed to maturity. (For full-length works on the topic of hydrogen, see Spiegel, 2007; Sperling and Cannon, 2004; or Larminie and Dicks, 2000)

Our primary focus is on the use of hydrogen in fuel cells rather than internal combustion engines for energy efficiency reasons. The hydrogen fuel cell is in fact one of a family of fuel cell designs that convert a range of fuels, such as methanol, solid oxides, or phosphoric acid, into electricity through a chemical reaction. Christian Schonbein developed the principle of the fuel cell in 1838, and in 1843 William Grove assembled the first working prototype. The first experimental application in a vehicle was the Allis-Chalmers Tractor Co.'s use of an alkaline fuel cell in a prototype farming tractor unveiled in 1959. The first commercial application was in the 1960s, when NASA used

[7]Calculation assumes that one divides the \$325/year capital cost for the control equipment across the 3,650 kWh for \$0.089/kWh, and includes this amount along with \$0.125/kWh for battery depreciation in the price per kWh.

a fuel cell developed by General Electric in the Gemini 5 mission, demonstrating the availability of a working alternative to batteries for electricity supply on board space flight. At about the same time, General Motors tested a fuel cell in a motor vehicle for the first time in its "Electrovan" prototype, retrofitting an ICE-powered van to run on a fuel cell powered by hydrogen and oxygen.

Since the time of the Gemini missions, fuel cells have been used primarily in space flight, where they are capable of providing power over long durations, in a weightless environment, without the noise or emissions of other types of generators. In recent years, both vehicle manufacturers and energy companies have been taking an increasing interest in the fuel-cell powered vehicle as an alternative to the ICEV powered by petroleum-derived liquid fuels. To this end, makers including Daimler, Ford, GM, Honda, and Toyota have been retrofitting production ICEVs to run on hydrogen fuel cells and also developing prototype hydrogen fuel cell vehicles from the ground up, so as to take advantage of new design possibilities made available by the fuel cell platform. Energy companies have also begun installing demonstration hydrogen filling stations, dispensing liquid and/or compressed hydrogen, in cities such as Washington and Tokyo. In 2005, Honda became the first automaker to lease an HFCV to a private individual for continuous use, providing a retrofitted version of a subcompact hatchback to a family in southern California.

Compared to passenger vehicles, operation of urban public buses may provide a more advantageous application for demonstrating the use of hydrogen. Public urban buses are operated, refueled, and maintained by professional staff; they can more easily accommodate the extra volume required for on-board hydrogen storage than passenger cars; they are usually refueled exclusively in a single location (typically the depot or maintenance facility where they are stored when not in service), and on many urban bus routes, they are not required to travel great distances in a single day's service (typically not more than 160 to 240 km, or 100 to 150 miles). A number of bus vendors including Daimler, El Dorado National (United States), and Toyota-Hino have placed prototype HFCV buses in service in various cities, including Madrid, Stuttgart, Stockholm, Nagoya, Vancouver, Palm Springs, California, and Newark, Delaware (Fig. 15-16), among others. One possible pathway for developing a successful HFCV is to perfect the technology on buses first, and then transfer the mature technology to passenger cars with appropriate modifications.

In the remainder of this section, the basic thermodynamics governing the function of the fuel cell is first presented, since this technology is relatively new (e.g., compared to the internal combustion engine or electric motor) and may not be as familiar. However, the technology is not as mature as that of the EV or PHEV, and predictions of a future hydrogen economy are speculative at this point. The latter parts of this section discuss practical aspects of implementing fuel cells in vehicles, and prospects for successfully commercializing them.

15-6-1 Function of the Hydrogen Fuel Cell and Measurement of Fuel Cell Efficiency

The function of the individual fuel cell is shown in Fig. 15-17. In the diagram, the fuel cell is supplied with pure hydrogen on the right and ambient air containing approximately 21% oxygen on the left. At the anode, the hydrogen separates into protons and electrons (in the case of a deuterium atom, the neutron remains joined

FIGURE 15-16 Prototype hydrogen fuel cell bus at the University of Delaware, Newark, DE, United States.

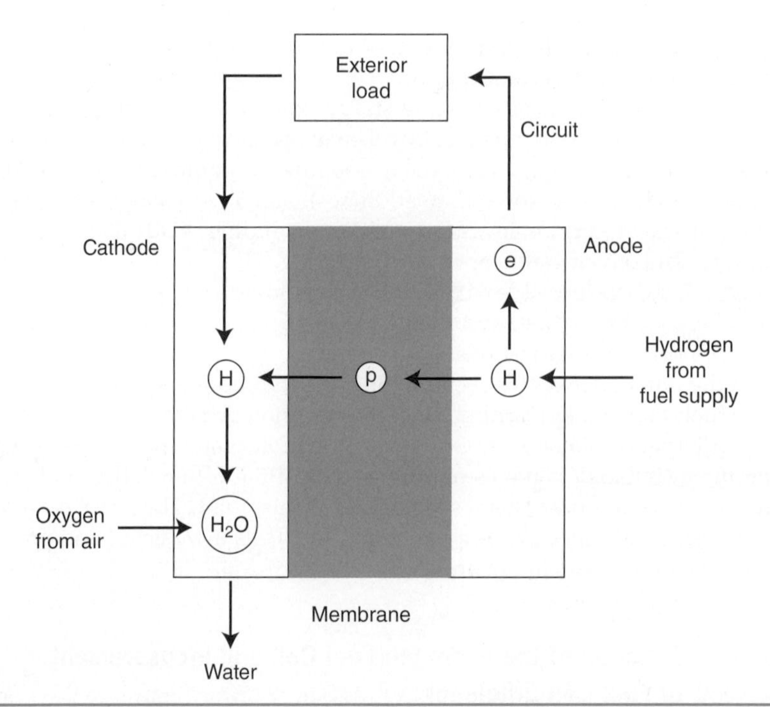

FIGURE 15-17 Schematic of hydrogen fuel cell function, showing anode, cathode, and proton exchange membrane (PEM).

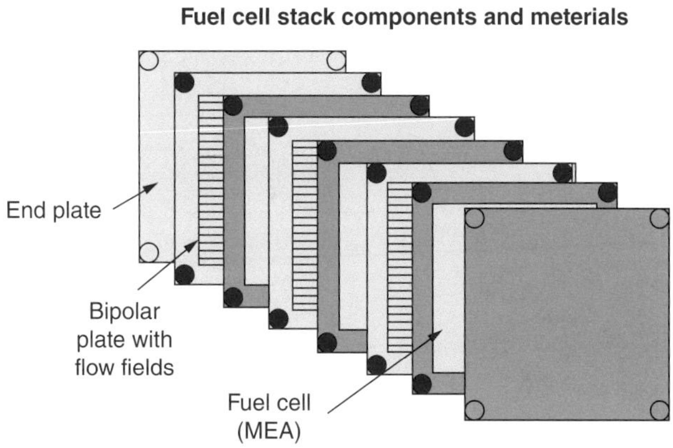

Fuel cell stack components and meterials

End plate
Bipolar plate with flow fields
Fuel cell (MEA)

FIGURE 15-18 Fuel cell stack. Exploded view of a fuel cell stack, made up of alternating units of fuel cells and bipolar plates. The connection is made to exterior loads from the end plates. *Source:* Spiegel (2007). Reprinted with permission.

to the proton), and then the proton exchange membrane (PEM) allows only the proton to pass. This leaves an excess of electrons on the anode side, which creates a voltage difference that can do work across an exterior load (e.g., electric motor in a vehicle). Since one fuel cell by itself does not create a sufficient voltage difference for most applications, multiple fuel cells are typically combined into a *fuel cell stack* to reach the desired voltage (Fig. 15-18). On the cathode side of the membrane, electrons and protons recombine into hydrogen, and then the hydrogen combines with the oxygen in the air to form water, which is the main component of the fuel cell's exhaust.

Next, we focus on estimation of the maximum efficiency of the fuel cell, using a theoretical model that ignores the effect of imperfect function and losses. We begin by considering the *Gibbs free energy* g_f of a compound, which is a measure of its ability to do external work, given temperature and pressure conditions. Because the value of Gibbs free energy changes with the reference point chosen, we focus on the change in Gibbs free energy, or Δg_f, when a chemical reaction takes place, since value of Δg_f is not affected by choice of reference point. For example, at ambient conditions of 298K and 100 kPa pressure, the reaction of one mole of H_2 molecules with ½ mole of O_2 molecules to form 1 mole of water molecules has a change in Gibbs free energy value of –237.2 kJ/mole. (We use gram-moles and not kilogram-moles throughout this section.) In other words, the decrease in Gibbs free energy due to the transformation of hydrogen and oxygen into water is energy that is available to do other work, such as moving electrons in a circuit.

Next, we can compute the voltage difference created by the release of energy if all of the electrons available in the reaction flow from the anode through the load to the cathode in the fuel cell. Recall that *Faraday's constant F* is the amount of charge in 1 mole of electrons, or $(1.602 \times 10^{-19}$ coulomb/electron$) \times (6.02 \times 10^{23}$ electron/mole$) = 9.64 \times 10^4$C/mole. Furthermore, since 1 C (coulomb) is equivalent to 1 J/V, dividing Δg_f among the total charge, as represented by the number of moles of electrons n in the reaction multiplied by F gives the potential E in units of volts, i.e.,

$$E = \frac{\Delta g_f}{nF} \tag{15-4}$$

Taking the case of the reaction in the HFC, 1 mole of hydrogen molecules is equal to 2 moles of hydrogen atoms. Plugging known values at ambient conditions into Eq. (15-4) gives

$$E = \frac{237{,}200 \text{ J}}{(2)(96{,}400 \text{ J/V})} = 1.23 \text{ V}$$

Thus the theoretical maximum voltage from this fuel cell, assuming all energy is converted to the flow of electrons and that all the hydrogen is consumed in the reaction, is 1.23 V. Note that this value for maximum voltage is specific to the HFC; fuel cells that run on other fuels may have different maximum voltages.

We can now evaluate the theoretical maximum efficiency of the fuel cell relative to the energy available from combustion with oxygen of the same quantity of hydrogen. Using the higher heating value (HHV) for combustion, which includes the condensation of steam to liquid, the combustion of 1 mole of H_2 molecules yields $\Delta h_f = -285.8$ kJ/mole.[8] In practice, not all of the hydrogen fuel will be reacted in the fuel cell, so we take account of this limitation by factoring in the consumption rate of fuel μ_f, which is the ratio of mass of fuel consumed to mass of fuel input. Comparing energy released in the fuel cell to energy available in combustion, and including losses due to unreacted fuel, the theoretical maximum efficiency of the cell ηFC is

$$\eta_{\text{FC}} = \mu_f \frac{\Delta g_f}{\Delta h_f} \tag{15-5}$$

For a well-designed fuel cell in normal operation, a value of $\mu_f = 95\%$ may be achievable. At an operating temperature of 80°C (353°K), which is current practice with typical prototype HFCVs, change in Gibbs free energy is decreased slightly, to $\Delta g_f = -228.2$ kJ/mole. Thus the maximum achievable efficiency is

$$\eta_{\text{FC}} = 0.95 \frac{-228.2}{-285.4} = 0.759 = 75.9\%$$

This simple calculation explains much of the interest in HFCVs. If efficiency values close to the value shown can be achieved in vehicular fuel cells in the future, then the HFC will have a clear efficiency advantage over the ICE, which is limited by the Carnot limit. Note that the operation at relatively low temperature (less than the boiling point of water) is advantageous for the fuel cell, Temperature and pressure have a strong effect on the value of Δg_f, and as the temperature increases, the theoretical maximum efficiency decreases. For example, at 1,000°C, the maximum with $\mu_f = 95\%$ is $\eta_{\text{FC}} = 59\%$. Also, this comparison of HFC and ICE efficiency does not consider balance of system requirements for loads that are required to operate either technology, such as pumps for

[8]Alternatively, the lower heating value (LHV) of −241.8 kJ/mole could be used, which is the value prior to condensation. We use HHV throughout this section.

controlling oxygen and air flows in the fuel cell. A more accurate comparison would require an evaluation of the net energy available for propulsion after taking into account all required loads.

Along with the proportion of fuel consumed, the partial pressures of both reactants and products in the fuel cell has an effect on the performance of the cell. Changes to partial pressures affect the value of E as calculated in Eq. (15-4). This effect is evaluated by first setting a reference value for E, called E^0, that is equal to the ideal change in potential at ambient conditions, that is, $E^0 = 1.23$ V at 25°C, 0.1 MPa. Next, the partial pressures of reactants are introduced, P_{H2} for hydrogen, P_{O2} for oxygen, and P_{H2O} for water, each in units of atmospheres (1 atm. = 100 kPa = 0.1 MPa), for example, at ambient conditions and assuming the operation of the fuel cell as in Fig. 15-17, $P_{H2} = 1$, $P_{O2} = 0.21$, and so on. The adjusted potential E taking into account partial pressures can now be written

$$E = E^0 + \frac{RT}{2F}\ln\left(\frac{P_{H_2}P_{O_2}^{\ 0.5}}{P_{H_2O}}\right) \tag{15-6}$$

This equation is known as the *Nernst equation*. In the equation, R is the ideal gas constant, and T is the temperature of the fuel cell reaction in degrees Kelvin. Also, the partial pressure of a compound is raised to same power as the number of molecules in the reaction. Therefore, $P_{O2}^{\ 0.5}$ appears in Eq. (15-6) because there is ½ molecule of O_2 reacted per mole of H_2.

For some situations, it is useful to be able to rewrite the partial pressures of compounds as follows:

$$P_{H_2} = \alpha P$$
$$P_{O_2} = \beta P$$
$$P_{H_2O} = \delta P$$

Here P is the total pressure in atmospheres, and α, β, and δ are factors corresponding to the relative value of the partial pressures of hydrogen, oxygen, and water, respectively, compared to the total pressure. Equation (15-6) can now be rewritten to separate the natural log term into separate components:

$$\begin{aligned} E &= E^0 + \frac{RT}{2F}\ln\left(\frac{\alpha P \beta^{0.5} P^{0.5}}{\delta P}\right) \\ &= E^0 + \frac{RT}{2F}\ln\left(\frac{\alpha\beta^{0.5}}{\delta}\right) + \frac{RT}{2F}\ln(P^{0.5}) \\ &= E^0 + \frac{RT}{2F}\ln\left(\frac{\alpha\beta^{0.5}}{\delta}\right) + \frac{RT}{4F}\ln(P) \end{aligned} \tag{15-7}$$

Equation (15-7) facilitates the evaluation of the effect of changes to individual fuel cell operating parameters on the value of E. We can write the value of the change ΔE as the difference between E^{new} and E^{old}, or the value of E after and before the change in parameter, respectively, that is,

$$\Delta E = E^{new} - E^{old} = E(\alpha^{new}, \beta^{new}, \delta^{new}, P^{new}) - E(\alpha^{old}, \beta^{old}, \delta^{old}, P^{old}) \tag{15-8}$$

since the terms that do not change will cancel out. The evaluation of E using the Nernst equation and the effect of changing operating parameters is demonstrated in Example 15-7.

Example 15-7 A hydrogen fuel cell operates at 25°C and pressure of 100 kPa, using a supply of pure hydrogen and oxygen from ambient air. The partial pressure of the water that forms from the hydrogen and oxygen is 30 kPa. (a) What is the voltage drop across the fuel cell? (b) if the supply of air is replaced with a supply of pure oxygen, and nothing else changes, what is the new voltage drop?

Solution

(a) Given the initial data, partial pressures for each of the components of the reaction are $P_{H2} = 1$, $P_{O2} = 0.21$, and $P_{H2O} = 0.3$. Substituting into Eq. (15-6) gives

$$E = 1.23 + \frac{8.314(298)}{2(96,400)}\ln\left(\frac{1(0.21)^{0.5}}{0.3}\right) = 1.23 + 0.0054 = 1.24 \text{ V}$$

(b) If the fuel cell is operated with pure oxygen, $P_{O2} = 1$. Thus $\beta^{new} = 1$ while $\beta^{old} = 0.21$. Rewriting Eq. (15-8) to include only nonzero terms gives

$$\Delta E = E^{new} - E^{old} = \frac{RT}{2F}[\ln((\beta^{new})^{0.5}) - \ln((\beta^{old})^{0.5})]$$

$$= (0.5)\frac{RT}{2F}[\ln(1) - \ln(0.21)] = 0.01$$

Thus $E^{new} = E^{old} + \Delta E = 1.24 + 0.01 = 1.25$ V

15-6-2 Actual Losses and Efficiency in Real-World Fuel Cells

Real-world fuel cells in current use do not achieve the 75% efficiency value predicted above based on Eq. (15-5); instead, they achieve on the order of 50% in the best circumstances. Some losses come from incomplete fuel consumption as mentioned above. In addition, in an ideal fuel cell, constant partial pressure of hydrogen is maintained despite continuing consumption of the fuel as it is reacted across the PEM membrane. In an actual fuel cell, the amount of fuel available and hence partial pressure declines as the fuel is consumed, especially in regions close to the outlet from the cell. In this area, current will decline, so that the overall power output of the cell declines as well. Also, some of the energy released in reacting the fuel is released to the surrounding as heat, rather than being converted to electrical energy. While the theoretical upper bound on fuel cell voltage may be on the order of 1.2 V, in practice a value of less than 1 V is more realistic.

15-6-3 Implementing Fuel Cells in Vehicles

The core component of an HFCV is the *fuel cell stack* introduced above in Fig. 15-18, in which a large number of fuel cells (on the order of 120 to 200) are connected in series so as to achieve a sufficient voltage to serve the propulsion load. Once the voltage is fixed, current flow from each cell must be adequate to meet overall power requirements. Once the best possible efficiency has been achieved, the stack designer can increase current flow from a single cell by increasing its cross-sectional area, so the stack is designed both with enough cells (for voltage) and with enough area per cell (for current) to provide sufficient power. From the FC stack, electricity is delivered to the drivetrain to turn the wheels. One approach is to have the stack connected by wires to electric motors at each of the four wheels. This layout provides four-wheel drive, and also gives maximum design flexibility, since the motors are positioned close to the wheels, the stack takes up relatively little space, and the wire harnesses connecting stack to motors can be designed around other elements in the vehicle frame (e.g., dimensions of the passenger compartment). A regenerative braking system with secondary battery system for short-term charge storage like that of an HEV is also possible.

Systems for storage of hydrogen on board the vehicle are quite different from those for gasoline or diesel in an ICEV. One kilogram of pure hydrogen has approximately the same energy equivalent as 2.8 kg of gasoline (1 gal or 3.8 L at ambient temperature and pressure). Assuming efficiency gains in mature HFCVs in the future, stored mass of 5 to 6 kg hydrogen might achieve sufficient range per refueling, on the order of 400 to 500 km. At low pressures, this amount of hydrogen would take up far too much space on the vehicle, so to achieve a realistic storage volume, the hydrogen must either be stored in a liquid form at cryogenic temperatures, or compressed to 34 to 68 MPa (5,000 to 10,000 psi). These storage specifications require a mixture of efficient insulation and strong containers that can withstand the shock of accidents. They also require careful design of the interface between the vehicle and the refueling facility. In terms of primary energy source, most hydrogen would initially come from steam reforming of natural gas, so it would not provide a carbon-free energy alternative to petroleum at first. However, if hydrogen takes root as a transportation energy source, production could be shifted to carbon-free sources.

Ancillary systems on the HFCV also have different requirements. For example, with the fuel cell operating at low temperatures (on the order of 80 to 100°C), there is no longer a source of high-temperature exhaust gases for use in heating the passenger compartment as in the ICEV. Another significant issue is that the low temperature complicates heat rejection, requiring large heat exchangers. Even a 60% efficient FC delivering 20 kW has to throw away 8 kW of low-grade heat, and at peak vehicle loads the heat rejection task is much greater.

15-6-4 Advantages and Disadvantages of the Hydrogen Fuel Cell Vehicle

Assuming all technological hurdles can be bridged, the HFCV may be able to achieve a number of advantages as an alternative for vehicle propulsion. First, like the electricity grid, it can make use of a wide range of energy sources, including some that might in the future be redesigned to produce hydrogen directly, such as nuclear or solar-thermal. Next, because the refueling process resembles refueling with petroleum products, it may be faster and easier to refuel than the EV, and the hydrogen filling station of the future may resemble the liquid fuel station of today. Also, transmission and storage of hydrogen in a mature distribution network, with large intercity transmission lines and smaller urban-region distribution networks, might in the long run prove cheaper and more robust than distribution as electricity (all distribution can be carried out at low pressures). As discussed above, at low temperatures, FCs have a much higher theoretical efficiency limits than Carnot-limited ICEs. In addition, since the HFCV dispenses with the mechanical driveshaft, there is a new flexibility in locating the hydrogen storage, fuel cell stack, and network of electric motors that might enable automotive designers to radically reshape the passenger car so as to create new design possibilities.

Lastly, the presence of a high-quality electricity source on-board the vehicle in the form of the FC stack may expand the role of the vehicle as a mobile power generator.

There are also substantial challenges with the HFCV, as follows:

1. *Cost-competitive and reliable fuel cell:* At present prices per kW of capacity, fuel cell stacks are too expensive to compete with ICEs, to the point that they make the entire HFCV too expensive to market competitively with current ICEVs or HEVs. Typical ICEs cost on the order of $25 to $35/kW of capacity, or $3,000 to $4,000 per engine. Fuel cell costs dropped from $500,000/kW for NASA in the 1960s to roughly $500/kW by the year 2000, and $80 to $200 per kW in large-scale production in 2010, depending on the estimate used, but parity with ICEs has not yet been reached. Current-generation fuel cell stacks also require an occasional complete overhaul in order to function properly. Over the course of a 160,000 km (100,000 mile) vehicle lifetime, this might amount to several times having the entire fuel cell stack completely removed and refurbished. Customers would likely not tolerate this inconvenience. Platinum loading (total platinum or similar precious metal requirements) in current-generation are also higher than desired. Although major gains have been made in reducing the amount of platinum required per kW of capacity from ~100g/kW in the 1960s to 2g/kW today, at current rates, further reductions to values on the order of ~0.1g/kW are needed to make the technology sustainable in terms of platinum availability.
2. *Safe and cost effective storage of hydrogen on board:* Cryogenic cooling and storage of hydrogen incurs a large energy penalty (on the order of 30% of the initial energy in the hydrogen), so compressed hydrogen storage is the preferred option at present. However, storage at very high pressures requires containers that will not rupture in an accident and that are not vulnerable to fire. Dispensing the fuel, while faster than taking hours to charge an EV, is a riskier and more complex task than with an ICEV, perhaps requiring professional attendants at all hydrogen refueling facilities.
3. *New infrastructure needed to distribute and dispense hydrogen:* Although energy companies have experience with making and handling hydrogen, as a society we have very few of the infrastructure pieces in place to use hydrogen as a transportation fuel on a large scale. This includes an adequate number of manufacturing plants, a distribution grid, and dispensing stations. Although some demonstration filling stations have been opened, energy companies and vehicle makers have been criticized for launching public refueling stations too soon, when the basic viability of the vehicles themselves has not yet been proven (Fig. 15-19).

Although the experience of launching the hydrogen station provided proof-of-concept that a prototype fleet of HFCVs could be refueled on a continuing basis at a commercial outlet, it did not by itself launch a local market for HFCV purchases among the general public.

Given the extent of these challenges and the emergence of competing technologies such as the PHEV or cellulosic ethanol, it is possible that we will create an end-game technology that runs on stored electric charge from non-fossil sources whenever possible, and on advanced biofuels when not, so that there will be no place for the HFCV in the future. On the other hand, if other technologies such as PHEVs cannot completely solve the sustainability problem, the HFCV may yet emerge as the endpoint technology

Figure 15-19 Co-author Vanek at hydrogen dispenser at Shell's Benning Road station in Washington, D.C., November 2006.

that is technically superior to all other options, and eventually becomes the permanent solution for sustainable transportation energy. Such an evolution would likely take a long time, on the order of decades. In the meantime, we can expect that vehicle makers and others will continue R&D at a steady pace to improve fuel cells and other components of this system.

One potential outcome in the short to medium term is that, rather than either the PHEV or HFCV emerging as the single solution to replace the gasoline-fueled ICEV, the two incoming technologies might exist side by side, each creating a niche around its respective advantage. Vehicle owners who rarely travel long distances might opt for PHEVs and use primarily the charging capability, while only occasionally using the liquid fuel option (possibly in the form of a biofuel) to travel long distances. HFCVs would then enter the market by appealing to drivers who frequently travel long distances: if a hydrogen refueling network can grow up alongside the market penetration of these vehicles, HFCV drivers would rely entirely on hydrogen for both short and long trips, simplifying their fuel purchasing requirements. For example, General Motors is developing a fuel cell version of its Chevrolet Equinox crossover sport utility vehicle (ICEV version curb weight approximately 3,800 lb, depending on the model) with a focus on this niche.

15-7 Concluding Discussion of Options: EVs, PHEVs, V2G, and HFCVs

The electrification of the LDV fleet faces several interesting options and tradeoffs. One of these is the choice of EVs versus PHEVs, and whether the extra complexity and cost of the PHEV brings gains that justify it. In the short run, it is consumers who will decide which option grows faster, with EVs like the Nissan Leaf and PHEVs like the Chevy Volt selling in the thousands of units per year. Longer term, though, it is relevant to transportation energy policy to evaluate the strengths and weaknesses of either option.

In the preceding example of V2G deployment, the EV was used as a basic building block, since it is simpler and clearly delineates the role of the grid as unique original provider of energy, and the vehicle as buyer and seller of electricity, depending on conditions in the moment. Under this scenario, the EV cannot call on an on-board ICE to provide propulsion, so it must always retain a sufficient amount of charge to meet the driver's basic driving range needs. This constraint does not affect the PHEV, assuming it has liquid fuel available (which in the V2G system the ISO might monitor as well as available charge). On the other hand, the EV is simpler (having a drivetrain with just battery, motor, controller, and transmission) and as a result could be expected to be cheaper: as examples, in 2013 the Leaf listed for about $30,000 but the Volt for about $38,000.[9] PHEVs incur a cost penalty despite having smaller battery packs (the necessary range may be only 50 miles for a PHEV versus 80 to 100 miles for an EV) because of the added cost of the fuel-delivery system, the ICE, and a more complex drivetrain that integrates both ICE and motor power. More equipment on board may also imply less space for passengers and storage: the Volt has only four seats versus five for the Leaf.

PHEVs also have advantages in the long run that may give them an edge in the alternative fuel vehicle market. First, for the extra price, PHEVs give the driver the capability for both urban and long-distance intercity travel; the latter is not available to the EV driver unless a swapping station system becomes widespread. To overcome this problem, Tesla Motors is installing a network of fast charging stations that would allow Tesla drivers to stop for a 20-minute recharge every 200 miles to make long-distance travel possible. Drivers may find even this relatively convenient arrangement for recharging too time consuming, which would limit its ability to compete with PHEVs, HEVs, or ICEVs. The PHEV is also more resilient in times when both the grid and the liquid fuel supply system face uncertainties: if either liquid fuel from refineries or charge from the grid are not available, the vehicle can access the other energy source.

Lastly, under V2G, PHEVs might be able to sell more electricity to the grid because if the demand for energy from vehicles were high, PHEVs could always drain batteries more and then call on the liquid fuel and ICE for power until the next opportunity to recharge the battery. However, this scenario must make economic sense from the vehicle owner's perspective; that is, the price paid per kWh of electricity must be higher than the equivalent cost of driving the vehicle using liquid fuel, which is a high economic hurdle given that liquid fuels for transportation are a relatively expensive per unit of energy.[10] The PHEV owner would therefore likely only sell to the grid at times when the cost paid per kWh were high, such as during peak demand periods. Also, the smaller storage capacity of the PHEV battery system might work against V2G participation.

---

[9]Prices for 2013 models with mid-level options taken from Kelley Blue Book (*www.kbb.com*).

[10]Suppose a PHEV owner buys either electricity at $0.14/kWh or gasoline at $4/gal ($1.06/L). If the PHEV has a fuel economy of either 3 miles/kWh or 35 miles/gallon, the owner will forego the opportunity to drive on electric charge at 4.7 cents/mile and instead drive on gasoline at 11.4 cents/mile, unless the owner can sell the electricity at a premium price much higher than $0.14/kWh.

15-7-1 Advantages and Disadvantages of Adding V2G Capability

In an analogy with the discussion of EVs versus PHEVs, the V2G concept adds complexity compared to a grid-vehicle system where charge flows only to the system, but also provides advantages. It is assumed here that the starting point for any system is a modernized smart grid capable of optimally charging EVs in a way that the charging occurs without straining the maximum capacity of the grid, preferably when other types of demand are low, such as at night. Thus large-scale rollout of electrified LDVs would go hand in hand with the upgrading of the grid, and events like the 2003 Northeast blackout would be extremely unlikely to occur.

In terms of disadvantages, the adoption of V2G would add a layer of complexity to the grid upgrade beyond the level required for large-scale charging of electrified LDVs without V2G. For example, one-way charging of vehicles requires charge control at each charge point and wireless communication with the ISO to regulate the flow of energy so that overall grid load does not increase excessively. V2G requires an additional inverter to take DC charge out of the battery system in the vehicle and transmit it to the grid at the correct AC oscillation to match the grid AC; thus the no-V2G option might be substantially cheaper. Indeed, smart charging without V2G might achieve many of the same benefits as V2G, because the ISO could detect either peaking in intermittent renewable source availability (solar and/or wind), or else peaks in available capacity at large dispatchable conventional plants due to fall-off in demand, and accelerate the charging of the network of connected vehicles to absorb this extra capacity. V2G might also raise security concerns, since the need to coordinate both flow to the vehicle when charging and flow to the grid when discharging might make the system more vulnerable to malfeasance intended to disrupt the normal function of electricity and transportation systems.

In its favor, V2G's main advantage is that it takes an expensive, underutilized resource, namely the battery systems of the vehicles in the network, and wrings more utilization out of them.[11] This change benefits first the individual vehicle owner, who has the opportunity to generate a revenue stream to offset the substantial additional cost of the battery system. Alternatively, the vehicle owner might not take ownership of the batteries at all, with the grid operator or some other entity retaining ownership of the battery system, so that the cost of the vehicle would resemble that of an ICEV, potentially with lower energy cost per unit of distance driven.

There is also a system-wide benefit from the use of the batteries in the V2G system. In effect, they become part of the network of storage systems along with pumped storage, flywheels, etc., discussed earlier, thus reducing the need for investment in expensive stationary energy storage systems that do not have the second purpose of providing propulsion to vehicles.

15-7-2 Comparison of Electricity and Hydrogen as Alternative Energy Sources

Although at present, the vehicle market emphasizes the launch of EVs and PHEVs, it is entirely possible that in the future a similar launch of commercial hydrogen vehicles will arise, with many of the same systems-level characteristics of electrification as

[11]The battery system is underutilized in the sense that it adds nearly $10,000 to the cost of a mass-market EV like a Nissan Leaf, yet over a 24-hour period an LDV is typically unoccupied and turned off for 90 to 95% of the time.

described in the preceding section. Like electrification, hydrogen might be derived from a wide range of primary energy sources, leading to guaranteed supply and stable prices. Primary energy sources could be converted to hydrogen and transported, possibly over long distances, to transportation energy markets, or else the energy could be transmitted as electricity and converted to hydrogen close to point of consumption. With hydrogen, there would be no tailpipe emissions except water. Regarding CO_2 emissions, currently most hydrogen comes from steam reformation of natural gas and is therefore not carbon-free, but in the future there could be carbon-free hydrogen from various sources, including nuclear, renewables, and fossil with CCS.

The main limitations for hydrogen in competition with electrification mentioned above include (1) cost-effective and reliable fuel cell stacks, (2) safe and cost-effective on-board hydrogen storage, and (3) the infrastructure needed to deliver hydrogen to dispensing stations. These disadvantages dictate that in the short run, electrification will be the alternative energy source of choice, along with biofuels such as ethanol and biodiesel (see Chap. 16). Focusing on the third limitation, the petrochemical industry has some amount of hydrogen production capacity for use in industrial processes, but it is not positioned to deliver hydrogen to widely dispersed LDV fleets, especially in terms of delivering hydrogen long distances from plants to urban population centers. The addition of hydrogen-dispensing capacity at filling stations around the United States as well as in other countries would cost in the billions of dollars, whereas an EV or PHEV owner can charge at home today with a 240-V charger that costs on the order of $1,000 installed. Furthermore the existing transmission and distribution network already exists for the grid; some modifications to the distribution network close to residence might be needed to handle a large number of EVs or PHEVs receiving and sending high currents simultaneously during peak charging and discharging times, but this adaptation is not on the same scale as building a complete national hydrogen grid.

In longer terms, electrification has disadvantages and limitations that may create an opening for hydrogen to compete. Lack of range at an affordable price is an issue for EVs even with the most advanced batteries that have the highest specific energy (kWh storage per kg of mass), especially for vehicles on the heavier end of the spectrum such as light trucks. Limits on the availability of lithium mentioned above may eventually limit the growth of EVs. Lastly, hydrogen distribution and storage is expensive at present, but if perfected on a large scale might one day be cheaper in terms of cost to the vehicle operator than using electricity.

What may eventually emerge from the evolution of alternatives to petroleum is a system where both electricity and hydrogen play a role. For small, lightweight vehicles that move primarily in urban areas, the electric system once mature may prove an ideal resource for which no further conversion to hydrogen is necessary. For larger LDVs or those that primarily travel long distances, hydrogen when and if it matures may be the more suitable fuel.

15-8 Summary

This chapter has presented the battery-electric vehicle platform and discussed recent developments, including the launch of a new generation of both mainstream and luxury EVs and plug-in hybrid electric vehicles (PHEVs) that seek to grow the market for electrified vehicles alongside the incumbent ICEV and HEV technologies. Since driving range per charge is such an important concern for EVs, a model of range as a function of capacity was presented, which showed that although any EV will have difficulty competing with the ICEV/HEV on range per charge, the charge density of lithium-based battery technology over other alternatives (e.g., lead-acid) can make a significant difference. Because of concern about EV range, PHEVs are proposed as an alternative that blends the low energy cost per mile and lack of emissions of the EV for short trips with the longer range of the HEV for intercity trips.

This chapter has also reviewed electrification at a systems level, specifically regarding light-duty highway vehicles. Benefits include a diversity of primary energy sources and relative price stability. They can be obtained by different means, such as the deployment of either EVs or PHEVs as the chargeable vehicles, or the development of a smart grid for charging vehicles either with or without vehicle-to-grid capability. For electrification to succeed, the components of the existing grid system, including the different types of either intermittent or dispatchable electricity generating assets and electrical energy storage systems, would require coordination that takes into account EV presence. The grid components, along with the home- and workplace-based charge points, network of electrical transmission and distribution lines, and the independent system operator together form the complete system which provides energy to both vehicles and other types of electrical loads. The electrification of the vehicle fleet provides both near-term opportunities for using existing extra capacity at night for recharging vehicles, and longer-term opportunities to develop sustainable, carbon-free electricity sources such as renewable energy in tandem with the rollout of electric vehicles. In addition, V2G capability would allow two-way flow of energy between vehicles and the grid so that V2G-enabled vehicles can act as energy storage units as well as grid energy purchasers. Lastly, many of the benefits of electrification would apply to a transition to hydrogen, such as the diversification of energy sources, since the hydrogen would be produced from a variety of sustainable, carbon-free sources.

References

Albertus, P., J. Couts, and V. Srinivasan (2008). "A Combined Model for Determining Capacity Usage and Battery Size for Hybrid and Plug-in Hybrid Electric Vehicles." *Journal of Power Sources* 183:771–782.

Grainger, J. and W. Stevenson (1994). *Power Systems Analysis*. McGraw-Hill, New York.

Hoerig, C., D. Grew, and H. Munedzimwe (2010). *New York State Wind Energy Study: Final Report.* Cornell University Master of Engineering team thesis project. Electronic resource, available at *www.lightlink.com/francis/*. Accessed Sep. 27, 2013.

Husain, I. (2010). *Electric and Hybrid Vehicles: Design Fundamentals, 2d ed.* CRC Press, Boca Raton, FL.

Kannan, G., P. Sasikumar, and K. Devika (2010). "A Genetic Algorithm Approach for Solving a Closed Loop Supply Chain Model: A Case of Battery Recycling." *Applied Mathematical Modeling* 34: 655–670.

Kempton, W. and J. Tomic (2005*a*). "Vehicle-to-Grid Power Fundamentals: Calculating Capacity and Net Revenue." *Journal of Power Sources* 144, 268–279.

Kempton, W. and J. Tomic (2005*b*). "Vehicle-to-Grid Power Implementation: From Stabilizing the Grid to Supporting Large-Scale Renewable Energy." *Journal of Power Sources* 144:280–294.

Larminie, J. and J. Dicks (2000). *Fuel Cell Systems Explained*. John Wiley, Chichester, West Sussex.

Sperling, D. and J. Cannon, eds. (2004). *The Hydrogen Energy Transition: Moving toward the Post-Petroleum Age.* Elsevier, Amsterdam.

Spiegel, C. (2007). *Designing and Building Fuel Cells.* McGraw-Hill, New York.

USDOE (2009). "Hydropower Resource Potential." Office of Energy Efficiency and Renewable Energy, U.S. Department of Energy, Washington, DC.

USEIA (2012). *Annual Energy Outlook.* U.S. Energy Information Administration, Washington, DC.

Wood, A. and B. Wollenberg (1996). *Power Generation, Operation, and Control, 2d ed.* John Wiley & Sons, New York.

Further Readings

Braess, H. and U. Seiffert, eds. (2004). *Handbook of Automotive Engineering. Society of Automotive Engineers,* Warrendale, PA.

Cengel, Y. and M. Boles (2010). *Thermodynamics: An Engineering Approach, 7th ed.* McGraw-Hill, Boston.

Commission on Engineering and Technical Systems (2000). *Review of the Research Program of the Partnership for a New Generation of Vehicles,* 6th Report. National Academies Press, Washington, DC.

Davis, S. and S. Diegel (2013). *Transportation Energy Data Book, 33d ed.* Oak Ridge National Labs, Oak Ridge, TN.

DeCicco, J. (2004). "Fuel Cell Vehicles." *Encyclopedia of Energy,* Vol. 2. Elsevier, Amsterdam.

Doerffel, D. and S. Abu-Sharkh (2007). "System Modeling and Simulation as a Tool for Developing a Vision for Future Hybrid Electric Vehicle Drivetrain Configurations." *Proceedings of 2006 IEEE Vehicle Power and Propulsion Conference,* Winsor, UK.

Fay, J. and D. Golomb (2002). Energy and the Environment. Oxford University Press, New York.

Galloway, R. and S. Haslam. (1999). "The ZEBRA Electric Vehicle Battery: Power and Energy Improvements." *Journal of Power Sources,* 80:164–170.

Gao, Y. and M. Ehsani (2010). "Design and Control Methodology of Plug-in Hybrid Electric Vehicles." *IEEE Transactions on Industrial Electronics,* 57(2):633–640.

General Motors Corp. (2001). *Well-to-Wheel Energy Use and Greenhouse Gas Emissions of Advanced Fuel/Vehicle Systems,* North American Analysis. Published by Argonne National Laboratories, Argonne, Ill. Available at: *http://www.ipd.anl.gov/anlpubs/2001/08/40409.pdf.* Accessed Sep. 14, 2007.

Gillespie, T. D. (1999). *Fundamentals of Vehicle Dynamics.* Society of Automotive Engineers, Warrendale, PA.

Kessels, J., B. Rosca, and H. Bergveld (2011). "On-Line Battery Identification for Electric Driving Range Prediction. *IEEE Explore Library.*

Kroeze, R. and P. Kreine (2008). "Electrical Battery Model for Use in Dynamic Electric Vehicle Simulations." *IEEE Transactions*:1336–1342.

NYSERDA (2009). *Annual Energy Data Report,* New York State Energy Research & Development Authority, Albany, NY.

Randolph, J. and G. Masters (2008). *Energy for Sustainability.* Island Press, Washington, DC.

Romm, J. (2004). *The Hype about Hydrogen: Fact and Fiction in the Race to Save the Climate.* Island Press, Washington, DC.

Somayajula, D., A. Meintz, and M. Ferdowsi (2008). "Study on the Effects of Battery Capacity on the Performance of Hybrid Electric Vehicles." *IEEE Vehicle Power and Propulsion Conference.*

Song, K., J. Zhang, and T. Zhang (2011). "Design and Development of a Pluggable PEMFC Extended Range Vehicle." *IEEE Explore Library.*

Sorensen, B. (2002). *Renewable Energy: Its Physics, Engineering, Use, Environmental Impacts, Economy and Planning Aspects, 2d ed*. Academic Press, London.

Sperling, D. and D. Gordon (2008). *Two Billion Cars: Driving toward Sustainability*. Oxford: Oxford University Press.

Stone, R. and J. Ball (2004). *Automotive Engineering Fundamentals.* SAE International, Warrendale, PA.

Tester, J., E. Drake, and M. Driscoll (2012). *Sustainable Energy: Choosing among Options, 2d ed.* MIT Press, Cambridge, MA.

Vanderburg, W. (2006). "The Hydrogen Economy as a Technological Bluff." *Bulletin of Science, Technology, and Society* 26(4):299–302.

Vanek, F., S. Galbraith, and I. Shapiro (2006). *Final Report: Alternative Fuel Vehicle Study for Suffolk County*, New York. Technical Report, New York State Energy Research and Development Authority (NYSERDA), Albany, NY.

Vanek, F., L. Albright, and L. Angenent (2012). *Energy Systems Engineering: Evaluation and Implementation, 2d ed.* McGraw-Hill, New York.

Wark, K. (1983). *Thermodynamics, 4th ed*. McGraw-Hill, New York.

Exercises

15-1. A future midsize EV has a production cost and weight, before adding any batteries, of $18,000 and 750 kg, respectively. The batteries are available with a charge density of 120 Wh/kg and a cost of $350/kWh. The energy intensity of the vehicle is 0.11 Wh/kg-km, and the maximum depth of discharge is 90%. (a) Produce two figures for this vehicle, one plotting vehicle range as a function of total mass, and the other plotting total cost as a function of range. (b) Compare this vehicle to an ICEV in the same class that costs $30,000 and has a typical range per tank of 350 km. What would the EV cost if it were to have the same range? By what percent would the range be reduced compared to the ICEV if it had the same cost?

15-2. A fleet manager is considering replacing a fleet of gasoline ICEVs with equivalent EVs. One factor in the decision is the CO_2 emissions from the vehicles. Each vehicle in the fleet drives an average of 10,000 miles/year, and the ICEVs have an average fuel economy of 20 mpg. The comparable EV consumes 274 Wh of electricity per mile. Use U.S. national average emissions factor from Table 15-2. You can assume that upstream CO_2 emissions (i.e., either from extracting petroleum and converting it to gasoline provided at the pump, or from extracting and transporting coal or gas to the electric plant) are even between the two options and that therefore they are outside the scope of your analysis. (a) How many pounds of CO_2 does each vehicle emit per year? (b) Does the EV reduce CO_2 emissions? (c) Repeat the calculation for your own location, based on local emissions per kWh of electricity. Does the result in (b) change?

15-3. You are considering two alternatives, either an ICEV, which includes initial cost of the car and gasoline cost over its lifetime, or an EV, which includes the cost of the car without batteries, the cost of the batteries, and the cost of electricity. For either option, assume a discount rate of 7%, that any capital investment in cars or battery systems will have $0 salvage value at the end of their investment lifetime, that vehicles are owned for 10 years, and that the demand is

8,000 miles per year. Ignore maintenance, insurance, and other ownership costs. For the EV, assume a battery life of 40,000 miles, and a required available capacity of 25 kWh. The maximum depth of discharge is 85%. The car without the battery system costs $18,000. Electricity costs $0.16/kWh, and the EV drives 2.7 mile/kWh consumed. The charging efficiency is 90%, meaning that there are 10% losses between the amount of electricity purchased and delivered to the battery system. For the ICEV, the purchase cost is $25,000, the fuel economy is 25 mpg, and the cost of gas is $4/gal. How much does the battery have to cost per kWh of storage capacity to make the EV cost the same as the ICEV?

15-4. A fuel cell stack consists of 75 fuel cells arranged in series, each one 12 cm^2 in size. The stack operates at 80°C and the amount of current generated is 4.257 A/cm^2. The fuel cell operates at a system pressure of 1 MPa using a supply of pure hydrogen and oxygen in ambient air. The partial pressure of the water in the outflow is 0.5 MPa. Assume any losses to be negligible and that the voltage at which it operates is the open circuit voltage; in reality, the voltage would be lower due to the current flow. What is the output of the fuel cell stack when operating at full capacity, in kW?

15-5. An electric vehicle has a battery system with rated capacity of 24 kWh and maximum depth of discharge of 80%. For comparison, a hatchback ICEV has a 12-gal tank and averages 31 mpg overall fuel economy. (a) Suppose the system is discharge to its maximum depth and I wish to fully recharge this system in 20 minutes, with a 480-Vcharger. Assuming the battery is charged at a constant rate from start to finish (i.e., linearly; this is a simplification since batteries actually have a variable charging rate as the battery approaches fully charged), what is the energy flux in kilowatts and the current in amperes? (b) Now suppose instead the battery charges at typical household wall outlet conditions, 20 amperes current at 120 V. How long does it take to charge in the same situation? (c) Next, suppose the hatchback starts on a full tank, drives 285 miles, and then stops to refuel. If the gas station pump averages 5 gal/minute, how long does it take to fill up the tank? Give your answer to the nearest tenth of a minute. (d) Short answer: What conclusion could you draw in comparing the refueling of EVs versus ICEVs from the answers in parts a to c? Comment in one or two sentences.

15-6. Consider total CO_2 emissions from the PHEV car in Example 15-2, compared to an HEV which is identical except that it uses only gasoline. The CO_2 emissions from electricity generation are the same as in Table 15-2. Assume 88% well-to-tank efficiency for gasoline usage, 11% losses in transmission and distribution of electricity from the generating plants to the vehicle, and 5% average energy losses upstream from the various electric power plants. Furthermore, assume that CO_2 emissions are proportional to energy losses. (a) Does the PHEV reduce energy consumption or CO_2 emissions compared to the HEV? (b) If the electricity is generated 1/3 from gas and 2/3 from coal, does the answer in (a) change?

CHAPTER 16

Bioenergy Resources and Systems

16-1 Overview

This chapter focuses on bioenergy generation from biomass (mostly plant material) or bioenergy generation with a bioprocessing step that uses microbes, including yeast, algae, or bacteria, to convert inorganic or organic materials into an energy carrier. We predominantly discuss technologies that are already used at an industrial scale or that are very close to be implemented. It is important to note that we cannot be comprehensive—there are simply too many technologies that generate bioenergy. On the other hand, we do discuss anaerobic gasification (anaerobic means without oxygen), because of the possibility to generate liquid biofuels as an energy carrier from one of the gasification products. We explain several bioprocessing systems with yeast or bacteria that generate biofuels. We also describe the production of the gaseous energy carriers—methane and hydrogen. We conclude with a discussion of the integration of bioenergy resources into the energy supply for specific transportation modes.

16-2 Introduction

Biofuels for transportation include any products derived from living organisms, ranging from food crops to plant and tree material to microbes, such as algae, which can be processed into substitutes for liquid transportation fuels. The use of biofuels for transportation dates back to the early years of the modern transportation vehicles, when Rudolf Diesel, who is the inventor of the diesel engine [i.e., compression-ignition (CI) engine], used peanut oil as a fuel in his early engine prototypes. One of Diesel's motivations was to create a source of mechanical power that could utilize a wide variety of fuels, so that small businesses of the day would not be captives of the coal industry for their energy supplies. In recent years, interest in biofuels has surged as nations, such as Brazil, Germany, and the United States, have sought to reduce petroleum imports. Also, many countries see biofuels as a way to prepare for anticipated dwindling petroleum reserves by developing a renewable alternative. For more information on biofuels, see, for example, Drapcho et al. (2008); Nag (2008); Olsson (2007); Sorensen (2002) Chap. 4; or Tester et al. (2005) Chap. 10.

To provide real benefit to society, a biofuel must achieve at least the following two objectives:

1. It must measurably reduce emissions of CO_2 and other pollutants compared to petroleum when taking into account life-cycle energy inputs and emissions. These inputs include energy and emissions resulting from growing and harvesting crops, processing crops into finished fuels, and transportation of raw materials and finished fuels.
2. It must be available in sufficient quantities to displace a measurable fraction of the fuel currently derived from petroleum without curtailing sufficient grain for nutritional purposes. It is especially important to safeguard the survival of the poorest people in the world, who depend on food imports at times when locally produced food supplies fall short. Thus, even if it were economically attractive to shift a large part of the grain harvest of countries such as the United States from exports to domestic biofuel production, it would be morally untenable to pursue such a policy.

The presence of a positive net energy balance (NEB) ratio is often used as a criterion for evaluating a biofuel, meaning that a biofuel that delivers more energy to the vehicle than it requires in the production life cycle. A biofuel with a considerably positive NEB ratio is considered sustainable. While useful as a surrogate for a more complete analysis of environmental benefit and impact, the NEB ratio is imperfect because it does not consider the relative CO_2 emissions of the energy source (e.g., coal vs. natural gas for process heat). It also does not consider land use changes that might be necessary for producing the biofuel; for example, clearing forests to grow crops for biofuels releases a large quantity of CO_2. Thus, if possible, measures of the target environmental concern (CO_2, air pollutants, water usage, and pollution) should be sought to evaluate the net environmental benefits of a biofuel. NEB ratios are also subject to change over time due to improvements in technology and agricultural or industrial practices, so it is important to work with the most up-to-date values available when assessing biofuels. For example, the NEB ratios for the corn-to-ethanol industry in the United States have been a moving target and are now more favorable than 10 years ago due to improved processing technologies at the ethanol plant, including lower energy needs for distillation of ethanol—the most energy-consuming process for ethanol production.[1]

16-2-1 Policies

Developing a bioenergy industry cannot be uncoupled from local, national, and/or international policies. Communities, societies, and intergovernmental organizations may have very different reasons why they chose to develop bioenergy system rather than to rely on fossil sources for energy carriers. These reasons may include:

1. *Environmental:* To tackle problems, such as air pollution of fossil fuel use (including greenhouse gas release), or impact from oil exploration.

[1]A recent report by Shapouri et al (2010) has estimated the most current NEB ratios for the United States corn-to-ethanol industry for 2008.

2. *Socio-economical:* To equalize the difference between the rich and the poor by focusing on decentralized renewable energy systems rather than centralized fossil fuel exploration by multinational organizations.
3. *Geostrategical:* To prevent national dependency on fossil fuels from other nations or intergovernmental organizations, such as the organization of petroleum exporting countries (OPEC). A society could also chose a laissez-faire policy to have the economic market dictate whether energy should come from fossil or renewable sources.

It is important to note that bioenergy research, development, and production occurs within a complex and dynamic policy environment. Policies shape the bioenergy industry mainly through allocation of government funds and renewable energy mandates. Government programs, such as the U.S. Department of Energy Biomass Program, provide direct funding to biofuel companies, amounting to federal investment in bioenergy. Other bioenergy investments are made in the form of feed-in tariffs, which are essentially long-term government contracts for renewable energy producers that include cost-based compensation (e.g., in the form of a set price for a biofuel for a certain length of time). This directly decreases the pressure of competition and allows renewable energy companies to grow and develop. Feed-in tariffs also encourage technological advancement and cost reduction through a planned and gradual decrease in the price paid for the energy. The feed-in tariff system has been particularly successful in Germany and has resulted in dramatic increases in renewable energy production (Mitchell et al., 2006).[2]

In contrast, mandates do not directly support bioenergy development, although they encourage investment from private interests and other government programs. For example, the European Union has passed a directive requiring each member state to obtain at least 10% of total energy used from renewable sources by 2020. Some European Union members have also made more ambitious renewable energy goals, such as Denmark, which has mandated a minimum of 30% renewable energy by 2020. These mandates will make renewable energy companies more attractive to investors and more competitive in markets. However, not all policies are favorable toward bioenergy development. On a broad scale, subsidies and tax benefits for fossil fuel companies make it more difficult for biofuels to compete. Indeed, the Organization for Economic Co-operation and Development has estimated that removal of fossil fuel subsidies in some countries could lower global greenhouse gas emissions by up to 10% through a combination of energy conservation and increased use of renewable energy.[3] In addition, the bioenergy industry is subject to all zoning and local regulations just like any other industry. This means that before embarking on any bioenergy enterprise it is always important to consider the current policy situation to determine the technologies and locations that make sense not only scientifically and industrially, but also politically.

16-2-2 Net Energy Balance Ratio and Life-Cycle Analysis

The *NEB ratio* is a metric used to evaluate bioenergy systems by comparing the energy available for consumption in the produced energy carrier (e.g., ethanol) to the energy

[2]Mitchell, Bauknecht, and Connor (2006) discuss the effectiveness of the German feed-in tariff policy.
[3]This OECD paper describes the effects of removing subsidies for fossil fuels for bioenergy production.

(e.g., petrodiesel and natural gas) required for growing, harvesting, and processing. For renewable energy systems, the available energy is seen as renewable while the consumption of energy is often nonrenewable. A NEB ratio greater than one indicates that more energy is available for consumption than is required to produce the biofuel, while a NEB ratio less than one indicates that more energy is required to produce the fuel than is available for consumption, which is unattractive. This metric has been widely used in the biofuel debate, but as we mentioned earlier, cannot stand alone. *Life-cycle assessment (LCA)* is a method of product assessment that considers all aspects of product's life cycle. It is also referred to as cradle-to-grave analysis, and evaluates a product from raw material extraction through disposal and/or recycling. Life-cycle assessment was developed to account for additional factors missing from conventional economic analysis, such as environmental impacts of the production, use, and disposal of a product. For example, think about a grocery store in Virginia that needs to decide whether to purchase oranges from a farm in Florida or a farm in California. Using traditional economic analysis, the grocery store would evaluate the cost of the oranges (including transportation costs), while a life-cycle assessment would include additional factors, such as increased greenhouse gas emissions associated with shipping the oranges across a larger distance (California vs. Florida). As our society becomes increasingly aware of environmental issues, life-cycle assessment provides a tool for a holistic comparison of alternative processes. Example 16-1 explores life-cycle assessment considerations.

Example 16-1 A corn-to-ethanol plant is located 40.2 km (25 miles) from farm A and 160.9 km (100 miles) from farm B. Farm A will sell corn to the plant for $289.36/metric ton ($6.30/bushel), while farm B will sell corn at a price of $248.02/metric ton ($5.40/bushel). Each truckload can carry 10.9 metric tons (500 bushels), and the truck emits 212.3 g CO_2eq/metric ton-km (310 g CO_2eq/ton-mile). The plant needs 130.6 metric tons/year (6,000 bushels/year), and the truck weighs 9.1 metric tons empty. Examine the two farms, considering both economics and greenhouse gas emissions. From which plant is it better to buy corn grain?

Solution The corn-to-ethanol plant needs

$$130.6 \text{ metric tons/year @ } 10.9 \text{ metric tons} = 12 \text{ truckloads/year}$$

The economic return for farm A is

$$130.6 \text{ metric tons/year} \times \$289.36\text{/metric ton} = \$37{,}800\text{/year}$$

The total length for transportation between farm A and the plant for one way is

$$40.2 \text{ km/trip (one-way)} \times 12 \text{ trips/year} = 482.4 \text{ km/year full (same amount empty to return)}$$

Greenhouse gas emissions for farm A are then a combination of transportation with an empty truck

$$482.4 \text{ km/year} \times 9.1 \text{ metric tons} \times 212.3\ CO_2\text{eq/metric ton-km} = 0.93 \text{ Mg } CO_2\text{eq}$$

and a full truck

$$482.4 \text{ km/year} \times 20 \text{ metric tons} \times 212.3\ CO_2\text{eq/metric ton-km} = 2.05 \text{ Mg } CO_2\text{eq}$$

The economic return for farm B is

$$130.6 \text{ metric tons/year} \times \$248.02\text{/metric ton} = \$32{,}400\text{/year}$$

The total length for transportation between farm B and the plant for one way is

$$160.9 \text{ km/trip (one way)} \times 12 \text{ trips/year} = 1{,}930.8 \text{ km/year full (same amount empty to return)}$$

Greenhouse gas emissions for farm B are again a combination of transportation with an empty truck

$$1{,}930.8 \text{ km/year} \times 9.1 \text{ metric tons} \times 212.3 \text{ CO}_2\text{eq/metric ton-km} = 3.72 \text{ Mg CO}_2\text{eq}$$

and a full truck

$$1{,}930.8 \text{ km/year} \times 20 \text{ metric tons} \times 212.3 \text{ CO}_2\text{eq/metric ton-km} = 8.18 \text{ Mg CO}_2\text{eq}$$

Discussion Based purely on the corn price difference between farm A and B, choosing farm B over farm A will result in a 14% economic savings per year ($5,400), but will result in a 300% increase (9 Mg) in annual greenhouse gas emissions. Obviously it will also be more expensive to drive a longer distance, but the cost for that may not be as considerable as the increase in greenhouse gas emissions. Regardless, from this simple assessment it seems reasonable that most corn-to-ethanol plants are located in the growing areas of corn.

16-2-3 Productivity of Fuels per Unit of Crop Land per Year

Before choosing a regional crop for biofuel production it is important to understand the productivity of fuel per unit of cropland per year. This analysis does not only take into account the yields of the crop at the regional soil and climate conditions, it also includes the efficiencies of harvest and the crop-to-fuel conversion technology used. This is especially important because of a relatively low retention rate of solar energy in plants and trees for later conversion into biofuels. Of the incoming energy, which may average 100 to 250 W/m^2 year round, taking into account diurnal cycles and variations in regional climate and latitude, less than 1% is available in the starches or oils as a raw material for conversion to fuel. This limit affects the ability of agriculture or forestry to provide sufficient raw materials to meet the greater part of the world transportation energy demand, especially as our societies must also generate enough nutrition to feed the world. In addition, because the crops for biofuels accumulate energy at such low density, a large energy expenditure is required to gather the material together for processing, which hurts the balance of energy produced relative to energy input requirements (i.e., NEB ratio). Much research work is therefore under way to find ways to capture more of the solar energy in the biofuel crop and to use more of the biomass (i.e., lignocellulose) of the entire plant as a raw material. Breakthroughs in research along these lines are in fact critical if biofuels are to become a major source of energy for transportation; otherwise, using current crops and technology, they can play at best a moderate role in displacing petroleum consumption and reducing CO_2 emissions.

It is also important to note that simply integrating crop yields per acre with harvest and conversion efficiencies is not enough. The decision maker should include data on the changes to the land that must be made to sustain a large-scale biofuels program. Noteworthy is a 2008 study,[4] which showed that the conversion of nonagricultural land such as rainforests, peatlands, savannas, or grasslands to

[4]Fargione et al. (2008) estimated the carbon debt for land clearing activities.

produce crop-based biofuels in Brazil, Southeast Asia, and the United States creates much more CO_2 (i.e., carbon debt) than the annual greenhouse gas reductions that these biofuels would, for a very long time, provide by displacing fossil fuels (i.e., carbon credit). This vast release of CO_2 during land changes is from burning biomass or slow microbial decomposition of organic carbon stored in biomass (including roots) and soils. Such a CO_2 release does not occur when biomass is grown on degraded and abandoned agricultural lands. Similarly, the use of organic wastes to generate biofuels does also not have a carbon debt and offers immediate greenhouse gas credits. Besides the carbon debt from land conversion, the decision maker should also take into consideration data on the quantity of inputs of water, nutrients, and nonrenewable energy that is necessary to grow the bioenergy crops to account for the CO_2 release during the agricultural activity. Taking the carbon credit for both land changes and for agricultural activity into account, analyses may show that it is, from a greenhouse gas reduction standpoint, better to maintain low-intensity prairie grasses for biomass production than high-intensity monoculture crops, such as corn, on, for example, abandoned cropland.[5] It is therefore important to both remain critical and have an open mind when it comes to the development of bioenergy systems. Constant increases in yields and efficiencies and technological breakthroughs may change the outlook of a technology over time, and therefore periodic re-evaluation of bioenergy system alternatives is also important.

16-3 Biomass

Biomass comes in many different forms: plant materials, including their extracted sugars, starches, and oils and lignocellulosic stalks or wood, algae, and organic waste materials, including agricultural wastes, food wastes, and urban wood wastes. Currently, most ethanol is produced from either sugar from sugarcane or from pretreated cornstarch. The sugarcane and corn crops are specifically grown to produce ethanol. In tropical climates, growing sugarcane is attractive because of its high growth rates and for the sugars that are easily extractable without any other conversion needs (Fig. 16-1). Sugarcane can consist of up to 20% of sugar in the form of sucrose (i.e., a dimer[6] of a glucose and a fructose molecule, each with 6 carbon atoms). Yeast cells, which convert sugars with 6 carbon atoms into ethanol, can use sucrose in their metabolic pathways, and thus no conversion is needed before feeding it to yeast cells. However, the leftover lignocellulosic stalks from sugarcane (bagasse) are not used very efficiently, and this encompasses the largest part of the carbon in sugarcane. In most climates, the predominant quantities of biomass as a feedstock for bioenergy systems can come from lignocellulosic biomass, such as corn stover (i.e., leftover maize plant materials after removal of the corn grain), soft woods (e.g., poplar), and grasses, such as mixed prairie grasses or perennial grasses, such as switchgrass or miscanthus. In this chapter, we discuss three feedstocks for bioenergy systems in more depth: lignocellulose materials, organic wastes, and algae. This section ends with some basic information on physical, chemical, and thermal pretreatment steps

---

[5]Fargione et al. (2008) also compared the greenhouse gas reduction from prairie biomass vs. monoculture crops.

[6]i.e., combination of two molecules together.

FIGURE 16-1 Sugarcane plantation, which surrounds an ethanol plant, in the northeastern region of Brazil (state of Paraíba).
Source: Lars Angenent.

that are available to make lignocellulose materials accessible to hydrolysis (i.e., breakdown of polymers into monomers and oligomers).

16-3-1 Sources of Biomass

Lignocellulose Feedstocks

Lignocellulosic feedstocks are materials derived from plants, including trees, and form the most abundant biomass source on land. For the United States, estimates were made of 1.3 billion tons of biomass per year that could be harvested (a large majority consists of lignocellulose materials, such as wood, corn stover, and grasses) to offset 30% of the fossil fuel for the transportation sector.[7] These numbers were further verified and updated in 2011.[8] In very general terms, lignocellulose consists of three components—cellulose, hemicellulose, and lignin (Fig. 16-2). Cellulose is a polymer of glucose molecules, which is the reason so much attention has been placed on using lignocellulose for biofuel generation. Cellulose consists of crystalline and amorphous cellulose to form cellulose fibers. Hemicellulose is a complex material that consists of sugars with 5 carbon atoms (including xylan) and sugars with 6 carbon atoms. Due to its branched nature, hemicellulose is easier to break down than cellulose and its function is to form a connective network to keep the cellulose bundles and lignin together. Finally, lignin is an aromatic and nonsoluble compound that gives the plant its structure and protection against stresses. It is very hard to break down and is often toxic at high concentration to

[7]Perlack et al. (2005) used conservative estimates to investigate if enough harvestable biomass was present in the United States to sustain a biofuel industry that could offset fossil fuel use for transportation.
[8]Perlack et al. (2011) updated and nuanced the Perlack et al. (2005) report.

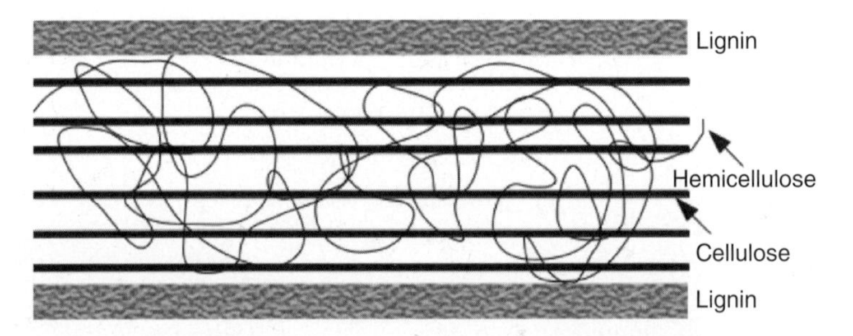

FIGURE 16-2 Simplified representation of the lignocellulosic matrix. Lignin forms a protective coat with hemicellulose acting as a connective network to keep the cellulosic fibers and lignin together.

microbes. The relative composition of these three major components can vary considerably between different lignocellulosic feedstocks and this will have important consequences for the choice of effective conversion processes. Hydrolysis of polymeric lignocellulosic compounds is seen as the rate-limiting step in bioprocessing of feedstocks. This makes evolutionary sense, because lignocellulosic materials form the scaffolds for plant metabolism and must resist microbial attack.

Organic Wastes

Organic wastes are diverse in composition and are abundant; often they include relatively large quantities of lignocellulosic materials. Examples of organic wastes are animal wastes, such as dairy waste; agricultural leftover materials, such as corn stover; and urban food and wood wastes. Because of the relatively easy breakdown of food waste, this source of biomass has been getting more attention for bioenergy systems, especially for renewable methane production in Europe. Data is available from a recent EPA report on the availability of food, yard, and urban wood wastes.[9] In this report, the authors estimated that in the United States alone 26 million dry tons of waste from urban sources could produce 8.7 billion L (2.3 billion gal) of ethanol-equivalent fuel per year. This is a considerable fraction of the required 61 billion L (16 billion gal) of fuels derived from cellulosic feedstocks as mandated by the U.S. government (more explanation on this fuel mandate follows). One currently mostly unexploited and growing renewable energy source is urban wood biomass residues (i.e., urban wood) from trees, yards, municipalities, and construction or demolition sites. In the United States, this biomass source is the fifth largest projected source of biomass. In certain areas of the United States, however, including the northeastern region (state of New York), this is the largest source of biomass (32%) with a total of 2 million tons produced per year (Milbrandt, 2005).

Algae

Microalgae are unicellular organisms capable of growing in inorganic salt media by fixing CO_2 and deriving energy from light harnessed by photosynthesis. Microalgae have high photosynthetic efficiencies (at least an order of magnitude higher than terrestrial

[9]The EPA (2010*a*) report has a section on organic wastes as a considerable source of biomass.

crop plants), and thus a faster growth rate compared to traditional terrestrial plants. The significance of their rapid biomass assimilation is seen in global element cycling, where microalgae have been found to be responsible for upwards of 30% of atmospheric carbon fixation. Unlike terrestrial biomass, problems with freshwater consumption can be averted because many strains of algae are capable of growth in saline or wastewater. The combination of rapid growth and low nutritional demands could make algae an attractive biomass feedstock option for sustainable biofuels.

Currently, of the three primary components in algal biomass (cellular lipids, carbohydrates, and proteins), biodiesel derived from algal lipids is the most popular studied form of biofuel being produced from algae. By-products of this biodiesel process include glycerol (derived from triglycerides), and the remaining algal carbohydrates and proteins. Anaerobic digestion, however, is capable of using all components from algal biomass for the production of energy, and thus can generate more renewable methane (bioenergy) with the left-over components after lipid removal. During life-cycle assessments it has become clear that the NEB ratio for biodiesel production from microalgae is much lower than 1, which makes this process unattractive as a bioenergy system. However, one of the life-cycle assessments also shows that a best-case scenario exists with

1. Special photobioreactors rather than algae ponds (Fig. 16-3);
2. Modern separation technology to extract lipids; and
3. Anaerobic digestion to convert left-over biomass into methane gas, and that this would result in a NEB ratio that is close to 1, albeit still lower than 1 (and thus still unattractive).[10]

This does show, however, that considerable improvements are possible during the maturation of the conversion technology.

16-3-2 Pretreatment Technologies

Yields and rates for converting biomass are greatly enhanced by pretreating the lignocellulosic material prior to enzyme hydrolysis and/or fermentation. Biomass is pretreated by subjecting it to physical, chemical, and/or thermal conditions that open up the cell wall structure, displace the xylan and lignin (when present) away from the cellulose polymers, and decrystallize (i.e., soften) the cellulose structure. Pretreatments thereby serve to give cellulase (enzymes to breakdown cellulose) access to the cellulose polymer, allowing for rapid and complete hydrolysis. A variety of pretreatments have been described which have been widely reviewed (Hendriks and Zeeman, 2009). Here, we mention four pretreatment schemes:

1. *Milling:* This is done to increase the surface area of the biomass feedstock and to decrease the level of polymeric structure and the shear of the biomass. Milling increases the speed of hydrolysis and the yields of production.
2. *Dilute-sulfuric acid:* Pretreating with dilute-sulfuric acid is very effective at releasing starch and xylan sugars as monomers (i.e., degrading polymeric

[10]Brentner et al. (2011) compared a base-case with a best-case scenario for algae-to-biodiesel conversion and found it to be unattractive as bioenergy systems at this point in time.

FIGURE 16-3 Photograph of a flat-plate photobioreactor, size approximately 50cm × 50cm.
Source: Lars Angenent.

hemicellulose sugars) and leaving the more digestible cellulose in an accessible form. Major shortcoming of this pretreatment method is the generation of toxic by-products from sugar decomposition products (e.g., furfural) that inhibit, for example, yeast cells. Often this method is combined with hot-water schemes. The produced toxic by-products have less of an effect on bioenergy systems with mixed consortia of microbes (e.g., anaerobic digesters) because by-products can be degraded.

3. *Hot-water:* Liquid hot-water pretreatment has been developed to solubilize the starch and xylan polymers and to produce digestible cellulose. This pretreatment only involves the use of process water and avoids the use of chemicals. However, this method does not completely degrade polymeric hemicellulose sugars.
4. *Alkaline:* Alkaline pretreatment can be performed by adding different types of chemicals. Examples include ammonia fiber expansion (AFEX) and lime addition. The chemical basis of these pretreatments is the dissolution of xylan at a pH of 10 and above, which makes it highly effective at producing digestible cellulose. Similar to the hot-water scheme, it does not completely degrade polymeric hemicellulosic sugars.

16-4 Platforms

Within the biorefinery concept of integrating biomass conversion processes and equipment to generate bioenergy and biochemicals, several different platforms have been identified. We will discuss four different platforms—sugar platform, syngas platform, bio-oil platform, and carboxylate platform. The difference between these four platforms is the method of hydrolysis to untangle the lignocellulosic matrix of biomass, and thus the platforms are classified based on the most difficult step of biomass conversion. Each of the different methods of hydrolysis results in different platform chemicals (i.e., sugar, syngas, bio-oil, and carboxylates), which gives the platforms their names. Subsequent conversion steps into bioproducts are interchangeable between platforms. For example, anaerobic fermentation is able to convert both sugars from the sugar platforms and syngas from the syngas platform into ethanol. It is important to note that all platforms may be compatible and should be used in different and variable combinations with the goal to maximize product formation and economic viability at the biorefinery. In the end, a biorefinery is the same as a refinery; the bottom line is to maximize the revenue from the conversion of a raw product into refined products.

16-4-1 Sugar Platform

The best-known biorefinery platform is the sugar platform. The hydrolysis is performed with special enzymes (including cellulases) to convert the polymers of sugars (cellulose and hemicellulose) into their derived 6-carbon and 5-carbon atom sugars after which bacterial or yeast fermentation can convert the sugars into alcohols. Before enzymes can be added, most often a pretreatment step is necessary to soften the feedstock and to make it accessible to the enzymes. Several companies in the United States are scaling this technology up for conversion of, for example, sustainable soft woods, such as poplar trees, into ethanol. The technology has been slow to develop due to technical and economical difficulties, including slow enzyme kinetics, costly bulk addition of extracted enzyme, and inhibiting effects of pretreatment by-products, among other factors. However, continued improvements may be making the sugar platform a promising technology.

16-4-2 Syngas Platform

When certain types of lignocellulosic biomass are hard or impossible to degrade by enzymatic processes due to their high lignin content and/or strong cellulosic matrix

Figure 16-4 A gasification system at a chicken farm in the Eastern region of the United States (state of West Virginia).

Source: Johannes Lehmann, Cornell University. Used with permission.

(Fig. 16-2), alternative strategies for biomass to liquid biofuel conversion must be used. A very promising strategy is to first convert these biomass feedstocks to synthesis gas (syngas, which is mostly H_2, CO, and CO_2) by gasification (Fig. 16-4) and then to use syngas as building blocks to synthesize a fuel. In many ways, gasification can be seen as a combustion process without oxygen. Besides syngas, the process also generates biochar (a char-like solid material) and bio-oil (a viscous solution). The gasification process can be operated in many different configurations and operating conditions, which changes the ratios of different products and their consistencies. Syngas can be used for fuel production by biologic or abiotic processes. Biologically, bacteria of the class Clostridia can convert CO and H_2 from syngas into alcohols. Several companies in the United States are currently scaling up bioreactors with an enhanced transfer of these gases into solution so that the bacteria can convert them at sufficient rates. In the past, syngas from coal has mostly been converted by metal catalysts (an abiotic process), and this is currently also under investigation for the conversions of syngas from biomass. Besides producing fuels, syngas can also be used to power engines or fuel cells for combined heat and power generation. Finally, organic wastes, such as agricultural wastes or urban wastes, can be converted to syngas, adding to the potential of the syngas platform.

16-4-3 Bio-Oil Platform

This platform also uses anaerobic gasification technology. To maximize bio-oil production from biomass feedstocks, rather than syngas and biochar production, the gasification process is operated at a temperature of 500°C and a relatively short biomass residence time of around 2 seconds. This process is often called fast pyrolysis due its speed of conversion. The main product is a viscous solution that is referred to as bio-oil. Some researchers have compared bio-oil to crude oil that can be further upgraded in biorefineries. However, problems exist—the bio-oil contains much water (20%), may be corrosive, and is very complex in composition. Numerous fast pyrolysis configurations have been developed to convert biomass into bio-oil, and this is a thriving area of research.

16-4-4 Carboxylate Platform

For the carboxylate platform, mixed consortia of microbes hydrolyze the biomass. These microbes produce enzymes themselves, which are needed for hydrolysis, and therefore the enzymes do not have to be purchased. A proxy for the carboxylate platform is the animal gut, in which complex biomass is converted to volatile fatty acids (short-chain carboxylic acids) that are taken up by the host. In bioenergy systems, the microbial process takes place in large-volume tanks. To speed up hydrolysis, pretreatment of the biomass may be necessary. The end products of this platform that can be generated with mixed consortia are either volatile fatty acids (acetic acid, propionic acid, lactic acid, or *n*-butyric acid) or the products from these acids. End products, then, can consist of methane, hydrogen, and medium-chain carboxylic acids. We will discuss the production of methane from volatile fatty acids later in this chapter (anaerobic digestion). The use of mixed consortia is advantageous because it has the genetic depth to handle a complex and variable feedstock and it is an open culture without the requirement to sterilize feedstock.

16-5 Alcohol

Ethanol, which is the best-known alcohol and often used synonymously with alcohol, is the name for a chemical compound that resembles ethane (C_2H_6) with 2 carbon atoms, but has 1 hydrogen atom replaced by a hydroxide ion (C_2H_5OH). Other alcohols have a shorter (i.e., methanol with 1 carbon atom) or longer carbon chain (e.g., *n*-butanol with 4 carbon atoms and *n*-hexanol with 6 carbon atoms) compared to ethanol. Most ethanol is produced at present by fermenting crop sugars (such as sugarcane in Brazil or sugar-derived cornstarch in the United States) with pure-cultures of yeasts. The ethanol is produced in highly aqueous solutions of up to 15% ethanol (150 g/L). A higher concentration of ethanol would not be possible because the yeasts that carry out the fermentation would not survive. To produce a transportation fuel that is 99% or more ethanol, the producer must distill the solution to remove water, which is a highly energy-intensive process. In modern plants, distillation produces a solution with 95% ethanol, and next a water-adsorbing system (i.e., molecular sieve) removes more water to produce a 99% ethanol solution. Ethanol is completely miscible, which means that at any concentration it remains in solution when water is present. The total energy consumption at any ethanol plant is high because of the need to distill—for a modern corn-to-ethanol plant in the United States, 20% of the energy content of ethanol itself may be required to separate it from the reaction solution (Shapouri et al., 2010).[11] When natural gas is used as a source for steam to power the distillation systems, the corn-to-ethanol technology in fact converts a gaseous fossil fuel in an easier-to-store liquid fuel. Fortunately, increasing amounts of biomass, including agricultural wastes, are used to power

[11]Shapouri et al. (2010) have estimated this percentage approximately.

the steam boilers to considerably decrease the use of nonrenewable fuels at the plant, and thus the NEB ratio.

The energy content per unit volume for ethanol is lower than that of gasoline (21.3 MJ/L or 75,700 Btu/gal net energy content vs. 32.5 MJ/L or 115,400 Btu/gal, respectively), so any comparisons of ethanol and gasoline must be conducted on an equivalent energy content basis. It is possible to combust pure ethanol in spark-ignition (SI) engines with appropriate modifications, or to blend it with gasoline and sell to consumers under the label "EXX," where XX stands for the percentage of ethanol in the blend. Spark-ignition engines can combust up to 10% ethanol without modification. Above this ratio, the engine must be modified to function well with the higher proportion of ethanol. In flex-fuel ethanol vehicles, the engine detects the proportion of ethanol and adjusts combustion accordingly. Different countries have taken different approaches. In Brazil, the passenger car fleet includes a mixture of vehicles that run on approximately 95% ethanol (~5% water): vehicles that use a fuel made up of a fixed ratio of gasoline to ethanol, and flex-fuel vehicles that can adapt to changing proportions (Fig. 16-5). In the United States, conventional vehicles can run on ethanol blends up to E10; vehicles that can run on over E10 ethanol are flex-fuel vehicles, and can combust ethanol mixtures up to E85.

While world ethanol production is currently only a small fraction of the total output of gasoline, there has been robust growth in recent years. United States production of ethanol grew from 7.6 to 53 billion L from 2001 to 2010 (2.0 to 14 billion gal). Brazil produced 26 billion L (6.9 billion gal) in 2010, making it the second largest producer in the world after the United States. There is room for further growth in output before limitations on the total size of sugarcane or corn crops would curtail production. However, most of the future production growth in the United States will have to come from ethanol that is produced from other biomass sources than cornstarch because this is

Figure 16-5 In Brazil, a car driver can choose between ethanol fuel or conventional gasoline based on their seasonal dependent costs, because their cars are flex-fuel vehicles.

Source: Lars Angenent.

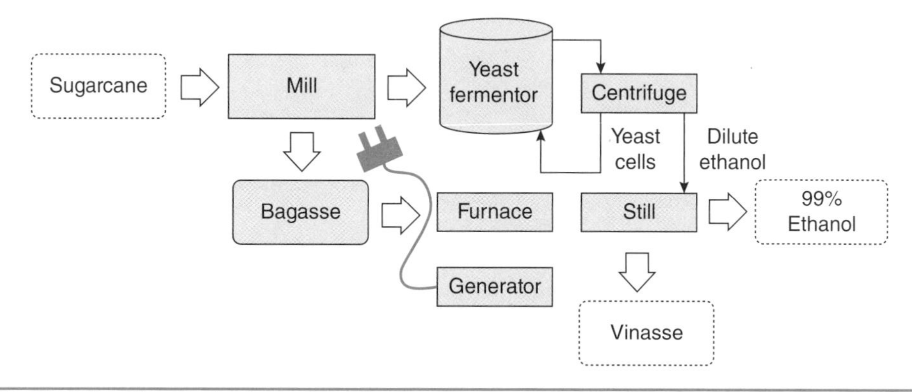

FIGURE 16-6 Simplified schematic of the sugarcane-to-ethanol process.

mandated by the Energy Independence and Security Act of 2007 (EISA). This Act specifies that in 2020 about 136 billion liters (36 billion gal) of ethanol-equivalent fuel should be produced from renewable sources of which 79 billion L (21 billion gal) from non-cornstarch sources. A total of 61 billion L (16 billion gal) is expected to come from cellulosic feedstocks. Thus, the United States has almost maximized the ethanol that can be produced from cornstarch (56 billion L or 15 billion gal). In 2010, this close-to-maximized value resulted in the use of approximately 30% of U.S. corn produced allocated to ethanol.[12]

16-5-1 Sugarcane-to-Ethanol

Sugarcane is a tall perennial grass with fibrous stalks that are rich in sugar (sucrose) and may reach up to 5-m (16-ft) tall (Fig. 16-1). In the Northeastern area of Brazil, which has a tropical climate, this grass is harvested three times a year. Some irrigation and nutrient addition may be necessary to increase the crop yields depending on the location. After harvesting, stalks of the sugarcane are transported to the ethanol plant (often the fields with sugarcane are surrounding the plant) and crushed by milling to recover the sugar (Fig. 16-6). The solution with high sugar content (i.e., molasses) is heated and transported to fermentation tanks with yeast to generate ethanol at concentrations of approximately 10%. Off gases from the yeast fermentation tanks are scrubbed for ethanol recovery.

The fermentation process may take 2 to 3 days, and yeast cells are recycled for a next batch (Fig. 16-6). The miscible ethanol from the fermentation tank is then pumped to distillation towers (stills) for ethanol recovery. Distillation to recover ethanol is energy intensive, and the required energy is generated in a furnace by burning the fibrous material (i.e., bagasse) that is left after crushing sugarcane (Fig. 16-6). The left-over solution without ethanol is called vinasse and can be partly recycled, used for irrigation, sold, or used to generate methane in anaerobic digesters. In the past, the furnace

[12]The U.S. EPA (2010*a*) Renewable Fuel Standard Program (RFS2) Regulatory Impact Analysis estimated the maximum of renewable fuel that may be generated from cornstarch.

FIGURE 16-7 Fermentation tanks (foreground) and bagasse furnace (left in background) at an ethanol plant in the Northeastern region of Brazil (state of Paraíba).
Source: Lars Angenent.

delivered just enough energy to power the ethanol plant, including pumps to irrigate the surrounding fields (Fig. 16-7). More modern and efficient energy-recovery systems have made the Brazilian ethanol industry a net electricity exporter, but more can be done to harvest energy from sugarcane, for example, by converting some of the bagasse into liquid fuels.

16-5-2 Corn Grain-to-Ethanol

In the United States, ethanol is produced from corn grain, which is also referred to as corn kernel. Two different processes are used for ethanol processing—dry milling and wet milling. The former process is less capital intensive and is simpler to operate, resulting in utilization of dry milling for many of the small- to mid-size ethanol plants that have been built over the last 5 to 10 years in the United States (Fig. 16-8). Wet milling, on the other hand, is more capital intensive but also generates more diverse and higher value products. Additional fractionation and separation technologies are used for wet milling to first completely separate all different compounds of corn grain. Only the purified cornstarch is used for ethanol production. Here, we discuss dry milling in more detail. More of the corn grain compounds than just cornstarch are used in dry milling, which increases the ethanol yield for dry milling versus wet milling.

Dry Milling

The entire corn grain is milled and then cooked during which added enzymes (α-amylases) liquefy the cornstarch into dextrins, which are polymers of glucose (Fig. 16-9). The solution with dextrin is further heated with steam and then cooled again to body temperature. In the fermentor, yeast cells, antibiotics (to control bacterial infections), and enzymes (glycoamylases) are combined for simultaneous saccharification (formation of glucose) and fermentation. After 2 to 3 days, a 15% ethanol concentration is achieved in the fermentation broth and the process is terminated. The off gases from

Figure 16-8 Four fermentation tanks at a dry-milling plant in the Northeastern region of the United States (state of New York). This plant is powered by natural gas and electricity (hydropower).

Source: Lars Angenent.

the fermentor are treated to recover ethanol, and possibly to harvest carbon dioxide for the bottling industry. The final solution in the fermentor is called beer and is pumped to the distillation towers (stills) to recover ethanol at a concentration of approximately 95%. Molecular sieves increase the concentration of ethanol to 99% after which the ethanol is stored on site.

The solution without ethanol from the liquid outflow of the stills is called whole stillage, and this is pumped to a centrifuge for recovery of the leftover corn grain solids and yeast cells. A wet cake is generated that is dried in large rotating dryers (Fig. 16-9). The liquid outflow of the centrifuges has still large quantities of solids, albeit at much

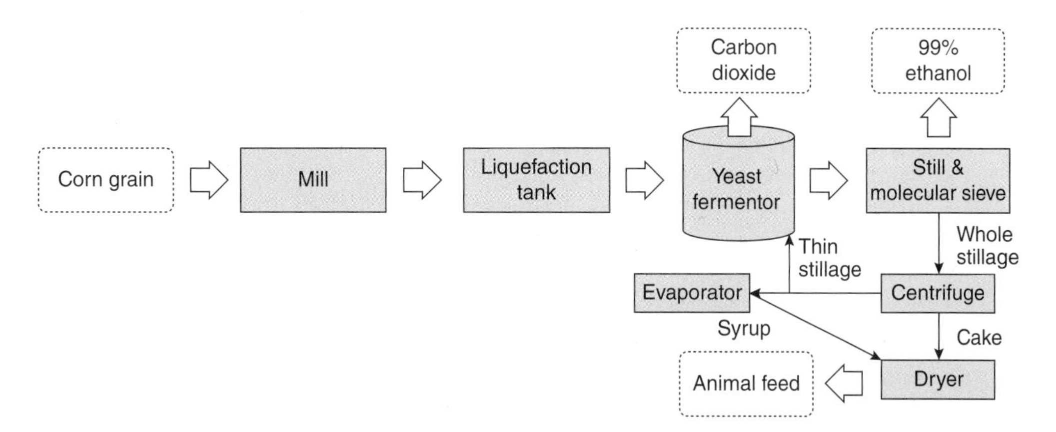

Figure 16-9 Simplified schematic of the corn-grain-to-ethanol process for dry milling.

lower concentrations, and this is pumped to evaporators that are powered with waste heat from the ethanol plant. Evaporators have two products—evaporated water and syrup. The evaporated water is recovered and recycled back to the start of the process to reduce the consumption of fresh water. Syrup is combined with the wet cake during drying. The combined dried material generated in the dryer is sold as animal feed and is called dried distillers grain (DDGS) in the ethanol industry. The quantities of DDGS are large and the economic viability of the ethanol industry is dependent on this co-product. This also shows that the discussion about fuel versus food is not as black and white as it may seem upon first introduction, because ethanol plants generate a large value of pretreated animal feed. In fact, in life-cycle assessments, the production of animal feed results in energy credit (increases the NEB ratio).

Life-Cycle Assessments

Many life-cycle assessments have been performed for the United States corn-to-ethanol industry. Several years ago, the average NEB ratio that was published was in the range of 1.25 and was a reason for criticism of the entire ethanol industry. The life-cycle assessments were broad (as they should be) and included production of ammonia for plant uptake, production of farm equipment, and farm and household energy use, among many other factors. It also, on the other hand, included a credit for the production of animal feed. The most recent assessments show an improved NEB ratio between 1.9 and 2.3,[13] by modernizing the ethanol plants, including more efficient ethanol-separation technology (stills and molecular sieves). Some plant locations are already incorporating agricultural-waste biomass for energy production rather than using natural gas or coal, and the estimated NEB ratio for such as plant would be 2.8 with 50% biomass power.[14] More can be done, though, and new ideas are emerging. For example, thin stillage can be converted to renewable biogas by integration of anaerobic digesters in the plant,[15] or miscible ethanol can be converted to a different chemical that does not need as much energy to be separated from solution.[16]

Even though the NEB ratio is not high in terms of sustainability, one attraction of ethanol production in the United States is that it serves as a petroleum multiplier because it takes a relatively small input of liquid fuel (petroleum) as part of its life cycle (e.g., for farming and for truck or rail transportation to move the corn crop or ethanol product), and converts this input into a larger quantity of liquid fuel (ethanol) in terms of energy equivalent. This process requires large inputs of other fossil energy sources, but since natural gas and coal are used for these inputs, and the United States primarily relies on domestic reserves for these fuels, ethanol helps the United States balance of trade. In addition, the 10% addition of ethanol to United States gasoline adds oxygen to the fuel mix and this results in a more efficient car engine. Already, this shows that the issues are more complex than just the discussion of what an appropriate NEB ratio

[13]Shapiro et al. (2010) evaluated several dry milling plants based on 2008 data gathering.
[14]This is the same study from Shapiro et al. (2010) but now by adding data from biomass energy inputs.
[15]Agler et al. (2008) performed a lab-based study by integrating anaerobic digestion into the dry milling process to estimate the NEB ratio to increase from 1.3 to 1.7 due to the formation of renewable methane—this included a correction for a lower production of animal feed, and lowered the animal feed credit.
[16]Agler et al. (2012) performed a lab-based study to convert beer (15% ethanol) from the corn-to-ethanol industry into *n*-caproic acid, which can be separated from water with a lower energy input than ethanol.

should be. Additional factors that are important to take into consideration in the debate about the U.S. ethanol industry are:

- The published NEB ratios are based on average numbers after evaluating different ethanol plants. There is a large amount of variability in ethanol production processes, and it would be difficult to create a study that truly represented the *average* process. Rather than generalizing, we should attempt to identify best practices.
- Even though the NEB ratio for ethanol is modest, producing ethanol should be seen as a steppingstone toward producing more effective biofuels in the future, rather than an end in itself. Different fuels and processes could be adopted when these emerge out of research. When that occurs, the United States has in the meantime gathered infrastructure and a trained work force that is able to design and build facilities to produce biofuels.
- The debate over energy inputs and outputs is a distraction from a much more important concern, namely, whether diverting too much corn to ethanol, and achieving modest environmental gains, will eventually have repercussions on basic food supplies that the United States as a country may come to regret.

Overall, it is clear that for ethanol production in temperate regions to have a real positive effect on the environmental bottom line, the ethanol production process will need to be changed to reduce energy input required and use nonfossil resources.

16-5-3 Cellulosic Ethanol

In the future, a relatively large part of the renewable fuel mix will have to come from cellulosic feedstocks as mandated by the U.S. government. Currently, technology is being scaled up that can convert cellulosic biomass into ethanol by either using sugars with 5- and 6-carbon atoms from the sugar platform or syngas from the syngas platform. Bacteria and yeast cells can convert sugars into ethanol. When bacteria produce ethanol, the maximum ethanol concentration they can achieve is approximately 4% to 5%. When wild-type yeast is used for sugar conversion, a higher concentration of ethanol may be achieved (15%), but the problem then is their inability to use the majority of sugars that stem from hemicellulose (5-carbon-atom sugars), which results in a lower lignocellulose-to-ethanol conversion efficiency. For syngas fermentation, only bacteria are being used with an anticipated ethanol concentration that is approximately 4%. Thus, for cellulosic ethanol a lower ethanol concentration of 4% to 5% is anticipated than the 15% that is achieved for the corn ethanol industry. With a lower concentration of ethanol, a larger amount of energy is needed to distill the product, which will result in a higher requirement for energy at the plant to distill ethanol. Estimations are that about 24% of the energy content in ethanol is used for distillation (Galbe and Zacchi, 2002), when cellulosic ethanol is distilled to 95%, while this is 20% for corn ethanol. Even though the energy requirement to distill ethanol for the corn-to-ethanol industry is lower, the NEB ratios for cellulosic ethanol are anticipated to be considerably higher than corn ethanol. We will have to wait until the cellulosic ethanol industry matures to find out what these NEB ratios will be.

16-5-4 *n*-Butanol

n-Butanol is an alcohol with a carbon length of 4 atoms in a straight carbon chain. Compared to ethanol, the chemical properties of *n*-butanol make it more similar to gasoline and a better fuel for SI engines. *n*-Butanol is produced from fermentation of sugars by bacteria from one class (Clostridia), which co-produce acetone and ethanol as by-products. The process to generate *n*-butanol is called acetone-ethanol-butanol (ABE) fermentation and was first publicized in 1861 by the microbiologist Louis Pasteur, who gave pasteurization its name. An industrial-scale fermentation process to supply acetone for gunpowder for the British army was developed during the First World War. The co-produced *n*-butanol was used as a solvent for lacquer in the car industry. Later, the ABE fermentation process was also used to produce *n*-butanol as a fuel. With the rise of the petroleum industry after the Second World War, the industrial production of *n*-butanol and acetone by fermentation gradually declined due to price competition, until the last commercial ABE fermentation plant in South Africa closed in 1982. Today, the ABE-fermentation process has seen a small revival due to higher fuel prices and concerns about the future availability of fossil fuels. But, although *n*-butanol is the better engine fuel, producing it in large quantities as a biofuel for transportation comes with even greater challenges than ethanol. *n*-Butanol is considerably more toxic to bacteria than ethanol is to yeast, resulting in much lower maximum concentrations that are achievable during ABE fermentation, and resulting higher costs for energy-intensive distillation. Another problem is that the sugars for fermentation should come from lignocellulosic feedstock rather than from starch (as mandated by the U.S. government). Even though *n*-butanol production from lignocellulosic feedstocks is studied in research labs, a widespread *n*-butanol industry is still missing.

16-6 Biodiesel

Both biodiesel and petrodiesel fuel consist of complex hydrocarbon strings with an average composition of $C_{12}H_{26}$, but with the number of carbon atoms varying around this average (10 to 15 carbons). The chemical composition of biodiesel is substantially different from that of petrodiesel, and it has lower energy content per unit volume than petrodiesel (33.0 MJ/L or 117,000 Btu/gal net energy content vs. 36.2 MJ/L or 128,700 Btu/gal, respectively). Most biodiesel in use today fits this description, although there is work ongoing to develop a bio-derived synthetic diesel that originates from plant or animal sources but is chemically much closer to petrodiesel. Total sales of biodiesel in the United States began rising rapidly in the late 1990s, increasing from 1.9 million L (500,000 gal) in 1999 to an estimated 3 billion L (0.8 billion gal; this is equivalent to 1.2 billion gal of ethanol) in 2011. Germany produced an estimated 5.7 billion L (1.5 billion gal) of biodiesel in 2010, mostly from rapeseed oil. Feedstocks for biodiesel production in Germany and the United States come primarily from agricultural crops. Biodiesel is also produced from local waste oil streams by individuals and small-scale businesses, which also produce straight vegetable oil or vegetable fat (i.e., not converted to diesel fuel using the process described above) that can be combusted in a CI engine when preheated so as to reduce viscosity.

The most common form of biodiesel is as a 20% biodiesel-80% petrodiesel blend known as "B20," since it requires very little adaptation of the vehicle or fuel supply system for use in existing vehicles. With adaptation, B100 (100% biodiesel) can also be used. Diesel (CI) engines are in general more sensitive to cold weather operation than gasoline (SI) engines, and with use of biodiesel, this sensitivity is increased. Therefore, biodiesel operators must take precautions to avoid gelling of fuel in supply lines, especially when using biodiesel at higher concentrations (greater than B20). Otherwise, no significant changes are needed to use biodiesel in CI engines. In some ways, biodiesel can be beneficial to the engine and fuel supply system; for example, adding 2% biodiesel to fuel increases the lubricating capability of the fuel inside the engine.

To eliminate the price gap between petrodiesel and biodiesel, a U.S. federal tax incentive went into effect in 2005, which provided a $0.20/gal incentive for B20 made by blending #2 diesel with virgin biodiesel and a $0.10/gal incentive for B20 made by blending #2 diesel with reused animal- or vegetable-based oil. Here, virgin biodiesel is made from either crops, such as soybeans, or animal fat from the meatpacking industry, while reused biodiesel comes from the purification and reacting of waste oil products, such as cooking oil. Improving efficiency in the biodiesel industry and rising prices of crude oil on the world market since 2005 have made biodiesel more competitive. Nevertheless, the industry continues to depend on price supports. In the United States, this became very apparent in 2010 to 2011, when their tax incentive was removed only to see a considerable decrease of biodiesel production. In 2011, production was back up due to reinstatement of the tax incentive.

16-6-1 Production Processes

Biodiesel is considered renewable because it is derived from agriculturally produced and industry-wasted plant oils, such as palm oil, soybean oil, rapeseed oil; from algae-extracted triglycerides; or from animal fats, such as tallow or lard. The biological starting materials for biodiesel production are triglycerides: an organic molecule consisting of three fatty acid chains attached to a common glycerine backbone at its three hydroxyl groups with an ester bond (Fig. 16-10). In the best-known biodiesel manufacturing process, the triglyceride molecules are cleaved into their constituent fatty acid molecules and glycerol molecules through a thermochemical reaction known as transesterification. This reaction involves adding a catalyzing agent (typically a strong base, such as sodium hydroxide or potassium hydroxide), which mediates the replacement of the fatty acid–glycerine bond with that of an alcohol, such as methanol or ethanol. This results in the formation of biodiesel and glycerol.

The mixture of biodiesel and glycerol is then allowed to settle and the denser glycerol is removed by gravity separation. On a weight-to-weight basis, 1 part of glycerol is generated for every 10 parts of biodiesel produced. This biodiesel process is relatively easy to operate and the diesel engines are so forgiving that many small and decentralized plants have emerged with homeowners performing the production in

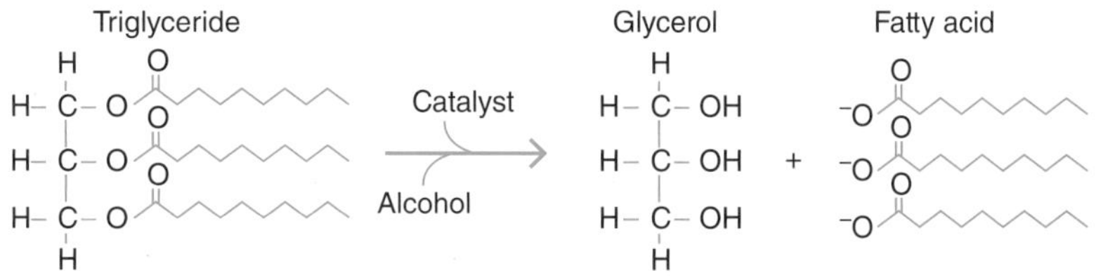

Triglyceride — Catalyst / Alcohol → Glycerol + Fatty acid

FIGURE 16-10 The reaction of the best-known biodiesel production process.

Note: This thermochemical reaction, which is known as transesterification, cleaves triglycerides in glycerol (by-product) and long-chain fatty acids (biodiesel). The carbon tails (right side of molecule) for the triglycerides and fatty acids can be of variable length.

their own garages. Glycerol, which is seen as a waste material from biodiesel production, is an ideal substrate to be mixed in with, for example, animal wastes for the production of renewable methane in anaerobic digesters.

Although less known, full-scale manufacturing plants exist with a different manufacturing process with the same feedstock as the transesterification process, but based on a thermochemical reaction known as dehydrogenation. This process yields significant volumes of biopropane as a by-product rather than glycerol (here the glycerine chain of the triglyceride is hydrogenated to produce propane). Biopropane is a small molecule with three carbons (C_3H_8) that can be used directly as liquid-petroleum gas for transportation. Liquid-petroleum gas is a generic name for mixtures of hydrocarbons [predominantly propane and butane (C_4H_{10})] that change from a gaseous to liquid state when compressed at moderate pressure or chilled. An alternative use for biopropane, which is especially important for rural areas that are not connected to the natural gas and/or electricity grid, is for cooking and heating of houses. This is the case for both developing and developed countries.

16-6-2 Life-Cycle Assessment

Most, though not all, recent studies of the life-cycle energy inputs and outputs from biodiesel production have estimated a substantially positive NEB ratio, which is greater than that of corn ethanol. One study performed the assessment for both corn grain ethanol and soybean biodiesel production as a comparative effort. Their study showed a NEB ratio of 1.9 for biodiesel (1.25 for ethanol) when co-products are included in the analysis (Hill et al., 2006). The apparent improvement in NEB ratio compared to corn ethanol comes from the reduced energy requirements in the processing stage, since the used vegetable oils are closer to biodiesel in their chemical formulation, and require processing with a chemical catalyst but do not require distillation. When greenhouse gases other than CO_2 are taken into account (e.g., N_2O from agriculture—note that N_2O has a much higher greenhouse gas equivalent than CO_2), biodiesel may release on the order of 60% of the greenhouse gas equivalent of petrodiesel over its life cycle per unit of energy delivered to the vehicle. Again, for life-cycle assessments, it is important to include all greenhouse gases during the entire life cycle of the biofuel even when this includes the impact from agriculture or waste management.

Although biodiesel from crops such as soybeans may have a considerably higher NEB ratio compared to corn ethanol, soybeans as a crop suffer from lower yields per land area compared to corn. Thus, the maximum soybean biodiesel production that could be achieved worldwide may be more limited in terms of total energy content. For example, in the United States, without substantial changes to the allocation of planting or noticeable reduction in soybean exports, the maximum biodiesel output is thought to be about 5% to 6% of current petrodiesel consumption on an energy content basis. Ethanol output from corn grain, on the other hand, is already generating approximately 10% of the gasoline consumption in the United States. Again, decision makers must take into account many factors besides just the NEB ratios.

16-7 Methane and Hydrogen (Biogas)

The gaseous fuels methane and hydrogen are often seen as fuels to power stationary power-generation systems, such as generators (SI engines) and fuel cells. However, methane is used to power cars, buses, and trains, while hydrogen powers cars and buses.

Renewable methane is produced from organic materials (often part of wastes) by anaerobic digestion. Hydrogen can be used as a fuel to generate electricity and some heat in a fuel cell, but because of the presence of expensive metal catalysts and their sensitivity to toxicity by many different molecules, the hydrogen gas has to be ultra clean to ensure a long lifetime of the fuel cell.

16-7-1 Anaerobic Digestion

Anaerobic digestion consists of a tank with an enclosed (air-tight) environment in which diverse consortia of microbes degrade organic material to generate biogas. The advantage of this technology is that besides mixing, no other power sources are necessary because the microbes work under strict anaerobic conditions and aeration is not necessary. The process is very different from, for example, the ethanol fermentation systems that operate with a pure culture. Anaerobic digesters are open systems, which means that diverse types of microbes can come in with the waste streams to circumvent the need to sterilize the inflow streams, and this eventually results in thousands of microbial species being present in a relatively stable microbial consortium. Within anaerobic digesters, microbes comprise a food web, which means that a product from one microbe is the substrate (food) for another one. The biogas, which consists typically of 60% to 70% out of methane, can be combusted to generate heat and/or electricity in a boiler or combined heat and electric power system.

Anaerobic digestion is a mature technology and, if operated under stable conditions, is a very efficient energy recovery system because the final products—methane and carbon dioxide—are automatically and constantly removed from solution by degassing (bubble formation) to increase the thermodynamic potential for the biological reactions. In contrast to ethanol production, no energy is needed to separate the fuel product. Another reason why this is an efficient process is because microbial cell yields for anaerobic microbial growth are low and all intermediate fermentation products are maintained at very low concentrations (i.e., no other product that lowers efficiency is formed). In the United States, anaerobic digestion technology is currently used on farms, at municipal wastewater treatment plants to treat biosolids, and in industrial wastewater treatment applications. In countries, such as Brazil and India, besides these three uses, anaerobic digester technology is also used for domestic wastewater treatment.

In 2010, the USEPA estimated that there were only 157 digesters operational at commercial-scale livestock facilities in the United States, while 8,000 farms across the country are good candidates for biogas installations with a combined capacity of 1,670 MW (megawatts).[17] In another estimate, the increased value of both methane reduction credits (currently, untreated manure in piles generates methane, which considerably adds to greenhouse gas release from the agricultural sector) and electricity prices will result in a competitive economic model for anaerobic digestion in 2025. Under these favorable conditions, anaerobic digesters on farms could generate 5.5% of the U.S. electricity consumption.[18] This may even be a conservative estimate of the total capacity, because about 7,000 farm-based digesters have already been built in Germany alone over the last 10 years due to the favorable feed-in tariff system, albeit the German

[17]AgStar through the EPA (2010*b*) has estimated the opportunity for anaerobic digesters on commercial live-stock facilities (farms).

[18]Zaks et al. (2011) have discussed under what circumstances farm-based digesters have an economic opportunity.

FIGURE 16-11 In-ground anaerobic digester on a 1,000-head dairy farm in the Northeastern region of the United States (state of New York).

Note: The biogas from this co-digestion process, which treats animal manure and food wastes, generated over 330 KW of electricity by two SI engines at the moment this picture was taken.

Source: Lars Angenent.

digesters are co-fed with animal manures and energy crops, such as ensilaged corn plants, to increase capacity to about 2,500 MW. Figure 16-11 shows one of the 157 farm-based digesters in operation in the United States.

Anaerobic digestion offers many benefits to farmers in addition to its use as an on-site energy source. Some examples include odor reduction and the production of a stabilized fertilizer that can be applied to cropland. Why then are not more digesters built in the United States? The combination of low natural gas or electricity prices and the high capital cost for digesters results in a too long return-of-investment period. Farmers can increase the revenue potential of their digesters, and therefore shorten the return-of-investment period, by charging tipping fees to treat other organic wastes from, for example, food industries in addition to their manure (co-digestion of combined wastes). There is a renewed interest in using anaerobic digestion at domestic wastewater treatment plants in the United States, because co-digestion of food waste from the community with the biosolids that these digesters were designed to treat originally may generate more methane (and thus electricity when a combined heat and power system is present). Such addition of organic materials can generate enough electricity to cover all energy needs at the wastewater treatment plant to make them carbon neutral.

Anaerobic digesters are typically run in either mesophilic (25 to 37°C) or thermophilic (55 to 65°C) temperature ranges, with each temperature range having its own advantages and disadvantages in terms of performance. The food web underlying the anaerobic digestion process can be broken down into four stages: hydrolysis, acidogenesis, acetogenesis, and methanogenesis. During the first stage, hydrolyzing bacteria mediate the degradation of large, complex organic molecules, such as proteins, carbohydrates, and lipids, into smaller, simpler molecules of amino acids, sugars, and

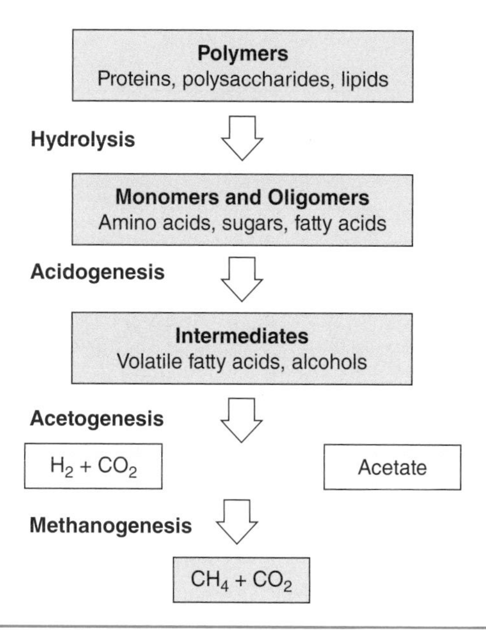

FIGURE 16-12 Simplified schematic of the anaerobic food web. The product from one group of microbes is the substrate for another group until methanogenens complete the conversion by producing methane and carbon dioxide.

fatty acids, respectively. Next, during acidogenesis, these compounds are further broken down by bacteria into volatile fatty acids and other intermediates. Acetogenesis constitutes the next degradation stage whereby the predominant intermediates (volatile fatty acids) are converted by bacteria to acetate, hydrogen, and carbon dioxide (CO_2). Finally, during methanogenesis two possible pathways exist [splitting acetate to methane (CH_4) and carbon dioxide (CO_2) or combining hydrogen (H_2) and CO_2 to produce CH_4] (Fig. 16-12). The group of microbes that perform these pathways are called methanogens and are not bacteria, but rather archaea. Archaea are the third domain of life, and are microscopic. Example 16-2 illustrates the function of anaerobic digestion.

Example 16-2 There are several digester configurations used for anaerobic digestion, including the continuously stirred anaerobic digester (CSAD), the plug-flow digester, the anaerobic sequencing batch reactor, and the upflow anaerobic sludge blanket reactor. One type of anaerobic digester system that is commonly used for animal waste treatment is CSAD. In an ideal CSAD system at steady state, the digester contents are assumed to be uniformly mixed, and the concentration of a compound coming out in the reactor effluent will equal the concentration of that compound within the reactor itself. Often, it is useful to estimate the final compound concentration leaving the reactor by performing a mass balance on the CSAD system. Sticks and Stones Farm has built a new anaerobic digester on their farm to process their dairy manure. They designed a digester with a volume (V) and flow rate (Q) to achieve a hydraulic residence time (θ) of 25 days. However, they realized that a chemical (used in the United States to increase the milk yields) at a concentration (C_o) of 6 mg/L in the dairy manure will also be entering the digester. The farmers want to ensure that the effluent (C) leaving the reactor has 0.5 mg/L or less of the chemical. Assume that the antibiotic undergoes first-order decay in the reactor with a decay rate constant (k) of 3.5×10^{-4} min^{-1}. Determine whether the final antibiotic concentration leaving the steady-state reactor would be less than the desired 0.5 mg/L (assume an ideal CSAD system).

Solution First, set up a mass balance on the system:

Rate (milligrams/day) of change of antibiotic concentration within digester	Rate (milligrams/day) of antibiotic entering the digester	Rate (milligrams/day) of antibiotic leaving digester	Rate (milligrams/day) of decay of antibiotic within digester

Therefore

$$V\frac{dC}{dt} = QC_0 - QC - kCV \tag{16-1}$$

Assume the digester is at steady state

$$V\frac{dC}{dt} = 0 \tag{16-2}$$

Also, we know that

$$\text{hydraulic residence time}, \theta = \frac{V}{Q} \tag{16-3}$$

Substituting Eqs. (16-2) and (16-3) into Eq. (16-1) and rearranging yields

$$C = C_0\left(\frac{1}{1+k\theta}\right) \tag{16-4}$$

Finally, putting in provided values into Eq. (16-4)

$$C = 6\ \text{mg/L}\left(\frac{1}{1+(3.5\times10^{-4}\ \text{min}^{-1})\left(\frac{1{,}400\ \text{min}}{d}\right)(25\ d)}\right) = 0.4\ \text{mg/L}$$

Note that this value is just below the maximum allowed concentration for Sticks and Stones Farm (!).

16-7-2 Anaerobic Hydrogen-Producing Systems

Researchers have explored the use of mixed consortia of microbes in engineered systems to produce hydrogen gas. When the methanogenic archaea are inhibited within the mixed consortia, hydrogen and volatile fatty acids—such as acetate, butyrate, and propionate—are the most abundant fermentation end products from organic material conversion. Most of the chemical energy is amassed in the volatile fatty acids, and unfortunately not in the produced hydrogen gas. Of the 12 hydrogen atoms that are present in, for example, glucose (i.e., $C_6H_{12}O_6$), only a maximum of 4 hydrogen atoms (one-third of the maximum yield) can be generated with bacteria, but in reality only approximately 2 hydrogen atoms (one-sixth) is generated. From a renewable energy point of view, focusing solely on hydrogen production with mixed consortia may therefore not have much future. However, two anaerobic digesters that are placed in series (outflow stream of the first digester is the feed stream of the second digester) to convert an energy-rich organic waste may generate some hydrogen in the off gas from the first digester. In the second digester, the soluble volatile fatty acids are then converted into methane. By mixing the off gases from both digesters, a biogas with some hydrogen gas is produced and this has a higher energetic value compared to biogas without hydrogen gas when burned in a generator (SI engine) or boiler. One problem with generating hydrogen that always needs to be considered, however, is the very diffusive nature of the small molecule hydrogen; it leaks through most plastics and therefore out of holding tanks, tubing, and connectors (special metal components do not leak hydrogen, but these add costs). This diffusive character of hydrogen gas is also a problem for other renewable energy systems that have been proposed, such as using algae in special photobioreactors to generate hydrogen gas from sunlight.

16-8 Bioenergy Integration into Transportation Energy Supply

The discussion in preceding sections has focused on biofuels in terms of the various pathways for conversion of biological raw materials into possible energy carriers to be used in vehicles (e.g., for combustion in reciprocating engines or jet turbines). However, biofuels may also be seen as an alternative to other options (e.g., electricity) for short-term energy storage for renewable energy derived from the sun (Fig. 16-13). From this perspective, solar energy on the left-hand side of the figure may be converted to electricity in a photovoltaic panel and delivered via the grid to a recharging electric vehicle (EV), where it is stored in the battery until used during the drive cycle. Also, since in a mature solar-to-EV system of the future, production of solar electricity might not exactly match demand for recharging in real time, an energy storage option is shown, for example, using pumped storage or conversion to hydrogen, for later reconversion back to electricity. Biofuels can achieve the same conversion-and-storage function, as shown on the right-hand side, through creation of oils and starches in plants using sunlight, water, and CO_2. Thus, the biofuel harvesting and processing system acts as a large battery bank, storing energy in biofuels and releasing it later through combustion in engines.

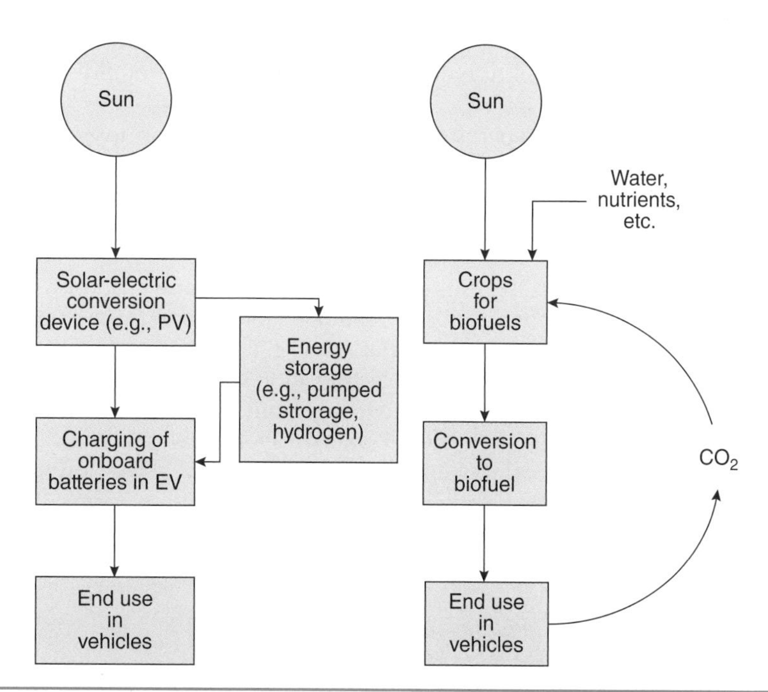

Figure 16-13 Comparison of pathways to solar-derived energy for vehicles from PV panels and EVs (left) and from biofuels and ICEVs (right).

When the biofuel is combusted, the CO_2 is released back to the atmosphere, completing the cycle.

In considering the possible niches for biofuels, it is important to consider the varying needs of different transportation applications, both in terms of sector (passenger, freight) and modes (road, rail, marine, air). As a transportation energy carrier, biofuels have advantages over electricity or hydrogen that make them especially suited to certain combinations of sector and mode. For example, liquid biofuels, such as ethanol or biodiesel, are relatively rapid to refuel. Biofuels are also relatively simple to carry long distances because the simple storage vessel (i.e., on-board fuel tank) adds relatively little tare weight compared to the net weight of the fuel. By contrast, electricity requires battery systems that are heavy, and therefore reduce total range between recharges, and hydrogen requires either compression or liquefaction, which add complexity and weight.

The advantages of biofuels point them in the direction of use in applications that require long travel distances between refueling stops, lightweight onboard energy storage, or both. One of the most obvious examples is aviation: using bioenergy feedstocks and chemical engineering techniques to produce a replacement for petroleum-based jet fuel that has equivalent performance. Also, many intercity freight applications, including long-distance trucking, rail, and marine transportation, are planned on the basis of long gaps between refueling stops. Although there is a small energy density penalty when biofuels are substituted for petroleum-based diesel, vehicles, locomotives, or vessels can travel nearly as far on a full tank, so that the function of the system would not be greatly affected by shifting to biodiesel assuming other adaptations could be made (e.g., modification of combustion engines). Lastly, for long-distance travel of light-duty vehicles (LDVs) beyond the maximum range of electric charge, biofuels are a leading contender for energy supply, for example, in plug-in hybrid electric vehicles (PHEVs) that might run on ethanol or biodiesel for the portion of a long-distance trip not powered by electricity.

A comparison of U.S. energy demand in the transportation applications mentioned in the previous paragraph and energy provided in biofuels in 2010 shows the extent of the challenge but also the opportunity for market growth when biofuel technology advances (Fig. 16-14). The total biofuel energy content of 1.2 quads is small compared to the total of 12.9 quads consumed by trucks, aviation, LDVs, marine, and rail.[19] The potential market for advanced biofuels in the future is, therefore, on the order of at least 10 quads, if the output can be adequately developed. The potential market for biofuels is large in part because electricity and hydrogen are not suitable at present to meet this demand. Of the loads shown, only rail is suitable for electrification since it can be connected to overhead catenary to deliver power from the grid to locomotives. However, this load is only a small fraction of the total (approximately 0.5 quads). Hydrogen is more flexible, as it can theoretically be applied to road, rail, marine, or air applications. The only one where significant R&D progress has been made, however, is LDV; in particular, aviation and heavy-truck applications are both very large energy loads where the technical challenges

---

[19]In Fig. 6-14, the measure of energy consumption shown for truck includes consumption for urban truck movements. Although the underlying data source from the U.S. Department of Energy does not provide information about energy consumed in urban versus intercity movements, the fraction consumed in urban movements is not negligible and could alternatively be powered by electricity in the form of battery-electric delivery vehicles.

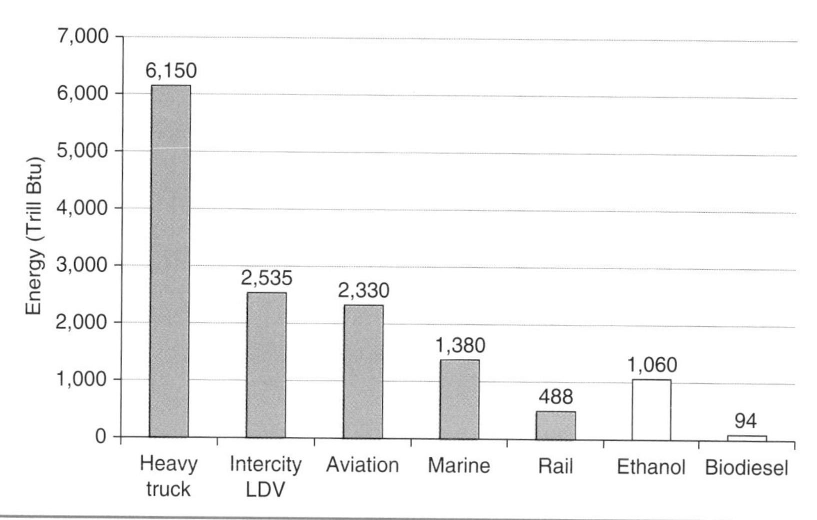

FIGURE 16-14 Comparison of biofuel output and transportation loads suited for bioenergy in the United States, 2010.

Source: Adapted from Davis et al (2013).

for hydrogen power are large. This situation leaves biofuels as the leading alternative energy source for each of the five loads in Fig. 16-14. In each case, the loads in question have been powered by biofuels to some degree, either at a test case level (aviation) or in actual routine use in biofuel-petroleum blends (e.g., truck, rail).

16-8-1 Early Examples of Prototype Applications

Along with growing use of ethanol and biodiesel in fuel blends in countries, such as Brazil, Germany, the United States, and elsewhere, biofuels are making inroads in other ways. The aviation industry is making progress in developing biofuels for jet-powered aviation, but substantial challenges remain. This is in part due to the difficult conditions under which jets operate at cruising altitude, where it is more difficult for alternative fuels to function well because of cold temperatures. Also, the potential higher cost of alternative fuel would have a large economic effect on the industry, since energy costs are an especially large part of overall aircraft operating cost. As a stepping-stone toward using biofuels in jets, in 2006 the U.S. Air Force for the first time used a mixture of traditional jet fuel and Fischer-Tropsch synthetic fuel derived from natural gas during a routine training flight. Other milestones include the first-time use of biofuels in a commercial airliner by Virgin Atlantic Airlines in February 2008, and the development and testing by Boeing in 2008 and 2009 of a biofuel-based aviation fuel that meets or exceeds jet fuel standards. By 2013, a number of airlines, including Virgin Atlantic and All-Nippon Airways, were using biofuels in small quantities to displace conventional jet fuel.

Biomethane used in place of compressed conventional natural gas is another pathway for bringing bioenergy into vehicles. A growing fraction of road vehicles are capable of running on natural gas, as is the case in Europe, where this figure was estimated at 600,000 units in 2008. Biomethane generated on farms, at wastewater treatment plants, or at other sources can be processed (CO_2 and water removed, gas compressed) and then used in place of conventional fossil-derived natural gas in vehicles. It can be delivered to vehicles in one of four ways: (1) by selling to the gas grid and extracting elsewhere;[20] (2) by refueling vehicles at the source where the gas is generated; (3) by using a dedicated pipeline to move gas from source to refueling stations; or (4) by transporting gas by truck to refueling stations. Although the compression (or in some cases liquefaction) stage and need for specialized nozzles and storage tanks add complexity, there is often a return on this investment because the price per unit of energy for biomethane is lower than that of gasoline or diesel. The Swedish biomethane sector has emerged as a world leader and, as an example, the city of Trollhattan, Sweden, has established a system for refueling transit buses and city LDV fleets from wastewater treatment by-products that supports more than 200 vehicles as of 2008 (Strehler et al., 2008).

16-9 Summary

In this chapter, we have discussed several different bioenergy systems to produce energy carriers. We have focused primarily on systems that are already being used at an industrial scale, including ethanol production from sugarcane or corn grain, or that have potential in the short term. We have not discussed novel biofuels that are developed by using synthetic biology (reshaping metabolic pathways on a cellular level), longer-chain carboxylic acids with mixed consortia of microbes, and drop-in biofuels that are similar in behavior to existing fossil fuels and that can be used without adaptation to fuel systems. Many new ideas are being investigated and hopefully the biofuel industry will look very different 20 years from now. This is especially pertinent because for most of the discussed bioenergy systems, the NEB ratio is only slightly above 1, while for a true sustainable bioenergy system this ratio may need to be 10 to 20. Of course, we must look further than just the NEB ratio and evaluate the land and water uses, and never forget that we must produce enough human food.

References

Agler, M. T., M. L. Garcia, E. S. Lee, M. Schlicher, and L. T. Angenent (2008). "Thermophilic Anaerobic Digestion to Increase the Net Energy Balance of Corn Grain Ethanol." *Environmental Science & Technology* 42:6723–6729.

Agler, M. T., C. M. Spirito, J. G. Usack, J. J. Werner, and L. T. Angenent (2012). "Chain Elongation with Reactor Microbiomes: Upgrading Dilute Ethanol to Medium-Chain Carboxylates."*Energy & Environmental Science* 5: 8189–8192.

Brentner, L. B., M. J. Eckelman, and J. B. Zimmerman (2011). "Combinatorial Life Cycle Assessment to Inform Process Design of Industrial Production of Algal Biodiesel." *Environmental Science & Technology* 45: 7060–7067.

Drapcho, C., J. Nghiem, and T. Walker (2008). *Biofuels Engineering Process Technology.* McGraw-Hill, New York.

Davis, S., S. Diegel, and R. Boundy (2013). *Transportation Energy Data Book, 32d ed.* Oak Ridge National Laboratories, Oak Ridge, TN.

[20]In this option, the gas extracted from the grid for use in vehicles would not be physically the same gas as was produced from biological sources, but the net system-wide benefits would be the same, since the same physical quantity of conventional natural gas would be displaced.

EPA (2010*a*). *Renewable Fuel Standard Program (RFS2) Regulatory Impact Analysis*, EPA-420-R-10-006. U.S. Environmental Protection Agency, Washington, DC.

EPA (2010*b*). *US Anaerobic Digester Status Report*, AgSTAR, Oct. U.S. Environmental Protection Agency, Washington, DC.

Fargione, J., J. Hill, D. Tilman, S. Polasky, and P. Hawthorne. (2008). "Land Clearing and the Biofuel Carbon Debt."*Science* 319: 1235.

Galbe M. and G. Zacchi (2002). "A Review of the Production of Ethanol from Softwood."*Applied Microbiology and Biotechnology* 59: 618.

Hendriks A. and G. Zeeman (2009). "Pretreatments to Enhance the Digestibility of Lignocellulosic Biomass." *Bioresource Technology* 100: 10

Hill, J., E. Nelson, D. Tilman, S. Polaskey, and D. Tiffany. (2006)."Environmental, Economic, and Energetic Costs and Benefits of Biodiesel and Ethanol Biofuels."*Proceedings of the National Academy of Sciences* 103: 11206.

Milbrandt A. (2005). *A Geographic Perspective on the Current Biomass Resource Availability in the United States.* National Renewable Energy Laboratory, Golden, CO.

Mitchell, C., D. Bauknecht, and P. M. Connor (2006). "Effectiveness through Risk Reduction: A Comparison of the Renewable Obligation in England and Wales and the Feed-in Tariff System in Germany." *Energy Policy* 34: 297.

Nag, A. (2008). *Biofuels Refining and Performance.* McGraw-Hill, New York.

Olsson, L., ed. (2007). *Biofuels (Advances in Biochemical Engineering/Biotechnology).* Springer-Verlag, Berlin.

Perlack R., L. Wright, A. Turhollow, et al. (2005). *Biomass as Feedstock for a Bioenergy and Bioproducts Industry: The Technical Feasibility of a Billion-Ton Annual Supply*, Sponsored by the U.S. DOE and USDA.

Perlack R.D. and B. J. Stokes (2011). *U.S. Billion-Ton Update: Biomass Supply for a Bioenergy and Bioproducts Industry.* Rep. ORNL/TM-2011/224, Oak Ridge National Laboratory, Oak Ridge, TN.

Sorensen, B. (2002). *Renewable Energy: Its Physics, Engineering, Use, Environmental Impacts, Economy and Planning Aspects, 2d ed.* Academic Press, London.

Shapouri, H., P. W. Gallagher, W. Nefstead, et al. (2010). "2008 Energy Balance for the Corn-Ethanol Industry." Washington, DC, USDA: 16.

Strehler, J., D. Parry, and R. Campbell (2008). "Fill 'Er Up … With Biomethane, Please!" *Biosolids Technical Bulletin* 13(2):1–16.

Tester, J., E. Drake, M. Driscoll, M. Golay, and W. Peters. (2012). *Sustainable Energy: Choosing among Options, 2d ed.* MIT Press, Cambridge, MA.

Zaks, D., N. Winchester, C. Kucharik, C. Barford, S. Paltsev, and J. Reilly (2011). "Contribution of Anaerobic Digesters to Emissions Mitigation and Electricity Generation under U.S. Climate Policy." *Environmental Science & Technology* 45: 6735–6742.

Further Readings

Andreoli, S. and P. De Souza (2006). "Sugarcane: The Best Alternative for Converting Solar and Fossil Energy into Ethanol."*Energy & Economy* 9(59) Electronic resource, available at: *http://ecen.com/*. Accessed Nov. 8, 2007.

Burniaux, J. and J. Chateau (2011). "Mitigation Potential of Removing Fossil Fuel Subsidies: A General Equilibrium Assessment," *OECD Economics Department Working Papers* 853, OECD Publishing. DOI:10.1787/5kgdx1jr2plp-en.

DeCicco, J. (2013). "Biofuels Carbon Balance: Doubts, Certainties, and Implications." *Climatic Change*, 121:801–817.

Pimentel, D. and T. Patzek (2005)."Ethanol Production Using Corn, Switchgrass, and Wood; Biodiesel Production Using Soybean and Sunflower." *Natural Resources Research*, 14(1):65–76.

Plevin, R., M. Delucchi, and F. Creutzig (2013). "Using Attributional Life Cycle Assessment to Estimate Climate-Change Mitigation Benefits Misleads Policy Makers." *Journal of Industrial Ecology*, 18:1, 73–84.

Smith, H., J. Finkle, M. Sharif, and T. Kanuparthy (2009). Biogas Generation from the Ithaca Area Wastewater Treatment Facility: Use for Facility CHP and Ithaca City Automobiles. Masters thesis, Dept. of Biological and Environmental Engineering, Cornell University, Ithaca, N.Y.

Wang, M., C. Saricks, and D. Santini (1999). *Effects of Fuel Ethanol Use on Fuel-Cycle Energy and Greenhouse Gas Emissions*. Report ANL/ESD-38, Argonne National Laboratories, Argonne, Ill. Available on internet at:*http://www.ethanol-gec.org/information/briefing/10.pdf*. Accessed Sep.12, 2007.

WestStart-CALSTART (2004). *Swedish Biogas Industry Education Tour 2004: Observations and Findings*. Technical report, WestStart-CALSTART, Pasadena, CA.

Exercises

16-1. As an exercise in exploring the impact of energy input in ethanol production on NEB, as discussed in the body of this chapter, consider the following process, in which each liter of ethanol requires 2.69 kg of corn as a feedstock. The corn must first be grown on arable land; per hectare (100m × 100m), the corn field has the following energy input requirements:

Input	Energy (GJ)
Machinery (embodied)	4.97
Energy products (diesel, gasoline, electricity)	6.04
Nitrogen	10.25
Balance of inputs	12.72

The hectare yields approximately 9000 kg of corn. The corn is then processed in an ethanol plant to make almost pure (99.5%) ethanol. Per 1000 liters of ethanol, the ethanol production process requires the energy inputs shown in the table below, in addition to the corn. Assume that a liter of ethanol contains 21.3 MJ net. For simplicity, ignore the energy impact of by-products resulting from corn ethanol production. Calculate the ratio of energy available in the resulting ethanol to the total energy input. Does the ethanol provide more energy than it consumes?

Input	Energy (GJ)
Transportation	1.35
Distillation steam	10.66
Electricity	4.87
Balance of inputs	0.59

16-2. Repeat Exercise 16-1, but this time for biodiesel production from soy. The energy requirement per hectare for soy is the following:

Component	Energy (GJ)
Fuels	2.67
Fertilizer	1.60
Embodied energy	0.53
All Other	0.53

Each hectare yields 508 kg of soy, and each liter of soy biodiesel requires 0.925 kg of soy as input. A liter of biodiesel contains 32.6 MJ of net energy content. Production energy requirements per 1,000 kg of soy beans processed are in the following table. Calculate ratio of output to input, ignoring by-product energy impact.

Component	Energy (GJ)
Process heat and electricity	4.14
Embodied energy	1.38
Transportation	0.69
All Other	0.69

16-3. Repeat the calculation of exercise 16-1 for corn ethanol for the advanced process with higher efficiency discussed in the chapter. Suppose that the net energy input in agriculture per hectare is cut by 50% due to advances in practices for growing corn. Suppose also that the "transportation" and "balance of inputs" figures for energy consumption remain fixed. What must the combined energy value per 1,000 L of production for distillation and electricity be to achieve the net energy balance ratio of 2:1, which is the average of the reported current values discussed in the chapter?

16-4. As an exercise in exploring the impact of energy input in ethanol production on NEB, as discussed in the body of this chapter, consider the following process, in which each gallon of ethanol requires 22.4 lb of corn as a feedstock. The corn must first be grown on arable land; per acre, the corn field has the following energy input requirements measured in millions of Btus (Million Btu):

Input	Energy [Million Btu]
Machinery (embodied)	1.906
Energy products (diesel, gasoline, electricity)	2.317
Nitrogen	3.932
Balance of inputs	4.879

The acre yields approximately 8,013 lb of corn. The corn is then processed in an ethanol plant to make almost pure (99.5%) ethanol. Per 1,000 gal ethanol, the ethanol production process requires the energy inputs shown in the table below, in addition to the corn. Assume that a gallon of ethanol contains 75.7 MBtu (1 MBtu = 1,000 Btu). For simplicity, ignore the energy impact of by-products resulting from corn ethanol production.

Input	Energy [Million Btu]
Transportation	4.84
Distillation steam	38.24
Electricity	17.47
Balance of inputs	2.12

Calculate the ratio of energy available in the resulting ethanol to the total energy input. Does the ethanol provide more energy than it consumes?

16-5. Repeat Exercise 16-4, but this time for biodiesel production from soy. The energy requirement per acre for soy is the following:

Component	MBtu
Fuels	1,025
Fertilizer	615
Embodied energy	205
All Other	205

Each acre yields 452 lb of soy, and each gallon of soy biodiesel requires 7.7 lb of soy as input. A gallon of biodiesel contains 117,000 Btu of net energy content. Production energy requirements per 1,000 lb of soy beans processed are in the following table. Calculate ratio of output to input, again ignoring byproduct energy impact.

Component	1000 Btu
Process heat and electricity	1,784
Embodied energy	595
Transportation	297
All Other	297

16-6. Repeat the calculation of Exercise 16-4 for corn ethanol for the advanced process with higher efficiency discussed in the chapter. Suppose that the net energy input in agriculture per acre is cut by 50% due to advances in practices for growing corn. Suppose also that the "transportation" and "balance of inputs" figures for energy consumption remain fixed. What must the combined energy value per 1,000 gal of production for distillation and electricity be to achieve the net energy balance ratio of 2.1, which is the average of the reported current values discussed in the chapter?

CHAPTER 17

Conclusion: Toward Sustainable Transportation Systems

17-1 Overview

The purpose of this chapter is to briefly survey transportation issues of general interest not covered in Chaps. 14 to 16, including air quality, transportation security, preparedness for extreme weather, and hazardous materials transportation. Thereafter, the chapter looks at some possible scenarios for meeting future transportation demand (either passenger-miles or ton-miles) in an environmentally sustainable way. Lastly, the chapter describes ways that the transportation professional can support the development of sustainable transportation, both in the professional arena and through extracurricular activities.

17-2 Introduction

The body of this book (Chaps. 3 to 16) has focused on several key components for sustainable transportation, including advanced technologies, multimodalism, and best practices for passenger and freight transportation, as well as sustainable, carbon-free energy for both sectors. A number of observations are appropriate for this concluding chapter. First, the use of information technology and electronic communication can substitute across the board for provision of additional physical capacity in transportation systems, since it often provides a less expensive alternative for using existing capacity more efficiently. Second, passenger transportation systems need multiple options, including not only ways of managing light-duty vehicle traffic more effectively but also other options such as public transportation, nonmotorized options, vehicle-sharing systems, and telecommuting opportunities. Third, differences between passenger and freight transportation should be acknowledged: unlike passenger transportation, freight is primarily an intercity and long-distance activity, so spatial patterns and long supply chains are important factors. Lastly, in terms of energy alternatives, there may be many technology options for engines, motors, drivetrains, means of storing energy, and other components, but the basic energy options are only three: electricity, hydrogen, or hydrocarbons.

In addition, there are several topics not covered in the previous chapters that comprise system-wide issues:

- *Transportation and air quality:* Environmental impact on human health and the natural environment of pollutants emanating from the exhaust of road vehicles, locomotives, ships, and aircraft
- *Transportation security management:* Precautionary steps taken to prevent malicious attacks on the transportation network and contingency plans in case of an attack, or preparedness for the effects of natural disasters
- *Extreme weather events and pre-event planning:* The impact of extreme weather events on the need for evacuation using the transportation network, and the development of evacuation plans including recommended evacuation routes and pre-positioning of shelters and supplies
- *Management of hazardous materials:* Planning of routes for the movement of dangerous materials, including both toxic and radioactive materials, with minimal risk, and contingency plans in case of accidents

These four topics are covered in turn in the following section.

17-3 Other Transportation System-Wide Issues

17-3-1 Transportation and Air Quality

One of the most important ecological impacts of the transportation sector is on the ambient air quality of both urban and rural regions. The transportation sector is one among several major economic sectors that are large contributors to air-quality problems. Different modes of transportation contribute to varying degrees to air pollution. One of the most important distinctions in level of impact is the difference between gasoline-powered engines [primarily in light-duty vehicles (LDVs)] and diesel-powered engines [primarily in heavy-duty vehicles (HDVs) such as trucks or buses as well as railroad locomotives and ship engines], since they emit air pollutants in different proportions. In this section, we focus on the U.S. situation, although many of the same characteristics and problems occur in other countries.

Like greenhouse gases (GHGs) such as CO_2, most transport-related air pollutants are emitted primarily from the tailpipes or stacks of road vehicles, locomotives, ships, and aircraft. Unlike GHGs, however, air pollution entails some additional complexities. First, the spatial distribution of air pollutant emissions is important, since emissions from a vehicle in an already-polluted urban area will have greater negative consequences in general than emissions from the same vehicle in a remote and sparsely populated rural region. (This assumes that prevailing winds do not carry rural emissions into urban areas, aggravating air-quality problems.) Second, the temporal distribution of emissions are important, since background levels of many air pollutants rise and fall depending on the time of day or season of year, so consequences of additional emissions may be more or less grave.

Emissions of pollutants are of concern in the first instance because of their potential negative impact on human health, contributing to both acute conditions such as respiratory distress and chronic ones such as lung disease. Degraded air quality also has a negative effect on other animals and plant life, some of which are agricultural crops, whose value can be reduced by lower health in grain crops and fruit and vegetable harvests. Even in cases where air pollution does not translate into lost economic value, the human community has a moral obligation to minimize harm to species with which we share the planet. Reduced air quality can also impact the tourism industry if tourists avoid a certain region due to foul air. Lastly, it can lead to premature maintenance or replacement costs on buildings and other structures if they are harmed by corrosive pollutants such as sulfur dioxide.

Measurement of Emissions and Steps to Reduce Transportation Contribution

Attainment and maintenance of satisfactory air quality requires the measurement of *criteria pollutant* levels and comparing their concentration against maximum levels of concentration or *criteria* above which the U.S. federal government has determined that human health is at risk. Concentration is measured in parts per million or per billion of volume (ppmv or ppbv) and may be measured in terms of transient peaks or of ongoing levels over time. Although many pollutants are emitted directly from transportation and other sources, some are by-products of the presence of other pollutants. Ground-level ozone, for example, is a measured pollutant which forms when oxides of nitrogen and volatile organic compounds react in the presence of sunlight.[1]

Table 17-1 shows national inventories of pollution in 2001 and 2011 for selected criteria pollutants. The list includes carbon monoxide (CO), which is toxic in high enough concentrations; oxides of nitrogen and sulfur (NO*x* and SO*x*), where the *x* represents different possible numbers of oxygen atoms that combine with either nitrogen or sulfur to form a pollutant; volatile organic compounds (VOCs), which represent a range of possible uncombusted hydrocarbons; and particulate matter (PM), which are

Year	CO	NOx	VOC	PM10	PM2.5	SOx
2001	99.5	12.41	7.5	0.53	0.45	0.7
Percent	82.4	55.5	41.7	2.2	6.1	4.4
2011	38.56	6.11	3.61	0.21	0.19	0.17
Percent	61.8	50.9	29.8	2.7	4.2	2.1

*For example, total emissions of CO from all sectors in 2001 were 120.8 million tons, of which 82.4%, or 99.5 million tons, were from transportation. Definitions: CO = carbon monoxide, NO*x* = oxides of nitrogen, VOC = volatile organic compounds, PM-10 = particulate matter 10 microns in diameter or less, PM-2.5 = particulate matter 2.5 microns in diameter or less, SO*x* = oxides of sulfur.
Source: Davis et al. (2013).

TABLE 17-1 Emissions of Criteria Pollutants from Transportation Sector in Millions of Tons in 2001 and 2011, and Percent Share of Total from All Sectors*

[1]Note that ground-level ozone is different from stratospheric ozone; the former is harmful to health when inhaled, whereas the latter protects the surface of the earth from harmful solar rays, and is therefore desirable to be maintained in sufficient quantities higher in the atmosphere in the form of the so-called *ozone layer*.

further classified into the size of 10 μm (microns) in diameter or less or 2. 5 μm in diameter or less. Regarding particulates, a distinction is made between larger and smaller particles because those that are under 2.5 μm are particularly adept at penetrating human and other respiratory systems, and therefore pose the greatest health risk.

The data in Table 17-1 show that across all pollutants, steady progress has been made since 2001 in reducing the total national inventory from transportation. Note, however, that since the figure is a national total, it does not take into account the location of the emissions and therefore the impact of transportation in specific urban regions. One could conclude from the data in the table that with the substantial reductions over 10 years shown, changes in the transportation sector contributed to improved air quality in many or most urban areas. For instance, transportation CO emissions were reduced by nearly two-thirds during this time. The other indicator in the table is the percent contribution of transportation to the total emissions inventory across all sectors. For the three criteria pollutants for which transportation was a significant contributor (>10%), namely, CO, NO*x*, and VOCs, not only was the absolute amount of pollution reduced, but transportation's percent contribution declined as well.

Efforts to reduce air pollution caused by transportation include reducing emissions from newly sold vehicles, inspection of vehicles already on the road, and retirement of vehicles sooner rather than later at the end of their useful life. Table 17-2 shows examples of current U.S. emissions standards for both LDVs and HDVs for a selection of criteria pollutants. On the LDV side, the values shown are for a particular *bin* of vehicles with specific maximum emissions, namely Bin 5, for their anticipated end-of-life emissions rate at the end of an expected 120,000 mile vehicle life. Emissions rates are given in maximum grams per mile. For HDVs, values shown are also anticipated emissions rates, in this case after 100,000 miles of operation, but because these vehicles experience such a wide range of operating conditions, it is not practical to evaluate emissions on a per-mile basis. Therefore, emissions are measured at the level of the engine instead, on the basis of grams per *brake horsepower-hour*, a measure of work done against a constant load by an engine in a laboratory setting.[2] For both types of vehicles, by gradually tightening maximum allowable emissions over time, new vehicles become cleaner compared to the existing fleet, and total emissions decline.

Turning to vehicles already in use, one concern is that at any time some fraction of the fleet that is active on the road will have malfunctioning emissions control equipment, such

Vehicle Type: Units	Light-Duty g/Mile	Heavy-Duty g/bhp-h
NOx	0.07	0.2
CO	4.2	15.5
PM	0.01	0.01

TABLE 17-2 Emissions Standards in Grams per Mile for Light-Duty Vehicles and Grams per Brake Horsepower Hour for Heavy-Duty Diesel Engine

[2]A horsepower hour is equivalent to a power output of one horsepower maintained for 1 hour.

that the emissions of that individual vehicle exceed the allowable emissions in Table 17-2 and contribute excessively to air pollution. A program of required regular inspection and maintenance (I&M) can identify vehicles that are in violation and eliminate those emissions by requiring the owner to make repairs if they want to continue to operate the vehicle. In some cases, regional or state governments may subsidize repairs to keep compliance affordable for vehicle owners on limited incomes. Another type of financial mechanism for improving air quality involves *super-emitters*, or aged vehicles with very high emissions rates per vehicle mile compared to new vehicles on sale in the market. Government agencies may offer cash incentives which allow the owner of a super-emitter who cannot afford vehicle replacement to purchase a new or newer vehicle.

17-3-2 Transportation Security Management

Transportation security management can be defined as the protection of transportation systems from the effects of malicious attacks or natural disasters through preemptive planning to avoid attacks or disasters, as well as the planning of recovery strategies in case an event cannot be avoided. Both private and public sectors have an interest in transportation security management. Whether disrupted by attacks or disasters, either sector can suffer losses if it is not prepared for security challenges. This section focuses on attacks, whereas the next section considers natural disasters, particularly extreme weather events, in more detail. (For a full-length work on the subject of transportation security, see Ritter et al., 2007.)

Three elements contribute to the heightened need for transportation security management at this time:

1. *Impact of globalization:* As both wireless communications and the efficiency of international freight transportation make it increasingly possible for firms to buy and sell goods and services between locations all over the globe, the task of keeping flows of information and goods secure becomes more complex.
2. *Dependence on critical infrastructure:* Global trade in goods and ideas depends in turn on transportation network linkages (highways, railroads, and canals), interchange points such as ports and airports, wireless communications, and electricity grids. Defects in one level of infrastructure can impact other levels.
3. *Possibility of disruption due to attacks:* Because connections in the global economy of today can be made over long distances and because key line-haul links or intermodal exchange points can be critical to thousands of global buyer-seller relationships, a single successful attack can cause widespread disruption.

The current circumstances are such that the effects of malicious attacks can be asymmetrical: a small group of individuals with malicious intent can have a disproportionate impact on a large, global system in which millions of people are directly involved in trading and millions more benefit either through employment opportunities or as consumers of globally manufactured products. The size of the impact comes in part because it affects not only the vehicles or infrastructure that are the target of the attack, but movements elsewhere in the network as the system struggles to recover afterwards. To give a concrete example, in the three days following the attacks of September 11, 2001, domestic aviation of all types in the United States was grounded while the system was secured. Once aircraft returned to the air, the air freight industry saw a precipitous decline in demand. According the U.S. Bureau of Transportation Statistics, total ton-miles moved fell from 14.9 billion in 2000 to 13.1 billion and 13.4 billion ton-miles in 2001 and 2002, respectively (see Fig. 11-1).

Transportation Security Management Application: Private Firm Perspective

In the example application that follows, we view transportation security management from the perspective of a private company engaged in the sale of products or services to the global marketplace. (In principle, many of the same goals and practices apply to a government agency seeking to protect vital infrastructure.) Success in transportation security management requires the pursuit of several key goals. First, a firm should pursue only those security measures whose return on investment can be anticipated in advance, and whose performance can be measured once implemented. In transportation security, even large budgets can be depleted rapidly on measures that aim blindly at preventing attacks and add little value, or worse encumber the smooth operation of business, leading to added cost instead of added value. For example, across a complex, multitiered global supply chain there are a multitude of locations where materials, components, or products could be inspected for malicious activity, so only the most cost-effective should be chosen.

Comprehensive security management also requires looking at the entire chain leading in to the firm's core product or service, not just the first-tier suppliers. (By definition, the first-tier supplier is the vendor that supplies directly to the firm, the second-tier supplies to the first-tier, and so on.) Since the firm relies on the first-tier supplier for provision of components, it has only indirect access to entities further up the supply chain. Disruptions at those earlier tiers can, however, work their way down and negatively impact the firm. Therefore, the firm should reach out to collaborate with second-, third-, and fourth or higher-tier firms involved in the supply chain on a comprehensive security solution. Of course, this activity requires enlisting the first-tier supplier in a collaborative partnership.

A third goal is for the firm to be proactive about anticipating possible security threats. If these threats can be anticipated, the firm can take preemptive steps to avoid being the victim of a successful attack. At the same time, it would be unrealistic to think that all threats can be anticipated and avoided all the time, so developing plans to survive and recover from disruptions is important as well.

Effective security management requires certain key practices. Security practices should be understood well throughout the organization to be effective, so the central management of the firm should educate the entire workforce about these practices. The firm can also improve the quality of practices by looking outside its own boundaries at other firms with which it interacts, and adopt new practices based on what it learns from their experience. Security-related information should flow clearly to key decision-makers in the firm, especially in an evolving security situation, where a rapid response to a threat can be crucial in limiting damage from a disruption. Firms can also prepare to respond to situations by practicing their response, either in a role play around a table in a "situation room," or, as a more elaborate activity involving greater outlay of resources, an actual field exercise in which a hypothetical scenario is played out.

In conclusion, the effect of heightened security threats in the early twenty-first century should not be overstated. The effect of increased global awareness and communications among peoples of the world is in many ways positive. Most citizens appear headed toward an era of increasing cooperation without resorting to violence, as they work as individuals, as either owners or employers of enterprises, or as voters or elected officials in government. They are becoming more united in solving global problems and raising living standards in an interconnected world no longer caught between two main competing political ideologies, namely, liberal democracy and communism. Nevertheless, the presence of a minority of malevolent actors who seek to reverse this shift to a more interconnected world or take action for other reason have left both national governments and major enterprises continuously on guard against threats in the short to medium term. Thus we can expect transportation security management to remain an important topic for the foreseeable future.

17-3-3 Extreme Weather Events and Pre-Event Planning

Transportation systems, including both physical transportation infrastructure and the networks of public and private entities that operate them, must be prepared for a wide range of possible natural disasters. Possible disasters include earthquakes, floods, volcanic eruptions, tidal waves, tornadoes, and hurricanes or cyclones. Preparedness includes building and maintaining robust, resilient systems that can withstand the effects of these disasters where possible, as well as contingency plans for situations where parts of networks fail under the stress of a disaster and people and goods require alternate solutions to meet their transportation needs. This section considers extreme weather events, and in particular one type of extreme weather, hurricanes, but many of the observations can be applied to other types of natural disasters. (For full discussion of this topic, see, for example, Li et al, 2012.)

Preparedness for extreme weather events is crucial because of the potential for great loss of life. Events such as hurricanes or cyclones are especially dangerous because they concentrate large amounts of power into strong winds and storm surges that can affect a wide area, both along the coast and inland. Unlike tornadoes, which can arise with little warning, hurricanes take days to develop, but their exact course is difficult to predict and they can also strengthen unexpectedly over short periods of time, making preparation difficult. In the case of Hurricane Katrina in the United States in August and September of 2005, winds and flooding from the hurricane combined with poor preparation led to more than 1,000 deaths along with more than $100 billion in damage.

Modeling of transportation networks can play a vital role in preparing for an extreme weather event. Models of an area at risk from a hurricane typically include the following elements:

- *Numbers and geographic distribution of populations at risk:* For a region susceptible to hurricanes, the distribution of the population that might need to be evacuated in advance of an approaching storm. Depending on the path and intensity of the storm, some fraction might be required to move to safer ground while others could remain in place, so a model must be able to handle different scenarios.
- *Geographic distribution of potential evacuation centers:* Designation of centers that are available as temporary shelters for evacuees, along with prepositioned supplies to be used for housing and feeding during the storm. One potential role of the model is to determine, for any given storm, a preferred subset of all available shelters that should be prepared and opened for evacuees as they arrive, so as to reduce the overall financial cost of the response.

- *Information about transportation networks by which populations will evacuate:* Primarily roads although possibly including other modes as well. The model of the network should include information about maximum capacity, distance, travel time, and interconnectivity for different links in the network. Models should consider both the movements of evacuees who own private vehicles and those who do not, as one of the major shortfalls of the Katrina response was the failure to consider residents who did not own a vehicle.

Preparation for a hurricane event involves creating and running the event model for a range of scenarios, and then developing a set of recommendations for how to prepare to share with elected officials and first responders. When a storm approaches, a preplanned response can be put in place to help residents evacuate efficiently and safely to shelters. Transportation professionals, elected officials, and first responders would then work with the public to implement the response.

A basic approach to the modeling effort is to optimize the selection of routes and shelters for an affected population, that is, run an optimization model that minimizes some combination of preparation and evacuee travel cost (see Chap. 3 for discussion of optimization). Residents would then be encouraged to use recommended routes and shelters so that the actual response during the event would resemble the model ideal as closely as possible.

Although conceptually relatively simple, the optimization approach is limited because during an evacuation most evacuees are given information about routes and shelters but generally they act autonomously. Therefore the choice of routes and shelters is likely to be far from the model output, especially given the duress of a full-scale evacuation. A more sophisticated model that better reproduces "real-world" flows is therefore desirable, even if it requires greater implementation effort. Such a model starts with a probabilistic assignment of evacuees among several candidate shelters. The combination of evacuation traffic and background vehicle movements (i.e., everyday movements that would take place with or without the presence of a storm) are then loaded into the transportation network model using a dynamic user equilibrium (DUE) approach, where each network user is assigned to an origin-destination path in such a way that no user can improve their travel time without increasing that of others. The DUE model therefore captures the tendency of growing congestion on links in the network to increase link travel time and make them less desirable so that travelers shift to other less congested links and routes. With a more realistic picture of how traffic is likely to assign itself as evacuees move from vulnerable areas to shelters, planning agencies can anticipate the network of shelters that is most likely to be needed in a given event, and focus their resources on preparing those shelters.

17-3-4 Transportation and Storage of Hazardous Materials

Hazardous materials constitute a range of by-products of industrial activities that possess one or more of the following characteristics:

1. *Explosive or flammable materials:* Materials that are at risk of either exploding or igniting if they are handled improperly or exposed to conditions in which they become volatile.
2. *Corrosive materials:* Materials with either a very low or very high pH level that are capable of corroding containing materials if they are not contained properly, that is, in a vessel made of materials that can withstand their corrosive capacity.

3. *Toxic materials:* Materials that can poison or sicken living organisms either through surface contact (for example, on the skin of people and animals or the bark of trees) or else if ingested in food or drink.
4. *Radioactive materials:* Materials that can harm surrounding organisms or damage inanimate objects by emitting radiation from different types of nuclear reactions.

From a transportation systems perspective, the management of hazardous materials involves the movement and intermediate storage of these materials from the point at which they are generated to the point at which they are either subject to some transformational process that renders them harmless or put into permanent disposal in a way that puts them beyond further access or risk (Turnquist and Nozick, 1997). Thus while it is clearly desirable to change manufacturing processes so that these materials are not generated in the first place, or advance the level of processes that can recycle them or render them harmless, these activities are outside the scope of hazardous materials management.

There are three levels of planning to consider in hazardous materials management. The first is the appropriate design of containers and structures for their transportation and storage. Thus containers must be robust to prevent any leakage during handling, and also to prevent exposure to conditions that would allow these materials to ignite, to corrode surrounding materials, or to sicken or poison surrounding organisms either through toxicity or radioactivity. Similarly, buildings and other large structures designed to store hazardous waste must be able to prevent impact on their surroundings. A key feature of this containment system is *redundancy,* with multiple layers of defense, so that if one layer fails, there is still a backup layer in place. For instance, hazardous waste might be kept in robust containers, which are then kept in a facility where should something untoward happen to one of the containers, the facility itself will prevent leakage into the surrounding soil, water, or air.

The second level of response is the planning of transportation routes and siting of storage sites in a way that minimizes risk to the surroundings. Thus transportation routes are chosen that avoid locations where an accident would pose a particularly high cost, and storage is sited away from locations where either the human population or natural ecosystems would be at the greatest risk. For example, in the United States, low-level radioactive waste storage is distributed to one or more storage sites in each of the states, but for high-level radioactive waste, a single long-term storage facility is contemplated, and high-level waste is currently being stored onsite at nuclear power plants until a final decision is made about permanent storage. The proposed permanent storage facility will likely be situated far from any population centers, so that the distinction between low- and high-level waste-disposal shows how a higher degree of risk requires greater site selectivity. Although the region around a high-level waste site may be sparsely populated, precautions must be taken nevertheless to protect populations in the vicinity of a site, especially in cases where these communities have limited incomes and do not have the economic resources to independently assess the impact of the site. In addition to populated settlements, it is also important to consider impact on sites of natural or cultural significance such as habitat of endangered species or sites that are significant for indigenous populations.

The final level of response is planning for accidents in the movement and storage of hazardous materials. This risk can be greatly minimized but not entirely eliminated. For instance, if a storage facility leaks, equipment and procedures must be at hand to deal with the response. Similarly, large quantities of petroleum products and chemicals are moved using the rail network because of the economic advantages of using rail over highway networks for these bulk products. Statistically, railcars are less likely to encounter accidents than the same freight moved by truck since they are not interacting frequently with LDVs, cross traffic, signalized intersections, and the like. Nevertheless, railcars carrying chemicals do occasionally derail and leak, sometimes with devastating results, such as the derailment in the town of Lac-Megantic, Quebec, Canada, in July 2013, which led to loss of life and destruction of many buildings by fire. Transportation professionals must work with first responders to react to such emergencies by evacuating nearby residents, controlling the spread of the hazardous materials in question, and managing the spread of fire or any other secondary impacts.

17-4 Pathways to a Sustainable Transportation Future: Urban, National, and Global Examples

This section looks at pathways to sustainable transportation through the lens of transportation demand (both passenger and freight), energy requirements, and greenhouse gas emissions. Three geographic scales are considered: (1) a metropolitan area, (2) the United States as an example country, and (3) the global demand for transportation.

At any geographic scale, a wide range of measures for improving the sustainability of transportation systems are possible. These measures can be broadly divided between the passenger and freight sectors, and also between *technological* measures and *operations and best practices* measures. Table 17-3 provides examples of each. In this context, measures are technological when they either bring into the marketplace a new technology that has been under development, or expand the use of an existing technology to exploit its benefits in regard to congestion, pollution, or climate change. Measures are operational when they leave unchanged the numbers of different technologies deployed (including both vehicles and transportation, energy, or telecommunications infrastructure) and instead use them in a more efficient way. Measures such as ITS or vehicle sharing system rely on technology but are viewed as primarily operational since they take an existing vehicle technology (such as passenger cars or light trucks) and use them in a more effective way.

17-4-1 An Example of Future Adaptation of a Metropolitan Area

In this section, a preliminary concept is presented for incorporating options available in the near to medium term (on the order of 10 to 30 years) into a sustainable transportation plan for a metropolitan region. This plan applies to a typical urban region (hereafter referred to as "the region") in an industrialized country in Europe, North America, or Asia, which seeks to reduce total transportation energy consumption, emissions, and impact across the board. For brevity, energy savings will be used as a representative measure for all types of sustainable transportation benefits. The term "metropolitan region" or "urban region" implies the urban core of the city and surrounding suburban and exurban developments out to the perimeter of the developed area. For many cities, this definition includes both the area within the politically-defined city limits, and the built-up areas surrounding the city limits. The plan focuses on cities in industrialized countries because of their high per-capita transportation demand (measured in passenger-miles and ton-miles) and high per-capita energy use: many cities in developing countries could also

Transportation Sector	Technological Measures	Operations and Best Practice Measures
Passenger	• New fuels & drivetrain platforms • Lightweight materials • Improved aerodynamics • New compact *city cars* (compact 2-seat car) • Self-driving cars • Expanded public transportation infrastructure • Expanded nonmotorized transportation (NMT) (bicycling, walking, etc.) infrastructure • Expanded multimodal intercity infrastructure • Integration of sustainable transportation into design of built environment	• ITS to improve traffic flow • Promotion of multimodal options • Vehicle sharing systems (cars, bicycles) • Carpooling and vanpooling (scheduled or dynamic) • *Active transportation* (promotion of NMT and using NMT to access public transportation, partially for personal health) • Promotion of telecommuting
Freight	• Alternative fuels and drivetrain platforms for trucks, trains, ships • Expanded rail and ship linehaul capacity • Expanded and upgraded intermodal transfer points • Improved truck technology: aerodynamics, materials • Alternative networks for urban freight	• Use of Commercial Vehicle Operations (CVO) measures from ITS • Truck efficiency measures: reduced speed and idling • Greater use of existing intermodal systems • Expanded use of regional and local suppliers for materials and products • More efficient loading practices to increase utilization

Table 17-3 Representative List of Technological and Operational Measures for Passenger and Freight Transportation

benefit from some parts of the plan. As a simplification, the plan is static in nature, suggesting end targets for reducing energy consumption from each option in percentage terms, but not considering how long implementation might take or what might happen to baseline transportation demand and energy consumption in the meantime.

The plan incorporates measures for both passenger and freight transportation. On the passenger side, options including the transformation of the light duty vehicle fleet to more efficient models of vehicles, expansion of the use of public transportation, development of the use of limited-use vehicles and nonmotorized modes, and changes to land use patterns that can be implemented in the short to medium term. On the freight side, the region works with firms that provide movement of goods to, from, and within the region to use more energy-efficient modes, and also works with businesses within and outside of the region to replace some of the distant sources of goods with others that are closer. The plan does not include longer-term technologies such as fuel cell vehicles that may not be available in significant quantities for some time.

The summary of the plan is shown in Table 17-4. In the table, the percentage values given are compared to the future baseline value of total transportation energy consumption in a business-as-usual (BAU) case. The potential energy savings shown are

Function	Efficiency Option	Potential Energy Savings	Rationale
Passenger	Improved efficiency of light-duty vehicle fleet	10%–25%	Current and proposed improvements in fuel efficiency standards in all major auto markets of the world; efficiency improvement and rate of market penetration by HEVs during the period 2000–2007.
	Expanded (quasi-)public transportation: new rapid transit systems, expanded bus service, support for carpooling and vanpooling*	5%–15%	Up to 50% reduction in energy use possible for U.S. public transportation compared to car per passenger-km. 5%–15% figure assumes limited ability to add transit and/or carpools to all areas of region. Ability to reduce energy consumption assumes sufficient load factors in alternatives to cars.
	Motorcycles, LUVs & nonmotorized modes: motorcycles, mopeds, scooters, limited-use EVs, expanded use of bicycling & walking through improved infrastructure	2%–10%	30%–40% of all trips in Amsterdam, Netherlands, by bicycle. Rate of bicycle trips in Melbourne, Australia, increases fourfold in 20 years. Untapped market for electric LUVs and "smart cars" in industrialized cities.
	Short- and medium-term land use changes: locally available shops and facilities, telecommuting centers, revitalizing traditional shopping districts	2%–15%	California "smart growth" plan could achieve 3%–10% reduction in vehicle-km and energy use compared to baseline by 2020.[3] Smart growth concept is widely applicable in the United States and other countries.
Freight	Encouraging modal shifting: greater use of intermodalism, supporting the development of intermodal terminals within the region	2%–10%	Projection based on savings per tonne-kilometers moved by intermodal vs. truck, and maximum applicability of intermodal (not all sources of goods are served by intermodal service)
	Rationalizing freight demand: substituting near for far sources, developing local resources, e.g., local food production	2%–10%	Types of products most easily substituted (foods, farm products, certain building products, etc.) and their relative contribution to freight energy consumption

Note: Percentage reduction figures are for total urban transportation energy consumption relative to future value for baseline ("do-nothing") scenario; see explanation in text.

*Note that carpools and vanpools typically rely on collaboration between private individuals to share vehicles, but can be supported by local and regional governments, and are therefore labeled quasi-public in this option.

TABLE 17-4 Range of Possible Percent Energy Savings from Options Available to a Metropolitan Region

[3]Estimated savings published by California Energy Commission (2001).

representative of the range of values thought to be plausible; individual statistics are provided in support, where available. Several points can be made about Table 17-4:

- The plan in the table does not address the potential increase in energy consumption in the BAU case due to growing passenger-miles and ton-miles that might occur during the 10 to 30 year implementation period. As was shown in Example 14-3 in Chap. 14, it is possible for an efficiency option to reduce energy consumption relative to the future baseline and still have the resulting future value be higher than the current value. In a region experiencing rapid population growth, increases in population and growing demand per capita might wash out all the improvements achieved by the plan. If this happened, the region would be better off than if no steps had been taken, but no closer to reducing the absolute value of its transportation energy consumption footprint.
- The implementation of most or all of the options, rather than just one, makes it more likely that the absolute amount of energy consumption will decrease, as well as total transportation demand.
- The upper bounds of the percentage range for each option are intentionally made to be ambitious. It is likely that the twin challenges of reducing CO_2 emissions and preserving petroleum resources will motivate cities to move toward unprecedented levels of reductions.
- Although the projected percent savings are higher for some options than others, none of them are mutually exclusive of one another, so all are worthy of pursuit.
- No attempt is made to calculate a total energy savings value on the basis of the percent values provided. Such an analysis would require a careful treatment of the interaction between different options, for example, once the fleet has been transformed into a more efficient one, the total energy savings available from modal shifting away from light duty vehicles is less than the baseline.

In conclusion, the table does not result in a calculation of exactly how much transportation demand or energy consumption could be reduced overall by the region. However, the values in the table illustrate potential overall reduction ranging from 20 to 40 percent. These savings would lead to substantial reductions in CO_2 emissions, even if most of the remaining demand for motorized transportation was powered by fossil fuels. Furthermore, over the longer term, the region might shift the remaining transportation energy consumption to a source that did not increase CO_2 in the atmosphere (using technologies discussed in Chaps. 15 and 16).

17-4-2 Example of U.S. National Scenario to 2050: Passenger and Freight Sectors

In this section, we look at possible scenarios for meeting CO_2 reduction targets while still meeting transportation demand measured in passenger-miles for passenger and ton-miles for freight. In the scenarios presented, demand, emissions rates, and emissions are tracked over the intervening years as the total emissions across all modes approach a set target. Note that we do not attempt to construct a model of possible policies that would map the interaction between various incentives and disincentives and the development of technologies or best practices that could in practical terms result in the desired emissions levels. Such models are complex and beyond the scope of this chapter, although they can be found in the literature (for example, NAS, 2013; Leighty, 2009). The goal here is to show the rate of change in emissions rates, total demand, or modal share necessary to achieve the targets.

U.S. Passenger Transformation to 2050

The U.S. passenger transportation scenario assumes a starting demand level of approximately 4.5 trillion passenger-miles spread across four main mechanized modes: combined car and light truck (also known as light-duty vehicles or LDVs), aviation, bus, and rail (Table 17-5). The bus and rail modes include both urban and intercity travel. Not included is a mix of other modes with small numbers of passenger miles (motorcycle, demand-response, ferryboat, etc.) that would add approximately 0.5% to the total number of passenger-miles. Because of their small volume, these modes can be omitted without loss of accuracy. Also omitted is nonmotorized transportation, since it does not contribute to CO_2 emissions at the end-use stage. Emissions rates per passenger-mile and total CO_2 emissions in 2010 from the four modal groups are also shown in Table 17-5.

Three possible pathways are considered for U.S. passenger transportation:

1. *Business-as-usual case:* Transportation demand continues to grow linearly to 2050 with constant modal share values.
2. *Alternative 1:* 2050 emissions reduced by 80% compared to 2010 by reducing emissions values per passenger-mile for all modes while keeping percent modal share fixed.
3. *Alternative 2:* Additional emissions reductions achieved beyond Alternative 1 by keeping Alternative 1 emissions rates fixed and further reducing and modal shifting ton-miles.

Business-As-Usual Case

Under a BAU assumption, total demand grows linearly from 4.54 trillion passenger-miles in 2010 to 6.35 trillion in 2050, or 40% total growth. Each mode grows linearly by a constant amount so that percent modal share values in 2010 and 2050. This growth rate is slower than historical rates for the period 1970 to 2010, based on the anticipated dampening effect of sustained higher energy costs, congestion in transportation networks, and the increasing competitiveness of telecommuting for physical travel.

Mode	Demand (bill. pmi)	Emit Rate (lb/pmi)	Total (mill. tons)	2050 Emit Rate (lb/pmi)
Car + LT	3,645	0.650	1,184.7	0.093
Air	565	0.553	156.2	0.079
Bus	292	0.429	62.7	0.061
Rail	36	0.462	8.3	0.066
Total	4,538		1,411.9	

Source: Adapted from Davis et al (2013).

Table 17-5 Passenger-Mile Demand, Emissions Rates, and Total Emissions by Mode in Alternative 1

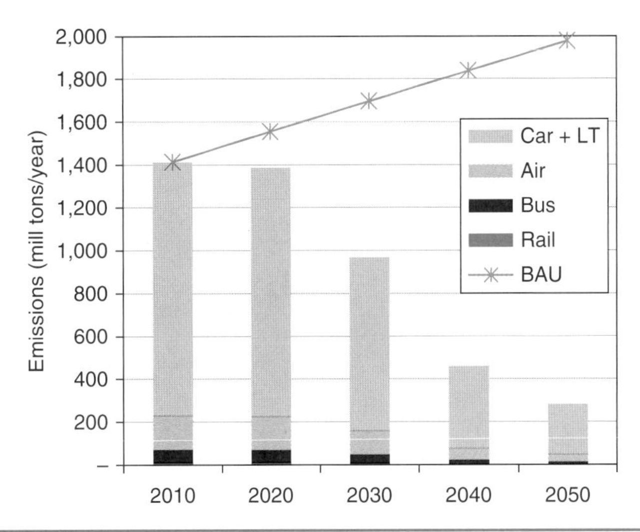

FIGURE 17-1 Pathway to 80% U.S. passenger CO_2 reduction over the 2010 to 2050 period via Alternative 1.

Note: "BAU" equals BAU case with baseline 40% increase in passenger-miles and constant emissions rates for all modes.

Because of constant modal split values, LDVs and air carry 80% and 12% of passenger miles, respectively, and the combination of bus and rail carry the remaining 8%. In a BAU case, emissions rates remain constant, so total CO_2 emissions also grow linearly by 40% from 1.4 billion tons in 2010 to 1.9 billion tons in 2050.

Alternative 1: Technological Improvement Scenario

In the first scenario, the goal is to achieve the emissions target by allowing the passenger-mile growth rates to remain unchanged and reducing emissions per passenger-mile for each mode. The target 2050 CO_2 per passenger-mile can be calculated from the known total passenger-miles, modal share for each mode, and 2050 target of 282 million tons. The resulting values are shown in the right-most column in Table 17-5 for each of the four modes. Each emissions rate is decreased by 85.7%, reflecting both the 80% reduction target and the 40% growth in demand.

Next, a triangle function is applied to model the change pathway of the emissions rates over the lifetime of the scenario, as a simple way of approximating an "S-shaped" transition that takes place far into the future where many parameters are uncertain (see Chap. 3).[4] This approach is used in subsequent scenario construction as well. The triangle function captures three stages of the transition scenario life cycle: (1) early ramp-up stage where efforts to reduce emissions are still gaining momentum, (2) maximum rate of change stage where efforts from the first stage result in the fastest rate of change, and (3) diminishing returns stage where the effort is nearing its target and further gains are the most difficult to secure.

Figure 17-1 shows emissions from each mode and the combined total as emission are reduced at first modestly by 2% from 2010 to 2020, and more rapidly after that.

[4]Law and Kelton (1991, p. 341) state that the triangle function is "used as a rough model in the absence of data."

Reductions over the period 2020 to 2040 are particularly rapid, amounting to a change of 928 million tons per year in the emissions rate, or an average of 46 million tons per year reduced. Note that the rail emissions figures are imperceptible in the figure due to their small values, but for four data points in the period 2020 to 2050 amount to 8.2, 5.7, 2.7, and 1.7 million tons per year. With emissions rates as the only means of reducing CO_2 (due to the exclusion of demand reduction and modal shifting), the emissions rates reduction are substantial, especially in the years after 2020. The LDVs is reduced from 0.58 lb CO_2 per passenger-mile in 2020 to 0.162 lb CO_2 per passenger-mile in 2040. We have a line-of-sight on the mix of technologies that might achieve this reduction, including electrified EVs and PHEVs (the latter possibly using part or all biofuels as liquid fuels), the launch of hydrogen vehicles, the improvement of vehicle efficiency across all transmission platforms using advanced aerodynamics and lightweight materials, and the deployment of sufficient generation assets to provide CO_2-free energy.

Alternative 2: Technological Improvement with Modal Shifting and Demand Reduction

As discussed in Alternative 1, we can theoretically achieve secure, low-carbon transportation energy use by shifting all modes (LDV, air, bus, rail) to a low CO_2 per passenger-mile solution. A possible weakness of this approach is its reliance on a single strategy, namely advanced technologies. Alternative 2 proposes the continued development of advanced technologies coupled with a transformation of our pattern of land use and modal split among passenger modes so that urban and suburban residents have the same access to amenities, but generate fewer passenger-miles of travel compared to the BAU case. Remaining passenger miles would then be shared more evenly among public transportation, NMT, and private LDVs, which would transition to EV, PHEV, or HFCV technology. Such a change would have the effect of reducing transportation energy demand and in some cases making it easier to provide the needed energy (e.g., electrified public transportation using catenary). For intercity travel, modal share for bus, rail, and especially high-speed rail would be increased, and these modes along with aviation would reduce CO_2 emissions by shifting to an appropriate mix of electrification (applicable only to rail) and biofuels. Note that Alternative 2 represents a substantial shift in direction from the 1970 to 2010 period in terms of per-capita demand and modal share figures for the bus and rail modes, which have relatively low modal shares. The figures used are therefore representative numbers to explore the impact of ambitious changes over a multidecade time horizon on emissions. Policies that might achieve these changes are outside the scope of the discussion—see discussion in the next section.

As a guiding tenet, Alternative 2 reduces energy consumption in urban and local travel, and hence the need for new sustainable energy infrastructure, by making it easier for the great majority of residents to choose modes other than light-duty vehicles for many of their routine urban and short-distance trips. Three major policy elements support this vision:

1. *Land use designed to facilitate use of modes other than private light-duty vehicles:* Housing, workplaces, educational centers, shopping, and other amenities are located to provide connectivity for either public transportation or NMT. This includes both geographic location of facilities in the urban region and

design of connections with the public transportation and NMT network. Over time, work and shopping amenities that have migrated to the fringes of the urban area shift back to public transportation hubs. Alternatively, edge developments are transformed into public transportation and work, school, retail, or educational hubs through appropriate planning and development of a balance of built environment elements.

2. *Prominent role for public transportation:* Both work- and nonwork trips are served by public transportation. Service is abundant, convenient, and provided in an appropriate mix of local and express service so that both short- and longer trips are feasible. Vehicles use a mixture of carbon-free electricity and biofuels for propulsion, eliminating both CO_2 emissions and air pollutants.
3. *Improved infrastructure for NMT modes:* By providing a comprehensive network of NMT infrastructure facilities (bike lanes, walk-cycle paths parallel to roadways or on separate trajectories, etc.), walking and bicycling become available to most or all residents for short- to medium-distance trips when weather conditions permit. Thus each residential area has NMT infrastructure that allows residents to walk or cycle within the neighborhood, and also to access nearby amenities (e. g., shopping, workplace, rapid transit stops). An interconnected network of NMT facilities throughout the region allows residents to occasionally make longer trips to more remote parts of the region by cycle or on foot if they desire.

In Alternative 2, LDVs using sustainable energy resources continue to have a role, but the balance has changed compared to the current situation, which private LDV dominates. Whereas in Alternative 1 the latter serve more than 80% of the passenger-miles for the entire duration between 2010 and 2050, the share of passenger-miles is shared more evenly between private LDV, public transportation, and nonmotorized modes in Alternative 2.

Alternative 2 treats urban and intercity passenger-miles differently. First, 72% of the 6.35 trillion passenger-miles, or 4.57 trillion, observed in Alternative 1 in the year 2050 are urban. The urban LDV figure of 4.1 trillion is reduced by 50% thanks to land use and multimodal changes, leaving 2.05 billion passenger-miles to be shared among all urban modes. The modal share is no longer dominated by LDV but divided evenly between a mix of other modes. The resulting shares are 5% for NMT, 27.5% for public bus systems, 27.5% for public rail systems, and the remaining 50% for LDV. An additional 0.47 trillion urban passenger-miles are carried by modes other than LDV. The remaining 28% of passenger-miles are intercity: 12% are air travel and an additional 16% are intercity LDV passenger-miles. Intercity travel is not affected by land use or multimodal efforts in the urban area.

The share for LDVs in 2050 is treated in the same way as in Alternative 1. The LDV fleet shifts to relying on electricity, hydrogen or liquid fuels, with low overall CO_2 per passenger-mile. Energy companies provide a comprehensive network of charge points and filling stations for the new fuels, and grid operators control the charging of vehicles to optimally use electricity made available from the grid. LDVs would have a prominent role in one-off trips with unusual origin-destination pairing, trips where it is convenient to move large numbers of people in a single vehicle, or other situations where public transport or NMT is not well-suited.

For intercity travel, the major mode in use in 2010 is aviation, with 565 billion passenger-miles, and intercity LDV, with 727 billion passenger-miles. Unlike the case of urban transportation, no large-scale policy to change intercity travel habits is assumed on

Mode	Demand 2010	Demand 2050
Car + LT	3,645	1,474
Air	565	452
Bus	292	912
Rail	36	912
Total	4,538	3,750

Table 17-6 Passenger-Mile Demand by Mode in 2010 and 2050 in Alternative 2

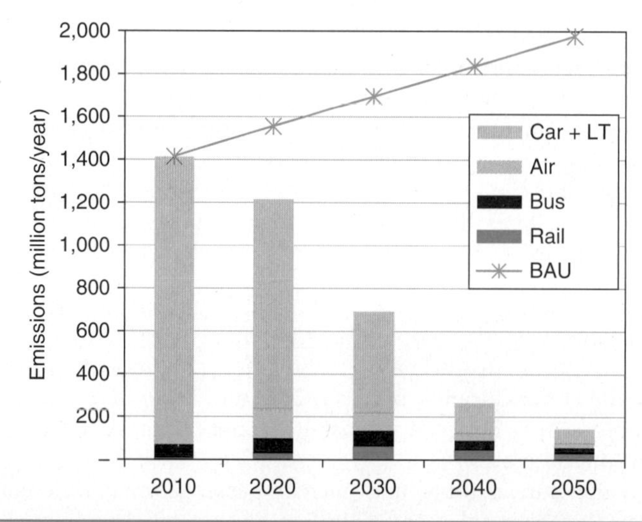

Figure 17-2 U.S. passenger CO_2 reduction pathway over the 2010 to 2050 period via Alternative 2, compared to BAU case.

the grounds that intercity destinations often meet unique needs (e.g., tourism, visiting loved ones) and are therefore more difficult to replace. Thus combined figure of 1.29 trillion passenger-miles grows by 40% to 1.81 trillion as in Alternative 1, but instead of being carried only by air and LDV, it is divided evenly between air, LDV, rail, and bus. The combined urban and intercity passenger-miles for the four major modes for both the start year (same as Alternative 1) and finish year are shown in Table 17-6. The 2050 figure for NMT is 102 billion passenger-miles; these are not shown in the table, however, since they do not contribute to CO_2 emissions.[5] Lastly, for simplicity the same emissions factors from Alternative 1 are used (Table 17-5), and the triangle function is used to model the change in modal passenger miles and emissions factors between 2010 and 2050 for all four modes.

Emissions for each of the four modes are shown in Fig. 17-2, where urban and intercity emissions are combined into a single amount. The emissions from LDV and air

[5]For reference, the 102 billion passenger-miles implies that with a population of 450 million in 2050, the average U.S. resident travels approximately 225 miles per year by NMT.

modes decrease moderately to 2020 and substantially thereafter, as rapidly declining emissions rates and decreasing modal share drive down emissions for these modes. Emissions from bus and rail at first increase due to rising modal share in the period 2010 to 2030, but eventually decrease due to the effect of declining emissions rates in the period 2030 to 2050. The result is emissions of 144 million tons CO_2 in the year 2050, a 90% reduction.

Although the emissions cuts are larger for Alternative 2, it should not be inferred that this approach is inherently more effective than Alternative 1. Both scenarios represent a sustained campaign that could eliminate most of the CO_2 emissions from passenger transportation by 2050. Furthermore, if the total passenger-mile and modal share changes in Alternative 2 could be achieved by 2050, it would be possible to reduce emissions factors less so that the passenger sector achieved but did not exceed the 80% target. The creation of an alternate scenario that achieves 80% reductions with the same modal shares and demand in 2050 as in Alternative 2 is beyond the scope of the current presentation. As a simple example, however, suppose the goal is to achieve emissions in 2050 for all modes of 282 million tons, equal to Alternative 1. If the LDV emissions factor were 0.28 lb CO_2per passenger-mile (instead of 0.093 lb as in Alternatives 1 and 2 as given), LDV emissions would be 206 million tons, and emissions for remaining modes would be 76 million tons, so that the objective would be met. Total LDV emissions per mile would still be reduced substantially (by 57%) compared to 2010.

Comparison of Alternatives 1 and 2: Advantages and Disadvantages

We discuss the advantages of Alternative 2 first, and then those of Alternative 1. From a systems perspective, the multipronged solution in Alternative 2 has several advantages over the Alternative 1 solution where technological changes are applied in isolation with no consideration given to reducing demand or shifting transportation modes. Because many fewer passenger-miles are moved by LDVs, the total energy requirement is less, so that the necessary investment in carbon-free energy sources is reduced. A substantial number of LDV passenger miles and hence LDVs still remain, however, and if these vehicles use electricity or hydrogen, they function as absorbers of intermittent renewable energy (and potentially suppliers back to the grid if the V2G component is applied as well). However, to the extent that passengers travel in *electrified* public transportation that uses electric catenary (i.e., overhead wires or third rails) for propulsion, including trolleybuses, light- and heavy rail, streetcars, and the like, the complexity of battery technology—manufacture, charging and discharging, disposal and recycling—can be avoided. Nonelectrified public transportation in the form of motor buses may also be able to use biofuels more efficiently, and may be better suited than private LDVs to hydrogen fuel cell propulsion as that technology matures.

The whole system solution can address other problems not related to energy and CO_2 emissions, including the following:

1. *Congestion from excessive auto dependency:* By redistributing trips away from urban expressways and arterial streets, the Alternative 2 solution can help to reduce congestion.
2. *Improved access to urban amenities not requiring an automobile:* At any given time, an urban area will invariably have some significant segment of the population that does not have access to a car. These residents may be too young, too old, or too poor; they may also have disabilities which prevent them from being able

to operate a private LDV. The whole systems solution improves access for this population segment.

3. *Improved health due to opportunities to use nonmotorized modes:* Improving facilities for walking and biking, and making connections to various urban amenities possible by these modes, gives resident opportunities to get exercise as they go about their daily travel routine.

For Alternative 2 to succeed, as opportunities arise, urban regions must develop supporting land use patterns, often at or near the urban core where urban renewal may be a possibility. Public transportation and NMT infrastructure must then be simultaneously put in place so that the new land use and transportation alternatives can work together. At the same time, the LDV fleet would be gradually converted to alternative energy sources.

The advantage of the Alternative 1 solution is that it builds on a direction that the transportation system has already taken in the period leading up to 2010, and therefore has the benefit of momentum. Like Alternative 2, the Alternative 1 solution requires sustained effort over many decades to transform the system and eliminate widespread ecological problems. Unlike Alternative 2, the needed change is limited to the transportation energy supply and vehicle fleet. A supply of some 15 to 20 quads per year might be needed by the year 2050 for a fleet of 300 to 400 million cars and light trucks, plus the required aircraft, buses, and rail vehicles. Further advantages of Alternative 1 are laid out as follows:

- *Alternative 1 avoids need for major restructuring of the U.S. urban and suburban built environment:* This transformation would avoid any major reshaping of popular patterns of living, working, and accessing amenities. Millions of residents would continue to live in suburban neighborhoods as they do now, with access to places of employment either in the urban core or in edge cities at the periphery of the metropolitan area.
- *Alternative 1 builds on technological directions currently being pursued:* Intelligent transportation systems (ITS) and advanced concepts such as self-driving vehicles might mitigate problems with congestion. In the 2010 to 2013 period, the research emphasis in the private sector is very much on the technological solutions that could make this vision possible in 2050 (ITS, electricity, advanced biofuels, and hydrogen as energy carriers, development of wind and solar technologies for primary energy).

Restructuring of the urban form is happening in places, such as examples of transit-oriented development around new metro stations (e.g., in the Washington, D.C. area) but the sum total of these efforts is small compared to what would be needed to profoundly shift modal share of passenger-miles or trips away from privately owned LDVs (see data presented in Chaps. 7 and 10). The most likely outcome of this transition is some mixture of Alternatives 1 and 2: wholesale transition by 2050 to some mix of electricity, hydrogen, and biofuels in LDVs, complemented by changes in the makeup of public transportation, NMT offerings, and land use patterns. If present trends continue, the majority of the change is likely to be of the former type and not the latter.

17-4-3 U.S. Freight Transformation to 2050

Like the preceding section on passenger transportation, the goal of the model of future freight demand and CO_2 emissions in this section is to create a baseline prediction of the growth in CO_2 emissions to the year 2050, and then to study a set of changes that might result in a substantial emissions reduction. Such a transformation would enable the freight sector to make a contribution to climate protection alongside passenger transportation, electric utilities, and other sectors. The freight scenario presented in this section is adapted from a study in Master of Engineering project at Cornell University (see Al-Jefri et al., 2008).[6]

For this scenario, the base year chosen is 2005 rather than 2010. The target is 50% reduction compared to 2005 levels. This target is chosen because the freight sector is already relatively efficient compared to passenger in terms of "mass-miles" moved, and the relatively embryonic state of efforts to develop carbon-free energy sources for freight.[7] Of the 4.2 trillion ton-miles of freight carried in 2005, about 1.7 trillion were carried by rail, 1.3 trillion by truck, and 600 million by marine, so these three modes together comprise 86% of the total market. The scope of the activity and emissions in the model is limited to truck, rail, and marine, due to the difficulty of obtaining emissions data for pipeline and air, the other two main modes.

Baseline 2005 and Business-as-Usual CO_2 Emissions to 2050

For the baseline year of 2005 used in the scenario, most of the surface freight CO_2 emissions come from the truck mode. Of the estimated 480 million tons of CO_2 emitted from surface freight excluding pipeline in 2005, 85% are from truck, with rail emitting 10% and marine emitting 5%. Therefore, many of the policy options in the study revolve around reducing emissions from truck, either by making intramodal changes (affecting truck only, such as implementing best efficiency practices or the use of biodiesel), or by shifting freight to less CO_2 intensive modes.

The further assumptions for the activity and emissions during the 2005 to 2050 timeframe are as follows:

1. Predictions for baseline demand for truck ton-miles in a BAU scenario are for an increase in response to growing GDP, but at rates slower than those of 1970 to 2005 and below the rate of growth of GDP. We predict that expansion of geographic distribution patterns will slow as globalization runs its course and freight grows as a result of increasing flows on existing supply chains, rather than the disruptive transition from a national to a global market. Also, congestion and higher energy costs become stronger factors in dampening the growth of physical movement. Truck freight growth has historically followed the GDP trend closely, but going forward to 2050, freight activity per unit of GDP is expected to decline. Thus truck ton-miles will continue to grow, but not as fast as GDP.
2. Baseline demand for rail continues to grow in line with recent historical trends, thus exhibiting steady growth through 2050. Water freight ton-miles had been in decline since the 1980s but are assumed to remain constant as the inland, coastal, and Great Lake networks provide an alternative for landside networks.

---

[6]Acknowledgment: Thanks to Hamzah Al Jefri, Jeff Bernstein, Richard Larin, Sunjeet Matharu, Tao Shi, and Selin Un for their contribution.

[7]If one considers the total mass moved multiplied by the distance traveled for the U.S. passenger sector, one can derive an estimate of ton-miles that is found to be of similar value to the total ton-miles moved by the freight sector, but consuming approximately three times as much energy.

3. CO_2 emissions per ton-mile include primarily tailpipe emissions from the combustion of diesel fuel, since this fuel is common to all three modes. There are additional small amounts of life-cycle emissions from the combustion of biodiesel, and electricity used in electrified railways in the case of freight. The baseline model includes a modest efficiency improvement factor for all three modes, in line with recent historical trends, so that emissions per ton-mile decline somewhat over the 2005 to 2050 timeframe. For truck and rail, the emissions rate does not fall as fast as ton-miles for each mode rise, so that overall emissions increase. For marine, ton-miles are constant per statement number 2, and emissions rates are decreasing, so total emissions decrease.

CO_2 emissions are computed on the basis of ton-miles of demand and CO_2 per ton-mile for each year of the model to create a BAU scenario for total emissions to the year 2050 (Fig. 17-3). For background, emissions are shown from the 2000 to 2005 period, during which time emissions increased from 420 to 480 million tons. The curved shape of truck emissions are due to the reduction in emissions per ton-mile dampening growth in total emissions up to approximately the year 2040 but accelerating economic growth becoming the dominant factor thereafter. The BAU outcome of 820 million tons of CO_2 in 2050 can be compared to the target of 50% of 480 million or approximately 240 million tons, which is 71% less than the BAU value.

Results from Individual and Combination Improvement Options

In this section, we focus first on four options for reducing CO_2 emissions from the modeling efforts: (1) implementation of truck best practices, (2) influx of biodiesel from all feedstock sources, (3) modal shifting from truck to rail, and (4) electrification of rail service. The impact of each option is illustrated in isolation to show its maximum potential impact. Combinations of options are presented later. In each case, the triangle

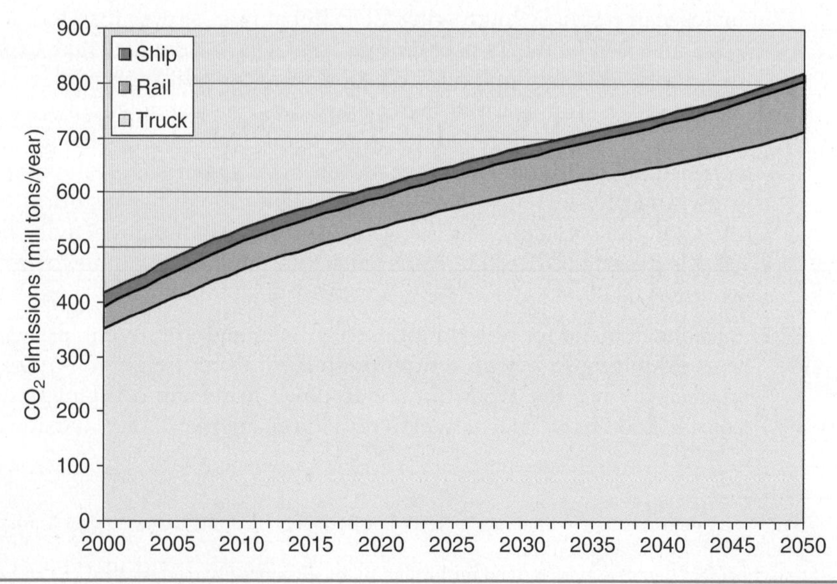

FIGURE 17-3 Predicted CO_2 emissions from U.S. domestic truck, rail, and ship modes in a BAU scenario through the year 2050.

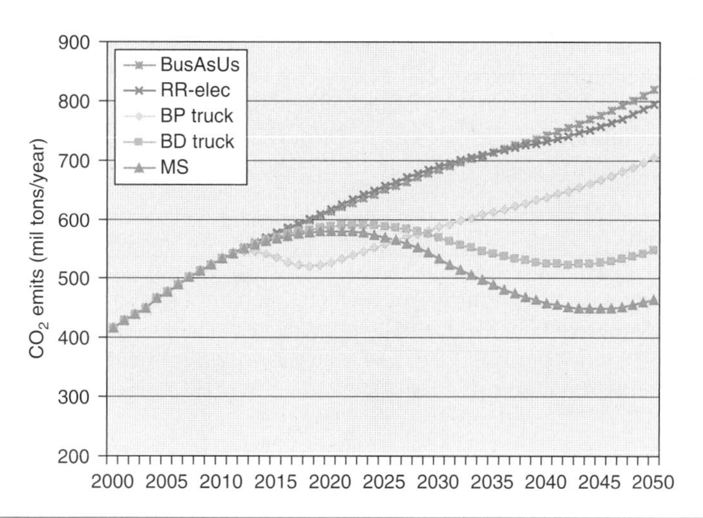

FIGURE 17-4 Pathway for CO_2 emissions from BAU and single policy scenarios (RR-elec = rail electrification, BP truck = best practice, BD truck = biodiesel, MS = modal shifting).

curve shape is used to chart a path for the new technology in the model for each year from a starting year in 2005 through the target year of 2050. The new technology is imposed on the freight transportation demand in ton-miles, which is held to a single predetermined pathway for all years and for all options. The effect of each individual option is shown in Fig. 17-4.

- *Implementing best practice in the truck fleet:* Best practices for energy efficiency in trucks are implemented between 2010 and 2020 because the technologies involved are commercially available, and because there is strong incentive to implement these technologies, due to the attraction of avoided fuel cost. Penetration reaches 80% of truck ton-miles in 2020 and remains at this level for the remainder of the time horizon. On this pathway, emissions decrease from a peak of 546 million tons in 2012 to 522 million tons in 2020. After 2020 emissions per ton-mile are constant and demand continues to grow, so that total emissions are driven upward to 706 million tons in the year 2050.
- *Influx of biodiesel from all feedstock sources:* In this scenario, biodiesel from crops such as soybeans is aggressively implemented and is eventually supplemented by biodiesel from algae and other noncrop biofuels, so that 50% of the petro-diesel sales in the BAU scenario in 2050 is replaced with biodiesel. In response, emissions first peak at 591 million tons in 2023, then fall to 525 million tons in 2042, then increase to 549 million tons. Emissions rise from 2042 to 2050 because during this late stage most of the transition to biodiesel has already occurred, but freight ton-miles are rising at their fastest rate.
- *Truck to Rail Freight Mode Shift:* In this scenario, a major investment in new rail infrastructure is expended to overcome the service disadvantages of rail and transfer a significant amount of freight from truck. The new program adds capacity along key linehaul corridors. It also adds and upgrades capacity at intermodal transfer points to make them more numerous and more convenient, so as to be more competitive with trucking. Assuming the investments can be

made, in this scenario the market share of rail increases so that, by 2050, 50% of the freight that would have been moved by truck in the BAU scenario is moved by rail. As a result, CO_2 emissions peak at 580 million tons in 2020, then fall to 451 million tons in 2044 as most of the modal shifting is completed, and then rise slightly to 466 million tons in 2050.

- *Rail shift to electrification:* In this scenario, the modal shares of truck and rail are held constant, and the rail share is shifted to 50% carried on electrified lines by the year 2050. The analysis assumes that enough of the intercity rail lines have been electrified and enough electric locomotives are available in the national locomotive fleet for this to be physically possible. Nevertheless, because rail has such a low percentage of CO_2 emissions to begin with, the effects on emissions of this scenario are small. Emissions in 2050 are just 3% lower than the base case, at 796 million tons rather than 820 million tons.

As shown in Fig. 17-4, all of the scenarios except rail electrification make significant reductions compared to the BAU scenario (labeled "BusAsUs" in the figure) but they do not approach the target of 240 million tons/year. The deepest reductions are achieved by the truck to rail shift option, at 466 million tons, but this level is only slightly less than the 2005 value of 480 million tons. We therefore look at two possible combined technology scenarios, in which most or all of the options are used in combination to further reduce emissions:

- *Combination 1—Multiple technology scenario that achieves the 50% reduction target:* In this scenario, 2050 targets of 80% penetration of best practices and 50% penetration of biodiesel are combined from the preceding section. Modal shifting to rail is set at 20%, which requires a substantial investment in new infrastructure but is less ambitious than the 50% target used above. Lastly, 20% of the rail freight is electrified by 2050, further reducing CO_2 emissions. Thanks to the combination of measures, 2050 emissions attain the target of 240 million tons/year. As in the comparison of individual options above, the contribution of rail electrification is modest. Of the components in the scenario, best practice, biodiesel, and modal shifting combined reduce emissions to 247 million tons, which comprises most of the reduction.
- *Combination 2: Multiple technology scenario based on technology available today:* The "Combination 1" scenario achieves the 50% target but also relies on the development of technologies that are not yet available in a mature, commercially viable form. Therefore, in the "Combination 2" scenario, we limit the influx of the new technologies to levels that are commercially available at present. Truck best practices still achieve a penetration of 80% in 2020, since this option relies on technologies that are at or close to the required level of maturity. However, the influx of biodiesel and modal shifting is limited to 10% in each case, for example, based on biodiesel exclusively from crops and an aggressive implementation of biodiesel manufacturing, a peak of 10% of the total energy content of the fuel is thought to be achievable. Also, modal shifting to truck is limited to 10% because each additional increment of modal shifting requires both more infrastructure and development of new services that can match the flexibility and high level of service of trucking across many different supply chain sectors. Rail electrification is not included at all in this scenario because

there is no present-day movement actively promoting its implementation (although some additional electrification may be forthcoming on the passenger rail side).

The result of Combination 2 is a substantial reduction from BAU in 2050, but a slight net increase in CO_2 compared to 2005. The resulting gap between what can be achieved with today's technology and what is required for achieving the 2050 goal is shown in Fig. 17-5. The difference between the two scenarios is a function of two main factors:

1. *Need to develop advanced biofuels beyond what is currently available:* These biofuels would be generated from plants or woody materials grown on marginal lands not suitable for growing crops (thus not competing with the food supply), or else from algae. Since the resulting advanced biofuel supply chain would be more productive than the current soybean-based one, the additional available fuel would allow the increases in biofuel supply in Combination 1.
2. *Broadly available, high level-of-service rail network:* Although rail is making inroads on the truck high-value freight market, especially with double-stack container service, it has not yet reached the point of rapid transition envisioned in Combination 1. Intermodal terminals would need significantly higher throughput, and more origins and final destinations for truck freight would need convenient linkages to these terminals.

Adequacy of supply for multiple transportation uses poses a long-term challenge for the development of advanced biofuels. Not only would the surface freight system depend on some form of biodiesel or similar fuel for long-distance haulage without the need to make frequent refueling stops, but aviation and intercity LDV travel may in the future depend on biofuels as well. Already, the aviation industry envisions biologically derived jet fuels to replace current petroleum-derived versions, and drivers may expect to have a liquid biofuel alternative for PHEVs when electric charge is exhausted on long trips. In the short run, biofuels appear more favorable than electrification for the rail network as a means of eliminating CO_2 emissions because of the high capital cost for

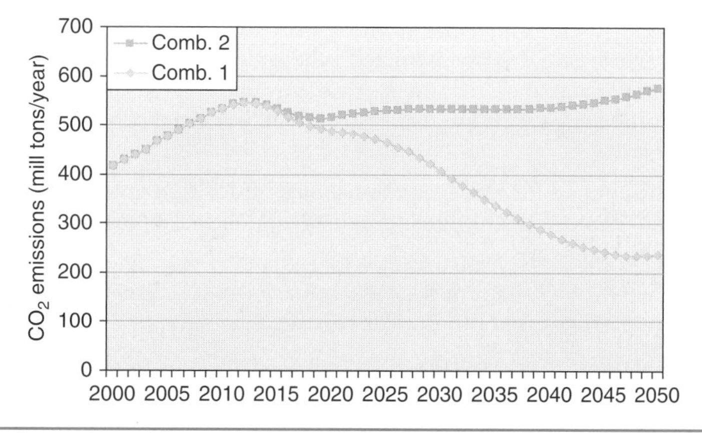

FIGURE 17-5 Comparison of CO_2 emissions from multiple-technology scenarios Combination 1 and Combination 2.

catenary and new locomotives compared to the current high energy efficiency and hence relatively low energy cost per ton-mile achieved by rail. However, if it is not possible to power intercity freight, intercity LDV travel, and aviation all on available biofuels, it may be necessary to increase the amount of electrification of rail despite the higher cost because rail is the only one of the three where electrification is even practical. At that point, liquid hydrogen stored onboard the train and used as a substitute for biodiesel in locomotives might become attractive, despite its high cost per unit of energy, so as to avoid even higher capital investment in electrification.

17-4-4 Example of Global Transformation to 2100: Transportation Growth with CO_2 Phase-Out

In this section, we continue the presentation of possible pathways to a reduction in transportation CO_2 in the twenty-first century by examining transportation demand at a global rather than U.S. national level. The projection of CO_2 emissions scenarios for the twenty-first century presented here is calculated based on the assumptions for the world transportation market already presented in Chap. 1: (1) world population stabilizing at 9 billion by 2050; (2) world transportation market divided into high-income high per-capita demand, high-income low per-capita demand, mid-income, and low-income; (3) separate presentation of passenger and freight demand scenarios; (4) stabilization in world transportation demand by 2100, with both total population and per-capita demand for passenger and freight transportation roughly constant after that (minor fluctuations year-on-year around an average value would be expected). The purpose of the scenarios is to examine changes in total CO_2 emissions from transportation from decade to decade as both demand (in passenger-kilometers and tonne-kilometers) and CO_2 emissions rates change. The CO_2 reduction targets are 100% reduction for both passenger and freight in this case, but the approach would be similar for 80% or 50% reductions.[8]

Assumptions Common to Both Passenger and Freight Sectors

For both passenger and freight sectors, the following assumptions hold:

- Annual passenger-kilometers and tonne-kilometers for each of the four segments are the same as those presented in Chap. 1. Although an alternate scenario could be constructed where these volumes are measurably reduced compared to the Chap. 1 base case (through changes in land use, telecommuting, shifting supply chains, etc.), this possibility is not considered here for reasons of brevity.
- The climate for transportation energy transition is favorable: leaders agree on and carry out strong and effective policies, and technologies evolve in a favorable way.
- The two high-income segments phase out CO_2 emissions from transportation by 2050 because of their relatively wealthy status.

[8]At the global level 100% reductions are chosen since there are many U.N. member countries, such as low-lying island nations, that view the need to eliminate CO_2 emissions with greater urgency than the U.S. targets of 50% to 80% would suggest.

- The mid- and low-income countries phase out of CO_2 emissions from transportation is extended to the year 2100 because of their relatively low per-capita economic activity, the rich countries' disproportionate contribution to climate change up to the present, and the anticipated economic hardship of growing their economies while at the same time transitioning transportation fuels. The Kyoto Protocol, in which industrialized countries were required to achieve reductions in the first round to 2012 while industrializing ones were not, provides a precedent for this delay in implementation of requirements.
- The decade-by-decade reduction in emissions rates per passenger-kilometers or tonne-kilometers is modeled using a triangle function, similar to the U.S. model above. Thus the rate of reduction of emissions reaches a maximum in the year 2025 for the high-income countries and 2050 for the low-income countries.

Assumptions Specific to Either Passenger or Freight Sectors

For passenger transportation, the starting emissions rate in the year 2000 is based on an average of emissions rates per passenger-kilometers in that year for France, Germany, Japan, the United Kingdom, and the United States of 0.162 kg CO_2 per pkm, which is in turn based on an energy consumption rate for those five countries of 1.8 MJ/pkm.[9] This emissions rate is then applied to the world as a whole. The calculation therefore assumes that per-capita transportation emissions in rich countries are high primarily due to high per-capita passenger-kilometers; energy efficiency per passenger-kilometer in these countries, excluding the United States, are already quite efficient, so energy efficiency and emissions rates per passenger-kilometer in mid- and low-income countries are not likely to be markedly better. For comparison, the U.S. rates are higher, at 2.0 MJ/pkm and 0.209 kg CO_2 per pkm. The result of the assumed average emissions rate is that in 2000 total emissions amount to 6.7 billion tonnes CO_2, on the basis of 41.4 trillion pkm worldwide.

On the freight side, more uniformity of emissions rates are assumed between countries in the high-income high-intensity segment (e.g., United States), the high-income low-intensity segment (e.g., France, Japan, etc.), and the other two segments. Therefore the starting emissions rate in the year 2000 is assumed to be equal to the U.S. rate in that year of 0.099 kg CO_2 per tonne-kilometer. This emissions rate is then applied to the world as a whole, resulting in 2000 emissions of 2.9 billion tonnes CO_2 on the basis of 29.3 trillion tkm worldwide.

The combined emissions for passenger and freight of 9.6 billion tonnes can be compared to the world total of 25.9 billion tonnes for all uses of energy. Thus transportation sources represent 37% of the total, and the combination of industrial, commercial, and residential emissions the remaining 63%. The transportation share is higher than would be observed in the share of total world energy consumption, but this is to be expected since transportation is almost entirely petroleum dependent whereas other sectors use measurable amounts of renewable and nuclear energy as primary sources, reducing CO_2 emissions per unit of energy consumed.

---

[9]Per-capita energy consumption per passenger-km obtained from Schafer et al. (2009).

Results of the Scenario Analysis

The trajectory of worldwide transportation CO_2 emissions for the twenty-first century is shown for passenger in Fig. 17-6 and freight in Fig. 17-7. On the passenger side, in 2050 when all CO_2 emissions in high-income countries have been eliminated, 7.4 billion tonnes from low- and middle-income countries remain, after which the emissions fall

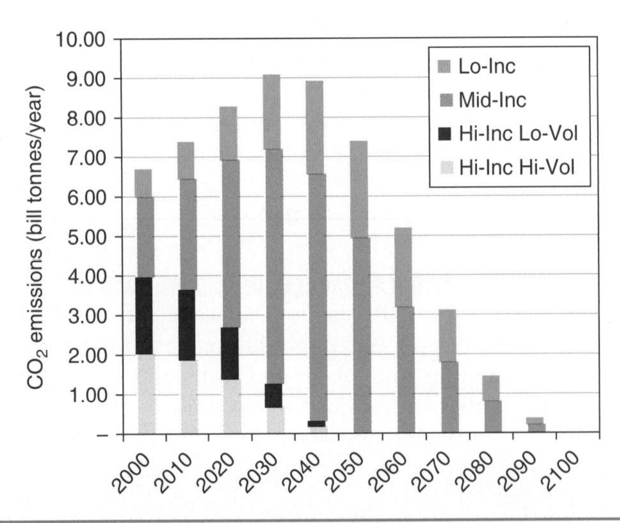

FIGURE 17-6 CO_2 emissions from world passenger transportation divided into four economic segments 2000 to 2100.

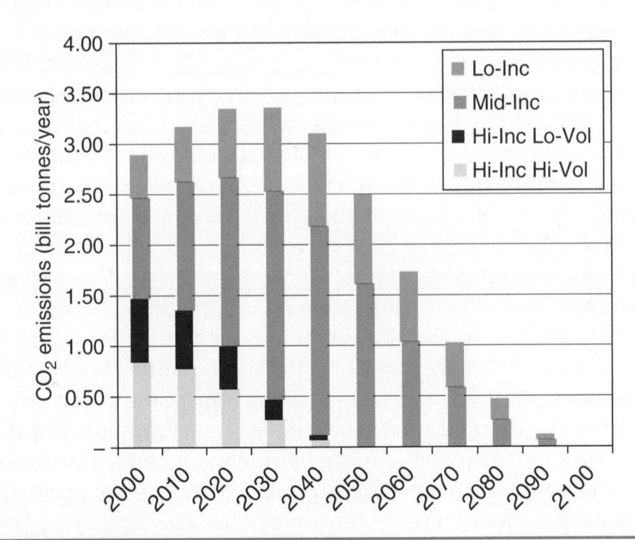

FIGURE 17-7 CO_2 emissions from world freight transportation divided into four economic segments 2000 to 2100.

Note: "Lo-Inc" = low income, "Mid-Inc" = middle income, "Hi-Inc Lo-Vol" = high-income low per-capita volume, "Hi-Inc Hi-Vol" = high-income high per-capita volume.

steadily to zero in 2100. On the freight side, the 2050 figure is 2.5 billion tonnes when emissions in the high-income countries have been eliminated. Total freight emissions peak sooner and fall more rapidly than passenger because total tonne-kilometers in mid- and low-income countries do not grow as much or as rapidly as passenger-kilometers on the passenger side.

Although many variations on the scenario presented are possible (see the end of the chapter for an exercise), the analysis shows that given the current world trajectory in the 2000 to 2100 period and likely rapid economic growth it will be challenging to reduce total annual CO_2 emissions, despite ambitious assumptions about the rate of technological change. For example, the model assumes 100% elimination of CO_2 emissions in the high-income countries by 2050, but studies such as that of the U.S. National Academy of Sciences or Leighty et al. target only 80% elimination by that point (NAS, 2013; Leighty, 2009), which is in itself a daunting task. As a result, using only the values at the beginning of each decade and not considering intervening years, the total annual emissions only drop below 2000 levels in the year 2050 for freight and 2060 for passenger. The rise for passenger is particularly marked, increasing from a 2000 value of 6.7 billion tonnes to a value in 2030 of 9.1 billion tonnes (a rise of 36%) before beginning to decline. It may be hoped that national governments working through bodies such as the Intergovernmental Panel on Climate Change (IPCC) would accelerate the transition process to avoid the initial rise in emissions shown, but the required rate of change makes the development and deployment of new technologies even more challenging.

17-4-5 Discussion of Future Scenarios

In the preceding case studies, we have looked at possible future scenarios for sustainable transportation at a regional, national, and global level. There are common conditions and challenges that run through all three levels, so a further discussion of the implications of these conditions and challenges is merited.

The underlying assumption of the scenarios at every level is that economic growth to 2050 or 2100 will generally move in a positive direction, and that the general strength of the global economy will allow a transition to sustainable transportation technologies and systems. Thus the fragility of the global economy in the 2008 to 2013 period is a temporary slowdown, to be replaced by stronger economic performance over the long term, occasionally punctuated by recessions. Economic strength for the majority of the time in the twenty-first century will allow national economies to sustain an investment in ITS, multimodal systems, alternative fuels and energy sources, and other ingredients of sustainability described in this book.

This positive outlook has a realistic probability of coming to pass, but it is by no means guaranteed, so it is wise to look at other possible outcomes and how they might shape the transition to sustainable transportation. One possibility is that the world economy enters a period of prolonged weakness. Such a phase might have two effects on efforts to transition to a more sustainable transportation system. On the one hand, poor economic performance might marginally benefit ecological sustainability, since the amount of transportation demand would decline, or at least growth would be dampened. On the other hand, the overriding factor might be the inability to make investments in the technologies and systems envisioned above in the scenarios for 2050 or 2100, as nations and international organizations such as the U.N. or World Trade Organization (WTO) would be preoccupied with short-term economic priorities.

A third possible outcome, equally pessimistic in its own way, is that increasing hardships generated by accelerating climate change would disrupt the function not only of natural ecosystems but also economic activity in the human-built environment, confounding efforts to transition to sustainable transportation systems. This disruption could come in the form of both prolonged changes in weather (either droughts or excess rainfall that disrupt agricultural production) or extreme weather events that severely damage transportation infrastructure such as happened in the United States in October 2012 (Superstorm Sandy on the east coast) or September 2013 (historic flooding in the state of Colorado). If such destructive events come too frequently, even the wealthiest countries on the planet will either become too economically burdened with reconstruction, or worse, be unable to afford to rebuild, so that capital resources are no longer available to invest in sustainable transportation. Ecological impact from everyday transportation activity would slow, but for very much the wrong reasons.

Either of these pessimistic scenarios could actually serve as a motivation for a robust transition to sustainable transportation. Investment in sustainable transportation executed successfully can create a *virtuous circle* in which the benefits from initial investments lead to increased profits and affluence, which can be reinvested in further infrastructure and vehicles. At the same time, they would help us to avoid the second or third outcomes above and the economic losses that come with them. Transportation would not be the only sector involved, as investments in sustainability in the electric utility, industrial, agricultural, commercial, and residential sectors would be necessary also. The transportation sector would, however, have a large and vital role to play.

Supply Chain Management Challenge to Infrastructure Transformation

The *robust transition* envisioned above would involve some mix of investments in LDVs and infrastructure (most likely the single largest area of investment by economic value), supported by investments in other mechanized modes, NMT, telecommuting, and changes to the built environment. Although the exact mix of investment in transportation modes and supporting technologies and systems is not known, any plausible mixture will suffer from the same problem: *building up infrastructure is a time-consuming process of many years, so the transition to energy secure, carbon-free, sustainable transportation will take time.* No matter what mixture of sustainable transportation technologies and systems eventually emerges, all of the new sources are starting from a small capacity and will take time to ramp up. For example, in the case of the electrification of LDVs resulting in a large fleet of EVs or PHEVs, new vehicle technology must be developed and rolled out, along with changes to the way the electric grid functions and new equipment for controlling the charging and discharging of vehicles while they are parked and connected to the grid. A similar time frame would be necessary to rebuild cities with comprehensive public transportation and NMT networks. It is evident that we confront a situation where the time horizon for completing the transition is on the order of decades.

The transition challenge becomes not just an R&D problem but a supply-chain management problem, sometimes also called a *scale-up* problem. From the field of logistics, a widely used definition of supply-chain management is "the ability to have the right product, in the right place, at the right time, in the right quality, at the right price, and with the right people" (Christopher, 2005). Based on each part of this definition, the following challenges arise as part of the supply chain challenge of introducing a large amount of new transportation infrastructure:

1. *Raw materials:* Especially for building large quantities of fixed infrastructure and large numbers of vehicles, sources are required for various metals, concrete, glass, etc.
2. *Manufacturing facilities:* The total amount of infrastructure can expand year on year only as fast as the maximum output from manufacturing facilities will allow.
3. *Locations for deployment:* New locations must be found for the new infrastructure, which have met local approval and contribute to the solution in a meaningful way.
4. *Financial capital:* The financial system must be mobilized to provide sufficient funding to invest in the necessary infrastructure.
5. *Human resources:* At all stages of the supply chain, there must be sufficient numbers of people with the right skills in the work force to carry out each activity: financial management, resource extraction, manufacturing, freight shipment, installation, and maintenance.

The supply-chain management problem implies that *shortfalls on any one of these points can greatly slow the evolution of the infrastructure*: if any piece is missing, the infrastructure will not be able to expand to its full necessary capacity. Fortunately, in a properly functioning market economy, if there is need for new infrastructure, the market will respond by attempting to fill the need. For any of the above five areas, investment will follow demand. Take human resources as an example: if manufacturers and other employers are having difficulty finding enough skilled employees to make the various components of the new infrastructure, both government and educational institutions will respond to this gap. Governments can publicize the need for the skills and encourage entry into the workforce, and educational institutions can launch training programs and new degrees to meet the perceived need. Nevertheless, the rate at which these projects can be carried out may become the limiting factor in how quickly the infrastructure can be transformed. Similar arguments could be articulated for points 1 to 4 in the list.

Government leadership can be critical in accelerating the rate of transition, and although unlikely in the short term based on the political climate in the 2010 to 2013 period, might nevertheless make a real difference in the future. An analogy can be made between the scope of investment in sustainable transportation systems envisioned and the output of the U.S. war effort during World War II (Brown, 2006). During a period from 1942 to 1945, the United States manufactured 230,000 aircraft of all descriptions for use in fighting the war, as well as many other record-setting production quantities for other types of war materials, a level of industrial output that was unprecedented up to that time. The government achieved this outcome by restricting certain types of industrial production (e.g., discontinuing the production of new private cars for the duration of the war, although allowing continued production of replacement parts) and diverting the factories and workforce into production of airplanes, military transport vehicles, and other equipment. Another example, although less aggressive, is the exploitation of the "peace dividend" at the end of the Cold War with the former Soviet Union and the U.S. government's encouragement of redundant defense contractors to begin developing products for civilian purposes. The greater the level of national government intervention in the economy, the greater the controversy and potential opposition. Nevertheless, if the need for sustainable transportation were perceived not just as a matter of ecological or economic necessity but of national security, there might be scope for steps in this direction.

Regardless of the path taken and the level of government activity as catalyst, we can expect a long time horizon for the upcoming transition. One benefit of a long time horizon is that we have broad latitude for looking at a range of options from the outset. The "whole-system solution" combining on the passenger side alternative fuel LDVs, changes to urban land use, public transportation, and nonmotorized modes, or on the freight side efficient vehicles, modal shifting, and rethinking of the distances and amounts of goods shipped, is just such a broad option. Looking at the case of passenger transportation in the United States, the dominant mode is the LDV, and because of this, there is already in place a supply chain to sell 10 to 12 million new vehicles to consumers each year as old vehicles reach the end of their useful lives. There is not yet a similarly large supply chain that could support other modes of passenger transportation at this level. Over the time frame to 2050 used in the scenarios above, however, there would be time to build one up.

17-5 Concluding Discussion of the Conflict between Development and Environment

Running deeply beneath all efforts to both deliver the required services of a passenger or freight transportation system and to protect the natural environment is a fundamental tension between the economic cost of foregoing business opportunities versus foregoing steps to protect the health and wellbeing of humans and other living beings. The following parable is indicative:

> A leather dealer builds a tannery just outside a small town that has no major business. Tanneries use chemicals and other materials that have unpleasant and penetrating odors, and the fumes can be irritating and dangerous to health. Following both law and common sense, the tannery was located downwind, to the east of town. The tannery became very successful and helped the town prosper. Eventually, as the town grew, houses were built to the east of the tannery, and the people who moved in found themselves breathing foul air. They brought a suit against the tannery, demanding that it either modify its operations or move further east. The tannery owner claimed that doing either would be too expensive and might force him to close his facility and move to another town.[10]

Several threads common to many challenges with sustainable transportation are apparent from the parable. The plaintiffs in the lawsuit against the tannery owner might represent all people everywhere who are critical of the negative effects of transportation, including congestion, air pollution, the negative effect of space occupied by transportation infrastructure in the built environment, species loss due to habitat destruction, and other issues. Naturally they demand that decision-makers with authority over the transportation system, represented in the parable by the tannery owner, take action to

[10]From Bentley, P. (1998),"Business and Environment: A Case Study," pp. 225–226, in *Ecology and the Jewish Spirit: Where Nature and the Sacred Meet*, copyright 1998 Ellen Bernstein (Woodstock, VT: Jewish Lights Publishing). Permission granted by Jewish Lights Publishing, P.O. Box 237, Woodstock, VT 05091, *www. jewishlights.com.*

protect their interests (including the benefits of a thriving environment as well as personal health). These decision-makers include elected officials as the most visible element but also career civil servants involved with transportation and leaders in the private sector who provide vehicles and infrastructure.

The paradox and dilemma is that the same tannery that is harming the plaintiffs is also providing their livelihood. Prior to the arrival of the tannery (which could be likened to the arrival of modern transportation, or the industrial revolution in general), the small town likely had only a subsistence economy, and many of the advantages of economic prosperity were not available to them. The risk as laid out by the owner is that steps taken to modify or move the tannery will lead to its economic undoing. In the same way, there is a risk that changes to the transportation system may be so expensive that the primary effect is not to improve health but to derail the economy.

An additional factor that would apply if the tannery were operated at any time in recent history is the impact of fossil fuel consumption and greenhouse gas emissions. Assuming the tannery uses coal or natural gas for heat, or is connected to the electric grid, or receives supplies or ships finished product using motor vehicles, it is contributing to the release of CO_2 into the atmosphere. One could conceivably operate a tannery using only renewable energy, but virtually any actual tannery operated around the world in the last 100 years would have used fossil fuels in one form or another. This situation becomes a further problem and challenge for both the tannery owner and the aggrieved townspersons: not only must they consider the direct pollution from the tannery in the form of noxious chemicals and fumes, but the tannery on which the town economy depends is contributing to climate change.

17-5-1 Overcoming the Conflict by Closing the Loop on Energy and Material Flows

The dilemma of modern technology represented by the tannery is not solved easily or quickly. In pursuit of sustainable transportation we can, however, look to the natural world for an example of a model that can be replicated in our manufacturing and recycling of products.

Natural systems use energy and physical materials to create and recreate organisms. At the heart of this system are the various transformative processes that build up matter into living organisms and then convert organisms back into raw matter. Most of the energy comes originally from the sun in a high-quality form, meaning that it has low entropy and is capable of a wide array of applications. The energy may arrive into processes on the earth's surface directly in the form of solar rays, but it may also arrive as wind energy, or the wind energy may in turn create waves, or solar energy can be stored in plants in the form of biomass energy. Energy is conserved as it is used to perform transformative processes (first law of thermodynamics), but even though the amount of energy remains constant, it exits in a lower quality form that is no longer as capable of acting as a catalyst (second law of thermodynamics). For instance, energy used once as a catalyst for photosynthesis in a leaf may eventually be stored in a plant and consumed by an animal when some part of the plant is eaten, but the same energy is no longer available for a second round of photosynthesis. However, because the sun continues to provide solar rays, the process can continue using newly provided high-quality energy.

At the level of life cycle of animals and plants, organisms may consume foods that provide both nutrients (proteins, fats, etc.) and energy (caloric content) for growth and maintenance of health. Organisms that are capable of locomotion also use the energy content in food to meet travel requirements of their daily life. At the end of their lives, energy is also used by the natural system for the reverse process of breaking down organisms into component materials, perhaps using the physical force of solar, wind, or water energy, or perhaps relying on microorganisms that have their own life cycle from matter to organism to matter. Over geologic time periods, some of the material is trapped under the earth's surface (hence the formation of fossil fuels), but most becomes available again for formation into new organisms.

Sustainable transportation systems can emulate the natural world on two levels, as shown in Fig. 17-8. First, they can use high-quality incoming energy as a source of propulsion to move vehicles of all types. The energy is dissipated as heat and friction as the vehicles move (thus becoming "degraded low-quality energy") but because a mechanism exists to continuously convert incoming energy in nature into transportation fuels (i.e., the sun continues to supply primary energy every day), the system continues to operate. Second, energy is used to assemble physical resources into the physical products needed for transportation: vehicles, infrastructure, and other supporting systems. Products are designed with recyclability in mind, so that at the end of their life, ideally 100% of the product can be broken down and returned to be used as a physical resource. At first, the economic cost of developing this system is prohibitive, but as it develops, society begins to reap the benefits of reusing materials and avoiding the creation of waste streams that pollute the environment and harm health.

Importance of Understanding Cultural Context as an Enabler of the Sustainability Transition

Although the parable in the preceding section and subsequent discussion of closed-loop energy and material systems dealt primarily with the economic and ecological dimensions of sustainability, Chap. 1 also introduced the social and cultural sustainability dimension alongside the other two. Social and cultural sustainability can be interpreted both as an end in itself and as an awareness of cultural context that enables the pursuit of economic and ecological sustainability. From this perspective, cultural norms must be accounted for so that plans for new or upgraded transportation systems are implemented effectively in their chosen location.

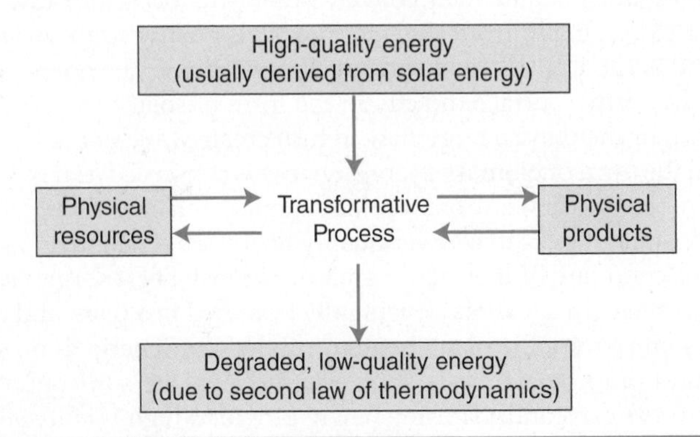

Figure 17-8 Transformative process that use high-quality energy to convert physical resources into physical products and vice versa.

(a)

(b)

FIGURE 17-9 "Cultural context matters": (a) Two signs, one in Michigan, United States and (b) the other in Edinburgh, Scotland, that both convey the same meaning but with different choice of words.

The two photographs in Fig. 17-9 symbolize this type of awareness: both of the streets signs shown seek to improve quality of life in a neighborhood by preventing the passage a certain type of vehicle, namely a "truck" in one country or a "heavy goods vehicle" in another. However, each uses language suitable to its regional and cultural surroundings. In this example, the adaptation of message to context is fairly simple: language must be chosen that conveys the meaning in a way that meets local expectations. More complex transportation technologies or systems must also be adapted to their situation. Local populations may have specific expectations around new types of vehicles or infrastructure projects, especially at the human-technology interface, such as the interior of a commercial passenger vehicle or passenger terminal.

The successful solution to the quest for sustainable transportation must find a balance between economic and ecological sustainability. It must neither create a crushing economic burden for populations that seek access to amenities and goods, nor must it allow ecological degradation to continue unimpeded. It must further be adapted to the many different cultures around the world that demand transportation, both in the tropical and temperate countries, in countries that have relatively modern built environments (Australia, Canada, New Zealand, United States) and those with older cities and towns (Africa, Asia, Europe, Latin America, Middle East). In principle, any transportation technology or system can in theory be adapted to any region. Economic, ecologic, and cultural factors in various regions will, however, tend to push development in different directions.

17-6 The Transportation Professional's Role in Creating Sustainable Transportation Systems

We now transit from the question of "What needs to be done in pursuit of sustainable transportation?" to the question of "Who is going to do this work?" The transportation professionals who design, operate, or manage transportation systems will shape our transportation future. They can serve in this role in both a professional and extracurricular context.

Turning first to the professional role, evaluation of the various technologies and systems in the book shows us where developments will take place over the next several decades as we adapt our transportation systems to make them more sustainable. Broadly speaking, these challenges can be categorized into the following three areas, which will provide career opportunities for transportation professionals interested in sustainability:

- *Research and development:* Challenges include, among others: (1) improvements to battery technology for use in electric vehicles, (2) development of all components necessary for integration of the electric vehicle fleet with the grid, (3) carbon-free solutions for aviation energy supply, (4) systems for bringing carbon-free energy to freight transportation, (5) the development of more cost-effective and less land-intensive biofuels, (6) the development of long-lived and cost-effective fuel cells and on-board hydrogen storage, and (7) the development of carbon-free hydrogen production and distribution networks, should hydrogen take root as a transportation fuel.
- *Identifying opportunities for improving efficiency:* Finding ways to use existing vehicles, systems, and infrastructure more efficiently. A major focus of this area is the further development of ITS.
- *Streamlining the design/build/operate process*: Standardizing the process for implementing new technologies and systems so that it is cheaper, faster, and more predictable. This task is especially for novel transportation systems, such as networks of electric cars and charging stations with access to CO_2-free electricity, or possibly networks for hydrogen distribution.

It is clear from this list that many disciplines within the fields of engineering, planning, and management will play a role in this process. For these efforts to succeed, decision makers in business and government must be well-informed. Therefore, transportation professionals with the relevant experience will be needed in executive positions in transportation businesses large and small, in policy decision-making positions in government, and as leaders in energy and environment-related nongovernmental organizations.

17-6-1 Roles for Transportation Professionals Outside of Formal Work

There are also many extracurricular roles for transportation professionals outside of their formal work in engineering, management, policy, or advocacy. For example, in many spheres of daily life the transportation professional can help to bridge the gap between efforts by government to provide sustainable transportation and the public's participation in these efforts. While there are numerous examples of new, state-of-the-art transportation technologies and systems in use, there are also many instances of transportation systems used poorly: for example, inadequately maintained motor vehicles; provision of excessive capacity in various vehicles that is mostly unused; time and fuel wasted in congested urban streets; or shipment of freight over very long distances when a local or regional alternative could save energy. The level of *sustainable transportation literacy* can be continually improved among transportation consumers—whose numbers, in the industrial countries, include almost the entire population—and this will be an important part of the project to create global transportation sustainability.

There are many ways in which a transportation professional can help those outside the profession understand transportation better and make better decisions about it. One way is to lead by example, by upgrading one's choice of personal vehicle to one that is as efficient as possible. Another way is to use alternative transportation sources, such as walking, bicycling, public transportation, carpooling, or carsharing. Explaining to friends and acquaintances what the changes are, and the benefits that are expected, expands these efforts into an opportunity to encourage others to do the same.

In the case of one's workplace, there may be scope for serving in an extracurricular capacity as a sustainable transportation champion who encourages efforts to use transportation more efficiently. For example, a medium or large employer may be required to reduce travel to work in single-occupant vehicles by encouraging carpooling or public transportation. Taking leadership in these efforts can help to ensure their success. Related to outreach in the workplace is outreach to tradespeople who fabricate and maintain transportation systems. Although the thrust of this book has been toward the design and financial management of transportation systems, none can be built or maintained without the skills of excavators, contractors, welders, electricians, plumbers, pipefitters, carpenters, telecommunications workers, and a host of other trades that contribute to transportation projects. Transportation professionals should nurture strong relations with the trades, generate buy-in and enthusiasm for projects, especially innovative or experimental ones, and invite practical input about how systems should be physically constructed and operated so that they work effectively and reliably.

Another important area is involvement in local government in an advisory capacity, helping public servants who do not have a transportation background to better understand technology, policy, and how local laws and design codes affect transportation systems. Many communities are interested in making their transportation systems more sustainable, both to reduce costs and to help protect the environment. In some cases, local governments may be able to bid out contracts to professional transportation service companies, but in other cases they may need to make decisions about transportation and sustainability without paid professional help. Often, when a transportation professional gives a clear explanation of technical aspects of a project or decision, it takes relatively little of her/his time but goes a long way toward assisting public servants to make better decisions. For instance, a community may be considering some new transportation service (e.g., public transit line, multiuse trail) that on the surface appears to be able to deliver savings in diminished congestion, reduced pollution, or other benefits. However, the technical and/or economic strengths and weaknesses of the technology or system may be difficult to discern for those without a technical background. The transportation professional can provide a valuable service by gathering and passing along this information. Of course, there are some design questions where volunteer work is inadequate and the services of a paid professional are required instead, so a balance must be struck between the two.

Another critical area affecting communities is the media coverage of transportation issues. The transportation professional can contribute by monitoring this coverage for accuracy and clarity. Members of the media are no different from other parts of society: some have a solid grasp of transportation issues and are able to present ideas accurately, but others may, even with the best of intentions, give incorrect information about transportation, leaving the readership or listeners with misconceptions. Transportation professionals can, when appropriate, respond to media reports and commentaries with letters to the editor or op-ed pieces. Not surprisingly, other community members often express their gratitude for these efforts to shed light on the vital yet confusing world of transportation technology, systems, and policy.

In addition to working individually in this way in one's community, it is also possible to join with others to support organizations that promote new transportation solutions, sustainable transportation, and the like. Along with financial contributions, one can also contribute to the outreach activities of these organizations, for example, by assisting with programming or joining a speakers' bureau. Also, community organizations such as Rotary, Soroptimist, or League of Women Voters, as well as religious organizations or retirement communities, may have regular meetings where the schedule occasionally includes a guest speaker on topics of interest. Given the awareness of transportation issues at this time, such organizations may be very interested in a talk about transportation solutions.

Lastly, and perhaps most importantly, the transportation professional can lend assistance by supporting young people's interest in the transportation field. They can find opportunities to talk to children and young adults about personal experience such as the types of projects on which transportation professionals work, or one's career path into the field. For example, it may be possible to visit a local elementary or secondary school as a guest speaker on sustainable transportation. Many high schools and colleges have student teams that work on design projects related to transportation [for example, designing an alternative fuel vehicle, or an all-weather human powered vehicle, (HPV)], and the services of a transportation professional who can act as an advisor are highly valued. Not only does her or his input help the students with the project at hand, but the entire interaction provides an opportunity for young people to understand the connection between the individual project and the larger world context in which it is set.

At the beginning of this book, it was stated that modern transportation systems reach almost every human being on the planet: residents of the industrial countries all the time, poor residents of less developed countries on a less frequent basis, but even the most isolated residents from time to time. Transportation professionals share the benefits of this system with a huge range of people all around the world. They can be seen as both leaders of society and partners with other users in creating the sustainable transportation systems of the future. They are leaders in the sense that they have the technical skills to create and deploy the systems, as well as to educate about and advocate for them. They are partners in the sense that they interact with the rest of the public to learn how changes in the systems are received and what adaptations are needed to make the systems function better. At this time, we cannot say exactly what mixture of systems will eventually emerge. But whichever specific technologies, systems, and aspects the transportation professional works on, this type of engagement with society is essential for the sustainability of the transportation systems of the future. The sustainability of these systems is, in turn, essential for the health and well-being of individuals in all walks of life, for the livability of our towns and cities, and for the protection of the natural environment on which all life on our planet depends.

17-7 Summary

Although the establishment of sustainable energy supplies and reduction of CO_2 emissions as discussed in Chaps. 14 to 16 are major transportation concerns, they are not the only system-wide considerations. All types of transportation contribute to air quality, and although the technology for controlling emissions from vehicles has greatly advanced in the last few decades in the industrialized countries, air-quality problems persist, and many developing countries do not yet have access to this technology. The need for transportation security in the face of natural disasters and malicious attacks is another universal concern, the latter especially after the attacks of Sept. 11, 2001. In areas prone to periodic natural disasters, such as coastal regions subject to hurricanes, pre-event planning for mass evacuations and temporary shelters can help to minimize loss of life. Lastly, the need to move and store hazardous materials as safely as possible affects many parts of the transportation system, including road, rail, and waterway networks.

A survey of sustainable transportation systems engineering as discussed in this book logically concludes with a view of how we might progress toward these sustainable systems as we move into the future. An individual urban region is limited in its ability to influence major changes in transportation energy supplies or vehicle fuels or drivetrain platforms. Steps at the regional level can, however, encourage different consumer choices, develop multimodal systems, and encourage more local sourcing of goods to reduce freight demand. At the national and global levels, single countries or groups of countries can work with manufacturers to roll out a range of alternative, zero-carbon fuels that can both power transportation systems for the long term and phase out greenhouse gas emissions to the atmosphere. This transition will happen over the course of decades rather than in just a few years. Therefore, there is latitude to not only transform the world's fleet of road vehicles (both passenger and freight) to new energy sources, but also to change land-use patterns and multimodal infrastructure so as to support a high quality of life with fewer passenger-miles and ton-miles required. Transportation professionals can make a career out of working on these problems, and they can also play an active role promoting sustainable transportation in the community outside of paid professional work.

References

Al-Jefri, H., J. Bernstein, R. Larin, S. Matharu, T. Shi, and S. Un (2008). *Toward Reduced-Carbon U.S. Freight Transportation in the Year 2050: Analysis of Policy Options for Technological Transition.* Technical Report: Master of Engineering in Engineering Management program, School of Civil and Environmental Engineering, Cornell University.

Brown, L. (2006). *Plan B 2. 0: Rescuing a Planet under Stress and a Civilization in Trouble.* Norton, New York.

Christopher, M. (2005). *Logistics and Supply Chain Management: Creating Value-Adding Networks.* Financial Times Press, London.

Davis, S., S. Diegel, and R. Boundy (2013). *Transportation Energy Data Book, 32d ed.* Oak Ridge National Laboratories, Oak Ridge, tn.

Law, A. and D. Kelton (1991). *Simulation Modeling and Analysis, 2d ed.* McGraw-Hill, New York.

Leighty, W. (2012). *Deep Reductions in Greenhouse Gas Emissions from the California Transportation Sector: Dynamics in Vehicle Fleet and Energy Supply Transitions to Achieve 80% Reduction in Emissions from 1990 Levels by 2050.* Technical report, Institute for Transportation Studies, U.C. Davis.

Li, A., L. Nozick, N. Xu, and R. Davidson (2012). "Shelter Location and Transportation Planning under Hurricane Conditions." *Transportation Research Part E* 48:715–729.

NAS. (2013) *Transitions to Alternative Fuels and Vehicles.* National Academies Press, Washington, D.C.

Ritter, L., J. Barrett, and R. Wilson (2007). *Securing Global Transportation Networks: A Total Security Management Approach.* McGraw-Hill, New York.

Schafer, A., H. Jacoby, and J. Heywood (2009). "The Other Climate Threat: Transportation." *American Scientist: The Magazine of Sigma Xi, the Scientific Research Society*:476–485.

Turnquist, M. and L. Nozick (1997). "Hazardous Waste Management." Ch. 6 in Revelle, C. and A. McGarity, Editors, *Design and Operation of Civil and Environmental Engineering Systems.* Wiley, New York.

Further Readings

Kamarianakis, Y., H. Gao, B. Holmen, and D. Sonntag (2011). "Robust Modeling and Forecasting of Diesel Particle Number Emissions Rates. *Transportation Research Part D* 16:435–443.

Nakicenovic, N. and M. Jefferson (1995). *Global Energy Perspective to 2050 and Beyond.* World Energy Council, London.

REN21 (2006). *Renewables Global Status Report: 2006 update.* Renewable Energy Policy Network for the 21st Century. Web resource, available at *http://www. ren21. net/globalstatusreport/download/RE_GSR_2006_Update.pdf.* Accessed Oct. 16, 2007.

Schafer, A., J. Heywood, H. Jacoby, and I. Waitz (2009). *Transportation in a Climate-Constrained World.* MIT Press, Cambridge, MA.

Tester, J., E. Drake, M. Driscoll, M. Golay, and W. Peters (2012). *Sustainable Energy: Choosing Among Options.* MIT Press, Cambridge, MA.

Vanek, F. (2010). Long-Term Solutions for Carbon-Free Urban Passenger Transportation: Vehicle Technology Transition or Whole-Systems Transformation? Paper presented at University of Pennsylvania Transportation Systems Engineering Alumni Council Conference, Philadelphia, PA, Jun. 10, 2010.

Von Weiszacker, U., A. Lovins, and H. Lovins (1997). *Factor Four: Doubling Wealth, Halving Resource Use.* Earthscan, London.

Vuchic, V. (1999). *Transportation for Livable Cities.* Center for Urban Policy Research, Rutgers University, New Brunswick, NJ.

Exercises

17-1. Consider the CO_2 reduction pathways for passenger and freight transportation presented in Figs. 17-6 and 17-7. (a) Reconstruct the emissions shown in the figures using population values given in Exercise 1-2, the passenger-kilometers and tonne-kilometers per capita figures given in Chap. 1, the CO_2 emissions factors given in this chapter, and the triangle function. (b) Use the model from part (a) to develop alternative pathways for emissions based on changes to the assumptions about each of the four country groups. Discuss the reasoning behind your changed assumptions, and the implications for the resulting total CO_2 emissions over the course of the twenty-first century.

17-2. Consider the two CO_2 reduction pathways presented in Figs. 17-1 and 17-2. (a) Use the given data and the triangle function to reconstruct the pathways shown. (b) The discussion in the chapter notes that Alternative 2 (as presented in Fig. 17-2) exceeds the 80% target because it uses the emission factor targets from Alternative 1 but also substantially reduces total passenger-miles. Develop an alternative set of emissions targets that exactly meet the 80% target when implemented according to the triangle function. Explain the reasoning behind your assumed targets.

APPENDIX A
Common Conversions

NOTE: *Where relevant, heat content values given below use the lower heating value.*

1 standard ton	= 2,000 pounds (lb)
1 metric ton (aka "tonne")	= 1,000 kilograms (kg)
1 kg	= 2.2 lb
1 tonne	= 1.1 standard ton
1 gallon (gal)	= 3.785 liters (L)
1 mile	= 1.61 kilometers (km)
1 meter (m)	= 3.28 feet (ft)
1 knot (nautical mile per hour)	= 1.15 miles per hour (mph)
1 British thermal unit (Btu)	= 1.055 kilojoules (kJ)
1 kilowatt (kW)	= 1.341 horsepower (hp)
1 hp	= 2,544 Btu
1 MBtu	= 1,000 Btu
1 MMBtu	= 1 million Btu
1 watt (W)	= 1 J/s
1 kilowatt-hour (kWh)	= 3.6 megajoules (MJ)
1 kWh	= 3,412 Btu
1 gal gasoline	= 115,400 Btu energy content
1 gal diesel	= 128,700 Btu
1 gal biodiesel (B100)	= 117,100 Btu
1 gal ethanol	= 75,700 Btu
1 L gasoline	= 32.2 MJ energy content
1 L diesel	= 35.9 MJ
1 L biodiesel (B100)	= 32.6 MJ
1 L ethanol	= 21.3 MJ
1 gal gasoline combusted	= 19.6 lb CO_2 resulting emissions
1 gal (petro-)diesel combusted	= 22.4 lb CO_2 resulting emissions
1 L gasoline combusted	= 2.35 kg CO_2 resulting emissions
1 L (petro-)diesel combusted	= 2.69 kg CO_2 resulting emissions

Metric unit prefixes:

Kilo-	$= 10^3$
Mega-	$= 10^6$
Giga-	$= 10^9$
Tera-	$= 10^{12}$
Peta-	$= 10^{15}$
Exa-	$= 10^{18}$

APPENDIX B

Online Appendices

The following appendices are available online at www.mhprofessional.com/STSE:

Appendix C—"Engineering Economy Tools." Includes discussion of the time value of money; discounted and nondiscounted cash flow analysis; present worth, annual worth, and benefit-cost ratio methods of analyzing engineering investments; and mechanisms for supporting investments to achieve social ends such as sustainability.

Appendix D—"Geometric Design." Considers the geometric design of transportation infrastructure such as roads, railroads, and runways, including horizontal alignment, vertical alignment, and the combination of horizontal and vertical dimensions of alignment into a geometric design solution.

Index

Note: Page numbers followed by *f* denote figures; page numbers followed by *t* denote tables.

B

F

G

H

J

N

S

U